BIOLOGICAL SCIENCE

ABOUT THE PHOTO:

Great tiger moth, *Arctia caja*, as photographed by
Joseph Scheer. Scheer, of New York's Alfred University,
is the author of *Night Visions: The Secret Designs of Moths*
(Prestel, 2003).

BIOLOGICAL SCIENCE
SECOND EDITION

SCOTT FREEMAN
University of Washington

CONTRIBUTORS

Healy Hamilton
*California Academy of Science,
Center for Biodiversity*

Sara Hoot
University of Wisconsin, Milwaukee

Greg Podgorski
Utah State University

James M. Ryan
Hobart and William Smith Colleges

Sally Sommers Smith
Boston University

Carol Trent
Western Washington University

Charles Walcott
Cornell University

D. Scott Weigle
University of Washington

PEARSON

Prentice
Hall

Upper Saddle River, New Jersey 07458

Library of Congress Cataloging-in-Publication Data

Freeman, Scott
  Biological science / Scott Freeman.— 2nd ed.
    p. cm.
  ISBN 0-13-140941-7 (hardcover)
  1. Biology. I. Title.

QH308.2.F73 2005
570—dc22
              2004027312

Publisher: Sheri L. Snavely
Senior Development Editor: Karen Karlin
Production Editor: Donna Young
Project Manager: Karen Horton
Senior Media Editor: Patrick Shriner
Manager of Electronic Composition: Allyson Graesser
Electronic Production Specialists/Electronic Page Makeup:
  Karen Stephens, Vicki Croghan, Julita Nazario,
  Richard Foster, Jim Sullivan
Project Art Director: Kenny Beck
Editor in Chief: John Challice
Editor in Chief of Development: Carol Trueheart
Senior Marketing Manager: Shari Meffert
Marketing Manager: Andrew Gilfillan
Director of Science Marketing: Linda Taft MacKinnon
Executive Managing Editor: Kathleen Schiaparelli
Director of Creative Services: Paul Belfanti
Managing Editor, Audio/Video Assets: Patricia Burns
Manufacturing Buyer: Alan Fischer
Editorial Assistants: Nancy Bauer, Lisa Tarabokjia

Marketing Assistant: Laura Rath
Assistant Managing Editor, Science Media: Nicole Jackson
Assistant Managing Editor, Science Supplements: Becca Richter
Media Production Editor: Aaron Reid
Copy Editor: Chris Thillen
Proofreader: Brian I. Baker
Cover and Interior Designer: Joseph Sengotta
AV Editor: Connie Long
Illustrators: Pearson Artworks, Quade Paul, Imagineering
National Sales Director for Key Markets: David Theisen
Photo Researcher: Yvonne Gerin
Director, Image Resource Center: Melinda Reo
Manager, Rights and Permissions: Zina Arabia
Manager, Visual Research: Beth Boyd-Brenzel
Manager, Cover Visual Research and Permissions: Karen
  Sanatar
Image Permission Coordinator: LaShonda Morris
Cover photo: © Joseph Scheer
Other image credits appear in the backmatter.

© 2005, 2002 Pearson Education, Inc.
Pearson Prentice Hall
Pearson Education, Inc.
Upper Saddle River, NJ 07458

Pearson Prentice Hall™ is a trademark of Pearson Education, Inc.

Printed in the United States of America
10 9 8 7 6 5 4 3 2

ISBN 0-13-140941-7 (Student Edition)
ISBN 0-13-141050-4 (Instructor's Edition)
ISBN 0-13-150293-X (Volume 1)
ISBN 0-13-150295-6 (Volume 2)
ISBN 0-13-150296-4 (Volume 3)

Pearson Education LTD., *London*
Pearson Education Australia PTY, Limited, *Sydney*
Pearson Education Singapore, Pte. Ltd
Pearson Education North Asia Ltd, *Hong Kong*
Pearson Education Canada, Ltd., *Toronto*
Pearson Educación de Mexico, S.A. de C.V.
Pearson Education—Japan, *Tokyo*
Pearson Education Malaysia, Pte. Ltd

Brief Contents

About the Author

Scott Freeman received his Ph.D. in Zoology from the University of Washington and was subsequently awarded an Albert Sloan Postdoctoral Fellowship in Molecular Evolution at Princeton University. His research publications explore a range of topics, including the behavioral ecology of nest parasitism and the molecular systematics of the blackbird family. Scott teaches the majors' general biology course as a Lecturer at the University of Washington. He assisted in the groundbreaking and influential redesign of the course, which emphasizes an inquiry-based approach and the logic of experimental design. With Jon Herron, Scott is co-author of the standard-setting *Evolutionary Analysis*, which over 50,000 students have used to explore evolution with the same spirit of inquiry. He is currently conducting research on how active learning and peer teaching techniques affect student learning.

CONTRIBUTORS

The author and publisher are grateful to the large number of teachers and experts in all fields of biology who shared their experiences and offered advice at every stage of planning and writing the second edition. In particular, we'd like to thank the following contributors for lending their expertise and providing new ideas and material for this revision.

Healy Hamilton
California Academy of Science, Center for Biodiversity

Sara Hoot
University of Wisconsin, Milwaukee

Greg Podgorski
Utah State University

James M. Ryan
Hobart and William Smith Colleges

Sally Sommers Smith
Boston University

Carol Trent
Western Washington University

Charles Walcott
Cornell University

D. Scott Weigle
University of Washington

ILLUSTRATOR

Kim Quillin combines training and experience in biology and art to create effective and scientifically accurate visual representations of biological principles. She received her B.A. in Biology at Oberlin College and her Ph.D. in Integrative Biology from the University of California, Berkeley, and has taught undergraduate biology at both schools. Students and instructors alike have praised Kim's illustration programs for *Biology: A Guide to the Natural World*, by David Krogh, and *Biology: Science for Life*, by Colleen Belk and Virginia Borden, for their success at clearly conveying complex biological ideas in a visually appealing manner.

Preface

Cultural evolution can be defined as a change in the frequency of ideas and practices over time. Introductory biology courses and *Biological Science* are textbook examples.

The courses we design for our majors are changing in response to selection pressure from two sources: the knowledge explosion in biology and dramatic advances in research on how introductory students learn. The knowledge explosion has made it less and less viable to teach an introductory biology course that emphasizes the memorization of facts. At the same time, research on student learning has shown that introductory students struggle to differentiate key unifying concepts from supporting details and that the greatest gains in understanding occur when students have to apply the facts and concepts they are learning to new situations. In a recent article, Handelsman et al.[1] noted that "There is mounting evidence that supplementing or replacing lectures with active learning strategies and engaging students in discovery and scientific process improves learning and knowledge retention."

Instead of being satisfied with memorization, instructors are training students to use facts. The goal is to have students demonstrate a mastery of content and concepts by applying them in new contexts.

The second edition of *Biological Science* is designed to make the transition to active, higher-level learning easier for both professors and students. Every sentence and figure in the text has been revised with that goal in mind.

Ease of Use

To make the transition to inquiry-based active learning easier, I made two major changes to the second edition: I increased the amount of content coverage to give you more flexibility in the topics you emphasize, and I added study aids to help students with the task of stepping up to a college-level biology course.

Increased Content Coverage

Compared with the first edition of *Biological Science*, this book contains much more content. Recommendations from well over 500 instructors guided decisions on which topics and terms to add. The goal was to provide students with more core coverage and vocabulary, and thereby provide instructors with more flexibility in designing a syllabus and better support for organizing lectures and labs. Experiments still play a central role in this edition, but I trimmed the overall number so that the re-

[1]Handelsman, J., D. Ebert-May, R. Beichner, P. Bruns, A. Chang, R. De-Haan, J. Gentile, S. Lauffer, J. Stewart, S. M. Tilghman, and W. B. Wood. April 23, 2004. Scientific teaching. *Science* 304(5670): 521–522.

maining experiments could be developed more thoroughly and with a clearer focus on the concept they illustrate. Throughout, the text retains its commitment to presenting topics in the context of questions, hypotheses, tests, and conclusions. Facts are tools for understanding—not ends in themselves.

New Study Tools

As introductory biology instructors, one of our most important jobs is to help our students become better students. As my colleague Mary Pat Wenderoth says, "We need to help them learn how to learn." The students in our courses are novices in biology. Like novices in any field, they have a difficult time distinguishing important points from unimportant points. They also struggle with self-diagnosis—to recognize that they do not understand something well. To help students get better at studying biology, and to take some of the burden for doing so off you, this edition offers several new features:

- **Key Concepts** are listed at the start of the chapter and then revisited in the Summary of Key Concepts. Each chapter's "big ideas" are laid out at the start, developed in detail, and then summarized.

- **Check Your Understanding** boxes appear at the ends of key sections within each chapter. These features briefly summarize one or two fundamental points and then present two to three tasks that students should be able to complete in order to demonstrate a mastery of the material. These boxes are checkpoints—a way for students to make sure that they understand what is going on before they move ahead.

- **Diversity Boxes** serve as the capstone for each of the chapters on biodiversity (Chapters 27–34). Their goal is to present a focused summary of features in key lineages. The detailed information about each group is tied to (1) where it occurs on the tree of life and (2) how and why the featured lineage diversified. Instead of swamping students with details during a traditional "march through lineages," the diversity boxes present selected information in a well-developed conceptual context.

The Forest and the Trees: Helping Students Synthesize and Unify

In addition to coping with an enormous amount of content in this course, instructors have to manage its diversity. In *Biological Science*, the emphasis on inquiry and experimentation provides a unifying theme from biochemistry through ecosystem ecology. In addition, the text highlights the fundamental how and why questions of biology. How does this event or process occur at the molecular level? In an evolutionary context, why does it exist?

Most chapters include at least one case history of an analysis done at the molecular level. Natural selection is introduced by exploring the evolution of antibiotic resistance via point mutations in the RNA polymerase gene of *Mycobacterium tuberculosis*. A box in the chapter on behavior features research on alleles that influence fruit-fly foraging behavior.

Similarly, evolutionary analyses do not begin or end with the unit on evolution. Concepts such as adaptation, homology, natural selection, and phylogenetic thinking are found in virtually every chapter. Unit 1, for example, presents traditional content in biochemistry—ranging from covalent bonding to the structure and function of macromolecules—in the context of chemical evolution and the origin of life. Meiosis is analyzed in terms of its consequences for generating genetic variation and hypotheses to explain the evolution of sex.

The overriding idea is that molecular and evolutionary analyses can help unify introductory biology courses, just as molecular tools and evolutionary questions are helping to unify many formerly disparate research fields within biology.

Supporting Visual Learners

Clear, attractive, and extensive graphics are critical to our success in the classroom. The second edition offers a major improvement in the visual presentation of the material. Kim Quillin has revised virtually every figure in the book to increase clarity, accuracy, and visual appeal and to tighten the focus on the central teaching point. Compared with the first edition, this book has 350 additional diagrams and 325 additional photographs.

To support active learning and conceptual understanding, the figures contain several important features:

- **Caption Questions and Exercises** challenge students to critically examine the information in the figure—not just absorb it.
- **Experiment Boxes** offer a standardized design to help students see how biologists answer questions by posing hypotheses and testing predictions, and to give students practice with interpreting data. In some experiment boxes, space is left blank for the null hypothesis, predicted outcomes, or conclusion. Students are challenged to fill them in.
- **Figure Pointers** act like your hand at the whiteboard so that students can easily find a figure's central teaching point.

Throughout the revision, the goal was to build an art program that supports the book's focus on thinking like a biologist. Color is used judiciously to distinguish important points from supporting details and general context. Layouts flow from top to bottom or left to right, and extensive labeling lets students work through each figure in a step-by-step manner. The overall look and feel of the art is clean, clear, accessible, and inviting.

Serving a Community of Teachers

As instructors, we have at least four major texts available that are essentially well organized, well written, and beautifully illustrated encyclopedias of the life sciences. *Biological Science* is different. By de-emphasizing the encyclopedic approach to learning biology and focusing more on the questions and experimental tools that make the science come alive, my aim is to offer a book that is more readable, attractive, and contemporary than traditional texts. Learning concepts well enough to apply them to new examples and data sets may be more challenging for some students than simply memorizing facts, but also it is more compelling. By motivating the presentation with questions and then using facts as tools to find answers, students of biology may come to think and feel more like the people who actually do biology.

Thank you for your devotion to biology, for your commitment to teaching, and for considering *Biological Science*.

Scott Freeman
University of Washington

Acknowledgments
Contributors

First and foremost, I'd like to acknowledge the second edition contributors for lending their expertise and for providing new ideas and material for this revision. Their commitment to scholarship and their passion for teaching resonated throughout the contributed chapters and had an enormous impact on the published version. In writing an introductory text, it is challenging to appeal to biologists of all specialties yet focus on what a student needs. The contributors were invaluable. They made the material more accurate and teachable and gave me much-needed partners in this endeavor.

Healy Hamilton, *California Academy of Science, Center for Biodiversity*
Sara Hoot, *University of Wisconsin, Milwaukee*
Greg Podgorski, *Utah State University*
James M. Ryan, *Hobart and William Smith Colleges*
Sally Sommers Smith, *Boston University*
Carol Trent, *Western Washington University*
Charles Walcott, *Cornell University*
D. Scott Weigle, *University of Washington*

Focus Group Participants

I have been fortunate to be the beneficiary of advice and inspiration from biology instructors attending a series of workshops at Sundance, Utah, since the inception of this book. The first and second editions are influenced by the experiences and wisdom of these visionaries. The focus group attendees read through chapters and helped me make countless critical decisions about content that should be added or deleted or handled differently.

Michel Bellini, *University of Illinois, Urbana-Champaign*
Peter Berget, *Carnegie Mellon University*
Jack Burk, *University of California, Fullerton*
Ruth Buskirk, *University of Texas, Austin*
Patrick Carter, *Washington State University*
Thomas Christianson, *University of Chicago*
Jim Colbert, *Iowa State University*
William Collins, *Stony Brook University*
Mark Decker, *University of Minnesota*
Kathryn L. Edwards, *Kenyon College*
Jeffrey Feder, *University of Notre Dame*
Ross Feldberg, *Tufts University*
Michael Gaines, *University of Miami*
Miriam Golbert, *College of the Canyons*
Harry Greene, *Cornell University*
Judith Heady, *University of Michigan*
Jean Heitz, *University of Wisconsin*
Carole Kelley, *Cabrillo College*
Kevin Kelley, *California State University, Long Beach*
Stephen Kelso, *University of Illinois, Chicago*
Judith Kjelstrom, *University of California, Davis*
Jeff Klahn, *University of Iowa*
Loren Knapp, *University of South Carolina*
Karen Koster, *University of South Dakota*
Dan Krane, *Wright State University*
Mary Rose Lamb, *University of Puget Sound*
Andrea Lloyd, *Middlebury College*
Jim Manser, *Harvey Mudd College*
Mike Meighan, *University of California, Berkeley*
Robert Newman, *University of North Dakota*
Harry Nickla, *Creighton University*
Shawn Nordell, *Saint Louis University*
Julie Palmer, *University of Texas, Austin*
Marc Perkins, *Orange Coast College*
Randall Phillis, *University of Massachusetts, Amherst*
Carol Reiss, *Brown University*
Amanda Schivell, *University of Washington*
Tom Sharkey, *University of Wisconsin*
Fred Singer, *Radford University*
Sally Sommers Smith, *Boston University*
Lori Stevens, *University of Vermont*
Briana Timmerman, *University of South Carolina*
Carol Trent, *Western Washington University*
Barbara Wakimoto, *University of Washington*
Charles Walcott, *Cornell University*
D. Scott Weigle, *University of Washington*
John Whitmarsh, *University of Illinois*
Susan Whittemore, *Keene College*
David Wilson, *Parkland College*
Dan Wivagg, *Baylor University*

Media and Supplements Contributors

The media and support materials that accompany the second edition were created by a team of talented and dedicated introductory biology instructors who brought an extraordinarily high level of creativity, experience, and ability to their re-spective projects. Our goal was to provide an innovative and tightly focused support package that addresses the unique challenges facing instructors and students in introductory biology today. I thank the instructors who attended workshops in which the critical roles of assessment and media in this course were carefully considered and discussed and that lead to guidelines inspiring the creation of content throughout the textbook's support package.

Media Contributors
Jennifer Butler, *Willamette University*
Carol Chihara, *University of San Francisco*
Cheryl Ingram-Smith, *Clemson University*
James M. Ryan, *Hobart and William Smith Colleges*
Eric Stavney, *DeVry University*
Mark D. Decker, *University of Minnesota*

Supplements Contributors
Marc Albrecht, *University of Nebraska, Kearney*
Brian Bagatto, *University of Akron*
Jay Brewster, *Pepperdine University*
Warren Burggren, *University of North Texas*
Sharon Eversman, *Montana State University*
Michelle Fay
Michael Gaines, *University of Miami*
Vanessa Handley, *University of California, Berkeley*
Harry Nickla, *Creighton University*
Laurel Hester, *University of South Carolina*
Christopher Keller, *Minot State University*
Marc Perkins, *Orange Coast College*
Randall Phillis, *University of Massachusetts, Amherst*
Greg Podgorski, *Utah State University*
Susan Rouse, *Brenau University*
Elena Shpak, *University of Washington*
Sally Sommers Smith, *Boston University*
Ellen Smith
Briana Timmerman, *University of South Carolina*
David Wilson, *Parkland College*
Cindy Wedig, *University of Texas, Pan American*

The Book Team

The production team for this edition brought a high level of experience and expertise to bear and was characterized by a single-minded focus on quality. The tenacity, work rate, and attention to detail of Senior Development Editor Karen Karlin were both instrumental and exemplary. Production Editor Donna Young held the reins with an expert's touch as a large team moved forward at high speed. Illustrator Kim Quillin acted as lead on the figure program and is a talent that comes along once in a generation. In addition to coordinating what is probably the largest review program in the history of textbook publishing, Project Manager Karen Horton managed focus groups, contributors, and media and supplements authors. Senior Media Editor Patrick Shriner worked tirelessly to revise and improve the quality of the media program—never losing sight of the fact that media must solve problems for professors and students. Photo Researcher Yvonne Gerin again provided superb photo research and was particularly effective at contacting scientists for images of research results that are not available from stock agencies. Formatters Karen Stephens, Vicki Croghan, and Julita Nazario worked patiently under relentless deadlines to create pages for the book and ensure that the layout works for students. Research Scientist Kathleen Hunt, of the University of Washington, is responsible for the dramatically improved Glossary. The entire team was assembled and inspired by Publisher Sheri Snavely, who has been the driving force behind this project since its inception. Her commitment to innovative biology publishing and devotion to meeting the needs of instructors are the reasons that this book exists.

The art program was executed by the talented crews at Pearson Artworks, Quade Paul, and Imagineering; a special note of thanks goes to Managing Editor Patricia Burns and AV Editor Connie Long. I'm particularly grateful to Lee Wilcox, of the University of Wisconsin, who worked with Kim Quillin to upgrade the photo and art program dramatically for all of the plant chapters, and to Robin Manasse, of RMBlueStudios, who worked with Kim to improve the illustrations in the anatomy and physiology unit. Project Art Director Kenny Beck managed multiple rounds of revision on the design and cover and added key creative input. Designer Joseph Sengotta is responsible for creating a clear and accessible text design and a striking cover.

A textbook can help students and professors only if it ends up on their desks. The marketing and sales efforts for this edition are directed by Director of Science Marketing Linda Taft MacKinnon, who was instrumental in making the first edition the most successful launch in the history of majors' biology. Sincere thanks to Senior Marketing Manager Shari Meffert and to Marketing Manager Andrew Gilfillan for their work on the thoughtful preview booklet and their continued tireless efforts in the field on behalf of *Biological Science*. I extend a special thank you to Director of ESM Sales Programs Meghan O'Donnell and to the Sales Directors—especially Don O'Neal, Rebecca Bersagel, Kate Brousseau, Brian Buckley, Megan Donnelly, Meghan Duffy, Christine Henry, Tom Johnson, and Michelle Renda—for their input, travel, and commitment to this effort. I'm also deeply grateful to Dave Theisen, National Sales Director for Key Markets, for his tactical skill and devotion to this book.

Finally, I thank my students at the University of Washington for inspiration, Barb Radin for invaluable help with cataloging and organizing reviewer comments, and Ben and Peter Freeman for love and support.

This book has two dedications. The first is to the memory of Bill Keeton and Neil Campbell, whose books inspired two generations of introductory biology students. As teachers and authors, they are the giants whose shoulders I try to stand on. The second is to my wife, Susan. After 24 years together, I have one thing to say: I am the luckiest man alive.

Reviewers

The review program for the second edition was even more rigorous than that for the first edition. The chapters were reviewed three times as they moved through the revision process. Reviewers included star instructors who addressed issues such as level, pacing, accuracy, and student comprehension. Other reviewers were experts in particular fields who focused primarily on making sure that the details are correct and that chapters are authoritative and current. In addition, all 55 chapters underwent a fourth and final review for accuracy just prior to publication. To a person, our reviewers supplied exemplary attention to detail, expertise, and empathy for students. I am deeply indebted to all of the colleagues who reviewed chapters of *Biological Science*; it is not possible to overstate how crucial these individuals are to the success of this book. Their effort reflects a deep commitment to excellence in teaching and a profound belief in the importance of introductory courses for training the next generation of professionals.

Julie Aires, *Florida Community College at Jacksonville*
Marc Albrecht, *University of Nebraska, Kearney*
Terry C. Allison, *University of Texas, Pan American*
Jorge E. Arriagada, *St. Cloud State University*
David Asai, *Harvey Mudd College*
David Asch, *Youngstown State University*
Karl Aufderheide, *Texas A & M University*
Christopher Austin, *Louisiana State University*
Ellen Baker, *Santa Monica College*
Christopher Beck, *Emory University*
Robert Beckman, *North Carolina State University*
Patricia Bedinger, *Colorado State University*
Peter Bednekoff, *Eastern Michigan University*
Michel Bellini, *University of Illinois*
Carl Bergstrom, *University of Washington*
John Bishop, *Washington State University, Vancouver*
Meredith Blackwell, *Louisiana State University*
Andrew Blaustein, *Oregon State University*
Dona F. Boggs, *Eastern Michigan University*
Barry Bowman, *University of California, Santa Cruz*
Jerry Brand, *University of Texas, Austin*
Angela Brown, *University of Idaho*
Albert Burchsted, *College of Staten Island*
Warren Burggren, *University of North Texas*
John Burr, *University of Texas, Dallas*
Scott Burt, *Truman University*
David Byres, *Florida Community College at Jacksonville*
Jeff Carmichael, *University of North Dakota*
Patrick Carter, *Washington State University*
David Champlin, *University of Southern Maine*
Jung Choi, *Georgia Institute of Technology*
Thomas Christianson, *University of Chicago*
Cynthia Church, *Metropolitan State College*
Vitaly Citovsky, *Stony Brook University*
Michael Clancy, *Boston University*

Anne B. Clark, *Binghamton University*
Jim Colbert, *Iowa State University*
Jerry L. Cook, *Sam Houston State University*
Scott Cooper, *University of Wisconsin, LaCrosse*
Erica Corbett, *Southeastern Oklahoma State University*
David Craig, *Williamette University*
Sarah Cunningham, *University of California, Berkeley*
Elizabeth Dahlhoff, *Santa Clara University*
David Dalton, *Reed College*
Sandra Davis, *University of Louisiana, Monroe*
Neta Dean, *Stony Brook University*
Lynda Delph, *Indiana University*
Charles F. Delwiche, *University of Maryland*
Jean DeSaix, *University of North Carolina, Chapel Hill*
Kathryn Dodd, *University of Texas, Pan American*
John Downing, *Iowa State University*
Marvin Druger, *Syracuse University*
Ernest F. Dubrul, *University of Toledo*
John Dudley, *University of Illlinois Champaign-Urbana*
Charles Duggins, Jr., *University of South Carolina*
William Eckberg, *Howard University*
Jean Everett, *College of Charleston*
Sharon Eversman, *Montana State University*
Stephanie Fabritius, *Southwestern University*
Scott Fay, *University of California, Berkeley*
Ross Feldberg, *Tufts University*
Siobhan Fennessy, *Kenyon College*
Anne Findley, *University of Louisiana, Monroe*
Jon Fischer, *Saint Louis University*
Teresa G. Fischer, *Indian River Community College*
Robert Fogel, *University of Michigan, Ann Arbor*
Don Fontes, *Community College of Rhode Island*
Larry J. Forney, *University of Idaho*
Irwin Forseth, *University of Maryland*
Marty Fox, *Edinboro University*
Krista Frankenberry, *Marshall University*
Michael Gaines, *University of Miami*
Gary Galbreath, *Northwestern University*
George Gilchrist, *College of William and Mary*
John Godwin, *North Carolina State University*
Miriam Golbert, *College of the Canyons*
Walter M. Goldberg, *Florida International University*
Sara V. Good-Avila, *Acadia University*
John Graham, *Bowling Green State University*
Eileen Gregory, *Rollins College*
Patricia A. Grove, *College of Mount Saint Vincent*
Alan Gubanich, *University of Nevada*
Cary Guffey, *Our Lady of the Lake University*
Bill Hamilton, *Washington and Lee University*
Samuel Hammer, *Boston University*
Vanessa Handley, *University of California, Berkeley*
Alice Harmon, *University of Florida*
Carla Hass, *Pennsylvania State University*
Stephen Hauschka, *University of Washington*
Judith Heady, *University of Michigan, Dearborn*
Albert Herrera, *University of Southern California*
Karen Hicks, *Kenyon College*

Jay Hirsch, *University of Virginia*
William Hoese, *California State University, Fullerton*
Mark Holbrook, *University of Iowa*
John Hoogland, *University of Maryland*
Sara Hoot, *University of Wisconsin, Milwaukee*
Margaret Horton, *University of North Carolina, Greensboro*
Kelly Howe, *University of New Mexico*
Lawrence E. Hurd, *Washington and Lee University*
Erin Irish, *University of Iowa*
Donald Jackson, *Brown University*
Rebecca Jann, *Queens University of Charlotte*
Lee Johnson, *Ohio State University*
Walter S. Judd, *University of Florida*
Thomas C. Kane, *University of Cincinnati*
Elizabeth A. Kellogg, *University of Missouri, St. Louis*
Stephen Kelso, *University of Illinois, Chicago*
Chris Kennedy, *Simon Fraser University*
Gwendolyn Kinnebrew, *John Carroll University*
John Kiss, *Miami University of Ohio*
Jeff Klahn, *University of Iowa*
Helen Koepfer, *Queens College*
John La Claire, *University of Texas at Austin*
Mary Rose Lamb, *University of Puget Sound*
Thomas Lehman, *Morgan Community College*
Paula Lemons, *Duke University*
Lynn Lewis, *Mary Washington College*
Andi Lloyd, *Middlebury College*
Paula Lovett, *University of Maryland*
John Lugthart, *Dalton State College*
Robert D. Lynch, *University of Massachusetts, Lowell*
Richard Malkin, *University of California, Berkeley*
Charles Mallery, *University of Miami*
Paul Manos, *Duke University*
James Manser, *Harvey Mudd College*
Diane Marshall, *University of New Mexico*
Steven L. Matzner, *Augustana College*
Michael Mazurkiewicz, *University of Southern Maine*
Richard E. McCarty, *John Hopkins University*
Kelly McLaughlin, *Tufts University*
Mona Mehdy, *University of Texas*
Eli Meier, *Simbiotic, Inc.*
Michael Meighan, *University of California, Berkeley*
Madeline Mignone, *Dominican College*
Molly Morris, *Ohio University*
Dale Mueller, *Texas A & M University*
Leann Naughton, *University of Wyoming*
Jacalyn S. Newman, *University of Pittsburgh*
Robert Newman, *University of North Dakota*
Karen Bushaw Newton, *American University*
Harry Nickla, *Creighton University*
Shawn Nordell, *Saint Louis University*
Amanda Norvell, *The College of New Jersey*
Deborah O'Dell, *Mary Washington College*
John Olsen, *Rhodes College*
John Osterman, *University of Nebraska*
Norman Pace, *University of Colorado, Boulder*
Julie Palmer, *University of Texas, Austin*

Matthew B. Parks, *University of Idaho*
C. O. Patterson, *Texas A & M University*
Andrew Pease, *Villa Julie College*
Craig L. Peebles, *University of Pittsburgh*
Curtis Pehl, *University of California, Berkeley*
Marc Perkins, *Orange Coast College*
Gary Peterson, *South Dakota State University*
Patricia Phelps, *Austin Community College*
Randall Phyllis, *University of Massachusetts*
Greg Podgorski, *Utah State University*
Frank Polanowski, *Elizabethtown College*
Donald Potts, *University of California, Santa Cruz*
F. Harvey Pough, *Rochester Institute of Technology*
Jerry Purcell, *San Antonio College*
Jonathan Reed, *Community College of Southern Nevada*
Stuart Reichler, *University of Texas, Austin*
Robin Richardson, *Winona State University*
Jared Rifkin, *Queens College*
Bruce Riley, *Texas A & M University*
John Romeo, *University of South Florida*
Peter Russell, *Reed College*
James M. Ryan, *Hobart and William Smith Colleges*
Brody Sandel, *University of California, Berkeley*
Amanda Schivell, *University of Washington*
Brian G. Scholtens, *College of Charleston*
Susan Schreier, *Villa Julie College*
Robert Seagull, *Hofstra University*
Eli Seigel, *Tufts University*
Marty Shankland, *University of Texas, Austin*
Thomas B. Shea, *University of Massachusetts, Lowell*
Allen Shearn, *Johns Hopkins University*
Tim Sherman, *University of South Alabama*
Richard M. Showman, *University of South Carolina*
Michele Shuster, *New Mexico State University*
Amanda Simcox, *Ohio State University*
Anne Simon, *University of Maryland*
Sally Sommers Smith, *Boston University*
William Smith, *Wake Forest University*
James Sniezek, *Montgomery College*
Phillip Sokolove, *University of Maryland, Baltimore County*
James Staples, *University of Western Ontario*
Eric Stavney, *University of Washington*

David Steen, *Andrews University*
William Stein, *State University of New York, Binghamton*
Lori Stevens, *University of Vermont*
Charles Stinemetz, *Rhodes College*
Sarah H. Swain, *Middle Tennessee State University*
Kevin Teather, *University of Prince Edward Island*
Ethan J. Temeles, *Amherst College*
Joshua Tewksbury, *University of Washington*
Briana Timmerman, *University of South Carolina*
Albert Torzill, *George Mason University*
Carol Trent, *Western Washington University*
John True, *Stony Brook University*
Nancy Trun, *Duquesne University*
Stephen Turnbull, *University of New Brunswick*
Elizabeth Van Volkenburgh, *University of Washington*
Neal J. Voelz, *St. Cloud State University*
Susan Waaland, *University of Washington*
Charles Walcott, *Cornell University*
Margaret Wallace, *John Jay College of Criminal Justice*
Jennifer Warner, *University of North Carolina, Charlotte*
Cindy Wedig, *University of Texas, Pan American*
D. Scott Weigle, *University of Washington*
Elizabeth A. Weiss, *University of Texas, Austin*
Barry Welch, *San Antonio College*
Sue Simon Westendorf, *Ohio University*
Susan Whittemore, *Keene College*
David Wilkes, *Harvey Mudd College*
Judy Williams, *Southeastern Oklahoma State University*
David Wilson, *Parkland College*
Eric Winkler, *Blinn College*
Bob Winning, *Eastern Michigan University*
Dan Wivagg, *Baylor University*
C. B. Wolfe, *University of North Carolina, Charlotte*
Lorne Wolfe, *Georgia Southern University*
Denise Woodward, *Pennsylvania State University*
Shawn Wright, *Albuquerque TVI*
Richard P. Wunderlin, *University of South Florida*
Peter H. Wyckoff, *University of Minnesota, Morris Campus*
Todd Christian Yetter, *Cumberland College*
Stephen Yezerinac, *Reed College*
Donna Young, *University of Winnipeg*
Gregory Zimmerman, *Lake Superior State University*

Print and Media Resources for Instructors and Students

For the Instructor

Lecture Presentation Tools

Instructor Resource Center on CD/DVD

As the demands of teaching introductory biology continue to grow, it becomes increasingly important for instructors that the right textbook be accompanied by the right tools to aid them during every stage—in lecture preparation and presentation, in assessment, and in the lab, as well as in the overall management of the course. The Freeman *Biological Science* Instructor Resource Center, provided in both CD and DVD formats, offers a rich suite of electronic tools designed to help instructors make the most of their limited time. Features include:

- As JPEG files, the entire textbook illustration program: all line drawings with labels individually enhanced for optimal projection results (as well as unlabeled versions), all tables, and all photos, as well as additional photos not found in the textbook

- The entire textbook illustration program pre-loaded into comprehensive PowerPoint presentations for each chapter

- A second set of PowerPoint presentations consisting of a thorough lecture outline for each chapter, augmented by key text illustrations

- An impressive series of concise instructor animations, Web Tutorials, and video clips adding depth and visual clarity to the most important topics and dynamic processes described in the text

- These same illustrative animations, Web Tutorials, and video clips pre-loaded into PowerPoint presentations for each chapter

- PowerPoint presentations containing a comprehensive set of in-class CRS questions for each chapter

- In Word files, a complete set of the assessment materials and study questions and answers from the Test Bank for Assessment, the textbook's in-chapter questions, and the student media practice questions, as well as files containing the entire Instructor's Guide

- Finally, to help instructors keep track of all that is available in this media package, a printable Media Integration Guide consisting of a PDF file that lists the media offerings for each chapter

Transparency Package and Instructor Resource Kit

Transparencies are an effective way to reinforce your lecture presentation visually. Every illustration from the text is available on four-color transparency acetates. The transparency set is three-hole-punched and organized in manila folders, which are stored in an Instructor Resource Kit file box along with the printed lecture tools from the Instructor's Resource Guide. Labels and images have been enlarged and modified to ensure optimal readability in a large lecture hall.

Instructor's Guide

Susan Rouse (Brenau University) with Brian Bagatto (University of Akron)

The Instructor's Guide includes traditional instructor support tools—lecture outlines, answers to end-of-chapter questions, vocabulary lists—as well as innovative activities to help motivate students and reinforce their understanding of the material. All of the content in the Instructor's Guide is available electronically on the Instructor Resource Center on CD/DVD.

Assessment Tools

Test Bank for Assessment and TestGen EQ

Edited by Harry Nickla (Creighton University) and Marc Perkins (Orange Coast College)

Contributors: Marc Albrecht (University of Nebraska, Kearney), Brian Bagatto, (University of Akron), Warren Burggren (University of North Texas), Sharon Eversman (Montana State University), Michelle Fay, Michael Gaines (University of Miami), Vanessa Handley (University of California, Berkeley) Laurel Hester (University of South Carolina), Christopher Keller (Minot State University), Randall Phillis (University of Massachusetts, Amherst), Greg Podgorski (Utah State University), Elena Shpak (University of Washington), Sally Sommers Smith (Boston University), Briana Timmerman (University of South Carolina), David Wilson (Parkland College), and Cindy Wedig (University of Texas, Pan American)

Our assessment materials were guided by a team of faculty who participated in an assessment workshop. The group developed guidelines for writing effective test questions for students and set our standards for quality and accuracy. The Test Bank features questions that:

- are based on the major biological themes and processes covered in each chapter rather than the minutiae

- require critical thinking or application/integration of pre-existing knowledge, not rote memorization and recall
- are biologically accurate, with only one clearly correct answer choice
- are relatively resistant to test-taking strategy
- have been peer reviewed and student tested

To make this test bank more useful, we have made a few significant changes compared with the typical test bank, including:

- *In-class questions*: For each chapter, we feature three to five questions that could be used in class to help motivate discussion and formative assessment during class sessions.
- *Literature questions*: Questions based on data or ideas contained in a relevant peer-reviewed journal article. These questions extrapolate from material covered in the chapter.

Laboratory Support

Symbiosis

We have assembled a set of inquiry-based experiments to complement the approach of *Biological Science*. Symbiosis, the state-of-the-art Pearson Custom Publishing service, allows you to choose from hundreds of other experiments and/or include your own. The result is a well-integrated lab manual customized to your needs.

Course Management Tools

Prentice Hall's **OneKey** online course management content offers instructors using *Biological Science* everything they need to plan and administer their course. It includes the best teaching and learning resources all in one place. Organized by textbook chapter, the compiled resources help you save time and help your students reinforce and apply what they have learned in class. All resources from the Instructor Resource Center on CD/DVD and the Companion Website are included. OneKey is available for institutions using **WebCT** or **Blackboard** and is also available in Prentice Hall's nationally hosted **Course Compass** course management system. If desired, WebCT and Blackboard cartridges containing only the Test Bank for Assessment are available for download. Visit http://cms.prenhall.com for details.

For The Student

Companion Website (www.prenhall.com/freeman/biology)

Student study time is an increasingly valuable resource in introductory biology courses, and the greatest challenge many students face is making sure they use it wisely. The Companion Website has been designed to allow students using *Biological Science* to zero in quickly on the chapter sections and topics where they need review or further explanation. Its **Online Study Guide** offers concise lesson summaries punctuated with key illustrations, as well as probing review questions that offer hints and feedback on correct and incorrect responses. **Web Tutorials** offer students the opportunity to visualize complex topics and dynamic processes. The media's strict adherence to both the principles and specific lessons of the textbook means that study time is not being wasted. Tabs like the one in the margin of this page alert students that a Web Tutorial exists on a topic related to the coverage on that page.

Additional features of the Companion Website include an **Art Notebook**, hundreds of section-specific **Weblinks** to aid in online research; and **Textbook Answers** for all end-of-chapter and figure-caption questions.

Accelerator CD-ROM

The Accelerator CD-ROM that accompanies all student editions of *Biological Science*, Second Edition, offers the same **Web Tutorials** that are found on the Companion Website but in a stand-alone CD format.

Student Study Guide

Edited by Warren Burggren (University of North Texas)
Contributors: Jay Brewster (Pepperdine University), Laurel Hester (University of South Carolina), and Brian Bagatto (University of Akron)

The Student Study Guide helps students focus on the fundamentals chapter by chapter. Each chapter presents a breakdown of key biological concepts, difficult topics, and quizzes. In addition, the Study Guide features four introductory, stand-alone chapters: Introduction to Experimentation and Research in the Biological Sciences, Presenting Biological Data, Understanding Patterns in Biology and Improving Study Techniques, and Reading and Writing to Understand Biology.

Contents

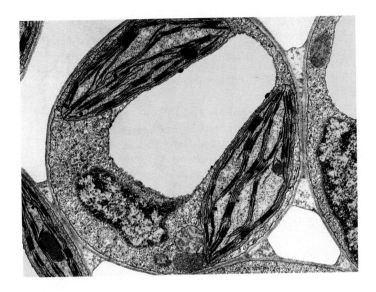

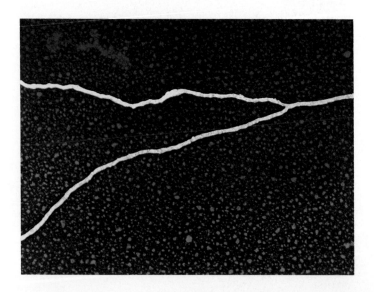

Biology and the Tree of Life

1

A biologist measuring an Atlantic puffin chick on an island off the coast of Norway. The data he's collecting will help him test hypotheses about factors that influence the growth and survival of seabirds.

KEY CONCEPTS

▨ Biological science was founded with the development of (1) the cell theory, which proposes that all organisms are made of cells and that all cells come from preexisting cells, and (2) the theory of evolution by natural selection, which maintains that the characteristics of species change through time—primarily because individuals with certain heritable traits produce more offspring than do individuals without those traits.

▨ A phylogenetic tree is a graphical representation of the evolutionary relationships among species. Phylogenies can be estimated by analyzing similarities and differences in traits. Species that share many traits are closely related and are placed close to each other on the tree of life.

▨ Biologists ask questions, generate hypotheses to answer them, and design experiments that test the predictions made by competing hypotheses.

In essence, biology is a search for ideas and observations that unify the incredible diversity of life. Chapter 1 is an introduction to this search. Its goals are to introduce the amazing variety of life-forms alive today, consider some fundamental traits shared by all organisms, and explore how biologists go about answering questions about life. Appreciating the diversity of life, understanding its underlying unity, and learning how to think like a biologist are themes that will resonate throughout this book.

We begin by examining two of the greatest unifying ideas in all of science: the cell theory and the theory of evolution by natural selection. When these concepts emerged in the mid-1800s, they revolutionized the way that biologists understand the world. The cell theory proposed that all organisms are made of cells and that all cells come from preexisting cells. The theory of evolution by natural selection maintained that species have changed through time and that all species are related to one another through common ancestry. The theory of evolution by natural selection established that bacteria, mushrooms, roses, and robins are all part of a family tree, similar to the genealogies that connect individual people.

A **theory** is an explanation for a very general class of phenomena or observations. The cell theory and the theory of evolution provided a foundation for the development of modern biology because they focused on two of the most general questions possible: How are organisms structured? Where did they come from? Let's begin by tackling the first of these two questions.

1.1   The Cell Theory

The initial conceptual breakthrough in biology—the cell theory—emerged after some 200 years of work. In 1665 Robert Hooke used a crude **microscope** to examine the structure of cork (a bark tissue) from an oak tree. The instrument magnified objects to just 30 times (30×) their normal size, but it allowed Hooke to see something extraordinary. In the cork he observed small, pore-like compartments that were invisible to the naked eye (**Figure 1.1a**). These structures came to be called cells.

Soon after Hooke published his results, Anton van Leeuwenhoek succeeded in developing much more powerful microscopes, some capable of magnifications up to 300×. With these instruments, Leeuwenhoek inspected samples of pond water and made the first observations of single-celled organisms like the *Paramecium* in **Figure 1.1b**. He also observed and described the structure of human blood cells and sperm cells.

In the 1670s a researcher who was studying the leaves and stems of plants with a microscope concluded that these large, complex structures are composed of many individual cells. By the early 1800s enough data had been accumulated for a biologist to claim that *all* organisms consist of cells. This conclusion was a classic example of inductive reasoning. Scientists made a broad generalization only after making thousands of supporting observations.

(a) The first view of cells: Robert Hooke's drawing from 1665

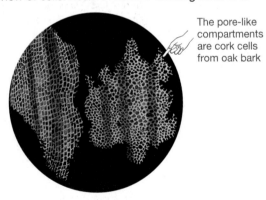

The pore-like compartments are cork cells from oak bark

(b) Anton van Leeuwenhoek was the first to view single-celled "animalcules" in pond water.

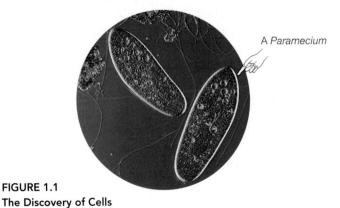

A *Paramecium*

FIGURE 1.1
The Discovery of Cells

Are *All* Organisms Made of Cells?

The smallest organisms known today are bacteria that are barely 200 nanometers wide, or 200 *billionths* of a meter. (See the Appendix at the back of this book to review the metric system and its prefixes.) Lining up bacteria end to end would take 5000 of these organisms to span a millimeter—the distance between the smallest hash marks on a metric ruler. In contrast, sequoia trees can be over 100 meters tall—the equivalent of a 20-story building. Bacteria and sequoias are composed of the same fundamental building block, however—the cell. Bacteria are unicellular (one-celled) organisms; sequoias are multicellular (many-celled) organisms.

Biologists have become increasingly dazzled by the diversity and complexity of cells as advances in microscopy have made it possible to examine cells at higher magnifications. The basic conclusion made in the 1800s is intact, however: As far as is known, all organisms are made of cells. Today, a **cell** is defined as a highly organized compartment that is bounded by a thin, flexible structure called a plasma membrane and that contains concentrated chemicals in an aqueous (watery) solution. The chemical reactions that sustain life take place inside cells. Most cells are also capable of reproducing by dividing—in effect, by making a copy of themselves.

The realization that all organisms are made of cells was fundamentally important, but it formed only the first part of the cell theory. In addition to understanding what organisms are made of, scientists wanted to understand how cells come to be.

Where Do Cells Come From?

Most scientific theories have two components: The first describes a pattern in the natural world, and the second identifies a mechanism or process that is responsible for creating that pattern. The early workers had articulated the pattern component of the cell theory. In 1858 Rudolph Virchow added a process component by stating that all cells arise from preexisting cells. The complete **cell theory**, then, can be stated as follows: All organisms are made of cells, and all cells come from preexisting cells.

This claim was a direct challenge to the prevailing explanation, called **spontaneous generation**. At the time, most biologists believed that organisms arise spontaneously under certain conditions. For example, the bacteria and fungi that spoil foods such as milk and wine were thought to appear in these nutrient-rich media of their own accord—meaning they spring to life from nonliving materials. Spontaneous generation was a **hypothesis**: a proposed explanation. The all-cells-from-cells hypothesis, in contrast, maintained that cells do not spring to life spontaneously but are produced only when preexisting cells grow and divide.

Soon after the all-cells-from-cells hypothesis appeared in print, Louis Pasteur set out to test its predictions experimentally. Pasteur wanted to determine whether microorganisms could arise spontaneously in a nutrient broth or whether they appear only when a broth is exposed to a source of preexisting cells.

To address the question, he created two treatment groups: a broth that was not exposed to a source of preexisting cells and a broth that was. The spontaneous generation hypothesis predicted that cells would appear in both treatments. The all-cells-from-cells hypothesis predicted that cells would appear only in the treatment exposed to a source of preexisting cells.

Figure 1.2 shows Pasteur's experimental setup. Note that the two treatments are identical in every respect but one. Both used glass flasks filled with the same amount of the same nutrient broth. Both were boiled for the same amount of time to kill any existing organisms such as bacteria or fungi. But because the flask pictured in Figure 1.2a had a straight neck, it was exposed to preexisting cells after sterilization by the heat treatment. These preexisting cells are the bacteria and fungi that cling to dust particles in the air. They could drop into the nutrient broth because the neck of the flask was straight. In contrast, the flask drawn in

Figure 1.2b had a long swan neck. Pasteur knew that water would condense in the crook of the swan neck after the boiling treatment and that this pool of water would trap any bacteria or fungi that entered on dust particles. Thus, the swan-necked flask was isolated from any source of preexisting cells even though it was still open to the air. The experimental setup was effective because there was only one difference between the two treatments and because that difference was the factor being tested—in this case, a broth's exposure to preexisting cells.

The result? As Figure 1.2 shows, the treatment exposed to preexisting cells quickly filled with bacteria and fungi. This treatment was important because it showed that the heat sterilization step had not altered the nutrient broth's capacity to support growth. But the treatment in the swan-necked flask remained sterile. Even when the broth was left standing for months, no organisms appeared in it.

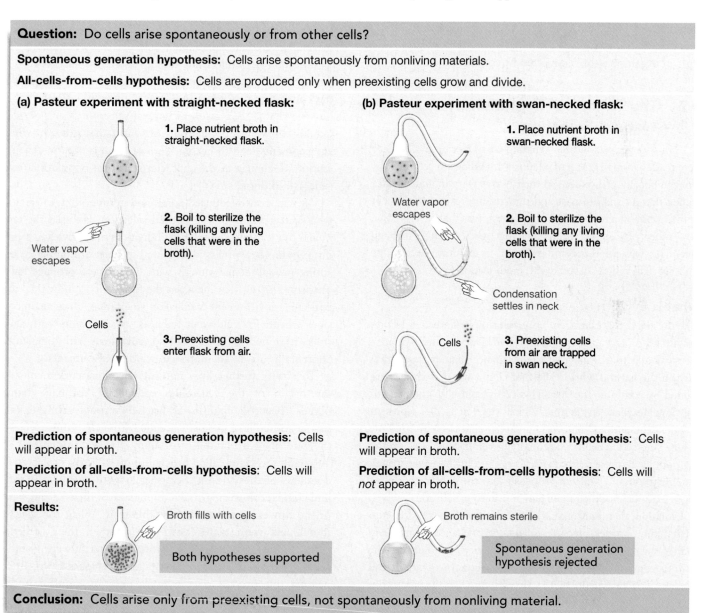

Question: Do cells arise spontaneously or from other cells?

Spontaneous generation hypothesis: Cells arise spontaneously from nonliving materials.

All-cells-from-cells hypothesis: Cells are produced only when preexisting cells grow and divide.

(a) Pasteur experiment with straight-necked flask:

1. Place nutrient broth in straight-necked flask.

2. Boil to sterilize the flask (killing any living cells that were in the broth).

Water vapor escapes

Cells

3. Preexisting cells enter flask from air.

(b) Pasteur experiment with swan-necked flask:

1. Place nutrient broth in swan-necked flask.

Water vapor escapes

2. Boil to sterilize the flask (killing any living cells that were in the broth).

Condensation settles in neck

Cells

3. Preexisting cells from air are trapped in swan neck.

Prediction of spontaneous generation hypothesis: Cells will appear in broth.

Prediction of all-cells-from-cells hypothesis: Cells will appear in broth.

Prediction of spontaneous generation hypothesis: Cells will appear in broth.

Prediction of all-cells-from-cells hypothesis: Cells will *not* appear in broth.

Results:

Broth fills with cells

Both hypotheses supported

Broth remains sterile

Spontaneous generation hypothesis rejected

Conclusion: Cells arise only from preexisting cells, not spontaneously from nonliving material.

FIGURE 1.2 The Spontaneous Generation Hypothesis Was Tested Experimentally

Because Pasteur's data were in direct opposition to the predictions made by the spontaneous generation hypothesis, the results persuaded most biologists that the all-cells-from-cells hypothesis was correct.

The success of the cell theory's process component had an important implication: If all cells come from preexisting cells, then all individuals in a population of single-celled organisms must be related by common ancestry. Similarly, in a multicellular individual such as you, all of the cells present are descended from preexisting cells, tracing back to a fertilized egg. A fertilized egg is a cell created by the fusion of sperm and egg—cells that formed in individuals of the previous generation. In this way, all of the cells in unicellular and multicellular organisms are connected by common ancestry.

The second great founding idea in biology is similar, in spirit, to the cell theory. It also happened to be published the same year as the all-cells-from-cells hypothesis. This was the realization, made independently by Charles Darwin and Alfred Russel Wallace, that all *species*—meaning all distinct, identifiable types of organisms—are connected by common ancestry.

1.2 The Theory of Evolution by Natural Selection

In 1858 short papers written separately by Darwin and Wallace were read to a small group of scientists attending a meeting of the Linnean Society of London. In their essays Darwin and Wallace argued that all species, past and present, are related by descent from a common ancestor. Their hypothesis was that species come from other, preexisting species and that species change through time. This was the theory of evolution by natural selection, or what Darwin called "descent with modification."

What Is Evolution?

Like the cell theory, the theory of evolution by natural selection has a pattern and a process component. Darwin and Wallace's theory made two important claims concerning patterns that exist in the natural world. The first claim was that species are related by common ancestry. This contrasted with the prevailing view in science at the time, which was that species represent independent entities that were created separately by a divine being. The second claim was equally novel. Instead of accepting the popular hypothesis that species remain unchanged through time, Darwin and Wallace proposed that the characteristics of species can be modified from generation to generation.

Evolution, then, means that species are not independent and unchanging entities, but are related to one another and can change through time. This part of the theory of evolution—the pattern component—was actually not original to Darwin and Wallace. Several scientists had already come to the same conclusions about the relationships among species. The great insight by Darwin and Wallace was in proposing a process, called **natural selection**, that explained *how* evolution occurs.

What Is Natural Selection?

Natural selection occurs whenever two conditions are met. The first condition is that individuals within a population vary in characteristics that are heritable. A **population** is defined as a group of individuals of the same species living in the same area at the same time. Darwin and Wallace had studied natural populations long enough to realize that variation among individuals is almost universal. In wheat, for example, some individuals are taller than others. **Heritable** traits are characteristics that can be passed on to offspring. As a result of work by wheat breeders, Darwin and Wallace knew that short parents tend to have short offspring. Subsequent research has shown that heritable variation exists in most traits and populations. The second condition of natural selection is that in a particular environment, certain versions of these heritable traits help individuals survive better or reproduce more than do other versions. For example, if tall wheat plants are easily blown down by wind, then in windy environments shorter plants will tend to survive better and leave more offspring than tall plants will.

If certain heritable traits lead to increased success in producing offspring, then those traits become more common in the population over time. In this way, the population's characteristics change as a result of natural selection acting on individuals. In wheat, populations of wheat that grow in windy environments tend to become shorter from generation to generation. A change in the characteristics of a population, over time, is evolution.

Darwin also introduced some new terminology to identify what is happening during natural selection. He used the term **fitness** to mean the ability of an individual to survive and reproduce. In biology, fitness is measured in units of number of offspring produced. Individuals with high fitness produce many offspring. A trait that increases the fitness of an individual in a particular environment is called an **adaptation**. Once again consider wheat: In windswept habitats, wheat plants with short stalks have higher fitness than do individuals with long stalks. Short stalks are an adaptation to windy environments.

To clarify further how natural selection works, consider the origin of the vegetables called the "cabbage family plants." Broccoli, cauliflower, Brussels sprouts, cabbage, kale, savoy, and collard greens descended from the same species—the wild plant in the mustard family pictured in **Figure 1.3a**. To create the plant called broccoli, horticulturists selected individuals of the wild mustard species with particularly large and compact flowering stalks. In mustards, the size and shape of the flowering stalk is a heritable trait. When the selected individuals were mated with one another, their offspring turned out to have larger and more compact flowering stalks, on average, than the original population (**Figure 1.3b**). By repeating this process over many generations, horticulturists produced a population with extraordinarily large and compact flowering stalks. The derived population has been artificially selected for the size and shape of the flowering stalk;

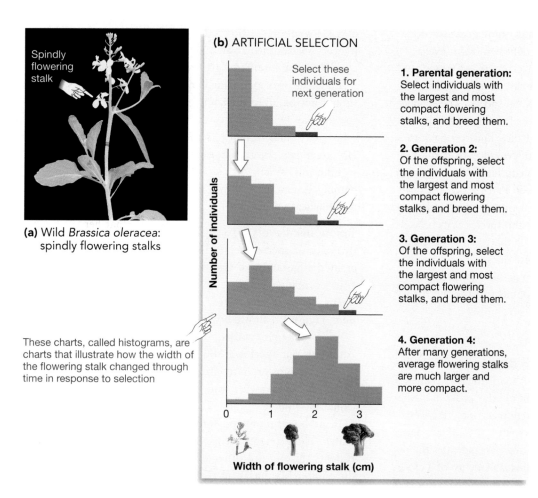

(a) Wild *Brassica oleracea*: spindly flowering stalks

Spindly flowering stalk

(b) ARTIFICIAL SELECTION

Select these individuals for next generation

Number of individuals

1. Parental generation: Select individuals with the largest and most compact flowering stalks, and breed them.

2. Generation 2: Of the offspring, select the individuals with the largest and most compact flowering stalks, and breed them.

3. Generation 3: Of the offspring, select the individuals with the largest and most compact flowering stalks, and breed them.

4. Generation 4: After many generations, average flowering stalks are much larger and more compact.

These charts, called histograms, are charts that illustrate how the width of the flowering stalk changed through time in response to selection

Width of flowering stalk (cm)

(c) Broccoli: extremely large, compact flowering stalks

Large, compact flowering stalks

FIGURE 1.3 Artificial Selection Can Produce Dramatic Changes in Organisms

as **Figure 1.3c** shows, it barely resembles the ancestral form. Note that during this process, the size and shape of the flowering stalk in each individual plant did not change within its lifetime—the change occurred in the characteristics of the population over time.

Darwin pointed out that natural selection changes the characteristics of a wild population over time, just as the deliberate manipulation of "artificial selection" changes the characteristics of a domesticated population over time. But no horticulturist is involved in the case of natural selection. Natural selection occurs naturally, simply because certain individuals in wild populations have heritable traits that allow them to leave more offspring than do individuals without those traits. Evolution, or change in the population over time, is the outcome of this process.

Since Darwin and Wallace published, biologists have succeeded in measuring hundreds of examples of natural selection in wild populations and have accumulated a massive body of evidence documenting that species have changed through time.

Together, the cell theory and the theory of evolution provided the young science of biology with two central, unifying ideas:

1. The cell is the fundamental structural unit in all organisms.

2. All species are related by common ancestry and have changed over time in response to natural selection.

✓ CHECK YOUR UNDERSTANDING

Natural selection occurs when heritable variation in certain traits leads to improved success in reproduction. Because individuals with these traits produce many offspring with the same traits, these traits increase in frequency and evolution occurs. Evolution is simply a change in the characteristics of a population over time. Although these ideas appear simple, they are often misunderstood. Research has shown that some biology students think that evolution is progressive, meaning that species always get larger, more complex, or "better" in some sense. In addition, it is common for students to think that *individuals* as well as populations change when natural selection occurs, or that individuals with high levels of fitness are stronger or bigger or "more dominant." None of these ideas are correct. Using the example of selection on height of wheat stalks, you should be able to explain why each of these three common misconceptions is wrong.

1.3 The Tree of Life

In Section 1.2 we focused on how individual populations change through time in response to natural selection. But over the past several decades, biologists have also documented dozens of cases in which natural selection has caused populations of one species to diverge and form new species. This divergence process is called **speciation**. In several instances, biologists are documenting the formation of new species right before our eyes.

Research on speciation supports a claim that Darwin and Wallace made over a century ago—that natural selection can lead to change *between* species as well as within species. The broader conclusions are that all species come from preexisting species and that all species, past and present, trace their ancestry back to a single common ancestor. If the theory of evolution by natural selection is valid, biologists should be able to reconstruct a **tree of life**—a family tree of organisms. If life on Earth arose just once, then such a diagram would describe the genealogical relationships among species with a single, ancestral species at its base.

Has this task been accomplished? If the tree of life exists, what does it look like? To answer these questions, we need to step back in time and review how biologists organized the diversity of organisms *before* the development of the cell theory and the theory of evolution.

Linnaean Taxonomy

In science, the effort to name and classify organisms is called **taxonomy**. This branch of biology began to flourish in 1735 when a botanist named Carolus Linnaeus set out to bring order to the bewildering diversity of organisms that were then being discovered.

The building block of Linnaeus' system is a two-part name unique to each type of organism. The first part indicates the organism's **genus** (plural: **genera**). A genus is made up of a closely related group of species. For example, Linnaeus put humans in the genus *Homo*. Although humans are the only living species in this genus, several extinct organisms, all of which walked upright and made extensive use of tools, were later also assigned to *Homo*. The second term in the two-part name identifies the organism's species. In Section 1.1 we defined a species as a distinct, identifiable type of organism. More formally, a **species** is made up of individuals that regularly breed together or have characteristics that are distinct from those of other species. Linnaeus gave humans the specific name *sapiens*.

An organism's genus and species designation is called its scientific name or Latin name. Scientific names are always italicized. Genus names are always capitalized, but species names are not: for instance, *Homo sapiens*. Scientific names are based on Latin or Greek word roots or on "Latinized" words from other languages (see **Box 1.1**). Linnaeus gave a scientific name to every species then known. (He also latinized his own name—from Karl von Linné to Carolus Linnaeus.)

Linnaeus also maintained that different types of organisms should not be given the same genus and species names. Other species may be assigned to the genus *Homo*, and members of other genera may be named *sapiens*, but only humans are named *Homo sapiens*. As a result, each scientific name is unique.

Linnaeus' system has stood the test of time. His two-part naming system, or **binomial nomenclature**, is still the standard in biological science.

Taxonomic Levels To organize and classify the tremendous diversity of species being discovered in the 1700s, Linnaeus created a hierarchy of taxonomic groups: From the most specific grouping to the least specific, the levels are **species, genus, family, order, class, phylum** (plural: **phyla**), and **kingdom**. **Figure 1.4** shows how this nested, or hierarchical, classification scheme works, using humans as an example. Although our species is the sole living member of the genus *Homo*, humans are now grouped with the orangutan, gorilla, common chimpanzee, and pygmy chimpanzee in a family called Hominidae. Linnaeus grouped members of this family with gibbons, monkeys, and lemurs in an order called Primates. The Primates are grouped in the class Mammalia with rodents, bison, and other organisms that have fur and produce milk. Mammals, in turn, join other animals with structures called notochords in the phylum Chordata, and all other animals in the kingdom Animalia. Each of these named groups—primates, mammals, or *Homo sapiens*—can be referred to as a **taxon** (plural: **taxa**). The essence of Linnaeus' system is that lower-level taxa are nested within higher-level taxa.

Aspects of this hierarchical scheme are still in use. As biological science matured, however, several problems with Linnaeus' original proposal emerged.

How Many Kingdoms Are There? Linnaeus proposed that species could be organized into two kingdoms: plants and animals. According to Linnaeus, organisms that do not move and that produce their own food are plants; organisms that move and acquire food by eating other organisms are animals.

Scientific names and terms are often based on Latin or Greek word roots that are descriptive. For example, *Homo sapiens* is derived from the Latin *homo* for "man" and *sapiens* for "wise" or "knowing." The yeast that bakers use to produce bread and that brewers use to brew beer is called *Saccharomyces cerevisiae*. The Greek root *saccharo* means "sugar," and *myces* refers to a fungus. *Saccharomyces* is aptly named "sugar fungus" because yeast is a fungus and because the domesticated strains of yeast used in commercial baking and brewing are often fed sugar. The specific name of this organism, *cerevisiae*, is Latin for *beer*. Loosely translated, then, the scientific name of brewer's yeast means "sugar fungus for beer."

Most biologists find it extremely helpful to memorize some of the common Latin and Greek roots. To aid you in this process, new terms in this text are often accompanied by a reference to their Latin or Greek word roots in parentheses.

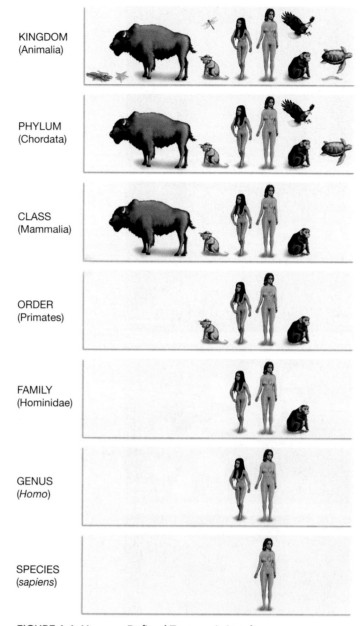

KINGDOM
(Animalia)

PHYLUM
(Chordata)

CLASS
(Mammalia)

ORDER
(Primates)

FAMILY
(Hominidae)

GENUS
(*Homo*)

SPECIES
(*sapiens*)

FIGURE 1.4 Linnaeus Defined Taxonomic Levels
In the Linnaean system, each animal species is placed in a taxonomic hierarchy with seven levels. Lower levels are nested within higher levels. Linnaeus proposed that these levels reflected the natural order of organisms.

Not all organisms fall neatly into these categories, however. Molds, mushrooms, and other fungi survive by absorbing nutrients from dead or living plants and animals. Even though they do not make their own food, they were placed in the kingdom Plantae because they do not move. The tiny, single-celled organisms called bacteria also presented problems. Some bacteria can move, and many can make their own food. Initially they, too, were thought to be plants. Eventually, though, it became clear that the two-kingdom system was simply inadequate.

Further, the development of the theory of evolution suggested a new goal for taxonomy. As evidence for evolution mounted, biologists concentrated on understanding how classification sys-

tems such as the one invented by Linnaeus could be modified to reflect the genealogical relationships among organisms—not just how similar they are in appearance or other broad characteristics. The goal of taxonomy became an attempt to reflect **phylogeny** (meaning "tribe-source")—the true historical relationships among types of organisms.

Linnaeus proposed the two-kingdom system because he thought it reflected a fundamental pattern in the natural world. His two-kingdom system was a hypothesis—a proposed explanation for an observed phenomenon. In this case, the pheno-menon that Linnaeus and other biologists were trying to explain was the relationship among the diverse forms of life on Earth.

For example, Linnaeus proposed that the fundamental division in organisms was between plants and animals. But when advances in microscopy allowed biologists to study the contents of individual cells in detail, a different fundamental division emerged and his hypothesis was rejected. In plants, animals, and the organisms that taxonomists call protists, cells contain a prominent component called a nucleus (**Figure 1.5a**). But in bacteria, cells lack this kernel-like structure (**Figure 1.5b**). Organisms with a nucleus are called **eukaryotes** ("true-kernel"); organisms without a nucleus are called **prokaryotes** ("before-kernel"). The vast majority of prokaryotes are unicellular ("one-celled"), many eukaryotes; are multicellular ("many-celled").

(a) Eukaryotic cells have a membrane-bound nucleus.

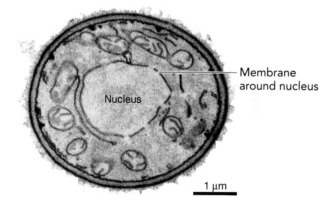

Membrane around nucleus

Nucleus

1 µm

(b) Prokaryotic cells do *not* have a membrane-bound nucleus.

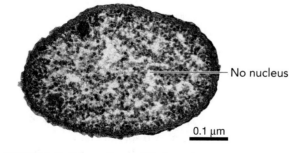

No nucleus

0.1 µm

FIGURE 1.5 Eukaryotes and Prokaryotes
Cross sections of **(a)** a eukaryotic and **(b)** a prokaryotic cell.
EXERCISE Study the scale bars; then draw two ovals that accurately represent the relative sizes of a eukaryotic cell and a prokaryotic cell.

When data began to conflict with Linnaeus' original scheme, biologists proposed alternative hypotheses. In the late 1960s one researcher suggested that a system of five kingdoms best reflects the patterns observed in nature. This five-kingdom system is depicted in **Figure 1.6**. Although the scheme has been widely used, it represents just one proposal out of many. Other biologists proposed that organisms are organized into three, four, six, or eight kingdoms. But it was still not clear which of these schemes, if any, accurately described the phylogeny of organisms.

About the time that the five-kingdom proposal was published, however, Carl Woese (pronounced "woes") and colleagues began working on the problem from a radically different angle. Instead of assigning organisms to kingdoms on the basis of characteristics such as the presence of a nucleus or the ability to move or to manufacture food, these researchers attempted to understand the relationships among organisms by analyzing their chemical components.

Using Molecules to Understand the Tree of Life

Woese and his co-workers had an explicit goal: to estimate where major branches occurred on the tree of life by analyzing the molecular components of cells. To accomplish this goal they needed to study a molecule that is found in all organisms. The molecule they selected is called small subunit rRNA. It is an essential part of the machinery that all cells use to grow and reproduce.

Although rRNA is a large and complex molecule, its underlying structure is simple. The rRNA molecule is made up of sequences of four smaller chemical components called ribonucleotides. Ribonucleotides are symbolized by the letters A, U, C, and G. In rRNA, ribonucleotides are connected to one another linearly, like boxcars of a freight train (**Figure 1.7**).

Why might rRNA be useful for understanding the relationships among organisms? The answer is that the ribonucleotide sequence in rRNA is a trait, similar to the height of wheat stalks or the size of flowering stalks of broccoli, that can change during the course of evolution. Although rRNA performs the same function in all organisms, the sequence of ribonucleotide building blocks in this molecule is not identical among species. In land plants, for example, the molecule might start with the sequence A-U-A-U-C-G-A-G. In green algae, which are closely related to land plants, the same section of the molecule might contain A-U-A-U-*G*-G-A-G. But in brown algae, which are not closely related to green algae or to land plants, the same part of the molecule might consist of A-*A*-A-U-*G*-G-A-G.

The research program that Woese and co-workers pursued was based on a simple premise: If the theory of evolution is correct, then rRNA sequences should be very similar in closely related organisms but less similar in organisms that are less closely related. The rRNA sequences of two species that diverged from each other long ago should be quite different, while rRNA sequences from two species that diverged from each other more recently should be much more alike.

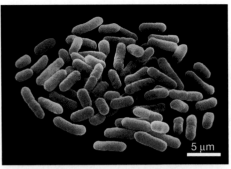
KINGDOM MONERA (includes all prokaryotes)

5 μm

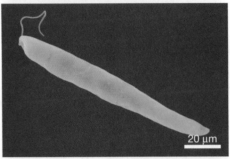
KINGDOM PROTISTA (includes several groups of unicellular eukaryotes)

20 μm

KINGDOM PLANTAE

3 cm

KINGDOM FUNGI

1 cm

KINGDOM ANIMALIA

1 mm

FIGURE 1.6 The Five-Kingdom Scheme
For decades, most biologists accepted the hypothesis that organisms naturally fall into the five kingdoms illustrated here. QUESTION How many times bigger is a fruit fly than one of the prokaryotic cells pictured here?

In rRNA four types of ribonucleotides (A, U, C, and G) are arranged in a linear sequence. The complete molecule contains about 2000 ribonucleotides; just eight are drawn here.

The sequence of ribonucleotides may vary among species. If the above sequence is observed in land plants, the sequence below might be found at the same location in the rRNA molecule of green algae.

FIGURE 1.7 RNA Molecules Are Made Up of Smaller Molecules
The four smaller molecules that make up an rRNA molecule are symbolized A, U, C, and G. The sequence of A, U, C, and G subunits in rRNA varies among species. QUESTION Suppose that in the same portion of rRNA, molds and other fungi have the sequence A-U-A-U-G-G-A-C. According to these data, are fungi more closely related to green algae or to land plants? Explain your logic.

To put this insight to work, the researchers determined the sequence of ribonucleotides in the rRNA of a wide array of species. Then they considered what the similarities and differences in the sequences implied about relationships among the species. The goal was to produce a diagram that described the phylogeny of the organisms in the study. A diagram that depicts evolutionary history in this way is called a **phylogenetic tree.** Just as a family tree shows relationships among individuals, a phylogenetic tree shows relationships among species. On a phylogenetic tree, branches that are close to one another represent species that are closely related; branches that are farther apart represent species that are more distantly related.

The rRNA Tree To construct a phylogenetic tree, researchers use a computer to find the arrangement of branches that is most consistent with the similarities and differences observed in the data—in this case, in the sequences of ribonucleotides observed in rRNA. The tree produced by comparing these sequences is shown in **Figure 1.8.** Because this tree includes species from many different kingdoms and phyla, it is often called the universal tree, or the tree of life.

The tree of life implied by rRNA data astonished biologists. According to data from this molecule:

- The fundamental division in organisms is not between plants and animals or even between prokaryotes and eukaryotes. Rather, *three* major groups occur: (1) the Bacteria; (2) another group of prokaryotic, single-celled organisms called

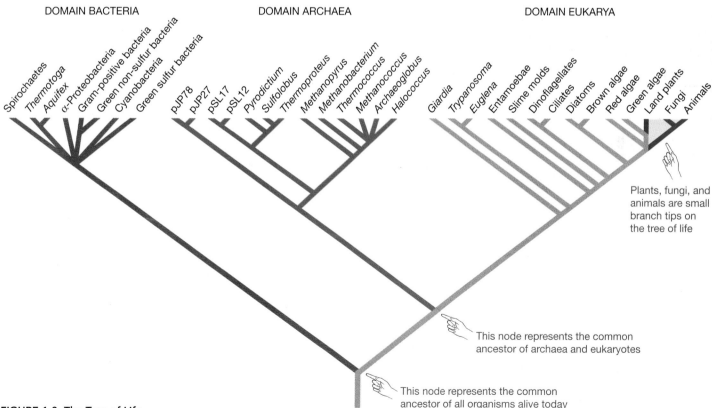

FIGURE 1.8 The Tree of Life
"Universal tree" estimated from rRNA sequence data. The three domains of life revealed by the analysis are labeled. Common names are given for most lineages in the domains Bacteria and Eukarya. Genus names are given for members of the domain Archaea, because most of these organisms have no common names. Archaean species with labels such as "pSL17" have not yet been given scientific names. EXERCISE Circle the branches and tips that represent prokaryotes.

the Archaea; and (3) the eukaryotes. To accommodate this new perspective on the diversity of organisms, Woese created a new taxonomic level called the **domain**. As Figure 1.8 indicates, the three domains of life are now called the Bacteria, Archaea, and Eukarya.

- Some of the kingdoms that had been defined earlier do not reflect how evolution actually occurred. For example, recall that Linnaeus grouped the multicellular eukaryotes known as fungi with plants. But the rRNA data indicate that fungi are much more closely related to animals than they are to plants.

- Bacteria and Archaea are much more diverse than anyone had imagined. If the differences among animals, fungi, and plants warrant placing them in separate kingdoms, then dozens of kingdoms exist among the prokaryotes.

The Tree of Life Is a Work in Progress   The rRNA tree was a hypothesis that inspired a flurry of research. Biologists in laboratories all over the world tested the conclusions by determining the sequences of other molecules found in cells and by reanalyzing older data in light of the new findings. In general, these studies have confirmed the major features of the tree in Figure 1.8. The discovery of the Archaea and the placement of lineages such as the fungi qualify as exciting breakthroughs in our understanding of life's diversity.

Work on the tree of life continues at a furious pace, however, and the location of certain branches on the tree is hotly debated. As databases expand and as techniques for analyzing data improve, the shape of the tree of life presented in Figure 1.8 will undoubtedly change.

✓CHECK YOUR UNDERSTANDING

A phylogenetic tree is analogous to a family tree. A phylogenetic tree shows the evolutionary relationships among species, just as a family tree shows the genetic relationships among individuals.

To infer where species belong on a phylogenetic tree, biologists examine the characteristics of the species involved. Closely related species should have similar characteristics, while less closely related species should be less similar. A characteristic that has been used extensively in phylogenetic analyses is the sequence of subunits in a molecule called rRNA, which is found in all species. You should be able to examine the sequences given below and draw a phylogenetic tree showing the relationships among Species A, B, and C that these data imply.

Species A: A   A   C   T   A   G   C   G   C   G   A   T

Species B: A   A   C   T   A   G   C   G   C   C   A   T

Species C: T   T   C   T   A   G   C   G   G   T   A   T

1.4   Doing Biology

This chapter has introduced some of the great ideas in biology. The development of the cell theory and the theory of evolution by natural selection provided cornerstones when the science was young; the tree of life is a relatively recent insight that has revolutionized the way researchers understand the diversity of life on Earth.

These theories are considered great because they explain fundamental aspects of nature, and because they have consistently been shown to be correct. They are considered correct because they have withstood extensive testing. How do biologists test ideas about the way the natural world works? The answer is that they test the predictions made by alternative hypotheses, often by setting up carefully designed experiments. To illustrate how this approach works, let's consider two questions currently being addressed by researchers.

Why Do Giraffes Have Long Necks? An Introduction to Hypothesis Testing

If you were asked why giraffes have long necks, you might say that long necks enable giraffes to reach food that is unavailable to other mammals. This hypothesis is expressed in African folktales and has traditionally been accepted by many biologists. The food competition hypothesis is so plausible, in fact, that for decades no one thought to test it. Recently, however, Robert Simmons and Lue Scheepers assembled data suggesting that the food competition hypothesis is only part of the story. Their analysis supports an alternative hypothesis—that long necks allow giraffes to use their heads as effective weapons for battering their opponents.

How did biologists test the food competition hypothesis? What data support their alternative explanation? Before we answer these questions, it's important to recognize that hypothesis testing is a two-step process. The first step is to state the hypothesis as precisely as possible and list the predictions it makes. The second step is to design an observational or experimental study that is capable of testing those predictions. If the predictions are accurate, then the hypothesis is supported. If the predictions are not met, then researchers do further tests, modify the original hypothesis, or search for alternative explanations.

The Food Competition Hypothesis: Predictions and Tests

Stated precisely, the food competition hypothesis claims that giraffes compete for food with other species of mammals. When food is scarce, as it is during the dry season, giraffes with longer necks can reach food that is unavailable to other species and to giraffes with shorter necks. As a result, the longest-necked individuals in a giraffe population survive better and produce more young than do shorter-necked individuals, and average neck length of the population increases with each generation. To use the terms introduced earlier in this chapter, long necks are adaptations that increase the fitness of individual giraffes during competition for food. This type of natural selection has gone on so long that the population has become extremely long necked.

(a) Most feeding is done below neck height.

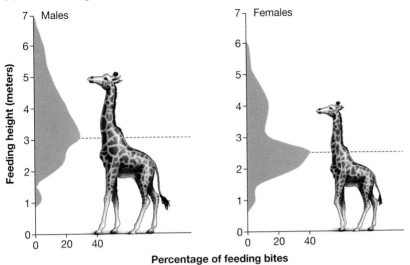

(b) Typical feeding posture in giraffes

FIGURE 1.9 Giraffes Do Not Usually Extend Their Necks to Feed
(a) The proportion of bites that male and female giraffes of average size take at different heights. **(b)** Although it is common to see photos of giraffes straining to reach leaves high in trees, giraffes usually feed at the heights shown here.

The food competition hypothesis makes several explicit *predictions*. A prediction is something that can be measured and that must be correct if a hypothesis is valid. For example, the food competition hypothesis predicts that (1) neck length is variable among giraffes; (2) neck length in giraffes is heritable; and (3) giraffes feed high in trees, especially during the dry season, when food is scarce and the threat of starvation is high.

The first prediction is clearly correct. Studies in zoos and natural populations confirm that neck length is variable among individuals.

The researchers were unable to test the second prediction, however, because they studied giraffes in a natural population and were unable to do breeding experiments. As a result, they simply had to accept this prediction as an assumption. In general, though, biologists prefer to test every assumption behind a hypothesis.

What about the prediction regarding feeding high in trees? According to Simmons and Scheepers, this is where the food competition hypothesis breaks down. Consider, for example, data collected by a different research team about the amount of time that giraffes spend feeding in vegetation of different heights. **Figure 1.9a** shows that in a population from Kenya, both male and female giraffes spend most of their feeding time eating vegetation that averages just 60 percent of their full height. Studies on other populations of giraffes, during both the wet and dry seasons, are consistent with these data. Giraffes usually feed with their necks bent (**Figure 1.9b**).

These data cast doubt on the food competition hypothesis, because one of its predictions does not appear to hold. Biologists have not abandoned this hypothesis completely, though, because feeding high in trees may be particularly valuable during extreme droughts, when a giraffe's ability to reach leaves far above the ground could mean the difference between life and death. Still,

Simmons and Scheepers have offered an alternative explanation for why giraffes have long necks. The new hypothesis is based on the mating system and social behavior of giraffes.

The Sexual Competition Hypothesis: Predictions and Tests
Giraffes have an unusual social system. Breeding occurs year round rather than seasonally. To determine when females are coming into estrus (or "heat") and are thus receptive to mating, the males nuzzle the rumps of females. In response, the females urinate into the males' mouths. The males then tip their heads back and pull their lips to and fro, as if tasting the liquid. Biologists who have witnessed this behavior have proposed that the males taste the females' urine to detect whether estrus has begun.

Once a female giraffe enters estrus, males fight among themselves for the opportunity to mate. Combat is spectacular. The bulls stand next to one another, swing their necks, and strike thunderous blows with their heads. Researchers have seen males knocked unconscious for 20 minutes after being hit and have cataloged numerous instances in which the loser died. Giraffes are the only animals that fight in this way.

These observations inspired a new explanation for why giraffes have long necks. The sexual competition hypothesis is based on the idea that longer-necked giraffes are able to strike harder blows during combat than can shorter-necked giraffes. In engineering terms, longer necks provide a longer moment arm. A long moment arm increases the force of the impact. (Think about the type of sledge hammer you'd use to bash down a concrete wall—one with a short handle or one with a long handle?) Thus, longer-necked males should win more fights and, as a result, father more offspring than do shorter-necked males. If neck length in giraffes is inherited, then the average neck length in the population should increase over time. Under the sexual

WEB TUTORIAL 1.2 Introduction to Experimental Design

competition hypothesis, long necks are adaptations that increase the fitness of males during competition for females.

Although several studies have shown that long-necked males are more successful in fighting and that the winners of fights gain access to estrous females, the question of why giraffes have long necks is not closed. With the data collected to date, most biologists would probably concede that the food competition hypothesis needs further testing and refinement and that the sexual selection hypothesis appears promising. It could also be true that both hypotheses are correct. But clearly, more work needs to be done.

In many cases in biological science, "more work" involves experimentation. Experimenting on giraffes is difficult. But in the case study considered next, biologists were able to test an interesting hypothesis experimentally.

Why Are Chili Peppers Hot? An Introduction to Experimental Design

Experiments are a powerful scientific tool because they allow researchers to test the effect of a single, well-defined factor on a particular phenomenon. Experiments that test the effect of neck length on food and sexual competition in giraffes have yet to be done. Instead, as an example of how experiments are designed, let's consider a different phenomenon: Why do chili peppers taste so spicy?

The jalapeño, anaheim, and cayenne peppers used in cooking descended from a wild shrub that is native to the deserts of the American Southwest. As **Figure 1.10a** shows, wild chilies produce fleshy fruits with seeds inside, just like their domesticated descendants. In both wild chilies and the cultivated varieties, the "heat" or pungent flavor of the fruit and seeds is due to a molecule called capsaicin. In humans and other mammals, capsaicin binds to heat-sensitive cells in the tongue and mouth.

In response to this binding, signals are sent to the brain that produce the sensation of burning. Similar signals would be transmitted if you drank boiling water. Asking why chilies are hot, then, is the same as asking why chilies contain capsaicin.

Josh Tewksbury and Gary Nabhan proposed that the presence of capsaicin is an adaptation that protects chili fruits from being eaten by animals that destroy the seeds inside. To understand this hypothesis, it's important to realize that the seeds inside a fruit have one of two fates when the fruit is eaten. If the seeds are destroyed in the animal's mouth or digestive system, then they never germinate (sprout). In this case, "seed predation" has occurred. But if seeds can travel undamaged through the animal, then they are eventually "planted" in a new location along with a valuable supply of fertilizer. In this case, seeds are dispersed. Here's the key idea: Natural selection should favor fruits that taste bad to animal species that act as seed predators. But these same fruits should not deter species that act as seed dispersers. This proposal is called the *directed dispersal hypothesis*.

Does capsaicin deter seed predators, as the directed dispersal hypothesis predicts? To answer this question, the researchers captured some cactus mice (**Figure 1.10b**) and birds called curve-billed thrashers (**Figure 1.10c**). These species are among the most important fruit- and seed-eating animals in habitats where chilies grow. Based on earlier observations, the biologists predicted that cactus mice destroy chili seeds but that curve-billed thrashers disperse them effectively.

To test the directed dispersal hypothesis, the biologists offered both cactus mice and curve-billed thrashers three kinds of fruit: hackberries, chilies from a strain of plant that can't synthesize capsaicin, and pungent chilies that have lots of cap-

(a) Chilies produce fruits that contain seeds.

(b) Cactus mice are seed eaters.

(c) Curve-billed thrashers eat chili fruits.

FIGURE 1.10 Chilies ... and Chili Eaters?
(a) Chili fruits are hot because they contain a molecule called capsaicin. **(b)** Cactus mice and **(c)** curve-billed thrashers are common in habitats where chilies grow.

saicin. The non-pungent chilies are about the same size and color as normal chilies and have similar nutritional value. The hackberries don't look anything like chilies, however, and contain no capsaicin. The three fruits were present in equal amounts. For each animal tested, the researchers recorded the percentage of hackberry, non-pungent chili, and pungent chili that was eaten during a specific time interval. Then they calculated the average amount of each fruit that was eaten by five test individuals from each species.

The directed dispersal hypothesis predicts that seed dispersers will eat the pungent chilies readily but that seed predators won't. Recall that a prediction specifies what we should observe if a hypothesis is correct. Good scientific hypotheses make testable predictions—predictions that can be supported or rejected by collecting and analyzing data. If the directed dispersal hypothesis is wrong, however, then there shouldn't be any difference in what various animals eat. This latter possibility is called a **null hypothesis**. A null hypothesis specifies what we should observe when the hypothesis being tested doesn't hold. These predictions are listed in **Figure 1.11.**

Do the predictions of the directed dispersal hypothesis hold? To answer this question, look at the results plotted in Figure 1.11. View the data for each type of fruit to see if different amounts were eaten by different types of predators. Then fill in the following table:

Type of fruit	Did the two predators eat about the same amount or very different amounts of this fruit?	If the amount eaten was very different, comment on the nature of the difference.
Hackberry		
Non-pungent chilies (no capsaicin)		
Pungent chilies (lots of capsaicin)		

Based on your analysis of the data, decide whether the results support the directed dispersal hypothesis or the null hypothesis. Use the conclusion stated in Figure 1.11 to check your answer.

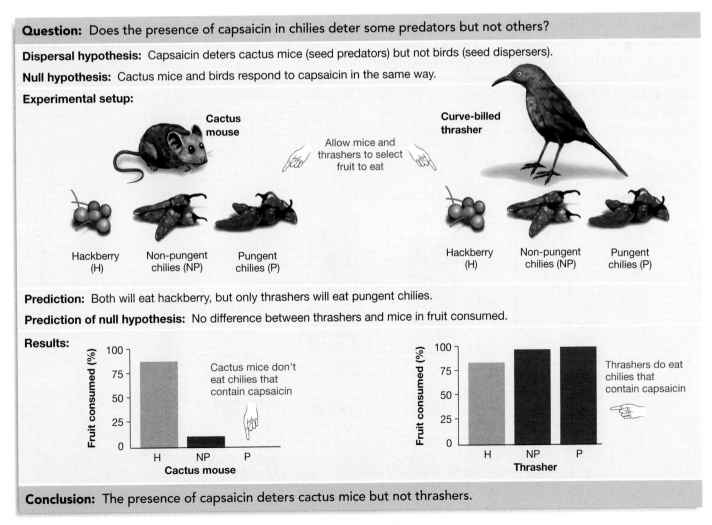

Question: Does the presence of capsaicin in chilies deter some predators but not others?

Dispersal hypothesis: Capsaicin deters cactus mice (seed predators) but not birds (seed dispersers).

Null hypothesis: Cactus mice and birds respond to capsaicin in the same way.

Experimental setup:

Cactus mouse

Curve-billed thrasher

Allow mice and thrashers to select fruit to eat

Hackberry (H) Non-pungent chilies (NP) Pungent chilies (P)

Hackberry (H) Non-pungent chilies (NP) Pungent chilies (P)

Prediction: Both will eat hackberry, but only thrashers will eat pungent chilies.

Prediction of null hypothesis: No difference between thrashers and mice in fruit consumed.

Results:

Cactus mice don't eat chilies that contain capsaicin

Cactus mouse

Thrashers do eat chilies that contain capsaicin

Thrasher

Conclusion: The presence of capsaicin deters cactus mice but not thrashers.

FIGURE 1.11 An Experimental Test: Does Capsaicin Deter Some Fruit Eaters?

In relation to designing effective experiments, this study illustrates several important points:

- *It is critical to include control groups.* A **control** checks for factors, other than the one being tested, that might influence the experiment's outcome. For example, if hackberries had not been included as a control, it would have been possible to claim that the cactus mice in the experiment didn't eat pungent chilies simply because they weren't hungry. But the not-hungry hypothesis can be rejected because all of the animals ate hackberries.

- *The experimental conditions must be carefully controlled.* The investigators used the same feeding choice setup, the same time interval, and the same definitions of predator response in each test. Controlling all of the variables except one—the types of fruits presented—is crucial because it eliminates alternative explanations for the results. For example, what types of problems could arise if the cactus mice were given less time to eat than the thrashers, or if the test animals were always presented with hackberries first and pungent chilies last?

- *Repeating the test is essential.* It is almost universally true that larger sample sizes in experiments are better. For example, suppose that the experimenters had used just one cactus mouse instead of five, and that this mouse was unlike other cactus mice because it ate almost anything. If so, the resulting data would be badly distorted. By testing many individuals, the amount of distortion or "noise" in the data caused by unusual individuals or circumstances is reduced.

To test the assumption that cactus mice are seed predators and curve-billed thrashers are seed dispersers, the researchers did a follow-up experiment. They fed fruits of the non-pungent chili to each type of predator. When the seeds had passed through the animals' digestive systems and were excreted, the researchers collected and planted the seeds—along with 14 uneaten seeds. Planting uneaten seeds served as a control treatment, because it tested the hypothesis that the seeds were viable and would germinate if they were not eaten. **Figure 1.12** shows the percentage of seeds that germinated. The data indicate that seeds pass through curve-billed thrashers unharmed but are destroyed when eaten by cactus mice.

Based on the outcomes of these two experiments, the researchers concluded that curve-billed thrashers are efficient seed dispersers and are not deterred by capsaicin. The cactus mice, in contrast, refuse to eat chilies. If they ate chilies, the mice would kill the seeds. These are exactly the results predicted by the directed dispersal hypothesis. The biologists concluded that the presence of capsaicin in chilies is an adaptation that keeps their seeds from being destroyed by mice. In habitats that contain cactus mice, the production of capsaicin increases the fitness of individual chili plants.

These experiments are a taste of things to come. In this text you will encounter hypotheses and experiments on questions ranging from how water gets to the top of 100-meter-tall sequoia trees to why the bacterium that causes tuberculosis has become resistant to antibiotics. A commitment to tough-minded hypothesis testing and sound experimental design is a hallmark of biological science. Understanding their value is an important first step in becoming a biologist.

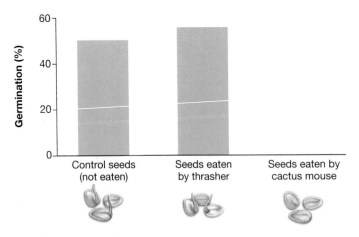

FIGURE 1.12 Mice Destroy Seeds, but Thrashers Do Not
Histograms showing the percentage of seeds that germinated after being planted directly into the ground versus being planted after passing through the digestive system of a curve-billed thrasher or cactus mouse. QUESTION Seven thrashers and five cactus mice were used in this experiment. Why wan't just one individual of each species used?

ESSAY Where Do Humans Fit on the Tree of Life?

Given the vast diversity of organisms that make up the tree of life, where do humans fit in? Most of the major branches on the tree diagrammed in Figure 1.8 represent unicellular organisms. Familiar multicellular species such as grasses and mushrooms are part of the twigs labeled "Plants" and "Fungi," respectively. At the tree's tip, the label "Animals" identifies the kingdom that appeared most recently among the eukaryotes, about 800 million years ago. Humans are found on this branch, along with millions of other species ranging from sponges and corals to insects and fish.

Our species is a tiny new twig on an enormous and ancient tree of life.

Just how recently did our species appear? In Chapter 2 we examine a technique called radiometric dating that allows geologists to estimate the age of rocks containing bones, teeth, shells, and other traces of organisms that lived in the past. According to this technique, the first traces left by our species appear about 100,000 years ago. To clarify just how recent this is in the sweep of Earth history, consider the calendar in **Figure 1.13**. The 12 months shown are scaled to represent the 4.6 billion years of Earth history. At this scale, 7 seconds make up a millennium; each hour denotes a span of 525,000 years; and each day represents an interval of 12.6 million years. Note that the first unicellular organisms appear in late March and the first multicellular life in early October. Hominids (members of the family Hominidae) walk upright for the first time on mid-afternoon of New Year's Eve. *Homo sapiens* appears about 15 minutes before the stroke of midnight.

The message of these analyses is clear: Our species is a tiny new twig on an enormous and ancient tree of life.

Given that our species evolved so recently, some other interesting questions arise. For example, a typical species of mammal lasts about 2.5 million years in the fossil record. In your opinion, how likely is it that our species will continue to leave fossil evidence for another 2.4 million years? What factors might contribute to the long-term survival of *Homo sapiens*? What factors might contribute to our demise?

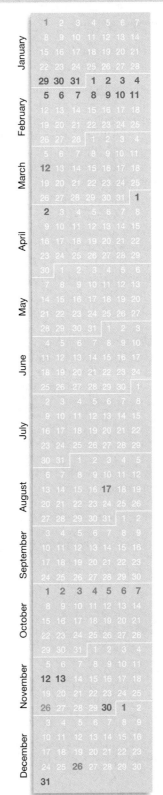

1: Earth forms

29–11: Oldest known rocks

12: Oldest chemical evidence of life

1–2: Oldest fossil cells

17: First eukaryotes

1–7: First multicellular organisms (algae)

12–13: First animals with shells and limbs

26: First animals with vertebrae

30: First land plants

1: First land animals

31: *Homo sapiens* appears one hour before midnight. Humans set foot on Moon $\frac{1}{4}$ second before midnight.

26: Extinction of dinosaurs

FIGURE 1.13 Earth's History as a Calendar

QUESTION If the length of Earth's history (4.6 billion years) were to correspond to the length of a football field (100 yards), where on the football field does the first fossil evidence of humans (100,000 years ago) appear?

1 day = 12.6 million years
1 second = 143 years

CHAPTER REVIEW

Summary of Key Concepts

Biologists have been discovering traits that unify the spectacular diversity of living organisms for over two hundred years.

▦ **Biological science was founded with the development of (1) the cell theory, which proposes that all organisms are made of cells and that all cells come from preexisting cells, and (2) the theory of evolution by natural selection, which maintains that the characteristics of species change through time—primarily because individuals with certain heritable traits produce more offspring than do individuals without those traits.**

The cell theory is an important unifying principle in biology, because it identified the fundamental structural unit common to all life. The theory of evolution by natural selection is another key unifying principle, because it states that all organisms are related by common ancestry. It also offered a robust explanation for why species change through time and why they are so well adapted to their habitats.

Web Tutorial 1.1   Artificial Selection

▦ **A phylogenetic tree is a graphical representation of the evolutionary relationships among species. Phylogenies can be estimated by analyzing similarities and differences in traits. Species that share many traits are closely related and are placed close to each other on the tree of life.**

The cell theory and the theory of evolution predict that all organisms are part of a genealogy of species, and that all species trace their ancestry back to a single common ancestor. To reconstruct this phylogeny, biologists have analyzed the sequence of components in a molecule called rRNA, which is found in all cells. A tree of life, based on similarities and differences in these sequences, has recently been constructed. According to the information contained in rRNA, the tree of life has three major lineages: the Bacteria, Archaea, and Eukarya. Analyses of other molecules and traits have largely supported the conclusions made from rRNA data.

▦ **Biologists ask questions, generate hypotheses to answer them, and design experiments that test the predictions made by competing hypotheses.**

Another unifying theme in biology is a commitment to hypothesis testing and to sound experimental design. Analyses of neck length in giraffes and the capsaicin found in chilies are case studies in the value of testing alternative hypotheses and conducting experiments. Biology is a hypothesis-driven, experimental science.

Web Tutorial 1.2   Introduction to Experimental Design

Questions

Content Review

1. Anton van Leeuwenhoek made an important contribution to the development of the cell theory. How?
   a. He articulated the pattern component of the theory—that all organisms are made of cells.
   b. He articulated the process component of the theory—that all cells come from preexisting cells.
   c. He invented the first microscope and saw the first cell.
   d. He invented more powerful microscopes and was the first to describe the diversity of cells.

2. Suppose that a proponent of the spontaneous generation hypothesis claimed that cells would appear in Pasteur's swan-necked flask eventually. According to this view, Pasteur did not allow enough time to pass before concluding that life does not originate spontaneously. Which of the following is the best response?
   a. The spontaneous generation proponent is correct: Spontaneous generation would probably happen eventually.
   b. Both the all-cells-from-cells hypothesis and the spontaneous generation hypothesis could be correct.
   c. If spontaneous generation happens only rarely, it is not important.
   d. If spontaneous generation did not occur after weeks or months, it is not reasonable to claim that it would occur later.

3. What does the term *evolution* mean?
   a. The strongest individuals produce the most offspring.
   b. The characteristics of an individual change through the course of its life, in response to natural selection.
   c. The characteristics of populations change through time.
   d. The characteristics of species become more complex over time.

4. What does it mean to say that a characteristic of an organism is heritable?
   a. The characteristic evolves.
   b. The characteristic can be passed on to offspring.
   c. The characteristic is advantageous to the organism.
   d. The characteristic does not vary in the population.

5. In biology, what does the term *fitness* mean?
   a. how well trained and muscular an individual is, relative to others in the same population
   b. how slim an individual is, relative to others in the same population
   c. how long a particular individual lives
   d. the ability to survive and reproduce

6. Could *both* the food competition hypothesis and the sexual selection hypothesis explain why giraffes have long necks? Why or why not?
   a. No. In science, only one hypothesis can be correct.
   b. No. Observations have shown that the food competition hypothesis cannot be correct.
   c. Yes. Long necks could be advantageous for more than one reason.
   d. Yes. All giraffes have been shown to feed at the highest possible height and fight for mates.

Conceptual Review

1. The Greek roots of the term *taxonomy* can be translated as "arranging rules." Explain why these roots were an appropriate choice for this term.

2. It was once thought that the deepest split among life-forms was between two groups: prokaryotes and eukaryotes. Draw and label a phylogenetic tree that represents this hypothesis. Then draw and label a phylogenetic tree that shows the actual relationships among the three domains of organisms.

3. Why was it important for Linnaeus to establish the rule that only one type of organism can have a particular genus and species name?

4. What does it mean to say that an organism is adapted to a particular habitat?

5. Compare and contrast natural selection with the process that led to the divergence of a wild mustard plant into cabbage, broccoli, and Brussels sprouts.

6. The following two statements explain the logic behind the use of molecular sequence data to estimate evolutionary relationships:

   "If the theory of evolution is true, then rRNA sequences should be very similar in closely related organisms but less similar in organisms that are less closely related."

   "On a phylogenetic tree, branches that are close to one another represent species that are closely related; branches that are farther apart represent species that are more distantly related."

   Is the logic of these statements sound? Why or why not?

Group Discussion Problems

1. A scientific theory is a set of propositions that defines and explains some aspect of the world. This definition contrasts sharply with the everyday usage of the word *theory*, which often carries meanings such as "speculation" or "guess." Explain the difference between the two definitions, using the cell theory and the theory of evolution by natural selection as examples.

2. Turn back to the tree of life shown in Figure 1.8. Note that Bacteria and Archaea are prokaryotes, while Eukarya are eukaryotes. On the simplified tree below, draw an arrow that points to the branch where the structure called the nucleus originated. Explain your reasoning.

3. The proponents of the cell theory could not "prove" that it was correct in the sense of providing incontrovertible evidence that all organisms are made up of cells. They could state only that all organisms examined to date were made of cells. Why was it reasonable for them to conclude that the theory was valid?

4. How do the tree of life and the taxonomic categories created by Linnaeus (kingdom, phylum, class, order, family, genus, and species) relate to one another?

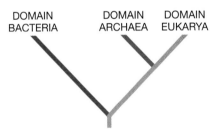

DOMAIN BACTERIA   DOMAIN ARCHAEA   DOMAIN EUKARYA

Answers to Multiple-Choice Questions   1. d; 2. d; 3. c; 4. b; 5. d; 6. c

The Origin and Early Evolution of Life

When life began, Earth's surface and oceans were roiled by volcanoes; the atmosphere lacked oxygen. This volcanic eruption in Hawaii can only hint at conditions that prevailed when the first organisms evolved.

2

The Atoms and Molecules of Ancient Earth

KEY CONCEPTS

- Molecules form when atoms bond to each other. Chemical bonds are based on electron sharing. The degree of electron sharing varies from nonpolar covalent bonds, to polar covalent bonds, to ionic bonds.

- Chemical reactions tend to be spontaneous if they lead to lower potential energy and higher entropy (more disorder). An input of energy is required for nonspontaneous reactions to occur.

- During chemical evolution, (1) energy in sunlight was converted to chemical energy and (2) carbon atoms were reduced.

- Water is a small, highly polar molecule. As a result, it is an extremely efficient solvent and has a high capacity for absorbing energy.

This chapter explores the earliest events in Earth's history, from the formation of the planet through the development of the first oceans and atmosphere.

Chapter 1 introduced experiments on the hypothesis of spontaneous generation, which tested the idea that life arises from nonliving materials. This work helped build a consensus that spontaneous generation does not occur. But for life to exist now, spontaneous generation must have occurred at least once, early in Earth's history. How did life begin? This simple query has been called "the mother of all questions."

In this chapter we examine a hypothesis, called chemical evolution, that attempts to answer this question. In its simplest sense, the term *evolution* means change through time. **Chemical evolution** is the proposition that early in Earth's history, simple chemical compounds in the atmosphere and ocean combined to form larger, more complex substances. As a result, the chemistry of the oceans and atmosphere changed over time. The hypothesis maintains that product chemicals then reacted with one another to produce even more complicated compounds and that continued chemical evolution eventually led to the origin of life.

More specifically, the hypothesis is that one of these complex compounds was able to make a copy of itself, or *self-replicate*. As this molecule multiplied, the process of evolution by natural selection took over—there was a switch from chemical evolution to biological evolution. Eventually, a self-replicating molecule became surrounded by a membrane, and cellular life began.

At first glance, the chemical evolution hypothesis seems wildly implausible. But is it? What evidence do biologists have that chemical evolution occurred? Let's start with the first steps—the formation of Earth and the chemical reactions that could have gotten chemical evolution started.

2.1 The Ancient Earth

Advocates of the chemical evolution hypothesis contend that when this process took place, Earth was a radically different place than it is today. To understand whether chemical evolution is plausible, then, it is important to understand what these claims about the nature of the early Earth are, and what evidence researchers use to support their argument. Let's take a closer look.

Studying the Formation of Planets

Astronomers and geologists pursue two strategies to study how Earth formed: They perform computer simulations to model conditions that led to the development of Earth, and they analyze how planets are being created elsewhere in the universe.

In biology and the other sciences, computer simulation is an important tool for analyzing processes that are difficult to study directly. Researchers write a computer program that specifies a set of starting conditions and the properties of the entities involved. The computer then records the outcome of interactions among these entities as the simulation proceeds.

For example, consider the results of simulations based on the condensation theory of planet formation. This theory maintains that our solar system originated when many billions of dust-sized particles accumulated around the forming Sun. To test this idea, researchers have created mathematical models that describe how small particles behave in orbit around the Sun. The properties of the particles are based on the types of atoms found in the solar system and their relative abundance. Then these equations are incorporated into computer simulations of planetary formation. In the initial stages of these simulations, dust and gas particles start colliding. Particles frequently stick together after the collisions, due to electrical and gravitational attraction. As the simulation continues to run, the collision-and-sticking events begin producing small bodies of rock called planetesimals. Collisions between these rocks eventually create larger bodies, called proto-planets, whose gravity attracts still more planetesimals. In this process of accumulation, called accretion, planet-sized bodies form.

Recently, these types of computer simulations have been augmented by direct studies of planet formation. Prior to 1992, astronomers had no firm evidence that planets existed anywhere outside our solar system. But numerous stars that appear to be surrounded by orbiting planets or planetary systems have since been identified. At least some of these planet-sized bodies are located in the middle of enormous clouds of orbiting dust and rock. These observations are consistent with the condensation theory, which predicts that particularly young planets should still be surrounded by dust and planetesimals.

Based on the success of these simulations and observational studies, the condensation theory is now widely accepted as the most plausible explanation for the formation of Earth. This theory also makes important predictions about the nature of the young Earth. Because immense heat would have been generated by the larger impacts during the "early accretion" phase of planet formation, most scientists conclude that Earth was molten initially. As Earth began to cool enough to form a crust of solid rock, two important events would have occurred. First, water would have rained out of the cooling atmosphere to form the first ocean. Second, volcanoes would have proliferated, as molten rock from below punctured the developing crust.

Several lines of evidence argue that heavy bombardment from space continued, however. For example, most astronomers believe that a particularly spectacular blow—a collision with a proto-planet the size of Mars—knocked Earth into its present tilt and blasted debris into space that formed the Moon. The extensive cratering on the Moon's surface suggests that Earth was also bombarded long after this giant impact.

When Did Chemical Evolution Take Place?

Geologists and astronomers estimate when these events occurred using a procedure called **radiometric** ("ray-measure") **dating**. To understand this technique, it is essential to recognize that all **atoms** share the same basic structure.

Atomic Components and Isotopes Extremely small particles called **electrons** orbit an atomic **nucleus** made up of larger particles called **protons** and **neutrons** (**Figure 2.1**). Protons have a positive electric charge, neutrons are electrically neutral, and

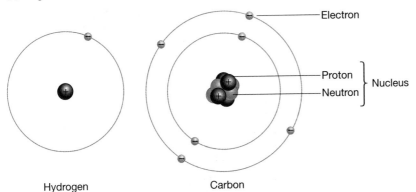

(a) Diagrams of atoms

Electron

Proton
Neutron } Nucleus

Hydrogen   Carbon

(b) Most of an atom's volume is empty space.

If an atom occupied the same volume as this stadium, the nucleus would be about the size of a pea

FIGURE 2.1 Parts of an Atom
(a) These crude diagrams of the hydrogen and carbon atoms show how the nucleus, made up of protons and neutrons, is surrounded by orbiting electrons. In reality, electrons do not orbit the nucleus in circles; their actual orbits are complex. **(b)** As the labels on this photograph show, the cartoons in part (a) are not to scale.

electrons have a negative electric charge. Opposite charges attract; like charges repel. When the number of protons and the number of electrons in an atom (or molecule) are the same, the charges balance and the atom is electrically neutral.

Each of the **elements** contains a characteristic number of protons. The number of neutrons present can vary, however. Forms of an element with different numbers of neutrons are known as **isotopes** ("equal-places"). For example, all atoms of the element uranium have 92 protons. But naturally occurring isotopes of uranium can have either 143 or 146 neutrons, giving them a total of 235 or 238 protons and neutrons, respectively. The sum of the protons and neutrons in an atom is called its **mass number**. Although the masses of protons, neutrons, and electrons can be measured in grams, the numbers involved are so small that chemists and physicists prefer to use a special unit called the **atomic mass unit (amu)**, or the dalton. The masses of protons and neutrons are virtually identical and are routinely rounded to 1 amu. A carbon atom that contains 6 protons and 6 neutrons has a mass of 12 amu and a mass number of 12. Uranium occurs in two isotopes with different mass numbers, which represent their masses in amu. They are symbolized ^{235}U and ^{238}U (or U-235 and U-238, respectively).

Radioactive Decay   One of the most striking observations about atomic structure is that certain numerical combinations of protons and neutrons make the nucleus that contains them inherently unstable. That is, these nuclei break down. Atoms of ^{235}U and ^{238}U, for example, are not stable (**Figure 2.2**). Like other **radioactive isotopes**, their nuclei tend to change spontaneously by emitting radiation or a particle. These changes are called **radioactive decay**. Most types of radioactive decay result

in the formation of a different isotope or element. The original isotope is called the parent, and the resulting isotope or element is called the daughter.

Radioactive decay has an important feature: The proportion of atoms that decay in a sample of an isotope over a given time period is constant. More specifically, the proportion of the sample that decays per unit of time does not change as a function of either the isotope's age or the quantity found in a particular sample. The rate at which decay occurs varies widely from isotope to isotope, however, and is usually reported in terms of a unit called a half-life. A **half-life** is the time it takes for half of the parent isotope to decay to a particular daughter isotope (**Figure 2.3**). The half-lives of many long-lived isotopes, such as ^{235}U and ^{238}U, have been measured. The half-life of ^{235}U decaying to lead-207 (^{207}Pb), for example, is 713 million years.

How Old Is the Earth?   To use radiometric dating to determine when a rock formed, geologists must estimate two quantities:

1. *The relative amounts of parent and daughter isotopes in a rock sample.* This can be measured directly, using instruments that identify the isotopes in the sample and record their abundance.

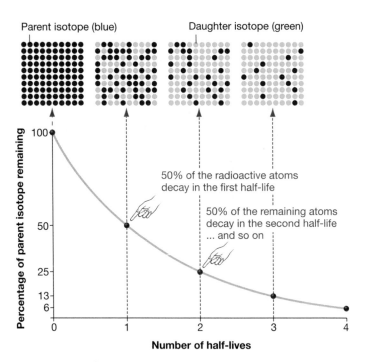

FIGURE 2.3 **Radioactive Decay**
A graph showing how a sample of radioactive atoms decays with time. When a sample forms, only the parent isotope exists. After the time represented by one half-life passes, the parent and daughter isotopes are present in equal proportions; after two half-lives, the ratio of parent to daughter isotopes is 1:3; after three half-lives, it is 1:7. **QUESTION** What percentages of the original parent isotope and daughter isotope exist after four half-lives? Five? Six? **EXERCISE** Suppose that a rock contains 3 percent parent isotope and 97 percent daughter isotope, and that the half-life of the parent isotope is 10 million years. How old do you estimate this rock to be?

FIGURE 2.2 **Some Isotopes Are Unstable**
A technician loads pellets of the radioactive isotope ^{235}U into a long rod. The isotope will be used as fuel to generate electricity in a nuclear reactor.

2. *The relative amounts of parent and daughter isotopes that existed when the sample formed.* This ratio is estimated from information about how the rock formed. For example, when molten rock containing uranium atoms cools, the uranium forms crystals that do not contain lead. Based on this observation, geologists assume that ^{235}U-containing rocks that formed during Earth's early accretion phase started out with crystals containing ^{235}U but no ^{207}Pb. Over time (that is, as half-lives passed), the crystals accumulated more and more ^{207}Pb until they reached their present quantities of ^{235}U and ^{207}Pb.

Once the ratio of parent to daughter isotopes has been determined in a sample, geologists can calculate the number of half-lives that have passed and multiply that number by the length of a half-life of the parent isotope to estimate the rock's age.

To determine the age of Earth, geologists would need to find a uranium-bearing rock that formed during the planet's early accretion phase. Unfortunately, this is impossible—the planet was molten when it formed, so no rocks existed. As an alternative, researchers assume that all components of our solar system formed at about the same time. The uranium:lead ratios in meteorites, for example, indicate that meteorites formed 4.58 billion years ago, or 4.58 Ga. (The phrase "billion years ago" is often abbreviated as Ga, for "giga-years ago.") The timeline presented in **Figure 2.4a** starts with these data, which coincide with the start of Earth's formation. The ages of the oldest Moon rocks indicate that the gigantic impact responsible for the Moon's creation occurred at 4.51 Ga. Because the most ancient rock crystals found thus far on Earth date to about 4.40 Ga, it is clear that some solid crust had begun to form by this time. Indeed, most astronomers agree that Earth's formation was largely complete by 4.47 Ga. Based on the ages of the most recent craters of the Moon, it is estimated that the era of heavy bombardment—when both Earth and Moon were regularly pummeled by large objects from space—had ended by about 3.9 billion years ago.

When did the oceans originate? The oldest sedimentary rocks on Earth date to 3.85 Ga. Because sedimentary rocks form from deposits of sand, mud, or other particles carried by water, geologists are confident that an ocean existed by this time. Other evidence suggests that oceans may have been present as early as 4.28 Ga. But the heat generated when Earth was smashed by asteroids and other objects would have repeatedly vaporized most or all of this early ocean.

When Did Life Begin?   If chemical evolution took place, it occurred after Earth formed but before life began. To estimate when the origin of life took place, biologists examine fossils—traces of living organisms left in rocks. The most ancient fossils that have been found to date are in rocks that formed about 3.85 Ga. As **Figure 2.4b** shows, these fossils consist of tiny grains of carbon. Although the grains do not contain cells or other direct evidence of life, a recent analysis suggests that the carbon atoms present were once part of organisms. Specifically, the isotope ^{12}C is much more abundant in the grains than expected, given the normal abundance of ^{12}C relative to heavier carbon isotopes. This is significant because living organisms that take up carbon in the form of CO_2 use ^{12}C much more readily than they use the heavier isotopes of carbon. Researchers hypothesize that even though heat and pressure have eliminated most traces of the original cells, the carbon isotope "signature" of life remains.

If these carbon grains actually do represent the remains of organisms, they suggest that life started just after the end of massive bombardment. The current consensus is that chemical evolution occurred during a fairly narrow window of time, perhaps 300 million years or less. Almost as soon as conditions on Earth stabilized enough for life to begin, it did.

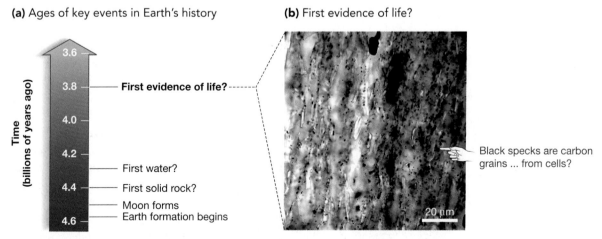

(a) Ages of key events in Earth's history

(b) First evidence of life?

First evidence of life?

First water?

First solid rock?

Moon forms
Earth formation begins

Black specks are carbon grains ... from cells?

20 μm

FIGURE 2.4 Early Events in the History of Earth
(a) Radiometric dating has assigned ages to key events in early Earth's history. The dates with question marks are still under investigation. **(b)** The black material is carbon that was found inside a 3.8-billion-year-old rock. The ratio of ^{13}C to ^{12}C inside is identical to the ratio observed in today's organisms, but very different from the ratios observed in the nonliving environment (air, water, and soil). Researchers hypothesize that this carbon is the remains of some of Earth's earliest organisms.

2.2 The Building Blocks of Chemical Evolution

Just four types of atoms—hydrogen, carbon, nitrogen, and oxygen—make up 96 percent of all matter found in organisms today. Many of the compounds found in living cells contain thousands, or even millions, of these atoms. But early in Earth's history, it is likely that these elements existed only in simple substances such as water and carbon dioxide, which contain just three atoms apiece.

The chemical evolution hypothesis maintains that simple compounds in the ancient atmosphere and ocean combined to form the larger, more complex substances found in living cells. To understand how this process could have begun, we need to consider the following questions:

- What is the physical structure of hydrogen, carbon, nitrogen, oxygen, and other atoms found in living cells?

- How do these atoms combine to form simple molecules such as water and carbon dioxide—the building blocks of chemical evolution?

What Atoms Are Found in Organisms?

Figure 2.5 shows one way that chemists summarize the structure of atoms. Notice that each atom has a symbol, such as the C shown for carbon, along with a superscript and a subscript. The subscript is the element's **atomic number**, meaning the number of protons in its nucleus. The superscript is the element's mass number (Section 2.1), which is the sum of the number of protons and neutrons in the nucleus. The mass number given for each element is that of the most common isotope of the element. For example, the most abundant isotope of carbon, ^{12}C, has six protons and six neutrons in its nucleus.

If an element is uncharged, its protons are matched by an equal number of electrons. Electrons carry a negative electric charge but

have very little mass compared with protons. To understand how the atoms involved in chemical evolution behave, it is critical to understand how electrons are arranged around the nucleus.

Electrons move around atomic nuclei in specific regions called **orbitals**. Each electron orbital has a distinctive shape, and each orbital can hold up to two electrons. Orbitals, in turn, are grouped into levels called **electron shells**. These are numbered 1, 2, 3, and so on, to indicate their relative distance from the nucleus. The electrons of an atom fill the innermost shells first, before filling outer shells.

To understand how the structures of atoms differ, take a moment to study **Figure 2.6**. This chart, which is a subset of the periodic table of the elements, highlights those atoms that are most abundant in living cells. The diagrams in each box of the table show how electrons are distributed in the shells of carbon and other key elements. The outermost shell of an atom is called the atom's **valence shell**, and the electrons found in that shell are referred to as **valence electrons**. Two observations are important:

1. In each of the highlighted elements, the outermost electron shell is not full. The highlighted elements have at least one unpaired valence electron.

2. The number of unpaired electrons in the valence shell varies among elements. Carbon has four unpaired electrons in its outermost shell; hydrogen has one. The number of unpaired electrons found in an atom is called its **valence**. Carbon's valence is four; hydrogen's valence is one.

These observations are significant because an atom is most stable when its valence shell is filled. One way that shells can be filled is through the formation of **chemical bonds**—strong attractions that bind atoms together.

How Does Covalent Bonding Hold Molecules Together?

To understand how atoms can become more stable by bonding to another atom, consider hydrogen. The hydrogen atom has just one electron, which resides in a shell that can hold two electrons. The hydrogen atom is most stable when this shell is full. When two atoms of hydrogen approach each other, the two electrons present become shared by the two nuclei (**Figure 2.7**). Both atoms now have a completely filled shell, and they have formed a chemical bond. Together, the bonded hydrogen atoms are more stable than the two individual hydrogen atoms. The shared electrons "glue" the atoms together in a **covalent bond**. Substances that are held together by covalent bonds are called **molecules**. In this case of two hydrogen atoms, the bonded atoms form a single molecule of hydrogen, written as H–H or H_2.

To visualize what is happening when a covalent bond forms, imagine two children who want the same toy. If both children grasp the toy tightly, through it they become attached to each

The highlighted elements are the most abundant elements found in organisms

Mass number (number of protons + neutrons)

Atomic number (number of protons)

^1_1H							^4_2He
^7_3Li	^9_4Be	$^{11}_5B$	$^{12}_6C$	$^{14}_7N$	$^{16}_8O$	$^{19}_9F$	$^{20}_{10}Ne$
$^{23}_{11}Na$	$^{24}_{12}Mg$	$^{27}_{13}Al$	$^{28}_{14}Si$	$^{31}_{15}P$	$^{32}_{16}S$	$^{35}_{17}Cl$	$^{40}_{18}Ar$

FIGURE 2.5 Chemical Symbols of Atoms Found in Organisms A segment of the periodic table, highlighting just the elements commonly found in organisms.

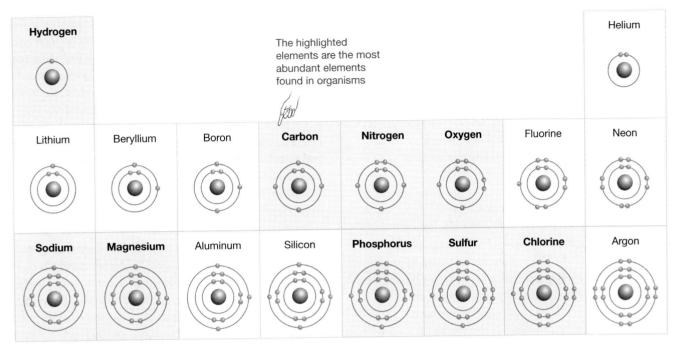

FIGURE 2.6 The Structure of Atoms Found in Organisms
The atomic nucleus is represented by a closed sphere. The first (inner), second, and third electron shells are shown as rings. The dots on the rings represent electrons. Electrons are drawn in pairs if they occupy filled orbitals within the same electron shell; they are drawn singly if they occupy unfilled orbitals. EXERCISE For each highlighted atom, write down the number of bonds it is capable of forming based on the number of unpaired electrons in its outermost electron shell.

other. For as long as they hold onto the toy, neither child can leave the other's side.

Another way to think about covalent bonding is in terms of electrical attraction and repulsion. As two hydrogen atoms move closer together, their positively charged nuclei repel each other and their negatively charged electrons repel each other. But each proton attracts both electrons, and each electron attracts both protons. Covalent bonds form when the attractive forces overcome the repulsive forces. This is the case when hydrogen atoms interact to form the hydrogen molecule (H_2).

To summarize, covalent bonding is based on electron sharing. In a hydrogen molecule, the electrons involved in the covalent bond are shared between the two nuclei. This situation is illustrated in **Figure 2.8a**. The covalent bond between hydrogen atoms is represented by a dash, and the electrons are drawn as dots that are between the two nuclei. An equal sharing of electrons, as in Figure 2.8a, results in a **nonpolar covalent bond**—a covalent bond that is symmetrical.

It's important to note, though, that the electrons participating in covalent bonds are not always shared equally between the atoms involved. Asymmetric sharing of electrons results in **polar covalent bonds**. Some atoms hold electrons in covalent

(a) Nonpolar covalent bond in hydrogen molecule

H—H   Electrons are shared equally

(b) Polar covalent bonds in water molecule

Electrons are not shared equally, so partial charges exist on the O and H atoms

FIGURE 2.8 Electron Sharing and Bond Polarity
(a) In a hydrogen molecule, electrons are shared equally by the two hydrogen nuclei. **(b)** In a water molecule, oxygen's high electronegativity causes that atom to hold electrons more tightly than do the hydrogen atoms. As a result, different parts of the molecule have partial charges. The delta (δ) symbols in the polar covalent bonds refer to these partial positive and negative charges.

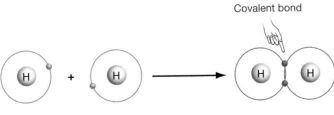

Hydrogen atoms each have one unpaired electron

H_2 molecule has two shared electrons

FIGURE 2.7 Covalent Bonds Result from Electron Sharing
When two hydrogen atoms come into contact, their electrons are attracted to the positive charge in each nucleus. As a result, their orbitals overlap, the electrons are shared by each nucleus, and a covalent bond forms.

bonds much more tightly than do other atoms, so the degree of sharing varies by element. Chemists call this property an atom's **electronegativity**. Oxygen is among the most electronegative of all elements: It attracts covalently bonded electrons more strongly than does any other atom commonly found in organisms. Nitrogen's electronegativity is somewhat lower than oxygen's. Carbon and hydrogen, in turn, have relatively low and approximately equal electronegativities. The electronegativities of the four most abundant elements in organisms are related as follows: $O \gg N > C \cong H$.

To put these ideas into practice, consider the water molecule. Water consists of oxygen bonded to two hydrogen atoms and is written H_2O. As **Figure 2.8b** illustrates, the electrons involved in the covalent bonds in water are not shared equally but are held much more tightly by the oxygen nucleus than by the hydrogen nuclei. Hence, water has two polar covalent bonds. Because electrons are shared unequally, the oxygen atom has a partial negative charge and the hydrogen atoms have a partial positive charge. Notice that the partial charges on the molecule are symbolized by the lowercase Greek letter delta, δ.

How Does Ionic Bonding Hold Ionic Compounds Together?

Ionic bonds are similar in principle to covalent bonds, but instead of sharing electrons between two atoms, the electrons in ionic bonds are completely transferred from one atom to the other. The electron transfer occurs because it gives the resulting atoms a full outermost shell. Sodium atoms (Na), for example, tend to lose an electron, leaving them with a full second shell. This is a much more stable arrangement, energetically, than having a lone electron in their third shell (**Figure 2.9a**). The atom

that results has a net electric charge of $+1$, because it has one more proton than it has electrons. An atom or molecule that carries a charge is called an **ion**. The sodium ion is written Na^+ and, like other positively charged ions, is called a **cation**. Chlorine atoms (Cl), in contrast, tend to gain an electron. When this occurs, the atom's outermost shell is full (**Figure 2.9b**). The atom has a net charge of -1, because it has one more electron than protons. This negatively charged ion, or **anion**, is written Cl^- and is called chloride. When sodium and chlorine combine to form table salt (sodium chloride, NaCl), the atoms pack into a crystal structure consisting of sodium cations and chloride anions (**Figure 2.9c**). The electrical attraction between the ions is so strong that salt crystals are difficult to break apart.

This discussion of covalent and ionic bonding supports an important general observation: The degree to which electrons are shared in chemical bonds forms a continuum. As the left-hand side of **Figure 2.10** shows, covalent bonds between atoms with exactly the same electronegativity—for example, between the atoms of hydrogen in H_2—represent one end of the continuum. The electrons in these nonpolar bonds are shared equally. In the middle of the continuum are bonds where one atom is much more electronegative than the other. In these asymmetric bonds, substantial partial charges exist on each of the atoms. These types of polar covalent bonds occur when a highly electronegative atom such as oxygen or nitrogen is bound to an atom with a lower affinity for electrons, such as carbon or hydrogen. Water (H_2O) and ammonia (NH_3) contain polar covalent bonds. At the right-hand side of the continuum are molecules made up of atoms with extreme differences in their electronegativities. In this case, electrons are transferred rather than shared, the atoms have full charges, and the bonding is

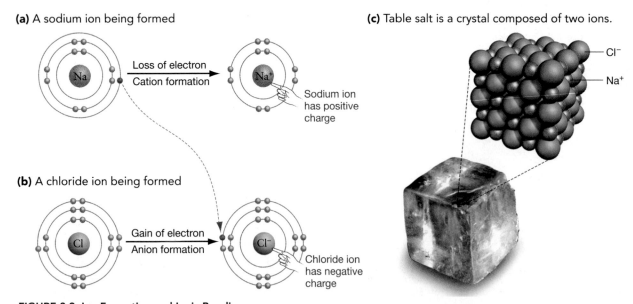

(a) A sodium ion being formed

Loss of electron
Cation formation

Sodium ion has positive charge

(b) A chloride ion being formed

Gain of electron
Anion formation

Chloride ion has negative charge

(c) Table salt is a crystal composed of two ions.

Cl^-
Na^+

FIGURE 2.9 Ion Formation and Ionic Bonding
(a) When an atom of sodium (Na) loses an electron, it forms a sodium ion (Na^+). The sodium ion is stable because it has a full valence shell. **(b)** When an atom of chlorine (Cl) gains an electron, it forms a chloride ion (Cl^-), which also has a full valence shell. **(c)** In table salt (NaCl), sodium and chloride ions pack into a crystal structure held together by electrical attraction.

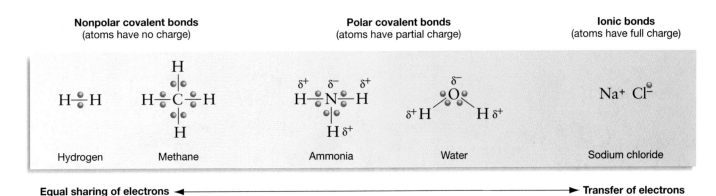

FIGURE 2.10 **The Electron-Sharing Continuum**
The degree of electron sharing in chemical bonds can be thought of as a continuum, from equal sharing in nonpolar covalent bonds to no sharing in ionic bonds.

ionic. Common table salt, NaCl, is a familiar example. In biology, ionic bonding is rare. The chemical bonds that occur in biological molecules are on the left-hand side of this continuum.

Some Simple Molecules Formed from H, C, N, and O

Look again at Figure 2.6 and count the number of unpaired electrons in the valence shells of carbon, nitrogen, and oxygen atoms: Carbon has four, nitrogen has three, and oxygen has two. Each unpaired electron can make up half of a covalent bond. As a result, a carbon atom can form a total of four covalent bonds; nitrogen can form three; and oxygen, two. When each of the four unpaired electrons of a carbon atom covalently bonds with a hydrogen atom, the molecule that results is written CH_4 and is called methane (**Figure 2.11a**). This is the most common molecule found in natural gas. When a nitrogen atom's three unpaired electrons bond with three hydrogen atoms, the result is NH_3, or ammonia. Similarly, an atom of oxygen can form covalent bonds with two atoms of hydrogen, resulting in a water molecule (H_2O).

Double and Triple Bonds In addition to forming multiple single bonds, atoms with more than one unpaired electron in the valence shell can sometimes produce double or triple bonds. **Figure 2.11b** shows how carbon forms double bonds with oxygen atoms to produce the molecule called carbon dioxide (CO_2). Triple bonds occur when three pairs of electrons are shared. **Figure 2.11c** shows the structure of molecular nitrogen (N_2), which forms when two nitrogen atoms establish a triple bond.

Bond Angles and the Shape of Molecules Proponents of the chemical evolution hypothesis argue that simple molecules such as CH_4, NH_3, H_2O, CO_2, and N_2 were important components of Earth's ancient atmosphere and ocean. Their reasoning is based on the observation that these molecules are found in volcanic gases on Earth and in the atmospheres of nearby planets. If so, then such molecules provided the building

(a) Single bonds

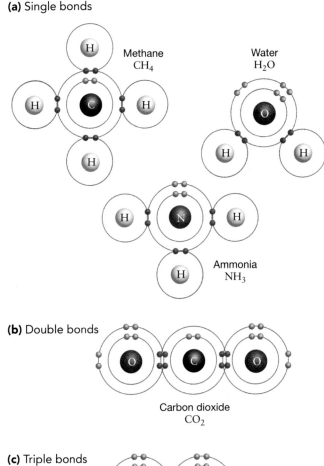

Methane
CH_4

Water
H_2O

Ammonia
NH_3

(b) Double bonds

Carbon dioxide
CO_2

(c) Triple bonds

Molecular nitrogen
N_2

FIGURE 2.11 **Unpaired Electrons in the Valence Shell Participate in Covalent Bonds**
Covalent bonding is based on sharing of electrons in the outermost shell. Covalent bonds can be **(a)** single, **(b)** double, or **(c)** triple.
QUESTION How do the electron configurations of H, C, O, and N in these molecules compare with their configurations shown in Figure 2.6?

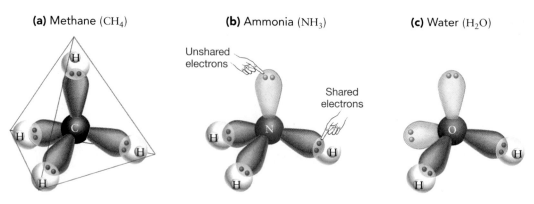

(a) Methane (CH_4)   **(b)** Ammonia (NH_3)   **(c)** Water (H_2O)

Unshared electrons

Shared electrons

FIGURE 2.12 The Geometry of Methane, Ammonia, and Water
Diagrams showing the shapes of the orbitals in the outermost shells of **(a)** methane, **(b)** ammonia, and **(c)** water. The dots represent electrons. Even though some of the same orbitals are involved in bonding, the overall shape of each molecule is radically different. **EXERCISE** Label which molecule is bent and planar, which forms a tetrahedron (having four identical faces), and which forms a pyramid. In parts (b) and (c), draw lines to indicate the overall shape of ammonia and water, similar to those drawn for methane.

blocks for chemical evolution. Before analyzing how these molecules could have combined to form more complex compounds, however, we need to consider their shape and how scientists represent their composition and shape. This is a crucial point. The overall shape of a molecule often dictates its behavior and function.

The shape of a simple molecule is governed by the geometry of its bonds. Bond angles, in turn, are determined by electrical repulsion that occurs between pairs of electrons. For example, **Figure 2.12a** shows the shape of methane (CH_4). The four pairs of electrons in carbon's outer shell form a tetrahedron when bonded to hydrogen because repulsion among the shared electron pairs orients them at an angle of 109.5° to one another. Ammonia (NH_3), in contrast, is shaped like a pyramid. As **Figure 2.12b** shows, the four orbitals in nitrogen's outermost shell form a tetrahedron, just as they do in the carbon atom in methane. But because one of the four orbitals in the nitrogen atom's outer shell is already filled with a pair of electrons, only the three orbitals with unpaired electrons bond with hydrogen. In this case, the geometry of the orbitals and the geometry of the molecule are different. In the oxygen atom, two of the four orbitals present in the outermost shell are filled with electron pairs. When hydrogen atoms bond to the remaining two electrons to form water (H_2O), the resulting molecule is bent and two-dimensional, or planar (**Figure 2.12c**).

Representing Molecules   Molecules can be represented in a variety of ways. The simplest representation is a **molecular formula**, which indicates the numbers and types of atoms in a molecule. Water has the molecular formula H_2O; methane has the molecular formula CH_4 (**Figure 2.13a**).

Molecular formulas are a compact way of stating a molecule's composition, but they contain no information about how the molecule is put together. **Structural formulas** indicate which

atoms are bonded together. In a structural formula, single, double, and triple bonds are represented by single, double, and triple dashes, respectively. **Figure 2.13b** gives structural formulas for molecules that were important components of ancient Earth's atmosphere and ocean.

The limitation of structural formulas is that they are two-dimensional, while molecules are three-dimensional. More complex diagrams, such as *ball-and-stick models* and *space-filling models*, are based on the three-dimensional structure of molecules. **Figures 2.13c** and **2.13d** show how ball-and-stick and space-filling models indicate a molecule's geometry as well as the relative sizes of the atoms involved.

Quantifying the Concentration of Key Molecules   Scientists quantify the number of molecules present in a sample by using a unit called the **mole**. A mole refers to the number 6.022×10^{23} (just as the unit called the *dozen* refers to the number 12). The mole is a useful unit because the mass of one mole of any molecule is the same as its molecular weight expressed in grams. **Molecular weight** is the sum of the mass numbers of all of the atoms in a molecule. For example, summing the mass numbers of two atoms of hydrogen and one atom of oxygen gives $1 + 1 + 16$, or 18, which is water's molecular weight. It follows that if you weighed a sample of 18 grams of water, it would contain 6.022×10^{23} water molecules, or one mole of water molecules.

Many important biological substances are normally dissolved in liquid. A **solution** is a homogenous (uniform) mixture of one or more substances dissolved in a liquid. When substances are dissolved in a liquid, their concentration is expressed in terms of molarity (symbolized by "M"). **Molarity** is the number of moles of the substance present per liter of solution. A 1-molar solution of carbon dioxide gas (CO_2) in water, for instance, means that 1 mole of CO_2—that is, 6.022×10^{23} CO_2 molecules—is contained in 1 liter of water.

	Methane	**Ammonia**	**Water**	**Oxygen**
(a) Molecular formulas:	CH_4	NH_3	H_2O	O_2

(b) Structural formulas:

$$H-\overset{\overset{\displaystyle H}{|}}{\underset{\underset{\displaystyle H}{|}}{C}}-H \qquad H-\overset{}{\underset{\underset{\displaystyle H}{|}}{N}}-H \qquad \overset{O}{H\quad H} \qquad O=O$$

(c) Ball-and-stick models:

(d) Space-filling models:

FIGURE 2.13 Molecules Can Be Represented Several Ways
Each method of representing a molecule has particular advantages. **(a)** Molecular formulas are compact and indicate the number and identity of atoms involved. **(b)** Structural formulas show which atoms are bonded to one another. **(c)** Ball-and-stick models take up more space than do structural formulas but include information about bond geometry. **(d)** Space-filling models are not as easy to read as ball-and-stick models but more accurately depict the spatial relationships between atoms.

✓CHECK YOUR UNDERSTANDING

Covalent bonds are based on electron sharing, while ionic bonds are based on electrical attraction between ions with opposite charges. Covalent bonds can be polar or nonpolar, depending on whether the electronegativities of the two atoms involved are the same or different. You should be able to (1) draw the structural formulas of methane (CH_4) and ammonia (NH_3) and add dots to indicate the relative locations of the covalently bonded electrons, and (2) draw the electron shells around sodium ions (Na^+) and chloride ions (Cl^-) and explain why table salt ($NaCl$) exists.

2.3 Chemical Reactions, Chemical Evolution, and Chemical Energy

Proponents of the chemical evolution hypothesis contend that simple molecules present in the atmosphere and ocean of ancient Earth participated in chemical reactions that produced larger, more complex molecules. A **chemical reaction** is an event in which one substance is combined with others or broken down into another substance. According to one hypothesis, the early atmosphere was made up of gases ejected from volcanoes (**Figure 2.14a**). Carbon dioxide, water, and nitrogen are the dominant gases ejected from volcanoes today; a small amount

(a) Volcanoes eject an array of gases, including water vapor.

(b) Liquid water forms when warm water vapor cools.

FIGURE 2.14 What Molecules Were Present at the Start of Chemical Evolution?
(a) When volcanoes erupt, they eject enormous quantities of dust, N_2, CO_2, and water into the atmosphere. **(b)** When water vapor cools, condensation occurs and liquid water forms. This process was responsible for the formation of the world's first lakes and oceans.

of molecular hydrogen (H_2) and methane (CH_4) may also be present. But when these molecules are placed in a glass tube together and allowed to interact, very little happens. Water vapor condenses to liquid water as the mixture cools (**Figure 2.14b**), but the simple molecules do not suddenly link together to create large, complex substances like those found in living cells. Instead, their bonds remain intact. How, then, did chemical evolution occur?

To answer this question we must explore two topics: (1) how chemical reactions occur, and (2) how conditions on ancient Earth made certain reactions possible.

How Do Chemical Reactions Happen?

Chemical reactions are written in a format similar to mathematical equations, with the initial, or **reactant**, atoms or molecules shown on the left and the resulting reaction **product**(s) shown on the right. For example, the most common reaction in the mix of gases and water that emerge from volcanoes is

$$CO_2(g) + H_2O(l) \rightleftharpoons H_2CO_3(aq)$$

This expression indicates that carbon dioxide (CO_2) reacts with water (H_2O), forming carbonic acid (H_2CO_3). The state of each reactant and product is indicated as gas (g), liquid (l), in aqueous solution (aq), or solid (s).

Note that the previous expression is balanced; that is, one carbon, three oxygen, and two hydrogen atoms are present on each side of the expression. Note also that the expression contains a double arrow, meaning that the reaction is reversible. When the forward and reverse reactions proceed at the same rate, the quantities of reactants and products remain constant, although not necessarily equal. A dynamic but stable state such as this is termed a **chemical equilibrium**. A chemical equilibrium can be disturbed by changing the quantity (*concentration*) of reactants or products. For example, adding CO_2 to the mixture would drive the reaction to the right, creating more H_2CO_3 until the equilibrium proportions of reactants and products are reestablished. Adding H_2CO_3 instead would drive the reaction to the left. Removing CO_2 would also drive the reaction to the left; removing H_2CO_3 would drive it to the right.

A chemical equilibrium can also be altered by changes in temperature. For example, the water molecules in this set of interacting elements, or **system**, would be present as a combination of liquid water and water vapor:

$$H_2O(l) \rightleftharpoons H_2O(g)$$

If liquid water molecules absorb enough heat, they transform to the gaseous state. This is called an **endothermic** ("within heating") process because heat is absorbed during the process. In contrast, the transformation of water vapor to liquid water releases heat and is called **exothermic** ("outside heating"). Raising the temperature of this system drives the

equilibrium to the right; cooling the system drives it to the left. In endothermic and exothermic processes, however, the system must be *isolated*, meaning one with no external source of energy. The components of the system can exchange energy (such as heat), but not matter, with their surroundings. If water vapor is lost to the environment, the system is not isolated and the equilibrium may not proceed as described earlier.

In relation to chemical evolution, though, these reactions and changes of state are not particularly interesting. Carbonic acid is not an important intermediate in the formation of more complex molecules. According to models developed by a series of researchers, however, interesting things do begin to happen when larger amounts of energy are added to the system.

What Is Energy?

Energy can be defined as the capacity to do work or to supply heat. This capacity exists in one of two ways—as a stored potential or as an active motion.

Stored energy is called **potential energy**. An object gains or loses its ability to store energy as a consequence of its position. An electron that resides in an outer electron shell will, if the opportunity arises, fall into a lower electron shell closer to the positive charges on the protons in the nucleus. As a result, an electron in an outer electron shell has more potential energy than does an electron in an inner shell (**Figure 2.15**).

Kinetic energy is the energy of motion. Molecules have kinetic energy because they are constantly in motion. This form of kinetic energy—the kinetic energy of molecular motion—is called **thermal energy**. The **temperature** of an object is a measure of how much thermal energy its molecules possess. If an object has a low temperature, its molecules are moving slowly (we perceive this as "cold"). If an object has a high temperature, its molecules are moving rapidly (we perceive this as "hot"). When two objects with different temperatures come into contact, thermal energy is transferred between them. We call this transferred energy **heat**.

There are many forms of potential and kinetic energies, and energy can change from one form into another. To drive this

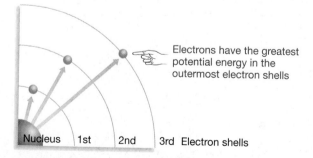

Electrons have the greatest potential energy in the outermost electron shells

Nucleus | 1st | 2nd | 3rd | Electron shells

FIGURE 2.15 Potential Energy as a Function of Electron Shells
Electrons in outer shells have more potential energy than do electrons in inner shells, because negative charges in outer shells are farther from the positive charges in the nucleus. Each shell represents a distinct level of potential energy.

point home, consider a water molecule sitting at the top of a waterfall, as in **Figure 2.16a**. This molecule has potential energy as a result of its position. If the molecule passes over the waterfall, its potential energy is converted to the kinetic energy of motion. When the molecule has reached the rocks below, it has experienced a change in potential energy because it has changed position. Panel 3 in Figure 2.16a shows that this change in potential energy is transformed into an equal amount of energy in other forms: mechanical energy, which tends to break up the rocks; heat (thermal energy), which raises the temperature of the rocks and the water itself; and sound.

An electron in an outer electron shell is analogous to the water molecule at the top of a waterfall (**Figure 2.16b**). If the electron falls to a lower shell, its potential energy is converted to the kinetic energy of motion. After the electron occupies the lower electron shell, it experiences a change in potential energy. As panel 3 in Figure 2.16b shows, the change in potential energy is transformed into an equal amount of energy in other forms—usually thermal energy, but sometimes light.

These examples illustrate the **first law of thermodynamics**, which states that energy is conserved. Energy cannot be created or destroyed, but only transferred and transformed.

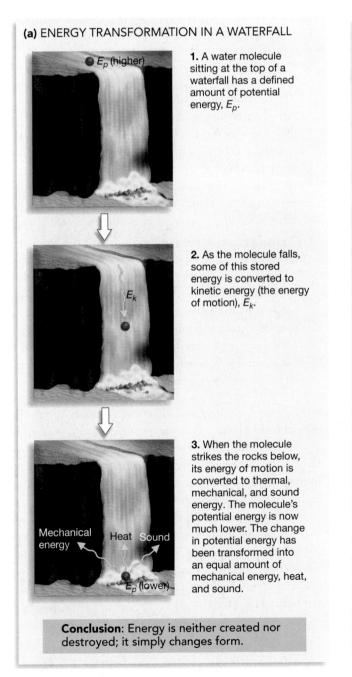

(a) ENERGY TRANSFORMATION IN A WATERFALL

E_p (higher)

1. A water molecule sitting at the top of a waterfall has a defined amount of potential energy, E_p.

E_k

2. As the molecule falls, some of this stored energy is converted to kinetic energy (the energy of motion), E_k.

Mechanical energy   Heat   Sound

E_p (lower)

3. When the molecule strikes the rocks below, its energy of motion is converted to thermal, mechanical, and sound energy. The molecule's potential energy is now much lower. The change in potential energy has been transformed into an equal amount of mechanical energy, heat, and sound.

Conclusion: Energy is neither created nor destroyed; it simply changes form.

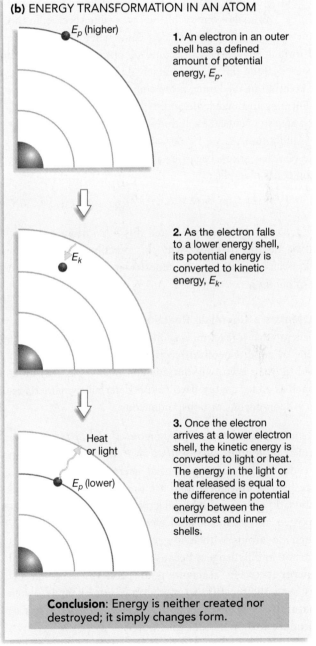

(b) ENERGY TRANSFORMATION IN AN ATOM

E_p (higher)

1. An electron in an outer shell has a defined amount of potential energy, E_p.

E_k

2. As the electron falls to a lower energy shell, its potential energy is converted to kinetic energy, E_k.

Heat or light

E_p (lower)

3. Once the electron arrives at a lower electron shell, the kinetic energy is converted to light or heat. The energy in the light or heat released is equal to the difference in potential energy between the outermost and inner shells.

Conclusion: Energy is neither created nor destroyed; it simply changes form.

FIGURE 2.16 Energy Transformations
During an energy transformation, the total amount of energy in the system remains constant.

If the Earth, as it cooled from a molten state, was being bombarded with large objects from space, then the simple molecules present in the ancient Earth's atmosphere and ocean would have been exposed to massive amounts of thermal energy from volcanic eruptions and asteroid impacts, in addition to high-energy ultraviolet radiation from the Sun. How would the application of large amounts of energy affect the course of chemical evolution?

Chemical Evolution: A Model System

To assess the impact of energy inputs on the simple molecules present in the early oceans and atmosphere, researchers have constructed computer models to simulate the reactions that can occur among carbon dioxide, water, nitrogen, and hydrogen molecules. The goal of one such study was quite specific: In this case the researchers wanted to determine whether a molecule called formaldehyde (H_2CO) could be produced. Along with hydrogen cyanide (HCN), formaldehyde is a key intermediate in the creation of the larger, more complex molecules found in cells. Forming formaldehyde and hydrogen cyanide is the critical first step in chemical evolution—a trigger that could set the process in motion.

The research group began by proposing that the following reaction could take place:

$$CO_2(g) + 2\,H_2(g) \longrightarrow H_2CO(g) + H_2O(g)$$

Before we explore how they tested this hypothesis, however, it will be helpful to know why this reaction doesn't occur spontaneously—that is, why doesn't it occur *without* an input of energy?

What Makes a Chemical Reaction Spontaneous?   When chemists say that a reaction is spontaneous, they have a precise meaning in mind: Chemical reactions are spontaneous if they proceed on their own, without any continuous external influence such as added energy. Two factors determine whether a reaction is spontaneous or nonspontaneous:

1. *Reactions tend to be spontaneous if the products have lower potential energy than the reactants.* Reaction products have lower potential energy if their electrons are held more tightly than are the electrons of the reactants. Recall that highly electronegative atoms such as oxygen and nitrogen hold electrons much more tightly than do atoms with a lower electronegativity, such as carbon and hydrogen. Because the difference in potential energy between reactants and products is given off as heat, the reaction is exothermic. For example, when natural gas burns, methane reacts with oxygen gas to produce carbon dioxide and water:

$$CH_4(g) + 2\,O_2(g) \longrightarrow CO_2(g) + 2\,H_2O(g)$$

FIGURE 2.17 Reactants and Products May Differ in Entropy
The chemical reaction called burning results in a system that is much less ordered than the original system. Stated another way, the reaction results in an increase in entropy.

The electrons involved in the C–O and H–O bonds of carbon dioxide and water are held much more tightly than they were in the C–H and O–O bonds of methane and oxygen. In chemical reactions, the difference in potential energy between the products and the reactants is symbolized by ΔH. (The uppercase Greek letter Δ, delta, is often used in chemical and mathematical notation to represent change.) When a reaction is exothermic, ΔH is negative.

2. *Reactions tend to be spontaneous when the product molecules are less ordered than the reactant molecules.* Wood is a highly ordered structure. But when wood burns, the molecules that result are much less ordered (**Figure 2.17**). The amount of disorder in a group of molecules is called its **entropy**, which is symbolized by S. When the products of a chemical reaction are less ordered than the reactant molecules, entropy increases and ΔS is positive. Reactions tend to be spontaneous if they increase entropy.

In general, physical and chemical processes proceed in the direction that results in lower potential energy and increased disorder (**Figure 2.18**). The **second law of thermodynamics**, in fact, states that entropy always increases in an isolated system. In the case of burning, the reaction is exothermic *and* results in higher entropy—less-ordered products.

Because reactions tend to be spontaneous when ΔH is negative and ΔS is positive, it's necessary to assess the *combined* contributions of changes in heat and disorder to determine whether a chemical reaction is spontaneous. To do this, chemists define a quantity called the **Gibbs free-energy change**, symbolized by ΔG:

$$\Delta G = \Delta H - T\Delta S$$

Here T stands for temperature measured on the Kelvin scale. The $T\Delta S$ term simply means that entropy becomes more im-

(a) Water has a higher potential energy at the top of a waterfall than at the bottom.

High potential energy

Low potential energy

(b) A sugar molecule has higher potential energy and more order (lower entropy) than carbon dioxide and water.

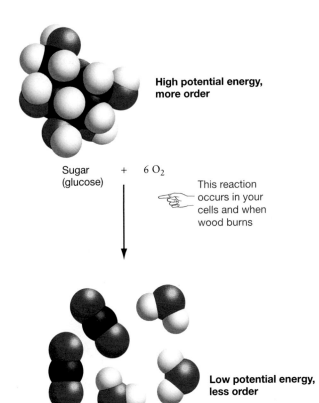

High potential energy, more order

Sugar   +   6 O_2
(glucose)

This reaction occurs in your cells and when wood burns

Low potential energy, less order

6 CO_2 +   6 H_2O

(not all product molecules are shown)

FIGURE 2.18 Spontaneous Processes Result in Lower Potential Energy or Increased Disorder, or Both
Examples of spontaneous processes include **(a)** a water molecule going over a waterfall, and **(b)** a sugar molecule undergoing a chemical reaction that forms smaller molecules.

portant in determining free-energy change as the temperature of the molecules increases.

Chemical reactions are spontaneous when ΔG is less than zero. Such reactions are said to be **exergonic**. Reactions are nonspontaneous when ΔG is greater than zero. Such reactions are termed **endergonic**. When ΔG is zero, reactions are at equilibrium. The reaction between carbon dioxide and hydrogen gas that forms formaldehyde and water is endergonic. It is nonspontaneous because it is endothermic and because it results in a decrease in entropy. For the reaction to occur, a large input of energy is required.

Energy Inputs and the Start of Chemical Evolution To explore how carbon dioxide (CO_2) and hydrogen gas (H_2) could have reacted to form formaldehyde (H_2CO) and trigger chemical evolution, a research group constructed a computer model of the ancient atmosphere. The model consisted of a list of all possible chemical reactions that can occur among CO_2, H_2O, N_2, NH_3, CH_4, and H_2 molecules. In addition to

the spontaneous reactions, they included reactions that occur when these molecules are struck by sunlight. This was crucial because sunlight represents a source of energy.

The sunlight that strikes Earth is made up of packets of light energy called photons. The amount of light energy contained in a photon can vary widely. Today, most of the higher-energy photons in sunlight never reach Earth's lower atmosphere. Instead, they are absorbed by a molecule called ozone (O_3) in the upper atmosphere. But if Earth's early atmosphere was filled with volcanic gases, it is extremely unlikely that appreciable quantities of ozone existed. As a result, we can infer that when chemical evolution was occurring, large quantities of high-energy photons bombarded the planet.

Why was the energy in photons important? Recall that the atoms in hydrogen and carbon dioxide have full outermost shells. As a result, these molecules are largely unreactive. But energy from photons can break molecules apart by knocking electrons away from the outer shells of atoms. The atoms that result, called **free radicals**, have unpaired electrons and are

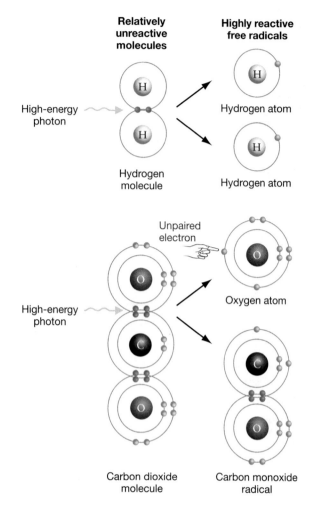

Relatively unreactive molecules

Highly reactive free radicals

High-energy photon

Hydrogen molecule

Hydrogen atom

Hydrogen atom

Unpaired electron

Oxygen atom

High-energy photon

Carbon dioxide molecule

Carbon monoxide radical

FIGURE 2.19 Free Radicals
When a high-energy photon strikes a hydrogen or carbon dioxide molecule, free radicals can be created. Formation of free radicals is thought to be responsible for some key reactions in chemical evolution.

extremely unstable (**Figure 2.19**). To mimic the conditions on early Earth more accurately, the computer model included several reactions that produce highly reactive free radicals.

To understand which of the long list of possible reactions would actually occur, and to estimate how much formaldehyde could have been produced in the ancient atmosphere, the researchers needed to consider the effects of two additional factors: temperature and concentration.

The Roles of Temperature and Concentration in Chemical Reactions Even if a chemical reaction occurs spontaneously, it may not happen quickly. For most reactions to proceed, one chemical bond has to break and another one has to form. For this to happen, the substances involved must collide in a specific orientation that brings the electrons involved near each other.

The number of collisions occurring among the substances in a mixture depends on the temperature and the concentrations of the reactants. When the concentration of reactants is high, more collisions occur and reactions proceed more quickly. When their temperature is high, reactants move faster and col-

lide more frequently. Higher concentrations and higher temperatures tend to speed up chemical reactions.

To model the behavior of simple molecules in the ancient atmosphere, then, the researchers needed to specify both the concentration of each molecule and the temperature. Then they were able to assign a rate to each of the reactions listed in their model based on the actual reaction rates observed in experiments conducted at controlled temperatures and concentrations.

Their result? They calculated that, under temperature and concentration conditions accepted as reasonable approximations of early Earth conditions by most atmospheric scientists, appreciable quantities of formaldehyde would have been produced. Using a similar model, other researchers have shown that significant amounts of hydrogen cyanide (HCN) could also have been produced in the ancient atmosphere. According to this research, large quantities of the critical intermediates in chemical evolution would have formed in the ancient atmosphere.

How Did Chemical Energy Change during Chemical Evolution?

The initial products of chemical evolution are important, for a simple reason: They have more potential energy than do the reactant molecules. When formaldehyde is produced, an increase in potential energy occurs because the electrons that bond CO_2 and H_2 together are held more tightly than they are in H_2CO or H_2O. This form of potential energy—the potential energy stored in chemical bonds—is called **chemical energy**.

This observation gets right to the heart of chemical evolution: The energy in sunlight was converted to chemical energy—meaning potential energy in chemical bonds. This energy transformation explains how chemical evolution was possible. When small, simple molecules absorb energy, chemical reactions can occur that transform the external energy into potential energy stored in chemical bonds. More specifically, the energy in sunlight was converted to chemical energy in the form of formaldehyde and hydrogen cyanide. The complete reaction that results in the formation of formaldehyde is written as

$$CO_2(g) + 2 H_2(g) + \text{sunlight} \longrightarrow H_2CO(g) + H_2O(g)$$

This reaction is balanced in terms of the atoms *and* the energy involved. An increase in chemical energy occurred that made the production of larger, more complex molecules possible.

✓CHECK YOUR UNDERSTANDING

Chemical reactions tend to be spontaneous if they lead to lower potential energy and higher entropy (more disorder). The combined effects of potential energy and entropy changes are summarized in the equation for the Gibbs free-energy change. You should be able to (1) write out the Gibbs equation and define each of the components, and (2) explain why potential energy might decrease as the result of a chemical reaction and why entropy might increase.

2.4 The Composition of the Early Atmosphere: Redox Reactions and the Importance of Carbon

In the models of the early atmosphere we just reviewed, chemical evolution did not begin until energy in the form of sunlight was added to the system. This conclusion makes sense at a very basic level. Energy is the capacity to do work, and it seems logical that building larger, more complex molecules requires work to be done. More specifically, the reactions involved in chemical evolution are endergonic, so inputs of energy were required.

It's important to recognize, though, that the start of chemical evolution also depended on the types of building-block molecules that were present in the atmosphere. In models, key compounds such as formaldehyde (H_2CO) and hydrogen cyanide (HCN) are produced only if molecular hydrogen (H_2), ammonia (NH_3), and methane (CH_4) are present in the atmosphere. No matter what type of energy is added to the model systems, chemical evolution does not take place unless these molecules are present. What evidence do biologists have that these molecules existed in the ancient atmosphere?

The earliest hypotheses for the composition of Earth's first atmosphere, developed in the 1920s and 1940s, proposed that H_2, NH_3, and CH_4 were abundant. Investigators based this claim on the idea that intense gravity and lack of volcanic activity had kept the atmospheres of Jupiter and Saturn unchanged since the founding of the solar system. If so, then ancient Earth's atmosphere was similar to the current atmosphere of Jupiter and Saturn. But in 1951 other researchers began arguing that volcanic gases such as CO_2, N_2, and H_2O dominated Earth's original atmosphere. Who is correct?

This controversy is difficult to resolve, because there is no direct evidence about the composition of the ancient atmosphere. The current consensus is based on recent models of the volcanic gases produced as Earth's crust formed. These emissions were undoubtedly very different from the volcanic gases produced today. These models support the hypothesis that small but significant amounts of H_2, NH_3, and CH_4 were in the early atmosphere, along with abundant CO_2, N_2, and H_2O.

Why are these particular molecules essential to chemical evolution? They trigger the most important chemical reactions in biology: reduction-oxidation, or redox, reactions.

What Is a Redox Reaction?

Reduction-oxidation reactions, or redox reactions, are a class of chemical reactions that involve the loss or gain of an electron. In a redox reaction, the atom that loses one or more electrons is said to be **oxidized**, and the atom that gains one or more electrons is said to be **reduced**. To help keep these terms straight, chemists use the mnemonic "LEO the lion goes GER"—Loss of Electrons is Oxidation; Gain of Electrons is Reduction. (An alternative is OIL RIG—Oxidation Is Loss; Reduction Is Gain.) Oxidation events are always coupled with a reduction; if one atom loses an electron, another has to gain it. Stated another way, a reactant that acts as an **electron donor** is always paired with a reactant that acts as an **electron acceptor**.

The gain or loss of an electron can be relative, however. That is, during a redox reaction, electrons can be transferred completely from one atom to another, or the electrons can simply shift their positions in covalent bonds. For example, consider the burning of methane diagrammed in **Figure 2.20**. The dots in the illustration represent the electrons involved in covalent bonds. Compare the position of the electrons in the reactant, methane, with their position in the product, carbon dioxide. Note that the electrons have moved farther from the carbon nucleus in the carbon dioxide product. This means that carbon has been oxidized. It has "lost" electrons. The change occurred because the carbon and hydrogen in CH_4 share electrons equally, while the carbon and oxygen in CO_2 do not. In carbon dioxide, the high electronegativity of oxygen pulls electrons away from the carbon atom. Now compare the position of the electrons in the oxygen reactant and their position in the product, water. These electrons have moved closer to the oxygen in the water molecule, meaning that the oxygen has been reduced. In this reaction, oxygen has "gained" electrons.

These shifts in electron position change the amount of chemical energy in the reactants and products. When methane burns, electrons are held more tightly in the product molecules than in

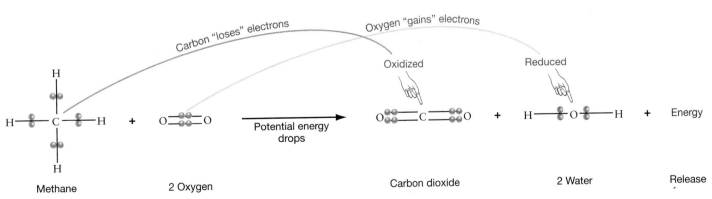

FIGURE 2.20 Redox Reactions Involve a Transfer of Electrons
Equation of the burning of methane. Burning is a redox reaction. Lines between atoms indicate covalent bonds; dots represent the relative positions of the electrons involved in those bonds. EXERCISE Circle the most electronegative atoms in this reaction.

BOX 2.1   Some Other Approaches to Understanding Redox Reactions

Redox reactions are absolutely fundamental to life, so it is critical to gain a solid understanding of redox chemistry. To further your understanding of how these reactions work, here are additional descriptions of redox dynamics:

- The electrons and bonds involved in redox reactions can be likened to a seesaw. In this analogy, an electron pair that is being shared equally by two atoms is like a seesaw balanced on its fulcrum.

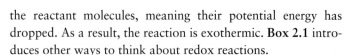

Because this state is inherently unstable, it has high potential energy. A redox reaction occurs, and an electron is transferred toward a more electronegative atom, like a seesaw with one end lowered to the ground.

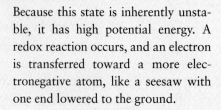

In this state, the seesaw is stable—meaning that it has much lower potential energy. In a redox reaction, the difference in potential energy between the two states is released as heat or transferred to another atom.

- During the redox reactions that occur in cells, electrons (e^-) are often transferred from an atom in one molecule to an atom in a different molecule. When this occurs, the electron is usually accompanied by a proton (H^+). As a result, the molecule that contains the reduced atom gains a hydrogen (H) atom. In many redox reactions in biology, understanding where oxidation and reduction have occurred becomes a matter of following hydrogen atoms. In many biological systems, reduction means "adding H's" and oxidation means "removing H's."

the reactant molecules, meaning their potential energy has dropped. As a result, the reaction is exothermic. **Box 2.1** introduces other ways to think about redox reactions.

Now consider **Figure 2.21**, which illustrates one of the key reactions in chemical evolution. The covalently bonded electrons in the hydrogen molecules (H_2) are shared equally and have relatively high potential energy. As a result, H_2 is a relatively unstable molecule and can act as an electron donor. The reactant, carbon dioxide (CO_2), in contrast, has covalently bonded electrons that are tightly held by oxygen. Consequently, CO_2 is a relatively stable molecule and can act as an electron acceptor. The interaction between H_2 and CO_2 is a redox reaction that couples an electron donor and an electron acceptor. Using Figure 2.20 as a guide, you should be able to add the electron positions for each of the bonds involved in this reaction. Then you should be able to determine whether carbon is reduced or oxidized as a result of this reaction. Based on this analysis, the following claim should make sense: The key step in launching chemical evolution was the reduction of carbon.

What Happens When Carbon Is Reduced?

Life has been called a carbon-based phenomenon, and with good reason. With the exception of water, almost all of the molecules found in organisms contain this atom.

Carbon is so important in biology because it is the most versatile atom on Earth. Due to its four valence electrons, it can form a large number of covalent bonds. With different combinations of single and double bonds, an almost limitless array of molecular shapes is possible. You have already examined the tetrahedral structure of methane and the linear shape of carbon dioxide. When molecules contain more than one carbon atom, these atoms can be bonded to one another in long chains, as in the component of gasoline called octane (C_8H_{18}; **Figure 2.22a**), or in a ring, as in the sugar glucose ($C_6H_{12}O_6$; **Figure 2.22b**). Molecules that contain carbon are called **organic molecules**. Other types of molecules are referred to as *inorganic compounds*.

Linking Carbon Atoms Together   The formation of carbon-carbon bonds was an important event in chemical evolution. It represented a crucial step toward the production of the types of

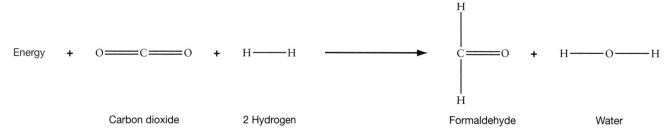

FIGURE 2.21 Tracking Electron Transfer during a Redox Reaction
A reaction that occurred during chemical evolution. EXERCISE Using Figure 2.20 as a guide, add dots to each covalent bond to show the relative positions of the electrons involved. Is the carbon atom reduced or oxidized in this reaction? Label the reactant molecules that act as the electron donor and the electron acceptor.

(a) Carbons linked in a linear molecule

C_8H_{18} Octane

(b) Carbons linked in a ring

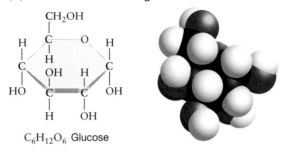

$C_6H_{12}O_6$ Glucose

FIGURE 2.22 The Shapes of Carbon-Containing Molecules
(a) Octane is one of the primary ingredients in gasoline. It is a linear molecule with carbon atoms that are highly reduced. **(b)** Glucose is a sugar that can form the ring-like structure illustrated here.

molecules found in living organisms. Once compounds with reduced carbon atoms such as formaldehyde and hydrogen cyanide had formed, continued chemical evolution could occur by the addition of heat alone. For example, when molecules of formaldehyde are heated, they react with one another to form a molecule called acetaldehyde. Acetaldehyde contains a carbon-carbon bond. With continued heating, reactions between formaldehyde and acetaldehyde molecules can produce the larger carbon-containing compounds called sugars.

In sum, advocates of the chemical evolution hypothesis propose that two sources of energy—the potential energy in inorganic compounds such as H_2 and the energy in sunlight—made the production of reduced carbon-containing compounds possible. Subsequently, the potential energy in these carbon-containing molecules made the production of the first complex organic compounds possible. **Figure 2.23** details these steps. In studying the figure, be sure to note two key messages: (1) Chemical evolution got under way because molecules with high potential energy were

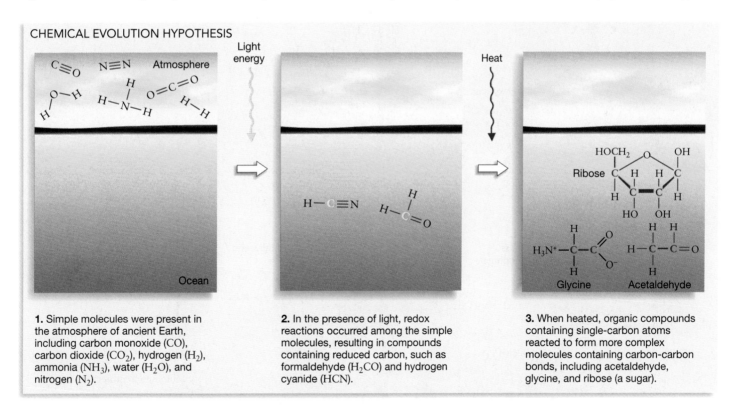

CHEMICAL EVOLUTION HYPOTHESIS

1. Simple molecules were present in the atmosphere of ancient Earth, including carbon monoxide (CO), carbon dioxide (CO_2), hydrogen (H_2), ammonia (NH_3), water (H_2O), and nitrogen (N_2).

2. In the presence of light, redox reactions occurred among the simple molecules, resulting in compounds containing reduced carbon, such as formaldehyde (H_2CO) and hydrogen cyanide (HCN).

3. When heated, organic compounds containing single-carbon atoms reacted to form more complex molecules containing carbon-carbon bonds, including acetaldehyde, glycine, and ribose (a sugar).

FIGURE 2.23 The Start of Chemical Evolution—an Overview
Chemical evolution is a process whereby simple molecules containing C, H, O, and N react to form molecules with reduced carbon atoms, which then react to form organic compounds with carbon-carbon bonds. The process is triggered by an energy source such as sunlight or the heat released in a volcanic eruption.

available to act as electron donors and because abundant sources of outside energy were available to trigger endergonic reactions; and (2) some of these molecules produced by chemical evolution are found in organisms living today.

Functional Groups Once chemical evolution was under way, a wide variety of small organic molecules would have accumulated in the ancient Earth. Although the molecular and structural formulas of small organic compounds are enormous-

ly variable, their chemical functions are reasonably predictable. In general, the carbon atoms in an organic molecule furnish a skeleton that gives the molecule its shape. The chemical behavior of the compound, however, is dictated by groups of H, N, or O atoms that are bonded to one of the carbon atoms in a specific way. These H-, N-, and O-containing groups of atoms are called **functional groups**. The composition and properties of six prominent functional groups recognized by organic chemists are summarized in **Table 2.1**.

TABLE 2.1 Six Functional Groups Commonly Attached to Carbon Atoms

EXERCISE Add notes to the table, predicting whether each functional group is polar or nonpolar, based on the electronegativities of the atoms involved.

Functional Group	*Formula	Family of Molecules	Example		Properties of Functional Group
Amino	$R-N\begin{smallmatrix}H\\H\end{smallmatrix}$	Amines		Glycine (an amino acid)	Acts as a base—tends to attract a proton to form $R-{}^+N\begin{smallmatrix}H\\H\\H\end{smallmatrix}$
Carbonyl	$R-C\begin{smallmatrix}O\\\\H\end{smallmatrix}$	Aldehydes		Acetaldehyde	Aldehydes, especially, react with compounds of form HR_2 to produce larger molecules with form $R_1-\underset{R_2}{\overset{OH}{C}}-H$
	$R-\underset{O}{\overset{}{C}}-R$	Ketones		Acetone	
Carboxyl	$R-C\begin{smallmatrix}O\\\\OH\end{smallmatrix}$	Carboxylic acids		Acetic acid	Acts as an acid—tends to lose a proton to form $R-C\begin{smallmatrix}O\\\\O^-\end{smallmatrix}$
Hydroxyl	$R-OH$	Alcohols		Ethanol	Highly polar, so makes compounds more soluble through hydrogen bonding with water
Phosphate	$R-O-\overset{O}{\underset{O^-}{P}}-O^-$	Organic phosphates		3–Phosphoglyceric acid	When several phosphate groups are linked together, breaking O–P bonds between them releases large amounts of energy
Sulfhydryl	$R-SH$	Thiols		Cysteine	When present in proteins, can form disulfide (S–S) bonds that contribute to protein structure

*In these structural formulas, "R" stands for the rest of the molecule.

The *carbonyl group*, for example, is found on aldehyde molecules such as formaldehyde and acetaldehyde. This functional group is the site of the reaction that links these molecules into larger, more complex compounds. The other functional groups listed in Table 2.1 confer equally distinctive properties, ranging from a propensity to participate in acid-base reactions to increasing the solubility of the compound in water.

Reactions among organic molecules launched the next phase of chemical evolution: the formation of the larger and more complex molecules found in living cells. Advocates of the chemical evolution hypothesis propose that these reactions occurred in water rather than in the atmosphere. Why?

2.5 The Early Oceans and the Properties of Water

Although it is accurate to say that life is based on carbon, it is even more meaningful to say that it is based on water. In a typical living cell, over 75 percent of the volume consists of this molecule. Nearly 70 percent of your body, by weight, is water (**Figure 2.24**). Virtually all researchers agree that most of the important steps in chemical evolution, including the origin of life itself, occurred in water. This hypothesis is logical because, as a dissolving agent, or **solvent**, water can dissolve more types of substances than any other molecule known. The hypothesis is also important because chemical reactions depend on direct, physical interaction between the reactants and because substances are most likely to collide when they are dissolved. Thus, life is based on water primarily because of water's solvent properties.

The formation of Earth's first ocean was a turning point in chemical evolution. It gave chemical evolution a place to happen.

Why Is Water Such an Efficient Solvent?

To understand why water is such an effective solvent, recall that the molecule contains two hydrogen atoms bonded to an oxygen atom. Then recall from Section 2.2 that oxygen and hydrogen differ in their ability to attract the electrons involved in covalent bonds, or their electronegativity. Because oxygen is among the most electronegative of all elements, it attracts covalently bonded electrons much more strongly than does the hydrogen nucleus. As a result, the electrons that participate in the bonds are not shared equally. The covalent bonds in water are polar.

If water were a linear molecule like carbon dioxide, the polarity of its covalent bonds would not matter very much. But recall from Section 2.2 that the water molecule is bent. The simple fact that water is a bent molecule with polar covalent bonds has an enormously important consequence: The molecule as a whole is **polar**, meaning that the overall distribution of charge is asymmetrical. As **Figure 2.25a** shows, the side of the molecule containing the oxygen atom is slightly more negative, and the side with the hydrogen atoms is slightly more positive. As before, the partial charges on the molecule are symbolized by the lowercase Greek letter delta, δ.

Figure 2.25b illustrates how the polarity of water affects its interactions with other substances in solution. When two liquid water molecules approach one another, the partial positive charge on hydrogen attracts the partial negative charge on oxygen. This weak electrical attraction forms a **hydrogen bond** between the molecules.

In a water-based, or aqueous, solution, hydrogen bonds also form between water molecules and other polar molecules. Similar interactions occur between water and ions. Ions and polar molecules stay in solution because of their interactions with

(a) Water is polar.

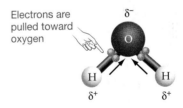

Electrons are pulled toward oxygen

(b) Hydrogen bonds form between water molecules.

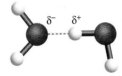

FIGURE 2.25 Water Is Polar and Participates in Hydrogen Bonds
(a) Because of oxygen's high electronegativity, the electrons that are shared when hydrogen and oxygen form a covalent bond are pulled toward the oxygen nucleus. The electrons spend more time close to the oxygen nucleus, so the oxygen atom has a slight negative charge and the hydrogen atom a partial positive charge.
(b) The electrical attraction that occurs between the partial positive and negative charges on water molecules forms a hydrogen bond.
EXERCISE Label the hydrogen bond in part (b).

FIGURE 2.24 Water Is the Most Abundant Molecule in Organisms
Fruits shrink when they are dried because they consist primarily of water.

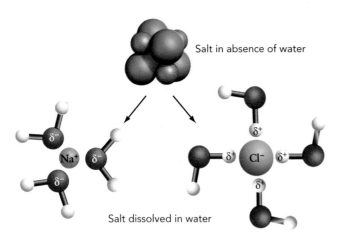

FIGURE 2.26 Polar Molecules and Ions Dissolve Readily in Water
Water's polarity makes it a superb solvent for polar molecules and ions.

water's partial charges (**Figure 2.26**). Hydrogen bonding makes it possible for almost any charged or polar molecule to dissolve in water. Although individual hydrogen bonds are not nearly as strong as covalent or ionic bonds, they are much more common. Hydrogen bonding is extremely important in biology because of the sheer number of hydrogen bonds that form in solution and between key molecules.

How Does Water's Structure Correlate with Its Properties?

Water's structure is unusual. Its small size, bent shape, highly polar covalent bonds, and overall polarity are unique among molecules. Because the structure of molecules routinely correlates with their function, it's not surprising that water has some remarkable properties. We've already reviewed the most important of water's chemical attributes—its ability to form hydrogen bonds with ions and polar compounds and its capacity to act as a solvent.

Water also has several striking physical properties that are the direct result of hydrogen bonding. In relation to chemical evolution, the most important of these are that water expands as it changes from a liquid to a solid, and that it has an extraordinarily large capacity for absorbing heat. Let's analyze why hydrogen bonding explains these properties, and connect water's physical and chemical behavior with its role in chemical evolution.

Water Is Denser as a Liquid than as a Solid   When factory workers pour molten metal or plastic into a mold and allow it to cool to the solid state, the material shrinks. When molten lava pours out of a volcano and cools to solid rock, it shrinks. But when you fill an ice tray with water and put it in the freezer to make ice, the water expands.

Unlike most substances, water is denser as a liquid than it is as a solid. In other words, there are more molecules of water in a given volume of liquid water than there are in the same

volume of solid water. **Figure 2.27a** illustrates why this is so. Note that in ice, each water molecule participates in four hydrogen bonds. These hydrogen bonds cause the water molecules to form a regular and repeating structure, or crystal. The crystal

(a) In ice, water molecules form a crystal lattice.

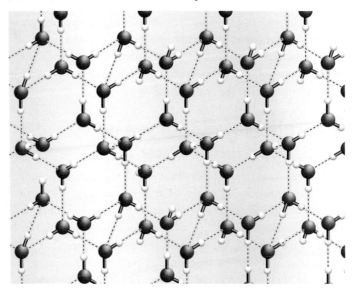

(b) In liquid water, no lattice forms, so liquid water is denser than ice.

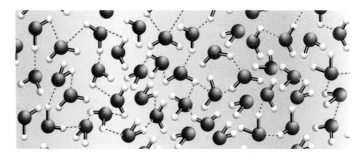

(c) As a result, ice floats.

FIGURE 2.27 Hydrogen Bonding Forms the Crystal Structure of Ice
EXERCISE In liquid water, each molecule can form four hydrogen bonds at one time. (Each oxygen atom can form two; each hydrogen atom can form one.) Choose two molecules in part (b), and draw in all four H bonds.

structure of ice is fairly open, meaning that there is a relatively large amount of space between molecules.

Now compare the extent of hydrogen bonding and the density of ice with that of liquid water, illustrated in **Figure 2.27b**. Although the exact structure of liquid water is not fully understood, it is clearly dynamic. Hydrogen bonds are constantly being made and broken. When hydrogen bonds form, energy is released; when hydrogen bonds break, energy is absorbed.

Overall, the extent of hydrogen bonding in liquid water is much less than that found in ice. As a result, molecules in the liquid phase are packed much more closely together than are molecules of solid water, even though their temperature is higher. Normally, heating a substance causes it to expand as molecules move faster and collide more often and with greater force. But heating ice causes hydrogen bonds to break and the open crystal to collapse. In this way, hydrogen bonding explains why water is denser as a liquid than as a solid.

This property of water has an important result: Ice floats (**Figure 2.27c**). If it did not, ice would sink to the bottom of lakes, ponds, and oceans soon after it formed. The ice would stay frozen in the cold depths. If water wasn't so unusual, it is almost certain that Earth's oceans would have frozen solid before life had a chance to start.

Water Has a High Capacity for Absorbing Energy

Hydrogen bonding is also responsible for another of water's remarkable physical properties: Water has a high capacity for absorbing energy. For example, water has an extraordinarily high specific heat. **Specific heat** is the amount of energy required to raise the temperature of 1 gram of a substance by 1°C. Water's specific heat is high because, when a source of energy such as sunlight or a flame strikes it, hydrogen bonds must be broken before heat can be transferred and the water molecules begin moving faster. As a result, it takes an extraordinarily large amount of energy to change the temperature of water (**Table 2.2**). Similarly, it takes a large amount of energy to break the hydrogen bonds in liquid water and change the molecules from the liquid phase to the gas phase. Water's **heat of vaporization**—the energy required to change 1 gram of it from a liquid to gas—is higher than that of most molecules that are liquid at room temperature. As a result, water has to absorb a great deal of energy to evaporate. Water's high heat of vaporization is the reason that sweating or dousing yourself is an effective way to cool off on a hot day. Because water molecules have to absorb a great deal of energy from your body in order to evaporate, you lose heat.

These properties of water are important to the chemical evolution hypothesis. Sources of energy trigger the formation of reduced carbon compounds such as formaldehyde, but energy inputs can also break them apart. Water's high specific heat insulates dissolved substances from sources of energy including asteroid bombardment, sunlight, and volcanism. Because formaldehyde and hydrogen cyanide dissolve readily in water,

TABLE 2.2 Some Specific Heats of Common Substances

The specific heats reported in this table were measured at 25°C (except for ice, which was measured at −11°C) and are given in units of joules per gram of substance per degree Celsius. (The **joule** is a unit of energy.) Notice that the specific heat of water, in both solid and liquid form, is extremely high. This means that water must absorb a great deal of thermal energy before it begins to change temperature.

Substance	Specific Heat
Air (dry)	1.01
Aluminum	0.90
Copper	0.39
Gold	0.13
Iron	0.45
Mercury	0.14
Table salt	0.86
Water (s)	2.03
Water (l)	4.18

Source: Table 8.1, p. 312 in John McMurry and Robert C. Fay, *Chemistry*, 4th Edition, ©2004. Reprinted by permission of Pearson Education, Inc., Upper Saddle River, NJ

they would have rained out of the atmosphere into the ocean—an environment where they were better protected. Similarly, if chemical evolution took place at the edges of ponds or on beaches, the evaporation of water would have kept the surfaces relatively cool. Hydrogen bonding gives water a moderating influence in terms of temperature.

Acid-Base Reactions and pH

One other aspect of water's chemistry influenced its role in the origin of life. Water is not a completely stable molecule. In reality, water molecules continually undergo a chemical reaction with themselves. This "dissociation reaction" can be written as follows:

$$H_2O \rightleftharpoons H^+ + OH^-$$

The double arrows indicate that the reaction proceeds in both directions.

The molecules on the right-hand side of the expression are the **hydrogen ion** (H^+) and the **hydroxide ion** (OH^-). A hydrogen ion is simply a proton. Substances that give up protons during chemical reactions are called **acids**; molecules or ions that acquire protons during chemical reactions are called **bases**. A chemical reaction that involves a transfer of protons is called an **acid-base reaction**. Acid-base reactions require a proton donor and a proton acceptor, just as redox reactions require an electron donor and an electron acceptor.

In reality, however, protons never exist by themselves. In water, for example, protons associate with water molecules to form the hydronium ion, H_3O^+. Thus, the dissociation of water is more accurately written

$$H_2O + H_2O \rightleftharpoons H_3O^+ + OH^-$$

One of the water molecules on the left-hand side of the expression has given up a proton and acted as an acid, while the other water molecule has accepted a proton and acted as a base. This illustrates another important property of water: Most acids act only as acids, and most bases act only as bases, but water can act both as an acid and as a base.

In a solution, the tendency for acid-base reactions to occur is largely a function of the number of protons present. How many protons are present in water?

Chemists answer this question by measuring the concentration of protons directly. In a sample of pure water at 25°C, the concentration of H^+ is 1.0×10^{-7} M (recall that M represents molarity, or moles per liter). Because this is such a small number (1 ten-millionth), the exponential notation is cumbersome. So chemists and biologists prefer to express the concentration of protons in a solution with a logarithmic notation called the **pH scale**. (The term *pH* is derived from the French *puissance d'hydrogène*, or "power of hydrogen.") By definition, the pH of a solution is the negative of the base-10 logarithm, or log, of the hydrogen ion concentration:

$$pH = -\log[H^+]$$

(The square brackets are a standard notation for indicating "concentration of" a substance in solution.) Taking antilogs gives

$$[H^+] = \text{antilog}(-pH) = 10^{-pH}$$

Figure 2.28 shows the pH scale and reports the pH of some common solutions. The pH of pure water at 25°C is 7. Pure water is used as a standard, or point of reference, on the pH scale. Solutions that contain acidic molecules (molecules that act as acids) have a proton concentration larger than 1×10^{-7} M and thus a pH < 7, because acidic molecules tend to release protons into solution. In contrast, solutions that contain basic molecules (molecules that act as bases) have a proton concentration less than 1×10^{-7} M and thus a pH > 7, because basic molecules tend to accept protons from solution. Solutions with a pH of 7 are considered neutral solutions, neither acidic nor basic. Rainwater is almost pure water, meaning that its pH is close to 7.

What Was Water's Role in Chemical Evolution?

Water's unusual structure and properties gave it a key role in chemical evolution. When the Earth cooled enough for rain to fall and oceans to form, the planet's surface came to be dominated by liquid water. As rain and wind slowly eroded rocks and rivers carried dissolved ions to the ocean, the oceans gradually became saline enough to resemble today's seawater. Organic compounds produced by redox reactions in the atmosphere dissolved in rainwater and accumulated in the ocean.

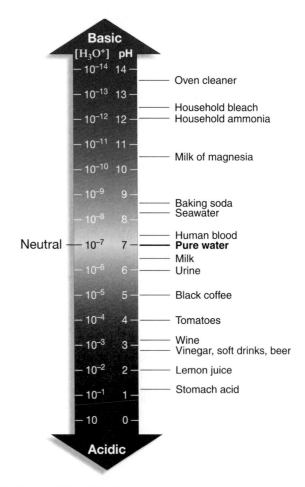

FIGURE 2.28 The pH Scale
Because the pH scale is logarithmic, a change in one unit of pH represents a change in the concentration of hydrogen ions equal to a factor of 10. Coffee has a hundred times more H^+ than pure water has.

Based on the arguments presented in this section, then, the environment from which chemical evolution proceeded was a salty solution of nearly neutral pH. The solution contained at least some compounds with reduced carbon atoms that resulted from redox reactions in the atmosphere, along with simple organic molecules produced when these carbon-containing molecules were heated. Thanks to water's high specific heat, however, these first products of chemical evolution were relatively well protected from further inputs of energy.

What happened next? For chemical evolution to continue, two things had to happen. First, reactions among relatively small and simple organic compounds had to produce the building blocks of the large molecules found in living cells. Second, these building blocks had to link together to form proteins, nucleic acids, and carbohydrates—the large, complex compounds found in organisms. Our task in the next three chapters is to analyze how these events occurred and how proteins, nucleic acids, and carbohydrates function in organisms today. As far as we know, the jump from nonlife to life has occurred only once in the history of the universe. According to the chemical evolution hypothesis, it happened as these molecules began to accumulate in the waters of the ancient Earth.

ESSAY The Search for Extraterrestrial Life

Astronomers who search for life on other planets have recently been invigorated by two findings: (1) evidence that liquid water is present under the ice-crusted surface of Jupiter's moon Europa (**Figure 2.29**); and (2) the discovery of cells living in rock formations hundreds of meters below Earth's surface.

Confirming that liquid water exists elsewhere in the solar system is exciting because life is probably impossible in the absence of liquid water. Finding evidence for liquid water on Europa narrows the search for extraterrestrial life—it gives scientists a promising place to look.

Finding organisms in the deep subsurface of Earth suggested that life could also be found under the surface of Mars, the Moon, or other bodies. Cells have recently been found in Earth rocks located up to 860 meters (about half a mile) below the surface. These discoveries have extended a general realization—that organisms are found in a wide variety of extreme environments on Earth. Species of the single-celled organisms called bacteria and archaea, and sometimes even multicellular (many-celled) animals, can thrive within glaciers, in extremely high pressure water near superheated steam vents deep on the ocean floor, in hot springs, and in ponds that are nearly saturated with salt. The presence of life in extreme environments on Earth has encouraged astronomers to suggest that life may also exist in extreme environments in space. Planets and moons that appear to be lifeless could actually be teeming with organisms just below the surface.

As biologists gain a better understanding of chemical evolution and the diversity of life on ancient Earth, the search for extraterrestrial life has become more focused. Few astronomers expect to find the sophisticated types of life-forms favored by science fiction writers. Instead, astronomers are looking for evidence of water, molecules with reduced carbon atoms, and traces of single-celled life-forms similar to Earth's bacteria and archaea.

Planets and moons that appear to be lifeless could actually be teeming with organisms below the surface.

Finding life elsewhere in the solar system would instantly qualify as a tremendous scientific advance. It would refute the hypothesis that chemical evolution occurred just once and that life is unique to Earth. It would also open up the possibility of finding organisms in many other locations throughout the universe—perhaps in the recently discovered planetary systems that are currently forming around young stars.

(a) Europa (moon of Jupiter)

(b) Closer view of ice crust on Europa

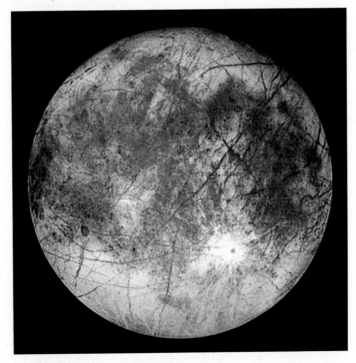

FIGURE 2.29 Europa—a Moon of Jupiter
The surface of Europa is frozen, but data suggest that liquid water exists underneath the ice.

CHAPTER REVIEW

Summary of Key Concepts

Chemical evolution is a hypothesis about the origin of the essential chemical components of life. It maintains that early in Earth's history, large and complex molecules formed from simple precursor compounds in the ancient atmosphere and ocean, as chemical reactions converted the energy in sunlight and other sources into chemical energy. Chemical energy is a form of potential energy and can be stored in bonds between atoms. Based on radiometric dating, advocates of the hypothesis suggest that chemical evolution took place over a span of some 300 million years, beginning about 3.85 billion years ago.

■ **Molecules form when atoms bond to each other. Chemical bonds are based on electron sharing. The degree of electron sharing varies from nonpolar covalent bonds, to polar covalent bonds, to ionic bonds.**

When atoms participate in chemical bonds to form molecules such as those produced during chemical evolution, the shared electrons give the atoms full valence shells and thus contribute to the atoms' stability. The electrons may be shared equally or unequally, depending on the electronegativities of the two atoms involved. Nonpolar covalent bonds result from equal sharing; polar covalent bonds are due to unequal sharing; and ionic bonds form when an electron is completely transferred from one atom to another.

■ **Chemical reactions tend to be spontaneous if they lead to lower potential energy and higher entropy (more disorder). An input of energy is required for nonspontaneous reactions to occur.**

The reactions involved in chemical evolution resulted in products that had higher potential energy and lower entropy than the reactants. As a result, these reactions were nonspontaneous. They could occur only because there was an input of kinetic energy in the form of heat, sunlight, and lightning.

■ **During chemical evolution, (1) energy in sunlight was converted to chemical energy and (2) carbon atoms were reduced.**

Researchers have developed models for how the first step in chemical evolution occurred. This step was the formation of compounds with reduced carbon atoms, such as formaldehyde and hydrogen cyanide, from molecules such as CO_2. The models show that these reactions could occur only if a source of energy, such as sunlight, *and* a source of inorganic compounds with high potential energy, such as ammonia (NH_3), methane (CH_4), and molecular hydrogen (H_2), were present in the ancient atmosphere. Although the composition of the ancient atmosphere is controversial, most investigators now agree that sources of energy were abundant and at least some inorganic molecules with high potential energy existed.

Web Tutorial 2.1   Redox Reactions

■ **Water is a small, highly polar molecule. As a result, it is an extremely efficient solvent and has a high capacity for absorbing energy.**

If compounds with reduced carbon atoms formed in the ancient atmosphere, they would have dissolved in water droplets and rained down into the early ocean. Advocates of the chemical evolution hypothesis propose that most subsequent chemical evolution took place in water, primarily because it is such an efficient solvent. Water is the most effective solvent known, because it is polar, meaning that it has partial positive and negative charges. Water is polar because it is bent and has two polar covalent bonds. As a result, polar molecules and charged substances, including ions, interact with water and stay in solution.

The general message of this chapter can be stated as follows: Chemical evolution was possible because the atmosphere and oceans of ancient Earth contained inorganic compounds with high potential energy, abundant and diverse sources of energy, and a large supply of water.

Web Tutorial 2.2   The Properties of Water

Questions

Content Review

1. Which of the following occurs when a covalent bond forms?
   a. The potential energy of electrons drops.
   b. Electrons in valence shells are shared between nuclei.
   c. Ions of opposite charge interact.
   d. Polar molecules interact.

2. If a reaction is exothermic, then which of the following statements is true?
   a. The products have lower potential energy than the reactants.
   b. Energy must be added for the reaction to proceed.
   c. The products have higher entropy (are more disordered) than the reactants.
   d. It occurs extremely quickly.

3. What is thermal energy?
   a. a form of potential energy
   b. the temperature increase that occurs when any form of energy is added to a system
   c. mechanical energy
   d. the kinetic energy of molecular motion, measured as heat

4. What determines whether a chemical reaction is spontaneous?
   a. if it increases the disorder, or entropy, of the substances involved
   b. if it decreases the potential energy of the substances involved
   c. the temperature only—reactions are spontaneous at high temperatures and nonspontaneous at low temperatures
   d. the combined effect of changes in potential energy and entropy

5. Which of the following is *not* an example of an energy transformation?
   a. A shoe drops, converting potential energy to kinetic energy.
   b. A chemical reaction converts the energy in sunlight into the chemical energy in formaldehyde.
   c. The electrical energy flowing through a light bulb's filament is converted into light and heat.
   d. Sunlight strikes a prism and separates into distinct wavelengths.

6. When an atom is reduced, it
   a. gains entropy
   b. loses entropy
   c. gains an electron
   d. loses an electron

Conceptual Review

1. Section 2.2 describes the reaction between carbon dioxide and water, which forms carbonic acid:

$$CO_2(g) + H_2O(l) \rightleftharpoons H_2CO(aq)$$

In aqueous solution, carbonic acid immediately dissociates to form a proton and the bicarbonate ion, as follows:

$$H_2CO_3(aq) \rightleftharpoons H^+(aq) + HCO_3^-(aq)$$

Does this reaction raise or lower the pH of the solution? Does the bicarbonate ion act as an acid or a base? If an underwater volcano bubbled additional CO_2 into the ocean, would this sequence of reactions be driven to the left or the right? How would this affect the pH of the ocean?

2. When chemistry texts introduce the concept of electron shells, they emphasize that shells represent distinct potential energy levels. In introducing electron shells, this chapter also emphasized that they represent distinct distances from the positive charges in the nucleus. Are these two points of view in conflict? Why or why not?

3. In using radiometric dating to estimate when a rock formed, geologists must know three quantities. What are they? How are they used to estimate the age of a rock sample? What sources of uncertainty are involved in measuring these three quantities?

4. Why does ice float?

5. Hydrogen bonds form because the opposite, partial electric charges on polar molecules attract. Covalent bonds form as a result of the electrical attraction between electrons and protons. Covalent bonds are much stronger than hydrogen bonds. Explain why, in terms of the electrical attractions involved.

6. Draw a ball-and-stick model of the water molecule, and explain why this molecule is bent. Indicate the location of the partial electric charges on it. Why do these partial charges exist?

Group Discussion Problems

1. Suppose you wanted to use radiometric dating to estimate the age of an archaeological site that was thought to be about 1500 years old. Would you use the uranium-lead system described in Section 2.1, or would you evaluate a radioisotope that had a shorter half-life? Explain your answer.

2. Oxygen is extremely electronegative, meaning that its nucleus pulls in electrons shared in covalent bonds. Because these electrons are close to the oxygen nucleus, they have lower potential energy. Explain the changes in electron position that are illustrated in Figure 2.20 in terms of oxygen's electronegativity.

3. When nuclear reactions take place, some of the mass in the atoms involved is converted to energy. The energy in sunlight is created during nuclear fusion reactions on the Sun. Explain what astronomers mean when they say that the Sun is burning down and that it will eventually burn out.

4. Why do coastal regions tend to have climates with moderate temperatures and lower annual variation in temperature than do inland areas at the same latitude?

Answers to Multiple-Choice Questions   **1.** b; **2.** a; **3.** b; **4.** d; **5.** d; **6.** c

3 Protein Structure and Function

KEY CONCEPTS

- Proteins are made of amino acids. Amino acids vary in structure and function because their side chains vary in composition. As a result, proteins vary widely in structure and function.

- In cells, most proteins are enzymes that function as catalysts. Chemical reactions occur much faster when they are catalyzed by enzymes. During enzyme catalysis, the reactants bind to an enzyme's active site in a way that allows the reaction to proceed efficiently.

- In cells, endergonic reactions occur in conjunction with an exergonic reaction involving ATP.

A biologist wearing 3-D goggles studies the structure of a protein, represented by gray ribbons. The researcher is using a computer model to test the hypothesis that a drug can bind to the protein at specific locations..

Chapter 2 introduced the hypothesis that chemical reactions in the atmosphere and ocean of ancient Earth led to the formation of complex carbon-containing compounds. This idea, called chemical evolution, was first proposed in 1923 by Alexander I. Oparin. The hypothesis was published again—independently and six years later—by J. B. S. Haldane. Today, the Oparin-Haldane proposal can best be understood as a formal scientific theory. As noted in Chapter 1, scientific theories typically have two components: a statement about a pattern that exists in the natural world and a proposed mechanism or process that explains the pattern. In the case of chemical evolution, the pattern is that increasingly complex carbon-containing molecules formed in the atmosphere and ocean of ancient Earth. The process responsible for this pattern was the conversion of energy, from sunlight and other sources, into chemical energy in the bonds of large, complex molecules.

Scientific theories are continuously refined as new information comes to light, and many of Oparin and Haldane's original ideas about how chemical evolution occurred have been extensively revised. In its current form, the theory can be broken into four steps, each requiring an input of energy:

1. Chemical evolution began with the production of small compounds with reduced carbon atoms, such as formaldehyde (H_2CO) and hydrogen cyanide (HCN).

2. These simple compounds reacted to form the mid-sized molecules called amino acids, sugars, and nitrogenous bases. These building-block molecules accumulated in the shallow waters of the ancient ocean, forming a complex solution called the **prebiotic soup**.

3. Mid-sized, building-block molecules linked to form the types of large molecules found in cells today, including proteins, nucleic acids, and complex carbohydrates. These large molecules are each made up of different chemical

46

subunits: Proteins are composed of amino acids; nucleic acids are composed of nucleotides; and complex carbohydrates are composed of sugars.

4. Life became possible when one of these large, complex molecules acquired the ability to make a copy of itself. This self-replicating molecule began to multiply by means of chemical reactions that it controlled. At that point, life had begun. Chemical evolution gave way to biological evolution.

Analyzing how the final three steps in chemical evolution occurred is the subject of this chapter and the next three chapters. Each of these chapters focuses on one of the four primary classes of biological molecules found in cells living today: proteins, nucleic acids, carbohydrates, and lipids. What do the subunits of these large molecules look like, and how could they have been created by chemical evolution? How are the subunits linked to form a protein or nucleic acid or complex carbohydrate or lipid? What do these large molecules look like, and what do they do in living cells? Finally, which type of molecule was responsible for the origin of life? This question is particularly intriguing, because researchers around the world are racing to synthesize a self-replicating molecule—to produce life in a test tube. Which type of molecule are they working with, and why?

Let's begin our analysis of these questions with the workhorse molecules called proteins, and the most famous experiment ever performed on the origin of life.

3.1 Early Origin-of-Life Experiments

In 1953 a graduate student named Stanley Miller performed a breakthrough experiment in the study of chemical evolution. Miller wanted to answer a simple question: Can complex organic compounds be synthesized from the simple molecules present in Earth's early atmosphere and ocean? In other words, is it possible to recreate the first steps in chemical evolution by simulating ancient Earth conditions in the laboratory?

Miller based his experimental design on the assumption that Earth's early atmosphere was dominated by molecules with high free energy when chemical evolution occurred. These molecules are likely to give up electrons, so their presence means that redox reactions can occur.

Miller's experimental setup (**Figure 3.1**) was designed to produce a microcosm of ancient Earth. The large glass flask represented the atmosphere and contained the gases methane (CH_4), ammonia (NH_3), and hydrogen (H_2), all of which have high free energy. This large flask was connected to a smaller flask by glass tubing. The small flask held a tiny ocean—200 milliliters (mL) of liquid water. Miller boiled this water constantly so that water vapor was added to the mix of gases in the large flask. As the vapor cooled and condensed, it flowed back into the smaller flask, where it boiled again. In this way, water vapor circulated continuously through the system. This was important: If the molecules in the simulated atmosphere reacted with one anoth-

Question: Can simple molecules and kinetic energy lead to chemical evolution?

Hypothesis: If kinetic energy is added to a mix of simple molecules with high free energy, redox reactions will occur that produce more complex molecules, perhaps including some with carbon-carbon bonds.

Null hypothesis: If kinetic energy is added to a mix of simple molecules with high free energy, more complex molecules will not be produced.

Experimental setup:

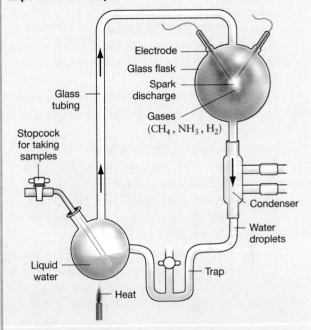

Prediction: Complex organic compounds will be found in the liquid water.

Prediction of null hypothesis: Only the starting molecules will be found in the liquid water.

Results:

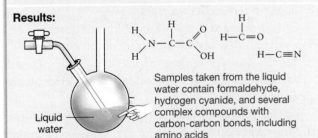

Samples taken from the liquid water contain formaldehyde, hydrogen cyanide, and several complex compounds with carbon-carbon bonds, including amino acids

Conclusion: Chemical evolution occurs readily if simple molecules with high free energy are exposed to a source of kinetic energy.

FIGURE 3.1 Miller's Spark-Discharge Experiment
The arrows in the "Experimental setup" diagram indicate the flow of water vapor or liquid, starting with the 200 milliliters in the small, boiling flask. The large glass flask can contain any mixture of gases desired; when a voltage is applied across the electrodes in that flask, a spark jumps across the gap between them. The condenser consists of a jacket with cold water flowing through it. **EXERCISE** Label the parts of the apparatus that mimic the ocean, the atmosphere, rain, and lightning.

BOX 3.1   Was Miller Correct about Conditions on Ancient Earth?

Miller's work is considered classic, primarily because he showed that hypotheses about chemical evolution can be tested experimentally. But today few researchers consider his experiment an accurate simulation of chemical evolution.

To understand why, recall from Chapter 2 that atmospheric scientists now propose that the atmosphere of ancient Earth was dominated by volcanic gases. If so, then it is likely that most carbon occurred in molecules with low free energy, such as carbon dioxide (CO_2) and carbon monoxide (CO), instead of molecules with high free energy, such as CH_4.

Why is this important? The answer lies in the position of the electrons in these molecules. As explained in Chapter 2, the electrons in the C–O bonds of CO_2 are held tightly by oxygen because of its high electronegativity. In contrast, the electrons in the C–H bonds of CH_4 are shared equally by the carbon and hydrogen atoms, which have relatively low

electronegativities. Thus the C–O bonds in CO_2 and CO are much stronger than the C–H bonds of CH_4. As a result, molecules such as CO_2 and CO are much less reactive than forms such as CH_4. This prediction has been confirmed experimentally. When CO_2, N_2, and H_2O are substituted for methane, ammonia, and hydrogen in Miller's spark-discharge experiment, virtually no chemical evolution takes place.

In response, atmospheric scientists began doing different types of experiments with the components of an oxidized atmosphere. For example, when water vapor and carbon monoxide or other volcanic gases are put into a glass flask and exposed to the types of high-energy radiation found in sunlight, a wide variety of compounds with reduced carbon atoms forms, including methane (CH_4), formaldehyde, and acetaldehyde. Similar experiments have altered the ratio of water vapor to carbon monoxide,

changed the temperature, extended the length of time the mixture was exposed to light, or included other molecules found in volcanic gases, such as hydrogen (H_2), nitrogen (N_2), or carbon dioxide (CO_2). In each case, a wide variety of organic compounds with high free energy formed.

In addition, some researchers recently asserted that the key events in chemical evolution did not begin in the atmosphere at all. Instead, they suggest that the process also occurred near hydrothermal ("hot-water") vents similar to the volcanic "black smokers" found at the bottom of oceans today (**Figure 3.2**). Black smokers form where molten rock occurs just below the ocean floor. The thermal energy in the liquid rock heats water in the surrounding rocks to temperatures as high as 450°C. Because there is intense pressure at these depths, this "superheated" water does not boil. Instead, it rises up through the crust. As it passes through the crustal rocks, it dissolves enough iron-, sulfur-, nickel-,

---

er, the "rain" would carry them into the simulated ocean, forming a simulated version of the prebiotic soup.

Had Miller stopped there, however, little or nothing would have happened. Even at the boiling point of water (100°C), the molecules involved in the experiment are stable. They do not undergo spontaneous chemical reactions, even at high temperatures.

Something did start to happen in the apparatus, however, when Miller sent electrical discharges across the electrodes he'd inserted into the atmosphere. These miniature lightning bolts added a crucial element to the reaction mix: pulses of intense electrical energy. After a day of continuous boiling and sparking, the solution in the boiling flask began to turn pink. After a week, it was deep red and cloudy. When Miller analyzed the molecules dissolved in the solution, he found that several complex carbon-containing compounds were present. The experiment, driven by the energy in the electrical discharges, had recreated the start of chemical evolution.

To find out exactly which products resulted from the initial reactions in the simulated atmosphere, Miller drew samples from the apparatus at intervals. In these samples he found large

quantities of hydrogen cyanide (HCN) and formaldehyde (H_2CO). The data were exciting because HCN and H_2CO are required for reactions that lead to the synthesis of more complex organic molecules. Indeed, some of these more complex compounds were actually present in the miniature ocean. The sparks and heating had led to the synthesis of compounds that are fundamental to life: amino acids.

3.2   Amino Acids and Polymerization

Based on the presence of amino acids, Miller claimed that his experiment simulated the second stage in chemical evolution—the formation of a prebiotic soup. Although the assumptions behind his experiments, and hence the results, eventually came under fire, follow-up studies have confirmed that the second step in chemical evolution occurred early in Earth's history. On the basis of the data reviewed in **Box 3.1** there is now a strong consensus that amino acids and other components of the prebiotic soup are readily produced under conditions that accurately simulate the atmosphere and oceans of ancient Earth.

and carbon-containing compounds to form a blackened solution that jets out into the surrounding water. At the bottom of the ocean, this surrounding water is frigid—typically 4°C.

Because a carbon- and sulfur-containing molecule called methanethiol (CH_3SH) is often found near black smokers, it appears that chemical evolution is occurring in these environments even today. A research group has shown that, under the temperature and pressure conditions found near black smokers, CO and CH_3SH react to form acetic acid (CH_3COOH), which contains a carbon-carbon bond.

Based on these and other experiments simulating early Earth conditions, most researchers now agree that early in Earth's history, quantities of formaldehyde, hydrogen cyanide, acetic acid, and other carbon-containing molecules were raining down from the skies and bubbling up from hydrothermal vents.

Chemical evolution takes place where superheated water emerges from vents in the otherwise frigid ocean floor

FIGURE 3.2 Black Smokers
The water that erupts from black smokers contains large amounts of iron, carbon, and nickel from the rocks below. This water is superheated, meaning that its temperature is above 100°C. It does not boil because black smokers are found deep in the ocean, where water is under extremely high pressure.

Researchers have also realized that large quantities of amino acids may have rained onto the planet from outer space during the early bombardment phase of Earth's history, described in Chapter 2. This conclusion is based on analyses of space debris that contains carbon. The Murchison meteorite that struck Australia in 1969, for example, contains as many as 18 different amino acids (**Figure 3.3**). These data suggest that about 4 billion years ago, chemical evolution on Earth could have been supplemented by similar reactions occurring in outer space.

Now let's look at the molecules themselves. What are amino acids, and how are they linked to form proteins?

The Structure of Amino Acids

The bacterial cells that live on your skin contain several thousand different proteins; as a group, the cells in your body produce tens of thousands of distinct proteins. But most of these proteins are composed of just 20 different building blocks, called **amino acids**. All 20 amino acids have a common structure.

To understand the structure of an amino acid, recall that carbon atoms have four unpaired valence electrons and

FIGURE 3.3 Some Meteorites Contain Complex Organic Molecules
A 23-gram fragment of the Murchison meteorite. The Murchison meteorite is one of many pieces of space debris that contain abundant and diverse types of complex organic molecules, including amino acids.

(a) Non-ionized form of amino acid

Amino group   Side chain   Carboxyl group   Non-ionized

(b) Ionized form of amino acid

Amino group   Side chain   Carboxyl group   Ionized

FIGURE 3.4 Amino Acid Structure
All amino acids have the same general structure: a central carbon, shown in red, bonded to an amino functional group, a carboxyl functional group, a hydrogen atom, and a side chain, or R-group.

participate in four bonds. Every amino acid has a carbon that makes the same four bonds. The first bond attaches this carbon to NH_2—the amino functional group (**Figure 3.4a**). The second bond links this carbon to COOH—the carboxyl functional group. The carboxyl group is acidic because its two oxygen atoms are highly electronegative. They pull electrons away from the hydrogen atom, which means that it is relatively easy for this group to lose a proton. The combination of the amino and carboxyl groups inspired the name *amino acid*. A third bond links the highlighted carbon to a hydrogen atom. In all amino acids, then, a carbon atom is bonded to an amino group, a carboxyl group, and hydrogen. As we'll see in a moment, the nature of the fourth attachment is what makes each amino acid unique.

The presence of an amino group and a carboxyl group in amino acids is important. **Figure 3.4b** shows what happens to these functional groups in solution. In water at pH 7, the concentration of protons causes the amino group to act as a base. It attracts a proton to form NH_3^+. The carboxyl group, in contrast, loses a proton to form COO^-. The charges on these functional groups help amino acids stay in solution and add to their chemical reactivity.

The Nature of Side Chains In Figure 3.4, the highlighted carbon atom in the amino acid forms a fourth bond with an atom or a group of atoms abbreviated as "R." Chemists use this symbol to indicate additional atoms called a side chain. In every amino acid, a carbon atom is linked to a hydrogen atom, an amino group, a carboxyl group, and an R-group. The 20 amino acids found in organisms are different because their R-groups are different. These R-groups vary from a single hydrogen atom to large structures containing carbon atoms linked into rings. Several of the side chains found in amino acids contain functional groups (see Chapter 2, Table 2.1).

Figure 3.5 shows the 20 amino acids found in cells and sorts them according to whether their side chain is nonpolar, polar, or electrically charged. This analysis is important because amino acids with nonpolar side chains do not have charged or electronegative atoms capable of forming hydrogen bonds with water. These R-groups are said to be **hydrophobic** ("water-fearing") because water does not interact with them. As a result, hydrophobic side chains tend to coalesce in aqueous solution. In contrast, amino acids with polar or charged side chains interact readily with water and are termed **hydrophilic** ("water-loving"). Hydrophilic amino acids dissolve in water easily. **Table 3.1** ranks the 20 amino acids according to how readily they interact with water.

In addition to affecting the solubility of amino acids, the side chain influences their chemical reactivity. In Figure 3.5, notice that some amino acids contain side chains consisting entirely of carbon and hydrogen atoms. These R-groups rarely participate in chemical reactions. As a result, the behavior of these amino acids depends primarily on their size and shape rather than reactivity. In contrast, amino acids that have hydroxyl, amino, or carboxyl functional groups in their side chains are more reactive. Soon we'll see that amino acids with sulfur atoms (S) in their side chains can help link different parts of large proteins. The point is that variation in the behavior of amino acids results from variation in the structure of their R-groups.

TABLE 3.1 How Amino Acids Interact with Water

The 20 amino acids are ranked according to how likely they are to interact with water. Color codes are based on Figure 3.5.

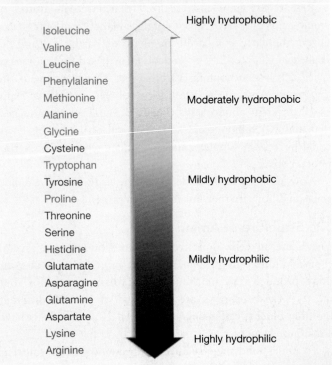

Amino acid	
Isoleucine	Highly hydrophobic
Valine	
Leucine	
Phenylalanine	
Methionine	Moderately hydrophobic
Alanine	
Glycine	
Cysteine	
Tryptophan	
Tyrosine	Mildly hydrophobic
Proline	
Threonine	
Serine	
Histidine	
Glutamate	Mildly hydrophilic
Asparagine	
Glutamine	
Aspartate	
Lysine	
Arginine	Highly hydrophilic

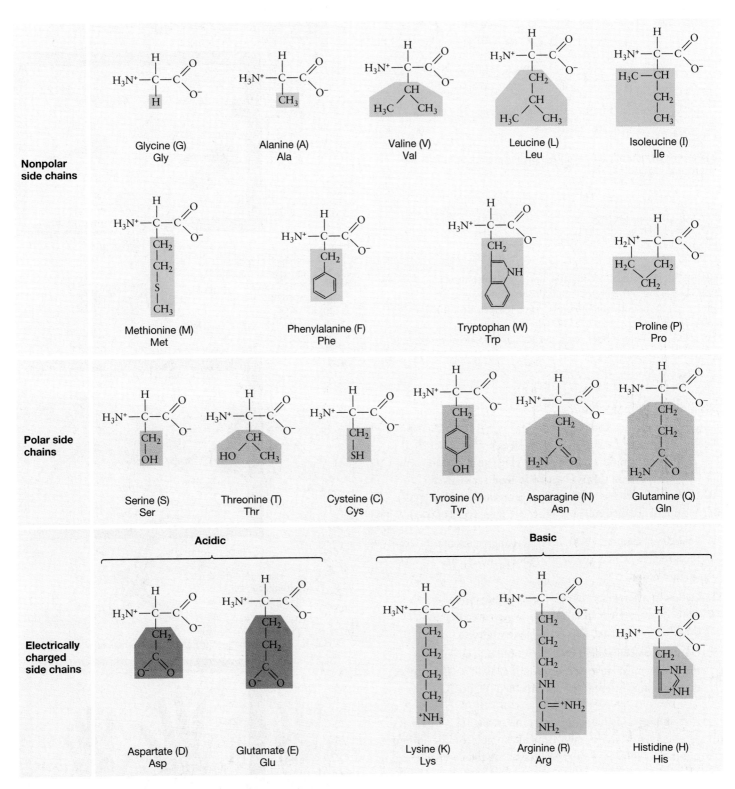

Nonpolar side chains

Glycine (G) Gly

Alanine (A) Ala

Valine (V) Val

Leucine (L) Leu

Isoleucine (I) Ile

Methionine (M) Met

Phenylalanine (F) Phe

Tryptophan (W) Trp

Proline (P) Pro

Polar side chains

Serine (S) Ser

Threonine (T) Thr

Cysteine (C) Cys

Tyrosine (Y) Tyr

Asparagine (N) Asn

Glutamine (Q) Gln

Electrically charged side chains

Acidic

Basic

Aspartate (D) Asp

Glutamate (E) Glu

Lysine (K) Lys

Arginine (R) Arg

Histidine (H) His

FIGURE 3.5 The 20 Major Amino Acids Found in Organisms

The structural formulas of the 20 major amino acids found in organisms, at the pH (about 7.0) found in cells. The side chains are highlighted, and standard single-letter and three-letter abbreviations for each amino acid are given. For clarity, the carbon atoms in the ring structures of phenylalanine, tyrosine, tryptophan, and histidine are not shown; each bend in a ring is the site of a carbon atom. The hydrogen atoms in these structures are also not shown. A double line inside a ring indicates a double bond. **EXERCISE** Label amino acid side chains that are hydrophilic versus those that are hydrophobic. To check your work, examine the data in Table 3.1.

(a) Structural isomers differ in the order which their atoms are attached.

Ethanol (C_2H_6O)   Dimethyl ether (C_2H_6O)

(b) Geometric isomers differ in the arrangement of atoms around a double bond.

trans-2-butene (C_4H_8)   *cis*-2-butene (C_4H_8)

FIGURE 3.6 Structural, Geometric, and Optical Isomers
(a) Ethanol is the active ingredient in alcoholic beverages; dimethyl ether, a gas at room temperature, is used in refrigeration. **(b)** The molecules *trans*-2-butene and *cis*-2-butene are used in the production of gasoline, synthetic rubber, and solvents. **(c)** All amino acids except glycine have optical isomers.

Explaining Optical Isomers   Experiments have shown that many of the amino acids in Figure 3.5 could have been produced by chemical evolution. But these experiments created a dilemma: Every amino acid except glycine exists in two forms, but only one of these forms is found in living cells. Why?

Molecules that have different structures but the same molecular formula are called **isomers**. There are three types of isomers:

1. **Structural isomers**, which have the same atoms but differ in the order in which covalently bonded atoms are attached (**Figure 3.6a**).

2. **Geometric isomers**, which have the same atoms but differ in the arrangement of atoms or groups on either side of a double bond or ring structure (**Figure 3.6b**).

3. **Optical isomers**, which have the same atoms but differ in the arrangement of atoms or groups around a carbon atom that has four different groups attached (**Figure 3.6c**).

Most amino acids have optical isomers. Figure 3.6c shows the arrangements in the two optical isomers of the amino acid alanine. Note that the two forms of the molecule are mirror images of one another, just as your left and right hands are mirror images of each other. Like your left and right hands, the left-handed and right-handed forms of alanine cannot be exactly superimposed. They also do not have a plane of symmetry. Carbon atoms with this feature exist in every amino acid except glycine. In fact, any carbon atom that has four different atoms or groups attached to it has an optical isomer.

(c) Optical isomers are mirror images of one another—they cannot be exactly superimposed.

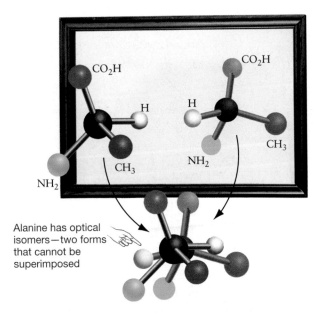

Alanine has optical isomers—two forms that cannot be superimposed

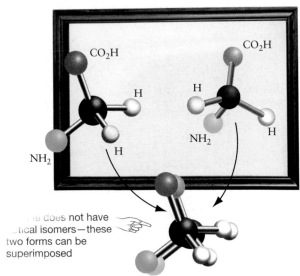

...ne does not have ...tical isomers—these two forms can be superimposed

Hands ... ...rror images, just as optical isomers are.

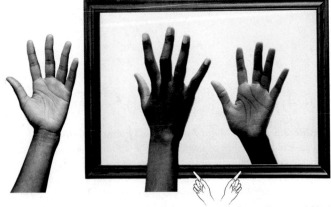

These two hands cannot be superimposed (that is, with both thumbs facing the same direction and both palms facing the same direction)

The existence of optical isomers is an important issue in biology. Because the structures of each optical isomer of a molecule are different, they have different functions. In cells, only the "left-handed" forms of amino acids exist. If the right-handed form of an amino acid is introduced into cells experimentally, it does not function normally. As in most other molecules, an amino acid's function is determined by its structure.

These observations are a challenge to the chemical evolution hypothesis, because no plausible mechanism has been proposed to explain how the process resulted in the production of only left-handed optical isomers. Was it simply a matter of chance? Or was there something unusual about the chemistry of early Earth that we still don't understand? To date, the issue is unresolved.

How Do Amino Acids Link to Form Proteins?

In the introduction to this chapter, we saw that amino acids link to form proteins. Similarly, the molecular building blocks called nucleotides attach to one another to form nucleic acids, and simple sugars connect to form complex carbohydrates. In general, a molecular subunit such as an amino acid, a nucleotide, or a sugar is called a **monomer** ("one-part"). When monomers bond together, the resulting structure is called a **polymer** ("many-parts"). The process of linking monomers is called **polymerization** (**Figure 3.7a**). Thus, amino acids polymerize to form proteins (**Figure 3.7b**). Biologists also use the word **macromolecule** to denote a very large molecule that is made up of smaller molecules joined together. A **protein** is a linear macromolecule—a polymer—that consists of linked amino acid monomers.

The theory of chemical evolution states that monomers in the prebiotic soup polymerized to form proteins and other types of macromolecules found in organisms. This is a difficult step, because monomers such as amino acids do not spontaneously self-assemble into macromolecules such as proteins. Based on the second law of thermodynamics reviewed in Chapter 2, this is not surprising. Complex and highly organized molecules are not expected to form spontaneously from simpler constituents, because polymerization organizes the molecules involved into a more complex, ordered structure. Stated another way, polymerization decreases the disorder, or entropy, of the molecules involved. In addition, polymers are energetically much less stable than their component monomers. In sum, the ΔH term of the Gibbs free-energy equation is positive and the $T\Delta S$ term is negative, making ΔG positive at all temperatures. Polymerization reactions are nonspontaneous. Monomers must absorb energy in order to link together. How could this have happened during chemical evolution?

(a) Monomers polymerize to form polymers.

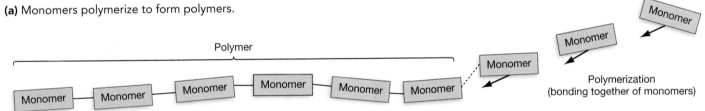

Polymer

Monomer — Monomer — Monomer — Monomer — Monomer — Monomer — Monomer — Monomer — Monomer

Polymerization
(bonding together of monomers)

(b) Proteins are polymers that are made up of amino acid monomers.

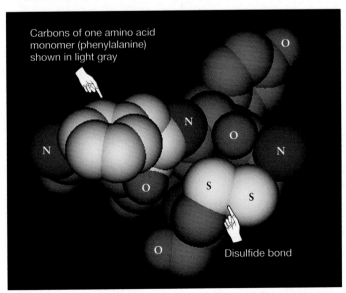

Carbons of one amino acid monomer (phenylalanine) shown in light gray

Disulfide bond

FIGURE 3.7 Monomers Are the Building Blocks of Polymers
(a) Monomers can be linked together to form polymers. (b) A polypeptide called ADH, which helps regulate urine formation in your body. One of the nine amino acids in the peptide chain is highlighted.

Could Polymerization Occur in the Energy-Rich Environment of Early Earth?

Researchers who added heat or electrical discharges to solutions of either amino acids or nucleic acids found that polymerization reactions proceed slowly, if at all. This is because monomers polymerize through **condensation reactions**, also known as **dehydration reactions**. These reactions are aptly named because the newly formed bond results in the loss of a water molecule (**Figure 3.8a**). The reverse reaction, called **hydrolysis**, breaks polymers apart by adding a water molecule (**Figure 3.8b**). The water molecule reacts with the bond linking the monomers, separating one monomer from the polymer chain.

In a solution, such as the prebiotic soup, condensation and hydrolysis represent the forward and reverse reactions of a chemical equilibrium. Hydrolysis dominates because it increases entropy and because it is energetically favorable—it lowers the potential energy of the electrons involved.

According to recent experiments, the key to overcoming hydrolysis during chemical evolution was, quite literally, as common as mud. Researchers have been able to create stable polymers by incubating monomers with tiny mineral particles—the size found in clay or mud. These experiments were based on the hypothesis that growing macromolecules would be protected from hydrolysis if they clung, or adsorbed, to the mineral surfaces.

More specifically, the experiments were designed to simulate events that could have occurred in the prebiotic soup. In one experiment, researchers put amino acids in a solution with tiny mineral particles and allowed them to react. After a day, the researchers separated the mineral particles from the solution. They then put the particles into a fresh solution containing amino acids and a source of energy. After repeating this procedure for several days, they analyzed the mineral particles and found polymers up to 55 amino acids long. These results support the hypothesis that adsorption to mineral particles protects polymers from hydrolysis. Because the experimental procedure was designed to mimic coastal environments where beaches are repeatedly washed with fresh waves or tidal flows, the results make it reasonable to claim that at least some muddy tide pools and beaches became covered with small proteins early in Earth's history.

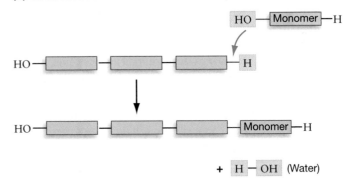

(a) Condensation reaction: monomer in, water out

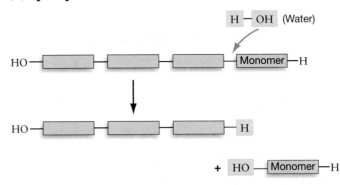

(b) Hydrolysis: water in, monomer out

FIGURE 3.8 Polymers Can Be Extended or Broken Apart
(a) In a condensation reaction, a monomer is added to a polymer to make a longer polymer. The new bond that forms results in the formation of a water molecule. **(b)** In hydrolysis, a water molecule reacts with the bond linking the monomers. A monomer is broken off the chain, resulting in a shorter polymer.

The Peptide Bond Exactly how do amino acids polymerize? As **Figure 3.9** shows, amino acids polymerize when a bond forms between the carboxyl group of one amino acid and the amino group of another. The C–N bond that results from this condensation reaction is called a **peptide bond**. This bond is particularly stable because electrons are partially shared between the neighboring carbonyl functional group and the peptide bond. The degree of electron sharing is great enough that peptide bonds ac-

FIGURE 3.9 Peptide Bond Formation
When the carboxyl group of one amino acid reacts with the amino group of a second amino acid, a peptide bond forms. **QUESTION** Is a peptide bond a hydrogen bond, a nonpolar covalent bond, a polar covalent bond, or an ionic bond?

(a) Polypeptide chain

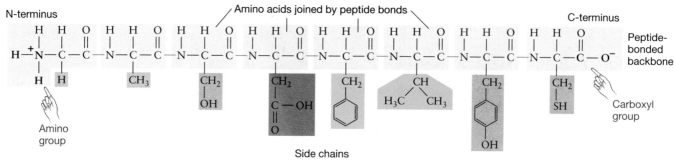

(b) Numbering system

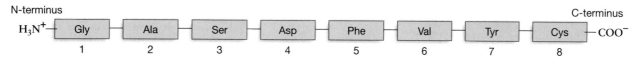

FIGURE 3.10 Amino Acids Polymerize to Form Polypeptides
(a) Amino acids can be linked into long chains, called polypeptides, by peptide bonds. **(b)** By convention, the sequence of amino acids in a polypeptide chain is numbered from the N-terminus to the C-terminus.

tually have some of the characteristics of a double bond. For example, the peptide bond is planar.

When a series of amino acids are linked by peptide bonds into a chain, the amino acids are referred to as *residues* and the resulting molecule is called a **polypeptide. Figure 3.10a** shows how the chain of peptide bonds in a polypeptide gives the molecule a structural framework, or a "backbone." Three points are important to note about this backbone: (1) The side chains present in each residue extend out from it, (2) it has directionality, and (3) it is flexible. The backbone is directional because there is an amino group $(-NH_3^+)$ on one end of every polypeptide chain and a carboxyl group $(-COO^-)$ on the other. By convention, biologists always write amino acid sequences in the same direction. The end of the sequence that has the free amino group is placed on the left and is called the N-terminus, or amino-

terminus, and the end with the free carboxyl group appears on the right-hand side of the sequence and is called the C-terminus, or carboxy-terminus. The amino acids in the chain are always numbered starting from the N-terminus (**Figure 3.10b**), because the N-terminus is the start of the chain when proteins are synthesized in cells. Although the peptide bond itself cannot rotate because of its double-bond nature, the single bonds on either side of the peptide bond can rotate. As a result, the structure as a whole is flexible (**Figure 3.11**).

When fewer than 50 amino acids are linked together in this way, the resulting polypeptide is called an **oligopeptide** ("few peptides") or simply a peptide. Polypeptides that contain 50 or more amino acids are formally called proteins. Proteins may consist of single polypeptides or multiple polypeptides that are bonded to each other.

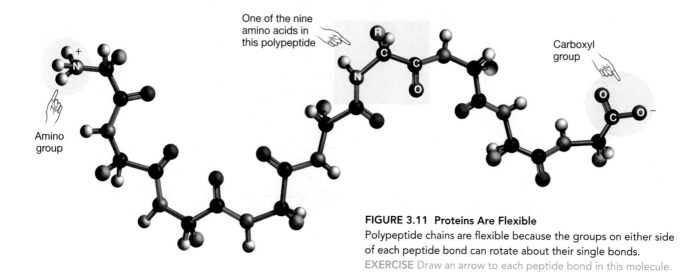

FIGURE 3.11 Proteins Are Flexible
Polypeptide chains are flexible because the groups on either side of each peptide bond can rotate about their single bonds.
EXERCISE Draw an arrow to each peptide bond in this molecule.

Proteins are the stuff of life. Let's take a look at how they are put together, and then at what they do.

✓ CHECK YOUR UNDERSTANDING

Amino acids are small molecules with a carbon atom bonded to a carboxyl group, an amino group, a hydrogen atom, and a side chain called an R-group. Each amino acid has distinctive chemical properties because each has a unique R-group. You should be able to draw the general form of an amino acid.

When the carboxyl group of one amino acid reacts with the amino group of another amino acid, a strong covalent bond called a peptide bond forms. Polypeptides are polymers made up of peptide-bonded amino acids. Small polypeptides are called oligopeptides, and large polypeptides are called proteins. You should be able to draw and label two amino acids linked by a peptide bond.

3.3   What Do Proteins Look Like?

With respect to their structure, proteins may be the most diverse class of molecules known. To drive this point home, consider the structures illustrated in **Figure 3.12**. The hormone glucagon, which is regulating the concentration of sugar in your blood right now, is shown in Figure 3.12a. This molecule is just 29 amino acids long, and it consists of a single polypeptide that folds into a simple coil. In contrast, a protein called cytochrome *c* oxidase, which is critical to energy production in your cells, contains over 3600 amino acids. It is made up of 13 distinct polypeptides and is roughly rectangular (Figure 3.12b). Aspartate transcarbamoylase has two types of subunit (shown in light and dark gray in (Figure 3.12c). The complete protein consists of two of the dark subunits and three of the light subunits and is triangular. In many cases, the shape of a protein correlates closely with its function. The TATA-box binding protein (Figure 3.12d) has a groove where DNA molecules fit. Porin is doughnut-

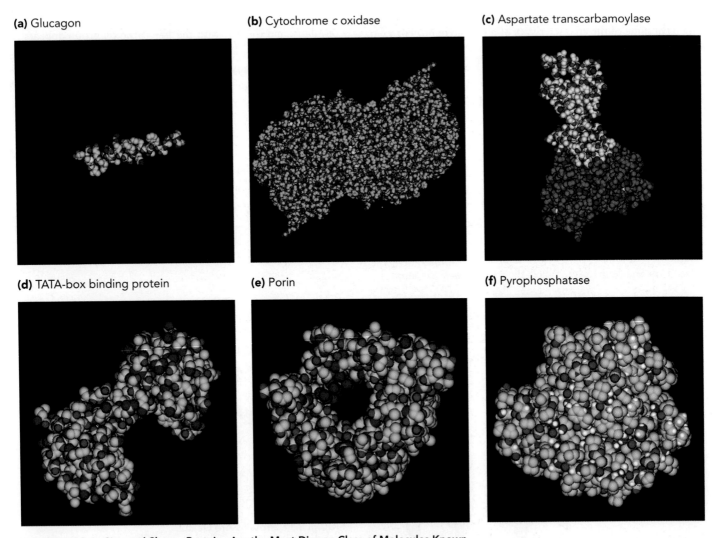

(a) Glucagon

(b) Cytochrome *c* oxidase

(c) Aspartate transcarbamoylase

(d) TATA-box binding protein

(e) Porin

(f) Pyrophosphatase

FIGURE 3.12 In Size and Shape, Proteins Are the Most Diverse Class of Molecules Known
The proteins found in cells may be (a) small or (b) large. In addition, they range from (a) coils to (b) rectangular, (c) triangular, (d) butterfly, (e) doughnut, and (f) globular shapes..

shaped, with a hole that forms a pore (Figure 3.12e). Most of the proteins found in cells function as enzymes and are globular (Figure 3.12f).

How can biologists make sense of this diversity of protein size and shape? Initially, the amount of variation seems overwhelming. Fortunately, it is not. No matter how large or complex a protein may be, its underlying structure can be broken down into just four basic levels of organization.

Primary Structure

Every protein has a unique sequence of amino acids. That simple message was the culmination of 12 years of study by Frederick Sanger and co-workers during the 1940s and 1950s. Sanger's group worked out the first techniques for determining the amino acid sequence of a protein and published the completed sequence of the hormone insulin, a protein that works in conjunction with glucagon to regulate sugar concentrations in the blood of humans and other mammals. When other proteins were analyzed, it rapidly became clear that each protein has a definite and distinct amino acid sequence.

Biochemists call the unique sequence of amino acids in a protein the **primary structure** of that protein. The sequence of amino acids in Figure 3.10, for example, defines that polypeptide's primary structure. With 20 types of amino acids available and a variation in size from two amino acid residues to tens of thousands long, the number of primary structures that are possible is practically limitless. There are, in fact, 20^n different polypeptides of length n. For a polypeptide that is just 10 amino acids long, 20^{10}, or 10,000 billion, primary sequences are possible.

Recall that the R-groups present on each amino acid affect its solubility and chemical reactivity. Based on this observation, it is reasonable to predict that the R-groups present in a polypeptide will affect the polypeptide's properties and function. In some cases, even a single change in the sequence of amino acids can cause radical changes in the way the molecule as a whole behaves. As an example, consider the hemoglobin protein of humans, which is carrying oxygen in your blood right now. In some individuals, hemoglobin has a valine instead of a glutamate at the amino acid numbered 6 in a strand of 146 amino acids (**Figure 3.13a**). Valine's side chain is very different from the R-group in glutamate. The change results in a protein that tends to crystallize instead of staying in solution when oxygen concentrations in the blood are low (**Figure 3.13b**). People whose hemoglobin contains this single amino acid change suffer from the debilitating illness called sickle-cell disease. A protein's primary structure is fundamental to its function.

Secondary Structure

Even though variation in the amino acid sequence of a protein is virtually limitless, it is only the tip of the iceberg in terms of generating structural diversity. The next level of organization in proteins is known as **secondary structure**, which is created by hydrogen bonding. More specifically, secondary structure results from hydrogen bonding that occurs between the carboxyl oxygen of one amino acid residue and the hydrogen on the amino group of another. The oxygen atom in the carboxyl group has a partial negative charge due to its high electronegativity, while the

(a) Normal amino acid sequence

Thr	Pro	Glu	Glu
4	5	6	7

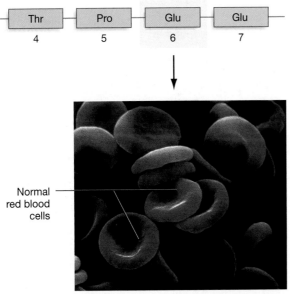

Normal red blood cells

(b) Single change in amino acid sequence

Thr	Pro	Val	Glu
4	5	6	7

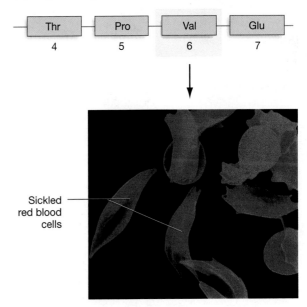

Sickled red blood cells

FIGURE 3.13 Changes in Primary Structure Affect Protein Function
Compare the primary structure of **(a)** normal hemoglobin with that of **(b)** hemoglobin molecules of people with sickle-cell disease. The single amino acid change causes red blood cells to change from their normal disc shape in (a) to a sickled shape in (b) when oxygen concentrations are low. Each red blood cell contains about 300 million hemoglobin molecules.

hydrogen atom in the amino group has a partial positive charge because it is bonded to nitrogen, which has high electronegativity (**Figure 3.14a**).

Note that secondary structure is not created by interactions among side chains but instead by interactions between atoms that are part of a protein's peptide-bonded backbone. This is a key point, because hydrogen bonding between sections of the backbone is possible only when different parts of the same polypeptide bend in a way that puts carboxyl and amino groups close together. The bending that aligns parts of the backbone and allows these bonds to form occurs in several distinct ways. **Figure 3.14b** shows two of the most important configurations that allow formation of hydrogen bonds: (1) an

(a) Hydrogen bonds form between peptide chains.

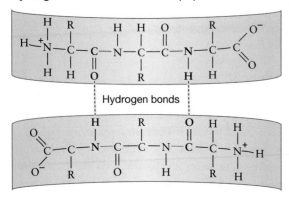

(b) Secondary structures of proteins result.

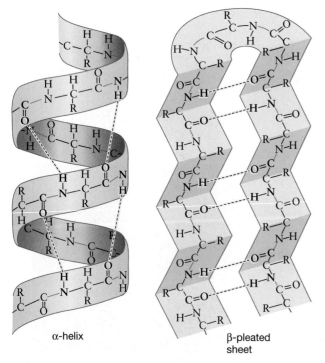

α-helix   β-pleated
                           sheet

FIGURE 3.14 Secondary Structures of Proteins
(a) The peptide-bonded backbone of a protein can coil or fold in on itself when hydrogen bonds form between amino groups and carboxyl groups. **(b)** The detailed structure of the coils called α-helices and the folds called β-pleated sheets.

α-helix (alpha helix), in which the polypeptide's backbone is coiled, and (2) a **β-pleated sheet** (beta-pleated sheet), in which segments of a peptide chain bend 180° and then fold in the same plane. In many cases, secondary structure consists of α-helices and β-pleated sheets. Which one forms, if either, is a product of the molecule's primary structure—specifically, the geometry of the amino acids in the sequence. Methionine and glutamic acid, for example, are much more likely to be involved in α-helices than in β-pleated sheets. The opposite is true for valine and isoleucine. Proline, in contrast, is unlikely to be involved in either type of secondary structure.

Although each of the hydrogen bonds in an α-helix or a β-pleated sheet is very weak relative to a covalent bond, the large number of hydrogen bonds in these structures makes them highly stable. As a result, they increase the stability of the molecule as a whole and help define its shape. In terms of overall shape and stability, though, a protein's tertiary structure is even more important.

Tertiary Structure

Alpha helices and β-pleated sheets form due to interactions between components of a protein's peptide-bonded backbone. In contrast, most of the overall shape, or **tertiary structure**, of a polypeptide results from interactions between R-groups or between R-groups and the peptide backbone. As **Figure 3.15a** shows, side chains can be involved in a wide variety of bonds and interactions. Because each of these contacts causes the peptide-bonded backbone to bend and fold, each contributes to the distinctive three-dimensional shape of a polypeptide.

Four types of interactions that involve side chains are particularly important:

1. Covalent bonds can form between sulfur atoms when a redox reaction occurs between sulfur-containing R-groups, such as those found in cysteine. These **disulfide** ("two-sulfur") **bonds** are frequently referred to as bridges because they create strong links between distinct regions of the same polypeptide.

2. In an aqueous solution, water molecules interact with hydrophilic side chains and force hydrophobic side chains to coalesce. Once the hydrophobic side chains are close to each other, they are stabilized by electrical attractions known as **van der Waals interactions**. These weak attractions occur because the constant motion of electrons gives molecules a tiny asymmetry in charge that changes with time. If molecules get extremely close to each other, the minute partial charge on one molecule induces an opposite partial charge in the nearby molecule and causes an attraction. Although the attraction is very weak relative to covalent bonds or even hydrogen bonds, a large number of van der Waals interactions can occur in a polypeptide when many hydrophobic residues congregate. The result is a significant increase in stability. Such hydrophobic interactions are responsible for folding that results in the globular shape of many proteins.

3. Ionic bonds form between groups that have full and opposing charges, such as the ionized amino and carboxyl functional groups highlighted on the right in Figure 3.15a.

4. Hydrogen bonds form in several ways: between hydrogen atoms and the carboxyl group in the peptide-bonded backbone, and between hydrogen and atoms with partial negative charges in side chains.

With so many interactions possible between side chains and peptide-bonded backbones, it's not surprising that polypeptides vary in shape from rod-like filaments to ball-like masses (**Figure 3.15b**).

Quaternary Structure

The first three levels of protein structure involve individual polypeptides. But many proteins contain several distinct polypeptides that interact to form a single structure. The combination of polypeptides as subunits gives proteins a **quaternary structure**. The individual polypeptides may be held together by bonds or other interactions among R-groups or sections of their peptide backbones.

In the simplest case, a protein with quaternary structure can consist of just two subunits that are identical. The Cro protein found in a virus called bacteriophage λ ("lambda")

(a) Interactions that determine the tertiary structure of proteins

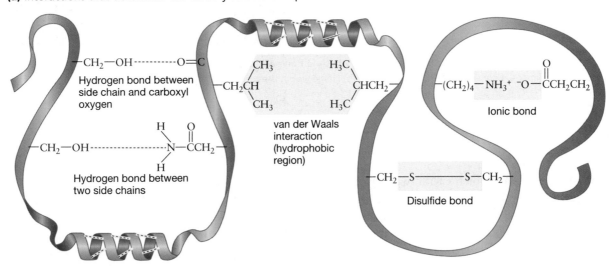

Hydrogen bond between side chain and carboxyl oxygen

Hydrogen bond between two side chains

van der Waals interaction (hydrophobic region)

Ionic bond

Disulfide bond

(b) Tertiary structures are diverse.

A tertiary structure composed mostly of α-helices

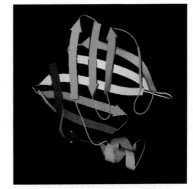

A tertiary structure composed mostly of β-pleated sheets

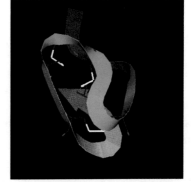

A tertiary structure rich in disulfide bonds

FIGURE 3.15 Tertiary Structure of Proteins Results from Interactions Involving R-Groups
(a) Each protein has a unique overall shape called its tertiary structure. Tertiary structure is created by bonds and other interactions that cause proteins to fold in a precise way. **(b)** The tertiary structure of these proteins includes interactions between α-helices (indicated by coils) and β-pleated sheets (indicated by flat arrows, with the arrowhead at the carboxyl end). The polypeptide chains are color-coded so that you can follow the chain from one end (red) to the other (dark blue). **EXERCISE** In part (b): (Left) Label the four α-helices; (center) label the α-helix and one of the β-pleated sheets; (right) label the disulfide bonds, shown in yellow.

(a) Cro protein, a dimer

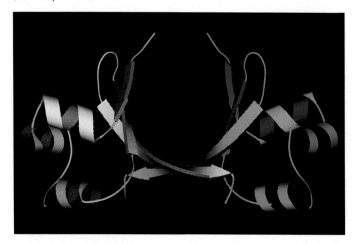

(b) Hemoglobin, a tetramer

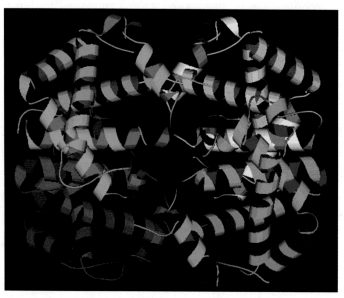

FIGURE 3.16 Quaternary Structures of Proteins Is Created by Multiple Polypeptides
(a) The Cro protein is a dimer—it consists of two polypeptides. **(b)** Hemoglobin is a tetramer—it consists of four polypeptides. **QUESTION** What elements of secondary structure can you identify in each of the two proteins shown here?

is an example (**Figure 3.16a**). Proteins with two polypeptide subunits are called dimers ("two-parts"). More than two polypeptides can be linked into a single protein, however, and the polypeptides involved may be distinct in primary, secondary, and tertiary structure. Hemoglobin, for example, is a tetramer ("four-parts"). As **Figure 3.16b** shows, the completed protein consists of two copies of each of two different polypeptides.

Table 3.2 summarizes the four levels of protein structure, using hemoglobin as an example. The table and preceding discussion have two important messages. First, the combination of primary, secondary, tertiary, and quaternary levels of structure is responsible for the fantastic diversity of sizes and shapes observed in proteins. Second, most elements of protein structure are based on folding of polypeptide chains. Does this folding occur spontaneously? What happens if normal folding is disrupted?

Folding and Function

If you were able to synthesize one of the polypeptides in hemoglobin from individual amino acids, and if you placed the resulting chain in water, it would spontaneously fold into the shape of the tertiary structure shown in Table 3.2. This result seems counterintuitive. Because an unfolded protein has many more ways to move about, it has much higher entropy than the folded version. Folding is spontaneous in some cases, however, because the bonds and van der Waals interac-

tions that occur make the folded molecule more stable energetically than the unfolded molecule. Thus, folding may release free energy.

Folding is also crucial to the function of a completed protein. This point was hammered home in a set of classic experiments by Christian Anfinson and colleagues during the 1950s. Anfinson studied a protein called ribonuclease that is found in many organisms. Ribonuclease breaks ribonucleic acid polymers apart. Anfinson found that the ribonuclease could be unfolded, or **denatured**, by treating it with compounds that break hydrogen bonds and disulfide bonds (**Figure 3.17**). The denatured ribonuclease was unable to function normally—it could no longer break apart nucleic acids. But when Anfinson removed the denaturing agents, the molecule refolded and began to function normally again. These experiments confirmed that ribonuclease folds spontaneously and that folding is essential for normal function.

More recent work has shown that in cells, folding is often facilitated by specific proteins called **molecular chaperones**. Many molecular chaperones belong to a family of molecules called the heat-shock proteins. These compounds are produced in large quantities after cells experience high temperatures or other treatments that make proteins lose their tertiary structure. Heat-shock proteins speed the refolding of other proteins into their normal shape after denaturation has occurred.

To summarize, a protein's function depends on its shape. In most cases, the final shape of a protein depends on folding. To

TABLE 3.2 A Summary of Protein Structure

Level	Description	Stabilized by	Example: Hemoglobin
Primary	The sequence of amino acids in a polypeptide	Peptide bonds	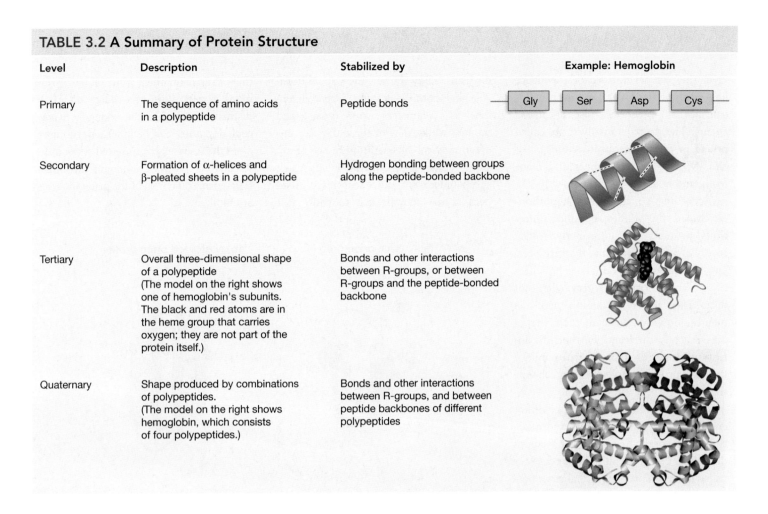
Secondary	Formation of α-helices and β-pleated sheets in a polypeptide	Hydrogen bonding between groups along the peptide-bonded backbone	
Tertiary	Overall three-dimensional shape of a polypeptide (The model on the right shows one of hemoglobin's subunits. The black and red atoms are in the heme group that carries oxygen; they are not part of the protein itself.)	Bonds and other interactions between R-groups, or between R-groups and the peptide-bonded backbone	
Quaternary	Shape produced by combinations of polypeptides. (The model on the right shows hemoglobin, which consists of four polypeptides.)	Bonds and other interactions between R-groups, and between peptide backbones of different polypeptides	

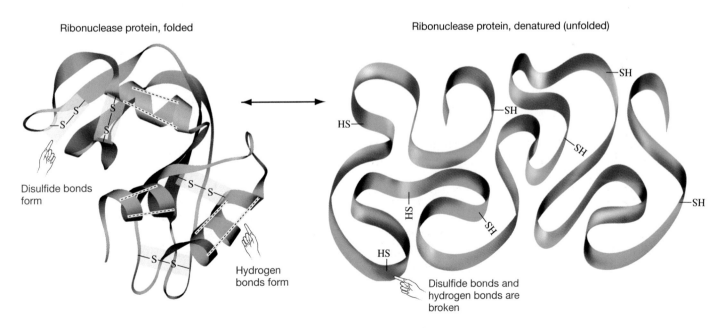

FIGURE 3.17 Proteins Fold into Their Normal, Active Shape
(a) The tertiary structure of ribonuclease is defined primarily by four disulfide bonds, shown in yellow.
(b) When the disulfide bonds and various non-covalent bonds are broken, the protein denatures (unfolds).

BOX 3.2 Prions

Over the past several decades, evidence has accumulated that certain proteins can act as infectious, disease-causing agents. The proteins involved are called **prions**, or proteinaceous infectious particles. Where do these molecules come from, and how do they work?

According to the prion hypothesis developed by Stanley Prusiner, infectious proteins are improperly folded forms of normal proteins that are present in healthy individuals. The infectious and normal forms do not necessarily differ in amino acid sequence, however. Instead, their shapes are radically different. Further, the infectious form of a protein can induce normal protein molecules to change their shape to the altered form.

Figure 3.18 illustrates the shape differences observed in the normal and infectious forms of the first prion to have been described. The molecule in Figure 3.18a is called the prion protein (PrP) and is a normal component of mammalian cells. Mutant versions of this protein, like the one in Figure 3.18b, are found in a wide variety of species and cause a family of diseases known as the spongiform encephalopathies—literally, "sponge-brain-illnesses." Hamsters, cows, goats, and humans afflicted with these diseases undergo massive degeneration of the brain. Cattle suffer from "mad cow disease"; sheep and goats acquire scrapie (so called because the animals itch so badly that because the animals itch so badly that they scratch off their wool or hair); humans develop kuru or Creutzfeldt-Jakob disease. Although some spongiform encephalopathies can be inherited, in many cases the disease is transmitted when individuals eat tissues containing the infectious form of PrP. All of the prion illnesses are fatal.

(a) Normal prion protein   **(b)** Misfolded prion protein

FIGURE 3.18 Prions Are Improperly Folded Proteins
Ribbon model of **(a)** a normal prion protein and **(b)** the misfolded form that causes mad cow disease in cattle. Secondary structure is represented by coils (α-helices) and arrows (β-pleated sheets).

drive this point home, **Box 3.2** investigates the way a change in the shape of the prion ("PREE-on") protein of mammals converts normal molecules into abnormally folded proteins. Misfolded prions cause the brains of humans and other mammals to disintegrate.

Given how proteins are put together and how they are folded, let's explore next what they do. In terms of diversity in function, no other type of molecule even begins to compare with proteins.

✓CHECK YOUR UNDERSTANDING

Proteins have up to four levels of structure. Primary structure consists of the sequence of amino acids. Secondary structure results from interactions between atoms in the peptide-bonded backbone of the same polypeptide, and yields structures such as α-helices and β-pleated sheets. Tertiary structure is a consequence of bonds or other interactions between R-groups or between R-groups and the peptide-bonded backbone of a polypeptide, and produces distinctive folds. Quaternary structure occurs when multiple polypeptides interact to form a single protein. You should be able to draw a hypothetical protein and label the parts that produce each level of structure.

3.4 What Do Proteins Do?

Proteins are the workhorse molecules in organisms. You've already been introduced to hemoglobin, which transports oxygen molecules around the human body. Other proteins protect the body by attacking disease-causing bacteria and viruses; still others make movement possible or form structures that give cells their shape. Let's take a closer look.

Proteins Have Diverse Functions in Cells

Table 3.3 provides a brief overview of protein functions. The table's message is simple: Proteins are crucial to most tasks required for cells to exist. The diverse functions of proteins include the following:

- *Defense.* Proteins called antibodies and complement proteins attack and destroy viruses and bacteria that cause disease.
- *Movement.* Motor proteins and contractile proteins are responsible for moving the cell itself, or for moving large molecules and other types of cargo inside the cell.

- *Catalysis.* Proteins called enzymes catalyze, or speed up, chemical reactions, many of which would not otherwise be able to proceed.

- *Signaling.* Proteins called peptide hormones bind to receptor proteins on particular cells. In response, the activity of the cell with the receptor protein changes. In this way, proteins are involved in carrying and receiving signals from cell to cell inside the body.

- *Structure.* Structural proteins give cells mechanical support. They also create structures such as fingernails and hair.

- *Transport.* Proteins are responsible for allowing particular molecules to enter or exit cells, and for carrying specific compounds throughout the body.

Most of these aspects of protein function will be explored in much greater detail elsewhere in the text. In many if not all cases, the function of the protein involved is closely correlated with its structure. Here let's focus on what may be the most fundamental of all protein functions: **catalysis**, which speeds the chemical reactions that allow cells to stay alive, grow, and reproduce.

An Introduction to Catalysis

In Chapter 2 we reviewed several fundamental ideas about chemical reactions, including how changes in potential energy and entropy determine whether a reaction is spontaneous. Even spontaneous chemical reactions are not necessarily fast, however. Oxidation of iron—the reaction known as rusting—is exothermic and spontaneous, but it occurs slowly. To appreciate why proteins are so important in the life of a cell, it is essential to get a deeper understand-ing of the factors that limit the speed at which chemical reactions occur.

Reaction rates depend on how the chemical bonds involved are broken and re-formed. Chapter 2 introduced the idea that molecules must collide with a specific orientation and a specific amount of energy for old bonds to break and new bonds to form. This is because the electrons involved in the reactions must interact. But these electrons repel one another as they come into contact. In many cases, for the reaction to proceed, the kinetic energy of the collision must be large enough to overcome this repulsion. If the energy is sufficiently large, then the collision creates a combination of old and new bonds called a **transition state**. The amount of free energy required to reach the transition state is called the **activation energy** of the reaction. The more unstable the transition state, the higher the activation energy and the less likely the reaction is to proceed quickly.

Figure 3.19 pulls these ideas together by showing the changes in free energy that take place during the course of a chemical reaction. In this graph, ΔG indicates the overall change in free energy in the reaction—that is, the energy of the products minus the energy of the reactants. In this case the products have lower potential energy than the reactants, meaning that the reaction is exothermic. But because the activation energy for this reaction, symbolized by E_a, is high, the reaction would proceed slowly.

Reaction rates, then, depend on both the kinetic energy of the reactants and the activation energy of the particular reaction—meaning the free energy of the transition state. If the kinetic energy of the participating molecules is high, then molecular collisions are likely to result in completed reactions. The kinetic energy of molecules, in turn, is a function

TABLE 3.3 A Summary of Protein Functions

Protein type	Role in cell or organism
Antibodies and complement proteins	Defense—destruction of disease-causing viruses and bacteria
Contractile proteins and motor proteins	Movement
Enzymes	Catalyze chemical reactions
Peptide hormones	Act as signals that help coordinate the activities of many cells
Receptor proteins	Receive chemical signals from outside cell and initiate response
Structural proteins	Provide support for cells and tissues; form structures such as hair, feathers, cocoons, and spider webs
Transport proteins	Move substances across cell membrane; move substances throughout body

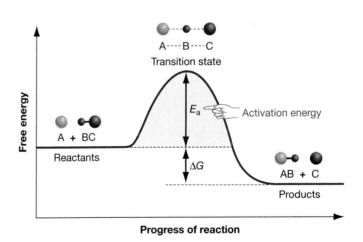

FIGURE 3.19 Changes in Free Energy during a Chemical Reaction
The changes in free energy that occur over the course of a hypothetical reaction between an atom A and a molecule containing atoms B and C. The overall reaction would be written as A + BC → AB + C. E_a is the activation energy of the reaction, and ΔG is the overall change in free energy. **EXERCISE** This graph illustrates an exergonic reaction. Draw the same type of graph for an endergonic reaction.

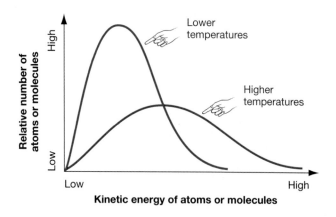

FIGURE 3.20 Atoms amd Molecules Have More Kinetic Energy at High Temperatures than at Low Temperatures
EXERCISE Add a vertical line about two-thirds of the way along the horizontal axis. Label this line "Activation energy for reaction without catalyst." At low and high temperatures, which molecules have enough kinetic energy for this reaction to proceed?

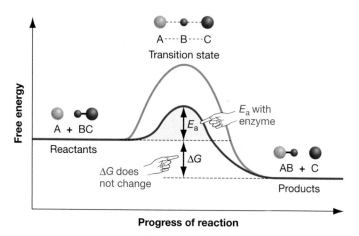

FIGURE 3.21 A Catalyst Changes the Activation Energy of a Reaction
The energy profile for the same reaction diagrammed in Figure 3.19, but with a catalyst present. Even though the energy barrier to the reaction, E_a, is much lower, ΔG does not change. QUESTION Can a catalyst make a nonspontaneous reaction occur spontaneously? EXERCISE Go back to Figure 3.20 and add another vertical line about one-third of the way along the horizontal axis. Label this line "Activation energy for reaction with catalyst." Label the molecules that have enough energy to undergo a reaction with a catalyst but not without a catalyst.

of their temperature (**Figure 3.20**). This is why chemical reactions tend to proceed faster at higher temperatures. But if the activation energy of a particular reaction is also high, then collisions are less likely to result in completed reactions.

In many cases, the electrons in the transition-state molecule can be stabilized when they interact with another ion, atom, or molecule. When this occurs, the activation energy required for the reaction drops and the reaction rate increases. A substance that lowers the activation energy of a reaction and increases the rate of the reaction is called a **catalyst**. A catalyst is not consumed in a chemical reaction, even though it participates in the reaction. The composition of a catalyst is exactly the same after the reaction as it was before.

Figure 3.21 diagrams how catalysts lower the activation energy for a reaction by lowering the free energy of the transition state. Note that the presence of a catalyst does not affect the overall energy change, ΔG, or change the energy of the reactants or the products. A catalyst changes only the free energy of the transition state.

Proteins that catalyze reactions are called **enzymes**. Most enzymes are quite specific in their activity—they catalyze just a single reaction by lowering the activation energy that is required. It's not difficult to appreciate why this is important. Most of the important reactions in biology do not occur at all, or else proceed at imperceptible rates, without a catalyst. In contrast, a single molecule of the enzyme carbonic anhydrase can catalyze over 1,000,000 reactions *per second*. It's not unusual for enzymes to speed up reactions by a factor of a million; some enzymes make reactions go many *trillions* of times faster than they would without a catalyst.

Enzymes are important because they speed up the chemical reactions that are required for life. Now the question is, how do enzymes do what they do?

How Do Enzymes Work?

The initial hypothesis for how enzymes work was proposed by Emil Fischer in 1894. According to Fischer's **lock-and-key model**, enzymes are rigid structures analogous to a lock. The keys are reactant molecules, called **substrates**, that fit into the lock and then react.

Several important ideas in this model have stood the test of time. For example, Fischer was correct in proposing that enzymes bring substrates together in a precise orientation that makes reactions more likely. His model also accurately explained why most enzymes can catalyze only one specific reaction. Enzyme specificity is a product of the geometry and chemical properties of the sites where substrates bind.

As researchers began to test and extend Fischer's model, the locations where substrates bind and react became known as the enzyme's **active site**. When techniques for solving the three-dimensional structure of enzymes became available, it turned out that enzymes tend to be very large relative to substrates and roughly globular, and that the active site is in a cleft or cavity within the globular shape. The enzyme hexokinase, which is at work in most cells of your body now, is a good example. (Many enzymes have names that end with –*ase*.) As the left-hand side of **Figure 3.22** shows, the active site in hexokinase is a small notch in an otherwise large, crescent-shaped enzyme.

As knowledge of enzyme action grew, however, Fischer's model was modified. Perhaps the most important change was based on the realization that enzymes are not rigid and static, but flexible and dynamic. In fact, many enzymes undergo a

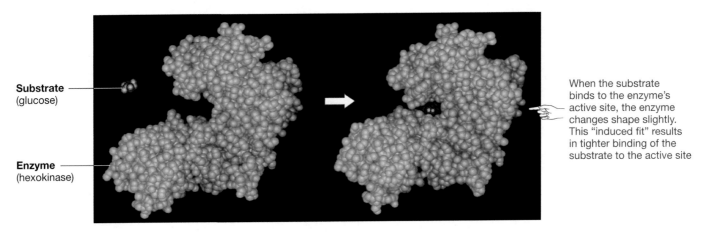

Substrate
(glucose)

Enzyme
(hexokinase)

When the substrate binds to the enzyme's active site, the enzyme changes shape slightly. This "induced fit" results in tighter binding of the substrate to the active site

FIGURE 3.22 Reactant Molecules Bind to Specific Locations in an Enzyme
The reactant molecule, shown in red, fits into a precise location, called the active site, in the green enzyme. In the enzyme shown here and in many others, the binding event causes the protein to change its shape.
EXERCISE Label the active site in the enzyme on the left.

significant change in shape, or conformation, when reactant molecules bind to the active site. This conformational change, called an **induced fit**, is noticeable in the hexokinase molecule on the right of Figure 3.22. As hexokinase binds its substrate—the sugar glucose—the enzyme rocks forward over the active site.

In addition, recent research has clarified the nature of Fischer's key. When one or more substrate molecules enter the active site, they are held in place through hydrogen bonding or other electrical interactions with amino acids in the active site. Once the substrate is bound, one or more R-groups in the active site come into play. The degree of interaction between the substrate and enzyme increases and reaches a maximum when the transition state is formed. Thus, Fischer's key is actually the transition state.

Recall a key point: Interactions with R-groups stabilize the transition state and thus lower the activation energy required for the reaction to proceed. At the atomic level, R-groups that line the active site may form short-lived covalent bonds that assist with the transfer of atoms or groups of atoms from one reactant to another. More commonly, the presence of acidic or basic R-groups allows the reactants to lose or gain a proton more readily.

Figure 3.23 provides a specific example of how substrates interact with an active site. The panels show a substrate fitting into the active site of the enzyme called ribonuclease. Notice that once the substrate has bound, R-groups in the active site interact with the substrate to form the transition state and get the reaction under way.

R-GROUPS IN AN ENZYME'S ACTIVE SITE STABILIZE THE TRANSITION STATE OF A SUBSTRATE.

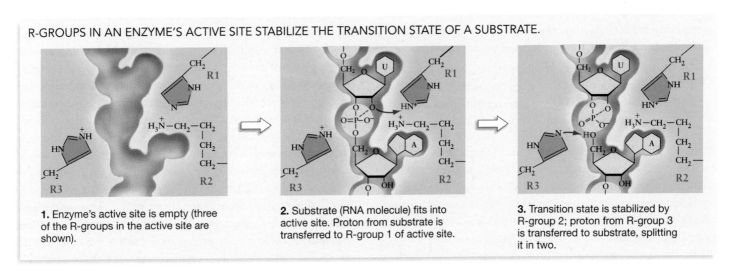

1. Enzyme's active site is empty (three of the R-groups in the active site are shown).

2. Substrate (RNA molecule) fits into active site. Proton from substrate is transferred to R-group 1 of active site.

3. Transition state is stabilized by R-group 2; proton from R-group 3 is transferred to substrate, splitting it in two.

FIGURE 3.23 R-groups in an Enzyme's Active Site Stabilize the Transition State
Part of the reaction sequence that occurs when the enzyme ribonuclease catalyzes a reaction that cuts a substrate in two. Note how precisely the substrate fits into the active site and how specific R-groups in the enzyme are involved in key steps in the reaction.

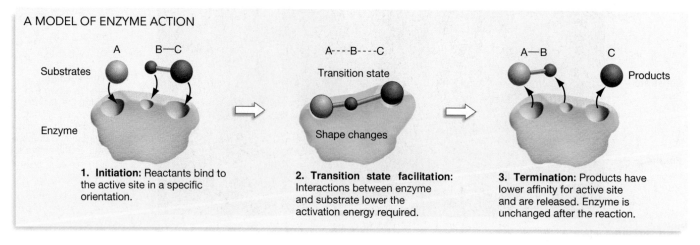

A MODEL OF ENZYME ACTION

1. Initiation: Reactants bind to the active site in a specific orientation.

2. Transition state facilitation: Interactions between enzyme and substrate lower the activation energy required.

3. Termination: Products have lower affinity for active site and are released. Enzyme is unchanged after the reaction.

FIGURE 3.24 Enzyme Action Can Be Analyzed as a Three-Step Process

Whatever the mechanism, enzymes induce the formation of the transition state and thereby increase reaction rates. The products of the reaction have a much lower affinity for the active site than do either the reactants or the transition state, however. As a result, they are released from the enzyme once they form.

Figure 3.24 summarizes these principles. Enzyme catalysis can be analyzed as a three-step process:

1. **Initiation.** Instead of reactants occasionally colliding in a random fashion, enzymes orient reactants precisely as they bind at specific locations within the active site.

2. **Transition state facilitation.** The act of binding induces the formation of the transition state. In some cases the transition state is stabilized by a conformational change in the enzyme. The interaction between the substrate and R-groups in the enzyme's active site lowers the activation energy required for the reaction. Inside a catalyst's active site, more reactant molecules have sufficient kinetic energy to reach this lowered activation energy. Thus, the catalyzed reaction proceeds much more rapidly than the uncatalyzed reaction.

3. **Termination.** The reaction products have considerably less affinity for the active site than does the transition state. As a result, binding ends. The enzyme returns to its original conformation, and the products are released.

Fischer's model inspired intensive and fruitful research into the mechanism of enzyme action. Enzymes speed the rate of reactions by orienting reactants and decreasing the reaction's activation energy. Activation energies drop because enzymes destabilize bonds in the reactant, stabilize the transition state, make acid-base and redox reactions more favorable, and/or change the reaction mechanism through a covalent bonding interaction. Enzyme specificity is a function of the active site's shape and the chemical properties of the R-groups present in the active site. In enzymes, as in many molecules, structure dictates function.

Do Enzymes Act Alone? Atoms or molecules that are not part of the enzyme's primary structure are often required for the enzyme to function normally. These **enzyme cofactors** can be either (1) metal ions such as Zn^{2+} (zinc) or Mg^{2+} (magnesium), or (2) small organic molecules called **coenzymes**. In many cases, the cofactor binds to the active site and is thought to play a key role in stabilizing the transition state during the reaction. As a result, the presence of the cofactor is essential for catalysis. To drive this point home, consider that many of the vitamins in your diet are required for the production of enzyme cofactors. Vitamin deficiencies cause enzyme-cofactor deficiencies. Lack of cofactors, in turn, disrupts normal enzyme function and causes disease. For example, thiamine (vitamin B_1) is required for the production of an enzyme cofactor called thiamine pyrophosphate, which is required by three different enzymes. Lack of thiamine in the diet dramatically reduces the activity of these enzymes and causes an array of nervous system and heart disorders collectively known as beriberi.

In addition, most enzymes are regulated by molecules that are not part of the enzyme itself. An enzyme is active or inactive, depending on the presence or absence of molecules that change the protein's structure in some way. In some cases, catalysis is inhibited when a molecule that is similar in size and shape to the substrate(s) binds to the active site. This event is called **competitive inhibition**, because the molecule involved competes with the substrate(s) for access to the enzyme's active site (**Figure 3.25a**). In other cases, a regulatory molecule binds at a location other than the active site. This type of regulation is called **allosteric** ("different-structure") **regulation**, because the molecule involved does not affect the active site directly. Instead, the binding event changes the shape of the enzyme in a way that makes the active site accessible or inaccessible (**Figure 3.25b**). Later chapters will provide detailed examples of how regulatory molecules change the activity of specific enzymes.

(a) Competitive inhibition prevents the substrate from binding to the enzyme.

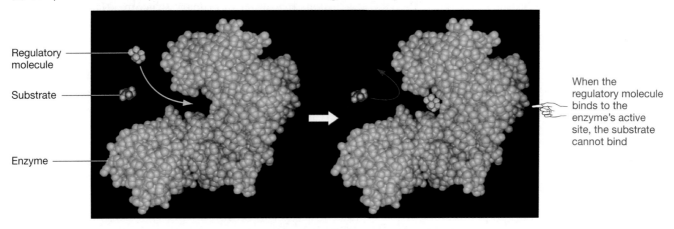

Regulatory molecule

Substrate

Enzyme

When the regulatory molecule binds to the enzyme's active site, the substrate cannot bind

(b) Allosteric regulation changes the enzyme's shape to activate or inactivate it.

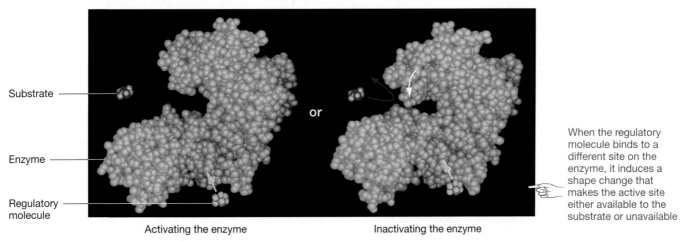

Substrate

Enzyme

Regulatory molecule

or

Activating the enzyme   Inactivating the enzyme

When the regulatory molecule binds to a different site on the enzyme, it induces a shape change that makes the active site either available to the substrate or unavailable

FIGURE 3.25 An Enzyme's Activity Is Precisely Regulated
Enzymes are turned on or off when specific molecules bind to them. **(a)** In competitive inhibition, binding occurs at the active site—the regulatory molecule competes with substrate(s). **(b)** In allosteric regulation, the regulatory molecule binds elsewhere on the enzyme and causes a shape change that activates or inactives the protein.

What Limits the Rate of Catalysis? For several decades after Fischer's model was published, most research on enzymes focused on rates of enzyme action, or what biologists call enzyme kinetics. Researchers observed that, when the amount of product produced per second—meaning the speed of the reaction—is plotted as a function of substrate concentration, a graph like that shown in **Figure 3.26** results. When the substrate concentrations are low, the reaction speed increases in a linear fashion. As substrate concentrations increase, however, the increase in speed begins to slow. Eventually the reaction rate plateaus at a maximum speed. This pattern is in striking contrast to the situation for uncatalyzed reactions, in which reaction speed tends to show a continuing linear increase with substrate concentration. The "saturation kinetics" of enzyme-catalyzed reactions were taken as strong evidence that the enzyme-substrate complex postulated by Fischer actually exists. The idea was that at some point, active sites cannot accept substrates any faster, no matter how large the concentration

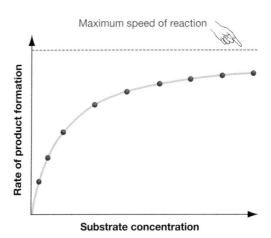

Maximum speed of reaction

Rate of product formation

Substrate concentration

FIGURE 3.26 Kinetics of an Enzyme-Catalyzed Reaction
The general shape of this curve is characteristic of enzyme-catalyzed reactions.

of substrates gets. Stated another way, reaction rates level off because all the available enzyme is being used.

It's important to recognize, however, that even though the general shape of the curve is similar for all enzymes, its position varies. In enzymes that have a high affinity for the substrate, the curve is shifted to the left. This means that substrate concentrations do not have to be very high for these enzymes to function at maximum speed. In enzymes with a lower affinity for the substrate, though, the curve is shifted to the right. This means that substrate concentrations have to be much higher before an enzyme is able to catalyze reactions at its maximal rate.

The processing of ethanol—the active ingredient in alcoholic beverages—in the human liver furnishes an example of these concepts. Ethanol is a poison broken down in two enzyme-catalyzed steps. First ethanol is converted to acetaldehyde (CH_3CHO). Then acetaldehyde is converted to acetic acid (CH_3COOH)—a harmless molecule that gives vinegar its bite. Humans have two versions of the enzyme that catalyzes the second reaction. One version of the enzyme works quickly and the other much more slowly. In some people, the fast-working version of the enzyme has a single amino acid change in its primary sequence that affects the enzyme's active site. In these individuals, the fast-working version of the enzyme is inactivated and only the slow-working version functions normally. Acetaldehyde must be at a high concentration before the slow-working version of the enzyme works at maximum speed. In these individuals, acetaldehyde builds up to high enough concentrations in the bloodstream that uncomfortable symptoms result, including a rapid heartbeat and pronounced flushing of the skin. People who lack the fast-acting version of the enzyme have virtually zero tolerance for alcoholic beverages. Imbibing even a small amount of ethanol makes them feel ill.

How Do Physical Conditions Affect Enzyme Function?

Given that an enzyme's structure is critical to its function, it's not surprising that an enzyme's activity is sensitive to conditions that alter its structure. In particular, the activity of an enzyme often changes drastically as a function of temperature and pH. Temperature affects the movement of the enzyme; pH affects the makeup and charge of amino acid side chains with carboxyl or amino groups, and the active site's ability to participate in proton-transfer or electron-transfer reactions.

Do data support these assertions? **Figure 3.27a** shows how the activity of an enzyme changes as a function of temperature. Data for the enzyme glucose-6-phosphate, which is actively producing energy in your cells right now, are shown for two species of bacteria. Note that in both of the bacterial species illustrated, the enzyme has a distinct optimum—a temperature at which it functions best. One of the bacterial species lives inside your gut, where the temperature is about 40°C, while the other lives in hot springs, where temperatures can be close to 100°C. The temperature optimum for the enzyme reflects these environments. The two

types of bacteria have different versions of the enzyme—versions that differ in primary structure. Natural selection has favored different structures that have different functions. The enzymes are adaptations that allow each species to thrive at different temperatures.

Figure 3.27b makes the same point for pH. The enzyme in this graph, called chitinase, protects bacterial cells by digesting a molecule found in the cell walls of fungi that eat bacteria. The data come from a species of bacterium that lives in acidic pools and a bacterial species that lives in the soil under palm trees. The organism that thrives in an acidic environment has a version of the enzyme that performs best at low pH; the organism that lives near palms has a version of the enzyme that functions best near neutral pH. Each enzyme is sensitive to changes in

(a) Enzymes from different organisms may function best at different temperatures.

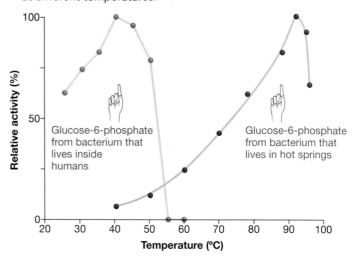

(b) Enzymes from different organisms may function best at different pHs.

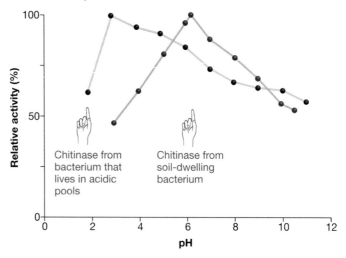

FIGURE 3.27 Enzymes Have an Optimal Temperature and pH
Enzymes are sensitive to changes in temperature and pH. Further, the structures of the enzymes found in a particular organism allow that organism to function well at the **(a)** temperature and **(b)** pH of its environment.

pH, but each species' version of the enzyme has a structure that allows it to function best at the pH of its environment.

How Does ATP Drive Endergonic Reactions?

Enzyme catalysis explains why the reactions required for life can proceed quickly. But it is critical to realize that the presence of an enzyme does not change the potential energy or entropy of the reactants or products. Recall from Figure 3.19 that enzymes change only the free energy of the transition state—not the ΔG for the reaction. How, then, do endergonic reactions occur? Answering this question is important for two reasons: (1) Most reactions that take place inside you and in the organisms around you are endergonic; and (2) the vast majority of reactions required for chemical evolution to proceed are endergonic.

The essence of chemical evolution is the production of larger and more complex molecules from simpler ones. The essence of life is the maintenance of highly ordered structures and the ability to grow and reproduce. To understand how these endergonic processes occur, let's return to a reaction that was summarized briefly earlier in the chapter: the production of polypeptides from amino acid monomers, by incubating the reactants on tiny mineral particles.

Polymerization reactions are endergonic because polymers have higher potential energy and lower entropy than do monomers. Recall from Section 3.2 that adsorption onto minerals protected the polypeptides from reactions that break polymers apart. But how was the reaction that first led to a higher free-energy state possible? How were researchers able to produce polypeptides from amino acids?

The answer is that the investigators added chemical energy to the reactants. More specifically, they added **phosphate groups** (PO_4^{3-}) to the amino acids in the reaction mix. The addition of a phosphate group adds two negative charges to an amino acid and raises its free energy enough to make the polymerization reaction proceed. This mechanism is occurring in your cells right now. Many reactions that would otherwise be endergonic, including the polymerization of amino acids into proteins, are being driven by the addition of a phosphate group to a reactant molecule. In your cells, the phosphate group comes from a molecule called **adenosine triphosphate**, or **ATP**. As **Figure 3.28a** shows, ATP contains three phosphate groups. The mechanism works as follows: ATP reacts with the substrate to produce a *phosphorylated* molecule—a molecule that contains a phosphate group—and an **adenosine diphosphate**, or **ADP** molecule. ADP has two phosphate groups rather than three. The reaction that adds a phosphate group to a substrate, called **phosphorylation**, is exergonic (**Figure 3.28b**). Next the "activated" substrate—activated in the sense that it now contains a phosphate group and has high free energy—reacts and forms the product molecule(s). In some cases, the enzyme that

(a) ATP has 3 phosphate groups and high free energy.

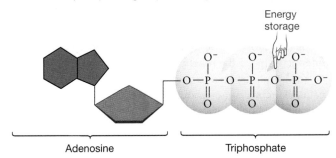

Adenosine   Triphosphate

FIGURE 3.28 Chemical Energy from ATP Can Drive Endergonic Reactions

(a) Adenosine triphosphate, or ATP, consists of a molecule called adenosine with three phosphate groups attached. **(b)** In cells, endergonic reactions occur if one reactant undergoes phosphorylation—the addition of a phosphate group from ATP. The phosphorylated reactant molecule has high enough free energy that the subsequent reaction is exergonic.

(b) Coupling of exergonic and endergonic reactions: How can monomer B be added to polymer A?

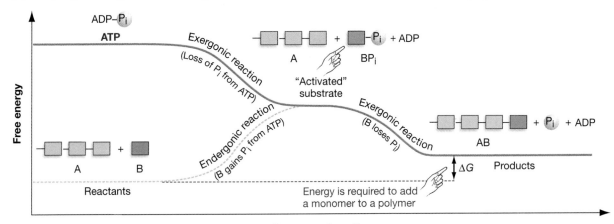

Progress of reaction

catalyzes the reaction is phosphorylated instead of a reactant. When either a substrate or an enzyme is phosphorylated, the exergonic phosphorylation reaction is said to be "coupled" to an endergonic reaction.

To summarize, two points are absolutely crucial to understanding how biological reactions occur in cells:

1. Phosphorylation makes biological reactions endergonic and thus spontaneous, and

2. Enzyme catalysis makes most biological reactions fast.

During chemical evolution, the phosphorylation of amino acids in the prebiotic soup could have been driven by solar, heat, or electrical energy. The phosphorylated amino acids could then have driven the endergonic polymerization reactions required for continued chemical evolution and the production of macromolecules. If small proteins were produced in this way 4 billion years ago, were they the molecules that triggered the origin of life?

Was the First Living Entity a Protein?

In the introduction to this chapter, we noted that life began with a molecule that could make a copy of itself. This self-replicating entity increased in number and formed a population of individuals. The population then began to evolve by natural selection—the process introduced in Chapter 1 that leads to change in populations of organisms over time. For such a molecule to make a copy of itself efficiently, it must have been a catalyst of some kind. This chapter's examination of protein structure and function raises a key question: Was the first self-replicating molecule on Earth a protein?

Several observations argue that the answer to this question is yes. Experimental studies have shown that amino acids were likely to be abundant in the prebiotic soup and that they could have polymerized to form small proteins. In addition, proteins are the most efficient catalysts known, and a self-replicating

molecule had to act as a catalyst during the assembly and polymerization of its copy. These observationss support the hypothesis that the self-replicating molecule was a polypeptide. Indeed, several laboratories that are currently working to create life have focused on synthesizing a self-replicating protein.

To date, however, attempts to simulate the origin of life with proteins have not been successful. Most origin-of-life researchers are increasingly skeptical about the hypothesis that life began with a protein. Their reasoning is that to make a copy of something, a mold or template is required. Proteins cannot furnish this information. Nucleic acids, in contrast, can. How they do so is the subject of Chapter 4.

✔CHECK YOUR UNDERSTANDING

Most proteins are enzymes that make specific chemical reactions occur rapidly. Enzymes catalyze reactions by means of a three-step mechanism: (1) Binding of reactants in a precise orientation inside the active site. (2) Facilitation of the transition state, thus lowering the activation energy required for the reaction. This step often involves a conformational change in the enzyme, resulting in an "induced fit" between active site and substrate. Enzyme cofactors are often required at this step. (3) Release of products, which do not bind tightly to the active site.

The activity of enzymes is controlled through allosteric regulation or competitive inhibition. Different enzymes work at different rates, and enzymes are sensitive to changes in temperature and pH.

Endergonic reactions occur in cells through the mechanism of phosphorylation. When a phosphate group from ATP binds to a substrate or an enzyme, its free energy is raised enough to make the reaction exergonic.

You should be able to diagram the three steps in enzyme catalysis. Your diagram should include (1) coupling with an exergonic reaction involving ATP and (2) regulatory processes.

ESSAY Molecular Handedness and the Thalidomide Tragedy

This chapter introduces the concept of molecular "handedness" in the context of research on chemical evolution. Most amino acids and many other molecules have different forms called optical isomers, which differ in the spatial arrangement of their atoms. The problem facing origin-of-life researchers is to explain why organisms have only one optical isomer of molecules such as the amino acids and the sugar called ribose. But the existence of optical isomers also has urgent practical consequences, especially for the pharmaceutical industry.

Most molecules that chemists synthesize in the laboratory are produced as a mixture of optical isomers in equal amounts. This was true for a sedative developed in the 1950s and marketed under the name thalidomide. People who used this tranquilizer were actually ingesting equal amounts of two mirror-image molecules.

In laboratory testing, thalidomide was considered to be extremely safe, because it had few side effects in experimental animals. As a result, it was widely prescribed in Europe; in Germany it was even available without a prescription. However, the drug was never approved for sale in the United States because

Dr. Frances O. Kelsey (an official with the U.S. Food and Drug Administration) interpreted the laboratory testing data differently. She believed the evidence indicated that the drug might be harmful.

Tragically, Dr. Kelsey was correct. One optical isomer of thalidomide acts as a safe and effective tranquilizer. But the mirror-image molecule can cause severe birth defects. Many women who took the drug during the first three months of a pregnancy bore children without arms or legs or with severely reduced limbs (**Figure 3.29**). Over 12,000 "thalidomide babies" were born before physicians were able to diagnose the cause and have the drug banned.

> *Over 12,000 "thalidomide babies" were born before physicians were able to diagnose the cause and have the drug banned.*

The tragedy made pharmaceutical companies acutely aware of the need to synthesize or isolate single optical isomers of molecules. It also underscores a very general point in molecular biology: Because chemical reactions depend on the orientation of the atoms involved, shape and geometry are fundamentally important characteristics. The function of a molecule is determined by its structure. Later research revealed that one optical isomer of thalidomide damages embryos because it inhibits blood vessel formation. Because cancerous tumors require extensive blood vessel development to sustain their growth, molecules related to thalidomide are now being investigated as possible anticancer drugs.

FIGURE 3.29 An Optical Isomer Caused Abnormal Development
Many mothers who took the drug thalidomide during pregnancy had children with reduced or no limbs.

CHAPTER REVIEW

Summary of Key Concepts

Chemical evolution begins when simple molecules such as methane, ammonia, water, carbon dioxide, and carbon monoxide react to form more complex organic molecules such as formaldehyde and hydrogen cyanide. Experiments using electrical discharges or ultraviolet radiation as energy sources have shown that these nonspontaneous reactions could have occurred on ancient Earth.

Chemical evolution continues when formaldehyde and hydrogen cyanide react to form amino acids, sugars, and nucleotides. These molecules are the building blocks, or monomers, needed to synthesize the complex macromolecules found in living organisms. Experiments have shown that amino acids and other monomers are easily produced under early Earth conditions. Many of these molecules are found in meteorites as well.

■ **Proteins are made of amino acids. Amino acids vary in structure and function because their side chains vary in composi-** tion. As a result, proteins vary widely in structure and thus in function.

The polymerization of monomers into macromolecules is the next key event in chemical evolution. Experiments have shown that these condensation reactions occur readily when growing polymers stick to clay minerals. Researchers have observed polypeptide formation from the polymerization of amino acids on clay particles.

Web Tutorial 3.1   Condensation and Hydrolysis Reactions

■ **In cells, most proteins are enzymes that function as catalysts. Chemical reactions occur much faster when they are catalyzed by enzymes. During enzyme catalysis, the reactants bind to an enzyme's active site in a way that allows the reaction to proceed efficiently.**

Proteins are far and away the most diverse and efficient class of catalysts known. A catalyst speeds the rate of that reaction by lowering the activation energy required, even though the catalyst itself is unchanged by the reaction. Enzymes are protein catalysts that lower activation energy by stabilizing the transition state of the reaction.

Enzyme-catalyzed reactions take place at a location called the active site of the enzyme. The active site in each type of enzyme has unique chemical properties and a distinctive size and shape. As a result, most enzymes are able to catalyze only one specific reaction. As a group, enzymes are able to catalyze many types of reactions because their chemical and physical structures are so diverse. This diversity is due to the variety of amino acids and the four levels of protein structure. A protein's primary structure, or sequence of amino acids, is responsible for most of its chemical properties. Interactions that take place between carboxyl and amino groups in the same peptide-bonded backbone, between R-groups found in the same polypeptide, and among different polypeptides all help determine the protein's overall shape.

Many enzymes function only with the help of cofactors. In addition, virtually all enzyme activity in cells is regulated by molecules that bind at the active site or at locations on the protein that induce a change in the size or shape of the active site.

Web Tutorial 3.2 Activation Energy and Enzymes

■ **In cells, endergonic reactions occur in conjunction with an exergonic reaction involving ATP.**

Almost all of the reactions that occur inside cells are endergonic. When ATP or a phosphate group from ATP is added to a substrate or enzyme that participates in an endergonic reaction, the potential energy of the substrate or enzyme is raised enough to make the reaction exergonic and thus spontaneous. In this way, ATP drives reactions that otherwise would not occur.

The final step in chemical evolution is the formation of a self-replicating molecule. A self-replicating molecule must be able to catalyze polymerization reactions, and it must furnish the template necessary to make a copy of itself. Although enzymes catalyze most of the polymerization reactions observed in cells today, only a handful of researchers currently support the hypothesis that the first living entity was a protein. Proteins are efficient catalysts, but there is no obvious way for them to self-replicate. Instead, most researchers now support the hypothesis that the first molecule to make copies of itself—a fundamental characteristic of life—was a nucleic acid. Chapter 4 explains why.

Questions

Content Review

1. What two functional groups are present on every amino acid?
   a. a carbonyl group and a carboxyl group
   b. an amino group and a carbonyl group
   c. an amino group and a hydroxyl group
   d. an amino group and a carboxyl group

2. Twenty different amino acids are found in the proteins of cells. What distinguishes these molecules?
   a. the location of their carboxyl group
   b. the location of their amino group
   c. the composition of their side chains, or R-groups
   d. their ability to form peptide bonds

3. What determines the primary structure of a polypeptide?
   a. its sequence of amino acids
   b. hydrogen bonds that form between carboxyl and amino groups on different residues
   c. hydrogen bonds and other interactions between side chains
   d. the number, identity, and arrangement of polypeptides that make up a protein

4. In a polypeptide, what is most responsible for the secondary structure called an α-helix?
   a. the sequence of amino acids
   b. hydrogen bonds that form between carboxyl and amino groups on different residues
   c. hydrogen bonds and other interactions between side chains
   d. the number, identity, and arrangement of polypeptides that make up a protein

5. What is a transition state?
   a. the complex formed as covalent bonds are being broken and re-formed during a reaction
   b. the place where regulatory molecules bind to an enzyme
   c. a reactant with high potential energy created by phosphorylation
   d. the shape adopted by an enzyme that has an inhibitory molecule bound at its active site

6. By convention, biologists write the sequence of amino acids in a polypeptide in which direction?
   a. carboxy- to amino-terminus
   b. amino- to carboxy-terminus
   c. polar residues to nonpolar residues
   d. charged residues to uncharged residues

Conceptual Review

1. Explain the lock-and-key model of enzyme activity. Be sure to comment on what the active site of an enzyme does.

2. Isoleucine, valine, leucine, phenylalanine, and methionine are amino acids with highly hydrophobic side chains. Suppose a section of a protein contains a long series of these hydrophobic residues. How would you expect this portion of the protein to behave when the molecule is in aqueous solution?

3. Compare and contrast competitive inhibition and allosteric regulation.

4. Figure 3.8a shows a generalized cartoon of monomers undergoing condensation reactions to form a polymer. Label the type of bond formed when amino acids polymerize to a polypeptide. Does it take energy for polymerization reactions to proceed, or do they occur spontaneously? Why or why not?

5. A major theme in this chapter is that the structure of molecules correlates with their function. Use this theme to explain why proteins can perform so many different functions in organisms and why proteins as a group are such effective catalysts.

6. Explain how endergonic reactions can be driven by phosphorylation.

Group Discussion Problems

1. The essay at the end of Chapter 2 discussed efforts to find life on other components of our solar system. Based on the information in this chapter, is it reasonable to expect that protein-based life forms could exist in extraterrestrial environments that are very hot or highly acidic? Why or why not?

2. Recently, researchers were able to measure movement that occurred in a single amino acid in an enzyme as reactions were taking place in its active site. The residue that moved was located in the active site, and the rate of movement correlated closely with the rate at which the reaction was taking place. Discuss the significance of these findings, using the information in Figures 3.22 and 3.23.

3. Researchers can analyze the atomic structure of enzymes during catalysis. In one recent study, investigators found that the transition state included the formation of a free radical, and that a coenzyme bound to the active site donated an electron to help stabilize the free radical. How would the reaction rate and the stability of the transition state change if the coenzyme were not available?

4. In a recent experiment, researchers found that if they took a mixture containing two optical isomers of a monomer and put a thin layer of the mixture on water, a crystal-like structure formed. Only one of the optical isomers—not both—was included in the crystal. The researchers suggested that a similar phenomenon might occur if both optical isomers of a monomer formed a thin coating on a clay mineral particle. How might this observation relate to the fact that proteins in cells contain only one isomer of each amino acid?

Answers to Multiple-Choice Questions   **1.** d; **2.** c; **3.** a; **4.** b; **5.** a; **6.** b

4

Nucleic Acids and the RNA World

KEY CONCEPTS

- Nucleotides are monomers that consist of a sugar, a phosphate group, and a nitrogen-containing base. Ribonucleotides polymerize to form RNA. Deoxyribonucleotides polymerize to form DNA.

- DNA's primary structure consists of a sequence of nitrogen-containing bases, which contain information in the form of a molecular code. DNA's secondary structure consists of two DNA strands running in opposite directions. The strands are held together by complementary base pairing and are twisted into a double helix.

- RNA's primary structure also consists of a sequence of nitrogen-containing bases that contain information in the form of a molecular code. Its secondary structure includes short double helices and structures called hairpins. RNA molecules called ribozymes catalyze important chemical reactions.

An artist's representation of deoxyribonucleic acid, or DNA. In today's cells, DNA functions as an information repository. But the nucleic acid called ribonucleic acid, or RNA, was probably responsible for the origin of life.

Chapter 3 began with experimental evidence that chemical evolution produced the monomers called amino acids and their polymers called proteins. But the chapter ended by stating that even though proteins are the workhorse molecules of today's cells, few researchers still favor the hypothesis that life began as a protein molecule. Instead, the vast majority of biologists now contend that life began as a polymer called a nucleic acid—specifically, a molecule of ribonucleic acid (RNA). This proposal is called the **RNA world hypothesis**. Just what is a nucleic acid? How could these types of molecules have been produced by chemical evolution? What do the nucleic acids RNA and deoxyribonucleic acid (DNA) do in cells today, and why isn't it plausible to propose that life began with DNA? Before attempting to answer these questions, let's consider an even more basic issue: How would we know when a molecule became alive? Stated another way, What *is* life?

Like many simple questions, the issue of defining life is not easy to answer. In reality, there is no precise definition of what constitutes life. Discussions about the definition of life are still taking place, but most biologists now point to two attributes to distinguish life from nonlife. The first attribute is the ability to reproduce. If something is alive, it can make a copy of itself. (In organisms that reproduce asexually, the copy is exact. In the case of sexually reproducing organisms, the copy is not exact. Instead, traits from male and female individuals are combined to produce an offspring.) In essence, this chapter is about the first molecule that could make a copy of itself.

For something to be alive, however, most biologists insist that it have a second attribute: the ability to acquire particular molecules and use them in controlled chemical reactions that maintain conditions suitable for life and that contribute to growth. In today's organisms, chemical reactions are precisely controlled because enzymes and reactants are bounded by a **plasma membrane**. To many biologists, then, the presence of a plasma membrane enclosing a large suite of coordinated chemical reactions, or **metabolism**, is also required for life.

According to the theory of chemical evolution, these two attributes of life did not emerge simultaneously. Instead, the theory predicts that chemical evolution first led to the existence of a molecule that could make copies of itself. Some researchers argue that this self-replicating molecule was also capable of metabolism, because it had to be able to catalyze the polymerization reactions that allowed copies of itself to form. Later, the theory holds, a descendant of this molecule became enclosed in a membrane. This event created the first cell. The formation of the cell allowed for reactants to be enclosed. As a result, it made precise control over reactions possible.

The key idea is that life started with replication and later became cellular. To distinguish a naked (membrane-less) self-replicator from later, cellular forms of "true" life, this chapter refers to the self-replicator as the first living entity or the first life-form—but *not* as the first organism. The term **organism** is reserved for cellular-based life. This distinction isn't trivial. Naked, self-replicating molecules undoubtedly existed early in

Earth's history, and researchers are almost certain to produce one in the laboratory within your lifetime. But it is much less likely that biologists will be able to create cellular life in the foreseeable future. Will humans be able to create life in a test tube? The answer depends, in part, on how you define life.

To understand how life began, it's helpful to delve into some more biochemistry. Let's begin with an analysis of nucleic acids as monomers and end with current research on creating a self-replicating molecule.

4.1 What Is a Nucleic Acid?

Nucleic acids are polymers, just as proteins are polymers. But instead of being made up of monomers called amino acids, **nucleic acids** are made up of monomers called nucleotides. **Figure 4.1a** diagrams the three components of a **nucleotide**: (1) a phosphate group, (2) a sugar, and (3) a nitrogenous (nitrogen-containing) base. The phosphate is bonded to the sugar molecule, which in turn is bonded to the nitrogenous base. A **sugar** is an organic compound with a carbonyl group ($>C=O$) and several hydroxyl (OH) groups. Notice that the "prime" symbols in Figure 4.1 indicate that the carbon being referred to is part of the sugar and not of the attached nitrogenous base.

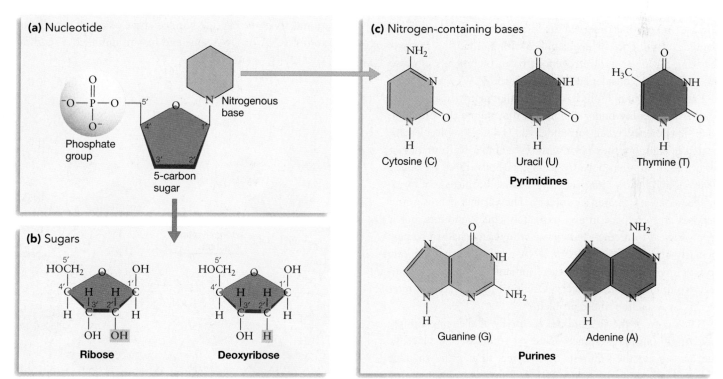

FIGURE 4.1 The General Structure of a Nucleotide
(a) The relationship between the phosphate group, the sugar, and the nitrogenous base of a nucleotide. The numbers indicate the position of each carbon in the sugar's ring. The nitrogenous base is bonded to carbon number 1 in the ring, while the phosphate is bonded to carbon number 5. The bond between the phosphate group and the sugar is called a 5′ linkage. Although hydrogen atoms are bonded to the carbon atoms in the ring (see part b), biologists routinely omit them for clarity. **(b)** Ribose and deoxyribose are similar sugars that are found in nucleotides. **(c)** Purines and pyrimidines are nitrogenous bases. A C–N bond links them to the sugar in a nucleotide. This bond forms at the nitrogen atom that is highlighted in each base. Purines are substantially larger than pyrimidines.

Although a wide variety of nucleotides are found in living cells, origin-of-life researchers concentrate on two types: ribonucleotides and deoxyribonucleotides. In **ribonucleotides**, the sugar is *ribose*; in **deoxyribonucleotides**, it is *deoxyribose*. As **Figure 4.1b** shows, these two sugars differ by a single atom. Ribose has an –OH group bonded to the second carbon in the ring. Deoxyribose has an H instead at the same location. Note that *deoxy-* means "lacking oxygen."

Cells today use four different ribonucleotides, each of which contains a different nitrogenous base. These bases, diagrammed in **Figure 4.1c**, belong to structural groups called **purines** and **pyrimidines**. Ribonucleotides include the purines adenine (A) and guanine (G), and the pyrimidines cytosine (C) and uracil (U).

Similarly, four different deoxyribonucleotides are found in cells today and are distinguished by the structure of their nitrogenous base. Like ribonucleotides, deoxyribonucleotides include adenine, guanine, and cytosine. But instead of uracil, a closely related pyrimidine called thymine (T) occurs in deoxyribonucleotides (**Figure 4.1c**).

Could Chemical Evolution Result in the Production of Nucleotides?

Based on data presented in Chapter 3, most researchers contend that amino acids were abundant early in Earth's history. As yet, however, no one has observed the formation of a nucleotide via chemical evolution. The problem lies with mechanisms for synthesizing the sugar and nitrogenous base components of these molecules. Let's consider each issue in turn.

Laboratory simulations have shown that many sugars can be synthesized readily under conditions that mimic the prebiotic soup. Specifically, when formaldehyde (H_2CO) molecules are heated in solution, they react with one another to form almost all of the sugars that have five or six carbons (these are called pentoses and hexoses, respectively). Ironically, the ease of forming these sugars creates a problem. The various pentoses and hexoses are produced in approximately equal amounts, but it seems logical to predict that ribose would have had to be particularly abundant for RNA or DNA to form in the prebiotic soup. How ribose came to be the dominant sugar during chemical evolution is still a mystery. Origin-of-life researchers refer to this issue as "the ribose problem."

The origin of the pyrimidines is equally challenging. Simply put, origin-of-life researchers have yet to discover a plausible mechanism for the synthesis of cytosine, uracil, and thymine molecules prior to the origin of life. Purines, in contrast, are readily synthesized by reactions among hydrogen cyanide (HCN) molecules. For example, both adenine and guanine have been found in the solutions recovered after spark-discharge experiments.

The ribose problem and the origin of pyrimidine bases are two of the most serious challenges that remain for the theory of chemical evolution. Research on how large quantities of ribose could have formed in the prebiotic soup, along with the pyrimidines, continues. But once formed, how did nucleic acids polymerize to form RNA and DNA? This question has an answer.

How Do Nucleotides Polymerize to Form Nucleic Acids?

Nucleic acids form when nucleotides polymerize. As **Figure 4.2** shows, the polymerization reaction involves the formation of a bond between the phosphate group of one nucleotide and the hydroxyl group of the sugar component of another nucleotide. The result of this condensation reaction is called a **phosphodiester linkage** or a phosphodiester bond. In Figure 4.2, a phosphodiester linkage joins the 5′ carbon on the ribose of one nucleotide to the 3′ carbon on the ribose of the other. When the nucleotides involved contain the sugar ribose, the polymer that is produced is called **ribonucleic acid**, or simply **RNA**. If the nucleotides contain the sugar deoxyribose instead, then the resulting polymer is **deoxyribonucleic acid**, or **DNA**.

Figure 4.3 shows how the chain of phosphodiester linkages in a nucleic acid acts as a backbone, analogous to the peptide-bonded backbone found in proteins. (Compare Figure 4.3 with Figure 3.10a.) The sugar-phosphate spine of a nucleic acid is directional, as is the peptide-bonded spine of a polypeptide. In a strand of RNA or DNA, one end has an unlinked 5′ carbon

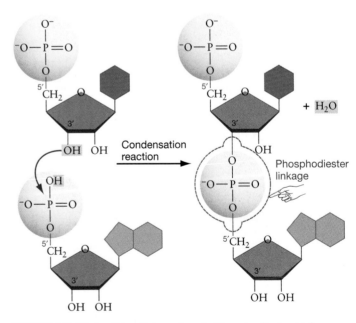

FIGURE 4.2 Nucleotides Polymerize via Phosphodiester Linkages
Ribonucleotides can polymerize via condensation reactions. The linkage that results, between the 3′ carbon of one ribonucleotide and the 5′ carbon of another ribonucleotide, is called a phosphodiester linkage.

while the other end has an unlinked 3′ carbon—meaning a carbon that is not linked to another nucleotide. By convention, the sequence of bases found in an RNA or DNA strand is always written in the 5′ → 3′ direction. (In cells, RNA and DNA are always synthesized in this direction. Bases are added at the 3′ end of the growing molecule.) The sequence of nitrogenous bases forms the primary structure of the molecule, analogous to the sequence of amino acids in a polypeptide.

In cells, the polymerization reactions that form nucleotides are catalyzed by enzymes. Like other polymerization reactions, the process is endergonic. Polymerization can take place in cells because the free energy of the nucleotide monomers is first raised by reactions that add two phosphate groups to the ribonucleotides or deoxyribonucleotides, creating nucleotide triphosphates (**Figure 4.4**). Recall from Chapter 3 that phosphorylation reactions also make the polymerization of amino acids to polypeptides possible. In both cases, the addition of one or more phosphate groups raises the potential energy of the substrate molecules enough to make the polymerization reaction possible. Researchers refer to the phosphorylated nucleotides or amino acids as "activated."

During chemical evolution, activated nucleotides probably polymerized on the surfaces of clay-sized (very fine-grained) mineral particles. In a suite of experiments analogous to those discussed in Chapter 3 for protein synthesis, researchers have produced RNA molecules by incubating activated ribonucleotides with tiny mineral particles. The hypothesis was that polymerization could occur without an enzyme if the ribonucleotides and growing RNA strands adhered to the clay particles. In one experiment, researchers isolated the clay particles after a day of incubation and then added a fresh batch of activated nucleotides. They repeated this reaction-isolation-reaction sequence for a total of 14 days, or fourteen additions of fresh ribonucleotides. At the end of the two-week experiment, they analyzed the mineral particles, using the techniques introduced in **Box 4.1** (page 78), and found RNA molecules up to 40 nucleotides long. Based on these results, there is a strong consensus that if ribonucleotides and deoxyribonucleotides were able to form during chemical evolution, they would be able to polymerize and form RNA and DNA. Now, what would these nucleic acids look like, and what could they do?

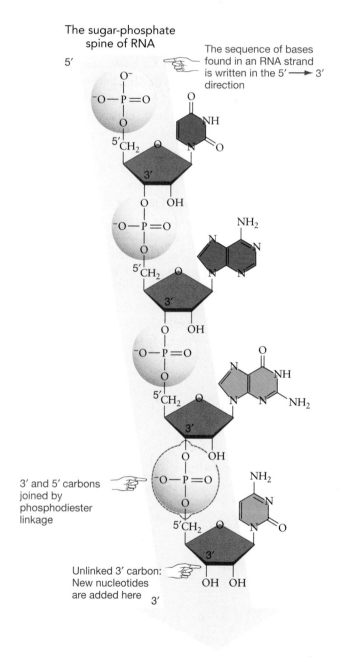

The sugar-phosphate spine of RNA

5′

The sequence of bases found in an RNA strand is written in the 5′ ⟶ 3′ direction

3′ and 5′ carbons joined by phosphodiester linkage

Unlinked 3′ carbon: New nucleotides are added here   3′

FIGURE 4.3 RNA Has a Sugar-Phosphate Backbone

EXERCISE Identify the four bases in this RNA strand, using Figure 4.1c as a key. Then write down the base sequence, starting at the 5′ end.

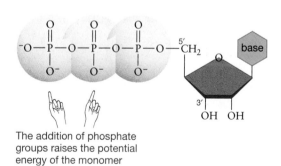

The addition of phosphate groups raises the potential energy of the monomer

FIGURE 4.4 Activated Monomers Drive Endergonic Polymerization Reactions

Polymerization reactions involving ribonucleotides are endergonic. But polymerization reactions involving ribonucleotide triphosphates are exergonic.

BOX 4.1   Gel Electrophoresis and Autoradiography

In molecular biology, the standard technique for separating and analyzing proteins and nucleic acids is called gel electrophoresis or, simply, electrophoresis ("electricity-moving"). This is the technique that was used to determine whether RNA could polymerize when ribonucleotides stuck to clay particles.

The principle behind electrophoresis is fairly simple. Both proteins and nucleic acids carry a charge. As a result, these molecules move when placed in an electric field. Negatively charged molecules move toward the positive electrode (the positive end of the field), and positively charged molecules move toward the negative electrode (the negative end). To separate a mixture of macromolecules so that each can be isolated and analyzed, researchers place the sample in a gelatinous substance. The "gel" consists of long molecules that form a matrix of fibers. This gelatinous matrix has pores through which molecules can pass. When an electrical field is applied across the gel, the molecules in the well move through the gel

toward an electrode. Molecules that are smaller or more highly charged for their size move faster than do larger or less highly charged molecules. As they move, the molecules separate by size and by charge.

Figure 4.5 shows the electrophoresis setup used in the polymerization experiment. Step 1 shows how investigators loaded samples of macromolecules, taken from different days during the experiment, into cavities or "wells" at the top of the gel slab. In step 2 the researchers immersed the gel in a solution that conducts electricity and applied a voltage across the gel. After the samples had run down the gel for some time (step 3), they removed the electric field. By then, molecules of different size and charge had separated from one another. In this case, small RNA molecules had reached the bottom of the gel. Above them were larger RNA molecules, which had run more slowly.

Once molecules have been separated in this way, they have to be detected.

Often, proteins or nucleic acids can be stained or dyed. In this case, however, the researchers had attached a radioactive atom to the monomers used in the experiment, so the polymers that resulted could be visualized by laying X-ray film over the gel. Because radioactive emissions expose film, a black dot appears wherever a radioactive atom is located in the gel. This technique for visualizing macromolecules is called autoradiography. If the samples are loaded into a rectangular well, as is commonly done, molecules of a particular size and charge form a band in an autoradiograph.

The autoradiograph that resulted from the polymerization experiment is shown in **Figure 4.6**. The samples, taken on days 2, 4, 6, 8, and 14 of the experiment, are labeled along the bottom. The far right lane contains macromolecules of known size; called a size standard or "ladder," this lane is used to estimate the size of the molecules in the experimental samples. The bands that appear in each sample lane

GEL ELECTROPHORESIS

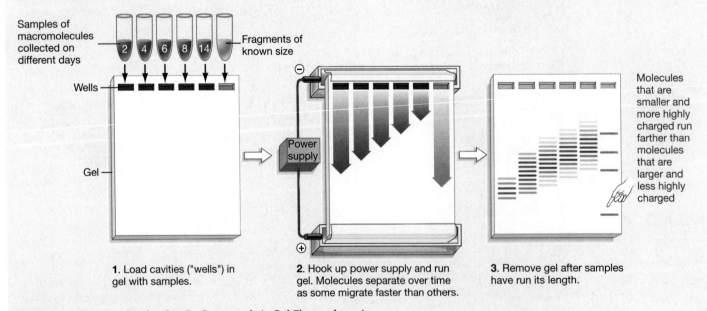

1. Load cavities ("wells") in gel with samples.

2. Hook up power supply and run gel. Molecules separate over time as some migrate faster than others.

3. Remove gel after samples have run its length.

FIGURE 4.5 Macromolecules Can Be Separated via Gel Electrophoresis

QUESTION DNA and RNA run toward the positive electrode. Why are these molecules negatively charged?

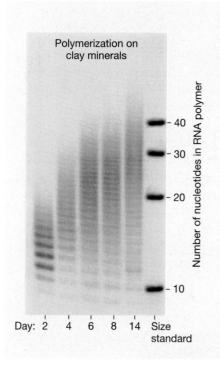

Polymerization on clay minerals

Number of nucleotides in RNA polymer

— 40
— 30
— 20
— 10

Day: 2 4 6 8 14 Size standard

FIGURE 4.6 Autoradiography Is a Technique for Visualizing Macromolecules
The molecules in a gel can be visualized in a number of ways. In this case, the RNA molecules in the gel exposed an X-ray film because they had radioactive atoms attached. When developed, the film is called an autoradiograph. EXERCISE Add labels to the photograph that read: "Top of gel—large molecules" and "Bottom of gel—small molecules."

represent the different polymers that had formed. Darker bands contain more radioactive marker, indicating the presence of many radioactive molecules. Lighter bands contain fewer molecules.

Several conclusions can be drawn from the data. First, a variety of polymers formed at each stage. After the second day, for example, polymers from 12 to 18 monomers long had formed on the clay particles. Second, the overall length of polymers produced increased with time. At the end of the fourteenth day, most of the RNA molecules were between 20 and 40 monomers long. Polymerization had occurred without the aid of an enzyme.

✓ CHECK YOUR UNDERSTANDING

Nucleotides are monomers that consist of a sugar, a phosphate group, and a nitrogen-containing base. Nucleotides polymerize to form nucleic acids through formation of phosphodiester linkages between the 3′ carbon on one nucleotide and the 5′ carbon on another. You should be able to draw a simplified cartoon of a nucleotide and a cartoon version of two nucleotides joined by a phosphodiester linkage.

4.2 DNA Structure and Function

Nucleic acids have a primary structure that is somewhat similar to the primary structure of proteins. Proteins have a peptide-bonded backbone with a series of R-groups that extend from it; DNA and RNA molecules have a sugar-phosphate backbone, created by phosphodiester linkages, and a sequence of four nitrogenous bases that extend from it.

Does DNA have secondary structure? The answer to this question, announced in 1953, ranks among the great scientific breakthroughs of the twentieth century. James Watson and Francis Crick presented a model for the secondary structure of DNA in a one-page paper published in *Nature*. At the time, Watson was a 25-year-old postdoctoral fellow and Crick was a 37-year-old graduate student.

Watson and Crick's finding was a hypothesis based on a series of results from other laboratories. They were trying to propose a secondary structure that could explain several important observations about the DNA found in cells:

- Chemists had worked out the structure of nucleotides and knew that DNA polymerized through the formation of phosphodiester linkages. Thus Watson and Crick knew that the molecule had a sugar-phosphate backbone.

- By analyzing the nitrogenous bases in DNA samples from different organisms, Erwin Chargaff had established two empirical rules: (1) The total number of purines and pyrimidines in DNA is the same; and (2) the numbers of T's and A's in DNA are equal, and the numbers of C's and G's in DNA are equal.

- By bombarding DNA with X rays and analyzing how it scattered the radiation, Rosalind Franklin and Maurice Wilkins had calculated the distances between groups of atoms in the molecule (see **Box 4.2**, page 80). The scattering patterns showed that three distances were repeated many times: 0.34 nanometer (nm), 2.0 nm, and 3.4 nm. Because the measurements repeated, the researchers inferred that DNA molecules had a regular and repeating structure. The pattern of X-ray scattering suggested that the molecule was helical or spiral in nature.

Based on this work, understanding DNA's structure boiled down to understanding the nature of the helix involved. What type of helix would have a sugar-phosphate backbone and explain both Chargaff's rules and the Franklin-Wilkins measurements?

Watson and Crick began by analyzing the size and geometry of deoxyribose, phosphate groups, and the nitrogenous bases. The bond angles and measurements suggested that the distance of 2.0 nm probably represented the width of the

BOX 4.2 An Introduction to X-Ray Crystallography

X-ray crystallography is the most widely used technique for reconstructing the three-dimensional structure of molecules. As its name implies, the procedure is based on bombarding crystals of a molecule with X rays. The crystals diffract (that is, scatter) the X rays, producing a diffraction pattern that can be recorded on X ray film or other types of detectors. (The approach is also called *X-ray diffraction analysis*.)

The basic principle is that X rays are scattered in precise ways when they interact with the electrons surrounding the atoms in a crystal (**Figure 4.7**). By varying the orientation of the X-ray beam that strikes a crystal and documenting the diffraction patterns that result, researchers can record a complex set of patterns that reflect the three-dimensional structure of the crystal.

The diffraction pattern can be used to construct a map representing the density of electrons in the crystal. By relating these electron-density maps to information about the primary structure of the nucleic acid or protein, a three-dimensional model of the molecule can be built.

It is important to appreciate the amount of work involved in producing an accurate three-dimensional representation of a nucleic acid or protein. Hundreds or thousands of failed attempts may occur before researchers produce crystals of sufficient quality for X-ray diffraction analysis. But the information contained in an accurate three-dimensional model makes the effort worthwhile.

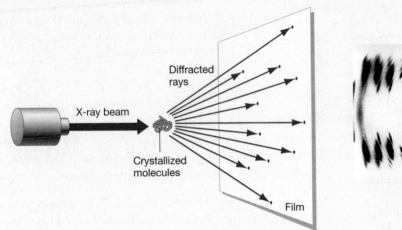
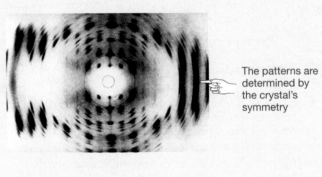

FIGURE 4.7 X-Ray Crystallography Provides Information on the Three-Dimensional Structure of Macromolecules
When crystallized molecules are bombarded with X rays, the radiation is scattered in distinctive patterns. The photograph at the right shows an X-ray film that recorded the pattern of scattered radiation from DNA molecules.

helix and that 0.34 nm was likely to be the distance between bases stacked in a spiral. Now they needed to make sense of Chargaff's rules and the 3.4-nm distance, which appeared to be exactly 10 times the distance between a single pair of bases.

To solve this problem, Watson and Crick constructed a series of physical models such as the one pictured with them in **Figure 4.8**. Building these models allowed them to tinker with different types of helical configurations. After many false starts, they hit on an idea that looked promising. They arranged two strands of DNA side by side and running in opposite directions—meaning that one strand ran in the $5' \rightarrow 3'$ direction while the other was oriented $3' \rightarrow 5'$. Strands with this orientation are said to be antiparallel. Watson and Crick found that if the antiparallel strands were twisted together to form a double helix, the coiled sugar-phosphate backbones ended up on the outside of the spiral and the nitrogenous

FIGURE 4.8 Building a Physical Model of DNA Structure
Watson (left) and Crick (right) did not know if the DNA helix was single, double, or triple stranded, or exactly how the sugar-phosphate backbones in the helix were oriented. To help sort out the possibilities, they built physical models of the four deoxyribonucleotides, using wires with precise lengths and geometries. Then they tested how the nucleotides fit together in different configurations.

(a) Only purine-pyrimidine pairs fit inside the double helix.

Purine-pyrimidine pair
JUST RIGHT

Purine-purine pair
NOT ENOUGH SPACE

Pyrimidine-pyrimidine pair
TOO MUCH SPACE

Space inside sugar-
phosphate backbones

FIGURE 4.9 Complementary Base Pairing Is Based on Hydrogen Bonding
(a) Only purine-pyrimidine pairs fit inside the double helix effectively.
(b) Hydrogen bonds form when guanine and cytosine pair, and when adenine and thymine pair.

(b) Hydrogen bonds form between G-C pairs and A-T pairs.

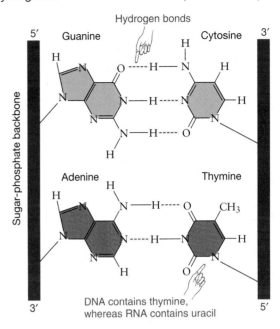

DNA contains thymine, whereas RNA contains uracil

bases on the inside. But for the bases from each backbone to fit in the interior of the 2.0-nm-wide structure, they had to form purine-pyrimidine pairs (see **Figure 4.9a**). With that came a fundamental insight: Inside the double helix, the bases lined up in a way that allowed hydrogen bonds to form between certain purines and pyrimidines. More specifically, adenine could form hydrogen bonds with thymine, and guanine could form hydrogen bonds with cytosine (**Figure 4.9b**). Because of this specificity, the A-T and G-C bases were said to be complementary. Two hydrogen bonds form when A and T pair, but three hydrogen bonds form when G and C pair. As a result, the G-C interaction is slightly stronger than the A-T bond.

Watson and Crick had discovered the phenomenon known as **complementary base pairing**. In fact, the term *Watson-Crick pairing* is now used interchangeably with the phrase *complementary base pairing*. **Figure 4.10** shows how antiparallel strands of DNA form when complementary bases line up and form hydrogen bonds. As you study the figure, notice that biochemists sometimes say that DNA is put together like a ladder whose ends have been twisted in opposite directions. The sugar-phosphate backbone forms the supports of the ladder; the base pairs represent the rungs of the ladder. The nitrogenous bases in the middle of the DNA helix stack tightly on top of each other. This tight packing forms a hydrophobic interior that is difficult to break apart. The molecule as a whole is water soluble, however, because the sugar-phosphate backbones, which face the exterior of the molecule, are negatively charged and thus are hydrophilic.

It is important to note that the outside of the helical DNA molecule forms two types of grooves and that the types differ in size. The larger of the two is known as the *major groove*, and the smaller one is known as the *minor groove*.

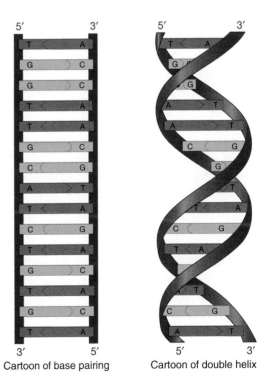

Cartoon of base pairing   Cartoon of double helix

FIGURE 4.10 The Secondary Structure of DNA Is a Double Helix
Complementary base pairing is responsible for twisting DNA into a double helix.

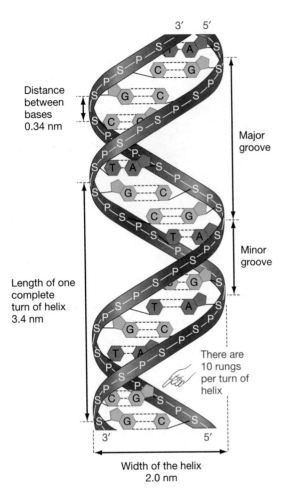

Distance between bases 0.34 nm

Major groove

Minor groove

Length of one complete turn of helix 3.4 nm

There are 10 rungs per turn of helix

Width of the helix 2.0 nm

FIGURE 4.11 Dimensions of DNA Secondary Structure
The double-helix hypothesis explains the measurements inferred from X-ray analysis of DNA molecules.

Figure 4.11 highlights these grooves and illustrates how DNA's secondary structure explains the measurements observed by Franklin and Wilkins.

Since the model of the double helix was published, experimental tests have shown that the hypothesis is correct in almost every detail. DNA's secondary structure consists of two antiparallel strands twisted into a double helix. The molecule is stabilized by hydrophobic interactions in its interior and by hydrogen bonding between the complementary base pairs A-T and G-C. How does this secondary structure affect the molecule's function?

DNA Is an Information-Containing Molecule

Watson and Crick's model created a sensation because it revealed how DNA could store and transmit biological information. In literature, information consists of letters on a page. In music, information is composed of the notes on a staff. But inside cells, information consists of a sequence of monomers in a nucleic acid. The four nitrogeneous bases function like letters of the alphabet. A particular sequence of bases is like the sequence of letters in a word—it has meaning. In all cells that have been examined to date, from tiny bacte-

ria to gigantic redwood trees, DNA carries the information required for the organism's growth and reproduction. Exploring how this information is encoded and translated into action is the heart of Chapters 14 through 17.

Here, however, our focus is on how life began. The theory of chemical evolution holds that life began as a self-replicating molecule—a molecule that could make a copy of itself. **Figure 4.12** shows how a copy of DNA could be made by complementary base pairing. Complementary base pairing allows each strand of a DNA double helix to be copied exactly, producing two daughter molecules. In this way, DNA contains the information required for a copy of itself to be made.

Watson and Crick ended their paper on the double helix with one of the classic understatements in scientific literature: "It has not escaped our notice that the specific pairing we have postulated immediately suggests a possible copying mechanism." The central idea is that DNA's primary structure serves as a mold or template for the synthesis of a complementary strand.

In today's cells, however, DNA does not self-replicate spontaneously. Instead, the molecule is copied through a complicated series of reactions catalyzed by a large suite of enzymes. Similarly, it is extremely unlikely that a DNA molecule began copying itself early in Earth's history. Let's explore why.

Is DNA a Catalytic Molecule?

The DNA double helix is a highly structured molecule that is much more stable than RNA and most proteins. DNA is regular and symmetric, with few chemical groups exposed that can participate in chemical reactions. In addition, the lack of a 2′ hydroxl group on deoxyribonucleotides (see Figure 4.1b) makes the polymer less reactive than RNA and highly resistant to chemical degradation. All of these features increase DNA's stability and thus its effectiveness as a reliable information-bearing molecule. Intact stretches of DNA have been recovered from fossils that are tens of thousands of years old. Despite death and exposure to a wide array of pH, temperature, and chemical conditions, the molecules have the same sequence of bases as the organisms had when they were alive.

The orderliness and stability that make DNA such a dependable information repository make it extraordinarily inept at catalysis, however. Recall from Chapter 3 that enzyme function is based on a specific binding event between a substrate and a protein catalyst. Thanks to the enormous diversity of primary through quaternary structures found in proteins, a wide array of binding events can occur. In comparison, DNA's primary and secondary structures are simple. It is not surprising, then, that DNA has never been observed to catalyze any reaction in any organism. Even though researchers have been able to construct single-stranded DNA molecules that can catalyze a few simple reactions in the laboratory, the number and diversity of reactions involved is a minute fraction of the activity catalyzed by proteins.

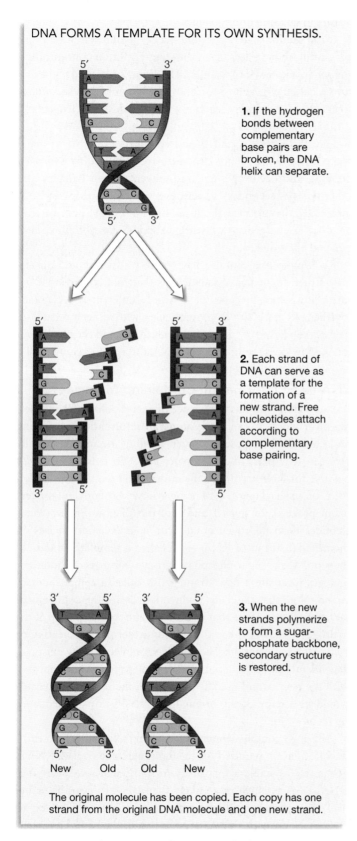

DNA FORMS A TEMPLATE FOR ITS OWN SYNTHESIS.

1. If the hydrogen bonds between complementary base pairs are broken, the DNA helix can separate.

2. Each strand of DNA can serve as a template for the formation of a new strand. Free nucleotides attach according to complementary base pairing.

3. When the new strands polymerize to form a sugar-phosphate backbone, secondary structure is restored.

The original molecule has been copied. Each copy has one strand from the original DNA molecule and one new strand.

FIGURE 4.12 Making a Copy of DNA
If new bases are added to each of the two strands of DNA via complementary base pairing, a copy of the DNA molecule can be produced.

In short, DNA furnishes an extraordinarily stable template for copying itself. But due to its inability to act as an effective catalyst, virtually no researchers support the hypothesis that the first life-form consisted of DNA. Most biologists who are working on the origin of life support the hypothesis that life began with RNA rather than DNA.

✓CHECK YOUR UNDERSTANDING

DNA's primary structure consists of a sequence of deoxyribonucleotides. Its secondary structure consists of two DNA molecules that run in opposite $5' \rightarrow 3'$ orientations to each other. The two strands are held together through hydrogen bonds between A-T and G-C pairs and hydrophobic interactions among atoms inside the helix. The sequence of deoxyribonucleotides in DNA contains information, much like the sequence of letters and punctuation on this page. Due to complementary base pairing, each DNA strand also contains the information required to form the complementary strand. You should be able to make a sketch of a DNA molecule that (1) uses different symbols to stand for the A, T, G, and C bases; (2) labels the molecule's sugar-phosphate backbone; and (3) indicates the $5' \rightarrow 3'$ orientation of each strand.

4.3 RNA Structure and Function

Like DNA, RNA has a primary structure that consists of a sugar-phosphate backbone formed by phosphodiester linkages and a sequence of four types of nitrogenous base that extend from that backbone. But it's important to recall two significant differences between these nucleic acids:

1. The pyrimidine base thymine does not exist in RNA. Instead, RNA contains the closely related pyrimidine base uracil.

2. The sugar in the sugar-phosphate backbone of RNA is ribose, not deoxyribose as in DNA.

The second point is critical because the hydroxyl (–OH) group on the 2'-carbon of ribose is much more reactive than the hydrogen atom on the 2'-carbon of deoxyribose. This functional group can participate in reactions that tear the polymer apart. The presence of this –OH group makes RNA much more reactive, while its absence makes DNA more stable.

A comparison of secondary structure in RNA and DNA is just as instructive. Like DNA molecules, most RNA molecules have secondary structure that results from complementary base pairing between purine and pyrimidine bases. In RNA, adenine forms hydrogen bonds only with uracil, and guanine again forms hydrogen bonds with cytosine. (Guanine can also bond with uracil, but it does so much less effectively than with cytosine.) Thus, the complementary base pairs in RNA are A-U and G-C. Three hydrogen bonds form between guanine and cytosine, but only two form between adenine and uracil.

How do the secondary structures of RNA and DNA differ? In the vast majority of cases, the purine and pyrimidine bases in RNA undergo hydrogen bonding with complementary bases on the *same* strand, rather than forming hydrogen bonds with complementary bases on a different strand, as in DNA. **Figure 4.13** shows how this works. The key is that when bases on one part of an RNA strand fold over and align with ribonucleotides on another segment of the same strand, the two sugar-phosphate strands are antiparallel. In this orientation, hydrogen bonding between complementary bases results in a stable double helix.

If the section where the fold occurs includes a large number of unbonded bases, then the stem-and-loop configuration shown in Figure 4.13 results. This type of secondary structure is called a **hairpin**. Several other types of RNA secondary structures are possible, each involving a different length and arrangement of base-paired segments.

Like the α-helices and β-pleated sheets observed in many proteins, RNA secondary structures are stabilized by hydrogen bonding and occur spontaneously. Even though hairpins and other types of secondary structures reduce the entropy of RNA molecules, they form without an input of energy because they are favored energetically. Hydrogen bond formation is exothermic and exergonic.

RNA molecules can also have tertiary structure and quaternary structure, owing to interactions that fold secondary structures into complex shapes or that hold different RNA strands together. As a result, RNA molecules with different base sequences can have very different overall shapes and chemical properties. How does the structure of RNA affect its function in cells?

RNA as an Information-Containing Molecule

Because RNA contains a sequence of bases analogous to the letters in a word, it can function as an information-containing molecule. And because hydrogen bonding occurs specifically between A-U pairs and G-C pairs in RNA, it is theoretically possible for RNA to furnish the information required to make a copy of itself. **Figure 4.14** shows how the copying process might proceed. In steps 1 and 2 of this diagram, free ribonucleotides form hydrogen bonds with complementary bases on the original strand of RNA—also called a **template strand**. As they do, their sugar-phosphate groups form phosphodiester linkages to create a new strand—also called a **complementary strand**. Note that the $5' \rightarrow 3'$ directionality of the complementary strand is opposite that of the template strand. In step 3, the hydrogen bonds between the strands are broken by heating or by a catalyzed reaction. The new RNA molecule now exists independently of the original strand. If these steps were repeated with the new strand used as a template, the resulting molecule would be a copy of the original. An RNA's primary sequence serves as a mold.

Although complementary base pairing allows RNA to carry the information required for the molecule to be copied, RNA is not nearly as stable a repository for particular sequences as is DNA. RNA is much more likely than DNA to degrade or hydrolyze in response to high temperatures, changes in pH, or attack by reactive chemicals—partly because it has 2′ hydroxyl groups and partly because its secondary structure is less extensive. For example, no one has yet found intact RNA in an ancient bone or pollen grain or other type of fossil. The molecule is simply too unstable. In today's cells, RNA molecules that are

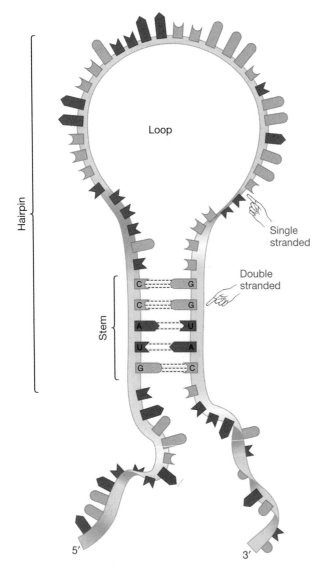

FIGURE 4.13 Complementary Base Pairing and Secondary Structure in RNA: Stem-and-Loop Structures
This RNA molecule has secondary structure. The double-stranded "stem" and single-stranded "loop" form a hairpin. The bonded bases in the stem are antiparallel, meaning that they are oriented in opposite directions.

RNA FORMS A TEMPLATE FOR ITS OWN SYNTHESIS

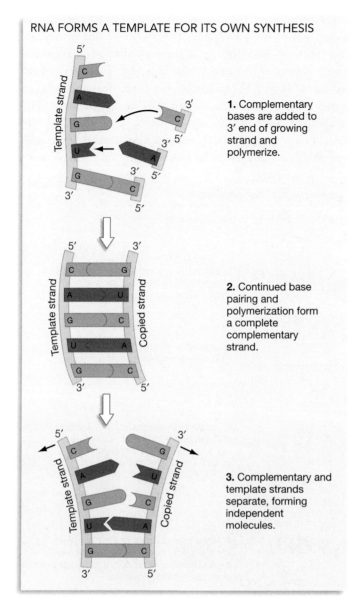

1. Complementary bases are added to 3′ end of growing strand and polymerize.

2. Continued base pairing and polymerization form a complete complementary strand.

3. Complementary and template strands separate, forming independent molecules.

FIGURE 4.14 RNA Molecules Can Be Copied Because They Furnish a Template
RNA molecules can be copied when complementary base pairing matches ribonucleotides on a template to form a complementary strand. If the three steps illustrated here were repeated using the copied strand as a template, the newly synthesized molecule would be identical to the original RNA.

shorter lived than DNA ferry the information required to make proteins and other compounds. But in the prebiotic soup, could an RNA molecule have made a copy of itself before it would have degraded?

Is RNA a Catalytic Molecule?

In terms of diversity in chemical reactivity and overall shape, RNA molecules do not begin to match proteins. The primary structure of RNA molecules is much more restricted, because there are only four types of nitrogenous base in RNA versus the 20 types of amino acid found in proteins, and RNA molecules cannot form the variety of bonds that give proteins their extensive tertiary structure. Yet RNA molecules have limited secondary and tertiary structure that allows distinct shapes to form, and their base sequences are variable enough to produce molecules with a certain amount of unique chemical behavior. Because RNA has a degree of structural and chemical complexity, it should be capable of stabilizing a few transition states and catalyzing at least a limited number of chemical reactions. Indeed, biologists have found that RNA *does* act as a catalyst in organisms today. Sidney Altman and Thomas Cech shared the 1989 Nobel Prize in chemistry for showing that RNA enzymes, or **ribozymes**, exist in organisms. The ribozymes they isolated, from a single-celled organism called *Tetrahymena*, could catalyze both the hydrolysis and condensation of phosphodiester linkages. Building on that accomplishment, researchers have discovered ribozymes that catalyze dozens of different reactions in cells. For example, ribozymes catalyze the formation of peptide bonds when amino acids polymerize to form polypeptides. Ribozymes are at work in your cells right now.

The discovery of ribozymes was a watershed event in origin-of-life research. Before Altman and Cech published their discovery, biologists thought that proteins were the only type of molecule capable of catalyzing chemical reactions in organisms. But if a ribozyme in *Tetrahymena* catalyzes the polymerization reactions similar to those diagrammed in steps 1 and 2 of Figure 4.12, it raises the possibility that an RNA molecule could also catalyze the reactions diagrammed in step 3. Such a molecule could copy itself and qualify as the first living entity. Does any experimental evidence support this hypothesis?

4.4 The First Life-Form

The theory of chemical evolution maintains that life began as a naked self-replicator. To make a copy of itself, the first living molecule had to provide a template that could be copied. It also had to catalyze polymerization reactions that would link monomers into a copy of that template. Because RNA is capable of both processes, most origin-of-life researchers propose that the first life-form was made of RNA. As noted in the introduction to this chapter, that proposal is called the *RNA world hypothesis*.

Because no self-replicating RNA molecules exist today, researchers test the hypothesis by trying to simulate the RNA world in the laboratory. The eventual goal is to create an RNA molecule that can catalyze its own replication. When this goal is achieved, a life-form will have been created in the laboratory.

To understand how researchers go about this work, consider a recent set of experiments by Wendy Johnston and others working in David Bartel's laboratory. This research group's goal was ambitious; they wanted to create, from scratch, a ribozyme that was capable of catalyzing the reactions shown in steps 1 and 2 of Figure 4.14. Such a ribozyme would be called RNA replicase. This research program is creating considerable excitement among biologists interested in the origin of life, because adding ribonucleotides to a growing strand is a key attribute of an RNA replicase.

The strategy Bartel's group pursued is outlined in **Figure 4.15**. As you study this figure, note that their procedure essentially mimics the process of natural selection. For example, the researchers began their experiments by establishing a large population of RNA molecules that were variable in structure and function, analogous to a large population of organisms with a variety of features. In this case, the researchers created millions of large RNA molecules that contained randomly generated primary sequences. When they incubated these newly synthesized, large RNAs with a small RNA template, they found that some of the large molecules could catalyze the addition of a few ribonucleotides that were complementary to the

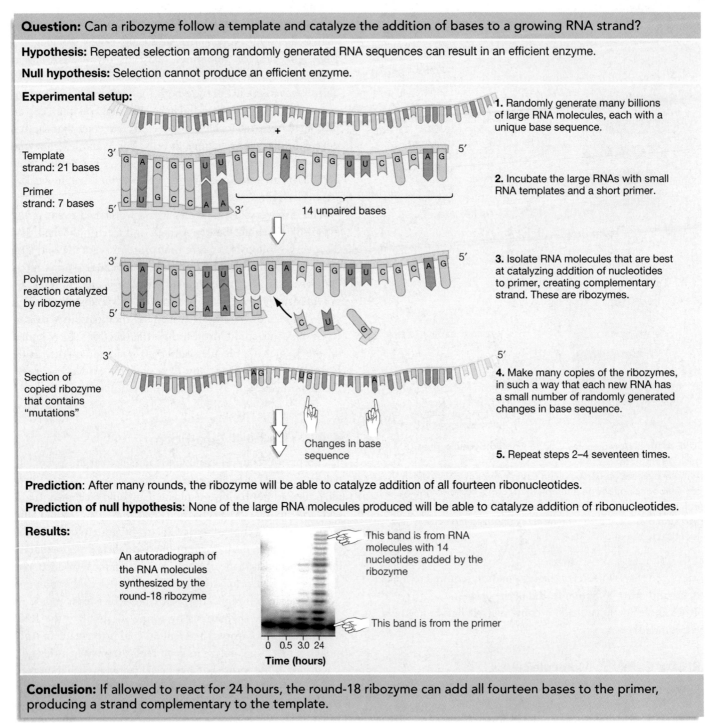

Question: Can a ribozyme follow a template and catalyze the addition of bases to a growing RNA strand?

Hypothesis: Repeated selection among randomly generated RNA sequences can result in an efficient enzyme.

Null hypothesis: Selection cannot produce an efficient enzyme.

Experimental setup:

1. Randomly generate many billions of large RNA molecules, each with a unique base sequence.

Template strand: 21 bases

Primer strand: 7 bases

14 unpaired bases

2. Incubate the large RNAs with small RNA templates and a short primer.

Polymerization reaction catalyzed by ribozyme

3. Isolate RNA molecules that are best at catalyzing addition of nucleotides to primer, creating complementary strand. These are ribozymes.

Section of copied ribozyme that contains "mutations"

4. Make many copies of the ribozymes, in such a way that each new RNA has a small number of randomly generated changes in base sequence.

Changes in base sequence

5. Repeat steps 2–4 seventeen times.

Prediction: After many rounds, the ribozyme will be able to catalyze addition of all fourteen ribonucleotides.

Prediction of null hypothesis: None of the large RNA molecules produced will be able to catalyze addition of ribonucleotides.

Results:

An autoradiograph of the RNA molecules synthesized by the round-18 ribozyme

This band is from RNA molecules with 14 nucleotides added by the ribozyme

This band is from the primer

0   0.5 3.0 24
Time (hours)

Conclusion: If allowed to react for 24 hours, the round-18 ribozyme can add all fourteen bases to the primer, producing a strand complementary to the template.

FIGURE 4.15 Selection of a Novel Ribozyme
Researchers use variations of this protocol to produce ribozymes that can catalyze a wide variety of reactions.

bases in the template. In this "round-1" experiment, the polymerization didn't happen very quickly or accurately, but it did occur. The researchers had discovered a ribozyme that could catalyze the synthesis of RNA.

To continue the experiment, the researchers isolated the round-1 ribozymes. This step was analogous to natural selection, because only certain variants survived to produce "offspring." To create these offspring, which would represent the next generation of ribozymes, Bartel's team copied the selected molecules in a way that introduced a few random changes to their sequence of bases. These changes were analogous to the types of mutations that occur each generation in natural populations. Mutation is a random process that produces new traits. When the researchers incubated the modified ribozymes with a new template RNA, they found that most of the modified ribozymes worked worse than the original ones. Some worked better, however.

The researchers selected the best of these round-2 ribozymes, copied them in a way that introduced a few additional, random changes in their base sequences, and allowed them to react with yet another template. The best ribozymes in this round 3 were isolated, and then the copying, reaction, and selection process was repeated … and repeated … and repeated. By round 18, the group had found a ribozyme that was a much better catalyst than the original molecule. Evolution had occurred. They had created a ribozyme that was reasonably proficient at adding ribonucleotides to a growing strand.

As this book goes to press, the round-18 ribozyme is the closest biologists have come to creating life. Thanks to similar efforts at other laboratories around the world, researchers have produced an increasingly impressive set of ribozymes—an array of molecules capable of catalyzing many of the key reactions responsible for replication and metabolism. Each result provides support for the RNA world hypothesis. Each result also brings research teams closer to the creation of an RNA replicase. If this goal is met, human beings will have created a living entity in a test tube.

ESSAY The Human Side of Research

The discovery of the double helix was a critically important advance in the early 1950s, because DNA had recently been established as the hereditary material. Stated another way, scientists had recently discovered that genes are made of DNA. Before Watson and Crick published, though, it was a complete mystery not only how genes are copied when new cells and offspring are produced but also how the information in genes is used to synthesize the enzymes and other molecules required for life. The discovery of complementary base pairing revealed how genetic information could be copied, and knowledge of DNA's secondary structure stimulated several decades of work aimed at understanding how cells use base sequences as information. It's no exaggeration to claim that the double helix triggered an explosion in research and knowledge. Molecular biology had come of age.

The field's early years were extraordinarily eventful. In the late 1940s and early 1950s, understanding DNA's structure was *the* question in biology. Researchers who were working on the problem knew that the greatest honor in science—the Nobel Prize—would almost certainly be awarded to the person or persons who were first to discover the correct structure. The race was on: Many of the brightest and most famous scientists of the day were involved, working in different laboratories around the world.

Arguably the most important event in the research was Wilkins' and Franklin's ability to produce X-ray pictures of DNA. Recall that the photographs made it possible to make crucial measurements and inferences about the molecule's structure. Although Wilkins and Franklin are justifiably given joint credit for this work, they actually did not cooperate closely. It was Franklin who took the photographs that led to the realization that DNA is helical, but it was Wilkins who showed her photographs to Watson and Crick. Wilkins did so without Franklin's knowledge or permission. Today, such an act would be considered a serious form of professional misconduct. And Linus Pauling—undoubtedly the century's greatest biochemist, who was also working on the problem at the time—was denied permission to travel from the United States to England to see the data. The United States was then in the throes of the Cold War, and the U.S. State Department had revoked Pauling's passport because his political views were considered too liberal.

Scientific research is a very human enterprise.

As expected, Watson and Crick, along with Wilkins, received the Nobel Prize. The award was made in 1962, after experimental work had confirmed that the double-helix model was correct. Tragically, Franklin died of ovarian cancer in 1958, when she was just 37. She never received the Nobel, because the honor is not awarded posthumously.

The moral of the story, if there is one, may be simply that scientific research is a very human enterprise. In addition, times have changed. Although competition can still be intense, biology is an increasingly cooperative and international enterprise; misconduct is usually detected and addressed promptly. Rosalind Franklin's professional life was made difficult at times because she was a woman. Today the majority of graduate students in molecular biology are female.

CHAPTER REVIEW

Summary of Key Concepts

◼ **Nucleotides are monomers that consist of a sugar, a phosphate group, and a nitrogen-containing base. Ribonucleotides polymerize to form RNA. Deoxyribonucleotides polymerize to form DNA.**

To understand the origin of life and how cells work today, it is critical to understand the structure and function of macromolecules—particularly proteins, nucleic acids, and carbohydrates. Each of these molecules is a polymer that is made up of monomers. In the case of nucleic acids, the monomers involved are nucleotides.

The origin of nucleotides remains an important challenge for the theory of chemical evolution. Experiments have shown that sugars and the nitrogenous bases called purines are easily produced under early Earth conditions. But it is not yet clear how the important sugar called ribose could have been produced in large quantities during chemical evolution or how the nitrogenous bases called pyrimidines could have been synthesized.

The polymerization of monomers into macromolecules is the next key event in chemical evolution. Experiments have shown that these condensation reactions occur readily when growing polymers stick to small mineral particles. For example, researchers have observed RNA formation from the polymerization of ribonucleic acids on clay-sized particles.

◼ **DNA's primary structure consists of a sequence of nitrogen-containing bases, which contain information in the form of a molecular code. DNA's secondary structure consists of two DNA strands running in opposite directions. The strands are held together by complementary base pairing and are twisted into a double helix.**

DNA is an extremely stable molecule that serves as a superb archive for information in the form of base sequences. DNA is stable because deoxyribonucleotides lack a reactive 2′ hydroxyl group and because antiparallel DNA strands form a secondary structure called a double helix. The DNA double helix is stabilized by hydrogen bonds that form between complementary purine and pyrimidine bases and by hydrophobic interactions between bases stacked on the inside of the spiral. This same structural stability and regularity make DNA ineffective at catalysis, however.

In addition to being stable, DNA is readily copied via complementary base pairing. Complementary base pairing occurs between A-T and G-C pairs in DNA, as well as between A-U and G-C pairs in RNA.

Web Tutorial 4.1   Nucleic Acid Structure

◼ **RNA's primary structure also consists of a sequence of nitrogen-containing bases that contain information in the form of a molecular code. Its secondary structure includes short double helices and structures called hairpins. RNA molecules called ribozymes catalyze important chemical reactions.**

Compared with proteins and DNA, RNA is exceptionally versatile. The primary function of proteins is to catalyze chemical reactions, and the primary function of DNA is to carry information. But RNA is an "all-purpose" macromolecule that can do both. Most origin-of-life researchers propose that the first self-replicating molecule was RNA, because RNA can catalyze a variety of chemical reactions and because complementary base pairing between nucleotides furnishes a mechanism for making a copy. Recent experiments have succeeded in producing a ribozyme that can catalyze the formation of phosphodiester linkages and the addition of ribonucleotides to a growing strand. Based on results to date, it is reasonable to predict that a self-replicating RNA will be created in the near future.

Questions

Content Review

1. What are the four nitrogenous bases found in RNA?
   a. uracil, guanine, cytosine, thymine (U, G, C, T)
   b. adenine, guanine, cytosine, thymine (A, G, C, T)
   c. adenine, uracil, guanine, cytosine (A, U, G, C)
   d. alanine, threonine, glycine, cysteine (A, T, G, C)

2. What determines the primary structure of an RNA molecule?
   a. the sugar-phosphate backbone
   b. complementary base pairing and the formation of hairpins
   c. the sequence of deoxyribonucleotides
   d. the sequence of ribonucleotides

3. DNA attains a secondary structure when hydrogen bonds form between the nitrogenous bases called purines and pyrimidines. What are the complementary base pairs that form in DNA?
   a. A-T and G-C
   b. A-U and G-C
   c. A-G and T-C
   d. A-C and T-G

4. By convention, biologists write the sequence of bases in RNA and DNA in which direction?
   a. $3' \rightarrow 5'$
   b. $5' \rightarrow 3'$
   c. N-terminal to C-terminal
   d. C-terminal to N-terminal

5. In RNA, when does the secondary structure called a hairpin form?
   a. when hydrophobic residues coalesce
   b. when hydrophilic residues interact with water
   c. when complementary base pairing between ribonucleotides on the same strand creates a stem-and-loop structure
   d. when complementary base pairing forms a double helix

6. The secondary structure of DNA is called a double helix. Why?
   a. Two strands wind around one another in a helical, or spiral, arrangement.
   b. A single strand winds around itself in a helical, or spiral, arrangement.
   c. It is shaped like a ladder.
   d. It stabilizes the molecule.

Conceptual Review

1. Explain how complementary base pairing makes the copying of RNA and DNA molecules possible. Include diagrams that provide examples.

2. Growing strands of nucleic acids are always extended in the $5' \rightarrow 3'$ direction. What do the $5'$ and $3'$ refer to? Draw the bond that forms at these locations to link nucleotides. Why can't nucleic acids polymerize in the $3' \rightarrow 5'$ direction?

3. Make a generalized cartoon of monomers undergoing condensation reactions to form a polymer. Label the type of bond that forms when nucleotides polymerize into RNA or DNA. Does it take energy for polymerization reactions to proceed, or do they occur spontaneously? Why or why not?

4. Why is DNA such a stable molecule?

5. Summarize the types of secondary structure found in RNA. Include a labeled diagram.

6. A major theme in this chapter is that the structure of molecules correlates with their function. Explain why DNA's secondary structure limits its catalytic abilities compared with that of RNA. Why is it logical that RNA molecules can catalyze a modest but significant array of reactions? Why are proteins the most effective catalysts?

Group Discussion Problems

1. Do you agree with the "definition" of life given at the start of this chapter? Why or why not? If you were looking for life on Jupiter's moon Europa, how would you know when you found it?

2. Suppose that experiments like those reviewed in Section 4.4 succeeded in producing a molecule that could make a copy of itself. Outline a one-page opinion piece for your local newspaper that explains the nature of the research and discusses the ethical and philosophical implications of the discovery.

3. Before Watson and Crick published their model of the DNA double helix, Linus Pauling offered a model based on a triple helix. Draw your conception of what a DNA triple helix might look like. What interactions would keep such a secondary structure together? How could such a molecule be copied?

4. Origin-of-life researcher Robert Crabtree maintains that experiments simulating early Earth conditions are a valid way to test the theory of chemical evolution. Crabtree claims that if scientists working in the field agree that an experiment is a plausible reproduction of early Earth conditions, it is valid to infer that its results are probably correct—that the simulation effectively represents events that occurred some 4 billion years ago. Do you agree? Do you find the models and experiments presented in this chapter and previous chapters to be convincing tests of the theory? Explain your answers.

Answers to Multiple-Choice Questions   **1.** c; **2.** d; **3.** a; **4.** b; **5.** c; **6.** a

5

An Introduction to Carbohydrates

KEY CONCEPTS

- Sugars and other carbohydrates are highly variable in structure. They perform a wide variety of functions in cells, ranging from energy storage to formation of tough structural fibers.

- Monosaccharides are monomers that polymerize to form polysaccharides, via different types of glycosidic linkages.

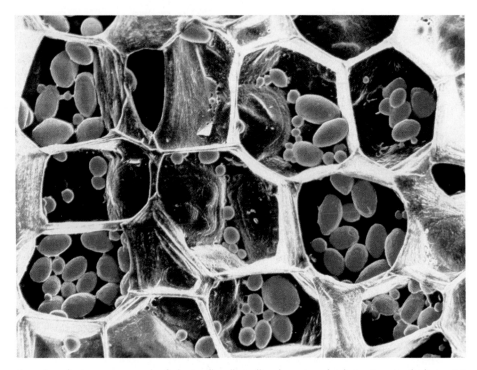

Scanning electron micrograph of plant cell walls (yellow honeycombed structures), which consist primarily of cellulose. The orange granules are stored starch. Both cellulose and starch are carbohydrates. Cellulose provides structural support for these cells, and starch supplies chemical energy.

This unit highlights the four types of macromolecules that are prominent in today's cells: proteins, nucleic acids, carbohydrates, and lipids. Understanding the structure and function of each of these macromolecules is a basic requirement for exploring how life began and how organisms work. In Chapters 3 and 4 we analyzed the way proteins and nucleic acids are put together and what they do. This chapter focuses on carbohydrates; Chapter 6 will introduce lipids.

The term **carbohydrate** encompasses both the monomers called monosaccharides ("one-sugar") and the polymers called polysaccharides ("many-sugars"). The name is logical because the chemical formula of many carbohydrates is $(CH_2O)_n$, where the n refers to the number of "carbon-hydrate" groups. The name is also misleading, though, because carbohydrates do not consist of carbon atoms bonded to water molecules. Instead, they are molecules with a carbonyl ($>C=O$) and several hydroxyl (–OH) functional groups, along with several to many carbon-hydrogen (C–H) bonds.

Let's begin by considering the structure of monosaccharides and continue by analyzing how these building blocks polymerize to form polysaccharides. The chapter closes with a look at what carbohydrates do in cells today and an analysis of their role in the origin of life.

5.1   Sugars as Monomers

Sugars are fundamental to life. They provide chemical energy in cells and furnish some of the molecular building blocks required for the synthesis of larger, more complex compounds. Monosaccharides were also important during chemical evolution, early in Earth's history. For example, the sugar called ribose would have been required for the formation of nucleotides. Laboratory simulations have shown that ribose and many other monosaccharides could have been produced in the prebiotic soup. What are these compounds, and how do they differ from one another?

Figure 5.1 illustrates the structure of the monomer called a **monosaccharide**, or simple sugar. Note that the carbonyl groups that serve as one of monosaccharides' distinguishing features can be found either at the end of the molecule, forming an aldehyde sugar (an *aldose*), or within the carbon chain, forming a ketone sugar (a *ketose*). The presence of a carbonyl group along with multiple hydroxyl groups gives sugars an array of functional groups. Based on this observation, it's not surprising that sugars are able to participate in a large number of chemical reactions.

The number of carbon atoms present also varies in monosaccharides. By convention, the carbons in a monosaccharide are numbered consecutively, starting with the end nearest the carbonyl group. Figure 5.1 features three-carbon sugars, or **trioses**. Ribose, which acts as a building block for nucleotides, has five carbons and is called **pentose**; the glucose that is cours-

ing through your bloodstream and being used by your cells right now is a six-carbon sugar, or **hexose**.

In addition to varying in the location of the carbonyl group and the total number of carbon atoms present, monosaccharides can vary in the spatial arrangement of their atoms. There is, for example, a wide array of pentoses and hexoses. Each is distinguished by the configuration of its hydroxyl functional groups. **Figure 5.2** illustrates glucose and galactose, which are six-carbon sugars that are optical isomers—they have the same chemical formula ($C_6H_{12}O_6$) but not the same structure. Although both are aldose sugars with six carbons, they differ in the spatial arrangement of the hydroxyl group at the carbon highlighted in Figure 5.2. Because their structures differ, their functions differ. In cells, glucose is used as a source of chemical energy in the production of ATP. But for galactose to be used as a source of energy, it first has to be converted to glucose via an enzyme-catalyzed reaction. A total of eight different hexoses exist, due to the different ways of configuring the hydroxyl groups in space. In addition, each hexose comes in two forms, each of which is a mirror image of the other. Thus sixteen distinct structures with the molecular formula $C_6H_{12}O_6$ are possible.

Sugars do not usually exist in the form of linear chains as illustrated in Figure 5.1 and Figure 5.2, however. Instead, in aqueous solution they tend to form ring structures. The chain and ring forms exist in equilibrium when simple sugars are in solution, with the ring forms predominating. Glucose serves

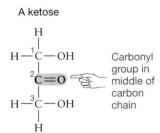

FIGURE 5.1 The Carbonyl Group in a Sugar Occurs in One of Two Configurations
The carbonyl group in a sugar can occur at the end of the molecule, forming an aldehyde sugar (an aldose), or within the carbon chain, forming a ketone sugar (a ketose).

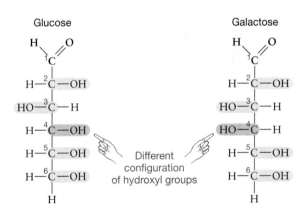

FIGURE 5.2 Sugars May Vary in the Configuration of Their Hydroxyl Groups
The two six-carbon sugars shown here vary only in the spatial orientation of their hydroxyl groups. These molecules are optical isomers, but they are not mirror images of each other. **EXERCISE** Next to these structures, draw the structural formula of mannose. Mannose is a six-carbon sugar that is identical to glucose, except that the hydroxyl (OH) group on carbon number 2 is switched in orientation.

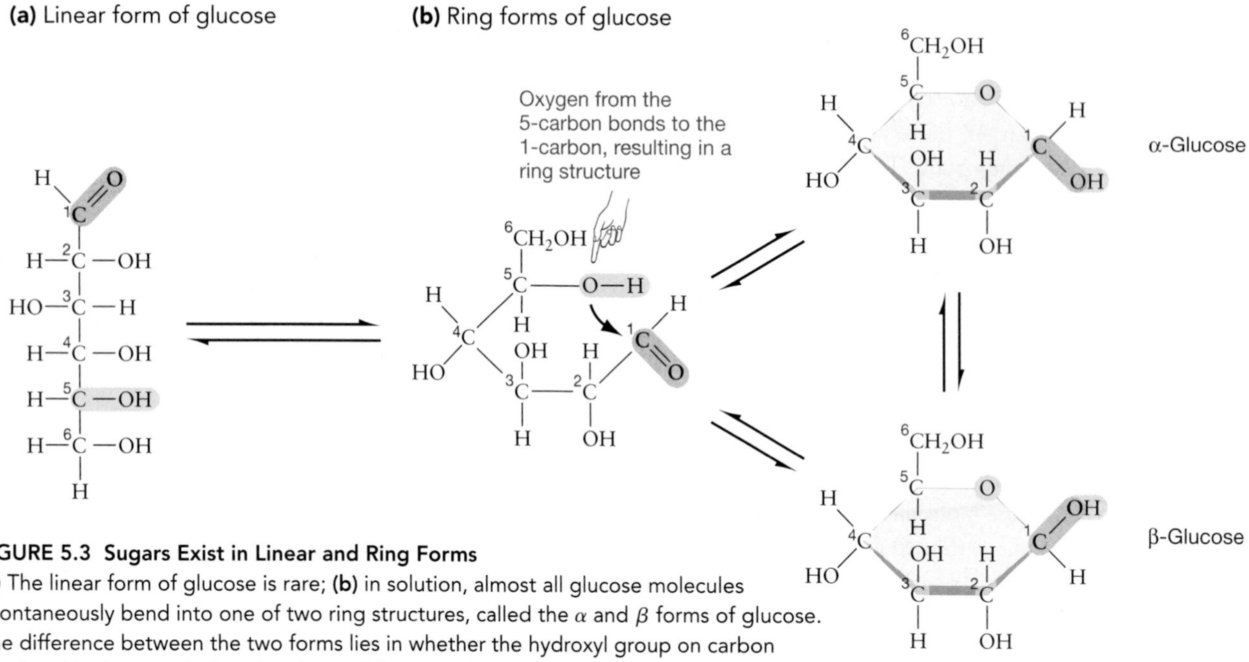

(a) Linear form of glucose   **(b)** Ring forms of glucose

FIGURE 5.3 Sugars Exist in Linear and Ring Forms
(a) The linear form of glucose is rare; **(b)** in solution, almost all glucose molecules spontaneously bend into one of two ring structures, called the α and β forms of glucose. The difference between the two forms lies in whether the hydroxyl group on carbon number 1 is above or below the plane of the ring. The two forms exist in equilibrium, but the β form is more common because it is slightly more stable than the α form.

as the example in **Figure 5.3**. When the cyclic structure forms in glucose, the carbon that is numbered 1 in the linear chain forms a bond with an oxygen atom and with a hydroxyl group that can be oriented in two distinct ways. As the right-hand side of Figure 5.3 shows, the different configurations result in the molecules α-glucose and β-glucose.

To summarize, the existence of alternate ring forms, optical isomers with different arrangements of hydroxyl groups in space and different mirror-image forms, variation in carbon number, and aldose or ketose placement of the carbonyl group makes a very large number of monosaccharides possible. Each has a unique structure and function.

Laboratory simulations have shown that most monosaccharides are readily synthesized under conditions that mimic the prebiotic soup. For example, when formaldehyde (H_2CO) molecules are heated in solution, they react with one another to form almost all of the pentoses and hexoses, as well as some seven-carbon sugars. In addition, researchers recently announced the discovery of the three-carbon ketose illustrated in Figure 5.1, along with a wide array of compounds closely related to sugars, on the Murchison meteorite (introduced in Chapter 3). Based on these observations, investigators suspect that sugars are synthesized on dust particles and other debris in interstellar space, and could have rained down onto Earth as the planet was forming. Most researchers interested in chemical evolution have become increasingly confident that a wide diversity of monosaccharides existed in the prebiotic soup. But it remains a mystery why ribose might have predominated and made the synthesis of nucleotides possible. It also appears highly unlikely that monosaccharides were able to polymerize to form the polysaccharides found in today's cells. Let's explore why.

✓ CHECK YOUR UNDERSTANDING

Simple sugars differ in three respects: (1) the location of their carbonyl group, (2) the number of carbon atoms present, and (3) the spatial arrangement of their atoms—particularly the relative positions of hydroxyl (OH) groups. You should be able to draw the structural formula of a monosaccharide in linear form and then draw other sugars that differ from this one in each of the three aspects listed.

5.2 The Structure of Polysaccharides

Polysaccharides are polymers that form when monosaccharides are linked together. They are also known as complex carbohydrates. The simplest polysaccharides consist of two sugars and are known as **disaccharides**. The two monomers involved may be identical, as in the two α-glucose molecules that link to form maltose. Or they may be different, as in the combination of a glucose molecule and a galactose molecule that forms lactose—the most important sugar in milk. (**Box 5.1** explains how problems with metabolizing lactose or galactose lead to disease in humans.)

Simple sugars polymerize when a condensation reaction occurs between two hydroxyl groups, resulting in a covalent bond called a **glycosidic linkage** (**Figure 5.4**). Glycosidic linkages are analogous to the peptide bonds that hold proteins together and to the phosphodiester bonds that connect the nucleotides in nucleic acids. There is an important difference, however. Peptide and phosphodiester bonds always form at the same location in their monomers. But because glycosidic linkages form between hydroxyl groups, and because every monosaccharide contains

BOX 5.1   Lactose Intolerance and Galactosemia

Most organisms have a large array of enzymes that catalyze reactions involving sugars. A different enzyme processes each type of sugar. This should not be surprising, considering that sugars differ in shape and that the ability of an enzyme's active site to catalyze a reaction depends on the shape of the substrate. As explained in Chapter 3, the specificity of enzyme action is a prime example of how a molecule's structure correlates with its function.

The human enzyme lactase, for example, catalyzes only one reaction, in which lactose is split into glucose and galactose. Another enzyme, called 1-phosphate uridyl transferase, is involved in converting galactose to glucose. When these enzymes do not function normally, lactose or galactose concentrations increase in the body and cause disease.

The condition known as lactose intolerance occurs in most adult humans to some extent. In lactose-intolerant individuals, the enzyme lactase is produced only during infancy—at a time when the primary source of nutrition is mother's milk. Lactase production ends as these individuals mature. This pattern has been observed in most human cultures, where traditional diets did not include cow's milk. Dairy cattle were unknown in Thai culture, for example, and 97 percent of Thai people are lactose intolerant as adults. This pattern is also logical, because it would be a waste of time and energy to produce an enzyme that has no substrate to work on. If these individuals drink milk once lactase production ends, though, the unprocessed lactose causes bloating, abdominal pain, and diarrhea. But in cultures where milk has been an important source of nutrition in adults for many centuries, lactase production continues throughout life. Among Danish people, for example, 97 percent are lactose tolerant throughout life.

To make sense of lactose tolerance and intolerance, recall the concept of adaptation introduced in Chapter 1. An adaptation is a trait that increases the fitness of individuals in a particular environment. Lactose intolerance is an adaptation to environments where milk is not available to adults. Lactose tolerance, in contrast, is an adaptation to environments where milk is readily available to adults.

Unfortunately, problems with galactose metabolism are completely nonadaptive. If a child lacks the enzyme that converts galactose to glucose, galactose concentrations build to high levels in the blood and cause the array of symptoms known as galactosemia ("galactose-sign"). Individuals who suffer from galactosemia are at risk for mental retardation or even death. The only cure is to exclude galactose from the diet. In large enough doses, milk, yogurt, cheese, and other dairy products are life-threatening substances for these individuals.

(a) Polymerization reaction

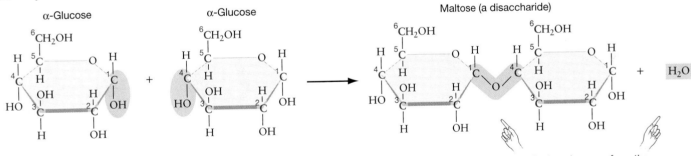

The hydroxyl groups from the 1-carbon and 4-carbon react to produce an α-1,4-glycosidic linkage and water

(b) Sugars are often drawn in simplified form for legibility.

FIGURE 5.4 Monosaccharides Polymerize through Formation of Glycosidic Linkages
(a) A glycosidic linkage occurs when hydroxyl groups on two monosaccharides undergo a condensation reaction to form the bond shown on the right. In this case, the hydroxyl groups involved were on carbons 1 and 4 of the two monomers, which were in the α-ring configuration. The resulting bond is called an α-1,4-glycosidic linkage. **(b)** Polysaccharides consist of a series of monosaccharides linked together by glycosidic bonds. **EXERCISE** Label the glycosidic linkages in part (b).

at least two hydroxyl groups, the location and geometry of glycosidic linkages can vary widely among polysaccharides. To drive this point home, consider the structures of the most common polysaccharides found in organisms today: starch, glycogen, cellulose, and chitin, along with a modified polysaccharide called peptidoglycan.

Each of these macromolecules can consist of a few hundred to many thousands of monomers, joined by glycosidic linkages at different locations.

Starch: A Storage Polysaccharide in Plants

In plant cells, monosaccharides are stored for later use in the form of starch. **Starch** consists entirely of α-glucose monomers that are joined by glycosidic linkages. As the top panel in **Table 5.1** shows, the angle of the linkages between carbons 1 and 4 causes the chain of glucose subunits to coil into a helix. Starch is actually a mixture of two such polysaccharides, however. One is an unbranched molecule called amylose. The other is a branched molecule called amylopectin. Branching occurs

TABLE 5.1 Polysaccharides Differ in Structure

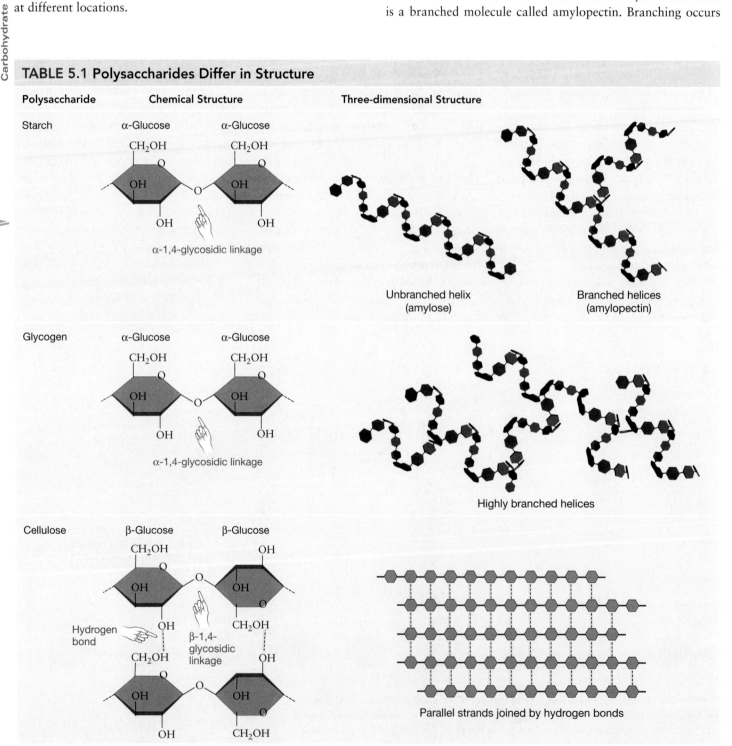

Polysaccharide	Chemical Structure	Three-dimensional Structure

Starch — α-Glucose, α-Glucose — α-1,4-glycosidic linkage

Unbranched helix (amylose) — Branched helices (amylopectin)

Glycogen — α-Glucose, α-Glucose — α-1,4-glycosidic linkage

Highly branched helices

Cellulose — β-Glucose, β-Glucose — Hydrogen bond — β-1,4-glycosidic linkage

Parallel strands joined by hydrogen bonds

TABLE 5.1 Polysaccharides Differ in Structure (continued)

Polysaccharide	Chemical Structure	Three-dimensional Structure

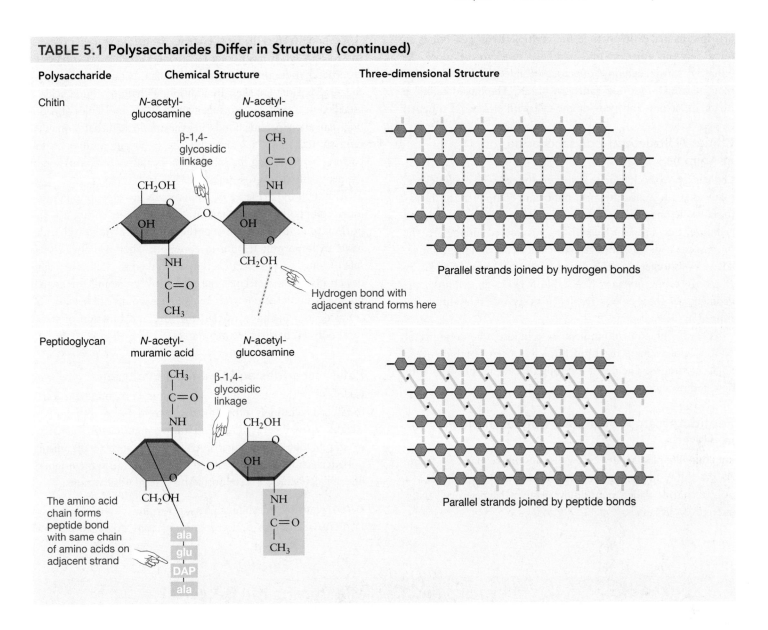

Chitin

Parallel strands joined by hydrogen bonds

Hydrogen bond with adjacent strand forms here

Peptidoglycan

The amino acid chain forms peptide bond with same chain of amino acids on adjacent strand

Parallel strands joined by peptide bonds

---

when glycosidic linkages form between carbon 1 of a glucose monomer on one strand and carbon 6 of a glucose monomer on another strand. In amylopectin, branches occur in about one out of every 30 monomers.

Glycogen: A Highly Branched Storage Polysaccharide in Animals

Glycogen performs the same storage role in animals that starch performs in plants. Glycogen is nearly identical to the branched form of starch. But instead of an α-1,6-glycosidic linkage occurring in about one out of every 30 monomers, a branch occurs in about one out of every 10 glucose subunits (see Table 5.1).

Cellulose: A Structural Polysaccharide in Plants

As we saw in Chapter 1, all cells are enclosed by a membrane. In most unicellular and multicellular organisms living today, the cell is also surrounded by a layer of material called a wall. A **cell wall** is a protective sheet that occurs outside the membrane. In algae, plants, bacteria, fungi, and many other groups, the cell wall is composed primarily of one or more polysaccharides.

In plants, cellulose is the major component of the cell wall. **Cellulose** is a polymer of β-glucose monomers, joined by β-1,4-glycosidic linkages. As Table 5.1 shows, the geometry of the bond is such that each glucose monomer in the chain is flipped in relation to the adjacent monomer. This arrangement increases the stability of cellulose strands, because the flipped orientation makes it possible for multiple hydrogen bonds to form between adjacent and parallel strands of cellulose. As Table 5.1 shows, cellulose often occurs in long, parallel strands that are joined by these hydrogen bonds. The linked cellulose fibers are strong and provide the cell with structural support.

Chitin: A Structural Polysaccharide in Animals

Chitin is a polysaccharide that stiffens the cell walls of fungi and many algae. It is also the most important component of the external skeletons of insects and crustaceans. Chitin is similar to cellulose, but instead of consisting of glucose monomers, the monosaccharide involved is one called N-acetylglucosamine. These monomers, abbreviated "NAc," are joined by β-1,4-glycosidic linkages (see Table 5.1). As in cellulose, the geometry of these bonds results in every other residue being flipped in orientation.

Like the glucose monomers in cellulose, the subunits of N-acetylglucosamine in chitin form hydrogen bonds between adjacent strands. The result is a tough sheet that provides stiffness and protection.

Peptidoglycan: A Structural Polysaccharide in Bacteria

Bacteria, like plants, have cell walls. But unlike plants, in bacteria the ability to produce cellulose is extremely rare. Instead, a polysaccharide called peptidoglycan gives bacterial cell walls strength and firmness.

Peptidoglycan is the most complex of the polysaccharides discussed thus far. It has a long backbone formed by two types of monosaccharides that alternate with each other and are linked by β-1,4-glycosidic linkages. In addition, a chain of amino acids is attached to one of the two sugar types. When molecules of peptidoglycan align, peptide bonds link the amino acid chains on adjacent strands. Box 5.2 describes how certain antibiotics kill bacteria by disrupting the enzymes that catalyze the formation of the peptide-bonded cross-links.

Although the presence of amino acids gives peptidoglycan a more complex structure than either cellulose or chitin, it is important to note that all three of these structural polysaccharides have an important feature in common: They usually exist as sets of long, parallel strands that are linked to each other. This design gives the molecules the ability to withstand forces that pull or push on them—what an engineer would call tension and compression. In this way, the structure and function of structural polysaccharides are correlated.

Polysaccharides and Chemical Evolution

Cellulose is the most abundant organic compound on Earth today, and chitin is probably the second most abundant by weight. Virtually every organism known manufactures glycogen or starch. But despite their current importance to organisms, polysaccharides probably played little to no role in the origin of life. This conclusion is supported by several observations:

- No plausible mechanism exists for the polymerization of monosaccharides under conditions that prevailed early in

---

BOX 5.2   How Do the Penicillins and Cephalosporins Kill Bacteria?

Antibiotics are molecules that kill bacteria. The first antibiotic to have been discovered and used to cure bacterial infections in humans is produced naturally, by a soil-dwelling fungus called *Penicillium chrysogenum*. Soil is packed with bacteria and fungi that compete for space, water, and nutrients. Some fungi reduce competition by producing and secreting antibiotics, which kill the bacteria surrounding them. The drug we call penicillin is such a molecule, and it was named in honor of the species that produces it.

Penicillin and the drug called cephalosporin, which is produced naturally by species of soil-dwelling fungi in the genus *Cephalosporium*, are closely related in structure and function. Both molecules are effective because they bind very tightly to the enzymes that catalyze the formation of cross-links between individual strands within peptidoglycan. In the absence of these cross-links, the bacterial cell wall begins to weaken. Eventually it fails and tears open, and the cell is destroyed. Penicillin and cephalosporin cause very few side effects in humans, because the binding event between these drugs and the cell-wall enzyme to which they bind is extremely specific. Human cells are not affected by the presence of penicillin or cephalosporin, because they lack the enzymes involved in cross-linking peptidoglycan.

In response to the widespread use of penicillin and cephalosporin, however, many bacterial populations have evolved an enzyme that breaks these drugs apart and renders them ineffective. (Chapter 23 will provide more detail on how drug resistance evolves.) In response, researchers have synthesized molecules that are very closely related to penicillin and cephalosporin structurally and have the same function, but that are not as severely affected by the newly evolved bacterial enzyme. You may, in fact, be familiar with some of these synthetic penicillins and cephalosporins from personal experience. They include widely prescribed molecules such as methicillin, oxacillin, ampicillin, and ceftriaxone.

Earth's history. In cells and in vitro, the glycosidic linkages illustrated in Figure 5.4 and Table 5.1 form only with the aid of specialized enzymes.

- It is highly likely that life began in the form of an RNA molecule. However, researchers have yet to discover a ribozyme that can join simple sugars by catalyzing the formation of glycosidic linkages. Thus, it appears extremely unlikely that polysaccharides were present in significant quantities during the RNA world.

- Even though monosaccharides contain large numbers of hydroxyl and carbonyl groups, they lack the diversity of functional groups found in amino acids. Polysaccharides also have very simple secondary structures, consisting of linkages between adjacent strands. Thus, they lack the structural and chemical complexity that makes proteins, and to a lesser extent RNA, effective catalysts. To date, no reactions have been discovered that are catalyzed by polysaccharides.

- The monomers in polysaccharides are not capable of complementary base pairing. Like proteins but unlike DNA and RNA, polysaccharides are not capable of providing the information required for themselves to be copied. As far as is known, no polysaccharides store information in cells. Thus, no one has proposed that the first living entity might have been a polysaccharide.

Even though polysaccharides probably did not play a significant role in the earliest forms of life, they became enormously important once cellular life evolved. Let's take a detailed look at how they function in today's cells.

✓ CHECK YOUR UNDERSTANDING

Polysaccharides form when enzymes catalyze the formation of glycosidic linkages between monosaccharides that are in the α- or β-ring form. Most polysaccharides are long, linear molecules, but some branch extensively. Among linear forms, it is common for adjacent strands to be linked by hydrogen bonding or other types of linkages. You should be able to make a rough sketch that shows the structures of glycogen and cellulose, and label at least two glycosidic linkages in each structure.

5.3 What Do Carbohydrates Do?

Chapter 4 introduced one of the three basic functions that carbohydrates perform in organisms: furnishing the building blocks of molecules needed by cells. Recall that both RNA and DNA contain sugars—the five-carbon sugars ribose and deoxyribose, respectively. In nucleotides, which consist of a sugar, a phosphate group, and a nitrogenous base, the sugar itself acts as a subunit of the larger molecule. But sugars frequently furnish only the raw "carbon skeletons" that are used in the synthesis of important molecules. Amino acids are being synthesized by your cells right now, for example, using sugars as a starting point.

Although the details of how sugars are used in synthesizing amino acids and other complex molecules are beyond the scope of this book, it will be productive to delve into the other three major roles of carbohydrates: indicating cell identity, storing chemical energy, and providing cells with fibrous structural materials.

The Role of Carbohydrates in Cell Identity

As we saw in Section 5.2, polysaccharides do not store information in cells. It's important to recognize, however, that polysaccharides can display important information. **Figure 5.5** shows how this information display happens, on the outer surface of a cell. Certain molecules project outward from this surface, into the environment surrounding the cell. These molecules are called glycoproteins. A **glycoprotein** is a protein that is covalently bonded to a carbohydrate—usually a relatively short chain of sugars.

Glycoproteins are key molecules in what biologists call cell-cell recognition and cell-cell signaling. Each distinct type of cell in a multicellular organism—for example, the nerve cells and immune system cells in your body—displays a different set of glycoproteins on its surface. The same is true for different species of unicellular organisms, such as the approximately 500 species of bacteria that live in your mouth as well as the cells that exist by the millions in a teaspoonful of soil or a drop of pond water. The role of the carbohydrates on the cell surface is to identify the type of cell that displays them. A glycoprotein is a "sugar coating" that acts like the magnetic stripe on the back of a credit card—it immediately identifies the individual that bears it.

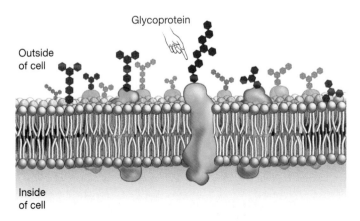

FIGURE 5.5 Carbohydrates Are an Identification Badge for Cells
Glycoproteins contain sugar groups that project from the surface of the plasma membrane. These sugar groups have distinctive structures that identify the type or species of the cell.

In later chapters we'll explore how the identification information displayed by glycoproteins is used by cells and cellular signals. The key point here is to recognize that the enormous number of structurally distinct monosaccharides makes it possible for an enormous number of unique, short polysaccharides to exist. As a result, each cell type and each species can display a unique identity.

The Role of Carbohydrates in Energy Production and Storage

Candy bar wrappers promise a quick energy boost, and ads for sports drinks claim to provide the "carbs" needed for peak activity. If you were to ask friends or family members what carbohydrates do in your body, they would probably say something like "They give you energy." And after pointing out that carbohydrates are also used to establish cell identity, as a structural material, and as a source of carbon skeletons for the synthesis of other complex molecules, you'd have to agree. Carbohydrates store and provide chemical energy in cells. What is it about carbohydrates that makes this possible?

The answer to this query lies in the discussion of redox reactions that began in Chapter 2. Recall that reduction-oxidation reactions involve the transfer of an electron. The atom that receives an electron is said to be reduced, while the atom that loses an electron is said to be oxidized. These points were introduced in Chapter 2 because they helped explain what was happening during the early stages of chemical evolution. You may remember that experimental and computer simulation studies have shown that if molecules with high free energy—such as molecular hydrogen (H_2), methane (CH_4), and ammonia (NH_3)—are available to act as electron donors, and if a source of intense kinetic energy—such as an electrical discharge or high-intensity solar radiation—is present, redox reactions occur. Those re-

actions result in the production of molecules with reduced carbon atoms, such as formaldehyde (H_2CO) and hydrogen cyanide (HCN). With continued inputs of kinetic energy, these molecules react to form sugars, amino acids, and other complex organic compounds.

Carbohydrates as Electron Donors Today, the key reaction that results in the production of sugar occurs in plants and other photosynthetic organisms. Photosynthesis is a complex process that can be summarized most simply as follows:

$$CO_2 + H_2O + sunlight \rightarrow (CH_2O)_n + O_2$$

where $(CH_2O)_n$ represents a carbohydrate. **Figure 5.6** shows the structural formulas of the molecules involved and represents the relative positions of their covalently bonded electrons. The figure has two key features:

1. Electrons are closer to the carbon atom in the product molecule $(CH_2O)_n$ than they are to the carbon atom in carbon dioxide (CO_2). Thus, photosynthesis results in the reduction of carbon.

2. The covalently bonded electrons in the C–H bonds of carbohydrates are shared more equally, and thus held less tightly, than they are in the C–O bonds of carbon dioxide. As a result, carbohydrates have much more free energy than carbon dioxide has.

The essence of photosynthesis, then, is that energy in sunlight is transformed into chemical energy that is stored in the C–H bonds of carbohydrates. Because the C–H bonds are unstable and have high free energy, carbohydrates can act as electron donors in redox reactions that lead to the production of chemical energy in the form of ATP—the phosphorylated molecule introduced in Chapter 4. The

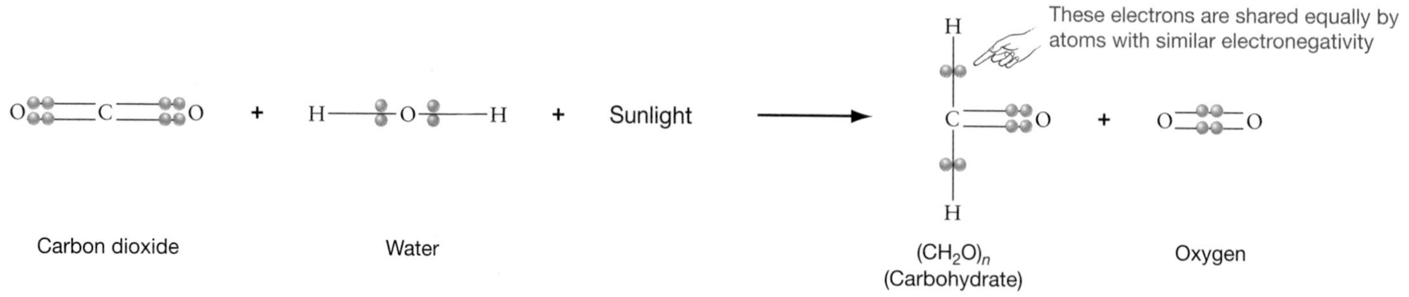

FIGURE 5.6 Carbohydrates Have High Free Energy
In these diagrams, the horizontal lines. indicate covalent bonds. The dots represent the relative positions of electrons in those bonds. QUESTION If carbon becomes reduced in this reaction, which atoms become oxidized? EXERCISE Circle the bonds in this diagram that have the highest free energy.

overall reaction for the oxidation of sugars in cells can be written as follows:

$$CH_2O + O_2 + ADP \rightarrow CO_2 + H_2O + ATP$$

Chemical energy stored in the C–H bonds of carbohydrate is transferred to chemical energy in the form of the third phosphate group in ATP. As indicated in Chapter 3, the free energy in ATP makes it possible for phosphorylation events to drive endergonic reactions, move your muscles, and perform other types of work in cells. Carbohydrates are like the water that piles up behind a dam; ATP is like the electricity, generated at a dam, that lights up your home.

Later chapters will analyze in detail how sugars and other carbohydrates are made in organisms, and how these carbohydrates are then oxidized to provide cells with usable chemical energy in the form of ATP. For now, the important thing is to recognize that carbohydrates store and provide chemical energy to cells because they contain a large number of C–H bonds and because C–H bonds have high free energy.

Figure 5.7 offers another way to think about carbohydrates as molecules that store energy. During chemical evolution and in today's plants, algae, and photosynthetic bacteria, the carbon atom in carbon dioxide is reduced. Stated another way, the carbon atoms in CO_2 molecules gain electrons. This occurs when a proton (H^+) replaces an oxygen in the CO_2 molecule. Compare the structures of carbon dioxide and the carbohydrate shown in Figures 5.7a and 5.7b, and you'll see that hydrogen atoms—an electron plus a proton—have been added to the carbohydrate. As Figure 5.7c shows, the fatty-acid subunits found in fats have even more free energy than carbohydrates have. Fatty acids are made up largely of the long hydrocarbon (C–H) chains introduced in Chapter 2. Compared with carbohydrates, fats contain many more C–H bonds and many fewer C–O bonds. C–H bonds have high free energy because the electrons are shared equally by atoms with low electronegativities. C–O bonds have low free energy because the highly electronegative oxygen atom holds the electrons so tightly. Both carbohydrates and fats are used as fuel in cells. Fats will be discussed in more detail in Chapter 6.

How Do Carbohydrates Store Energy?   Section 5.2 introduced the structure of the five most important polysaccharides found in organisms: starch, glycogen, cellulose, chitin, and peptidoglycan. Starch and glycogen serve as energy-storage molecules. They can do so because enzymes readily catalyze hydrolysis reactions that break off individual glucose subunits from these polysaccharides. The glucose subunits are then used as electron donors in redox reactions that result in the production of ATP.

The most important enzyme involved in catalyzing the hydrolysis of α-glycosidic linkages in glycogen is called **phosphorylase**. The enzymes involved in breaking these linkages in starch are called **amylases**. Most of your cells contain phosphorylase, so they can break down glycogen to provide glucose on demand. Your salivary glands and pancreas also produce amylases that are secreted into your mouth and small intestine, respectively. These amylases are responsible for digesting the starch that you eat.

Carbohydrates as Structural Molecules

Cellulose and chitin, along with the modified polysaccharide peptidoglycan, are not normally used as a source of chemical energy in cells. Instead, these molecules are structural. They form fibers that give cells and organisms strength and elasticity.

To appreciate why cellulose, chitin, and peptidoglycan are effective as structural molecules, recall that they form long strands and that bonds can form between adjacent strands. Table 5.1 detailed how these bonds form. In the cell walls of plants, a collection of about 80 cellulose molecules, cross-linked by hydrogen bonding, creates a tough fiber. These cellulose fibers, in

(a) Carbon dioxide

(b) A carbohydrate

(c) A fatty acid (a component of fat molecules)

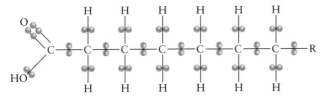

FIGURE 5.7 Carbon Atoms in Carbohydrates Are Reduced **(a)** The carbon atom in carbon dioxide is highly oxidized. **(b)** In carbohydrates such as the sugar shown here, carbon atoms are highly reduced compared with carbon dioxide. **(c)** The fatty acids found in fat molecules store even more free energy in the form of C–H bonds than carbohydrates do. **EXERCISE** Label the molecule with the highest amount of chemical energy, and the molecule with the lowest amount. Label which molecules are likely to act as electron donors and which are likely to act as electron acceptors in redox reactions.

turn, criss-cross to form a tough sheet (**Figure 5.8a**). Groups of chitin molecules are also organized into fibers held together by hydrogen bonding. These fibers stiffen the cell walls of fungi and algae, and they overlap in the external skeletons of insects to form a dense, waterproof sheet (**Figure 5.8b**). In the cell walls of bacteria, peptidoglycan molecules are linked by peptide bonds and layered in sheets (**Figure 5.8c**).

Almost all organisms have the enzymes required to break the α-1,4- and α-1,6-glycosidic linkages that hold starch and glycogen molecules together. But only a few have enzymes capable of hydrolyzing the β-1,4-glycosidic linkages in cellulose, chitin, and peptidoglycan. The shape and orientation of β-1,4-glycosidic linkages make them difficult to break. Few enzymes have active sites with the correct geometry and reactive groups to do so. As a result, the structural polysaccharides are resistant to degradation and decay. In fact, the cellulose that you ingest when you eat plant cells passes through your gut undigested. It is referred to as dietary fiber, and it helps digestion by adding moisture and bulk to the feces. Cellulose absorbs water, and the addition of mass to fecal material helps it move through the intestinal tract more quickly, avoiding constipation.

In short, then, the remarkable functional difference between the energy-storing polysaccharides and the structural polysaccharides is due to differences in their structures. One key distinction between storage polysaccharides and structural polysaccharides is the presence of α-glycosidic linkages versus β-glycosidic linkages. The α-linkages in storage polysaccharides are readily hydrolyzed to release sugars, while the β-linkages in structural polysaccharides resist enzymatic degradation. Another key distinction between the two types of polysaccharide is that the geometry of the glycosidic linkages in

cellulose and chitin make it possible for these macromolecules to form extensive hydrogen bonds. Although each individual hydrogen bond is relatively weak, the combination of many weak bonds allows the structural polysaccharides to form large fibers that are strong yet flexible. Similarly, the presence of amino acid residues in peptidoglycan allows adjacent fibers to cross-link into durable sheets.

Polysaccharides are essential to the formation of all the different types of cell walls observed in organisms today. But if polysaccharides were not produced by chemical evolution, and if all cell walls contain polysaccharides, then it is a virtual certainty that the first form of cellular life lacked a cell wall. It is equally certain, however, that the first cell was surrounded by a membrane. What molecules made up this structure, and how did it function? These are the questions taken up in Chapter 6, "Lipids, Membranes, and the First Cells."

✓CHECK YOUR UNDERSTANDING

Carbohydrates have several important functions in cells in addition to providing building blocks for the synthesis of more complex compounds. Glycoproteins project from the surface of cells and provide a molecular PIN that identifies the cell's type or species. Starch and glycogen store sugars for later use in reactions that produce ATP. Sugars contain large amounts of chemical energy because they contain reduced carbon atoms, meaning carbon atoms that are bonded to hydrogen. These bonds have high free energy because the electrons are shared equally by atoms with low electronegativity. Polysaccharides such as cellulose, chitin, and peptidoglycan form cell walls, which give cells structural strength. **You should be able to describe how the sugars you ate during breakfast today are functioning in your body right now.**

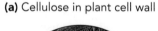 **(a)** Cellulose in plant cell wall

(b) Chitin in insect exoskeleton

(c) Peptidoglycan in bacterial cell wall

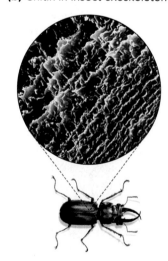

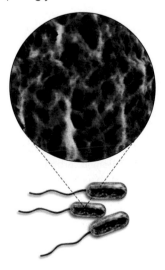

FIGURE 5.8 Cellulose, Chitin, and Peptidoglycan Form Tough Fibers or Sheets
High-magnification photographs of **(a)** cellulose fibers in the cell wall of a plant, **(b)** chitin from an insect's external skeleton, and **(c)** peptidoglycan from the cell wall of a bacterium.

ESSAY   Why Do We Have a Sweet Tooth?

If you look at your tongue closely in the mirror, you'll see tiny bumps scattered on the surface. Inside these bumps are small structures called taste buds; inside taste buds are cells called taste receptors. Taste receptors detect the presence of certain molecules and send messages to your brain about them. The sensed food molecules contribute to the tastes we call sweet, sour (acidic), salty, and bitter. Some taste buds respond specifically to amino acids, triggering the taste called umami, or "meaty."

If food in your mouth stimulates your bitter receptors, your brain will probably respond by initiating movements that lead you to spit the food out. But if a mouthful of something stimulates sweet or umami receptors, your brain will probably respond by directing movements that result in your eating more.

Instead of a sweet tooth, then, we actually have sweet receptors. These molecules respond more strongly to some sugars than to others. The six-carbon sugar fructose, for example, is the sweetest of all sugars. If you've read the list of ingredients on food labels, you may have noticed that many foods are sweetened with "high-fructose corn syrup." By adding fructose to foods, manufacturers can make them taste sweeter while lowering the total sugar content. The synthetic sweetener NutraSweet® is even more potent in its ability to stimulate our sweet receptors. It is about 200 times sweeter than sucrose (table sugar).

Why do these receptor molecules and cells exist? Carbohydrates serve so many vital roles in our cells that the ability to sense them and the desire to take in more should benefit individuals. To use the terms introduced in Chapter 1, responding to sweets should be an adaptation that increases the fitness of individuals in most environments. People who sense and respond to sweetness should survive better and reproduce more than do individuals who do not. But having a sweet tooth may have been more adaptive in the past than it is today. Historically, simple sugars were extremely rare in our environment. In many cultures, honey was the only source. But with the recent invention of techniques to refine large amounts of the disaccharide sucrose cheaply, the developed nations are flooded with it. Our taste receptors keep responding strongly, just as they did when our ancestors occupied sugar-poor environments. Now, overdoses of sugar are contributing to obesity and problems with regulating blood sugar levels. Humans love sweets for a reason, but in our current environment we probably love them a little too much.

Historically, simple sugars were extremely rare in our environment.

CHAPTER REVIEW

Summary of Key Concepts

■ **Sugars and other carbohydrates are highly variable in structure. They perform a wide variety of functions in cells, ranging from energy storage to formation of tough structural fibers.**

Carbohydrates are organic compounds that have a carbonyl group and several to many hydroxyl groups. They can occur as the simple sugars called monosaccharides or as complex polysaccharides. Many types of monosaccharide exist, each distinguished by one or more of the following features: (1) the location of their carbonyl group—either at the end of the molecule or within it; (2) the number of carbon atoms they contain—from three to many; and (3) the orientation of their hydroxyl groups in the linear chain and/or the ring form. In addition, most monosaccharides have optical isomers.

■ **Monosaccharides are monomers that polymerize to form polysaccharides, via different types of glycosidic linkages.**

Monosaccharides can be linked together by covalent bonds, called glycosidic linkages, that join hydroxyl groups on adjacent molecules. Unlike the peptide bonds that form proteins and the phosphodiester bonds of nucleic acids, glycosidic linkages may form at several locations on a monosaccharide. The different disaccharides ("two-sugars") and polysaccharides are distinguished by the type of monomers involved and the location of the glycosidic linkages between them. The most common polysaccharides in organisms today are starch, glycogen, cellulose, and chitin; peptidoglycan is an abundant polysaccharide that has short chains of amino acids attached.

Starch and glycogen are made up of glucose molecules that are in the ring form called α, and that are joined by glycosidic linkages between their first and fourth carbons. In some forms of starch, individual chains are occasionally joined to other chains; in glycogen these links are abundant enough to form a highly branched molecule. Both starch and glycogen function as energy-storage molecules. Because sugars contain many C–H bonds, they contain a significant amount of chemical energy. When cells need energy, enzymes hydrolyze the α-1,4-glycosidic linkages in starch or glycogen. The reaction releases individual glucose molecules, which then act as electron donors in redox reactions. These reactions result in the production of usable chemical energy in the form of ATP.

Cellulose, chitin, and peptidoglycan are made up of monosaccharides joined by β-1,4-glycosidic linkages. When individual molecules of these carbohydrates align side by side, bonds form between them. In cellulose and chitin, the individual molecules are joined by hydrogen bonds; in peptidoglycan, the intermolecular links consist of peptide bonds between amino acid chains that extend from the polysaccharide. Hydrogen bonding produces bundles of cellulose and chitin molecules that form strong, elastic fibers; when peptide bonding occurs between many peptidoglycan strands, the result is a tough fibrous or sheet-like substance. It is no surprise that cellulose, chitin, and peptidoglycan provide cells and organisms with support. In addition, few organisms have enzymes that can degrade β-1,4-glycosidic linkages, which makes cell walls made of cellulose, chitin, and peptidoglycan resistant to attack. In carbohydrates, as in proteins and nucleic acids, structure correlates with function.

Web Tutorial 5.1   Carbohydrate Structure and Function

Questions

Content Review

1. What is the difference between a monosaccharide, a disaccharide, and a polysaccharide?
   a. the number of carbon atoms in the molecule
   b. the type of glycosidic linkage between monomers
   c. the spatial arrangement of the various hydroxyl residues in the molecule
   d. the number of monomers in the molecule

2. What type of bond allows sugars to polymerize?
   a. glycosidic linkage
   b. phosphodiester bond
   c. peptide bond
   d. hydrogen bonds

3. What holds cellulose molecules together in bundles large enough to form fibers?
   a. the cell wall
   b. peptide bonds
   c. hydrogen bonds
   d. hydrophobic interactions between different residues in the cellulose helix

4. What are the primary functions of carbohydrates in cells?
   a. energy storage, cell identity, structure, and building blocks for synthesis
   b. catalysis, structure, and transport
   c. information storage and catalysis
   d. signal reception, signal transport, and signal response

5. Why is it unlikely that carbohydrates played a large role in the origin of life?
   a. They cannot be produced by chemical evolution.
   b. They have optical isomers.
   c. More types of glycosidic linkages are possible than are actually observed in organisms.
   d. They do not polymerize without the aid of enzymes.

6. What is a "quick and dirty" way to assess how much free energy an organic molecule has?
   a. Count the number of oxygen atoms it contains.
   b. Count the number of hydrogen atoms it contains.
   c. Determine whether it contains an amino group.
   d. Determine whether it contains a carbonyl group.

Conceptual Review

1. Draw a six-carbon sugar in linear and ring forms. Draw another six-carbon sugar that differs from the first one. Identify the aspects of these two molecules that are distinct.

2. Draw the ring structure of glucose in the α form. Now add another glucose molecule with an α-1,4-glycosidic linkage, and then a third glucose molecule with an α-1,6-glycosidic linkage.

3. Compare and contrast the structures and functions of starch and glycogen. How are these molecules similar? How are they different?

4. How is it possible for starch and glycogen to function as energy-storage molecules? Explain.

5. How is it possible for cellulose and chitin to function as molecules that provide cells with structural support? Explain.

6. Both glycogen and cellulose consist of glucose monomers that are linked end to end. How do the structures of these polysaccharides differ? How do their functions differ?

Group Discussion Problems

1. A weight-loss program for humans that emphasizes minimal consumption of carbohydrates has recently become popular. What is the logic behind this diet? (Note: This diet plan has caused controversy and is not endorsed by some physicians and researchers.)

2. To treat galactosemia, physicians exclude the monosaccharide galactose from the diet. Why does the disaccharide lactose also have to be excluded from the diet?

3. Amylase, an enzmye found in human saliva, catalyzes the hydrolysis of the α-1,4-glycosidic linkages in starch. If you hold a salty cracker in your mouth long enough, it will begin to taste sweet. Why?

4. Lysozyme, an enzyme found in human saliva, tears, and other secretions, catalyzes the hydrolysis of the β-1,4-glycosidic linkages in peptidoglycan. What effect does contact with this enzyme have on bacteria?

Answers to Multiple-Choice Questions   1. d; 2. a; 3. c; 4. a; 5. d; 6. b

Lipids, Membranes, and the First Cells

6

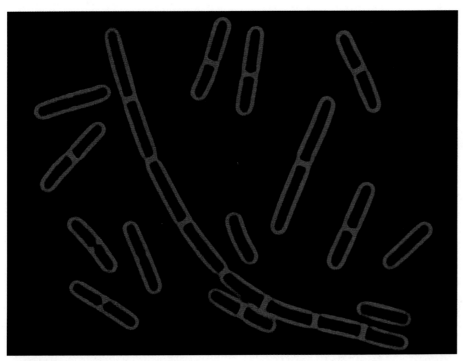

These bacterial cells have been stained with a red compound that inserts itself into the plasma membrane. The plasma membrane defines the basic unit of life. In single-celled organisms like those shown here, the membrane creates a physical separation between life on the inside and nonlife on the outside.

KEY CONCEPTS

▨ Phospholipids have a hydrophilic region and a hydrophobic region. In solution, they spontaneously form bilayers that are selectively permeable—meaning that only certain substances cross them readily.

▨ Ions and molecules spontaneously diffuse from regions of high concentration to regions of low concentration. In osmosis, a special case of diffusion, water moves across a selectively permeable membrane from regions of high concentration to regions of low concentration.

▨ In cells, membrane proteins are responsible for the passage of ions, polar molecules, and large molecules that do not readily cross phospholipid bilayers on their own.

The research discussed in previous chapters suggests that biological evolution began with an RNA molecule that could make a copy of itself. As the offspring of this molecule multiplied in the prebiotic soup, natural selection would have favored versions of the molecule that were particularly stable and efficient at catalysis. Another great milestone in the history of life occurred when a descendant of this replicator became enclosed within a membrane. This event created the first cell and thus the first organism.

The **cell membrane,** or **plasma membrane,** is a layer of molecules that surrounds the cell, separating it from the external environment and selectively regulating the passage of molecules and ions into or out of the cell. The evolution of the plasma membrane was a momentous development because it separated life from nonlife. Before plasma membranes existed, self-replicating molecules prob-

ably clung to clay-sized mineral particles, building copies of themselves as they randomly encountered the appropriate nucleotides in the prebiotic soup that washed over them. But the membrane made an internal environment possible—one that could have a chemical composition different from that of the external environment. This was important for two reasons. First, the chemical reactions necessary for life could occur much more efficiently in an enclosed area, because reactants could collide more frequently. Second, the membrane could serve as a selective barrier. That is, it could keep compounds out of the cell that might damage the replicator, but it might allow the entry of compounds required by the replicator. The membrane not only created the cell but also made it into an efficient and dynamic reaction vessel.

The goal of this chapter is to investigate how membranes behave, with an emphasis on how they distinguish an internal environment from the external environment. We begin the chapter by examining the structure and properties of the most abundant molecules in plasma membranes: the "oily" or "fatty" compounds called lipids. We then expand on this introduction by analyzing the way lipids behave when they form membranes. Which ions and molecules can pass through a membrane that consists of lipids? Which cannot, and why? We'll end the chapter by exploring how proteins that become inserted into a lipid membrane can control the flow of materials across the membrane.

6.1 Lipids

Most biochemists are convinced that the building blocks of membranes, called lipids, existed in the prebiotic soup. This conclusion is based on the observation that several types of lipid have been produced in experiments designed to mimic the

chemical and energetic conditions that prevailed early in Earth's history. For example, the spark-discharge experiments reviewed in Chapter 3 succeeded in producing at least two types of lipid.

An observation made by A. D. Bangham illustrates why this result is interesting. In the late 1950s, Bangham performed experiments to determine how lipids behave when they are immersed in water. But until the electron microscope was invented, he had no idea what these lipid-water mixtures looked like. As **Box 6.1** explains, **electron microscopes** allow investigators to magnify objects hundreds of thousands of times. When these instruments became available and Bangham was able to take high-magnification photographs of his experimental mixtures, he saw something astonishing: The lipids had spontaneously formed enclosed compartments filled

---

BOX 6.1   Electron Microscopy

Chapter 1 introduced the light microscope and explained how its invention and use stimulated the development of the cell theory. The electron microscope has had an equally important impact in biology since its invention in the 1950s. Two basic types of electron microscopy are now available: one that allows researchers to examine cross sections of cells at extremely high magnification, and one that offers a view of surfaces at somewhat lower magnification.

Transmission Electron Microscopy

The **transmission electron microscope**, or **TEM**, is an extraordinarily effective tool for viewing cell structure at high magnification. A TEM forms an image from electrons that pass through a specimen, just as a light microscope forms an image from light rays that pass through a specimen.

Biologists who want to view a cell under a TEM begin by "fixing" the cell, meaning that they kill it with a chemical agent that disrupts the cell's structure and contents as little as possible. Then they permeate the cell with an epoxy plastic that stiffens the structure. Once this hardens, the cell can be cut into extremely thin sections with a glass or diamond knife. Finally, the sec-

tioned specimens are impregnated with a metal—often lead. (The reason for this last step is explained shortly.)

Figure 6.1a shows a TEM, outlines how it works, and includes an image produced on a TEM. A beam of electrons is produced by a tungsten filament at the top of a column and directed downward. (All of the air is pumped out of the column so that the electron beam isn't scattered by collisions with air molecules.) The electron beam passes through a series of lenses and the specimen. The lenses are actually electromagnets, which alter the path of the beam much like a glass lens bends light. The lenses magnify and focus the image on a screen at the bottom of the column. There the electrons strike a coating of fluorescent crystals, which emit visible light in response—just like a television screen. When the microscopist moves the screen out of the way and allows the electrons to expose a sheet of black-and-white film, the result is a **micrograph**—a photograph of an image produced by microscopy.

The image itself is created by electrons that pass through the specimen. If no specimen were in place, all the electrons would pass through and the screen (and micrograph) would be uniformly bright. Unfortunately, cell materials by

themselves would also appear fairly uniform and bright. This is because the ability of an atom to deflect an electron depends on its density. An atom's density, in turn, is a function of its atomic number. The hydrogen, carbon, oxygen, and nitrogen atoms that dominate biological molecules have low atomic numbers. This is why cell biologists must saturate cell sections with lead solutions. Lead has a high atomic number and scatters electrons effectively. Different macromolecules take up lead atoms in different amounts, so the metal acts as a "stain" that produces contrast. In the TEM, areas of dense metal scatter the electron beam most, producing dark areas in micrographs.

Scanning Electron Microscopy

The **scanning electron microscope**, or **SEM**, is the most useful tool biologists have for looking at the surfaces of cells. Materials are prepared for scanning electron microscopy by coating their surfaces with a layer of metal atoms. This is in contrast to the TEM, which uses sectioned material impregnated with metal atoms. The SEM allows researchers to inspect the surfaces of objects such as cells; the TEM allows researchers to examine the interiors of objects instead.

"Scanning" describes how the SEM works. To create an image of a surface, the instrument scans the surface with a narrow beam of electrons (**Figure 6.1b**). Electrons that are reflected back from the surface or that are emitted by the metal atoms in response to the beam then strike a detector. The signal from the detector controls a second electron beam, which scans a TV-like screen and forms a magnified image of up to 50,000 times. Because the SEM records shadows and highlights, it provides images with a three-dimensional appearance. It cannot clearly magnify objects nearly as much as the TEM can, however.

(a) Transmission electron microscopy: High magnification of cross sections

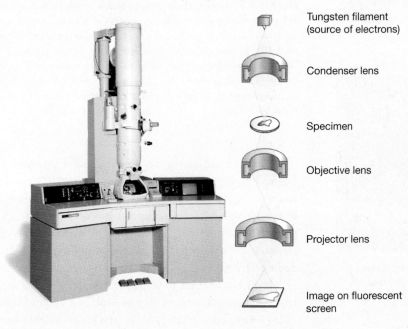

Tungsten filament (source of electrons)

Condenser lens

Specimen

Objective lens

Projector lens

Image on fluorescent screen

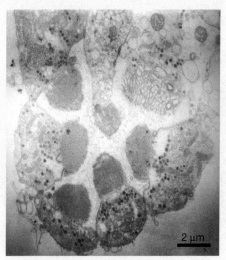

2 μm

Cross section of insect eye

(b) Scanning electron microscopy: Lower magnification of surfaces

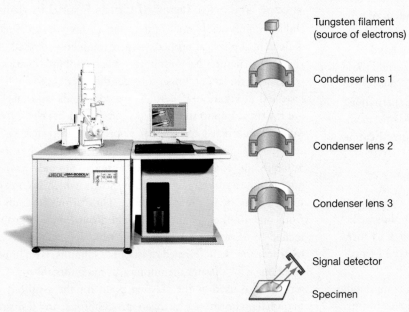

Tungsten filament (source of electrons)

Condenser lens 1

Condenser lens 2

Condenser lens 3

Signal detector

Specimen

20 μm

Surface view of insect eye

FIGURE 6.1 How Does Electron Microscopy Work?
(a) Transmission electron microscopes provide images of cross sections through cells, often at extremely high magnification. **(b)** Scanning electron microscopes offer images of the surfaces of objects.

(a) In solution, lipids form water-filled vesicles.

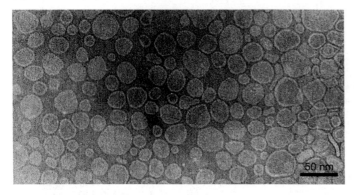

(b) Red blood cells resemble vesicles.

FIGURE 6.2 Lipids Can Form Cell-like Vesicles When in Water
(a) Transmission electron micrograph showing a cross section through the tiny, bag-like compartments that formed when a researcher shook a mixture of lipids and water. **(b)** Scanning electron micrograph showing red blood cells from humans. Note the scale bars in each photo.

with water (**Figure 6.2a**). He called these membrane-bound structures *vesicles* and noted that they were reminiscent of cells (**Figure 6.2b**). Bangham had not done anything special to the lipid-water mixtures; he had merely shaken them by hand.

The experiment raises a series of questions: How could these structures have formed? Is it possible that vesicles like these existed in the prebiotic soup? If so, could they have surrounded a self-replicating molecule and become the first plasma membrane? Let's begin answering these questions by investigating what lipids are and how they behave.

What Is a Lipid?

In earlier chapters we examined the structures of the organic molecules called amino acids, nucleotides, and monosaccharides and explored how these monomers polymerize to form macromolecules. Here we analyze another major type of mid-sized molecule found in living organisms: lipids.

Lipid is a catch-all term for carbon-containing compounds that are found in organisms and are largely nonpolar and hydrophobic—meaning that they do not dissolve readily in water. (Recall from Chapter 2 that water is a polar solvent.) Lipids do dissolve, however, in liquids consisting of nonpolar organic compounds.

(a) Isoprene   **(b)** Fatty acid

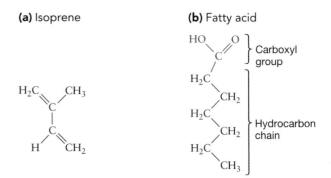

FIGURE 6.3 Hydrocarbon Groups Make Lipids Hydrophobic
A hydrocarbon is a molecule or a section of a molecule that is made up solely of carbon and hydrogen. Hydrocarbons are uncharged and nonpolar, so they do not interact with water. **(a)** Isoprenes are hydrocarbons. Isoprene subunits can be linked end to end to form long hydrocarbon chains. **(b)** Fatty acids have long hydrocarbon tails. Although only seven carbons are shown, fatty acids usually contain between 14 and 20. EXERCISE Circle the hydrophobic hydrocarbon components of both molecules.

To understand why lipids do not dissolve in water, examine **Figure 6.3**. Figure 6.3a shows a five-carbon compound called isoprene. It consists of a group of carbon atoms bonded to hydrogen atoms. Molecules that contain only carbon and hydrogen, such as isoprene, are known as **hydrocarbons**. Because electrons are shared equally in carbon-hydrogen bonds, hydrocarbons are nonpolar. This property makes them hydrophobic. Lipids do not dissolve in water, because they have a significant hydrocarbon component. Figure 6.3b is a type of compound called a **fatty acid**, which consists of a hydrocarbon chain bonded to a carboxyl (COOH) functional group. Isoprene and fatty acids are key building blocks of the lipids found in organisms.

A Look at Three Types of Lipids Found in Cells

Unlike amino acids, nucleotides, and carbohydrates, lipids are defined by a physical property—their solubility—instead of their chemical structure. As a result, the structure of lipids varies widely. To drive this point home, consider the three classes of lipids illustrated in **Figure 6.4**: steroids, phospholipids, and fats. These are the most important types of lipids found in cells. Note that each is constructed from either isoprene or fatty-acid subunits.

Figure 6.4a shows the cholesterol molecule, which belongs to a family of compounds called steroids. **Steroids** are distinguished by the four-ring structure shown, which is constructed from chains of isoprene subunits. The various steroids differ from one another by the functional groups or side groups attached to those rings. Cholesterol, which is distinguished by a hydrocarbon "tail" formed of isoprene subunits, is an important component of plasma membranes in many organisms. Cholesterol is also used as the starting point for the synthesis of the mammalian hormones estrogen, progesterone, and testosterone.

Figure 6.4b shows the basic structure of a phospholipid. A **phospholipid** consists of a three-carbon molecule called **glycerol** that is linked to a phosphate group (PO_4^{2-}) and to either two chains of isoprene or two fatty acids. In some cases, the phos-

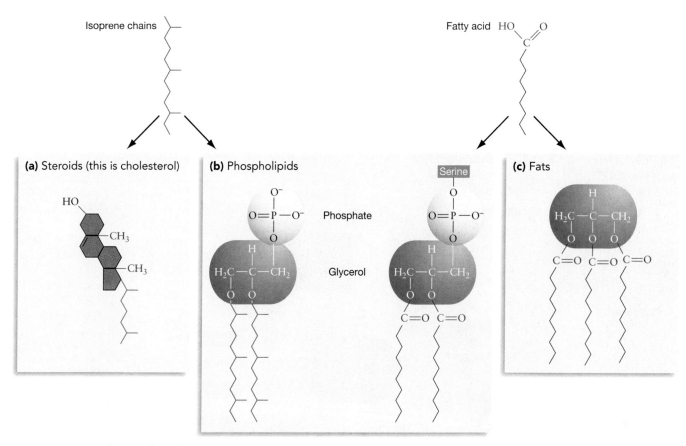

FIGURE 6.4 Major Classes of Lipids
Isoprenes and fatty acids are building blocks for **(a)** steroids, **(b)** phospholipids, and **(c)** fats. Many of the C and H symbols for carbon and hydrogen, respectively, have been omitted for clarity. The red or black lines indicate carbon-carbon bonds.

phate group is bonded to another small organic molecule, such as the amino acid serine in the figure. Phospholipids with isoprene tails are found in the domain Archaea introduced in Chapter 1; phospholipids composed of fatty acids are found in the domains Bacteria and Eukarya.

Fats, such as the molecule in Figure 6.4c, are composed of three fatty acids that are linked to a single glycerol molecule. Because of this structure, fats are also called *triacylglycerols* or *triglycerides*.

Both phospholipids and fats form when a condensation reaction occurs between the hydroxyl group of glycerol and the carboxyl group of a fatty acid (**Figure 6.5a**). The glycerol and fatty-acid molecules become joined by an **ester linkage**, which is analogous to the peptide bonds, phosphodiester bonds, and glycosidic linkages in proteins, nucleic acids, and carbohydrates, respectively. Phospholipids and fats are not polymers, however, and fatty acids are not monomers. As **Figure 6.5b** shows, fatty acids are not linked together to form a macromolecule in

(a) Dehydration reactions join fatty acids to glycerol.

(b) Fats consist of glycerol linked to three fatty acids.

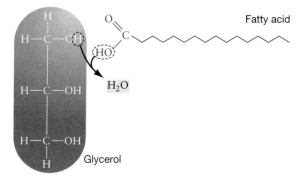

FIGURE 6.5 Fats Form via Dehydration Reactions
(a) When glycerol and a fatty acid react, a water molecule leaves. **(b)** The covalent bonds that result from this reaction are termed an ester linkage.

the way that amino acids, nucleotides, and monosaccharides are.

The classes of molecules in Figure 6.5 will reappear throughout this text, because lipids perform a wide variety of functions in cells. Lipids store chemical energy, act as pigments that capture or respond to sunlight, serve as signals between cells, form waterproof coatings on leaves and skin, and act as vitamins used in an array of cellular processes. The most important lipid function, however, is their role in the plasma membrane.

The Structures of Membrane Lipids

Not all lipids can form the artificial membranes that Bangham and his colleagues observed. In fact, just two types of lipids dominate plasma membranes. Membrane-forming lipids have a polar, hydrophilic region in addition to the nonpolar, hydrophobic region found in all lipids. For example, examine the phospholipid illustrated in **Figure 6.6a**. The molecule has a "head" region, containing covalent bonds that are highly polar, as well as positive and negative charges. The

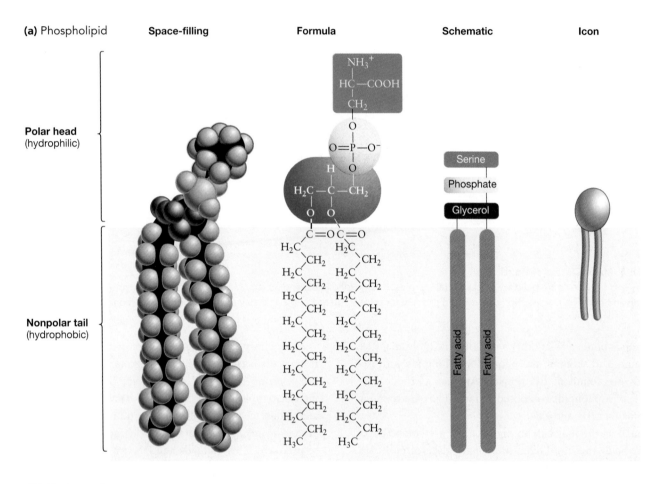

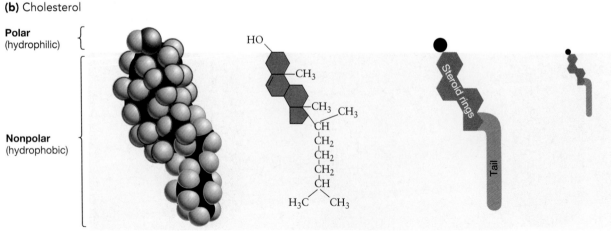

FIGURE 6.6 Amphipathic Lipids Contain Hydrophilic and Hydrophobic Elements
QUESTION If these molecules were in solution, where would water molecules interact with them?

charges and polar bonds in the head region interact with water molecules when a phospholipid is placed in solution. In contrast, the long isoprene or fatty-acid tails of a phospholipid are nonpolar. Water molecules do not interact with this part of the molecule.

Compounds that contain both hydrophilic and hydrophobic elements are **amphipathic** ("dual sympathy"). Phospholipids are amphipathic. As **Figure 6.6b** shows, cholesterol is also amphipathic. It has both hydrophilic and hydrophobic regions.

The amphipathic nature of phospholipids is far and away their most important feature biologically. It is responsible for their ability to form plasma membranes.

6.2   Phospholipid Bilayers

Phospholipids do not dissolve when they are placed in water. Water molecules interact with the hydrophilic heads of the phospholipids, but not with their hydrophobic tails. Instead of dissolving in water, then, phospholipids form one of two types of structures: micelles or lipid bilayers.

Micelles (Figure 6.7a) are tiny droplets created when the hydrophilic heads of phospholipids face the water and the

hydrophobic tails are forced together, away from the water. Phospholipid bilayers, or simply **lipid bilayers**, are created when two sheets of lipid molecules align. As **Figure 6.7b** shows, the hydrophobic heads in each layer face the solution while the tails face one another inside the bilayer. In this way, the hydrophilic heads interact with water while the hydrophobic tails interact with each other. Micelles tend to form from phospholipids with relatively short tails; bilayers tend to form from phospholipids with longer tails.

Energetically, the formation of micelles or lipid bilayers is highly favored. Phospholipids are much more stable in water when they form bilayers than they are when the individual molecules exist independently. As a result, these structures form spontaneously. No input of energy is required.

The key point here is that the amphipathic structure of phospholipids makes it possible for them to function as plasma membranes. The fact that bilayers form spontaneously also explains why Bangham's group was able to make vesicles so easily.

Artificial Membranes as an Experimental System

When lipid bilayers are agitated by shaking, the layers break and re-form as small, spherical structures. This is what happened in Bangham's experiment. The resulting vesicles had water on the inside as well as the outside because the hydrophilic heads of the lipids faced outward on each side of the bilayer.

Researchers have produced these types of vesicles by using dozens of different types of phospholipids. Artificial membrane-bound vesicles like these are called **liposomes**. The ability to create them supports an important conclusion: If phospholipid molecules accumulated during chemical evolution early in Earth's history, they almost certainly formed water-filled vesicles.

The discovery of liposomes sparked intense interest among biologists because lipid bilayers also form the basic structure of plasma membranes. Researchers began an effort to understand the properties of plasma membranes by creating and experimenting with artificial bilayers.

Some of the first questions posed by researchers concerned the permeability of lipid bilayers. The **permeability** of a structure is the structure's tendency to allow a given substance to diffuse across it. Once a membrane forms a water-filled vesicle, can other molecules or ions pass in or out? If so, is this permeability selective in any way? The permeability of membranes is a critical issue, because if certain molecules or ions pass through a lipid bilayer more readily than others, the internal environment of a vesicle can become different from the outside. This difference between exterior and interior environments is a key characteristic of cells.

To explore the permeability of phospholipid bilayers, biologists began performing experiments on artificial membranes.

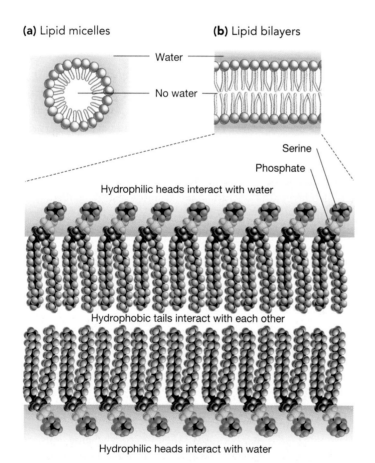

(a) Lipid micelles   **(b)** Lipid bilayers

Water

No water

Serine

Phosphate

Hydrophilic heads interact with water

Hydrophobic tails interact with each other

Hydrophilic heads interact with water

FIGURE 6.7 Phospholipids Form Bilayers in Solution
In **(a)** a micelle or **(b)** a lipid bilayer, the hydrophilic heads of phospholipids face out, toward water; the hydrophobic tails face in, away from water. Plasma membranes consist in part of lipid bilayers.

Figure 6.8 shows the two types of artificial membranes that are used to study the permeability of lipid bilayers. Figure 6.8a shows liposomes, roughly spherical vesicles. Figure 6.8b illustrates **planar bilayers**, which are lipid bilayers constructed across a hole in a glass or plastic wall separating two aqueous (watery) solutions.

(a) Liposomes: Artificial membrane-bound vesicles

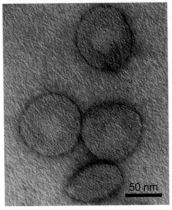
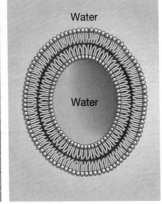

(b) Planar bilayers: Artificial membranes

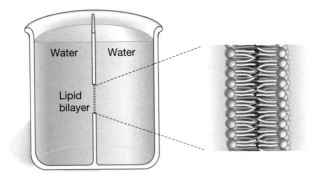

(c) Artificial-membrane experiments

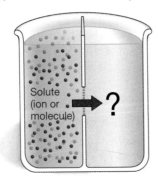

How rapidly can different solutes cross the membrane (if at all) when …

1. Different types of phospholipids are used to make the membrane?

2. Proteins or other molecules are added to the membrane?

FIGURE 6.8 Liposomes and Planar Bilayers Are Important Experimental Systems
(a) Transmission electron micrograph of liposomes in cross section (left) and a cross-sectional diagram of the lipid bilayer in a liposome. **(b)** The construction of planar bilayers across a hole in a glass wall separating two water-filled compartments (left), and a close-up sketch of the bilayer. **(c)** A wide variety of experiments is possible with liposomes and planar bilayers; a few are suggested here.

Using liposomes and planar bilayers, researchers can study what happens when a known ion or molecule is added to one side of a lipid bilayer (Figure 6.8c). Does the ion or molecule cross the membrane and show up on the other side? If so, how rapidly does the movement take place? What happens when a different type of phospholipid is used to make the artificial membrane? Does the membrane's permeability change when proteins or other types of molecules are added to it?

Biologists describe such an experimental system as elegant and powerful because it gives them precise control over which factor changes from one experimental treatment to the next. Control, in turn, is why experiments are such an effective way to explore scientific questions, as we learned in Chapter 1. A good experimental design allows researchers to alter one factor at a time and determine what effect, if any, each has on the process being studied.

Equally important for experimental purposes, liposomes and planar bilayers provide a clear way to determine whether a given change in conditions has an effect. By sampling the solutions on both sides of the membrane before and after the treatment and then analyzing the concentration of ions and molecules in the samples, researchers have an effective way to determine whether the treatment had any consequences.

Using such systems, what have biologists learned about membrane permeability?

Selective Permeability of Lipid Bilayers

When researchers put molecules or ions on one side of a liposome or planar bilayer and measure the rate at which the molecules arrive on the other side, a strong pattern emerges. Lipid bilayers are *highly* selective. **Selective permeability** means that some substances cross a membrane more easily than other substances can. Small, nonpolar molecules move across bilayers quickly. In contrast, most charged substances cross the membrane slowly, if at all. According to the data in **Figure 6.9**, small, nonpolar molecules such as oxygen (O_2) move across selectively permeable membranes more than a billion times faster than do chloride ions (Cl^-). Larger molecules can also move rapidly if they are nonpolar. Indole, for example, is a large, nonpolar compound that moves across membranes 100 million times faster than do potassium ions (K^+). Very small and uncharged molecules such as water (H_2O) can also cross membranes rapidly, even if they are polar.

The leading hypothesis to explain this pattern is that charged compounds and large, polar molecules can't pass through the nonpolar, hydrophobic tails of a lipid bilayer. The reasoning here is that because of their electrical charge, ions are more stable in solution than they are in the interior of membranes, which is electrically neutral. To test this hypothesis, researchers have manipulated the size and structure of the tails in liposomes or planar bilayers.

(a) Permeability scale (cm/sec)

(b) Size and charge affect the rate of diffusion across a membrane.

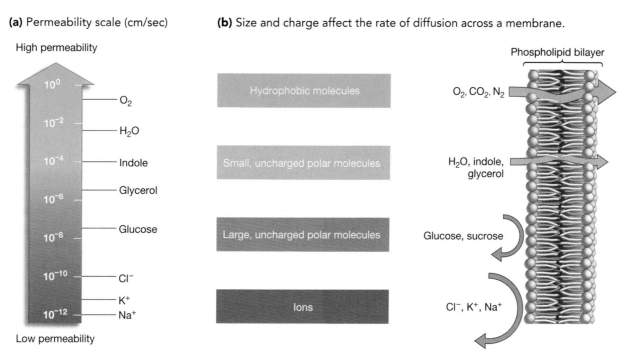

FIGURE 6.9 **Selective Permeability of Lipid Bilayers**
(a) The numbers represent "permeability coefficients," or the rate (cm/sec) at which an ion or molecule crosses a lipid bilayer. **(b)** The relative permeabilities of various molecules and ions, based on data like those presented in part (a). **QUESTION** About how fast does water cross the lipid bilayer?

Does the Type of Lipid in a Membrane Affect Its Permeability?

Theoretically, two aspects of a hydrocarbon chain could affect the way the chain behaves in a lipid bilayer: (1) the number of double bonds it contains and (2) its length. Recall from Chapter 2 that when carbon atoms form a double bond, the attached atoms are found in a plane instead of a (three-dimensional) tetrahedron. The carbon atoms involved are also locked into place; they cannot rotate freely, as they do in carbon-carbon single bonds. As a result, a double bond between carbon atoms produces a "kink" in an otherwise straight hydrocarbon chain (**Figure 6.10**).

When a double bond exists between two carbon atoms in a hydrocarbon chain, the chain is said to be **unsaturated.** Conversely, hydrocarbon chains without double bonds are said to be **saturated.** This choice of terms is logical, because if a hydrocarbon chain does not contain a double bond, it is saturated with the maximum number of hydrogen atoms that can attach to the carbon skeleton. If it is unsaturated, then fewer than the maximum number of hydrogen atoms are attached. Because they contain more C–H bonds, which have much more free energy than C=C bonds, saturated fats have much more chemical energy than unsaturated fats do. People who are dieting are often encouraged to eat fewer saturated fats. Foods that contain lipids with many double bonds are said to be *polyunsaturated* and are advertised as healthier than foods with more-saturated fats.

Why do double bonds affect the permeability of membranes? When hydrophobic tails are packed into a lipid bilayer, the kinks created by double bonds produce spaces among the tightly packed tails. These spaces reduce the strength of hydrophobic interactions among the tails. Because the interior of the membrane is "glued together" less

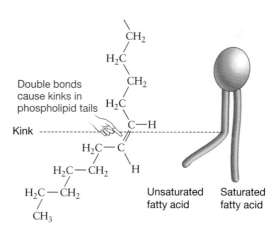

FIGURE 6.10 **Unsaturated Hydrocarbons Contain Carbon-Carbon Double Bonds**
A double bond in a hydrocarbon chain produces a "kink." The icon on the right indicates that one of the hydrocarbon tails in a phospholipid is kinked and therefore unsaturated. **EXERCISE** Draw figures analogous to those for an unsaturated fatty acid containing two double bonds.

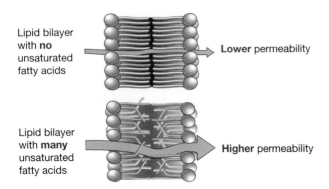

FIGURE 6.11 Fatty-Acid Structure Changes the Permeability of Membranes
Lipid bilayers that contain many unsaturated fatty acids contain more gaps and should be more permeable than are bilayers with few unsaturated fatty acids.

tightly, the structure should become more fluid and more permeable (**Figure 6.11**).

Hydrophobic interactions also become stronger as saturated hydrocarbon tails increase in length. Membranes dominated by phospholipids with long, saturated hydrocarbon tails should be stiffer and less permeable because the interactions among the tails are stronger.

A biologist would predict, then, that bilayers made of lipids with long, straight, saturated fatty-acid tails should be much less permeable than membranes made of lipids with short, kinked, unsaturated fatty-acid tails. Experiments on liposomes have shown exactly this pattern. Phospholipids with long, saturated tails form membranes that are much less permeable than membranes consisting of phospholipids with shorter, unsaturated tails.

The central point here is that the degree of hydrophobic interactions dictates the behavior of these molecules. This is another example in which the structure of a molecule—specifically, the number of double bonds in the hydrocarbon chain and its overall length—correlates with its function.

These data are also consistent with the basic observation that highly saturated fats are solid at room temperature (**Figure 6.12a**). Lipids that have extremely long hydrocarbon tails, as **waxes** do, form stiff solids at room temperature as a result of the extensive hydrophobic interactions that occur (**Figure 6.12b**). Birds, sea otters, and many other organisms synthesize waxes and spread them on their exterior surface as a waterproofing. In contrast, highly unsaturated fats are liquid at room temperature (**Figure 6.12c**). Liquid triacylglycerides are called **oils**.

In addition to exploring the role of hydrocarbon chain length and degree of saturation on membrane permeability, biologists have investigated the effect of adding cholesterol molecules. Because cholesterol is expected to fill spaces between phospholipids, adding cholesterol to a membrane should increase the density of the hydrophobic section. As predicted, researchers found that adding cholesterol molecules to liposomes dramatically reduced the permeability of the liposomes. The data behind this claim are presented in **Figure 6.13**. This graph makes another important point, however: Temperature has a strong influence on the behavior of lipid bilayers.

Why Does Temperature Affect the Fluidity and Permeability of Membranes?

At about 25°C, or "room temperature," the phospholipids found in plasma membranes are liquid and bilayers have the consistency of olive oil. This fluidity, as well as the membrane's permeability, decreases as temperature decreases. As temperatures drop, individual molecules in the bilayer move more slowly. As a result, the hydrophobic tails in the interior of membranes pack together

(a) Saturated lipids   **(b)** Saturated lipids with long hydrocarbon tails   **(c)** Unsaturated lipids

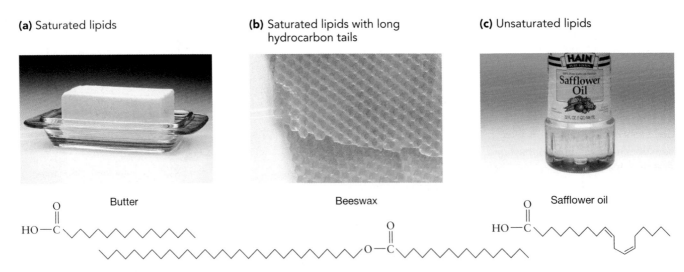

Butter   Beeswax   Safflower oil

FIGURE 6.12 The Fluidity of Lipids Depends on the Characteristics of Their Hydrocarbon Chains
The fluidity of a lipid depends on the length and saturation of its hydrocarbon chain. **(a)** Butter consists primarily of saturated lipids. **(b)** Waxes are lipids with extremely long hydrocarbon chains. **(c)** Oils are dominated by "polyunsaturates"—lipids with hydrocarbon chains that contain multiple double bonds.

more tightly. At very low temperatures, lipid bilayers begin to so-lidify. As the graphs in Figure 6.13 indicate, low temperatures can make membranes impervious to molecules that would nor-mally cross them readily.

The fluid nature of membranes also allows individual lipid molecules to move laterally within each layer, a little like a per-son moving about in a dense crowd. By tagging individual phospholipids and following their movement, researchers have

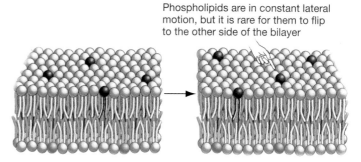

FIGURE 6.14 **Phospholipids Move within Membranes**
Membranes are dynamic—in part because phospholipid molecules move within each layer in the structure.

clocked average speeds of 2 micrometers (μm)/second at room temperature (**Figure 6.14**). At these speeds, phospholipids trav-el the length of a small bacterial cell every second.

These experiments on lipid and ion movement demonstrate that membranes are dynamic. Phospholipid molecules whiz around each layer, while water and small, nonpolar molecules shoot in and out of the membrane. How quickly molecules move within and across membranes is a function of tempera-ture and the structure of the hydrocarbon tails in the bilayer.

Given these insights into the permeability and fluidity of lipid bilayers, an important question remains: *Why* do certain molecules move across membranes spontaneously?

✓ CHECK YOUR UNDERSTANDING

In solution, phospholipids form bilayers that are selectively permeable—meaning that some substances cross them much more readily than others can. Permeability is a func-tion of temperature, the amount of cholesterol in the mem-brane, and the length and degree of saturation of the hydrocarbon tails in membrane phospholipids. You should be able to fill in a chart with (1) rows called "Temperature," "Cholesterol," "Length of hydrocarbon tails," and "Satura-tion of hydrocarbon tails" and (2) columns named "Factor," "Effect on permeability," and "Reason."

6.3 Why Molecules Move across Lipid Bilayers: Diffusion and Osmosis

A thought experiment can help explain why molecules and ions are able to move across membranes spontaneously. Suppose you rack up a set of blue billiard balls on a pool table containing many white balls and then begin to vibrate the table. Because of the vibration, the balls will move about randomly. They will also bump into one another. After these collisions, some blue balls will move outward—away from their original position. In fact, the overall (or net) movement of blue balls will be outward. This occurs because the random motion of the blue balls disrupts their original, nonrandom position—as they move at random, they are

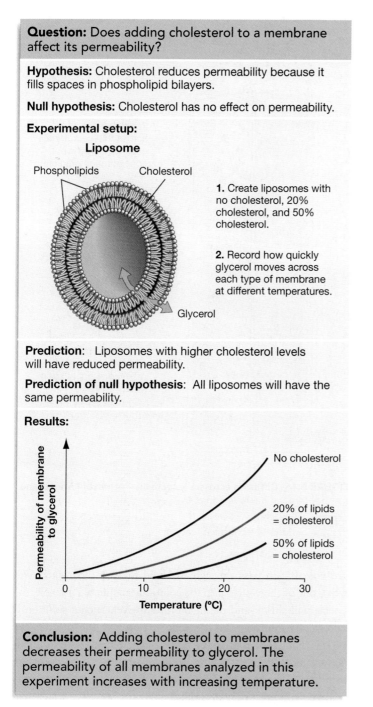

Question: Does adding cholesterol to a membrane affect its permeability?

Hypothesis: Cholesterol reduces permeability because it fills spaces in phospholipid bilayers.

Null hypothesis: Cholesterol has no effect on permeability.

Experimental setup:

Liposome

Phospholipids   Cholesterol

1. Create liposomes with no cholesterol, 20% cholesterol, and 50% cholesterol.

2. Record how quickly glycerol moves across each type of membrane at different temperatures.

Glycerol

Prediction: Liposomes with higher cholesterol levels will have reduced permeability.

Prediction of null hypothesis: All liposomes will have the same permeability.

Results:

No cholesterol

20% of lipids = cholesterol

50% of lipids = cholesterol

Permeability of membrane to glycerol

Temperature (°C)

0   10   20   30

Conclusion: Adding cholesterol to membranes decreases their permeability to glycerol. The permeability of all membranes analyzed in this experiment increases with increasing temperature.

FIGURE 6.13 **The Permeability of a Membrane Depends on Its Composition**
Experiments like this established that membrane permeability varies with the types of lipids found in the structure.

more likely to move away from each other than to stay together. Eventually, the blue billiard balls will be distributed randomly across the table. The entropy of the blue billiard balls has increased. Recall from Chapter 2 that entropy is a measure of the randomness or disorder in a system. The second law of thermodynamics states that in a closed system, entropy always increases.

This hypothetical example illustrates why molecules or ions located on one side of a lipid bilayer move to the other side spontaneously. The dissolved molecules and ions, or **solutes**, have thermal energy and are in constant, random motion. Because they change position randomly, they tend to move from a region of high concentration to a region of low concentration, along what biologists call a **concentration gradient**. This directed movement of molecules and ions is known as **diffusion**. Diffusion occurs along a concentration gradient because the random motion of molecules and ions causes a net movement from regions of high concentration to regions of low concentration. Diffusion is a spontaneous process because it results in an increase in entropy.

Once the molecules or ions are randomly distributed throughout a solution, equilibrium is established. For example, consider two solutions separated by a lipid bilayer. **Figure 6.15** shows how molecules that pass through the bilayer diffuse to the other side. At equilibrium, molecules continue to move back and forth across the membrane, but at equal rates—simply because each molecule or ion is equally likely to move in any direction. This means that there is no longer a net movement of molecules across the membrane.

What about water itself? As the data in Figure 6.9 showed, water moves across lipid bilayers extremely quickly. Like other substances that diffuse, water moves along its concentration gradient—from higher to lower concentration. The movement of water is a special case of diffusion that is given its own name: **osmosis**. Osmosis occurs only when solutions are separated by a membrane that is permeable to some molecules but not others—that is, a selectively permeable membrane.

The best way to think about water moving in response to a concentration gradient is to focus on the concentration of solutes in the solution. Let's suppose the concentration of a particular solute is higher on one side of a selectively permeable membrane than it is on the other side (**Figure 6.16**, step 1). Further, suppose that this solute cannot diffuse through the membrane to establish equilibrium. What happens? Water will move from the side with a lower concentration of solute to the side with a higher concentration of solute (step 2). It dilutes the higher concentration and equalizes the concentrations on both sides. This movement is spontaneous. It is driven by the increase in entropy achieved when solute concentrations are equal on both sides of the membrane.

This movement of water is important because it can swell or shrink a membrane-bound vesicle. Consider the liposomes illustrated in **Figure 6.17**. If the solution outside the membrane has a higher concentration of solutes than the interior, and the solutes

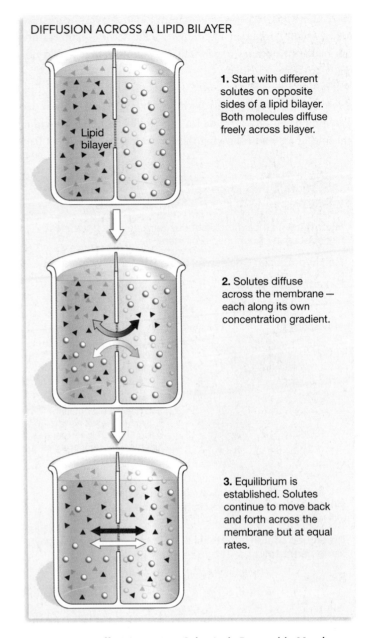

DIFFUSION ACROSS A LIPID BILAYER

1. Start with different solutes on opposite sides of a lipid bilayer. Both molecules diffuse freely across bilayer.

Lipid bilayer

2. Solutes diffuse across the membrane — each along its own concentration gradient.

3. Equilibrium is established. Solutes continue to move back and forth across the membrane but at equal rates.

FIGURE 6.15 Diffusion across a Selectively Permeable Membrane
EXERCISE If a solute's rate of diffusion increases linearly with its concentration difference across the membrane, write an equation for the rate of diffusion across a membrane.

are not able to pass through the lipid bilayer, then water will move out of the vesicle into the solution outside. As a result, the vesicle will shrink and the membrane shrivel. Such a solution is said to be **hypertonic** ("excess-tone") relative to the inside of the vesicle. The word root *hyper* refers to the outside solution containing more solutes than the solution on the other side of the membrane. Conversely, if the solution outside the membrane has a lower concentration of solutes than the interior, water will move into the vesicle via osmosis. The incoming water will cause the vesicle to swell or even burst. Such a solution is termed **hypotonic** ("lower-tone") relative to the inside of the vesicle. Here the word root *hypo* refers to the outside solution containing fewer

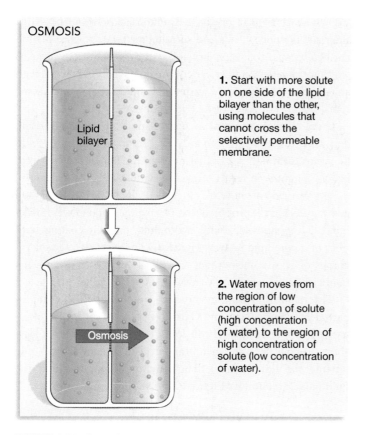

FIGURE 6.16 Osmosis

QUESTION Suppose you doubled the number of molecules on the right side of the membrane (at the start). At equilibrium, would the water level on the right side be higher or lower than what is shown here?

solutes than the solution inside. If solute concentrations are equal on either side of the membrane, the liposome will maintain its size. When the outside solution does not affect the membrane's shape, that solution is termed **isotonic** ("equal-tone").

In sum, diffusion and osmosis move solutes and water across lipid bilayers. What does all this have to do with the first membranes floating in the prebiotic soup? In effect, osmosis and diffusion tend to *reduce* differences in chemical composition between the inside and outside of membrane-bound structures. If liposome-like structures were present in the prebiotic soup, it's unlikely that their interiors offered a radically different environment from the surrounding solution. In all likelihood, the primary importance of the first lipid bilayers was simply to provide a container for self-replicating molecules. Experiments have shown that ribonucleotides can diffuse across lipid bilayers. Further, it is clear that cell-like vesicles grow as additional lipids are added and then divide if sheared by shaking or bubbling or wave action. Based on these observations, it is reasonable to hypothesize that once a self-replicating ribozyme had become surrounded by a lipid bilayer, this simple life-form and its descendants would continue to occupy cell-like structures that grew and divided.

We can now investigate the next great event in the evolution of life: the formation of a true cell. How can lipid bilayers become a barrier capable of creating and maintaining a specialized internal environment that is conducive to life? How could a simple cell become an effective plasma membrane—one that admits ions and molecules needed by the replicator while excluding ions and molecules that might damage it?

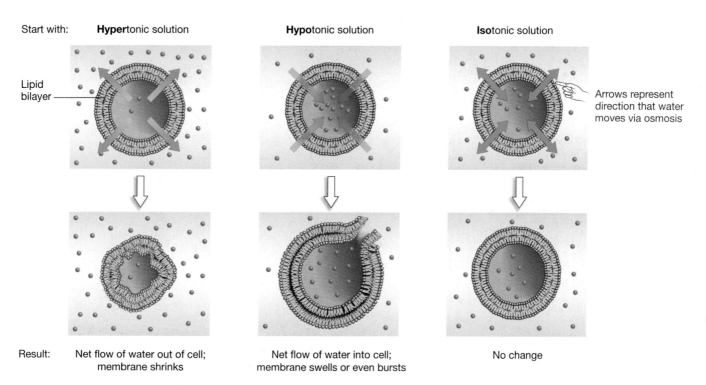

FIGURE 6.17 Osmosis Can Shrink or Burst Membrane-Bound Vesicles

QUESTION Some species of bacteria can live in extremely salty environments, such as saltwater-evaporation ponds. Is this habitat likely to be hypertonic, hypotonic, or isotonic relative to the interior of the cells?

✓CHECK YOUR UNDERSTANDING

Diffusion is the movement of ions or molecules from regions of high concentration to regions of low concentration. Osmosis is the movement of water across a selectively permeable membrane, from a region of low solute concentration to a region of high solute concentration. You should be able to (1) make a sketch analogous to Figure 6.17 and (2) predict what happens when the solution *inside* the cell is hypertonic, hypotonic, and isotonic relative to the outside.

6.4   Membrane Proteins

What sort of molecule could become incorporated into a lipid bilayer and affect the bilayer's permeability? The title of this section gives the answer away. Proteins that are amphipathic can be inserted into lipid bilayers.

Proteins can be amphipathic because they are made up of amino acids and because amino acids have side chains, or R-groups, that range from highly nonpolar to highly polar (or even charged; see Figure 3.5 and Table 3.1). It is conceivable, then, that a protein could have a series of nonpolar amino acids in the middle of its primary structure but polar or charged amino acids on both ends of its primary structure, as illustrated in **Figure 6.18a**. The nonpolar amino acids would be stable in the interior of a lipid bi-

(a) Proteins can be amphipathic.

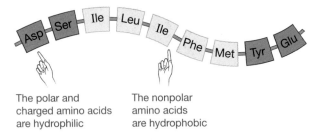

The polar and charged amino acids are hydrophilic

The nonpolar amino acids are hydrophobic

(b) Amphipathic proteins can integrate into lipid bilayers.

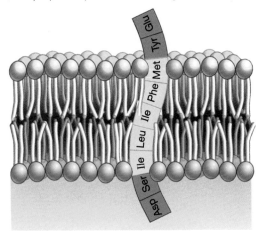

FIGURE 6.18 Proteins Can Be Amphipathic
Note that these drawings are conceptual and not to scale. It actually takes at least 15 amino acids to span a lipid bilayer.

layer, while the polar or charged amino acids would be stable alongside the polar heads and surrounding water (**Figure 6.18b**). Further, because the secondary and tertiary structures of proteins are almost limitless in their variety and complexity, it is possible to imagine that proteins could form tubes and thus function as some sort of channel or pore across a lipid bilayer.

Based on these theoretical considerations, it is not surprising that when researchers began analyzing the chemical composition of plasma membranes in organisms, they found that proteins were just as common, in terms of mass, as phospholipids. How were these two types of molecules arranged? In 1935 Hugh Davson and James Danielli proposed that plasma membranes were structured like a sandwich, with hydrophilic proteins coating both sides of a pure lipid bilayer (**Figure 6.19a**). Early electron micrographs of plasma membranes seemed to be consistent with the sandwich model, and for decades it was widely accepted.

The realization that membrane proteins could be amphipathic led S. Jon Singer and Garth Nicolson to suggest an alternative hypothesis, however. In 1972, they proposed that at least some proteins span the membrane instead of being found only outside the lipid bilayer. Their hypothesis was called the **fluid-mosaic model**. As **Figure 6.19b** shows, Singer and Nicol-

(a) Sandwich model

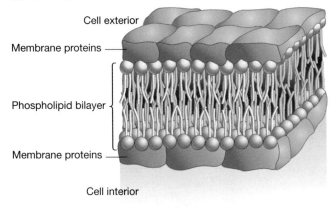

Cell exterior

Membrane proteins

Phospholipid bilayer

Membrane proteins

Cell interior

(b) Fluid-mosaic model

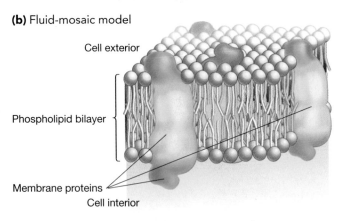

Cell exterior

Phospholipid bilayer

Membrane proteins

Cell interior

FIGURE 6.19 Past and Current Models of Membrane Structure
(a) The protein-lipid-lipid-protein sandwich model was the first hypothesis for the arrangement of lipids and proteins in plasma membranes. **(b)** The fluid-mosaic model was a radical departure from the sandwich hypothesis.

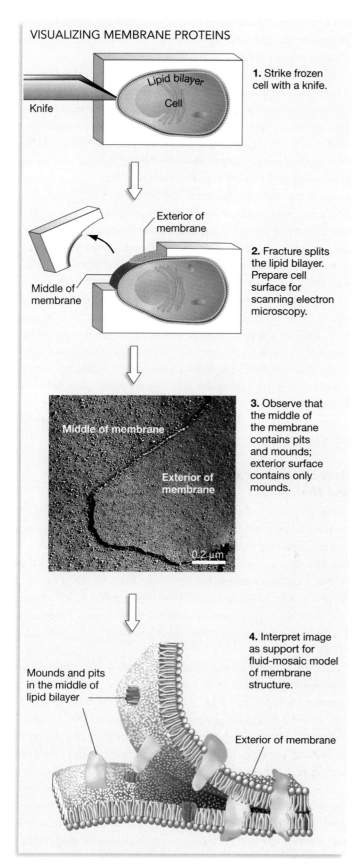

VISUALIZING MEMBRANE PROTEINS

1. Strike frozen cell with a knife.

Knife

Lipid bilayer

Cell

Exterior of membrane

Middle of membrane

2. Fracture splits the lipid bilayer. Prepare cell surface for scanning electron microscopy.

Middle of membrane

Exterior of membrane

0.2 μm

3. Observe that the middle of the membrane contains pits and mounds; exterior surface contains only mounds.

4. Interpret image as support for fluid-mosaic model of membrane structure.

Mounds and pits in the middle of lipid bilayer

Exterior of membrane

FIGURE 6.20 Freeze-Fracture Preparations Allow Biologists to View Membrane Proteins
EXERCISE Draw what the micrograph in step 3 would look like if the sandwich model of membrane structure were correct.

son suggested that membranes are a mosaic of phospholipids and different types of proteins. The overall structure was proposed to be dynamic and fluid.

The controversy over the nature of the plasma membrane was resolved in the early 1970s with the development of an innovative technique for visualizing the surface of plasma membranes. The method is called **freeze-fracture electron microscopy**, because the steps involve freezing and fracturing the membrane before examining it with a scanning electron microscope. As **Figure 6.20** shows, the technique allows researchers to split plasma membranes and view both the middle and the surface of the structure. The scanning electron micrographs that result show pits and mounds studding the inner surfaces of the lipid bilayer. Researchers interpret these structures as the locations of membrane proteins. As step 4 in Figure 6.20 shows, the pits and mounds are hypothesized to represent proteins that span the lipid bilayer.

These observations conflicted with the sandwich model but were consistent with the fluid-mosaic model. Based on these and subsequent observations, the fluid-mosaic model is now widely accepted. **Figure 6.21** summarizes where proteins and lipids are found in a plasma membrane. Note that some proteins span the membrane and have elements facing both the interior and exterior surfaces. Proteins such as these are called **integral membrane proteins**, or **transmembrane proteins**. Other proteins, called **peripheral membrane proteins**, are found only on one side of the membrane. Often these peripheral membrane proteins are attached to an integral membrane protein. The key point is that the arrangement of proteins makes the interior and exterior surfaces of the plasma membrane very different.

What do all these proteins do? In later chapters we'll explore how certain membrane proteins act as enzymes or are involved

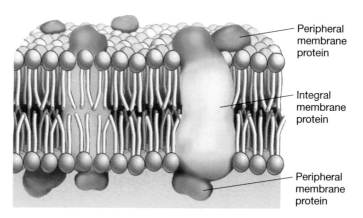

Peripheral membrane protein

Integral membrane protein

Peripheral membrane protein

FIGURE 6.21 Integral and Peripheral Membrane Proteins
Integral membrane proteins are also called transmembrane proteins because they span the membrane. Peripheral membrane proteins are often attached to integral membrane proteins. QUESTION Are the external and internal faces of a plasma membrane the same or different? Explain.

in cell-to-cell signaling or making physical connections between cells. Here, let's focus on how integral membrane proteins are involved in the transport of selected ions and molecules across the plasma membrane.

Systems for Studying Membrane Proteins

The discovery of integral membrane proteins was consistent with the hypothesis that proteins affect membrane permeability. The evidence was considered weak, though, because it was also plausible to claim that integral membrane proteins were structural components that influenced membrane strength or flexibility. To test whether proteins actually do affect membrane permeability, researchers needed some way to isolate and purify membrane proteins.

Figure 6.22 outlines one method that researchers developed to separate proteins from membranes. The key to the technique is the use of detergents. A **detergent** is a small, amphipathic molecule. As step 1 of Figure 6.22 shows, the hydrophobic tails of detergents clump in solution, forming micelles. When detergents are added to the solution surrounding a lipid bilayer, the hydrophobic tails of the detergent molecule interact with the hydrophobic tails of the lipids. In doing so, the detergent tends to disrupt the bilayer and break it apart (step 2). If the membrane contains proteins, the hydrophobic tails of the detergent molecules also interact with the hydrophobic parts of the membrane proteins. The detergent molecules displace the membrane phospholipids and end up forming water-soluble, detergent-protein complexes (step 3).

To isolate and purify these membrane proteins once they are in solution, researchers use the technique called gel electrophoresis, which was introduced in Box 4.1. **Gel electrophoresis** separates molecules based on their movement through a gelatinous substance (a gel) in an electric field applied across the gel. Molecules move through the gel—and thus are separated—according to their size and their charge. For example, if the detergent used to solubilize (dissolve) proteins carries a negative charge, then detergent-protein complexes will migrate toward the positively charged end of the gel. Larger detergent-protein complexes migrate more slowly than do smaller molecules, so the various proteins isolated from a plasma membrane separate from each other with time. After a voltage has been applied to a gel long enough for the protein molecules to become separated, the voltage is turned off and the gel is stained with a dye that adheres to proteins. Each of the proteins that was in the original solution has migrated to a specific position and appears as a band on the gel.

To obtain a pure sample of a particular protein, the appropriate band is cut out of the gel. The gel material is then dissolved to retrieve the protein. Once this protein is inserted into a planar bilayer or liposome, dozens of different experiments are possible.

How Do Membrane Proteins Affect Ions and Molecules?

In the 40 years since intensive experimentation on membrane proteins began, researchers have identified three broad classes of peptides or proteins—channels, transporters, and pumps—that affect membrane permeability. (Recall from Chapter 3 that peptides are proteins containing fewer than 50 amino acids.) What do these molecules do? Can plasma membranes that contain these proteins create an internal environment more conducive to life than the external environment is?

Facilitated Diffusion via Channel Proteins   One of the first membrane peptides investigated in detail is called gramicidin. Gramicidin is produced by a bacterium called *Bacillus brevis* and is used medicinally as an antibiotic ("against-life")—a substance that kills or inhibits the growth of bacteria. *Bacillus brevis* produces gramicidin as a weapon: *B. brevis* cells release the protein just before a resistant coating forms around their cell wall and membrane. The gramicidin wipes out competitors, giving *B. brevis* cells more room to grow when they emerge from the resistant phase. Gramicidin also affects human cells and is extremely toxic if taken internally. It can be used only to fight skin infections.

After observing that experimental cells treated with gramicidin seemed to lose large numbers of ions, researchers became interested in understanding how the molecule works. Could this protein alter the flow of ions across plasma membranes?

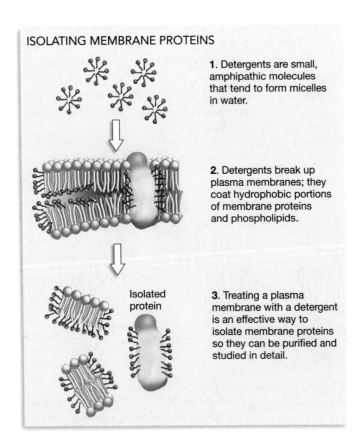

ISOLATING MEMBRANE PROTEINS

1. Detergents are small, amphipathic molecules that tend to form micelles in water.

2. Detergents break up plasma membranes; they coat hydrophobic portions of membrane proteins and phospholipids.

Isolated protein

3. Treating a plasma membrane with a detergent is an effective way to isolate membrane proteins so they can be purified and studied in detail.

FIGURE 6.22 Detergents Can Be Used to Get Membrane Proteins into Solution

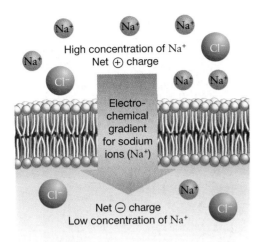

FIGURE 6.23 Electrochemical Gradients
When ions build up on one side of a membrane, they establish a combined concentration and electrical gradient. EXERCISE Add an arrow indicating the electrochemical gradient for chloride ions.

Biologists answered this question by inserting purified gramicidin into planar bilayers. The experiment they performed was based on an important fact about ion movement across membranes: Ions flow not only from regions of high concentration to regions of low concentration but also from areas of like charge to areas of unlike charge (**Figure 6.23**). Stated another way, ions move in response to a combined concentration and electrical gradient, or what biologists call an **electrochemical gradient**.

To determine whether gramicidin affected the membrane's permeability to ions, the researchers measured the flow of electric current across the membrane. Because ions carry a charge, the movement of ions produces an electric current. This property provides an elegant and accurate test for assessing the bilayer's permeability to ions—one that is simpler and more sensitive than taking samples from either side of the membrane and determining the concentrations of solutes present. If gramicidin facilitates ion movement, then an investigator should be able to detect an electric current across planar bilayers that contain gramicidin.

The result? The graph in **Figure 6.24** shows that when gramicidin was absent, no electric current passed through the membrane. But when gramicidin was inserted into the membrane, current began to flow. Based on this observation, biologists proposed that gramicidin is an **ion channel**. An ion channel is a peptide that makes lipid bilayers permeable to ions. Follow-up work corroborated that gramicidin is selective. It allows only positively charged ions, or cations, to pass. Gramicidin does not allow negatively charged ions, or anions, to pass through the membrane. It was also established that gramicidin is most permeable to hydrogen ions (or protons, H^+) and somewhat less permeable to other cations, such as potassium (K^+) and sodium (Na^+).

Researchers gained additional insight into the way gramicidin works when they determined its amino acid sequence

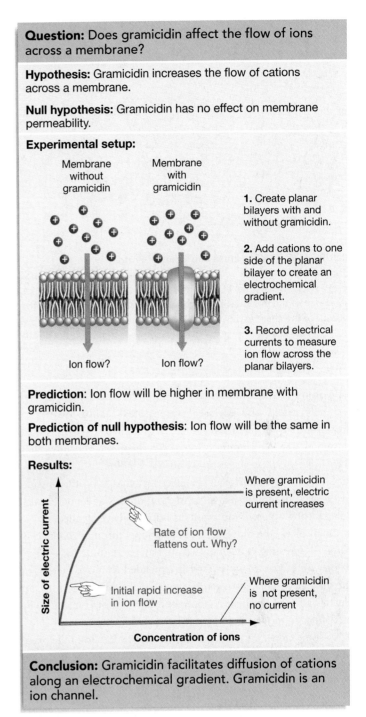

Question: Does gramicidin affect the flow of ions across a membrane?

Hypothesis: Gramicidin increases the flow of cations across a membrane.

Null hypothesis: Gramicidin has no effect on membrane permeability.

Experimental setup:

Membrane without gramicidin   Membrane with gramicidin

1. Create planar bilayers with and without gramicidin.

2. Add cations to one side of the planar bilayer to create an electrochemical gradient.

Ion flow?   Ion flow?

3. Record electrical currents to measure ion flow across the planar bilayers.

Prediction: Ion flow will be higher in membrane with gramicidin.

Prediction of null hypothesis: Ion flow will be the same in both membranes.

Results:

Where gramicidin is present, electric current increases

Rate of ion flow flattens out. Why?

Initial rapid increase in ion flow

Where gramicidin is not present, no current

Size of electric current (y-axis)
Concentration of ions (x-axis)

Conclusion: Gramicidin facilitates diffusion of cations along an electrochemical gradient. Gramicidin is an ion channel.

FIGURE 6.24 Measuring Ion Flow through the Channel Gramicidin
Experiment for testing the hypothesis that gramicidin is an ion channel.

(that is, primary structure) and tertiary structure. **Figure 6.25,** page 120, provides a view from the outside of a cell to the inside through gramicidin. The key observation is that the molecule forms a hole. The portions of amino acids that line this hole are hydrophilic, while regions on the exterior (in contact with the membrane phospholipids) are hydrophobic. The molecule's structure correlates with its function.

Cells have many different types of channels in their membranes, each with a structure that allows it to admit a particular

(a) Top view of gramicidin

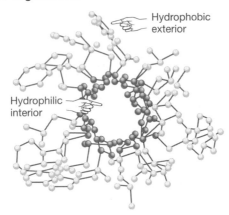

Hydrophobic exterior

Hydrophilic interior

(b) Side view of gramicidin

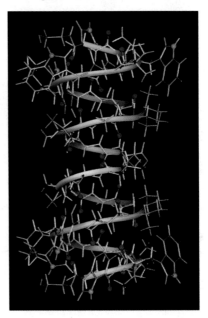

FIGURE 6.25 The Structure of a Channel Protein
Gramicidin is an α-helix consisting of only 15 amino acids. **(a)** In top view, the molecule forms a hole or pore. **(b)** In side view, a green helix traces the peptide-bonded backbone of the polypeptide. R-groups hang off the backbone to the outside. The interior of the channel is hydrophilic; the exterior is hydrophobic. EXERCISE In (a) and (b), add symbols indicating the locations of phospholipids relative to gramicidin in a plasma membrane.

type of ion or small molecule. For example, recently discovered channels called **aquaporins** ("water-pores") allow water to cross the plasma membrane rapidly. Recent research has also shown that most channels are **gated channels**—meaning that they open or close in response to the binding of a particular molecule or to a change in the electrical charge on the outside of the membrane. In this way, the flow of ions and small molecules through membrane channels is carefully controlled.

In all cases, the movement of substances through channels is passive—meaning it does not require an expenditure of energy. In **passive transport**, no energy is expended to move materials; the change in concentration is powered by diffusion along an electrochemical gradient. Channel proteins enable ions or polar molecules to move across lipid bilayers efficiently.

In some cases, however, passive transport does not take place through a channel. As **Figure 6.26** shows, the antibiotic valino-

mycin acts as an **ionophore** ("ion-mover") instead. The protein binds to a potassium (K^+) ion on the membrane's exterior, diffuses across the bilayer, and releases the cation on the interior.

Membrane proteins such as gramicidin and valinomycin circumvent the lipid bilayer's impermeability to small, charged compounds. As a result, their presence *reduces* differences between the interior and exterior. Water molecules and ions are not the only substances that move across membranes through membrane proteins, however. Larger molecules can, too.

Facilitated Diffusion via Transport Proteins Next to ribose, the six-carbon sugar glucose is the most important sugar found in organisms. Glucose was very likely important to the earliest cells simply because virtually all cells alive today use it as a building block for important macromolecules and as a source of chemical energy. But as we saw in Figure 6.9, lipid

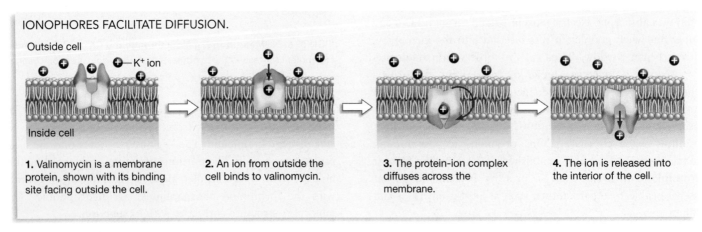

IONOPHORES FACILITATE DIFFUSION.

Outside cell

K^+ ion

Inside cell

1. Valinomycin is a membrane protein, shown with its binding site facing outside the cell.

2. An ion from outside the cell binds to valinomycin.

3. The protein-ion complex diffuses across the membrane.

4. The ion is released into the interior of the cell.

FIGURE 6.26 The Ionophore Valinomycin
Valinomycin is an ionophore that does not form a channel. It acts instead as a mobile ion carrier.

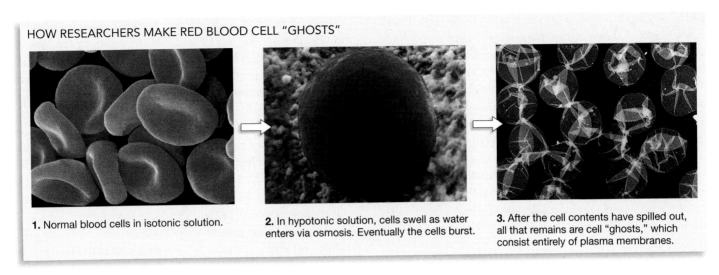

HOW RESEARCHERS MAKE RED BLOOD CELL "GHOSTS"

1. Normal blood cells in isotonic solution.

2. In hypotonic solution, cells swell as water enters via osmosis. Eventually the cells burst.

3. After the cell contents have spilled out, all that remains are cell "ghosts," which consist entirely of plasma membranes.

FIGURE 6.27 Red Blood Cell "Ghosts"
Red blood cell ghosts are simple membranes that can be purified and studied in detail.

bilayers are only moderately permeable to glucose. It is reasonable to expect, then, that plasma membranes have some mechanism for increasing their permeability to this sugar.

This expectation grew into a certainty when researchers compared the permeability of glucose across planar bilayers with its permeability across membranes from cells. The plasma membrane in this study came from human red blood cells, which are among the simplest cells known. Mature red blood cells contain a membrane, about 300 million hemoglobin molecules, and not much else (**Figure 6.27**, step 1). When these cells are placed in a hypotonic solution (step 2), water rushes into them by osmosis. As water flows inward, the cells swell. Eventually they burst, releasing the hemoglobin molecules and other cell contents. This leaves researchers with pure preparations of plasma membranes called red blood cell "ghosts" (step 3). Experiments have shown that these membranes are much more permeable to glucose than are pure lipid bilayers. Why?

After isolating and analyzing many proteins from red blood cell ghosts, researchers found one protein that specifically increases membrane permeability to glucose. When this purified protein was added to liposomes, the artificial membrane transported glucose at the same rate as a membrane from a living cell. This experiment confirmed that a membrane protein was indeed responsible for transporting glucose across plasma membranes. Follow-up work showed that this glucose carrier, a **transport protein** named GLUT-1, transports the "right-handed" optical isomer of glucose but not the left-handed form. This is logical because cells use only the right-handed form of glucose.

Exactly how GLUT-1 works is a focus of ongoing research. Because glucose transport by GLUT-1 is so specific, researchers presume that the mechanism resembles the action of enzymes. One hypothesis is illustrated in **Figure 6.28**. The idea is that glucose binds to GLUT-1 on the exterior of the membrane and that this binding induces a conformational change in the pro-

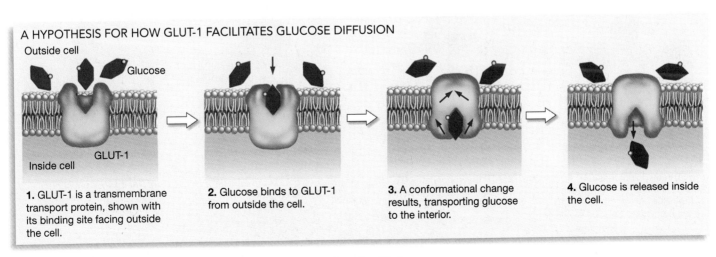

A HYPOTHESIS FOR HOW GLUT-1 FACILITATES GLUCOSE DIFFUSION

Outside cell

Glucose

GLUT-1

Inside cell

1. GLUT-1 is a transmembrane transport protein, shown with its binding site facing outside the cell.

2. Glucose binds to GLUT-1 from outside the cell.

3. A conformational change results, transporting glucose to the interior.

4. Glucose is released inside the cell.

FIGURE 6.28 A Hypothesis to Explain How Membrane Transport Proteins Work
This model suggests that the GLUT-1 transporter acts like an enzyme. It binds a substrate (in this case, a glucose molecule), undergoes a conformation change, and releases the substrate.

tein that transports glucose to the interior of the cell. Recall from Chapter 3 that enzymes frequently change shape when they bind substrates and that such conformational changes are often a critical step in the catalysis of chemical reactions.

Importing molecules into cells via transport proteins is still powered by diffusion, however. When glucose enters a cell via GLUT-1, it does so because it is following its concentration gradient. The movement is called **facilitated diffusion,** because it does not require an expenditure of energy for movement to occur and because it is facilitated—aided—by the presence of a transport protein or **carrier.** If the concentration of glucose is the same on both sides of the plasma membrane, then no net movement of glucose occurs even if the membrane contains GLUT-1.

Active Transport by Pumps   Whether diffusion occurs via channel proteins or is facilitated by carriers, it is a passive process that makes the cell interior and exterior more similar. But it is also possible for today's cells to import molecules or ions *against* their electrochemical gradient. Accomplishing this task requires energy, however, because the cell must counteract the entropy loss that occurs when molecules are concentrated. It makes sense, then, that transport against a concentration gradient is called **active transport.** The energy required to move substances against their electrochemical gradient is provided by phosphate groups in adenosine triphosphate, or ATP. In addition to energy, active transport requires a machine—a molecule that is capable of using energy to pump an ion or a molecule against its electrochemical gradient.

Figure 6.29 shows how chemical energy from ATP and a membrane protein called a pump accomplish this task. The figure highlights the first pump that was discovered and char-

acterized: a protein called the **sodium-potassium pump,** or more formally **Na$^+$/K$^+$-ATPase.** *Na$^+$/K$^+$* refers to the ions that are transported; *ATP* indicates that adenosine triphosphate is used; and *–ase* implies that the molecule acts like an enzyme. Unlike the situation with GLUT-1, the mechanism of action in Na$^+$/K$^+$-ATPase is now well known. When the protein is in the conformation shown in step 1 of Figure 6.29, three sodium ions from the inside of the cell bind to it. The protein then becomes phosphorylated by the addition of a phosphate group from ATP (step 2) and changes shape in a way that sends these ions to the exterior. This change in conformation also allows two potassium ions from the solution outside the cell to bind to the protein (step 3). When the phosphate group drops off inside the cell, the protein changes shape again and releases the K$^+$ to the interior (step 4). This movement of ions can occur even if an electrochemical gradient exists that favors the outflow of potassium and the inflow of sodium. By exchanging three sodium ions for every two potassium ions, the outside of the membrane becomes positively charged relative to the inside. In this way, the sodium-potassium pump sets up an electrical gradient as well as a chemical gradient across the membrane.

Similar pumps are specialized for moving protons (H$^+$), calcium ions (Ca^{2+}), or other ions or molecules. In this way, cells are capable of concentrating certain substances or setting up electrochemical gradients.

Taken together, the lipid bilayer and the proteins involved in passive and active transports enable cells to create an internal environment that is much different from the external one (**Figure 6.30**). When membrane proteins first evolved, then, the early cells acquired the ability to create an internal environment that was conducive to life—meaning that such

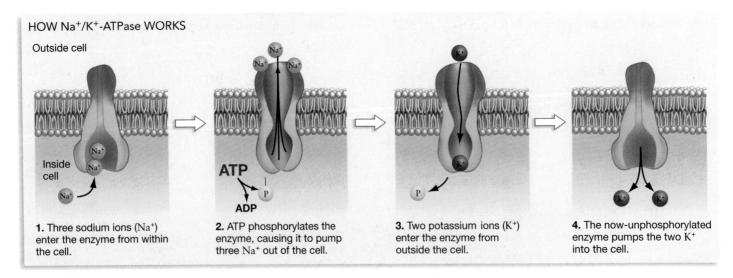

HOW Na$^+$/K$^+$-ATPase WORKS

Outside cell

Inside cell

ATP

ADP

P

P

1. Three sodium ions (Na$^+$) enter the enzyme from within the cell.

2. ATP phosphorylates the enzyme, causing it to pump three Na$^+$ out of the cell.

3. Two potassium ions (K$^+$) enter the enzyme from outside the cell.

4. The now-unphosphorylated enzyme pumps the two K$^+$ into the cell.

FIGURE 6.29 Active Transport Depends on an Input of Chemical Energy
If chemical energy in the form of a phosphate group is added to the sodium-potassium pump, the pump can transport ions against their electrochemical gradient. The mechanism involves a change in the protein's shape.

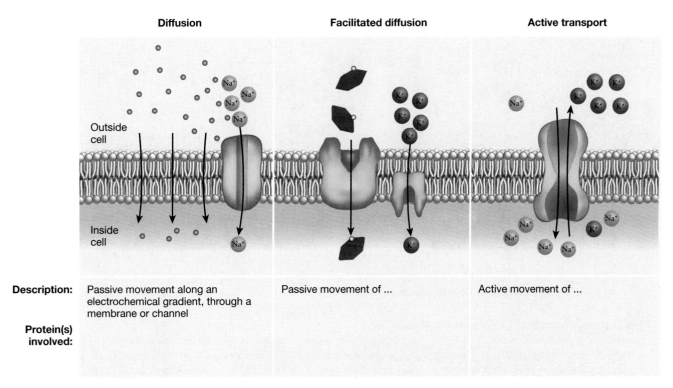

FIGURE 6.30 Mechanisms of Membrane Transport: A Summary
EXERCISE Complete the chart.

an environment contained the substances required for manufacturing ATP and copying ribozymes. Cells with particularly efficient and selective membrane proteins would be favored by natural selection and would come to dominate the population. Cellular life had begun.

Some 3.5 billion years later, cells continue to evolve. What do today's cells look like, and how do they produce and store the chemical energy that makes life possible? How do they use ATP to move pumps and channels and other molecules and machines where they're needed? Answering these and related questions is the focus of Unit 2.

✓ CHECK YOUR UNDERSTANDING

Membrane proteins allow ions and molecules that ordinarily do not readily cross lipid bilayers to enter or exit cells. Substances may move along an electrochemical gradient via facilitated diffusion through channel proteins or transport proteins, or they may move against an electrochemical gradient in response to work done by pumps. You should be able to (1) sketch a phospholipid bilayer and (2) indicate how ions and large molecules cross it via each major type of membrane transport protein.

ESSAY The Molecular Basis of Cystic Fibrosis

Cystic fibrosis (CF) is the most common genetic disease in humans of Northern European descent. In these populations, one in every 2500 infants born has CF. Children affected by CF suffer from a progressive deterioration of their lungs and gastrointestinal and reproductive tracts. Although the ability of physicians to manage the disease has improved dramatically in recent years, the median life expectancy for patients with severe forms of CF is still only 32.*

*Half of the numbers in a distribution are found above the median value, and half are found below. A median life expectancy of 32 years means that half of the people with CF live less than 32 years and half live more than 32 years.

CF produces a distinctive set of symptoms. Prominent among them are the production of salty sweat and the chronic development of thickened mucus in the linings of the lungs and associated air passages. The thickened mucus, which is extremely difficult to cough out, leads to two important complications: the direct obstruction of airways and the rapid growth of disease-causing bacteria.

Biologists now understand the molecular basis of cystic fibrosis.

(Continued on next page)

(Essay continued)

Ion movement in a
normal sweat duct:

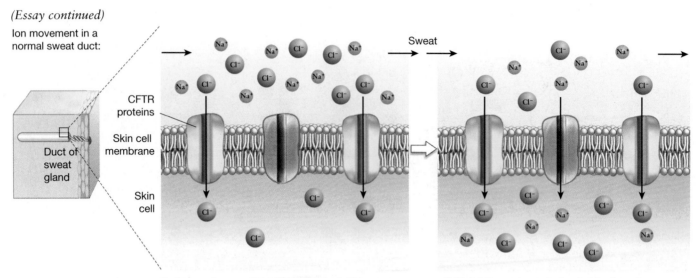

1. As sweat moves through sweat duct, Cl⁻ moves from sweat through CFTR proteins into skin cell.

2. Na⁺ follows along electrochemical gradient; sweat becomes less salty.

FIGURE 6.31 The CFTR Protein Is a Chloride Ion Channel
In the sweat ducts found in human skin, the CFTR protein acts as an ion channel that allows chloride ions to be reabsorbed from sweat. The movement of chloride ions sets up a charge gradient that also makes sodium ions move out of sweat before it reaches the skin surface. People with CF produce an abnormal form of the CFTR protein, so this movement cannot occur.
EXERCISE Cross out the CFTR protein in the right-hand drawing to represent the situation in an individual with CF. Then redraw the distribution of ions on both sides of the membrane.

What causes these symptoms? Understanding the molecular basis of CF frustrated researchers for decades. Then an important clue came to light in 1983, when P. M. Quinton showed that sweat-duct cells taken from individuals with CF are impermeable to chloride ions (Cl⁻). Sweat is produced as a salty solution. But as the solution moves through a sweat duct toward the surface of the skin, chloride ions pass through the membrane and are reabsorbed by the body (**Figure 6.31**). This movement establishes an electrochemical gradient that causes sodium ions (Na⁺) to follow. But in people with CF, both ions stay in the sweat duct and are excreted. As a result, these individuals have extraordinarily salty sweat.

Quinton's result suggested that the disease might be caused by a defect in a membrane protein—one that normally allows chloride ions to move across lipid bilayers. Nine years later, Christine Bear and colleagues confirmed this result in spectacular fashion. Bear's group was able to isolate and purify quantities of the normal protein, which had come to be called cystic fibrosis transmembrane conductance regulator (CFTR). The researchers inserted the protein into planar bilayers and showed that it did indeed selectively permit the passage of chloride ions. Subsequent studies showed that people with CF produce abnormal forms of the protein. As a result, their chloride channels do not function.

Biologists now understand the molecular basis of cystic fibrosis. Cells that line the respiratory tract of humans normally release a steady stream of chloride ions, followed by sodium ions. The high concentration of ions on the outside of the cell leads to an outflow of water. But chloride ions are not released

in CF individuals, and water remains inside the cells. The result is dramatically thickened mucus in the lungs and air passages, contributing to this fatal syndrome (**Figure 6.32**).

The research that identified the protein's mode of action unleashed a torrent of new studies aimed at finding a cure. As this book goes to press, that work is still in progress. One strategy being tested is to synthesize large quantities of normal protein, package the protein channels into liposomes, and then deliver them to the respiratory lining of CF patients via an inhalator. The hope is that the liposomes will fuse with the plasma membranes, introduce copies of CFTR that can function normally, and alleviate the symptoms of CF.

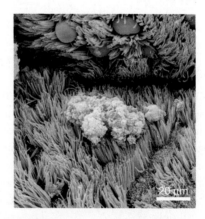

FIGURE 6.32 CF Patients Have Thickened Mucus in Their Lungs
Scanning electron micrograph of an airway in the lung of a CF patient. The hairlike structures are cilia; the pink mass is mucus.

CHAPTER REVIEW

Summary of Key Concepts

- **Phospholipids have a hydrophilic region and a hydrophobic region. In solution, they spontaneously form bilayers that are selectively permeable—meaning that only certain substances cross them readily.**

  The plasma membrane is a structure that forms a physical barrier between life and nonlife. Molecules called lipids, which do not dissolve in water, are a prominent component of plasma membranes. Lipids called fatty acids and isoprenes, both of which have long hydrocarbon tails, are often found in larger molecules called phospholipids. Phospholipids have a polar head and nonpolar tail. The basic structure of plasma membranes is created by a phospholipid bilayer.

  Small, nonpolar molecules tend to move across membranes readily; ions and other charged compounds cross rarely, if at all. The permeability and fluidity of lipid bilayers depend on temperature and on the types of phospholipids present. For example, because phospholipids that contain long, saturated fatty acids form a dense and highly hydrophobic membrane interior, they tend to be less permeable than phospholipids containing shorter, unsaturated fatty acids.

- **Ions and molecules spontaneously diffuse from regions of high concentration to regions of low concentration. In osmosis, a special case of diffusion, water moves across a selectively permeable membrane from regions of high concentration to regions of low concentration.**

  Molecules can move across membranes spontaneously because of diffusion. Diffusion is the directed movement of ions or molecules from a region of high concentration to a region of low concentration; it is driven by an increase in entropy. Water also moves across membranes spontaneously if a molecule or an ion that cannot cross the membrane is found in different concentrations on the two sides. In osmosis, water moves from the region with a lower concentration of solutes to the region of higher solute concentration. Osmosis is a passive process driven by an increase in entropy.

  Web Tutorial 6.1 Diffusion and Osmosis

- **In cells, membrane proteins are responsible for the passage of ions, polar molecules, and large molecules that do not readily cross phospholipid bilayers on their own.**

  The permeability of lipid bilayers can be altered radically by membrane transport proteins, which are scattered throughout the plasma membrane. Channel proteins, for example, are molecules that provide holes in the membrane and facilitate the diffusion of specific ions into or out of the cell. Transport proteins are enzyme-like proteins that allow specific molecules to diffuse into the cell. In addition to these forms of facilitated diffusion, membrane proteins that act as energy-demanding pumps actively move ions or molecules against their electrochemical gradient. In combination, the selective permeability of phospholipid bilayers and the specificity of transport proteins make it possible to create an environment inside a cell that is radically different from the exterior.

  Web Tutorial 6.2 Membrane Transport Proteins

Questions

Content Review

1. What does the term *hydrophilic* mean when it is translated literally?
   a. "oil loving"
   b. "water loving"
   c. "oil fearing"
   d. "water fearing"

2. If a solution surrounding a cell is hypotonic relative to the inside of the cell, how will water move?
   a. It will move into the cell via osmosis.
   b. It will move out of the cell via osmosis.
   c. It will not move, because equilibrium exists.
   d. It will evaporate from the cell surface more rapidly.

3. If a solution surrounding a cell is hypertonic relative to the inside of the cell, how will water move?
   a. It will move into the cell via osmosis.
   b. It will move out of the cell via osmosis.
   c. It will not move, because equilibrium exists.
   d. It will evaporate from the cell surface more rapidly.

4. When does a concentration gradient exist?
   a. when membranes rupture
   b. when solute concentrations are high
   c. when solute concentrations are low
   d. when solute concentrations differ on the two sides of a membrane

5. Which of the following must be true for osmosis to occur?
   a. Water must be at room temperature or above.
   b. Solutions with the same concentration of solutes must be separated by a selectively permeable membrane.
   c. Solutions with different concentrations of solutes must be separated by a selectively permeable membrane.
   d. Water must be under pressure.

6. Why are the lipid bilayers in cells called "selectively permeable"?
   a. They are not all that permeable.
   b. Their permeability changes with their molecular composition.
   c. Their permeability is temperature dependent.
   d. They are permeable to some substances but not others.

Conceptual Review

1. Cooking oil is composed of lipids that consist of long hydrocarbon chains. These are not amphipathic molecules. Would you expect these lipids to form membranes spontaneously? Why or why not? Describe, on a molecular level, how you would expect these lipids to interact with water. Your answer should explain the saying, "oil and water don't mix."

2. Ethanol, the active ingredient in alcoholic beverages, is a small, polar, uncharged molecule. Would you predict that this molecule crosses plasma membranes quickly or slowly? Explain your reasoning.

3. Why can osmosis occur only if solutions are separated by a selectively permeable membrane?

4. The text claims that the portion of membrane proteins that spans the hydrophobic tails of phospholipids is itself hydrophobic (see Figure 6.18b). Why does this make sense? Look back at Figure 3.5 and Table 3.1, and make a list of amino acids you would expect to find in these regions of transmembrane proteins.

5. Describe how the structure of phospholipids allows them to function as a plasma membrane. Explain how the amphipathic structure of some proteins allows them to function as integral membrane proteins.

6. Examine the membrane in the accompanying figure. Label the molecules and ions that will pass through the membrane as a result of osmosis, diffusion, and facilitated diffusion. Draw arrows to indicate where each of the molecules and ions will travel.

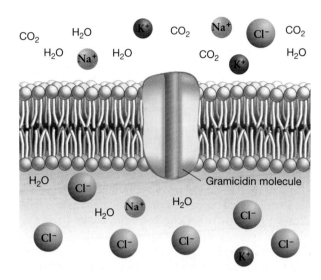

Gramicidin molecule

Group Discussion Problems

1. When phospholipids are arranged in a bilayer, it is theoretically possible for individual molecules in the bilayer to flip-flop. That is, a phospholipid could turn 180° and become part of the membrane's other surface. Sketch this process. Based on what you know about the behavior of polar heads and nonpolar tails, predict whether flip-flops are frequent or rare. Then design an experiment, using a planar bilayer made up partly of fatty acids that contain a dye molecule on their hydrophilic head, to test your prediction.

2. Unicellular organisms that live in extremely cold habitats have an unusually high proportion of unsaturated fatty acids in their plasma membranes. Some of these membranes even contain polyunsaturated fatty acids, which have more than one double bond in each hydrocarbon chain. Researchers have proposed that the lack of saturation helps these membranes maintain a semifluid state at low temperatures instead of becoming semisolid or solid. Draw a picture of this type of membrane, analogous to the sketch in Figure 6.11. Using this picture as a resource, comment on the hypothesis that membranes with unsaturated fatty-acid tails function better at cold temperatures than do membranes with saturated fatty-acid tails. How would you test this hypothesis? Make a prediction about the structure of fatty acids found in organisms that live in extremely hot environments.

3. When biomedical researchers design drugs that must enter cells to be effective, they sometimes add methyl (CH_3) groups to make the drug molecules more likely to pass through plasma membranes. Conversely, when researchers design drugs that act on the exterior of plasma membranes, they sometimes add a charged group to decrease the likelihood that the drugs will pass through membranes and enter cells. Explain why these strategies make sense.

4. Advertisements frequently claim that laundry and dishwashing detergents "cut grease." What the ad writers mean is that the detergents surround oil droplets on clothing or dishes, making the droplets water soluble. When this happens, the oil droplets can be washed away. Explain how this happens on a molecular level.

Answers to Multiple-Choice Questions **1.** b; **2.** a; **3.** b; **4.** d; **5.** c; **6.** d

UNIT 2
Cell Structure and Function

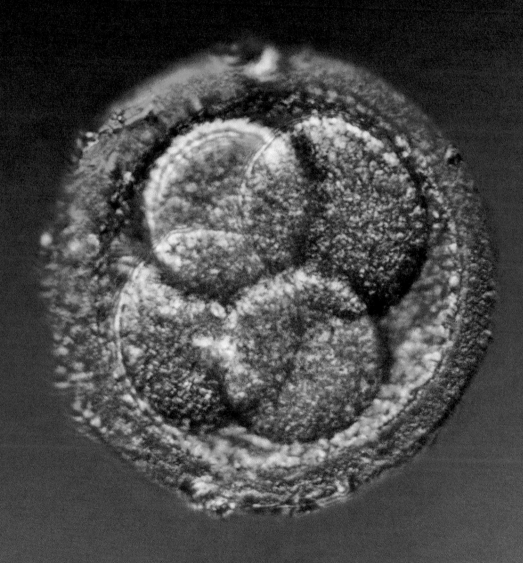

The four cells in this two-day-old human embryo will give rise to the trillions of cells in the adult body.

7 Inside the Cell

- The structure of cell components is closely correlated with their function.

- Inside cells, materials are transported to their destinations with the help of molecular "zip codes."

- Cells are dynamic. Thousands of chemical reactions occur each second within cells; molecules constantly enter and exit across the plasma membrane; cell products are shipped along protein fibers; and elements of the cell's internal skeleton grow and shrink.

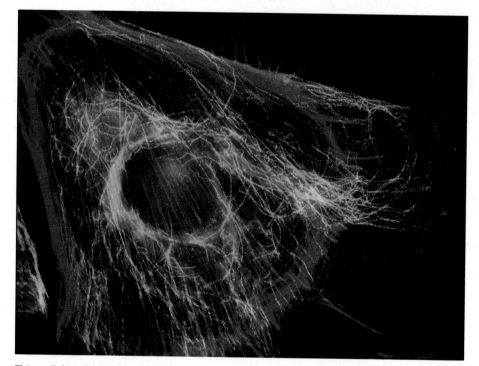

This cell has been treated with fluorescing molecules that bind to its fibrous skeleton. Microtubules (large protein fibers) are green; microfilaments (smaller fibers) are red. The cell's nucleus has been stained blue.

In Chapter 1 you were introduced to the cell theory, which states that all organisms consist of cells and that all cells are derived from preexisting cells. Since this theory was initially developed and tested in the 1850s and 1860s, an enormous body of research has confirmed that the cell is the fundamental structural and functional unit of life. Life on Earth is cellular.

In a very real sense, then, understanding how an organism works is a matter of understanding how cells are structured and how they function. To drive this point home, recall from Chapter 1 that many eukaryotic organisms and virtually all bacteria and archaea are unicellular. In number of individuals present, unicellular organisms dominate life on Earth. For researchers who study these species, understanding the cell is synonymous with understanding the organism as a whole. Even in plants, animals, and other multicellular eukaryotes, complex behavior originates at the level of the cell. For example, your

ability to read this page begins with changes in light-sensitive molecules located in cells at the back of your eyes. When these molecules change shape, they trigger changes in the membranes of nerve cells that connect your eyes to your brain. To understand complex processes such as vision, then, researchers often begin by studying the structure and function of the individual cells involved—the parts that make up the whole.

Chapter 6 introduced an essential part of the cell: the **plasma membrane**. Thanks to the selective permeability of phospholipid bilayers and the activity of membrane transport proteins, this structure creates an internal environment that is different from conditions outside the cell. Our task now is to explore the structures that are found inside the membrane and analyze what they do. We'll focus on several particularly dynamic structures and processes and introduce some of the experimental approaches that biologists use to understand them. Let's begin by surveying the basic types of cells, cell structures, and cell processes that biologists have documented to date.

7.1 What's Inside the Cell?

In Chapter 1 you read about the two fundamental types of cells observed in nature. Recall that eukaryotic cells have a membrane-bound compartment called a nucleus, while prokaryotic cells do not. In terms of *morphology* ("form"), then, species fall into the two broad categories: (1) prokaryotes and (2) eukaryotes. But in terms of *phylogeny*, or evolutionary history, organisms fall into the three broad groups called (1) Bacteria, (2) Archaea, and (3) Eukarya. Members of the Bacteria and Archaea are prokaryotic; members of the Eukarya—including algae, fungi, plants, and animals—are eukaryotic.

In the late seventeenth century, biologists began studying the structure of cells with microscopes. Over time, improvements in optics and cell preparation techniques allowed researchers to catalogue the structures reviewed in this section. When electron microscopes became widely available in the 1950s, investigators described the internal anatomy of these structures in more detail. More recent advances in microscopy have allowed investigators to videotape certain types of cell processes in living cells.

What have anatomical studies based on microscopy revealed? Let's look first at the general anatomy of prokaryotic cells and eukaryotic cells, and then consider how the structures that have been identified help cells function.

Prokaryotic Cells

Figure 7.1 shows the general structure of a prokaryotic cell. For most bacterial and archaeal species, the plasma membrane encloses a single compartment—meaning that the cell has few or no substructures delimited by internal membranes. Closer examination reveals a series of intricate structures, however. Let's take a look at a typical prokaryotic cell, starting at the outside and working in.

As Chapter 6 pointed out, the cell membrane, or plasma membrane, consists of a phospholipid bilayer and proteins that either span the bilayer or attach to one side. Inside the membrane, the contents of a cell are collectively termed the **cytoplasm** ("cell-formed"). Because the cytoplasm contains a high concentration of solutes, in most habitats it is hypertonic relative to the surrounding environment. When this is the case, water enters the cell via osmosis and makes the cell's volume expand. In virtually all bacteria and archaea, this pressure is resisted by a stiff **cell wall**. Bacterial and archaeal cell walls are a tough, fibrous layer that surrounds the plasma membrane. In many species the cell wall is made of a carbohydrate-protein complex called *peptidoglycan* or related substances (**Figure 7.2**). The pressure of the plasma membrane against the cell wall is about the same as the pressure in an automobile tire. The cell wall protects the organism and gives it shape and rigidity. In addition, many

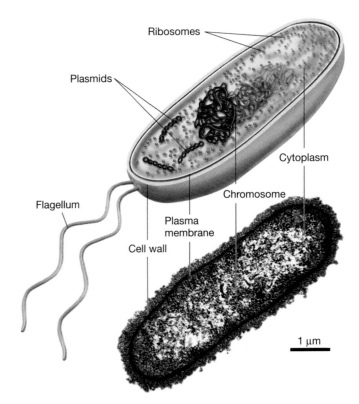

FIGURE 7.1 A Prokaryotic Cell
Prokaryotic cells are identified by a negative trait—the absence of a membrane-bound nucleus. Although there is wide variation in the size and shape of bacterial and archaeal cells, all such cells contain a plasma membrane, chromosomes, and ribosomes; almost all have a stiff cell wall. Some prokaryotes have flagella and/or inner membranes where photosynthesis takes place.

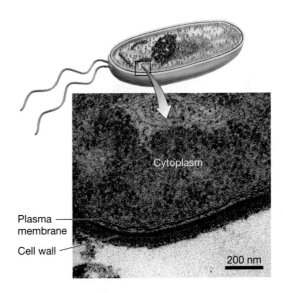

FIGURE 7.2 The Bacterial Cell Wall
In bacteria and archaea, the cell wall consists of peptidoglycan or similar polymers that are cross-linked into tough sheets. The inside of the cell wall contacts the plasma membrane, which pushes up against the wall. The outside of the cell wall makes direct contact with the outside environment, which is almost always filled with competitors and predators.

bacteria have an additional protective layer outside the cell wall that consists of lipids with polysaccharides attached. Lipid molecules that contain carbohydrate groups are termed **glycolipids**.

In the cytoplasm of a prokaryotic cell, the most prominent structure is the chromosome. The prokaryotic **chromosome** consists of a large DNA molecule associated with a small number of proteins. In most species there is a single, circular chromosome, but other species have several circular chromosomes, and a few species—including the bacterium that causes Lyme disease—have one to several linear chromosomes. In prokaryotes and all other organisms, the sequence of nitrogenous bases in DNA acts as a code that contains the genetic, or heredity, information. Stated another way, the primary structure of DNA contains the instructions for making the proteins and other molecules needed by the cell. A **gene** is a segment of DNA that contains the information for building an RNA molecule or a polypeptide. Genes are components of chromosomes.

Prokaryotic chromosomes are found in a localized area of the cell called the **nucleoid**. The genetic material is not separated from the rest of the cytoplasm by a membrane, however. In the well-studied bacterium *Escherichia coli*, the circular chromosome is 500 times longer than the cell itself. This situation is typical in prokaryotes. To fit into the cell, the DNA double helix coils on itself with the aid of enzymes to form the highly compact, "supercoiled" structure shown in **Figure 7.3**. Supercoiled regions of DNA resemble a rubber band that has been held at either end and then twisted.

Depending on the species and population being considered, prokaryotic cells may also contain one to about a hundred small, usually circular, supercoiled DNA molecules called **plasmids**. Plasmids contain genes but are physically independent of the main,

cellular chromosome. In most cases the genes carried by plasmids are not required under normal conditions; instead they help cells adapt to unusual circumstances, such as the sudden presence of a poison in the environment. As a result, plasmids can be considered auxiliary genetic elements. They are copied independently of the main chromosome and are passed along to daughter cells when the parent cell divides. Certain plasmid genes also allow a copy of the entire plasmid to be transferred from one cell to another. As a result, plasmids can spread through a population or even be passed between species. Plasmids have been studied intensively because some carry genes that confer resistance to antibiotics. One recently characterized plasmid carries genes that provide resistance to seven distinct antibiotics.

Two other prominent cell structures found in prokaryotes are **ribosomes**, which manufacture proteins, and **flagella** (singular: *flagellum*), which power movement. Ribosomes are observed in all prokaryotic cells and are found throughout the cytoplasm. Bacterial ribosomes are complex structures consisting of a total of three distinct RNA molecules and over 50 different proteins. These molecular components are organized into two major structural elements, called the large subunit and small subunit (**Figure 7.4**). It is not unusual for a single cell to contain 10,000 ribosomes. Both ribosomes and prokaryotic flagella lack a membrane. Not all bacterial species have flagella, however. When present they are usually few in number and are located on the surface of the cell. Over 40 different proteins are involved in building and controlling bacterial flagella. At top speed, flagellar movement can drive a bacterial cell through water at 60 cell lengths per second. In contrast, the cheetah qualifies as the fastest land animal but can sprint at a mere 25 body lengths per second.

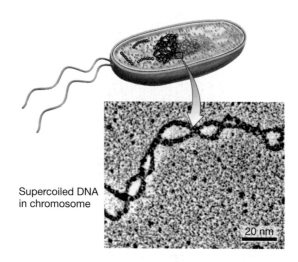

FIGURE 7.3 Bacterial DNA Is Supercoiled
The circular chromosomes of bacteria and archaea must be coiled extensively, into "supercoils," to fit in the cell. For example, the *Escherichia coli* chromosome would be more than 1 mm long if it were linear. This is about 500 times the length of the cell itself.

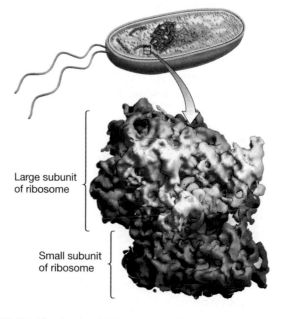

FIGURE 7.4 The Bacterial Ribosome
Bacterial ribosomes are made of RNA and protein molecules that are organized into large and small subunits.

Prokaryotes lack a nucleus, but it is not correct to say that no membrane-bound structures ever occur inside these cells. Many species contain membrane-bound storage containers, and extensive internal membranes occur in bacteria and archaea that perform photosynthesis. The photosynthetic membranes arise as invaginations of the plasma membrane. As the plasma membrane folds in, either vesicles pinch off or the types of flattened stacks shown in **Figure 7.5** form. The internal parts of this membrane contain the enzymes and pigment molecules required to convert the kinetic energy in sunlight into chemical energy in the form of sugar.

In addition, recent research indicates that at least one bacterial species has an internal compartment that qualifies as an organelle ("little organ"). An **organelle** is a membrane-bound compartment in the cytoplasm that contains enzymes specialized for a particular function. The bacterial organelle that was just discovered has proton pumps in its membrane and an acidic environment inside, where calcium ions (Ca^{2+}) are stored.

Recent research has also shown that bacteria and archaea contain long, thin fibers that serve a structural role in the cell. All bacterial species, for example, contain fibers made from the protein FtsZ. These filaments are essential for cell division to take place. Some species also have protein filaments that help maintain cell shape. Protein filaments such as these form the basis of the **cytoskeleton** ("cell skeleton").

Even though internal membranes and some cytoskeletal components are found in prokaryotic cells, their extent pales in comparison with that in eukaryotes. When typical prokaryotic and eukaryotic cells are compared side by side, three outstanding differences jump out: (1) Eukaryotic cells are usually much larger; (2) they contain extensive amounts of internal membrane; and (3) they feature a diverse and dynamic cytoskeleton.

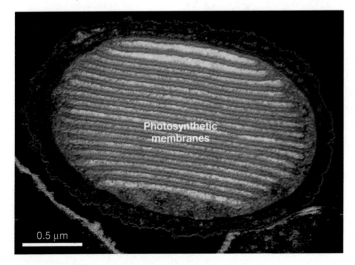

FIGURE 7.5 Photosynthetic Membranes in Bacteria
The green stripes in this photosynthetic bacterium are infoldings of the plasma membrane. They are green because they contain the pigments and enzymes required for photosynthesis.

Eukaryotic Cells

The lineage called Eukarya includes forms ranging from unicellular species to 100-meter-tall redwoods. Brown algae, red algae, fungi, amoebae, and slime molds are all eukaryotic, as are green plants and animals.

The first thing that strikes biologists about eukaryotic cells is how much larger they are on average than bacteria and archaea. Most eukaryotic cells range from about 5 to 100 μm in diameter, while most prokaryotic cells vary between 1 and 10 μm in diameter. A micrograph of an average eukaryotic cell, at the same scale as the bacterial cell in Figure 7.5, would fill this page. This difference in size inspired the hypothesis that when eukaryotes first evolved, they made their living by ingesting bacterial and archaeal cells whole. Stated another way, the evolution of large cell size is thought to have made it possible for eukaryotic cells to act as *predators*—organisms that kill and consume other organisms. Hundreds of eukaryotic species alive today still make their living by surrounding and taking in whole bacterial and archaeal cells.

The evolution of large cells has a downside, however. Ions and small molecules such as adenosine triphosphate (ATP), amino acids, and nucleotides cannot diffuse across a large volume quickly. If ATP is used up on one side of a large cell, ATP from the other side of the cell would take a long time to diffuse to that location. Prokaryotic cells are small enough that ions and small molecules arrive where they are needed via diffusion. In fact, the size of prokaryotic cells is probably limited by the distance that molecules must diffuse or be transported inside the cell.

How do eukaryotic cells solve the diffusion problem? The answer lies in the numerous organelles observed in eukaryotic cells. In effect, the huge volume inside a eukaryotic cell is compartmentalized into a large number of bacterium-sized parts. Because eukaryotic cells are subdivided, the molecules required for specific chemical reactions are often located within a given compartment and do not need to diffuse long distances to be useful. But solving the diffusion problem is not the only advantage conferred by organelles:

- Compartmentalization of the cell allows incompatible chemical reactions to be separated. For example, new fatty acids are synthesized in one organelle while excess or damaged fatty acids are degraded and recycled in a different organelle.

- Compartmentalization increases the efficiency of chemical reactions. First, the substrates required for particular reactions can be localized and maintained at high concentration within organelles. Second, groups of enzymes that work together can be clustered on internal membranes instead of floating free in the cytoplasm. Clustering these molecules increases the speed and efficiency of the reactions, because reactants have shorter distances over which to diffuse or be transported.

Based on their morphological differences, prokaryotic cells can be compared to small machine shops while eukaryotic cells

(a) Generalized animal cell

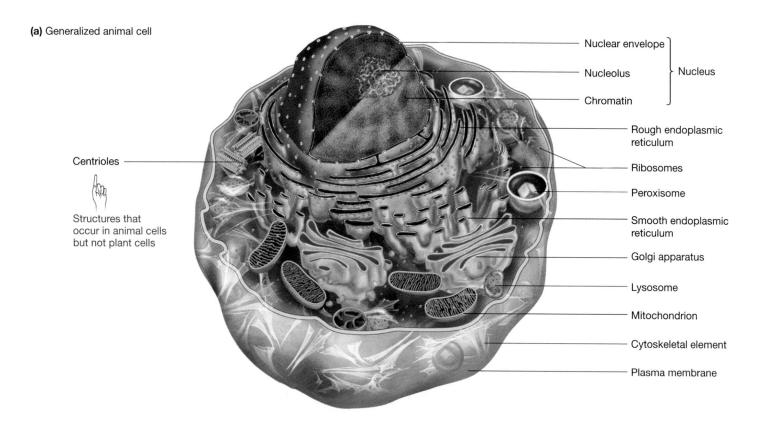

Nuclear envelope
Nucleolus — Nucleus
Chromatin

Rough endoplasmic reticulum

Ribosomes

Peroxisome

Smooth endoplasmic reticulum

Golgi apparatus

Lysosome

Mitochondrion

Cytoskeletal element

Plasma membrane

Centrioles

Structures that occur in animal cells but not plant cells

(b) Generalized plant cell

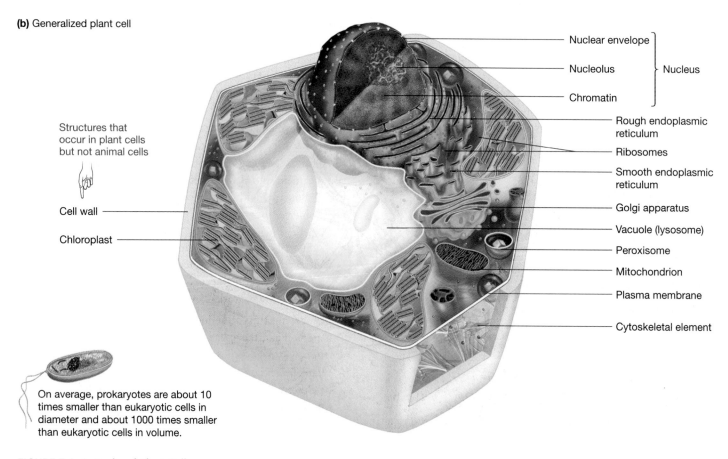

Nuclear envelope
Nucleolus — Nucleus
Chromatin

Rough endoplasmic reticulum

Ribosomes

Smooth endoplasmic reticulum

Golgi apparatus

Vacuole (lysosome)

Peroxisome

Mitochondrion

Plasma membrane

Cytoskeletal element

Structures that occur in plant cells but not animal cells

Cell wall

Chloroplast

On average, prokaryotes are about 10 times smaller than eukaryotic cells in diameter and about 1000 times smaller than eukaryotic cells in volume.

FIGURE 7.6 Animal and Plant Cells
Generalized or "typical" **(a)** animal and **(b)** plant cells. (Compare with the prokaryotic cell, shown at true relative size at bottom left.)
QUESTION Which organelles are unique to animal cells, and which organelles are unique to plant cells? Which are common to both?

resemble sprawling industrial complexes. The organelles and other structures found in eukaryotes are analogous to highly specialized buildings that act as factories, power stations, warehouses, transportation corridors, and administrative centers. **Figure 7.6** shows how organelles are arranged in a typical animal cell and plant cell. What are these structures, and what do they do?

The Nucleus   The **nucleus** is among the largest organelles and is highly organized (**Figure 7.7**). It is enclosed by a unique structure—a complex double membrane called the **nuclear envelope**, which is studded with openings called **nuclear pores**. The inside surface of the nuclear envelope is associated with fibrous proteins that form a lattice-like sheet called the **nuclear lamina**. The nuclear lamina stiffens the envelope and helps organize the chromosomes. Each chromosome occupies a distinct area inside the nucleus and is attached to the nuclear lamina and the inner surface of the nuclear envelope in at least one location. In eukaryotes, chromosomes are linear and consist of DNA that is tightly complexed with a series of ball-shaped *histone* proteins, forming a structure called **chromatin**. Some sections of each chromosome are condensed into a highly compact, supercoiled structure called **heterochromatin**; other sections are unwound into long, filamentous strands called **euchromatin**. The nucleus also includes a distinctive region called the **nucleolus**, where the RNA molecules found in ribosomes are manufactured and the large and small ribosomal subunits are assembled. Section 7.2

discusses the structure and function of the nucleus, and particularly the nuclear envelope, in more detail.

Ribosomes   In eukaryotes, the cytoplasm consists of everything inside the plasma membrane excluding the nucleus; the fluid portion of the cytoplasm is called the **cytosol**. Many of the cell's millions of ribosomes are scattered throughout the cytosol. The ribosomes shown in **Figure 7.8** are comprised of two subunits, one small and one large. Each subunit is composed of several different proteins and one large RNA molecule. In eukaryotes the large subunit also contains two small RNA molecules. (In prokaryotes the large subunit has just one small and one large RNA molecule.) Neither ribosomal subunit is enclosed by a membrane. When the two subunits come together, they form a complex molecular machine that synthesizes proteins.

Rough Endoplasmic Reticulum   In addition to the ribosomes found free in the cytosol, many ribosomes are associated with membranes. More specifically, hundreds of thousands of ribosomes are attached to a network of membrane-bound sacs and tubules called the **rough endoplasmic reticulum**, or **rough ER**. Translated literally, *endoplasmic reticulum* means "inside-formed network." Notice in Figure 7.6 that the ER is continuous with the outer membrane of the nuclear envelope. From there, the layers of sacs extend into the cytoplasm.

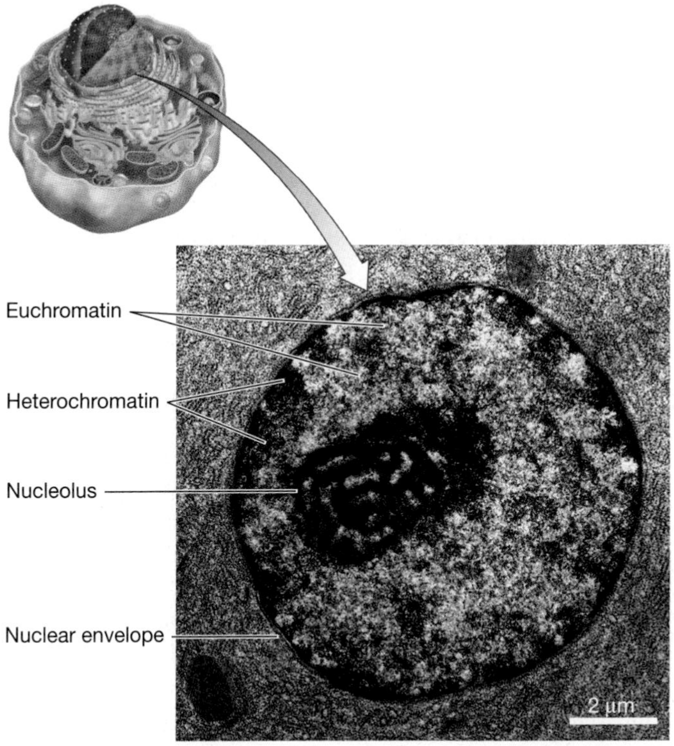

Euchromatin

Heterochromatin

Nucleolus

Nuclear envelope

2 μm

FIGURE 7.7 The Nucleus Is the Eukaryotic Cell's Information Storage and Retrieval Center
The genetic, or hereditary, information is encoded in DNA, which is a component of the chromosomes inside the nucleus.

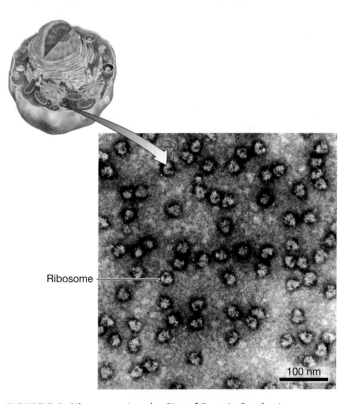

Ribosome

100 nm

FIGURE 7.8 Ribosomes Are the Site of Protein Synthesis
Eukaryotic ribosomes are similar in structure to bacterial and archaeal ribosomes—though not identical. They are comprised of large and small subunits, each of which contains both RNA molecules and proteins.

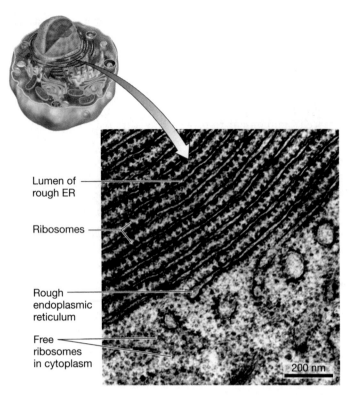

Lumen of rough ER

Ribosomes

Rough endoplasmic reticulum

Free ribosomes in cytoplasm

200 nm

FIGURE 7.9 Rough ER Is a Protein Synthesis and Processing Complex
Rough ER is a system of membrane-bound sacs and tubules with ribosomes attached. It is continuous with the nuclear envelope and with smooth ER.

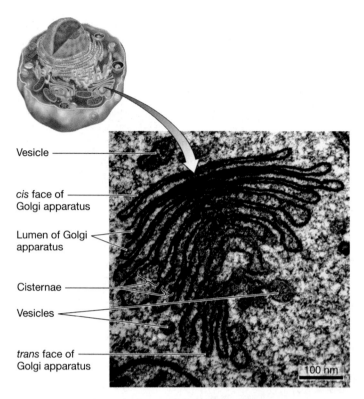

Vesicle

cis face of Golgi apparatus

Lumen of Golgi apparatus

Cisternae

Vesicles

trans face of Golgi apparatus

100 nm

FIGURE 7.10 The Golgi Apparatus Is a Site of Protein Processing, Sorting, and Shipping
The Golgi apparatus is a collection of flattened vesicles called cisternae. The organelle has a *cis* face oriented toward the rough ER and a *trans* face oriented toward the plasma membrane.

The ribosomes associated with the rough ER are responsible for synthesizing proteins that will be inserted into the plasma membrane, secreted to the cell exterior, or shipped to an organelle called the *lysosome*. As they are being manufactured by ribosomes, these proteins move to the interior of the sac-like component of the rough ER (**Figure 7.9**). The interior of any sac-like structure in a cell or body is called the **lumen**. In the lumen of the rough ER, newly manufactured proteins undergo folding and other types of processing.

The proteins produced in the rough ER have a variety of functions. Some carry messages to other cells; some act as membrane transport proteins or pumps; others are enzymes. The common theme is that rough ER products are destined for transport to a distant destination—often to the surface of the cell or beyond.

Golgi Apparatus In many cases, the products of the rough ER pass through the Golgi apparatus before they reach their final destination. The **Golgi apparatus** consists of flattened, membranous sacs called **cisternae** (singular: *cisternum*), which are stacked on top of one another (**Figure 7.10**). The organelle also has a distinct polarity, or sidedness. The *cis* ("this side") surface is closest to the rough ER and nucleus, and the *trans* ("across") surface is oriented toward the plasma membrane. The *cis* side receives products from the rough ER, and the *trans* side ships them out toward the cell surface. In between, within the cisternae, the rough ER's products are processed and packaged for delivery. Micrographs often show "bubbles" on either side of the Golgi stack. These are membrane-bound vesicles that carry proteins or other products to and from the organelle. Section 7.3 analyzes the intracellular movement of products in more detail.

Smooth Endoplasmic Reticulum Not all of the ER is associated with transport of material to Golgi sacs, and not all ER has ribosomes attached. While parts of the ER that contain ribosomes look dotted and rough in electron micrographs, the portions of the organelle that are free of ribosomes appear smooth and even. Appropriately enough, these parts of the ER are called **smooth endoplasmic reticulum** or **smooth ER** (**Figure 7.11**). The smooth ER membrane contains enzymes that are required for reactions involving lipids. Depending on the type of cell, these enzymes may be involved in synthesizing specialized types of lipids needed by the organism or in breaking down hydrophobic molecules that are poisonous to the cell. Smooth ER is the manufacturing site for phospholipids required for the plasma membrane, and smooth ER also functions as a reservoir for calcium ions (Ca^{2+}) that act as a signal inside the cell.

The structure of endoplasmic reticulum correlates closely with its function. Rough ER has ribosomes and functions primarily as a protein-manufacturing center; smooth ER lacks ribosomes and functions primarily as a lipid-processing center.

FIGURE 7.11 Smooth ER Is a Lipid-Handling Center and a Storage Facility
Smooth ER is a system of membrane-bound sacs and tubules that lacks ribosomes.

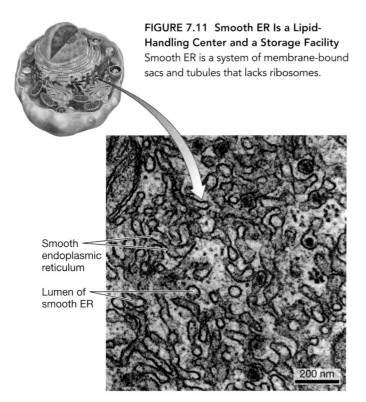

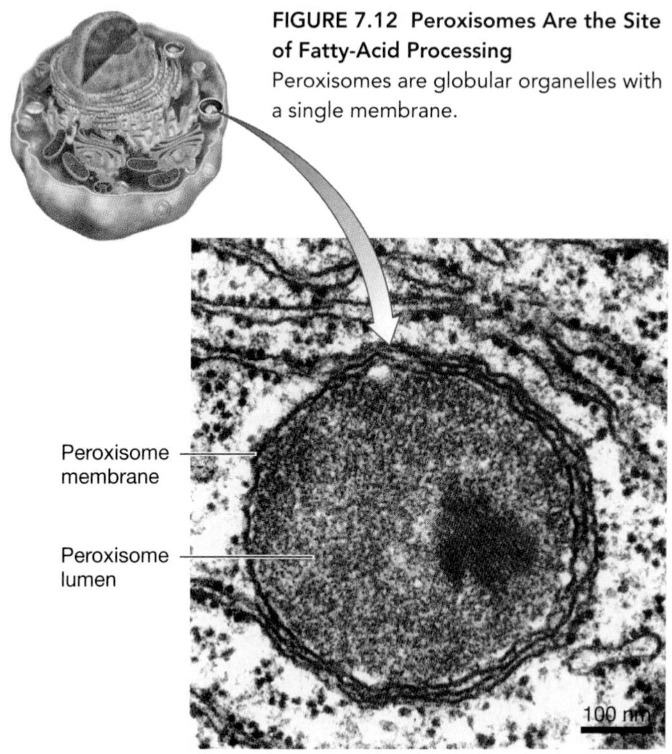

Smooth endoplasmic reticulum

Lumen of smooth ER

200 nm

FIGURE 7.12 Peroxisomes Are the Site of Fatty-Acid Processing
Peroxisomes are globular organelles with a single membrane.

Peroxisome membrane

Peroxisome lumen

100 nm

Together with the Golgi apparatus and lysosomes, the endoplasmic reticulum forms the **endomembrane system**. The endomembrane ("inner-membrane") system is the primary center for protein and lipid synthesis in eukaryotic cells.

Peroxisomes Peroxisomes are globular organelles that are found in virtually all eukaryotic cells (**Figure 7.12**). They have a single membrane and grow and divide independently of other organelles. Although different types of cells from the same individual may have distinct types of peroxisomes, these organelles are united by a common function: Peroxisomes are centers for oxidation reactions. In many cases the products of these reactions include hydrogen peroxide (H_2O_2), which is highly corrosive. If hydrogen peroxide escaped from the peroxisome, the H_2O_2 would quickly damage organelle membranes and the plasma membrane. This is rare, however. Inside the peroxisome, the enzyme catalase quickly converts hydrogen peroxide to water and oxygen.

The various types of peroxisomes that exist contain different suites of enzymes. As a result, each is specialized for oxidizing particular compounds. For example, the peroxisomes in your liver cells contain enzymes that oxidize an array of toxins, including the ethanol in alcoholic beverages. The products of these oxidation reactions are usually harmless and are either excreted from the body or used in other reactions. Other peroxisomes contain enzymes that catalyze the oxidation of fatty acids. These reactions result largely in the production of a molecule called *acetyl CoA*, which is used for the synthesis of important molecules elsewhere in the cell. In plant leaves, specialized peroxisomes called **glyoxisomes** are packed with enzymes that convert

one of the products of photosynthesis into a sugar that can be used to produce energy for the cell. Seeds do not perform photosynthesis, so they lack this type of peroxisome. Instead, they have peroxisomes with enzymes that oxidize fatty acids to yield glucose. The glucose is then used by the young plant as it begins to grow. In each case, there is a clear connection between structure and function: The enzymes found inside the peroxisome make a specialized set of oxidation reactions possible.

Lysosomes The major structures involved in solid-waste processing and materials storage in the cell are called **lysosomes**. The size and shape of these organelles vary widely, and in the cells of plants, fungi, and certain other groups they are referred to as **vacuoles**. In animal cells, lysosomes function as digestive centers (**Figure 7.13**). The organelle's interior, or lumen, is acidic

FIGURE 7.13 Lysosomes Are Recycling Centers
Lysosomes are usually oval or globular and have a single membrane.

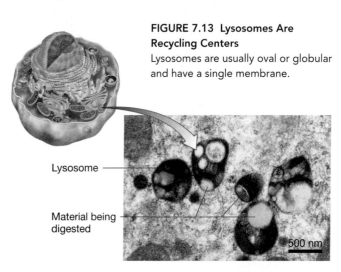

Lysosome

Material being digested

500 nm

because a proton pump in the lysosome membrane imports enough hydrogen ions to maintain a pH of 5.0. This organelle also contains about 40 different enzymes. Each of these proteins is specialized for breaking up a different type of macromolecule—proteins, nucleic acids, lipids, or carbohydrates—into its component monomers. These digestive enzymes are collectively called *acid hydrolases* because they hydrolyze macromolecules most efficiently at pH 5.0. In the cytosol, where the pH is about 7.2, these enzymes are less active.

Figure 7.14 illustrates three ways that materials are delivered to lysosomes in animal cells:

1. When **phagocytosis** ("eat-cell-act") occurs, the plasma membrane of a cell surrounds a smaller cell or a food particle and engulfs it. The resulting structure is delivered to a lysosome, where it is taken in and digested.

2. During **autophagy** ("same-eating"), damaged organelles are surrounded by a membrane and delivered to a lysosome. There the components are digested and recycled.

3. Materials can also find their way into lysosomes as a result of **receptor-mediated endocytosis**. This process begins when macromolecules outside the cell bind to membrane proteins that act as receptors. More than 25 distinct receptors have now been characterized, each specialized for responding to a different macromolecule. Once binding occurs, the plasma membrane folds in and pinches off to form a membrane-bound vesicle called an **early endosome** ("inside-body"). Early endosomes undergo a series of processing steps that include the receipt of digestive enzymes from the Golgi apparatus and the activation of proton pumps that gradually lower their pH. In this way, early endosomes undergo a gradual

maturation process that may lead to the formation of a **late endosome** and eventually a fully functioning lysosome.

Regardless of whether materials are delivered to lysosomes via phagocytosis, autophagy, or receptor-mediated endocytosis, the result is similar: Molecules are hydrolyzed. The amino acids, nucleotides, sugars, and other molecules that result from acid hydrolysis leave the lysosome via transport proteins in the organelle's membrane. Once in the cytoplasm, they can be reused.

It is important to note, however, that not all of the materials that are surrounded by membrane and taken into a cell end up in lysosomes. **Endocytosis** ("inside-cell-act") refers to any pinching off of the plasma membrane that results in the uptake of material from outside the cell. Endocytosis can occur in three ways: (1) phagocytosis, (2) receptor-mediated endocytosis, and (3) **pinocytosis** ("drink-cell-act"). Pinocytosis brings fluid into the cytoplasm via tiny vesicles that form from invaginations of the plasma membrane. The fluid inside these vesicles is not transported to lysosomes, but is used elsewhere in the cell. In addition, most of the macromolecules that collect in early endosomes are selectively removed and used long before the structure becomes a lysosome.

Compared with the lysosomes of animal cells, the vacuoles of plant and fungal cells are large—sometimes taking up as much as 80 percent of a plant cell's volume (**Figure 7.15**). Although some vacuoles contain enzymes that are specialized for digestion, most of the vacuoles observed in plant and fungal

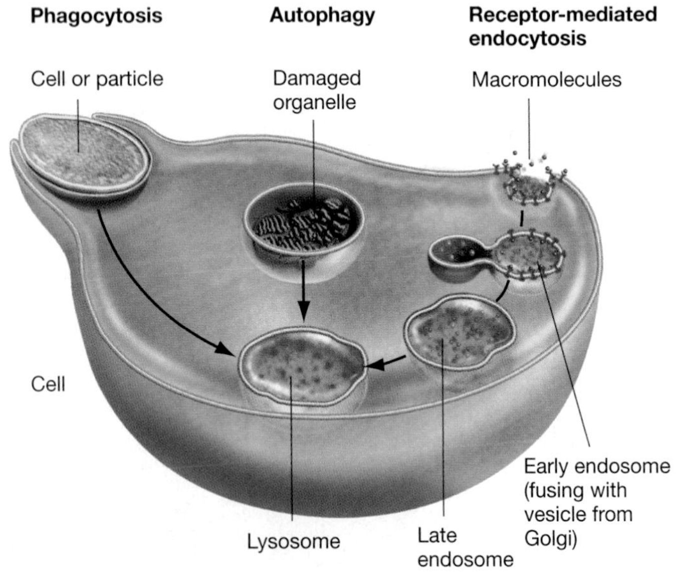

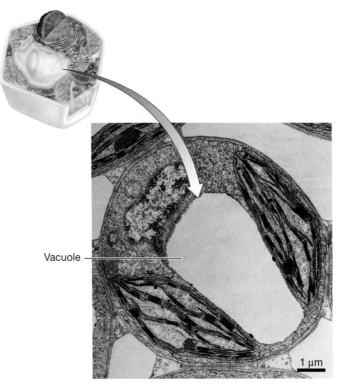

FIGURE 7.15 Vacuoles Are Storage Centers
Vacuoles are variable in size and function. Some contain digestive enzymes and serve as recycling centers; most are large storage containers.

Phagocytosis   **Autophagy**   **Receptor-mediated endocytosis**

Cell or particle   Damaged organelle   Macromolecules

Cell

Lysosome   Late endosome   Early endosome (fusing with vesicle from Golgi)

FIGURE 7.14 Three Ways to Deliver Materials to Lysosomes
Materials can be transported to lysosomes after phagocytosis or autophagy. Endosomes created by receptor-mediated endocytosis may mature into lysosomes.

cells act as storage depots. In many cases, the stored material is water, which maintains the cell's normal volume, or ions such as potassium (K^+) and chloride (Cl^-). But inside seeds, cells may contain a large vacuole filled with proteins. When germination occurs, enzymes begin digesting these proteins to provide amino acids for the growing individual. In cells that make up flower petals or fruits, vacuoles are filled with colorful pigments. Elsewhere in the plant, vacuoles may be packed with noxious compounds that protect leaves and stems from being eaten by predators. The type of chemical involved varies by species, ranging from bitter-tasting tannins to toxins such as nicotine, morphine, caffeine, or cocaine.

Mitochondria   The chemical energy required to build all of these organelles and do other types of work comes from adenosine triphosphate (ATP), most of which is produced in the cell's **mitochondria** (singular: *mitochondrion*). As **Figure 7.16** shows, each mitochondrion has two membranes. The outer membrane defines the organelle's surface, while the inner membrane contacts a series of sac-like **cristae**. The solution inside the inner membrane is called the **mitochondrial matrix.** In eukaryotes, most of the enzymes and molecular machines responsible for providing chemical energy in the form of ATP from food molecules are embedded in the membranes of the cristae or suspended in the matrix. Depending on the type of cell, from 50 to more than a million mitochondria may be present.

Each mitochondrion contains its own small chromosome, independent of the main chromosomes in the nucleus. **Mitochondrial DNA** is a component of a circular and supercoiled chromosome that is similar in structure to bacterial chromosomes. Mitochondria also manufacture their own ribosomes. Like most organelles, mitochondria can grow and divide independently of nuclear division and cell division.

Chloroplasts   Most algal and plant cells possess an organelle called the **chloroplast**, in which sunlight is converted to chemical energy during photosynthesis. The chloroplast has a double membrane around its exterior, analogous to the structure of a mitochondrion (**Figure 7.17**). Instead of featuring sac-like cristae that connect to the inner membrane, though, the interior of the chloroplast is dominated by hundreds of membrane-bound, flattened vesicles called **thylakoids**, which are independent of the inner membrane. Thylakoids are stacked like pancakes into piles called **grana** (singular: *granum*). Many of the pigments, enzymes, and molecular machines responsible for converting light energy into carbohydrates are embedded in the thylakoid membranes. Certain critical enzymes and substrates, however, are found outside the thylakoids in the region called the **stroma**.

The number of chloroplasts per cell varies from none to several dozen. Like mitochondria, each chloroplast contains a circular chromosome. **Chloroplast DNA** is independent of the main

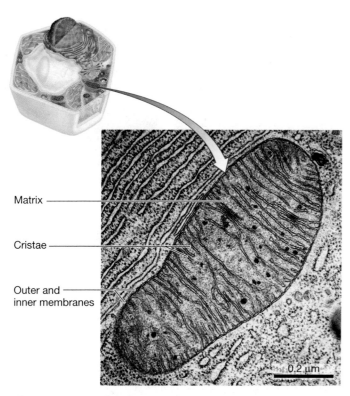

Matrix

Cristae

Outer and
inner membranes

0.2 µm

FIGURE 7.16 Mitochondria Are Power-Generating Stations
Mitochondria are variable in size and shape, but all have a double membrane with sac-like cristae inside.

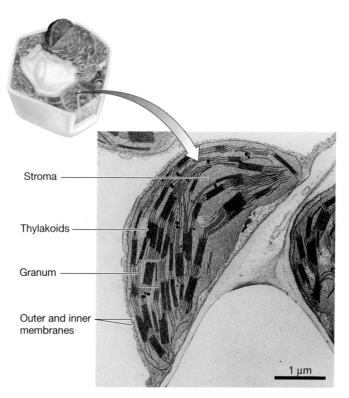

Stroma

Thylakoids

Granum

Outer and inner
membranes

1 µm

FIGURE 7.17 Chloroplasts Are Sugar-Manufacturing Centers
In photosynthesis, sunlight is converted to chemical energy in the form of sugar. The enzymes and other molecules required for photosynthesis are located in membranes inside the chloroplast. These membranes are folded into thylakoids and stacked into grana.

genetic material inside the nucleus. Chloroplasts also grow and divide independently of nuclear division and cell division.

Cytoskeleton   The final major structural feature that is common to all eukaryotic cells is an extensive system of protein fibers called the **cytoskeleton**. As we'll see in Section 7.4, the cytoskeleton contains several distinct types of proteins and fibers and has an array of functions. In addition to giving the cell its shape, cytoskeletal proteins are involved in moving the cell itself and in moving materials within the cell.

The Cell Wall   In fungi, algae, and plants, cells possess an outer **cell wall** in addition to their plasma membrane (**Figure 7.18**). Animals, amoebae, and other groups lack this feature. Although the composition of the cell wall varies among species and even between types of cells in the same individual, the general plan is similar: Rods or fibers composed of a carbohydrate run through a stiff matrix made of other polysaccharides and proteins. In addition, some plant cells produce a secondary cell wall that features a particularly tough molecule called lignin. Lignin forms a branching, cagelike network that is almost impossible for enzymes to attack. The combination of cellulose fibers and lignin in secondary cell walls makes up most of the material we call wood.

How Does Cell Structure Correlate with Function?

The preceding discussion emphasized how the structure of each organelle fits with its role in the cell. As **Table 7.1** indicates, an organelle's membrane and its complement of enzymes correlate closely with its function. The same connection between structure and function occurs at the level of the entire cell. Inside an individual plant or animal, cells are specialized for certain tasks and have a structure that correlates with those tasks. For example, the muscle cells in your upper leg are extremely long, tube-shaped structures. They are filled with protein fibers that slide past one another as the entire muscle flexes or extends. It is this sliding motion that allows your muscles to contract or extend as you run. Muscle cells are also jam packed with mitochondria, which produce the ATP required for the sliding motion to occur. In contrast, nearby fat cells are rounded, globular structures. They consist of little more than a plasma membrane, a nucleus, and a fat droplet. Neither cell bears a close resemblance to the generalized animal cell pictured in Figure 7.6a.

To drive home the correlation between the overall structure and function of a cell, examine the transmission electron micrographs in **Figure 7.19**. The animal cell in Figure 7.19a is from the pancreas and is specialized for the manufacture and export of digestive enzymes. It is packed with rough ER and

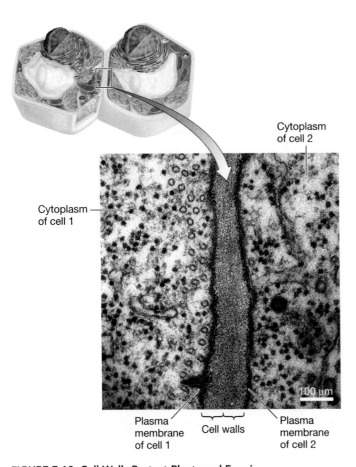

FIGURE 7.18 Cell Walls Protect Plants and Fungi
Plants have cell walls that contain cellulose; in fungi the major structural component of the cell wall is chitin.

(a) Animal pancreatic cell: exports digestive enzymes

(b) Animal testis cell: exports lipid-soluble signals

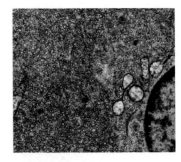

(c) Plant leaf cell: manufactures ATP and sugar

(d) Plant root cell: stores starch

FIGURE 7.19 Cell Structure Correlates with Function
EXERCISE In each cell, label ribosomes, rough ER, chloroplasts, the nucleus, smooth ER, mitochondria, vacuole, plasma membrane, and cell wall if they are visible.

TABLE 7.1 A Summary of Cell Components

	Structure		Function
	Membrane	Components	
Nucleus	Double ("envelope"); openings called nuclear pores	Chromosomes Nucleolus Nuclear lamina	Genetic information Assembly of ribosome subunits Structural support
Ribosomes	None	Large/small subunits Complex of RNA and proteins	Protein synthesis
Endomembrane system			
Rough ER	Single; contains receptors for entry of selected proteins	Network of branching sacs Ribosomes associated	Protein synthesis and processing
Golgi apparatus	Single; contains receptors for products of rough ER	Stack of flattened cisternae	Processing of proteins
Smooth ER	Single; contains enzymes for synthesizing phospholipids	Network of branching sacs Enzymes for synthesizing lipids	Lipid synthesis
Peroxisomes	Single; contains transporters for selected macromolecules	Enzymes that catalyze oxidation reactions Catalase (processes peroxide)	Processing of fatty acids
Lysosomes	Single; contains proton pumps	Acid hydrolases (catalyze hydrolysis reactions)	Digestion and recycling
Vacuoles	Single; contains transporters for selected molecules	Varies—pigments, oils, carbohydrates, water, or toxins	Varies—coloration; storage of oils, carbohydrates, water, or toxins
Mitochondria	Double; outer contains enzymes for processing pyruvate; inner contains enzymes for ATP production	Enzymes that catalyze oxidation-reduction reactions, ATP synthesis	ATP production
Chloroplasts	Double, plus membrane-bound sacs in interior	Pigments Enzymes that catalyze oxidation-reduction reactions	Production of ATP and sugars via photosynthesis
Cytoskeleton	None	Actin filaments Intermediate filaments Microtubules	Structural support Movement of materials In some species: movement of whole cell
Plasma membrane	Single; contains transport and receptor proteins	Phospholipid bilayer with transport and receptor proteins	Selective permeability—maintains intracellular environment
Cell wall	None	Carbohydrate fibers running through carbohydrate or protein matrix	Protection, structural support

Golgi, which make this function possible. The animal cell in Figure 7.19b is from the testis and synthesizes the lipid-soluble signaling molecule called testosterone. This cell is dominated by smooth ER, where lipid processing takes place. The plant cell in Figure 7.19c is from the leaf of a potato and is specialized for absorbing light and manufacturing sugar; the cell in Figure 7.19d is from a potato tuber (part of an underground stem) and functions as a starch storage container. The leaf cell contains hundreds of chloroplasts, while the tuber cell has a prominent storage vacuole filled with carbohydrate. In each case, the type of organelles in each cell and their size and number correlate with the cell's specialized function.

The Dynamic Cell

Biologists describe the structure and function of organelles and cells by a combination of tools and approaches. Light microscopes and transmission electron microscopes have allowed researchers to see cells at increasingly high magnification. Microscopy allowed biologists to characterize the basic size and shape of organelles and where they occurred in the cell; a technique called **differential centrifugation** made it possible to isolate particular cell components and analyze their chemical composition. As **Box 7.1** (page 140) explains, differential centrifugation is based on breaking cells apart to create a complex mixture and then separating components in a centrifuge. The individual parts of the cell can then be purified and studied in detail.

Although these techniques have led to an increasingly sophisticated understanding of how cells work, they have a limitation. Transmission electron microscopy is based on a fixed "snapshot" of the cell that is to be observed, and differential centrifugation is based on splitting cells into parts that are analyzed independently. Neither technique allows investigators to answer directly questions about how things move from place to place in the cell or how parts interact. The information gleaned from these techniques can make cells seem somewhat static. In reality, cells are dynamic.

BOX 7.1　How Does a Centrifuge Work?

For decades, the centrifuge was among the most common tools used by biologists who study life at the level of molecules and cells. It was vital to early studies of organelles and other cell structures because it can separate cell components efficiently. A centrifuge accomplishes this by spinning cells in a solution that allows molecules and other cell components to separate according to their density or size and shape.

The first step in preparing a cell sample for centrifugation is to release the organelles and cell components by breaking the cells apart. This can be done by putting them in a hypotonic solution, by exposing them to ultrasonic vibration, by treating cells with a detergent, or by grinding them up. Each of these methods breaks apart plasma membranes and releases the contents of the cells.

The pieces of plasma membrane broken up by these techniques quickly reseal to form small vesicles, often trapping cell components inside. The solution that results from the homogenization step is a mixture of these vesicles, free-floating macromolecules released from the cells, and organelles. A solution such as this is called a **cell extract** or **cell homogenate**.

When a cell homogenate is placed in a centrifuge tube and spun at high speed, the components that are in solution tend to move outward, along the dashed line in **Figure 7.20a**. The effect is similar to a merry-go-round, which seems to push you outward in a straight line away from the spinning platform. In response to this outward-directed force, the solution containing the cell homogenate exerts a centripetal ("center-seeking") force that pushes the homogenate away from the bottom of the tube. Larger, denser molecules or particles resist this inward force more readily than do smaller, less dense ones and so reach the bottom of the centrifuge tube faster.

To separate the components of a cell extract, researchers often perform a series of centrifuge runs. Steps 1 and 2 of **Figure 7.20b** illustrate how an initial treatment at low speed causes larger, heavier parts of the homogenate to move below smaller, lighter parts. The material that collects at the bottom of the tube

(a) How a centrifuge works

When the centrifuge spins from position A to B, the macromolecule tends to move along the dashed line. This motion pushes the molecule toward the bottom of the tube.

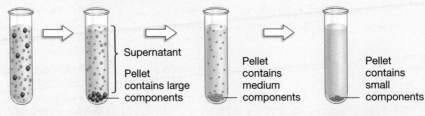

The solution in the tube exerts a centripetal force, which resists movement of the molecule to the bottom of the tube.

Very large or dense molecules overcome this centripetal force more readily than smaller, less dense ones. As a result, larger, denser molecules move toward the bottom of the tube faster.

(b) DIFFERENTIAL CENTRIFUGATION

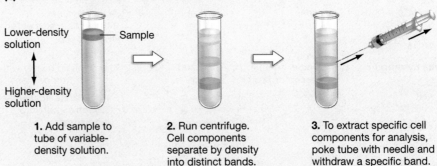

Supernatant

Pellet contains large components

1. Start with uniform cell homogenate in centrifuge tube.

2. Subject tube to low-speed centrifugation. Large components settle out below the supernantant.

Pellet contains medium components

3. Transfer supernatant to new tube, and subject it to medium-speed centrifugation.

Pellet contains small components

4. Transfer supernatant to new tube, and subject it to high-speed centrifugation.

(c) DENSITY GRADIENT CENTRIFUGATION

Lower-density solution

Sample

Higher-density solution

1. Add sample to tube of variable-density solution.

2. Run centrifuge. Cell components separate by density into distinct bands.

3. To extract specific cell components for analysis, poke tube with needle and withdraw a specific band.

FIGURE 7.20 Cell Components Can Be Separated by Centrifugation
(a) Overhead view of a centrifuge, illustrating why cell components separate by being spun.
(b) Through a series of centrifuge runs made at increasingly higher speeds, an investigator can separate fractions of a cell homogenate according to size by differential centrifugation.
(c) A high-speed centrifuge run can achieve extremely fine separation among cell components by density gradient centrifugation.

is called the **pellet**, and the solution and solutes left behind comprise the **supernatant** ("above swimming"). The supernatant is placed in a fresh tube and centrifuged at increasingly higher speeds and longer durations. Each centrifuge run continues to separate cell components based on their size and density.

To accomplish even finer separation of macromolecules or organelles, researchers frequently follow up with centrifugation

at extremely high speeds. One strategy is based on filling the centrifuge tube with a series of sucrose solutions of increasing density. The density gradient allows cell components to separate on the basis of small differences in size and shape. When the centrifuge run is complete, each cell component comprises a distinct band of material in the tube. A researcher can then collect the material in each band for further study.

The amount of chemical activity and the speed of molecular movement inside cells is nothing short of fantastic. Bacterial ribosomes add up to 20 amino acids per second to a growing polypeptide, and eukaryotic ribosomes typically add two per second. Given that there are about 15,000 ribosomes in each bacterium and possibly a million in an average eukaryotic cell, hundreds or even thousands of new protein molecules can be finished each second in every cell. In the same amount of time, a typical cell in your body uses an average of 10 million ATP molecules and synthesizes just as many. It's not unusual for a cellular enzyme to catalyze 25,000 or more reactions per second; most cells contain hundreds or thousands of enzymes. A minute is more than enough time for each membrane phospholipid in your body to travel the breadth of the organelle or cell where it resides. The hundreds of trillions of mitochondria inside you are completely replaced about every 10 days, for as long as you live. The plasma membrane is fluid, and its composition is constantly changing.

Because humans are such large organisms, it is impossible for us to imagine what life is really like inside a cell. At the scale of a ribosome or an organelle or a cell, gravity is inconsequential. Instead, electrostatic attractions between molecules and the kinetic energy of motion are the dominant forces. At this level, events take nanoseconds, and speeds are measured in micrometers per second. Contemporary methods for studying cells, including those featured in **Box 7.2**, capture this dynamism by tracking how organelles and molecules move and interact over time.

The rest of this chapter focuses on this theme of cellular dynamism and movement. To begin, let's look at how molecules move into and out of the cell's control center—the nucleus. Then we'll consider how proteins move from ribosomes into the lumen of the rough ER and then to the Golgi apparatus and beyond. The chapter closes by analyzing how cytoskeletal elements help transport cargo inside the cell or move the cell itself.

BOX 7.2 Techniques for Studying the Dynamic Cell

Contemporary methods for studying cells allow researchers to see specific molecules moving inside living cells. One of the most popular techniques for tagging molecules of interest relies on a fluorescent molecule called **green fluorescent protein**, or **GFP**. GFP is naturally synthesized in jellyfish that *fluoresce*, or emit light. By affixing GFP molecules to another protein and then inserting it into a cell, investigators can follow its fate over time and even videotape its movement. For example, recent studies have used GFP to tag proteins that are secreted from the cell. Control experiments show that GFP does not affect the behavior of these proteins. Researchers then videotaped the GFP-tagged protein's transport from the rough ER through the Golgi apparatus and out to the plasma membrane. This is cell biology: the movie.

To produce extremely high resolution still images of proteins that are tagged with GFP or other fluorescing tags, researchers often rely on **confocal microscopy**. This technique is based on mounting live cells on a microscope slide and then focusing a beam of ultraviolet light at a specific depth within the specimen. The fluorescing tag emits visible light in response. A detector for this light is then set up at exactly the position where the emitted light comes into focus. The result is a sharp image of a precise plane in the cell being studied (**Figure 7.21**).

(a) Conventional fluorescence image

(b) Confocal fluorescence image

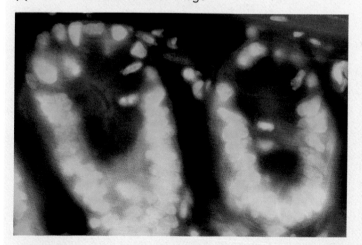
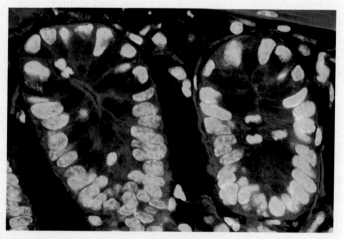

FIGURE 7.21 Confocal Microscopy Provides Sharp Images of Living Cells
The same living cells from the intestine of a mouse at the same magnification. **(a)** The conventional image is blurred, because it results from light emitted by the entire group of cells. **(b)** The confocal image is sharp, because it results from light emitted at a single plane inside the cells.

7.2 The Nuclear Envelope: Transport Into and Out of the Nucleus

The nucleus is the information center of eukaryotic cells. It is a corporate headquarters, design center, and library all rolled into one. Appropriately enough, its interior is highly organized. The organelle's overall shape and structure are defined by the mesh-like nuclear lamina, which also helps anchor each chromosome. The remainder of each chromosome occupies a well-defined region in the nucleus, and specific centers exist where the genetic information in DNA is decoded and processed. At these locations, large suites of enzymes interact to produce RNA messages from specific genes at specific times. Meanwhile, the nucleolus functions as the site of ribosome synthesis.

Consistent with its role as information repository and processing center, the nucleus is separated from the rest of the cell by the nuclear envelope. Biologists began to understand exactly how the nuclear envelope is structured when electron microscopy became available in the 1950s. As **Figure 7.22a** shows, the nuclear envelope has two membranes, each consisting of a lipid bilayer. The inner membrane and the outer membrane are separated by a space that is continuous with the lumen of the endoplasmic reticulum. Later, electron micrographs showed that the envelope contains thousands of openings called **nuclear pores (Figure 7.22b)**. Because these pores extend through both inner and outer nuclear membranes, they connect the inside of the nucleus with the cytoplasm. The pore itself consists of over 50 different proteins. These molecules form an elaborate structure called the **nuclear pore complex (Figure 7.22c)**.

A series of experiments in the early 1960s showed that molecules travel into and out of the nucleus through the nuclear pore complexes. The initial studies were based on injecting tiny gold particles into cells and then preparing them for electron microscopy. In electron micrographs, gold particles show up as black dots. One or two minutes after injection, the micrographs showed that most of the gold particles were in the cytoplasm. A few, however, were closely associated with nuclear pores. Ten minutes after injection, particles were inside the nucleus as well as in the cytoplasm. These data supported the hypothesis that the pores function as the doors to the nucleus. Follow-up work confirmed that the nuclear pore complex is the only gate between the cytoplasm and the nucleus and that only certain molecules go in and out. Passage through the nuclear pore is selective.

What substances traverse nuclear pores? DNA clearly does not—it never leaves the nucleus. But information coded in DNA is used to synthesize RNA inside the nucleus. Several distinctive types of RNA molecules are produced, each distinguished by size and function. For example, most ribosomal RNAs are manufactured in the nucleolus, where they bind to proteins to form completed ribosomal subunits. Messenger RNAs, in contrast, carry the information required to manufacture proteins out to the cytoplasm, where protein synthesis takes place. To perform their function, all of the various types of RNA move out of the nucleus. Traffic in the other direction is also impressive. Nucleotide

(a) The nuclear envelope has a double membrane.

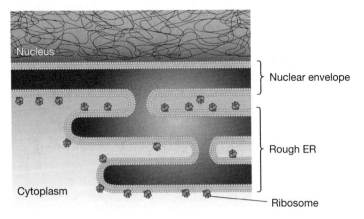

Nucleus

Nuclear envelope

Rough ER

Cytoplasm

Ribosome

(b) Surface view of nuclear envelope

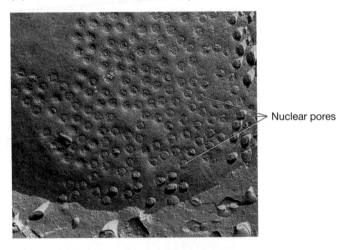

Nuclear pores

(c) Cross-sectional view of nuclear pore

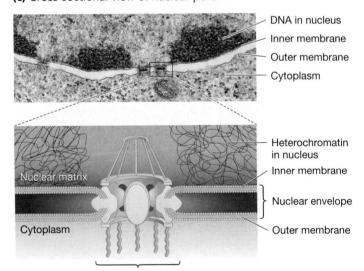

DNA in nucleus
Inner membrane
Outer membrane
Cytoplasm

Heterochromatin in nucleus

Inner membrane

Nuclear matrix

Nuclear envelope

Cytoplasm

Outer membrane

Nuclear pore complex

FIGURE 7.22 The Structure of the Nuclear Envelope
(a) The nuclear envelope is continuous with the endoplasmic reticulum. **(b)** Electron micrograph showing that the surface of the nuclear envelope is studded with nuclear pores. **(c)** The drawing (bottom) is based on electron micrographs of the nuclear pore complex.

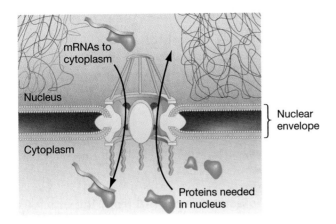

FIGURE 7.23 Molecules Move Into and Out of the Nucleus through the Nuclear Pores
Messenger RNAs are synthesized in the nucleus and must be exported to the cytoplasm. Proteins needed in the nucleus are synthesized in the cytoplasm and have to be imported to the nucleus.
EXERCISE Label the inner and outer membranes of the nuclear envelope. Label the nuclear pore.

triphosphates that act as building blocks for DNA and RNA have to enter the nucleus, as do the proteins responsible for copying DNA, synthesizing RNAs, extending the nuclear lamina, assembling ribosomes, or building chromosomes (**Figure 7.23**).

To summarize, ribosomal subunits and various types of RNAs exit the nucleus; proteins that are needed inside enter it. In a typical cell, over 500 molecules pass through each of the 3000–4000 nuclear pores every second. The traffic is intense. How is it regulated and directed?

How Are Molecules Imported into the Nucleus?

The first experiments on how molecules move through the nuclear pore focused on proteins that are produced by viruses. **Viruses** are parasites that use the cell's machinery to make copies of themselves. When a virus infects a cell, certain of its proteins enter the nucleus. Investigators noticed that if a particular amino acid in one of these proteins happens to be altered, the viral protein is no longer able to pass through the nuclear pore. This simple-sounding observation led to a key hypothesis: Proteins that are synthesized by ribosomes in the cytosol but are headed for the nucleus contain a "zip code"—a molecular address tag that marks them for transport through the nuclear pore complex. The idea was that viral proteins enter the nucleus if they have the same address tag as normal cellular proteins have. This zip code came to be called the **nuclear localization signal (NLS).**

A series of experiments on a protein called nucleoplasmin helped researchers better understand the nature of this signal. *Nucleoplasmin* plays an important role in the assembly of chromosomes and happens to have a distinctive structure: It consists of a globular protein core surrounded by a series of extended protein "tails." When researchers labeled nucleoplasmin with a radioactive atom and injected it into the cytoplasm of living cells, they found that the radioactive signal quickly ended up in the nucleus.

Figure 7.24 outlines how the nuclear localization signal in nucleoplasmin was found. Researchers began by using enzymes called proteases to separate the core sections of nucleoplasmin from the tails. After separating the two components, the researchers labeled each part with radioactive atoms and injected them into the cytoplasm of different cells. When they

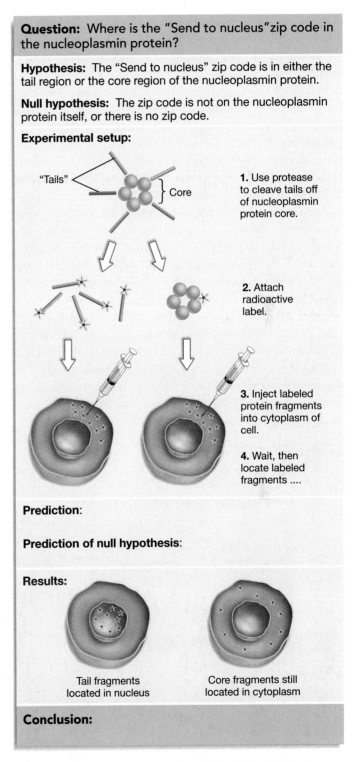

FIGURE 7.24 Where Is the "Send to Nucleus" Zip Code in the Nucleoplasmin Protein?
EXERCISE Without looking at the text, fill in the predictions and conclusion(s) in this experiment.

examined the experimental cells with the electron microscope, they found that tail fragments were transported to the nucleus. Core fragments, in contrast, remained in the cytoplasm. These data suggested that the zip code must be somewhere in the tail part of the protein.

By analyzing different stretches of the tail section, the biologists eventually found a 17-amino-acid-long section that had to be present to direct proteins to the nucleus. The biologists therefore concluded that instead of consisting of five numbers, the NLS zip code consisted of 17 specific amino acids in the tail.

Follow-up work confirmed that other proteins bound for the nucleus have similar localization signals, and that these signals interact with proteins called **importins**. **Figure 7.25** summarizes the current model for how nuclear import takes place. Several different molecules are involved: the protein being transported, an importin, ATP, *guanosine diphosphate* (GDP) or *guanosine triphosphate* (GTP), and a protein called Ran. GTP is similar in structure to ATP and has high potential energy. If you think of the protein as cargo, the importin as a delivery truck, and ATP as gas, then Ran is the unloading crew and GTP is their supervisor. More specifically, data suggest that when an importin in the cytoplasm binds to a molecule that has a nuclear localization signal, the importin/cargo complex enters the nucleus along with Ran that has GDP bound to it. The movement of the cargo requires ATP. Inside, an enzyme exchanges the GDP for GTP. When this reaction occurs, Ran's conformation changes. It binds to the importin/cargo complex, causing the cargo molecule to drop off. Ran then escorts the importin back out to the cytoplasm. There the import sequence starts anew.

Work on nuclear import carries two general messages:

1. Movement is highly regulated. Although small molecules can diffuse freely into and out of the nuclear pore complex, larger molecules can enter only if they contain a nuclear localization signal.

2. Movement of large molecules is an energy-demanding, active process.

Do the same general principles hold when RNAs and other materials move *out* of the nucleus?

How Are Molecules Exported from the Nucleus?

After several decades' worth of experiments, biologists have come to a satisfying conclusion: Export of ribosome subunits, proteins, and other materials from the nucleus is almost exactly the reverse of import. In almost all cases, Ran and GTP are involved, as are shuttle proteins called **exportins**. Further, proteins that leave the nucleus have a specific zip code—a nuclear export signal. Distinct types of exportins are specialized for binding to the different types of materials and ferrying them to the cytoplasm. The Ran molecules and exportins responsible for nuclear export cycle back and forth, into and out of the nucleus, just as they do during import. Like nuclear import, nuclear export is both highly regulated and energy demanding.

Currently, biologists are focused on understanding how Ran interacts with proteins inside the nuclear pore complex as cargo moves in and out. Investigators are also trying to unravel how traffic is regulated to avoid backups and head-on collisions. The goal is to understand the precise physical mechanisms responsible for moving cargo into and out of the nuclear pore complex.

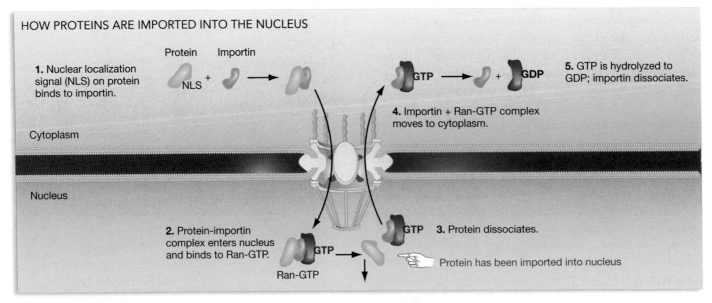

FIGURE 7.25 An Importin, Ran, and GDP Are Required to Import Proteins into the Nucleus
Importin, Ran, and GDP are recycled to the cytoplasm after they deliver cargo to the nucleus.

WEB TUTORIAL 7.2 A Pulse-Chase Experiment

7.3 The Endomembrane System: Manufacturing and Shipping Proteins

The nuclear membrane is not the only place in cells where cargo moves in a regulated and energy-demanding fashion. For example, Chapter 6 highlighted how specific ions and molecules are pumped into and out of cells or transported across the plasma membrane by specialized membrane proteins. In addition, proteins that are synthesized by ribosomes in the cytosol but are used inside mitochondria or chloroplasts contain special signal sequences, analogous to the nuclear localization signal, that target the proteins for transport to these organelles.

Perhaps the most intricate of all manufacturing and shipping systems, however, involves proteins that are synthesized in the rough ER and move to the Golgi apparatus for processing, and from there travel to the cell surface or other destinations. The idea that materials might move through the endomembrane system in an orderly way was inspired by a simple observation. According to electron micrographs, cells that secrete digestive enzymes, hormones, or other types of products have particularly large amounts of rough ER and Golgi. This correlation led to the idea that these cells have a "secretory pathway" that starts in the rough ER and ends with products leaving the cell (**Figure 7.26**). How does this hypothesized pathway work?

George Palade and colleagues did pioneering research on the secretory pathway with an experimental approach known as a pulse-chase experiment. The strategy is based on providing experimental cells with a large concentration of a labeled molecule for a short time. For example, if a cell receives a large amount of labeled amino acid for a short time, virtually all of the proteins synthesized during that interval will be labeled. This "pulse" of labeled molecule is followed by a chase—large amounts of an unlabeled version of the same molecule, provided for a long time. If the chase consists of unlabeled amino acid, then the proteins synthesized during the chase period will *not* be labeled. The general idea is to mark a population of molecules at a particular interval and then follow their fate over time. This approach is analogous to adding a small amount of dye to a stream and then following the movement of the dye molecules.

In testing the secretory pathway hypothesis, Palade's team focused on pancreatic cells that were growing in culture, or in vitro.[1] These cells are specialized for secreting digestive enzymes into the small intestine and are packed with rough ER and Golgi. The basic experimental approach was to supply the cells with a 3-minute pulse of the amino acid leucine, labeled with a radioactive atom, followed by a long chase with nonradioactive leucine. Because the radioactive leucine was incorporated into all proteins being produced during the pulse, it labeled them. Then the researchers

[1]The term *in vitro* is Latin for "in glass." Experiments that are performed outside living cells are done in vitro. The term *in vivo*, in contrast, is Latin for "in life." Experiments performed with living organisms are done in vivo.

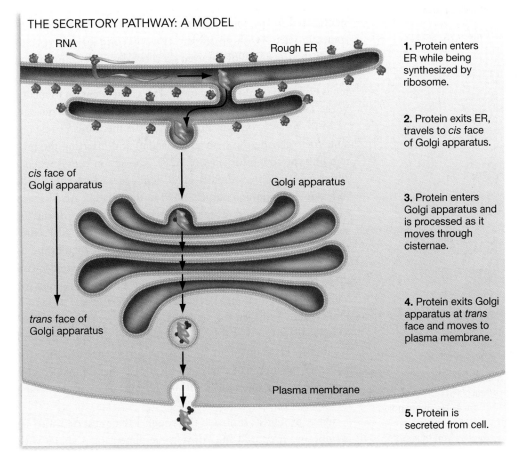

THE SECRETORY PATHWAY: A MODEL

RNA

Rough ER

cis face of Golgi apparatus

Golgi apparatus

trans face of Golgi apparatus

Plasma membrane

1. Protein enters ER while being synthesized by ribosome.

2. Protein exits ER, travels to *cis* face of Golgi apparatus.

3. Protein enters Golgi apparatus and is processed as it moves through cisternae.

4. Protein exits Golgi apparatus at *trans* face and moves to plasma membrane.

5. Protein is secreted from cell.

FIGURE 7.26 The Secretory Pathway Hypothesis
The secretory pathway hypothesis proposes that proteins intended for secretion from the cell are synthesized and processed in a highly prescribed set of steps.

prepared a sample of the cells for electron microscopy and autoradiography (see Chapter 4). When they examined cells immediately after the pulse, they found the newly synthesized proteins inside the rough ER (**Figure 7.27a**). Seven minutes later, most of the labeled proteins were in a Golgi apparatus or inside structures called secretory vesicles on the *trans* side of a Golgi apparatus (**Figure 7.27b**). After 80 minutes, most labeled proteins were in secretory vesicles or actually outside the cell (**Figure 7.27c**).

(a) Immediately after labeling

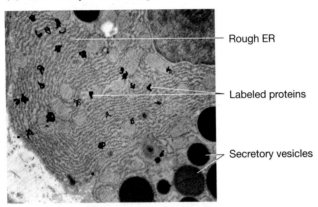

— Rough ER

— Labeled proteins

— Secretory vesicles

(b) 7 minutes after end of labeling

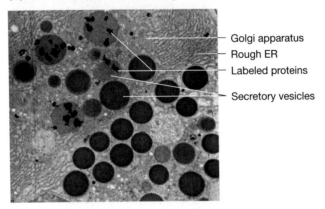

— Golgi apparatus
— Rough ER
— Labeled proteins
— Secretory vesicles

(c) 80 minutes after end of labeling

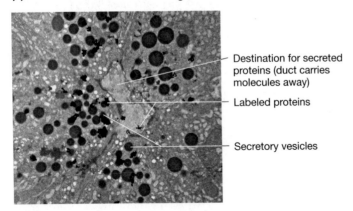

— Destination for secreted proteins (duct carries molecules away)

— Labeled proteins

— Secretory vesicles

FIGURE 7.27 Results of a Pulse-Chase Experiment
The position of labeled proteins immediately after the pulse of amino acids, then 7 and 80 minutes later. QUESTION Do the data support or contradict the secretory pathway hypothesis? Explain your answer.

These results were consistent with the hypotheses that a secretory pathway exists and that the rough ER and Golgi apparatus function as an integrated endomembrane system. Clearly, proteins produced in the rough ER don't float around the cytoplasm aimlessly or drift randomly from organelle to organelle. Instead, traffic through the endomembrane system is highly organized and directed. Now let's break the system down and examine four of the steps in more detail. The ribosomes in rough ER are bound to the outside of the membrane. How do the proteins that they manufacture get into the lumen of the ER? How do they move from the ER to the Golgi apparatus? Once they're inside the Golgi, what happens to them? And finally, how do the finished proteins get to their destination? Let's consider each question in turn.

Entering the Endomembrane System: The Signal Hypothesis

How do proteins enter the endomembrane system? The **signal hypothesis**, proposed by Günter Blobel and colleagues, predicted that proteins bound for the endomembrane system have a zip code analogous to the nuclear localization signal. The idea was that these proteins are synthesized by ribosomes that are attached to the outside of the ER and that the first few amino acids in the growing polypeptide act as a signal that brings the protein into the lumen of the ER.

This hypothesis received important support when researchers made a puzzling observation: When proteins that are normally synthesized in the rough ER are manufactured by naked ribosomes in vitro—with no ER present—they are 20 amino acids longer than usual. Blobel seized on these data. He claimed that the 20 amino acids are the "Send to ER" signal and that the signal is removed inside the organelle. His group went on to identify the exact sequence of amino acids in the **ER signal sequence**.

More recent work has documented the mechanisms responsible for receiving the send-to-ER signal and inserting the protein into the rough ER (**Figure 7.28**). The action begins when a ribosome synthesizes the ER signal sequence, which then binds to a **signal recognition particle (SRP)** in the cytosol. An SRP is a complex of RNA and protein that acts as a receptor for the ER signal sequence. The ribosome + signal sequence + SRP complex then attaches to an SRP receptor in the ER membrane itself. You can think of the SRP as a key that is activated by an ER signal sequence. The receptor in the ER membrane is the lock. Once the lock and key connect, the rest of the protein is synthesized, and then the signal sequence is removed. The finished polypeptide has one of two fates: (1) proteins that will eventually be shipped to an organelle or secreted from the cell enter the lumen of the rough ER; or (2) membrane proteins that remain in the rough ER membrane as they are being manufactured.

Once proteins are inside the rough ER or inserted into its membrane, they fold into their three-dimensional shape with

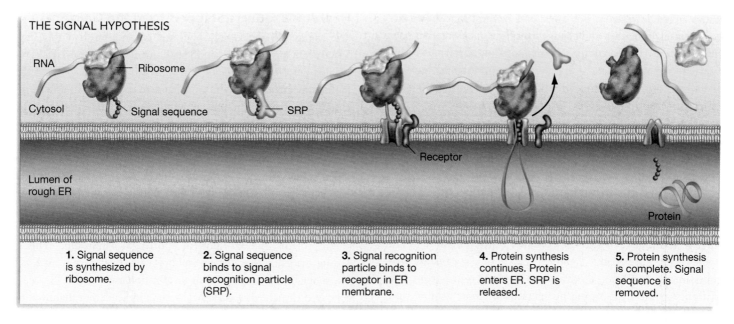

FIGURE 7.28 The Signal Hypothesis Explains How Proteins Destined for Secretion Enter the Endomembrane System
According to the signal hypothesis, proteins destined for secretion contain a short stretch of amino acids that interact
with a signal recognition particle (SRP) in the cytoplasm. This interaction allows the protein to enter the ER.

the help of chaperone proteins. In addition, proteins that enter the lumen are acted on by enzymes that catalyze the addition of carbohydrate side chains. Because carbohydrates are polymers of sugar monomers, the addition of one or more carbohydrate groups is called **glycosylation** ("sugar-together"). The resulting molecule is called a **glycoprotein** ("sugar-protein"). As **Figure 7.29** shows, proteins that enter the ER often gain a specific carbohydrate that consists of 14 sugar residues. Thus, proteins are not only synthesized in the rough ER, they are folded and modified by glycosylation. The completed glycoproteins are ready for shipment to the Golgi apparatus.

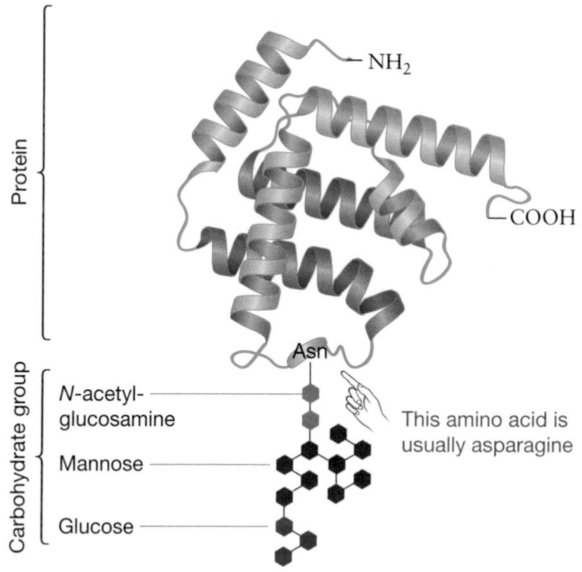

Getting from the ER to the Golgi

How do proteins travel from the ER to the Golgi apparatus? Palade's group thought they had the answer, based on data from the pulse-chase experiments that first confirmed the existence of the endomembrane system. When labeled proteins appeared in a region between the rough ER and the Golgi apparatus, they appeared to be inside small membrane-bound structures. Based on these observations, the biologists suggested that proteins are transported between the two organelles in vesicles. The idea was that vesicles bud off of the ER, move away, fuse with the membrane on the *cis* face of the Golgi apparatus, and dump their contents inside.

This hypothesis was supported when other researchers used differential centrifugation to isolate and characterize the vesicles that contained labeled proteins. Using this approach, investigators have established that distinctive types of vesicles carry proteins from the rough ER to the Golgi apparatus and from layer to layer within the Golgi apparatus.

What Happens Inside the Golgi Apparatus?

Recall from Section 7.1 that the Golgi apparatus consists of a stack of flattened vesicles called cisternae, and that cargo enters one side of the structure and exits the other. It is still not clear, however, exactly how material moves through the stack. There is strong evidence that at least some molecules move among the

FIGURE 7.29 Glycosylation Adds Carbohydrate Groups to Proteins
When proteins enter the ER, most acquire the 14 sugar residues shown here. Some of these sugars may be removed or others added as proteins pass through the Golgi apparatus.

cisternae inside vesicles. But other data suggest that the cisternae themselves mature and change over time, meaning that new cisternae are created at the *cis* face and old cisternae break apart at the *trans* face. If so, then cisternae would have to change in composition and activity over time. **Figure 7.30** illustrates these two hypotheses. Is each cisterna static except for occasional additions or subtractions via vesicle delivery and shipment, or is the entire structure dynamic? The answer is not yet known, and the dichotomy is not necessarily absolute—both processes may occur to some degree.

Although the structure of the Golgi apparatus is still somewhat uncertain, its function is not. By separating individual cisternae and analyzing their contents, researchers have found that each cisterna contains a different suite of enzymes that catalyze glycosylation reactions. As a result, proteins undergo further modification as they move from one cisterna to the next. Some proteins have sugar groups that are phosphorylated in a vesicle near the *cis* face. Later, the carbohydrate group that was added in the rough ER is removed. In other cisternae, various types of carbohydrate chains are attached that may protect the protein or help it attach to surfaces.

How Are Products Shipped from the Golgi?

The rough ER and Golgi apparatus are like an assembly line. Some of the products stay in the endomembrane system itself, replacing worn-out molecules. But if proteins are processed to the end of the line, they will be sent to one of several destinations, including lysosomes, the plasma membrane, or the outside of the cell. How are these finished products put into the right shipping containers, and how are the different containers addressed?

Studies on enzymes that are shipped to lysosomes have provided some answers to both questions. A key finding was that lysosome-bound proteins have a phosphate group attached to a specific sugar subunit on their surface, forming the compound mannose-6-phosphate. If mannose-6-phosphate is removed from these proteins, they are not transported to a lysosome. This is strong evidence that the phosphorylated sugar serves as a zip code, analogous to the nuclear localization and rough ER signals analyzed earlier. More specifically, data indicate that mannose-6-phosphate binds to a protein in membranes of certain vesicles. These vesicles, in turn, have proteins on their surface that interact specifically with proteins in the lysosomal membranes. In this way, the presence of mannose-6-phosphate targets proteins for vesicles that deliver their contents to lysosomes.

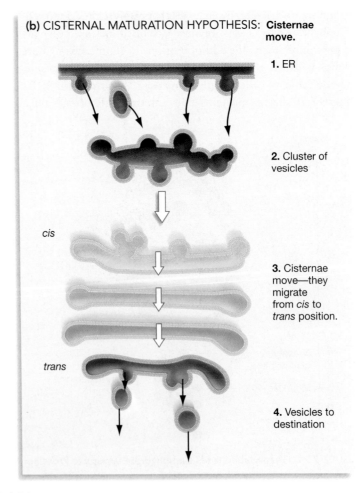

(a) VESICLE TRANSPORT HYPOTHESIS: Cisternae are fixed, vesicles move.

1. ER

2. Cluster of vesicles

cis

3. Cisternae don't move. Vesicles transport materials through cisternae.

trans

4. Vesicles to destination

(b) CISTERNAL MATURATION HYPOTHESIS: Cisternae move.

1. ER

2. Cluster of vesicles

cis

3. Cisternae move—they migrate from *cis* to *trans* position.

trans

4. Vesicles to destination

FIGURE 7.30 Two Hypotheses for How Materials Move through the Golgi Apparatus
QUESTION Are these hypotheses mutually exclusive? Explain your answer.

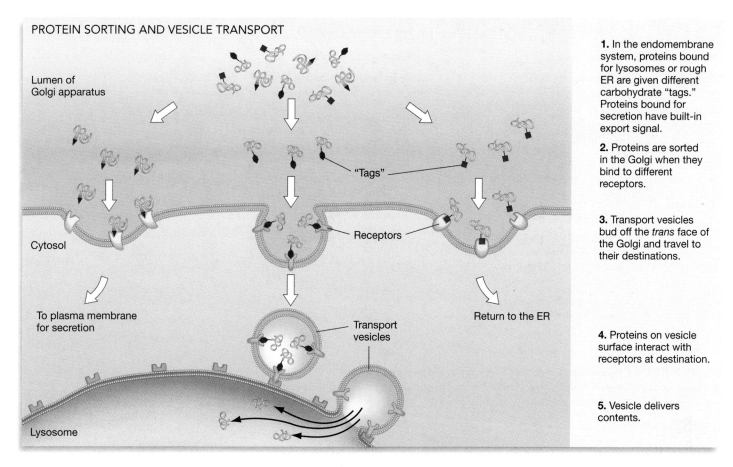

PROTEIN SORTING AND VESICLE TRANSPORT

Lumen of
Golgi apparatus

"Tags"

Receptors

Cytosol

To plasma membrane
for secretion

Transport
vesicles

Return to the ER

Lysosome

1. In the endomembrane system, proteins bound for lysosomes or rough ER are given different carbohydrate "tags." Proteins bound for secretion have built-in export signal.

2. Proteins are sorted in the Golgi when they bind to different receptors.

3. Transport vesicles bud off the *trans* face of the Golgi and travel to their destinations.

4. Proteins on vesicle surface interact with receptors at destination.

5. Vesicle delivers contents.

FIGURE 7.31 In the Golgi Apparatus, Proteins Are Sorted into Vesicles That Are Targeted to a Destination
Summary of the current model for how proteins are sorted into distinct vesicles in the Golgi apparatus and how these vesicles are then targeted to their correct destination.

Figure 7.31 pulls these observations together into a comprehensive model explaining how the products of the endomembrane system are loaded into specific vesicles and shipped to their correct destination. Notice that transport vesicles bound for the plasma membrane secrete their contents to the outside. This process is called **exocytosis** ("outside-cell-act"). When exocytosis occurs, the vesicle membrane and plasma membrane make contact and fuse. The two sets of lipid bilayers rearrange in a way that exposes the interior of the vesicle to the outside of the cell. The vesicle's contents then diffuse away from the cell into the space outside the cell. In this way, cells in your pancreas deliver the hormone insulin to your bloodstream.

The general message of this section is that cells have sophisticated cargo production, sorting, and shipping systems. Proteins that are synthesized in the cytoplasm also have zip codes directing them to mitochondria, chloroplasts, or other destinations.

If vesicles function like shipping containers for products that move between organelles, do they travel along some sort of road or track? What molecule or molecules function as the delivery truck, and does ATP or GTP supply the gas? In general, what physical mechanisms are responsible for moving vesicles to their destination?

✔CHECK YOUR UNDERSTANDING

Ions, ATP, amino acids, and other small molecules diffuse randomly throughout the cell, but the transport of proteins and other large molecules is energy demanding and tightly regulated. Proteins must have the appropriate molecular zip code to enter or leave the nucleus, enter the lumen of the rough ER, or become incorporated into vesicles destined for lysosomes or the plasma membrane. In many cases, proteins and other types of cargo are shipped in vesicles that contain molecular zip codes on their surface. You should be able to (1) propose a hypothesis for how proteins are targeted to chloroplasts, and (2) outline an experiment that would test your hypothesis.

7.4 The Dynamic Cytoskeleton

Based on early observations with light microscopes, biologists viewed the cytoplasm of eukaryotic cells as a fluid-filled space devoid of structure. As microscopy improved, however, researchers realized that the cytoplasm contains an extremely dense and complex network of fibers. This cytoskeleton helps maintain cell shape by providing structural support. It's

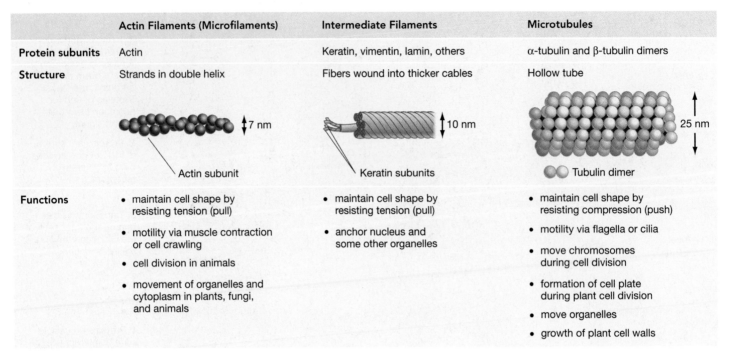

	Actin Filaments (Microfilaments)	Intermediate Filaments	Microtubules
Protein subunits	Actin	Keratin, vimentin, lamin, others	α-tubulin and β-tubulin dimers
Structure	Strands in double helix	Fibers wound into thicker cables	Hollow tube
	7 nm Actin subunit	10 nm Keratin subunits	25 nm Tubulin dimer
Functions	• maintain cell shape by resisting tension (pull) • motility via muscle contraction or cell crawling • cell division in animals • movement of organelles and cytoplasm in plants, fungi, and animals	• maintain cell shape by resisting tension (pull) • anchor nucleus and some other organelles	• maintain cell shape by resisting compression (push) • motility via flagella or cilia • move chromosomes during cell division • formation of cell plate during plant cell division • move organelles • growth of plant cell walls

FIGURE 7.32 The Cytoskeleton Comprises Three Types of Filaments
The three types of filaments found in the cytoskeleton are distinguished by their size and structure, and by the protein subunit of which they are made.

important to recognize, though, that the cytoskeleton is not a static structure like the scaffolding used at construction sites. The fibrous proteins that make up the cytoskeleton move and change to change the cell's shape, to move materials from place to place, and to move the entire structure. Like the rest of the cell, the cytoskeleton is dynamic.

As **Figure 7.32** shows, there are several distinct types of cytoskeletal elements: actin filaments (also known as microfilaments), intermediate filaments, and microtubules. Each of these elements has a distinct size, structure, and function. Let's look at each one in turn.

Actin Filaments

Actin filaments are sometimes referred to as **microfilaments** because they are the cytoskeletal element with the smallest diameter. As Figure 7.32 indicates, **actin filaments** are long, fibrous structures made of a globular protein called actin. In animal cells, actin is often the most abundant of all proteins—typically it represents 5–10 percent of the total protein in the cell. Each of your liver cells contains about half a billion of these molecules.

Actin filaments form when individual actin molecules polymerize. The completed structure resembles two strands that coil around each other. Because each actin monomer in the strand is asymmetrical, the structure as a whole has a distinct polarity. The two ends of an actin filament are different and are referred to as plus and minus ends. Actin filaments tend to grow at the plus end, because polymerization occurs fastest there.

Figure 7.33a shows a fluorescence micrograph of the actin filaments in a mammalian kidney cell. Note that groups of actin filaments are organized into long bundles or dense networks and that actin filaments are particularly abundant just under the plasma membrane. Whether they are arranged in parallel as part of bundles or crisscrossed in networks, individual actin filaments are linked to one another by other proteins. In combination, the bundles and networks of actin filaments help stiffen the cell and define its shape.

Although actin filaments are an important part of the cell's structural support, it would be a mistake to think that they are static. Instead, actin filaments grow and shrink as actin subunits are added or subtracted from each end of the structure. This phenomenon is called *treadmilling*, because the dynamics of the fibers resemble those of a treadmill.

In addition, many cells have actin filaments that interact with the specialized protein **myosin**. When ATP that is bound to myosin is hydrolyzed to ADP, the "head" region of the myosin molecule binds to actin and moves. The movement of this protein causes the actin filament to slide (**Figure 7.34a**). As **Figure 7.34b** shows, the (ATP-powered) interaction between actin and myosin is the basis for an array of cell movements:

• **Cell crawling** occurs in amoebae, slime molds, and certain types of human cells. Cell crawling is based on three processes: a directional extension of actin filaments that pushes the plasma membrane into bulges called **pseudopodia** ("false-feet"), adherence to a solid substrate, and a myosin-driven contraction of actin filaments at the cell's other end. In com-

(a) Fluorescence micrograph of actin filaments in mammalian cells

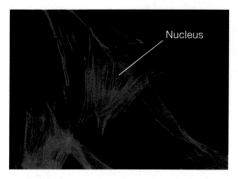

Nucleus

(b) Fluorescence micrograph of intermediate filaments in mammalian cells

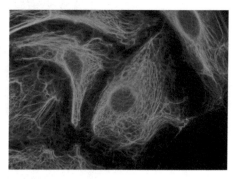

(c) Fluorescence micrograph of microtubules in mammalian cells

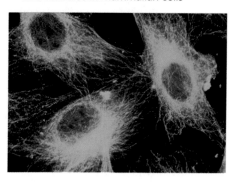

FIGURE 7.33 How Are Cytoskeletal Elements Distributed in the Cell?
To make these micrographs, researchers attached a fluorescent compound to **(a)** actin, the protein subunit of actin filaments, to **(b)** a protein found in intermediate filaments, and to **(c)** tubulin dimers.

bination, the three events result in directed movement by whole cells.

- **Cytokinesis** ("cell-moving") is the process of cell division in animals. For these cells to divide in two, actin filaments that are arranged in a ring under the plasma membrane must slide past one another. Because they are connected to the plasma membrane, the movement of the actin fibers pinches the cell in two.

- **Cytoplasmic streaming** is the directed flow of cytosol and organelles around plant and fungal cells. The movement occurs along actin filaments and is powered by myosin.

In addition, extension of actin filaments is responsible for the expansion of long, thin fungal cells into soil or rotting wood. The same mechanism causes structures called pollen tubes to grow toward the egg cells of plants, so sperm can be delivered prior to fertilization.

Intermediate Filaments

Unlike actin filaments and microtubules, **intermediate filaments** (**Figure 7.33b**) are defined by size rather than composition. Many types of intermediate filaments exist, each consisting of a different protein. In many cases, different types of cells in the same organism contain different types of intermediate filaments. This is in stark contrast to actin filaments and microtubules, which are made from the same protein subunits in all eukaryotic cells. In addition, intermediate filaments are not polar; instead, each end of these filaments is identical. As a result, intermediate filaments do not treadmill, and they are not

(a) Actin and myosin interact to cause movement.

Myosin

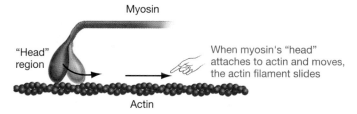

"Head" region

When myosin's "head" attaches to actin and moves, the actin filament slides

Actin

(b) Actin-myosin interactions produce several types of movement.

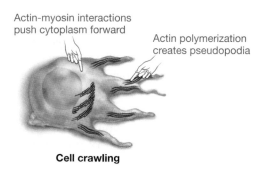

Actin-myosin interactions push cytoplasm forward

Actin polymerization creates pseudopodia

Cell crawling

Actin-myosin interactions pinch membrane in two

Cell division in animals

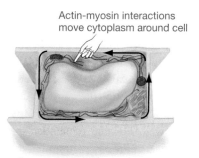

Actin-myosin interactions move cytoplasm around cell

Cytoplasmic streaming in plants

FIGURE 7.34 Many Cellular Movements Are Based on Actin-Myosin Interactions
(a) When the "head" region of the myosin protein binds to ATP or ADP, myosin attaches to actin and changes shape. The movement causes the actin filament to slide. **(b)** Actin-myosin interactions can move cells, divide cells, and move organelles and cytoplasm.

involved in directed movement driven by myosin or related proteins. Intermediate filaments serve a purely structural role in eukaryotic cells.

The intermediate filaments that you are most familiar with belong to a family of molecules called the keratins. The cells that make up your skin and that line surfaces inside your body contain about 20 types of keratin. The presence of these intermediate filaments provides the mechanical strength required for these cells to resist pressure and abrasion. Skin cells secrete another 10 distinct forms of keratin. Depending on the location of the skin cell and keratins involved, the secreted filaments form fingernails, toenails, or hair.

Nuclear lamins, which make up the nuclear lamina layer introduced in Section 7.1, also qualify as intermediate filaments. These fibers form a dense mesh under the nuclear envelope. Recall that in addition to giving the nucleus its shape, they anchor the chromosomes. They are also involved in the breakup and reassembly of the nuclear envelope when cells divide. Some intermediate filaments project from the nucleus through the cytoplasm to the plasma membrane, where they are linked to intermediate filaments that run parallel to the cell surface. In this way, intermediate filaments form a flexible skeleton that helps shape the cell surface and hold the nucleus in place.

Microtubules

Microtubules are composed of the proteins α-tubulin and β-tubulin and are the largest cytoskeletal components in terms of diameter (**Figure 7.33c**). Molecules of α-tubulin and β-tubulin bind to form **dimers** ("two-parts"), compounds formed by the joining of two monomers. Tubulin dimers then polymerize to form the large, hollow tube called a **microtubule**. Because each end of a tubulin dimer is different, each end of a microtubule has a distinct polarity. Like actin filaments, microtubules are dynamic and more likely to grow from one end than they are from the other. Microtubules grow and shrink in length as tubulin dimers are added or subtracted.

Microtubules are similar to actin filaments in function as well as structure. Both cytoskeletal elements provide structural support, and both are involved in cell division. Although microtubules are not involved in the physical division of the cell, they are essential for the directed movement of chromosomes to each of the two resulting cells. In animals and fungi, the microtubules involved in chromosome movement emanate from a structure called the **centrosome**. Distinctive structures called **centrioles** are found inside centrosomes (**Figure 7.35**). Centrioles may help organize microtubules; however, they are not essential for cell division to occur. In plants and many other eukaryotes, a region called the **microtubule organizing center** performs the same function as the centrosome.

Microtubules are involved in many other types of cellular movement as well. For the remainder of this chapter, we'll focus on how microtubules function in moving materials inside cells and in moving the entire cell.

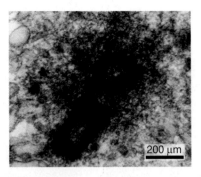

FIGURE 7.35 Centrosomes Are a Type of Microtubule Organizing Center
Microtubules emanate from microtubule organizing centers, which in animals are called centrosomes. The centrioles inside a centrosome are made of microtubules.

Studying Vesicle Transport Materials are transported to a wide array of destinations inside cells. To study how this movement happens, Ronald Vale and colleagues focused on a cell called the giant axon that is found in squid. The giant axon is an extremely large nerve cell that runs the length of a squid's body. If the animal is disturbed, the cell signals muscles to contract so the individual can jet away to safety.

The researchers decided to study this particular cell for three reasons. First, the giant axon is so large that it is relatively easy to see and manipulate. Second, signaling molecules are synthesized in the cell's ER and then transported in vesicles down the length of the cell. As a result, a large amount of cargo moves a long distance. Third, the researchers found that if they gently squeeze the cytoplasm out of the cell, vesicle transport still occurs in the cytoplasmic material.

In short, the squid giant axon provided a cell-free system that could be observed and manipulated efficiently. What did the biologists find out?

Microtubules Act as "Railroad Tracks" To watch vesicle transport in action, researchers mounted a video camera to a microscope. As **Figure 7.36** shows, this technique allowed them to document that vesicle transport occurred along a filamentous track. A simple experiment convinced the group that this movement was an energy-dependent process. If they depleted the amount of ATP in the cytoplasm, vesicle transport stopped.

To identify the type of filament involved, the biologists measured the diameter of the tracks and analyzed their chemical composition. Both types of data indicated that the tracks consisted of microtubules. Microtubules also appear to be required for movement of materials elsewhere in the cell. If experimental cells are treated with a drug that disrupts microtubules, the movement of vesicles from the rough ER to the Golgi apparatus is impaired.

The general message of these experiments is that transport vesicles move through the cell along microtubules. How? Do the tracks themselves move, like a conveyer belt, or are vesicles carried along on some sort of molecular truck?

(a) Micrograph image

(b) Video image (at higher magnification)

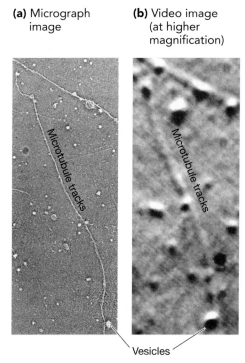

Microtubule tracks

Microtubule tracks

Vesicles

FIGURE 7.36 Vesicles Move along Microtubule Tracks
Transport vesicles moving along microtubules. The images are of extruded cytoplasm from a squid giant axon. **(a)** An electron micrograph that allowed researchers to measure the diameter of the filaments and confirm that they are microtubules. **(b)** A slightly fuzzy but higher-magnification videomicroscope image, in which researchers actually watched vesicles move.

A Motor Protein Generates Motile Forces To study the way vesicles move along microtubules, Vale's group set out to tear the squid axon's transport system apart and then put it back together. To begin, they assembled microtubule fibers from purified α-tubulin and β-tubulin. Then they used differential centrifugation to isolate transport vesicles. But when they mixed purified microtubules and vesicles with ATP, no transport occurred.

Something had been left out—but what?

To find the missing element or elements, the researchers purified one subcellular part after another, using differential centrifugation, and added it to the microtubule + vesicle + ATP system. Through trial and error, they found something that triggered movement. After further purification steps, the researchers finally succeeded in isolating a protein that generated vesicle movement. They named the molecule **kinesin**, from the Greek word *kinein* ("to move"). Like myosin, kinesin is a **motor protein** that converts chemical energy in ATP into mechanical work, just as a car's motor converts chemical energy in gasoline.

Biologists began to understand how kinesin works when X-ray diffraction studies similar to those that revealed the helical nature of DNA revealed the three-dimensional structure of kinesin. As **Figure 7.37a** shows, the protein consists of two intertwined polypeptide chains. It has three major regions: a head section with two globular pieces, a tail, and a stalk that connects the head and tail. Follow-up studies confirmed that the two globular components of the head bind to the microtubule. The tail region binds to the transport vesicle. The kinesin molecule is like a delivery person who carries transport vesicles along microtubule tracks. Cells contain a number of different kinesin proteins, each specialized for carrying a different type of vesicle.

How does kinesin move? More detailed studies of this protein's structure indicated that each of the globular components of the molecule's head has a site for binding ATP as well as a site that binds to the microtubule. To pull these observations together, biologists propose that kinesin transports vesicles by "walking" along a microtubule. The idea is that each part of the head region undergoes a conformational change when it binds ATP. As Chapter 3 showed, these types of shape changes often alter the activity of a protein. As **Figure 7.37b** shows, the ATP-dependent conformational change in kinesin results in a step forward. As each head

(a) Structure of kinesin

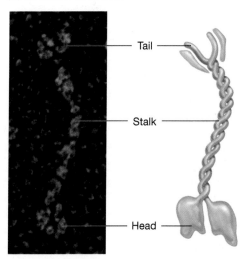

Tail

Stalk

Head

(b) Kinesin "walks" along a microtubule track.

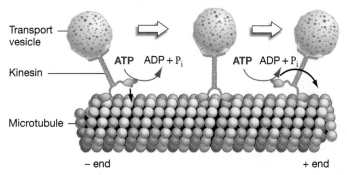

Transport vesicle

Kinesin

Microtubule

ATP   ADP + P_i

ATP   ADP + P_i

− end

+ end

FIGURE 7.37 A Motor Protein Moves Vesicles along Microtubules
(a) Kinesin has three major segments. **(b)** The current model depicting how kinesin "walks" along a microtubule track to transport vesicles. The two head segments act like feet that alternately attach and release in response to the gain or loss of a phosphate group.

alternately binds and hydrolyzes ATP, the protein and its cargo move down the microtubule track.

In short, kinesins move molecular cargo to destinations throughout the cell. They are not the only type of motor protein active inside cells, however. Recall that myosin causes actin filaments to slide, resulting in the movement of cells or cytoplasm. Myosin is also involved in the movement of organelles along tracks made of actin. And a third motor protein, *dynein*, powers the transport of certain organelles as well as swimming movements that move the entire cell. Let's take a closer look at how cells swim.

Cilia and Flagella: Moving the Entire Cell

Flagella are long hairlike projections from the cell surface that function in movement. Flagella are found in many bacteria and eukaryotes. The structure of flagella is completely different in the two groups, however. Bacterial flagella are made of a protein called flagellin; eukaryotic flagella are constructed from microtubules (tubulin). Bacterial flagella move the cell by rotating like a ship's propeller; eukaryotic flagella move the cell by undulating. Eukaryotic flagella are surrounded by plasma membrane; bacterial flagella are not. Based on these observations, biologists conclude that the two structures evolved independently—even though their function is similar.

To understand how cells move, we'll focus on eukaryotic flagella. Eukaryotic flagella are closely related to structures called **cilia** (singular: *cilium*), which are short filamentous projections that are also found in some eukaryotic cells. Unicellular eukaryotes may have either flagella or cilia, while some multicellular organisms have both. In humans, for example, the cells that line the respiratory tract have cilia; sperm cells have flagella.

Flagella are generally longer than cilia, and cells typically have just one or two flagella but many cilia (**Figure 7.38**). But when researchers examined the two structures with the electron microscope, they found that their underlying organization is identical.

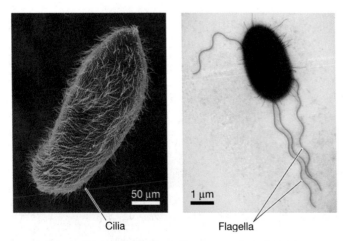

FIGURE 7.38 Cilia and Flagella Differ in Length and Number Cilia are relatively short and large in number; flagella are relatively long and few in number.

How Are Cilia and Flagella Constructed? In the 1950s, anatomical studies established that both cilia and flagella have a characteristic "9 + 2" arrangement of microtubules. As **Figure 7.39a** shows, nine microtubule pairs, or doublets, surround two central microtubules. The doublets, consisting of one complete and one incomplete microtubule, are arranged around the periphery of the structure. The entire 9 + 2 structure is called the **axoneme** ("axle-thread"). The axoneme attaches to the cell at a structure called the **basal body**. The basal body is derived from the centrioles found inside the centrosome. The basal body has a "9 + 0" arrangement of microtubules and plays a central role in the growth of the axoneme.

As electron microscopy improved, biologists gained a more detailed view of the structure. As the sketch in **Figure 7.39b** illustrates, spoke-like structures connect each doublet to the central pair of microtubules. In addition, molecular bridges connect the nine doublets to one another. Finally, each of the doublets has a set of arms that project toward an adjacent doublet. Microtubules are complex. How do their components interact to generate motion?

(a) TEM of axoneme

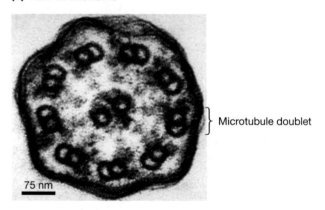

Microtubule doublet

75 nm

(b) Diagram of axoneme

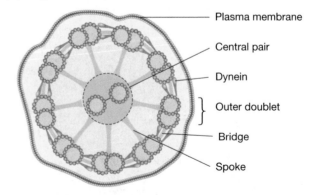

Plasma membrane

Central pair

Dynein

Outer doublet

Bridge

Spoke

FIGURE 7.39 The Structure of Cilia and Flagella
(a) Transmission electron micrograph of a cross section through an axoneme. **(b)** The major structural elements in cilia and flagella. The microtubules are connected by bridges and spokes, and the entire structure is surrounded by the plasma membrane. EXERCISE Label the "9 + 2" arrangement of microtubules

A Motor Protein in the Axoneme In the 1960s Ian Gibbons began studying the cilia of a common unicellular eukaryote called *Tetrahymena*, which lives in pond water. Gibbons found that by using a detergent to remove the plasma membrane that surrounds cilia and then subjecting the resulting solution to differential centrifugation, he could isolate axonemes. Further, the isolated structures would beat if Gibbons supplied them with ATP. These results confirmed that the beating of cilia is an energy-demanding process. They also provided Gibbons with a cell-free system for exploring the molecular mechanism of movement.

In an early experiment with isolated axonemes, Gibbons treated the structures with a molecule that affects the ability of proteins to bind to one another. The axonemes that resulted from this treatment could not bend or use ATP. When Gibbons examined them in the electron microscope, he found that the arms had fallen off. This observation led to the hypothesis that the arms are required for movement. Follow-up work showed that the arms are made of a large protein that Gibbons named **dynein** (from the Greek word *dyne*, meaning "force").

Like myosin and kinesin, dynein is a motor protein. Structural and chemical studies have shown that dynein undergoes a conformational change when a phosphate group from ATP attaches to it. More specifically, the end of a dynein molecule changes shape when it is phosphorylated. This shape change moves the molecule along the nearby microtubule. When the protein reattaches, it has succeeded in walking up the microtubule. This walking motion allows the microtubule doublets to slide past one another. But because each of the nine doublets in the axoneme is connected to the central pair of microtubules by a spoke, and because all of the doublets are connected to each other by molecular bridges, the sliding motion is constrained. So if dynein arms on just one side of the axoneme walk while those on the other side are at rest, the result of the constrained, localized movement is bending (**Figure 7.40**). The result of the bending of cilia or flagella is a swimming motion.

Scaled for size, flagellar-powered swimming can be rapid. In terms of the number of body or cell lengths traveled per second, a sperm cell from a bull moves faster than a human world-record-holder does when swimming freestyle. At the level of the cell, life is fast paced.

✓CHECK YOUR UNDERSTANDING

Each component of the cytoskeleton has a unique structure and set of functions. In addition to providing structural support, actin filaments and microtubules work in conjunction with motor proteins to move the cell or materials inside the cell. Intermediate filaments provide structural support. Most elements of the cytoskeleton are dynamic—they grow and shrink over time. You should be able to predict what will happen when experimental cells are treated with drugs that inhibit formation of each type of cytoskeletal filament.

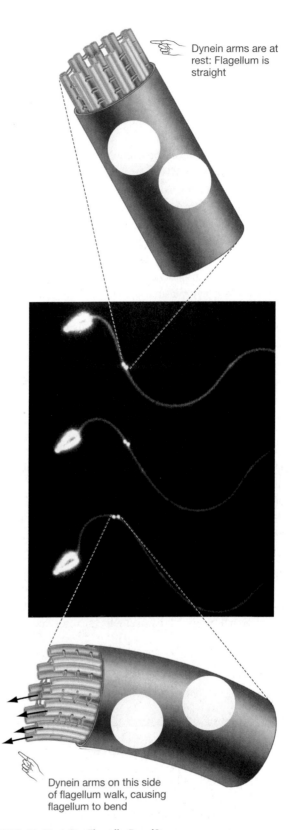

Dynein arms are at rest: Flagellum is straight

Dynein arms on this side of flagellum walk, causing flagellum to bend

FIGURE 7.40 How Do Flagella Bend?
Researchers attached a pair of gold beads to a flagellum and photographed its movement over a short time sequence. As the flagellum bends and beats back and forth, the sperm cell swims forward. When dynein arms walk along the microtubule doublets on one side of a flagellum, the structure bends.

ESSAY Organelles and Human Disease

What happens when organelles malfunction? Given the importance of organelles in the function of eukaryotic cells, it's not surprising that abnormalities in organelles cause disease. In humans, some of the best-understood organelle-based diseases involve lysosomes.

Lysosomes are the cell's recycling center. Defunct organelles and old or damaged RNA, protein, carbohydrate, and lipid molecules are transported to lysosomes. Enzymes inside these organelles then degrade the molecules to simpler components that can be reused. It is essential that these degradative enzymes be confined to the interior of the lysosome. If they escaped into the cytosol, they would threaten to destroy the cell's contents.

Several human diseases result from leakage of lysosomal enzymes into the cytoplasm or the cell exterior. Rheumatoid arthritis is an inflammatory joint disease that is partly caused by the leakage of lysosomal enzymes into the joint fluid. The debilitating lung disease called asbestosis also results from enzyme leakage, even though the population of cells involved is different. Asbestosis is observed in asbestos miners and in construction workers who handled large quantities of asbestos-containing insulation and fireproofing material during their careers. When these individuals breathed asbestos fibers into their lungs, the fibers were localized to the lysosomes of lung cells. But because lysosomal enzymes cannot degrade asbestos fibers, the fibers gradually accumulate inside the organelles. Eventually the buildup of fibers results in damage to the lysosomal membrane and leakage of enzymes into the cell cytoplasm. When many lung cells are harmed in this way, the result is severe coughing and shortness of breath.

In addition to disorders caused by enzyme leakage, about 40 human illnesses are caused by deficiencies in specific lysosomal enzymes. The most severe of these so-called lysosomal storage diseases is inclusion-cell disease, which causes facial and skele-

What happens when organelles malfunction?

tal abnormalities as well as mental retardation. Inclusion-cell disease occurs when most of the degradative enzymes found in lysosomes are missing. Their absence causes the organelles to swell with undigested materials, resulting in structures called inclusions. The swollen lysosomes ultimately cause cell damage, leading to disease symptoms.

Inclusion-cell disease is caused by a deficiency in a single enzyme—the one that is required for attaching mannose-6-phosphate to proteins. Recall from Section 7.3 that this phosphorylated sugar serves as the zip code that targets proteins to lysosomes from the Golgi apparatus. In the absence of the mannose-6-phosphate zip code, enzymes that are normally shipped to lysosomes are instead secreted from the cell. In fact, researchers originally discovered that mannose-6-phosphate serves as a zip code by comparing lysosomal proteins from healthy individuals with the same proteins secreted by individuals with inclusion-cell disease.

Tay-Sachs disease is another lysosomal storage disease. Unlike inclusion-cell disease, Tay-Sachs results from the absence of one particular enzyme from lysosomes. In normal individuals, this protein degrades a type of glycolipid that is abundant in brain cells. In Tay-Sachs patients, the glycolipid accumulates in these cells and disrupts their function. Symptoms include rapid mental deterioration after about 6 months of age, followed by paralysis and death within 3 years.

Although most organelle diseases cannot yet be cured, drug therapies are beginning to offer some hope. For example, the symptoms associated with certain lysosomal storage diseases show improvement when the normal enzyme is provided in pill form. Research continues on this and other strategies for alleviating the symptoms caused by dysfunctional organelles.

CHAPTER REVIEW

Summary of Key Concepts

■ **The structure of cell components is closely correlated with their function.**

Because all organisms consist of cells, many questions in biology can be answered by understanding the structure and function of cells and cell components. There are two basic cellular designs: prokaryotic and eukaryotic. Eukaryotic cells are usually much larger and more structurally complex than prokaryotic cells. Prokaryotic cells consist of a single membrane-bound compartment in which nearly all cellular functions occur. Eukaryotic cells contain numerous membrane-bound compart-

ments called organelles. Organelles allow eukaryotic cells to compartmentalize functions and grow to a large size.

Eukaryotic organelles are specialized for carrying out different functions, and their structure is often correlated closely with their function. Mitochondria and chloroplasts have extensive internal membrane systems, where the enzyme machines responsible for ATP generation and photosynthesis reside. Rough ER is named for the ribosomes that attach to it. Ribosomes are protein-making machines, and rough ER is a site for protein synthesis and processing.

Smooth ER lacks ribosomes because it is a center for lipid synthesis and processing.

■ **Inside cells, materials are transported to their destinations with the help of molecular "zip codes."**

The defining organelle of eukaryotic cells is the nucleus, which contains the cell's chromosomes and serves as its control center. For a cell to function properly, the movement of molecules into and out of the nucleus must be carefully controlled. Traffic across the nuclear envelope occurs through nuclear pores, which contain a multiprotein nuclear pore complex that serves as gatekeeper. Both passive and active transports of materials occur through these nuclear pore complexes. Active import and export of proteins and RNAs involves built-in signals that target cargo to the correct compartment.

The endomembrane system is an extensive, interconnected system of membranes and membrane-bound compartments that can extend from the nucleus to the plasma membrane. Two principal organelles in the endomembrane system are the endoplasmic reticulum (ER) and the Golgi apparatus. The ER is the site of synthesis for a wide array of proteins and lipids. Most ER products are shipped to the Golgi apparatus, which serves as a processing and dispatching station. In many proteins the major processing step is glycosylation, or the addition of carbohydrate groups.

The movement of materials through the endomembrane system is highly organized and takes place inside membrane-bound transport organelles called vesicles. Prior to products leaving the endomembrane system, they are sorted with molecular zip codes that direct them to vesicles headed for their final destination. The vesicles contain proteins that interact with receptor proteins on the surface of a target organelle or the plasma membrane, and allow the contents to be delivered.

■ **Cells are dynamic. Thousands of chemical reactions occur each second within cells; molecules constantly enter and exit across the plasma membrane; cell products are shipped along protein fibers; and elements of the cell's internal skeleton grow and shrink.**

The cell is a membrane-bound structure with a highly organized, dynamic interior. Inside the cell, thousands of different chemical reactions take place at incredible speeds. The products of these chemical reactions allow the cell to acquire resources from the environment, synthesize additional molecules, dispose of wastes, and reproduce.

The cytoskeleton is an extensive system of fibers that serves as a structural support for eukaryotic cells. Elements of the cytoskeleton also provide the machinery for moving vesicles inside cells and for moving the cell as a whole through the beating of flagella or cilia, or cell crawling. Both cell motility and the movement of vesicles inside cells depend on motor proteins, which can convert chemical energy stored in ATP into movement. Within the cell, movement of transport vesicles occurs as the motor protein kinesin "walks" along microtubule tracks. Cilia and flagella bend as the motor protein dynein "walks" along microtubule tracks. The bending motion allows these structures to beat back and forth, enabling cells to swim or generate water currents.

The data reviewed in this chapter provide a view of the cell as a dynamic reaction vessel that synthesizes and ships an array of products in a highly regulated manner. How does all this activity inside the cell relate to what is going on outside? This is the issue taken up in Chapter 8.

Questions

Content Review

1. Which of the following best describe the nuclear envelope?
   a. It is continuous with the endomembrane system.
   b. It is continuous with the nucleolus.
   c. It is continuous with the plasma membrane.
   d. It contains a single membrane and nuclear pores.

2. What is a nuclear localization signal?
   a. a stretch of amino acids that directs proteins from the nucleus to the ER
   b. a molecule that is attached to nuclear proteins so that they are retained inside the nucleus
   c. a signal built into a protein that directs it to the nucleus
   d. a component of the nuclear pore complex

3. Which of the following is not true of secreted proteins?
   a. They are synthesized in ribosomes.
   b. They are transported through the endomembrane system in membrane-bound transport organelles.
   c. They are transported from the Golgi apparatus to the ER.
   d. They contain a signal sequence that directs them into the ER.

4. To find the nuclear localization signal in the protein nucleoplasmin, researchers separated the molecule's core and tail segments, labeled both with a radioactive atom, and injected them into the cytoplasm. Why did the researchers conclude that the signal is in the tail region of the protein?
   a. The protein reassembled and folded into its normal shape spontaneously.
   b. Only the tail segments appeared in the nucleus.
   c. With a confocal microscope, tail segments were clearly visible in the nucleus.
   d. The tail and head segments appeared together in the nucleus.

5. Molecular zip codes direct molecules to particular destinations in the cell. How are these signals read?
   a. They bind to receptor proteins.
   b. They enter transport vesicles.
   c. They bind to motor proteins.
   d. They are glycosylated by enzymes in the Golgi apparatus.

6. The number and size of organelles in a cell correlates with that cell's function. Propose a function for cells that contain extensive rough ER.
   a. rapid cell division in growing bones or muscle tissues
   b. production and processing of fatty acids and other lipids
   c. movement via cell crawling
   d. production of proteins that are secreted from the cell

Conceptual Review

1. Compare and contrast the structure of a generalized plant cell, animal cell, and prokaryotic cell. Which features are common to all cells? Which are specific to certain lineages?

2. Draw a diagram that traces the movement of a secreted protein from its site of synthesis to the outside of a eukaryotic cell. Identify all of the organelles that the protein passes through, and indicate the direction of movement.

3. Describe how a motor protein such as kinesin can move a transport vesicle down a microtubule track. Include all necessary steps and components.

4. Describe the logic of a pulse-chase experiment. How was this approach used to document the pattern of protein transport through the endomembrane system?

5. Briefly describe how researchers use centrifugation to isolate particular cell components for further study.

6. Compare and contrast the structure and function of actin filaments, intermediate filaments, and microtubules. Why is it misleading to refer to the cytoskeleton as "scaffolding"?

Group Discussion Problems

1. In addition to delivering cellular products to specific organelles, eukaryotic cells can take up material from the outside and transport it to specific organelles. For example, specialized cells of the human immune system ingest bacteria and viruses and then deliver them to lysosomes for degradation. Suggest a hypothesis for how this material is tagged and directed to lysosomes. How would you test this hypothesis?

2. The leading hypothesis to explain the origin of the nuclear envelope is that a deep infolding of the plasma membrane occurred in an ancient prokaryote. Draw a diagram that illustrates this infolding hypothesis. Does your model explain the existence of the structure's inner and outer membranes? Explain.

3. Propose a function for cells that contain (a) a large number of lysosomes, (b) a particularly extensive cell wall, and (c) many peroxisomes.

4. Suggest a hypothesis or a series of hypotheses to explain why bacteria, archaea, algae, and plants have cell walls. Suppose that mutant individuals from each group lacked a cell wall. How could you use these individuals to test your idea(s)?

Answers to Multiple-Choice Questions: 1. a; **2.** c; **3.** c; **4.** b; **5.** a; **6.** d

Cell-Cell Interactions

8

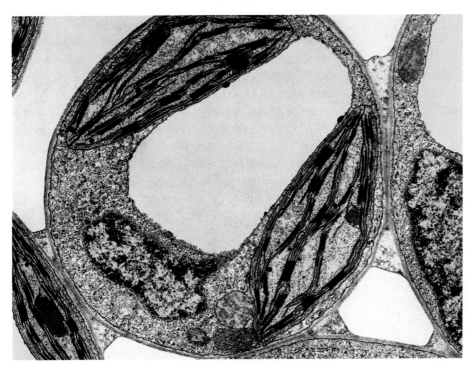

Transmission electron micrograph of the plasma membranes and cell walls (colored yellow) of several plant cells. The cell surface is where adjacent cells join and where signals from distant cells arrive.

KEY CONCEPTS

- Extracellular material strengthens cells and helps bind them together.

- Cell-cell connections help adjacent cells adhere. Cell-cell gaps allow adjacent cells to communicate.

- Intercellular signals are responsible for creating an integrated whole from many thousands of independent parts.

Chapter 6 introduced the structure and function of the plasma membrane, which is the defining feature of the cell. Chapter 7 surveyed the organelles, molecular machines, and cytoskeletal elements that fill the space inside that membrane and explored how cargo moves from sources to destinations within the cell. Both chapters highlighted the breathtaking speed and diversity of events that take place at the cellular level.

The cell is clearly a bustling enterprise. But it would be a mistake to think that cells are self-contained—that they are worlds in and of themselves. Instead, cells constantly interact with other cells. They continuously adjust their activities in response to stimulation from other cells and to changes in environmental conditions.

To understand the life of a cell thoroughly, it is critical to analyze how the cell interacts with the world outside. How do cells obtain information about the outside world and respond to that information? In particular, how do cells interact with nearby cells?

For most unicellular species, the outside environment is made up of soil or water that is teeming with other organisms, either of the same or different species. In your gut, hundreds of billions of bacterial cells are jostling for space and resources. Similar numbers of single-celled organisms are found in every tablespoon of good-quality soil. In addition to interacting with these individuals, each unicellular organism must contend with constantly shifting aspects of the physical environment, such as heat, light, ion concentrations, and food supplies. If unicellular organisms are unable to sense these conditions and respond appropriately, they die.

In multicellular species, the environment outside the cell is made up of other cells, both neighboring and distant. The cells that make up a redwood tree, an *Amanita* mushroom, or your body are intensely social. Although biologists often study cells in isolation, an individual tree, fungus, or person is actually an interdependent community of cells. If those cells do not communicate and cooperate, the whole will break into dysfunctional parts and die.

To introduce the ways that cells interact with each other, this chapter focuses on how cells in a multicellular organism communicate. Let's begin by looking at the plasma membrane and analyzing the molecules involved in cell-cell interactions.

8.1 The Cell Surface

Chapter 6 introduced the currently accepted model for the structure of the plasma membrane. The fluid-mosaic hypothesis contends that the plasma membrane is a phospholipid bilayer with interspersed proteins (**Figure 8.1**). Data from electron micrographs and chemical treatments suggest that these membrane proteins can be either *integral*, meaning that they are embedded in the bilayer, or *peripheral*, meaning that they are attached to one surface. Recall that if an integral protein spans the entire membrane, it is termed a transmembrane protein. Membrane proteins may float free in the lipid bilayer or be anchored by connections to the cytoskeleton or materials outside the cell.

Chapter 6 also analyzed the primary function of the plasma membrane: to create an environment inside the cell that is different from conditions outside. Ions and molecules move across plasma membranes by direct diffusion through the phospholipid bilayer or by means of several types of membrane proteins. Transport of materials across the membrane can be energy demanding or passive, but it is always selective.

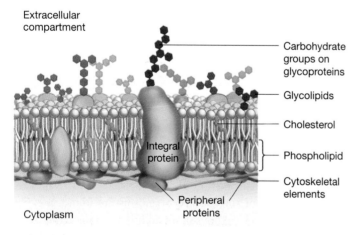

FIGURE 8.1 The Plasma Membrane Is a Mosaic of Phospholipids and Proteins
According to the fluid-mosaic model, the plasma membrane consists of integral and peripheral proteins scattered throughout a phospholipid bilayer.

This picture—of a dynamic, complex plasma membrane that selectively admits or blocks passage of specific substances—is accurate but not complete. The plasma membrane does not exist in isolation. Cytoskeletal elements introduced in Chapter 7 attach to the interior face of the bilayer, and a complex array of extracellular structures exists outside. Let's consider the nature of the material outside the cell first and then analyze how the cell interacts with other cells.

The Structure and Function of an Extracellular Layer

In species from across the tree of life, it is extremely rare for cells to be bounded simply by a plasma membrane. Most cells secrete a layer or wall that forms just beyond the membrane. The extracellular material helps define the cell's shape and either attaches it to another cell or acts as a first line of defense against the outside world. This observation holds for both unicellular organisms and the various types of cells in multicellular species.

Virtually all types of extracellular structures, in turn—from the cell walls of bacteria, algae, fungi, and plants to the extracellular matrix that surrounds most animal cells—follow the same fundamental design principle. Like reinforced concrete and fiberglass, they are "fiber composites": They consist of a cross-linked network of long filaments embedded in a stiff surrounding material, or ground substance (**Figure 8.2a**). The molecules that make up the rods and the encasing material vary from group to group, but the engineering principle is the same.

Fiber composites are a successful design because rods and filaments are extremely effective at withstanding stretching and straining forces, or tension. The steel rods in reinforced concrete and the cellulose fibers in a plant cell wall are unlikely to break as a result of being pulled or pushed lengthwise. Scaled for size, in fact, steel and cellulose are *equally* unlikely to break. Cellulose fibers have the same tensile strength as steel. In addition, the stiff surrounding substance is effective at withstanding the pressing forces called compression. Concrete performs this function in highways, and a gelatinous mixture of polysaccharides achieves the same end in plant cell walls.

Thanks to the combination of tension- and compression-resisting elements, fiber composites are particularly rugged. And in many living cells, the fiber and composite elements are flexible as well as sturdy. When this is the case, the extracellular material is both supple and strong.

What molecules make up the rods and the ground substance found on the surface of plant and animal cells? How are these extracellular layers synthesized, and what do they do?

The Plant Cell Wall

Most plant cells are surrounded by a layer of extracellular material called a cell wall. When new cells first form, they secrete an initial fiber composite designated a **primary cell wall**. As indicated in Chapter 5, the fibrous component of the cell wall consists of long strands of the polysaccharide cellulose, which is cross-linked by polysaccharide filaments and bundled into stout filaments termed microfibrils. Microfibrils form a crisscrossed network that becomes filled with gelatinous polysaccharides (**Figure 8.2b**). Chief among these gel-forming carbohydrates are **pectins**—the molecules that are used to thicken jams and jellies. Pectins are not as strong as cellulose, because the polysaccharides that make up their structure do not form tightly packed filaments strengthened with cross-linkages. Instead, the polysaccharides in pectin are hydrophilic. They attract and hold large amounts of water, forming a gel that helps keep the cell wall moist.

Pectins and other gelatinous polysaccharides that form the ground substance of the cell wall are synthesized in the rough endoplasmic reticulum (ER) and Golgi apparatus and secreted to the extracellular space. The cellulose microfibrils, in contrast, are synthesized by a complex of enzymes in the plasma membrane itself and "spun out" into the exterior.

The primary cell wall defines the shape of a plant cell. Under normal conditions, cytoplasm fills the entire volume of the cell and pushes the plasma membrane up against the wall. Because the concentration of solutes is higher inside the cell than outside, water tends to enter the cell via osmosis. The incoming water inflates the plasma membrane, exerting a force against the wall that is known as **turgor pressure**. Although plant cells experience turgor pressure throughout their lives, it is particularly important in young cells that are actively growing. Young plant cells secrete enzymes named *expansins* into their cell-wall matrix. Expansins catalyze reactions that allow the cellulose microfibrils in the matrix to slide past one another. Turgor pressure then forces the wall to elongate and expand. The result is cell growth.

As certain plant cells mature and stop growing, they secrete a layer of material—designated a **secondary cell wall**—inside the primary cell wall. The structure of the secondary cell wall varies from cell to cell in the plant and correlates with that cell's function. For example, cells on the surface of a leaf have cell walls that are impregnated with waxes that form a waterproof coating. Cells that furnish structural support for the plant's stem have secondary cell walls that contain a great deal of cellulose. In some species, the secondary cell walls of support cells include a complex and extremely tough substance named **lignin**. Lignin forms a rigid network that is exceptionally stiff and

(a) Fiber composites resist tension and compression.

(b) Primary cell walls of plants are fiber composites.

FIGURE 8.2 What Is the Nature of a Fiber Composite?
(a) Fiber composites consist of a massive ground substance that fills spaces between cross-linked rods. In reinforced concrete, the rods or fibers are made of steel, while the ground substance is concrete. The finished material is strong because the fibers resist tension and the ground substance resists compression. **(b)** In the primary plant cell wall, cellulose fibers are cross-linked by polysaccharide chains. The spaces between the fibers are filled with pectin molecules, which form a gelatinous solid.

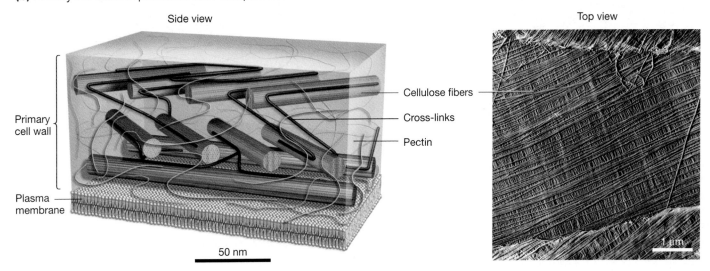

Side view

Primary cell wall

Plasma membrane

Cellulose fibers

Cross-links

Pectin

50 nm

Top view

1 μm

strong. Cells that form wood secrete a secondary cell wall that contains large amounts of lignin. Cells that have thick cell walls of cellulose and lignin help plants withstand the forces of gravity and wind.

Plant cell walls are also dynamic. If they are damaged by insect attack, they may release signaling molecules that trigger the reinforcement of walls in nearby cells. When a seed germinates (sprouts), the walls of cells that store oils or starch are actively degraded, releasing these nutrients to the growing plant. Cell walls are also degraded in a controlled way as fruits ripen, making the fruits softer and more digestible for the animals that disperse the seeds inside.

Although animal cells do not make a cell wall, they do form a fiber composite outside their plasma membrane. What is this substance, and what does it do?

The Extracellular Matrix in Animals

Virtually all animal cells secrete a fiber composite called the **extracellular matrix (ECM)**. Although the design of this substance follows the same principles observed in the cell walls of bacteria, archaea, algae, fungi, and plants, the fibrous component of the animal ECM consists of protein fibers instead of polysaccharide filaments. **Figure 8.3a** illustrates the most abundant protein found in the ECM—a fibrous, cable-like molecule termed **collagen**. As in cell walls, the matrix that surrounds collagen and the other fibrous components consists of gel-forming polysaccharides. But because collagen and the other common ECM proteins are much more elastic and bendable than cellulose or lignin, the structure as a whole is relatively pliable. Most ECM components are synthesized in the rough ER, processed in the Golgi apparatus, and secreted from the cell via exocytosis. Some of the polysaccharides that form the composite material are synthesized by membrane proteins, however.

The amount of ECM varies among different types of cells in the same organism. In human bone and cartilage, for example, the number of cells is relatively small and the amount of ECM very large (**Figure 8.3b**). Skin cells, in contrast, are packed together with a minimal amount of ECM between them. The composition of the ECM also varies among cell types. For example, the ECM of lung cells contains large amounts of a rubber-like protein called elastin, which allows the ECM to expand and contract during breathing movements. The structure of a cell's ECM correlates with that cell's function.

Wherever it is found in the body, one of the ECM's most important functions is structural support—a common theme in the extracellular materials found in other organisms. ECM provides support because the combination of protein filaments and surrounding polysaccharide gel creates a strong external layer, and because cells "grab" the layer via a chain of proteins. As **Figure 8.4** shows, actin filaments inside the plasma membrane are connected to transmembrane proteins called **integrins**. The integrins bind to nearby proteins in the ECM named **fibronectins**, which in turn bind to collagen fibers. This direct linkage between the cytoskeleton and ECM is critical. In addition to keeping individual cells in place, it helps adjacent cells adhere to each other via their common connection to the ECM.

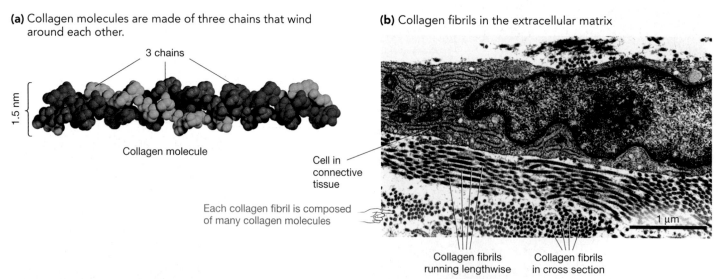

(a) Collagen molecules are made of three chains that wind around each other.

3 chains

1.5 nm

Collagen molecule

Each collagen fibril is composed of many collagen molecules

(b) Collagen fibrils in the extracellular matrix

Cell in connective tissue

1 µm

Collagen fibrils running lengthwise

Collagen fibrils in cross section

FIGURE 8.3 The Extracellular Matrix Is a Fiber Composite
(a) Although several types of protein filaments are found in the ECM of animal cells, the most abundant is collagen. A collagen molecule consists of three strands that wind around each other. Groups of collagen molecules coalesce to form collagen microfibrils, and bundles of microfibrils link up to form collagen fibers. **(b)** A cross section of ECM from monkey cartilage tissue, showing a cell surrounded by abundant ECM. The spaces between the collagen fibers are filled with gelatinous polysaccharides.

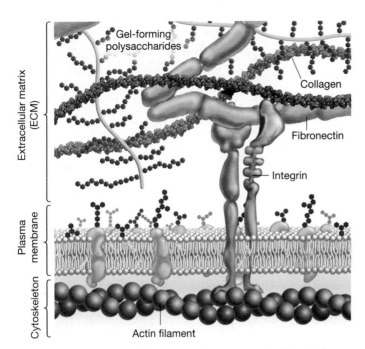

FIGURE 8.4 The Extracellular Matrix Connects to the Cytoskeleton
EXERCISE Circle the direct physical connection among actin filaments, integrins, fibronectin, and collagen fibers.

The cytoskeleton-ECM linkage is also important in the development of cancer. When the cell-ECM connection is lost—through mechanisms that are not well understood—cells that are growing in an uncontrolled fashion can leave the site of a tumor and begin migrating throughout the body. Eventually these cells settle in a new region of the body and seed the for-

mation of new tumors. The result is the development of full-blown cancer. **Box 8.1** details other types of maladies that result from breakdown of the ECM.

✔CHECK YOUR UNDERSTANDING

Most cells secrete a layer of structural material that stiffens the cell and helps define its shape. Although the types of molecules involved vary among species, in almost all cases the extracellular material is a fiber composite—a combination of cross-linked filaments surrounded by a ground substance. You should be able to (1) diagram the general structure of a fiber composite and (2) compare and contrast the molecular composition of a plant cell wall and the ECM of animal cells.

8.2 How Do Adjacent Cells Connect and Communicate?

To understand a cell fully, it is important to appreciate it as a social entity. Cells continually interact with other cells. This is true even for unicellular organisms, which usually live in habitats packed with other unicellular species. For example, the 500 species of bacteria and archaea that inhabit your mouth compete with each other for space and nutrients. Like many other prokaryotes that are adapted for life on a solid surface, certain oral bacteria secrete a hard, polysaccharide-rich substance that accumulates outside their cell walls. This substance creates a **biofilm** that encases the cells and attaches them to the surface.

BOX 8.1 What Happens When the Extracellular Matrix Is Defective?

Given the extent of ECM surrounding animal cells, it is not surprising that defects in this structure have serious consequences. Consider, for example, the effects of collagen breakdown. One of collagen's most common constituents is the amino acid proline. When collagen fibers first assemble in the ECM, many of the proline residues present in the newly synthesized structures are oxidized to hydroxyproline. This reaction depends on the molecule ascorbic acid, also called vitamin C. If vitamin C is lacking in the diet, hydroxyproline formation slows and collagen fibers begin to weaken or disintegrate. The ECM surrounding capillaries is particularly sensitive to this problem. Capillaries are the smallest blood vessels in the body. When collagen fibers in the ECM of capillary

cells weaken, the structure as a whole begins to lose its integrity. Blood cells begin leaking out of capillaries, forming bruises. If vitamin C deficiency continues, capillaries weaken to the point where extensive internal bleeding occurs. The potentially fatal disease called *scurvy* results. Scurvy is readily cured by eating citrus fruits or other foods rich in vitamin C.

Skin is the other major organ that is sensitive to collagen damage. Skin cells are layered over a thick, collagen-rich ECM called the **basal lamina**. For reasons that are not completely clear, exposure to intense sunlight signals the cells to produce and secrete the enzyme metalloproteinase-1, which catalyzes a reaction that cleaves collagen fibers. As a result, people who work for long periods in

bright sunlight or who cultivate a tanned skin are almost certain to develop what dermatologists called *photoaging*—a premature aging of the skin caused by the loss of collagen.

Defects in ECM components other than collagen can also cause problems. *Marfan syndrome* develops in people who inherit a genetic defect that prevents them from manufacturing the normal form of the protein fibrillin. Along with elastin, fibrillin is a key component of the elastic fibers in the ECM. People with Marfan syndrome tend to have exceptionally long fingers and be extremely tall and thin. They are also susceptible to severe heart problems. Based on his appearance, biomedical researchers have hypothesized that Abraham Lincoln may have had Marfan syndrome.

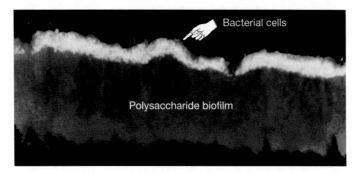

FIGURE 8.5 Unicellular Organisms May Form Aggregations
The yellow band consists of *Pseudomonas aeruginosa* cells—a common soil bacterium that also causes infections in cystic fibrosis patients. The red layer is a biofilm secreted by the cells.

Biofilms that form in the mouth are termed **dental plaque**, and they contain an array of species. In some cases, biofilm formation is actually a product of cooperation among members of the same species. When cells of the bacterium *Pseudomonas aeruginosa* settle on a lung cell, for instance, they secrete a signaling molecule that recruits other *P. aeruginosa* cells to the site. The cells then secrete a biofilm that attaches them to the surface of the lung and protects them from attack by your immune system cells or antibiotics (**Figure 8.5**).

Although unicellular species may live in close proximity and even communicate with each other, they do not make physical connections. Physical connections between cells are the basis of **multicellularity** (**Figure 8.6**). This should come as no surprise. The billions of cells that make up a rose bush, a bracket fungus, or a mountain gorilla must be held together somehow. And because cells in multicellular species are specialized for particular tasks, the cells must be able to communicate with each other to function as an integrated whole. It is logical, then, that rose and fungus and gorilla cells are organized into functional units, called tissues. **Tissues** consist of groups of similar cells that perform a similar function. In your body, individual muscle cells are grouped into muscle tissue that contracts and extends to make movement possible. Several tissues, in turn, combine to

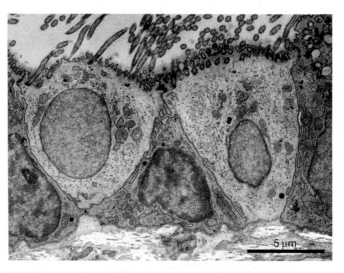

FIGURE 8.6 In Multicellular Organisms, Cells Are Connected
A cross section through cells that line the human trachea.
EXERCISE Label two types of cells: ciliated cells that sweep debris from the trachea and cells that secrete mucus, which traps bacteria and viruses.

make up the integrated structures called **organs**. Organs perform specialized functions such as reproduction, digestion, or support. Flowers, roots, and stems are organs; your gonads, small intestine, and bones also are organs.

A look at the phylogeny in **Figure 8.7** should convince you that multicellularity has evolved numerous times, in an array of independent groups. Based on this observation, it is logical to expect that each multicellular group has different types of intercellular ("between-cell") connections and communication mechanisms to organize their tissues and organs.

Many of the structures responsible for cell-cell adhesion and communication in these groups were first described in the 1950s, when transmission electron microscopy first made it possible to examine tissues at high magnification. In the decades that followed, researchers made exciting progress in understanding the molecular nature of cell-cell attachments and openings. Let's look first at the structures that physically

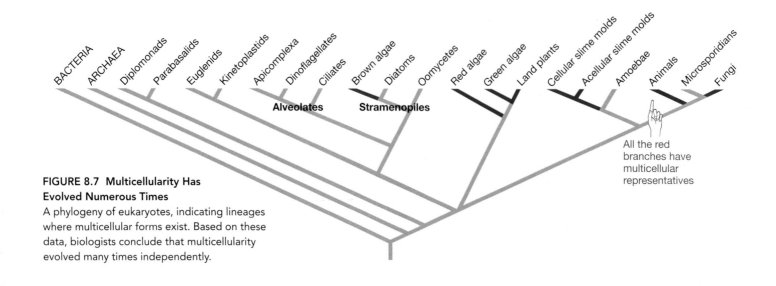

FIGURE 8.7 Multicellularity Has Evolved Numerous Times
A phylogeny of eukaryotes, indicating lineages where multicellular forms exist. Based on these data, biologists conclude that multicellularity evolved many times independently.

All the red branches have multicellular representatives

attach adjacent cells to each other, then at the openings that allow nearby cells to exchange materials and information.

Cell-Cell Attachments

The structures that hold cells together vary among multicellular organisms. To illustrate this diversity, consider the intercellular connections observed in the best-studied groups of organisms: plants and animals. Electron microscopy showed that a distinctive layer of material joins plant cells together. The same tool detailed the structure of cell-cell attachment points in animal cells.

As **Figure 8.8** indicates, the extracellular space between adjacent plant cells is composed of three layers. The primary cell walls of adjacent cells sandwich a central layer designated the **middle lamella**, which consists primarily of gelatinous pectins. Because this gel layer is continuous with the primary cell walls of the adjacent cells, it serves to glue them together. If enzymes degrade the middle lamella, as they do when flower petals and leaves detach and fall, the surrounding cells separate.

In many animal tissues, integrins connect the cytoskeleton of each cell to the extracellular matrix (see Section 8.1). A middle-lamella-like layer of gelatinous polysaccharides runs between adjacent cells, so cytoskeleton-ECM connections help hold individual cells together. But in certain animal tissues, the polysaccharide glue is reinforced by protein girders that span the ECM to connect adjacent cells. These structures are particularly important in **epithelia** (singular: **epithelium**)—tissues that form external and internal surfaces. Epithelial cells form a layer that separates parts of the body that contain different solutions—

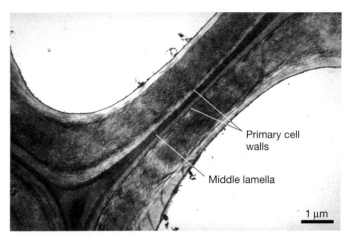

FIGURE 8.8 The Middle Lamella Connects Adjacent Plant Cells The middle lamella, which holds adjacent plant cells together, is dominated by pectins, a family of gelatinous polysaccharides.

urine versus blood, for example. Epithelial cells must be sealed to prevent mixing of solutions. Although a variety of cell-cell attachment structures exist in epithelia and other tissues, we'll analyze just two: tight junctions and desmosomes.

Tight Junctions   The electron micrograph in **Figure 8.9a** shows a cross section through cells that are linked by tight junctions. A **tight junction** is a cell-cell attachment composed of specialized proteins in the plasma membranes of adjacent animal cells. As the sketch in **Figure 8.9b** indicates, these proteins line up and bind to one another. The resulting structure resembles quilting, with the proteins acting as stitches.

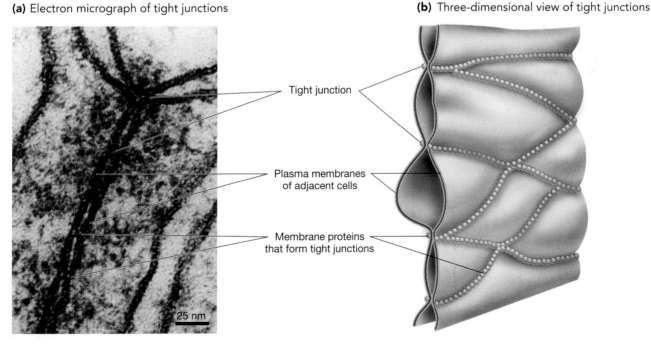

(a) Electron micrograph of tight junctions

(b) Three-dimensional view of tight junctions

Tight junction

Plasma membranes of adjacent cells

Membrane proteins that form tight junctions

FIGURE 8.9 In Animals, Tight Junctions May Form a Seal between Adjacent Cells **(a)** A cross section through adjacent cells that are stitched together by a tight junction. **(b)** Proteins in the membranes of each cell are arranged in a quilted pattern and bind to one another to form the junction.

Tight junctions form a watertight seal between membranes. Based on this observation, it is not surprising that tight junctions are common in cells that form a barrier, such as the cells lining your stomach and intestines. The presence of tight junctions keeps ions and molecules in your gut contents from leaking between stomach cells or intestinal cells and from diffusing into your body. Instead, only selected nutrients enter the cells. These ions and molecules are admitted via specialized transport proteins and channels in the plasma membrane.

Although tight junctions are tight, they are also variable and dynamic. Their structure and function differ in different types of epithelia—in skin versus the lungs, for example. Tight junctions are dynamic because they disassemble and reassemble to allow the passage of immune system cells between epithelial cells.

Desmosomes Figure 8.10a illustrates another type of cell-cell connection found in animals: desmosomes. **Desmosomes** are particularly common cell-cell attachments in epithelial cells and in certain types of muscle cells. The structure and function of a desmosome are analogous to the rivets that hold pieces of sheet metal together. As **Figure 8.10b** indicates, though, desmosomes are extremely sophisticated rivets. At their heart are proteins that form a physical link between the cytoskeletons of the adjacent cells. In addition to binding to each other, these proteins bind to larger proteins that anchor intermediate filaments in the cytoskeletons of the two types of cells. In this way, desmosomes bind together the cytoskeletons of the adjacent cells.

The proteins that form desmosomes may be involved in attaching animal cells long before a desmosome is complete, however. Early in the development of an animal, an embryo's first tissues form when selected cells begin to adhere to each other. Mature desmosomes and adult tissues and organs form much later. Which proteins are involved in these most basic types of cell-cell adhesion, early in development?

Selective Adhesion Long before electron micrographs revealed the presence of desmosomes, biologists realized that some sort of molecule must bind animal cells to each other. This insight grew out of experiments that H. V. Wilson conducted on the cells of sponges in the early 1900s. Sponges are aquatic animals, and the sponge species that Wilson worked with was composed of just two basic types of cells. When Wilson treated adult sponges with chemicals that made the cells dissociate from each other, the result was a jumbled mass of individual and unconnected cells. But when normal chemical conditions were restored, the cells gradually began to move and stick to other cells. With time, cells of each type began to aggregate and adhere to cells of the same tissue type. This phenomenon came to be called **selective adhesion**. In Wilson's experiment, the cells eventually re-formed complete and functional adult sponges.

In an even more dramatic experiment, Wilson dissociated the cells of adult sponges from two differently pigmented

(a) Micrograph of desmosome

(b) Three-dimensional view of desmosome

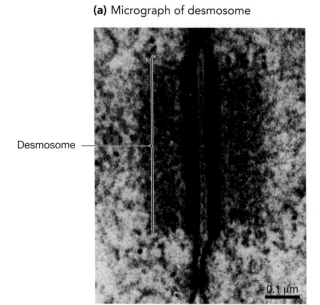

Desmosome

0.1 μm

Plasma membranes of adjacent cells

Anchoring proteins in each cell

Membrane proteins that link cells

Intermediate filaments

FIGURE 8.10 Adjacent Animal Cells Are Riveted Together by Desmosomes, Binding Together Cytoskeletons
(a) A cross section through a desmosome, showing the connections between adjacent cells and the intermediate filaments that link the desmosome to the cytoskeleton of each cell. **(b)** The major components of a desmosome. EXERCISE Add lines connecting the structures in part (a) to the labels in part (b).

sponge species and randomly mixed them together in a culture dish. As **Figure 8.11** shows, the cells eventually sorted themselves out into distinct aggregates containing cells from only one species and from only one cell type. These results implied that the cells had some way of physically linking to each other and that the linkage was specific to the species and cell type involved. Based on these observations, researchers hypothesized that there must be some sort of molecule on the surface of animal cells that attaches only to cells of the same species and cell or tissue type.

Question: Do animal cells adhere selectively?

Hypothesis: Cells of the same tissue type and from the same species have a mechanism for selectively adhering to each other.

Null hypothesis: Cells do not adhere or they adhere to each other randomly, not selectively.

Experimental setup:

1. Start with two adult sponges of different species.

2. Use a chemical treatment to dissociate the cells.

3. Mix cells of both tissues from the two species.

Prediction: Cells from the same tissue type will adhere to one another, and cells from the same species will adhere to one another.

Prediction of null hypothesis:

Results:

Cells spontaneously reaggregate into correct tissues and species.

Conclusion: Cells of the same tissue type and species have specific adhesion molecules or mechanisms.

FIGURE 8.11 Do Animal Cells Adhere Selectively?
QUESTION What is the prediction made by the null hypothesis?

The Discovery of Cadherins What is the molecular nature of selective adhesion? The initial hypothesis, proposed in the 1970s, was that certain membrane proteins might be involved. The idea was that if different types of cells produce different types of adhesion proteins in their membranes, the molecules might interact in a way that anchors cells of the same type to each other.

This hypothesis was tested through experiments that relied on molecules called antibodies. An **antibody** is a protein that binds specifically to a section of another protein. To test the hypothesis that cell-cell adhesion takes place via interactions between membrane proteins, researchers pursued the following strategy (**Figure 8.12**, page 168):

1. Isolate the membrane proteins from a certain cell type. Create pure preparations of each protein.

2. Inject one of the membrane proteins into a rabbit. The rabbit's immune system cells respond by creating antibodies to the membrane protein. Then purify antibodies that bind tightly to the injected membrane protein. Repeat this procedure for the other membrane proteins that were isolated. In this way, obtain a large collection of antibodies—each of which binds specifically to one (and only one) type of membrane protein.

3. Add one antibody type to the mixture of dissociated cells. Observe whether the cells are able to reaggregate normally. Repeat this experiment with the other antibody types, one type at a time. (Three different antibodies are shown in Figure 8.12, so three experimental treatments are required.)

4. If treatment with a particular antibody prevents the cells from attaching to each other, the antibody is probably bound to an adhesion protein. The hypothesis is that the interaction between antibody and adhesion protein blocks the adhesion protein and keeps it from functioning properly. In effect, the antibody "shakes hands" with the adhesion protein, which prevents the adhesion protein from "shaking hands" with other adhesion proteins to attach cells.

Based on studies such as this, researchers have identified several major classes of cell adhesion proteins. The molecules in desmosomes belong to a group that came to be named **cadherins**. Each major type of cell in the body has a different type of cadherin in its plasma membrane, and each type of cadherin can bind only to cadherins of the same type. In this way, cells of the same tissue type attach specifically to one another.

Animal cells attach to each other in a selective manner because only certain types of cell adhesion proteins can bind and rivet cells together. Cadherins provide the physical basis for most selective adhesion and are a critical component of the desmosomes that join mature cells.

Question: Do animal cells have adhesion proteins on their surfaces?

Hypothesis: Selective adhesion is due to specific membrane proteins.

Null hypothesis: Selective adhesion is not due to specific membrane proteins.

Experimental setup:

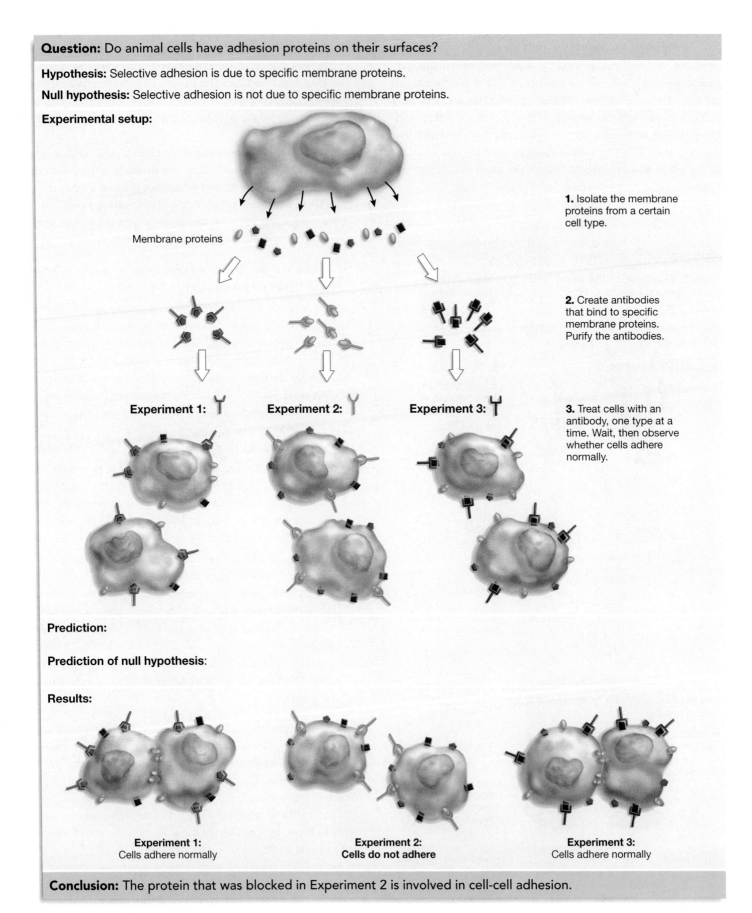

Membrane proteins

Experiment 1:

Experiment 2:

Experiment 3:

1. Isolate the membrane proteins from a certain cell type.

2. Create antibodies that bind to specific membrane proteins. Purify the antibodies.

3. Treat cells with an antibody, one type at a time. Wait, then observe whether cells adhere normally.

Prediction:

Prediction of null hypothesis:

Results:

Experiment 1:
Cells adhere normally

Experiment 2:
Cells do not adhere

Experiment 3:
Cells adhere normally

Conclusion: The protein that was blocked in Experiment 2 is involved in cell-cell adhesion.

FIGURE 8.12 Do Animal Cells Have Adhesion Proteins on Their Surfaces?
EXERCISE Fill in the prediction made by each hypothesis.

Cell-Cell Gaps

Once cells of the same tissue type are bound together, how do they communicate? Careful analysis of electron micrographs answered this question. In both plants and animals, direct connections between cells in the same tissue help the cells work in a coordinated fashion.

In plants, gaps in cell walls create direct connections between the cytoplasm of adjacent cells. At these connections, named **plasmodesmata** (singular: *plasmodesma*), the plasma membrane and the cytoplasm of the two cells are continuous. In addition, smooth endoplasmic reticulum (ER) runs through the hole (**Figure 8.13a**). Growing evidence suggests that plasmodesmata also contain proteins that regulate the passage of specific proteins, making the connections similar in function to the nuclear pore complex, which we introduced in Chapter 7.

At least some of the proteins that are transported through plasmodesmata are involved in coordinating the activity of adjacent cells.

In most animal tissues, adjacent cells are connected by holes in the ECM and adjacent plasma membranes. These holes, known as **gap junctions**, are created by specialized proteins that line each hole (**Figure 8.13b**). The proteins in gap junctions form channels that admit water, ions, and small molecules such as amino acids, sugars, and nucleotides. The flow of small molecules can help adjacent cells communicate and coordinate their activities by allowing the rapid passage of regulatory ions or molecules. In the muscle cells of your heart, for example, a flow of ions through gap junctions acts as a signal that coordinates contractions. Without this cell-cell communication, a normal heartbeat would be impossible.

(a) Plasmodesmata create gaps that connect plant cells.

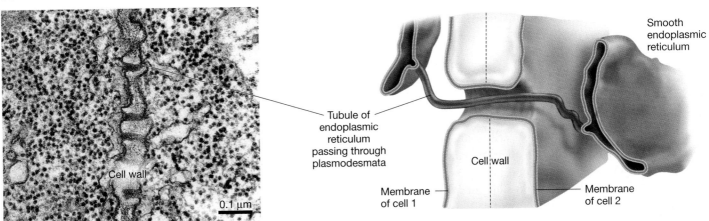

Smooth endoplasmic reticulum

Tubule of endoplasmic reticulum passing through plasmodesmata

Cell wall

Cell wall

Membrane of cell 1

Membrane of cell 2

0.1 μm

(b) Gap junctions create gaps that connect animal cells.

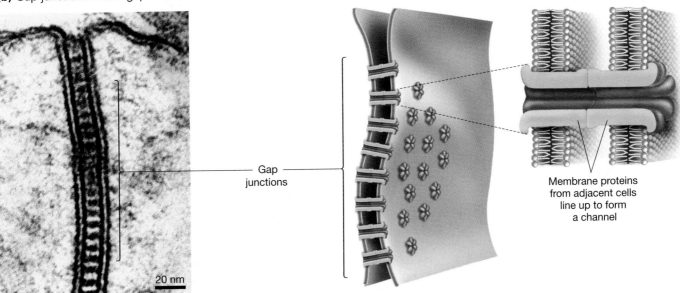

Gap junctions

Membrane proteins from adjacent cells line up to form a channel

20 nm

FIGURE 8.13 Adjacent Plant Cells and Adjacent Animal Cells Communicate Directly
(a) Plasmodesmata are holes in plasma membranes and cell walls of adjacent plant cells. A tubule of smooth ER runs through the hole. **(b)** Gap junctions are collections of protein-lined holes in the membranes of adjacent animal cells.

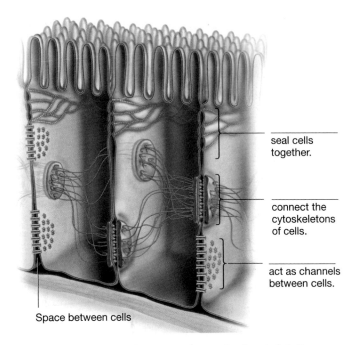

seal cells
together.

connect the
cytoskeletons
of cells.

act as channels
between cells.

Space between cells

FIGURE 8.14 An Array of Structures Is Involved in Cell-Cell Adhesion and Communication
EXERCISE In the blanks on the right, write the name of each type of cell-cell attachment.

In effect, plant cells and animal cells share most or all of the ions and small molecules in their cytoplasm but retain their own organelles, proteins, and nucleic acids. The existence of cell-cell gaps makes it possible for adjacent cells to communicate with each other efficiently. A tissue can act as an integrated whole if its component cells are connected by gaps in their extracellular material and in their plasma membranes (**Figure 8.14**).

How do more distant cells in a multicellular organism communicate? For example, suppose that leaf cells in a maple tree are being attacked by caterpillars or that the muscle cells in your arm are exercising so hard that they begin to run low on sugar. How do these cells signal tissues or organs elsewhere in the body to release materials that might be needed to fend off caterpillars or exhaustion?

✓ CHECK YOUR UNDERSTANDING

Adjacent plant cells and adjacent animals cells are physically connected. In addition, both plant cells and animal cells communicate with each other through gaps in their plasma membranes. You should be able to (1) compare and contrast the structure and function of the middle lamella of plants and the tight junctions and desmosomes of animals, (2) describe the structure and function of plasmodesmata and gap junctions, and (3) explain the molecular mechanism of selective adhesion.

8.3   How Do Distant Cells Communicate?

As you read this, molecules that carry information about conditions throughout your body are traveling through your bloodstream and arriving at cells from your head to your toes. These molecules might signal that viruses are attacking cells in your nose or that the amount of glucose in your blood is low. The signals are intended for tissues or organs that can respond appropriately. If the signal was triggered by the presence of a virus, then muscles in nearby blood vessels will relax—opening the vessels and increasing blood flow to the affected area. If the molecule indicates that glucose levels are low, cells in your liver that store glucose will release the sugar into your blood. Similar types of molecules course through the plants on your campus. They may carry signals about the direction and rate of growth in stems and roots, the direction of incoming sunlight and the current length of daylight, or the site and severity of attacks by insects or viruses. As in animals, these signals are meant for target cells that can respond in a way that increases the individual's chances of surviving and reproducing.

In a multicellular organism, it is logical that chemical signals convey information from one tissue type or organ to another. If the activities of different cells, tissues, and organs are to be coordinated, then information has to move back and forth among them. How do cells receive and respond to signals from other cells? Researchers began to get a handle on this question when they realized that intercellular signaling is based on a four-step process: (1) receiving a signal, (2) processing it, (3) responding to it, and (4) turning it off.

Signal Reception

Intercellular signals are chemical messengers. They deliver their message by binding to receptor molecules. Any molecule that binds to a specific site on a receptor molecule is a **ligand**. A ligand that acts as an intercellular signal is termed a **hormone**. Hormones are usually small molecules and are often present in minute concentrations, yet they significantly affect the activity of target cells. They are like a fleeting scent or whispered phrase from someone that you are attracted to—a tiny signal, but one that makes your cheeks flush and your heart pound.

As **Table 8.1** indicates, the chemical structure of plant and animal hormones varies widely. Hormones also have a wide array of effects. Despite this diversity in structure and function, all hormones have the same general role in the organism: They coordinate the activities of cells in response to information from outside or inside the body.

The messages carried by hormones are received by **signal receptors**—proteins that bind to a particular signal and change conformation or activity in response. If the signal receptors respond to steroid hormones, then these receptors are located inside the cell. This observation is not surprising, because steroid hormones are lipid soluble and thus readily diffuse through the plasma membrane. But because most hormones are not lipid soluble, most signal receptors are located in the plasma membrane.

TABLE 8.1 Hormones Have Diverse Structures and Functions

Hormone Name	Chemical Structure	Function of Signal
Nitric oxide	NO (a gas)	Produces short-lived messages in animal cells
Ethylene	C_2H_4 (a gas)	Stimulates fruit ripening, regulates aging
Insulin	Protein, 51 amino acids	Stimulates glucose uptake in animal bloodstream
Systemin	Peptide, 18 amino acids	Stimulates plant defenses against herbivores
Estrogen	Steroid	Stimulates development of female characteristics in animals
Brassinosteroids	Steroid	Stimulates plant cell elongation
Prostaglandins	Modified fatty acid	Perform a variety of functions in animal cells
Oligosaccharides	Carbohydrate	Triggers defensive reactions in plants
Thyroxine (T4)	Modified amino acid	Regulates metabolism in animals
FSH	Glycoprotein	Stimulates egg maturation, sperm production in animals
Auxin	Small organic compound	Signals changes in long axis of plant body

No matter where they are located, signal receptors have several important general characteristics:

- Each type of receptor is restricted to particular cell types. Your liver cells, for example, contain receptors for the hormone insulin, which signals that glucose levels in the blood are too high. The presence of insulin receptors on liver cells makes sense because liver cells store glucose. But insulin receptors are not present in bone cells or skin cells or other cell types that cannot help regulate glucose levels.

- Receptors are dynamic. The number of receptors in a particular cell may decline if hormonal stimulation occurs at high levels over a long period of time. The ability of a receptor molecule to bind tightly to a hormone may also decline in response to intensive stimulation.

- Receptors can be blocked. The drugs called beta-blockers, for example, bind to certain receptors for the chemical signal *epinephrine*. Because the binding of epinephrine to these receptors stimulates heart cells to contract, beta-blockers are prescribed to lower heart rate and thus blood pressure.

The most important and most general characteristic of signal receptors, though, is that their physical conformation changes when a hormone binds to them. This is a critical event in cell-cell signaling. The change in receptor structure means that the signal has been received. What happens next?

Signal Processing

Once a cell receives a signal, something has to happen to initiate the cell's response. In some cases the "something" occurs directly. For example, steroid hormones such as testosterone and estrogen diffuse through the plasma membrane and enter the cytoplasm, where they bind to a receptor protein. The hormone-receptor complex is then transported to the nucleus, where it triggers changes in the genes being expressed in the cell (**Figure 8.15**). Early in human development, the arrival of testosterone in cells leads to the expression of genes that trigger the development of male reproductive organs. (In the absence of testosterone, female reproductive organs develop.) Later in life, a surge of testosterone triggers the changes associated with puberty in boys, while the arrival of estrogen molecules spurs the development of adult female characteristics in girls.

The effects of hormones that *cannot* diffuse across the plasma membrane and enter the cytoplasm are not nearly so direct. In this case, the first event triggered by a hormone's arrival at the cell surface consists of **signal transduction**—the conversion of the signal from one form to another. A long and often complex series of events ensues, collectively called a **signal transduction pathway**.

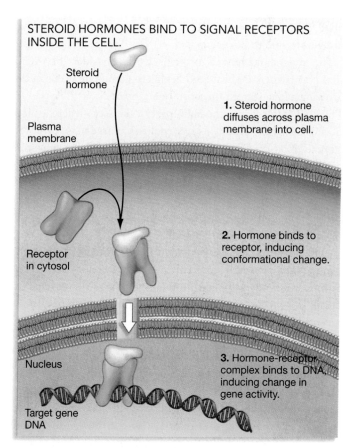

FIGURE 8.15 Some Cell-Cell Signals Enter the Cell and Bind to Receptors in the Cytoplasm
Because they are lipids, steroid hormones can diffuse across the plasma membrane. They then bind to signal receptors inside the cell. The hormone-receptor complex is transported to the nucleus and binds to genes, changing their activity.

Signal transduction is a common occurrence in everyday life. For example, the e-mail messages you receive are transmitted electronically from one computer to another over wires or cables. Software in your computer transduces the electronic signals to letters and words on the screen—a form that you can understand and respond to.

Signal transduction pathways work the same way. In a cell, signal transduction converts an extracellular signal to an intracellular signal. As in an e-mail transmission, a signal that is easy to transmit is converted to a signal that is easily understood and that triggers a response. In cells, signal transduction occurs at the plasma membrane and involves one of two major types of membrane proteins: (1) G proteins and (2) receptor protein kinases. Let's look at each in turn.

G Proteins Many signal receptors span the plasma membrane and are closely associated with peripheral membrane proteins inside the cell called **G proteins**. G proteins got their name because they bind guanosine triphosphate (GTP) and guanosine diphosphate (GDP).

G proteins are signal transducers that act like time-delayed light switches—after they are turned on, they later turn themselves off. They are turned on when they bind GTP, and they turn themselves off after they hydrolyze the bound GTP to GDP. Proteins that are on are said to be *activated*; proteins that are off are said to be *inactivated*.

To understand how G proteins work, consider the events that occur when a wheat seed begins to germinate. Cells in the growing plant secrete the hormone gibberellic acid 1 (GA_1). The GA_1 diffuses into the main part of the seed, where a sheet of cells called the *aleurone layer* encloses a large mass of stored starch. Upon reaching cells in the aleurone layer, the hormone binds to receptors in their plasma membranes and activates a G protein by inducing the G protein to release GDP and bind GTP. In response, the G protein splits into two parts. One of these parts activates a nearby enzyme that is embedded in the plasma membrane (**Figure 8.16a**).

The enzymes that are activated by G proteins catalyze the production of small molecules called second messengers. **Second messengers** are non-protein signaling molecules that increase in concentration inside a cell and elicit a response to the received signal. They are intracellular signals that spread the message carried by the extracellular signal—the hormone.

Several types of small molecules act as second messengers in cells. In wheat seed cells, for example, both calcium ions (Ca^{2+}) and cyclic guanosine monophosphate (cGMP) act as second messengers for GA_1. The G protein that is switched on by GA_1 activates enzymes that lead to the release of calcium ions from the smooth ER. In addition, the same G protein activates an enzyme that catalyzes the synthesis of cGMP in the cytoplasm. In response to a different hormone in a different type of cell, diacylglycerol (DAG), inositol triphosphate (IP_3), or cyclic adenosine monophosphate (cAMP) might serve as the second messenger.

Second messengers are effective because they are small and because they diffuse rapidly to spread the signal throughout the cell. In addition, they can be produced in large quantities in a short time. This characteristic is important. Because the arrival of a single hormone molecule can stimulate the production of many second messenger molecules, the signal transduction event amplifies the original signal.

Receptor Tyrosine Kinases Other receptor proteins have a completely different mode of action. Instead of activating a nearby G protein, **receptor tyrosine kinases** form dimers after they bind a signaling molecule. They then phosphorylate each other (**Figure 8.16b**). Recall from Chapter 3 that the addition of a phosphate group often changes a protein's conformation and activity. In many cases, phosphorylation activates a protein. A phosphorylated receptor tyrosine kinase is no exception. Once it is activated by phosphorylation, the receptor tyrosine kinase phosphorylates one or more specific proteins inside the cell, activating them. These phosphorylated proteins then catalyze the phosphorylation of still other proteins, which phosphorylate yet another population of proteins. This sequence of events is termed a **phosphorylation cascade**. Because each enzyme in the cascade catalyzes the phosphorylation of numerous "downstream" enzymes, the original signal is amplified many times over. One activated receptor tyrosine kinase can trigger the activation of thousands of proteins at the end of the cascade.

To summarize, signal transduction occurs via G proteins or receptor tyrosine kinases. The signal transduction event has two results: (1) It converts an easily transmitted extracellular message into an intracellular message, and (2) it amplifies the original message many times over. Although the hormone at the surface of the cell eventually disengages from its receptor and diffuses away or disintegrates, a "louder" intracellular signal—whether it is a large number of second messenger molecules or a large population of phosphorylated proteins—carries the information throughout the cell and triggers a response.

Signal Response

What is the ultimate response to the messages carried by hormones? The answer varies from signal to signal and from cell to cell. Recall that steroid hormones bind to intracellular receptors and directly change which genes are active in the cell. As a result, they change the cell's characteristics. Similar types of changes in gene expression can occur in response to second messengers or a cascade of protein phosphorylation events. In wheat seeds, for example, a rise in cellular cGMP concentrations triggers the production of a protein that activates the gene for a starch-digesting enzyme. As a result, large quantities of the starch-digesting enzyme α-amylase begin to enter the endomembrane system (introduced in Chapter 7). But recall that the G protein activated by GA_1 also triggers a rise in intracellular Ca^{2+}. Recent experiments have shown that the presence of

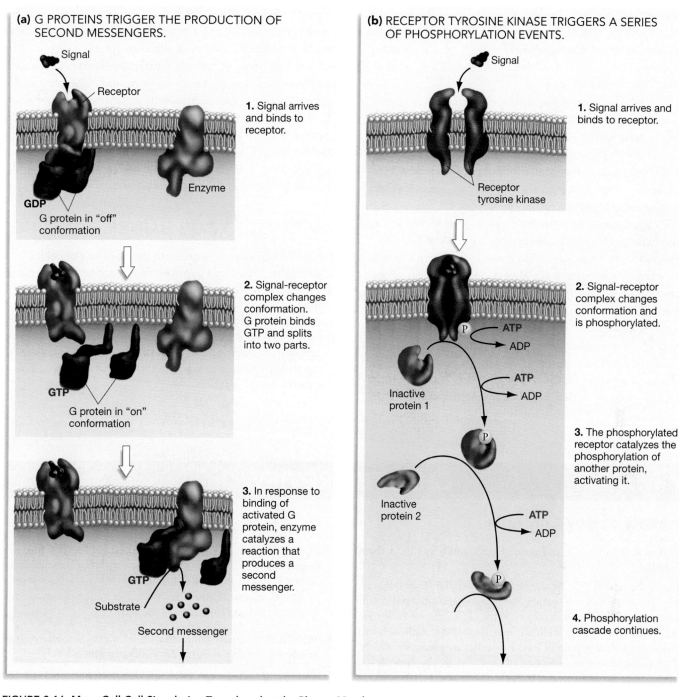

(a) G PROTEINS TRIGGER THE PRODUCTION OF SECOND MESSENGERS.

Signal

Receptor

1. Signal arrives and binds to receptor.

GDP

G protein in "off" conformation

Enzyme

2. Signal-receptor complex changes conformation. G protein binds GTP and splits into two parts.

GTP

G protein in "on" conformation

3. In response to binding of activated G protein, enzyme catalyzes a reaction that produces a second messenger.

GTP

Substrate

Second messenger

(b) RECEPTOR TYROSINE KINASE TRIGGERS A SERIES OF PHOSPHORYLATION EVENTS.

Signal

1. Signal arrives and binds to receptor.

Receptor tyrosine kinase

P   **ATP**
        ADP

ATP
ADP

Inactive protein 1

P

2. Signal-receptor complex changes conformation and is phosphorylated.

3. The phosphorylated receptor catalyzes the phosphorylation of another protein, activating it.

Inactive protein 2

ATP
ADP

P

4. Phosphorylation cascade continues.

FIGURE 8.16 Many Cell-Cell Signals Are Transduced at the Plasma Membrane
(a) If a receptor that is coupled to a G protein is activated by a signal, the G protein binds guanosine triphosphate (GTP) and splits into two parts. The change in the G protein activates a nearby enzyme and results in the production of a second messenger. **(b)** When a receptor tyrosine kinase is activated by a signal, it sets off a cascade of phosphorylation reactions.

Ca^{2+} allows vesicles that are packed with α-amylase to fuse with the plasma membrane and release their contents into the starch storage area. As a result of GA_1's arrival at the plasma membrane, cells in the aleurone layer begin secreting an enzyme that digests stored starch. When the enzyme begins to work, sugars are released. In this way, by sending a GA_1 signal, the germinating embryo has signaled for the release of the nutrients it needs to grow.

Many second messengers or phosphorylation cascades result in the activation or deactivation of a particular target protein that already exists in the cell—typically an enzyme. Either way, the activity of the target cell changes dramatically in response to the arrival of the signal.

But how is the signal turned off? Cells in the aleurone layer of wheat seeds do not need to secrete unlimited quantities of α-amylase, and abnormalities would result if the morphological changes induced by testosterone and estrogen continued past the early teenage years into adulthood. What keeps signals from being amplified indefinitely?

Signal Deactivation

Cells have built-in systems for turning off intracellular signals. For example, activated G proteins can hydrolyze GTP. This reaction converts GTP to GDP and releases a phosphate group. As a result, the G protein's conformation changes, activation of its associated enzyme stops, and production of the second messenger ceases. The messengers in the cytosol are also short lived. Pumps in the membrane of the smooth ER return calcium ions to storage, and enzymes called *phosphodiesterases* convert active cAMP and cGMP to inactive AMP and GMP.

Phosphorylation cascades wind down in a similar way. Enzymes called *phosphatases* are always present in cells. Phosphatases catalyze reactions that remove phosphate groups from proteins. If hormone stimulation of a receptor tyrosine kinase ends, phosphatases are able to dephosphorylate enough components of the phosphorylation cascade that the response begins to slow. Eventually it stops.

Although an array of specific mechanisms is involved, the general observation is that hormone-response systems are designed to shut down quickly. As a result, they are exquisitely sensitive to small changes in the concentration of hormones or in the number and activity of signal receptors.

The four steps of cell-cell signaling—reception, processing, response, and deactivation—allow hormone-secreting cells to elicit a specific response from cells in nearby or distant tissues. In this way, hormones coordinate the activity of cells throughout the body. In wheat plants and in your body, cell-cell signaling helps millions of individual cells function as an integrated whole. As a result, multicellular organisms can respond to changing conditions in an appropriate way.

✓ CHECK YOUR UNDERSTANDING

Intercellular signals coordinate the activities of cells throughout the body in response to changes in internal or external conditions. You should be able to (1) describe the four steps involved in signal transduction (signal reception, signal processing via the production of a second messenger or a phosphorylation cascade, signal response, and signal deactivation); (2) explain why only certain cells respond to particular signals; and (3) elucidate how signals are amplified.

CHAPTER REVIEW

Summary of Key Concepts

■ **Extracellular material strengthens cells and helps bind them together.**

The vast majority of cells secrete an extracellular layer. In bacteria, archaea, algae, and plants, the extracellular material is stiff and is called a cell wall. In animals, the secreted layer is flexible and is called the extracellular matrix (ECM). Although the types of molecules present in the external coating vary widely, the basic structure and function of the ECM are the same: It is a fiber composite that defines the cell's shape and helps protect it from mechanical damage. Fiber composites consist of cross-linked filaments that provide tensile strength and a ground substance that fills space and resists compression. In plants the extracellular filaments are cellulose microfibrils; in animals the most abundant filaments are made of the protein collagen. In both plants and animals, the ground substance is composed of gelatinous polysaccharides.

■ **Cell-cell connections help adjacent cells adhere. Cell-cell gaps allow adjacent cells to communicate.**

In both unicellular and multicellular organisms, cell-cell interactions are mediated by molecules in the extracellular layer and plasma membrane. Most interactions among unicellular species are competitive in nature, but the cells of multicellular organisms are intensely social and cooperate with each other. Many cells of multicellular organisms are physically bound to one another via the glue-like middle lamella that forms between plant cells or the tight junctions and desmosomes observed between animal cells. The cytoplasm of adjacent cells is in direct communication, via openings called plasmodesmata in plants and gap junctions in animals.

■ **Intercellular signals are responsible for creating an integrated whole from many thousands of independent parts.**

Distant cells in multicellular organisms communicate through signaling molecules that bind to receptors found on or in specific target cells. Once these signals are received, they are often transduced to a new type of intracellular signal, which is amplified. This internal signal triggers a sequence of events that leads to the activation of certain enzymes, the release or uptake of specific ions or molecules, or a change in the activity of target genes. Because enzymes inside the cell quickly deactivate the signal, the cell's response is tightly regulated. A continued response usually depends on continued stimulation by the signaling molecule. Thus cell-cell communication occurs by a four-step process: (1) signal reception, (2) signal processing, (3) signal response, and (4) signal deactivation. As a result, cells and tissues throughout the body can alter their activity in response to changing conditions, and do so in a coordinated way.

Web Tutorial 8.1 Cell-Cell Communication

ESSAY How Do Viagra®, Levitra®, and Cialis® Work?

In the United States alone, 30 million men are estimated to suffer from erectile dysfunction—the inability to produce and sustain an erection when sexually stimulated. In recent years, drugs that treat this disorder effectively have appeared on the market. Among them are molecules that are marketed under the trade names Viagra, Levitra, and Cialis. How do these drugs work?

The development of this drug therapy grew out of a detailed understanding of a signal transduction pathway.

The answer to this question is found in the cell-cell signaling system that leads to erection in males. **Figure 8.17** outlines the sequence of events. When a man is sexually stimulated, nerve impulses from the brain or direct physical contact lead to the release of the chemical messenger nitric oxide (NO) gas from blood-vessel cells in his penis. NO acts as an intercellular signal. Its targets are muscle cells that surround the blood vessels throughout the penis. When it arrives at one of these muscle cells, the NO diffuses through the plasma membrane and binds to the enzyme guanylate cyclase. The activated enzyme catalyzes the production of cyclic guanosine monophosphate (cGMP) from guanosine triphosphate (GTP). cGMP acts as a second messenger. It binds to and activates a protein kinase. This kinase then phosphorylates a protein in the membrane of the smooth ER.

When that protein is activated, it begins pumping calcium ions (Ca^{2+}) out of the cytosol and into the lumen of the smooth ER. When Ca^{2+} levels in the cytoplasm of the muscle cells are low, the muscle tissue that contains these cells relaxes. When muscles that surround blood vessels relax, the vessels widen. As a result, blood flow through the vessels increases. As muscles in the penis relax, blood flows into chambers in the structure and fills them, causing an erection.

How do the drugs Viagra, Levitra, and Cialis fit into this signaling pathway? The muscle cells of the penis contain an enzyme called phosphodiesterase 5 (PDE5), which turns off the NO-induced system. More specifically, PDE5 catalyzes the conversion of cGMP to an inactive form. Because it removes the second messenger, PDE5 diminishes the cells' response to the original signal. Viagra, Levitra, and Cialis inhibit PDE5. When these drugs are in the cytoplasm, the cell's mechanism for deactivating the signal is inhibited and cGMP levels tend to stay high. Because this second messenger is present at a high concentration, each cell's response to the signal tends to be stronger and of longer duration. The development of this drug therapy grew out of a detailed understanding of a signal transduction pathway.

CELL-CELL SIGNALING DURING ERECTIONS

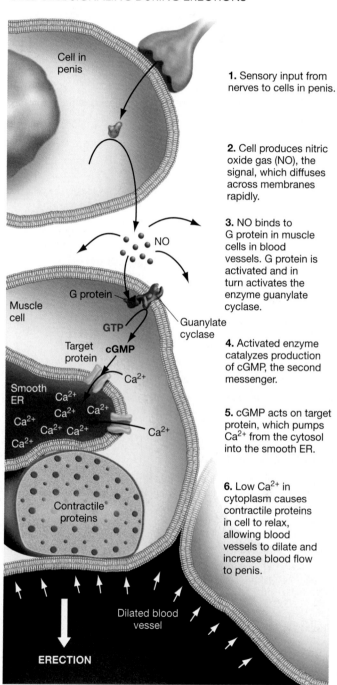

1. Sensory input from nerves to cells in penis.

2. Cell produces nitric oxide gas (NO), the signal, which diffuses across membranes rapidly.

3. NO binds to G protein in muscle cells in blood vessels. G protein is activated and in turn activates the enzyme guanylate cyclase.

4. Activated enzyme catalyzes production of cGMP, the second messenger.

5. cGMP acts on target protein, which pumps Ca^{2+} from the cytosol into the smooth ER.

6. Low Ca^{2+} in cytoplasm causes contractile proteins in cell to relax, allowing blood vessels to dilate and increase blood flow to penis.

FIGURE 8.17 A Signal Transduction Pathway in Male Reproductive Tissue
EXERCISE The enzyme PDE5 converts cGMP to an inactive molecule. Circle the step in the signal transduction pathway that is affected by PDE5. What effect does PDE5 activity have on the response?

Questions

Content Review

1. Which of the following represents a fundamental difference between the fibers found in the extracellular layers of plants and those of animals?
   a. Plant fibers are thicker; they are also stronger because they have more cross-linkages.
   b. Animal fibers consist of proteins; plant fibers consist of polysaccharides instead.
   c. Plant extracellular fibers never move; animal fibers can slide past one another.
   d. Cellulose microfibrils run parallel to each other; collagen filaments crisscross.

2. In plants and animals, where are most components of the extracellular material synthesized?
   a. smooth ER
   b. the rough ER and Golgi apparatus
   c. in the extracellular layer itself
   d. adjacent cells

3. Treating dissociated cells with certain antibodies makes the cells unable to reaggregate. Why?
   a. The antibodies bind to cell adhesion proteins called cadherins.
   b. The antibodies bind to the fiber component of the extracellular matrix.
   c. The antibodies bind to receptors on the cell surface.
   d. The antibodies act as enzymes that break down desmosomes.

4. What does it mean to say that a signal is transduced?
   a. The signal enters the cell directly and binds to a receptor inside.
   b. The physical form of the signal changes between the outside of the cell and the inside.
   c. The signal is amplified, such that even a single molecule evokes a large response.
   d. The signal triggers a sequence of phosphorylation events inside the cell.

5. Why are tight junctions found in only certain types of tissues, while desmosomes are found in a wide array of cells?
   a. Tight junctions are required only in cells where communication between adjacent cells is particularly important.
   b. Tight junctions are not as strong as desmosomes.
   c. Tight junctions have different structures but the same functions.
   d. Tight junctions are found only in epithelial cells that must be watertight.

6. What physical event represents the receipt of an intercellular signal?
   a. the passage of ions through a desmosome
   b. the activation of the first protein in a phosphorylation cascade
   c. the binding of a hormone to a signal receptor, which changes conformation in response
   d. the activation of a G protein associated with a signal receptor

Conceptual Review

1. Why is it difficult to damage a fiber composite?

2. Why does a phosphorylation cascade amplify an intercellular signal?

3. What is the difference between a tight junction and a desmosome?

4. Animal cells adhere to each other selectively. Summarize experimental evidence that supports this statement.

5. Make a sketch showing the reception, processing, and response steps in the signal transduction pathway featured in this chapter's essay.

6. Why do researchers use the term *pathway* in reference to signal transduction?

Group Discussion Problems

1. Suppose that an animal species and a plant species each lacked the ability to secrete an extracellular matrix. What would these organisms look like, and how would they live?

2. Suppose that a signal that is released from one cell triggered a response in another cell in the absence of signal transduction. Diagram the sequence of events that would occur between the arrival of the signal and the arrival of the response. Compare and contrast the response to a signal that does involve signal transduction.

3. According to the types of phylogenetic analyses introduced in Chapter 1, multicellular bodies evolved independently in plants and animals. Both plants and animals have structures that knit adjacent cells together and that form openings between cells. But the molecules in these structures differ markedly between the two groups. Are these observations logical? Explain.

4. Suppose you created an antibody that bound to one of the receptors illustrated in Figure 8.16. How would the signal transduction pathway be affected?

Answers to Multiple-Choice Questions **1.** b; **2.** b; **3.** a; **4.** b; **5.** d; **6.** c

Cellular Respiration and Fermentation

9

KEY CONCEPTS

▨ Cellular respiration produces ATP from molecules with high potential energy—often glucose.

▨ Glucose processing has three components: (1) glycolysis, (2) the Krebs cycle, and (3) electron transport coupled with oxidative phosphorylation.

▨ Fermentation pathways allow glycolysis to continue when the lack of an electron acceptor shuts down electron transport chains.

▨ Respiration and fermentation are carefully regulated.

When table sugar is heated, it undergoes the uncontrolled oxidation reaction known as burning. Burning gives off heat. In cells, the simple sugar called glucose is oxidized through a long series of carefully regulated reactions. Instead of being given off as heat, some of the energy produced by these reactions is used to synthesize ATP.

Cells are dynamic. Vesicles move cargo from the Golgi apparatus to the plasma membrane and other destinations; enzymes synthesize a complex array of macromolecules; and millions of membrane proteins pump ions and molecules to create an environment conducive to life. How does all this activity occur? The answer lies in the molecule **adenosine triphosphate (ATP)**. ATP has high potential energy and acts as the major energy currency in cells.

In earlier chapters we explored the role that ATP plays in driving endergonic reactions, in moving materials through cells, and in moving cells themselves through the environment. Recall that when a phosphate group is transferred to a molecule, the molecule often becomes more reactive. This action is particularly important in the case of proteins. When ATP or a phosphate group from ATP binds to a specific site on an enzyme, the enzyme's shape and activity can change. When ATP binds to a

motor protein such as kinesin or dynein or myosin, the protein responds by moving in a way that transports materials or changes the cell's shape. The transfer of a phosphate group to a protein pump changes its shape in a way that transports ions or molecules across the plasma membrane against an electrochemical gradient (see Chapter 6).

In short, ATP allows cells to do work. Because staying alive takes work, there is no life without ATP.

The goal of this chapter is to investigate how cells make adenosine triphosphate by oxidizing compounds such as sugars that have high potential energy. The energy released during the oxidation is used to transfer a phosphate group to **adenosine diphosphate (ADP)** and generate ATP. Section 9.1 introduces the events involved by reviewing the fundamental principles of reduction and oxidation and summarizing the reactions involved in oxidizing glucose, the most common fuel used by organisms.

Then Sections 9.2 through 9.4 delve into the glucose-oxidizing reactions in detail. Section 9.5 introduces an alternative route for ATP production—called fermentation—that occurs in many bacteria, archaea, and eukaryotes. In closing, the chapter examines how cells shunt certain carbon-containing compounds away from ATP production and into the synthesis of DNA, RNA, amino acids, and other molecules. This chapter is your introduction to **metabolism**—the chemical reactions that occur in cells.

9.1 An Overview of Cellular Respiration

Cells use chemical energy to move molecules, pump ions, complete endergonic reactions, and perform other types of work. Because the electrons in the carbon-hydrogen bonds of carbohydrates and fats contain a great deal of chemical energy, they make it possible for cells to do a great deal of work. But even if a cell has huge amounts of chemical energy stored in the form of carbohydrates and fats, no movement, pumping by membrane proteins, or endergonic reactions occur. Why? Just as the dollar, euro, and yen act as currencies in human economies, ATP acts as the major currency in the economy of a cell. A huge savings account at a bank won't do you any good when you walk up to a vending machine to get a snack—you need cash. Similarly, cells are not able to use the chemical energy stored in carbohydrates or fats to do work. To be useful, the chemical energy must first be transferred to ATP.

It's important to recognize, though, that cells don't hoard large wads of cash in the form of ATP. In general, a cell contains only enough ATP to last from 30 seconds to a few minutes. Like many other cellular processes, the production and use of ATP is dynamic. Because ATP makes every cell function possible, producing a steady supply of it is among the most fundamental of all cell processes.

If carbohydrates and fats act as a savings account for chemical energy, how do cells draw on this account to synthesize ATP? In plants, algae, animals, fungi, and many other organisms, the simple sugar glucose ($C_6H_{12}O_6$) is the key intermediary (**Figure 9.1**). Most glucose is produced by photosynthetic species, which use the energy in sunlight to reduce carbon dioxide (CO_2) to glucose and other carbohydrates. When photosynthetic species decompose or are eaten, they provide glucose to animals, fungi, and many bacteria and archaea. All organisms use glucose as a building block in the synthesis of carbohydrates, fats, and other energy-storage compounds. When a cell needs energy, glucose is used to produce ATP through processes called cellular respiration and fermentation.

Chapter 10 focuses on how plants and other photosynthetic organisms produce glucose from carbon dioxide, water, and sunlight. Here we focus on how glucose is used to produce ATP. Some of the chemistry involved in processing glucose actually predates the evolution of photosynthesis. Indeed, the reactions analyzed early in this chapter were probably occurring as the first cell sprang to life. The ancestry of some of the enzymes introduced in this chapter dates back 3.5 billion years.

Before analyzing what these enzymes are and what they do, though, let's briefly review the general nature of chemical ener-

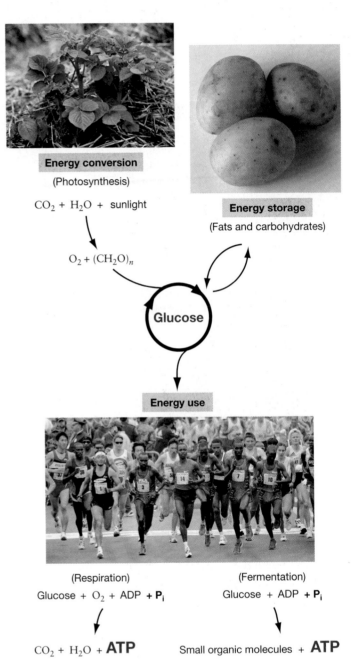

Energy conversion
(Photosynthesis)

$CO_2 + H_2O + $ sunlight

$O_2 + (CH_2O)_n$

Energy storage
(Fats and carbohydrates)

Glucose

Energy use

(Respiration)
Glucose + O_2 + ADP **+ P$_i$**

(Fermentation)
Glucose + ADP **+ P$_i$**

$CO_2 + H_2O +$ **ATP**

Small organic molecules + **ATP**

FIGURE 9.1 Glucose Is the Hub of Energy Processing in Cells
Glucose is the endpoint of energy production via photosynthesis. It is also the intermediary molecule in energy storage. In a cell, when glucose is abundant, it is used to synthesize fats and storage carbohydrates. When glucose is scarce, it is supplied by the breakdown of fats and storage carbohydrates. Glucose is also used to provide chemical energy in the form of ATP, via cellular respiration and fermentation. **EXERCISE** The CO_2 produced by respiration is used by plants during photosynthesis. The O_2 produced by photosynthesis is used during respiration. Add arrows to this diagram to indicate these relationships.

gy and reduction-oxidation (redox) reactions. Recall from Chapter 2 that redox reactions involve the transfer of an electron from one substance to another. The compound that receives the electron is reduced, while the atom or molecule that gives up the electron is oxidized.

The Nature of Chemical Energy and Redox Reactions

Recall from Chapter 2 that chemical energy is a form of potential energy. Potential energy is energy that is associated with position or configuration. In cells, electrons are the most important source of chemical energy—specifically the electrons in ATP. The amount of potential energy that an electron has is based on its position relative to other electrons and to the protons (positive charges) in the nuclei of nearby atoms. The potential energy of a molecule is a function of the way its electrons are configured.

ATP makes things happen in cells because it has a great deal of potential energy. To understand why this is so, study the mo-

lecular formula in **Figure 9.2a**. Note that the three phosphate groups in ATP contain a total of four negative charges and that these charges are confined to a small area. In part because these negative charges repel each other, the potential energy of the electrons in the phosphate groups is extraordinarily high.

When ATP reacts with water during a hydrolysis reaction, the bond between ATP's outermost phosphate group and its neighbor is broken, resulting in the formation of ADP and inorganic phosphate, P_i (**Figure 9.2b**). This reaction is highly exergonic. Under standard conditions of temperature and pressure in the laboratory, a total of 7.3 kilocalories (kcal) of energy, per mole of ATP, is released during the reaction. The hydrolysis of ATP is exergonic because the entropy of the product molecules is much higher than that of the reactants, and because there is a large drop in potential energy when ADP and P_i are formed from ATP. The change in potential energy occurs in part because the electrons from ATP's phosphate groups are now spread between two molecules instead of being clustered on one molecule, meaning that there is less electrical repulsion. In addition, the negative charges on ADP and P_i are stabilized much more efficiently by interactions with the partial positive charges on surrounding water molecules than are the charges on ATP.

For similar reasons, the transfer of a phosphate group from ATP to a protein is usually exergonic. When a phosphate group is transferred from ATP to a protein, two negative charges are added to the protein. In response to this phosphorylation event, the electrons in the protein change configuration and the molecule's shape, or conformation, is usually altered (**Figure 9.2c**).

(a) ATP consists of three phosphate groups, ribose, and adenine.

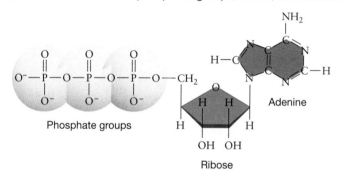

(b) Energy is released when ATP is hydrolyzed.

(c) Phosphorylation generates shape change in proteins.

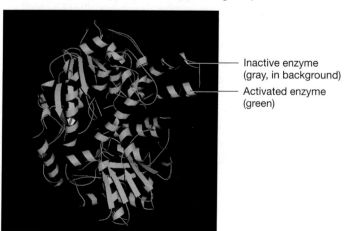

Inactive enzyme (gray, in background)

Activated enzyme (green)

FIGURE 9.2 Adenosine Triphosphate (ATP)
(a) ATP has high potential energy, in part because four negative charges are clustered in its three phosphate groups. The negative charges repel each other, raising the potential energy of the electrons. **(b)** When ATP is hydrolyzed to ADP and inorganic phosphate, a large free-energy change occurs. **(c)** When proteins are phosphorylated or an ATP molecule binds to them, they often change shape in a way that changes their activity. This protein is an enzyme called arsenite-transporting ATPase. **EXERCISE** Circle portions of the enzyme in part (c) that change conformation between the activated and inactive states.

Part of the protein moves. Movement in response to phosphorylation or to the addition of the entire ATP molecule is what transports materials, powers flagella or cilia, pumps ions, drives endergonic reactions, and completes other kinds of work in the cell.

If a great deal of energy is released when ATP loses a phosphate group, then a great deal of energy must be required to synthesize ATP from ADP by adding P_i. Where does this energy come from? The answer is redox reactions.

Using Redox Reactions to Produce ATP

Recall from Chapter 2 that redox reactions involve a transfer of electrons. When an atom or molecule is reduced, it gains an electron from another atom or molecule. The compound that donates the electron becomes oxidized in the process.

In cells, molecules that gain an electron and become reduced also often gain a proton (H^+) and thus a hydrogen atom. As a result, reduced compounds in cells tend to have many C–H bonds. The C–H bonds have high potential energy, because the electrons in these bonds are not held tightly—recall from Chapter 2 that carbon and hydrogen atoms have low and approximately equal electronegativities. Thus, molecules that have a large number of C–H bonds, such as carbohydrates and fats, store a great deal of potential energy. During photosynthesis, for example, the energy in sunlight is used to add electrons to carbon dioxide (CO_2), resulting in the formation of glucose ($C_6H_{12}O_6$). Glucose contains many C–H bonds and thus a great deal of potential energy.

Conversely, molecules that are oxidized in cells often lose a proton along with an electron. Instead of having many C–H bonds, oxidized molecules in cells tend to have many C–O bonds. Such molecules also have lower potential energy. To understand why, recall from Chapter 2 that oxygen atoms have extremely high electronegativity. Because oxygen atoms hold electrons so tightly, the electrons involved in bonds with oxygen atoms have low potential energy.

To put all of these ideas together, consider what happens when glucose undergoes the uncontrolled oxidation reaction called burning:

$$C_6H_{12}O_6 + 6\,O_2 \rightarrow 6\,CO_2 + 6\,H_2O + energy$$

In this reaction, electrons are transferred from glucose to oxygen, and protons follow. The carbon atoms in glucose are oxidized, forming carbon dioxide. The oxygen atoms in the oxygen molecule (O_2) are reduced, forming water. Glucose is the molecule that acts as an electron donor and becomes oxidized; oxygen is the molecule that acts as an electron acceptor and becomes reduced. A large drop in potential energy occurs during the oxidation of glucose because the electrons are much more tightly bound in the C–O and H–O bonds of the product molecules than they were in

the C–H bonds of the reactant molecules. Entropy also increases as a result of this reaction. When glucose undergoes the uncontrolled oxidation reaction called burning, the change in potential energy is converted to kinetic energy in the form of heat. When one mole of glucose burns, a total of 686 kcal of heat is released.

Glucose does not burn in cells, however. Instead, the glucose in cells is oxidized through a long series of carefully controlled redox reactions. These reactions are occurring, millions of times per minute, in your cells right now. Instead of being given off as heat, much of the energy released by the drop in potential energy is being used to make the ATP you need to think, move, and stay alive. In cells, the change in free energy that occurs during the oxidation of glucose is used to synthesize ATP from ADP and P_i.

How does the oxidation of glucose take place in a way that supports the production of ATP? In many organisms, cellular respiration is a three-step process: (1) Glucose is broken down to pyruvate; (2) pyruvate is oxidized to CO_2; and (3) compounds that were reduced in steps 1 and 2 are oxidized in processes that usually lead to the production of ATP. Let's get an overview of the way these steps work, and then pursue a more detailed analysis in Sections 9.2 through 9.4.

Processing Glucose: Glycolysis

An enormous diversity of organisms use glucose as their primary fuel. In virtually all species examined to date, the first step in the oxidation of glucose is a sequence of 10 chemical reactions that are collectively called **glycolysis** ("sugar breakdown"). These reactions occur in the cytosol of eukaryotic and prokaryotic cells. During glycolysis, one molecule of the six-carbon sugar glucose is broken into two molecules of the three-carbon compound pyruvate. Some of the potential energy released by this sequence of reactions is used to phosphorylate ADP molecules, forming ATP. In addition, one of the reactions in the sequence results in the reduction of a molecule called nicotinamide adenine dinucleotide, symbolized NAD^+. As **Figure 9.3** shows, the addition of two electrons and a proton to NAD^+ produces NADH. We'll soon see that NADH readily donates electrons to other molecules. As a result, it is called an **electron carrier** and is said to have "reducing power."

Figure 9.4a summarizes the results of the reactions of glycolysis. Notice that glucose goes in and that ATP, NADH, and pyruvate come out.

The Krebs Cycle

What happens to the pyruvate produced by glycolysis? If an electron acceptor such as oxygen is present in the cell, pyruvate enters a sequence of reactions called the **Krebs cycle**. In eukaryotic cells, the enzymes involved in the Krebs cycle are located inside mitochondria. Some of these enzymes are found in the

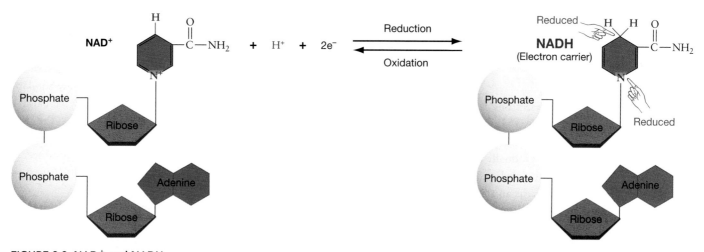

FIGURE 9.3 NAD⁺ and NADH
NADH is the reduced form of NAD^+. **QUESTION** Compare the structure of NAD^+ with
the structure of ATP in Figure 9.2. Which portions of the two molecules are identical?

matrix inside the organelle, while others are associated with
the surface of the mitochondrion's inner membrane.

In the reactions that lead up to and complete the Krebs
cycle, each pyruvate is oxidized to three molecules of carbon
dioxide. Some of the potential energy released by these reactions is used to (1) reduce NAD^+ to NADH; (2) reduce another electron carrier, called flavin adenine dinucleotide
(FAD), to $FADH_2$; and (3) phosphorylate ADP, yielding ATP

(**Figure 9.4b**). Pyruvate enters this sequence of reactions;
carbon dioxide, NADH, $FADH_2$, and ATP come out of the
sequence.

With the completion of the Krebs cycle, glucose has been completely oxidized to CO_2. This observation raises several questions, though. According to the overall reaction for the oxidation
of glucose $(C_6H_{12}O_6 + 6\ O_2 \rightarrow 6\ CO_2 + 6\ H_2O + energy)$,
molecular oxygen (O_2) is a reactant. Where does oxygen come

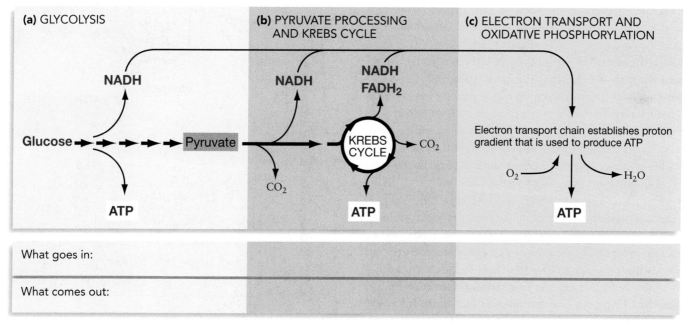

FIGURE 9.4 An Overview of Glucose Oxidation
Cells produce ATP from glucose via a series of processes: **(a)** glycolysis, **(b)** the Krebs cycle, and **(c)** electron
transport combined with oxidative phosphorylation. Each component produces at least some ATP. Because
the three components are connected, glucose oxidation is an integrated metabolic pathway. Glycolysis and
the Krebs cycle are connected by pyruvate, and glycolysis and the Krebs cycle are connected to electron
transport chains by NADH and $FADH_2$. **EXERCISE** Fill in the chart along the bottom.

into play, if not in glycolysis or the Krebs cycle? And what happens to all of the electrons that have been transferred from glucose to NADH and $FADH_2$?

Electron Transport

In cells, the potential energy of electrons carried by NADH and $FADH_2$ is gradually "stepped down" by molecules that participate in a series of redox reactions. Stated another way, the potential energy in the electrons is gradually decreased via a series of step-by-step reactions. The molecules involved in the redox reactions make up what is known as an **electron transport chain**. As electrons are passed along the electron transport chain, they gradually fall from a higher to a lower potential energy.

In eukaryotes, the components of the electron transport chain are located in the inner membrane and cristae of mitochondria; in prokaryotes, they are found in the plasma membrane. The transfer of an electron to each molecule in the electron transport chain changes the molecule's shape and activity. In several cases, the molecule that receives the electron responds by changing conformation in a way that transfers a proton across the mitochondrion's inner membrane. This is crucial. In effect, the movement of electrons through the chain results in the pumping of protons to the outside of the membrane. The subsequent buildup of protons results in a strong electrochemical gradient across the mitochondrial membrane. The gradient drives the protons through a membrane protein called ATP synthase. In eukaryotes, ATP synthase is located in the inner membrane and cristae of mitochondria. The force generated by the flow of protons through ATP synthase makes part of the protein spin. This change in conformation drives the phosphorylation of ADP. Indirectly, then, the sequence of redox reactions in the electron transport chain results in the production of ATP (**Figure 9.4c**).

Once the electrons donated by NADH and $FADH_2$ have passed through the electron transport chain, they are transferred to a final electron acceptor, which in many organisms is oxygen. A transfer of these electrons to oxygen, along with protons, results in the formation of water as an end-product. The four molecules in the overall reaction for the oxidation of glucose—glucose, oxygen, carbon dioxide, and water—are now accounted for. Chemical energy has been transferred from glucose to ATP, via NADH and $FADH_2$.

The vast majority of ATP production in cells results from the reactions in the electron transport chain. Hence, glycolysis and the Krebs cycle can be thought of as mechanisms for stripping electrons from glucose and feeding them to the electron transport chain. Biologists use the term **cellular respiration** for any process of ATP production that involves each of the following: a compound that acts as an electron donor, an electron transport chain, and an electron acceptor.

Methods of Producing ATP

The phosphorylation events that are catalyzed by ATP synthase are powered by a stream of protons. The gradient that drives these protons is set up by an electron transport chain that uses a molecule with low free energy—usually oxygen—as the final electron acceptor. Because this mode of ATP production links the phosphorylation of ADP with the oxidation of NADH and $FADH_2$, it is called **oxidative phosphorylation**. But when ADP is used as a substrate in enzyme-catalyzed phosphorylation reactions during glycolysis or the Krebs cycle, the event is called **substrate-level phosphorylation**. In substrate-level phosphorylation, the energy to produce ATP comes from a phosphorylated substrate and not from a proton gradient.

In many organisms, cellular respiration begins with the oxidation of glucose and ends with a flow of electrons through an electron transport chain to a final electron acceptor. The process starts with glycolysis and the Krebs cycle, which oxidize glucose to CO_2. As **Figure 9.5** shows, these reactions produce dramatic free-energy changes. Note that relative to the other compounds in glycoysis and the Krebs cycle, glucose has a great deal of potential energy and low entropy. During glycolysis and the Krebs cycle, the free energy of each subsequent compound involved drops—at times in small steps and at other times in large steps. The small drops in free energy are coupled with the production of ATP; the large drops are asso-

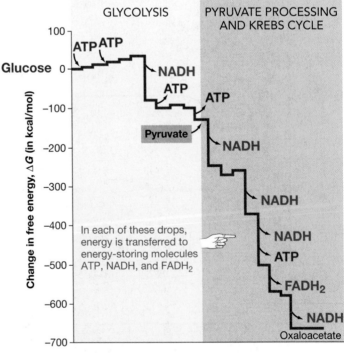

FIGURE 9.5 Free-Energy Changes as Glucose Is Oxidized
Graph of free-energy changes during glycolysis and the Krebs cycle. Drops in free energy are associated with the production of ATP, NADH, and $FADH_2$. Oxaloacetate is the endpoint of the Krebs cycle.
QUESTION Which is associated with larger changes in free energy: production of ATP or production of NADH and $FADH_2$? Add labels indicating where in the cell glycolysis and the Krebs cycle take place.

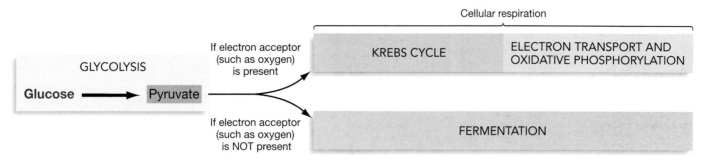

FIGURE 9.6 Cellular Respiration and Fermentation Are Alternative Pathways for Producing Energy
When oxygen or another electron acceptor is present in a cell, the pyruvate produced by glycolysis enters the Krebs cycle and the electron transport system is active. But if no electron acceptor is available, the pyruvate undergoes reactions known as fermentation.

ciated with the synthesis of NADH and FADH$_2$. In effect, then, the free energy present in glucose is transferred to ATP, NADH, and FADH$_2$. In the third (and final) stage of cellular respiration, NADH and FADH$_2$ carry these electrons to the electron transport chain.

If the molecule that functions as the final electron acceptor is unavailable, however, then cellular respiration cannot occur. In this case, alternative pathways called fermentation reactions take over (**Figure 9.6**). In eukaryotes, for example, fermentation occurs when oxygen is lacking. As we'll see in Section 9.5, **fermentation** consists of reactions that allow glycolysis to continue even if pyruvate is not drawn off to the Krebs cycle. In eukaryotes, by-products such as ethanol (the active ingredient in alcoholic beverages) or lactic acid result from various fermentation pathways. In bacteria and archaea, other products are possible.

This overview of cellular respiration is intended simply to introduce the processes responsible for most ATP production in cells. Now we need to explore glycolysis, the Krebs cycle, and electron transport chains in more detail. Let's take a closer look at each process in turn.

9.2 Glycolysis

Glycolysis may be the most fundamental of all metabolic pathways, but it was discovered by accident. In the late 1890s Hans and Edward Buchner were working out techniques for manufacturing extracts of baker's yeast for therapeutic use. (Yeast extracts are still added to some foods as a flavor enhancer or nutritional supplement.) In one set of experiments the Buchners added sucrose, or table sugar, to their extracts. Sucrose is a disaccharide consisting of glucose linked to another six-carbon sugar, called fructose. At the time, sucrose was commonly used as a preservative. But the Buchners found that instead of preserving the yeast extracts,

the sucrose was quickly broken down and fermented, with alcohol appearing as a by-product. This was a key finding because it showed that fermentation and other types of cellular metabolism could be studied in vitro—outside the cell. Until then, researchers thought that metabolism could take place only inside cells.

The Buchners and other researchers followed up on this observation by trying to determine how the sugar was being processed. An important early observation was that the reactions could be sustained much longer than normal if inorganic phosphate was added to the mixture. This finding implied that some of the compounds involved were being phosphorylated. Soon after, a molecule called fructose bisphosphate was isolated. (The *bis*–prefix means that two phosphate groups are attached to the molecule at distinct locations.) Subsequent work showed that all but two of the compounds involved in glycolysis—the starting and ending molecules, glucose and pyruvate—are phosphorylated.

A third major finding came to light in 1905, when researchers found that the processing of sugar by yeast extracts stopped if the reaction mix was boiled. Because enzymes were known to be inactivated by heat, this discovery suggested that enzymes were involved in at least some of the processing steps. Years later, investigators realized that each step in glycolysis is catalyzed by a different enzyme.

Over the next 35 years, each of the reactions and enzymes involved in glycolysis was gradually worked out by several different researchers. Because Gustav Embden and Otto Meyerhof identified many of the steps in glycolysis, this process is also known as the Embden-Meyerhof pathway.

A Closer Look at the Glycolytic Reactions

By breaking open cells, separating the cell components via differential centrifugation (introduced in Chapter 7), and testing which components could sustain glycolysis, biologists discovered that all 10 reactions of glycolysis occur in the cytosol.

FIGURE 9.7 Glycolysis Pathway

Glucose is oxidized to pyruvate through this sequence of 10 reactions. Each reaction is catalyzed by a different enzyme. The process begins with an investment of 2 molecules of ATP. Reactions later in the sequence result in the production of 2 molecules of NADH and 4 of ATP, for a net gain of 2 molecules of ATP.

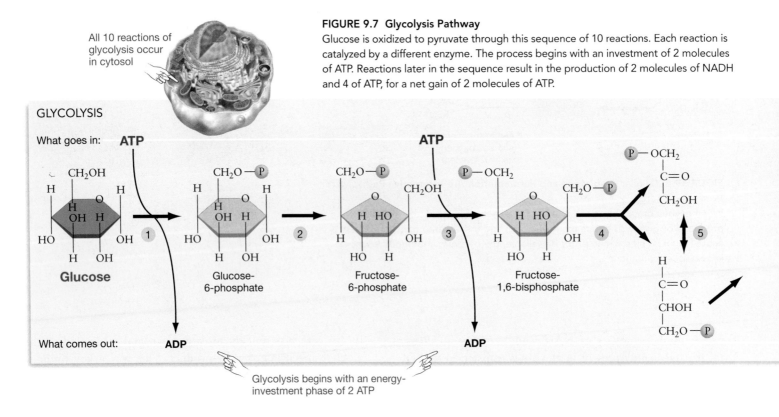

All 10 reactions of glycolysis occur in cytosol

GLYCOLYSIS

What goes in: ATP   ATP

What comes out: ADP   ADP

Glycolysis begins with an energy-investment phase of 2 ATP

Glucose • Glucose-6-phosphate • Fructose-6-phosphate • Fructose-1,6-bisphosphate

Figure 9.7 details the 10 reactions involved in glycolysis. There are three key points to note about the reaction sequence:

1. Contrary to what most researchers expected, glycolysis starts by *using* ATP, not producing it. In the initial step, glucose is phosphorylated to form glucose-6-phosphate. After an enzyme rearranges this molecule to fructose-6-phosphate in the second step, the third step in the reaction sequence adds a second phosphate group, forming the fructose-1,6-bisphosphate observed by early researchers. Thus, two ATP molecules are used up before any ATP is produced by glycolysis.

2. Once this energy-investment phase of glycolysis is complete, the subsequent reactions represent an energy payoff phase. The sixth reaction in the sequence results in the reduction of two molecules of NAD^+; the seventh produces two molecules of ATP and erases the energy "debt" of two molecules of ATP invested early in glycolysis. The final reaction in the sequence produces another two ATPs. For each molecule of glucose processed, the net yield is two molecules of NADH, two of ATP, and two of pyruvate.

3. The production of ATP during glycolysis occurs by substrate-level phosphorylation. In reactions 7 and 10 of Figure 9.7, an enzyme catalyzes the transfer of a phosphate group from a phosphorylated intermediate in glycolysis directly to ADP.

The discovery and elucidation of the glycolytic pathway ranks as one of the great achievements in the history of biochemistry. Because the enzymes involved have been observed in nearly every bacterium, archaean, and eukaryote examined, biologists infer that the pathway evolved very early in the history of life. It is very likely that the ancestor of all organisms living today made ATP by glycolysis. The reactions outlined in Figure 9.7 are among the most ancient and fundamental of life processes.

How Is Glycolysis Regulated?

Once the glycolytic pathway was worked out, researchers focused on the structures and functions of the enzymes involved and the way the sequence is regulated. It seemed logical to predict that glycolysis does not occur at the maximum rate at all times. Instead, researchers predicted that glycolysis would proceed only when cells needed fresh supplies of ATP. An important advance occurred when biologists observed that high levels of ATP inhibit a key glycolytic enzyme called phosphofructokinase. **Phosphofructokinase** catalyzes step 3 in Figure 9.7—the synthesis of fructose-1,6-bisphosphate from fructose-6-phosphate.

The discovery that ATP inhibits phosphofructokinase was important because this enzyme catalyzes a key step in the reaction sequence. Step 3 is highly exergonic. As a result, its product, fructose-1,6-bisphosphate, cannot easily be converted back into fructose-6-phosphate. In addition, an array of enzymes that aren't involved in glycolysis can convert the products of step 1 and step 2 to products used in other metabolic pathways. Before step 3, then, the sequence can be interrupted and the intermediates used elsewhere. But once fructose-1,6-bisphosphate is synthesized, it is difficult to turn back. Based on these observations, it is logical that the pathway is turned on or off at step 3.

The observation that ATP inhibits phosphofructokinase raises an important question, however: Why would a substrate that is required for the reaction in step 3 also inhibit the reac-

(Figure 9.7 continued)

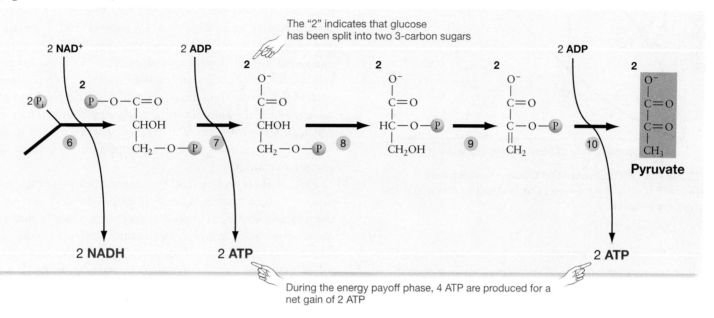

The "2" indicates that glucose has been split into two 3-carbon sugars

During the energy payoff phase, 4 ATP are produced for a net gain of 2 ATP

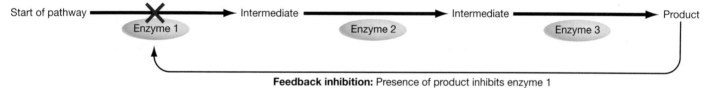

FIGURE 9.8 What Is Feedback Inhibition?
Feedback inhibition occurs when the product of a metabolic pathway inhibits an enzyme that is active early in the pathway. During glycolysis, feedback inhibition occurs when ATP binds to the regulatory site of phosphofructokinase.

tion? In the vast majority of cases, the addition of a substrate *speeds* the rate of a chemical reaction instead of slowing it.

To explain this unusual situation, biologists hypothesized that high levels of ATP are a signal that the cell does not need to produce more ATP. When an enzyme in a pathway is inhibited by the product of the reaction sequence, **feedback inhibition** is said to occur (**Figure 9.8**). The product molecule "feeds back" to stop the reaction sequence when the product is abundant. In glycolysis, cells that are able to stop the glycolytic reactions when ATP is abundant can conserve their stores of glucose for times when ATP is scarce. As a result, natural selection should favor individuals who have phosphofructokinase molecules that are inhibited by high concentrations of ATP.

This hypothesis provided a satisfying explanation for why ATP inhibits the key step of glycolysis, but it did not explain *how* high levels of the substrate inhibit the enzyme. The answer came later, when researchers were able to determine the three-dimensional structure of phosphofructokinase. As **Figure 9.9** shows, phosphofructokinase has two distinct binding sites for ATP. ATP can bind at the enzyme's active site or at a site that changes the

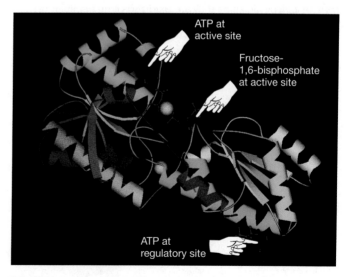

FIGURE 9.9 Phosphofructokinase Has Two Binding Sites for ATP
A model of one of the four identical subunits of phosphofructokinase. Notice the active site, where a phosphate group is transferred from ATP to form fructose-1,6-bisphosphate, and the regulatory site, where ATP binds.

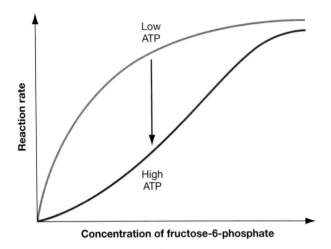

FIGURE 9.10 ATP Concentration Changes Reaction Rate
When ATP binds at the regulatory site, the reaction catalyzed at the active site is slowed.

enzyme's activity—a **regulatory site**. At the active site, ATP is converted to ADP and the phosphate group is transferred to fructose-6-phosphate. This reaction results in the synthesis of fructose-1,6-bisphosphate. But when ATP concentrations are high, the molecule also binds at a regulatory site. When ATP binds at this second location. The enzyme's conformation changes in a way that makes the reaction rate at the active site drop dramatically (**Figure 9.10**). To use the term introduced in Chapter 3, ATP acts as an allosteric regulator.

Thanks to advances such as these, glycolysis is among the best understood of all metabolic pathways. Now the question is, what happens to its product? How is pyruvate processed?

9.3 The Krebs Cycle

While Embden and Meyerhof and others were working out the sequence of reactions in glycolysis, biologists in several other laboratories were focusing on a different set of redox reactions that take place in actively respiring cells. These reactions involve small organic acids such as citrate, malate, and succinate. Because they have the form R-COOH, these molecules are called **carboxylic acids**. The reactions also result in the production of carbon dioxide. Recall from Section 9.1 that carbon dioxide is a highly oxidized form of carbon and is the endpoint of glucose metabolism. Thus, it was logical for researchers to propose that the oxidation of small carboxylic acids must be an important component of glucose oxidation.

Early workers made several key observations about these reactions. First, a total of eight small carboxylic acids are oxidized rapidly enough to imply that they are involved in glucose metabolism. Second, when one of these carboxylic acids is added to cells, the rate of glucose oxidation increases. The added molecules did not appear to be used up, however. Instead, virtually all of the carboxylic acid added seemed to be recovered later. Why this should be so was a puzzle. Even while this observation remained unexplained, however, the biochemists working on the reactions were able to determine the order in which the eight acids were oxidized. The result is the reaction pathway shown in **Figure 9.11a**.

The mystery of why intermediates in the pathway were not used up was resolved when Hans Krebs realized that the reaction sequence might occur in a cyclical fashion instead of a linear pathway. Krebs had another crucial insight when he suggested that the

(a) The carboxylic acids that are oxidized during cellular respiration

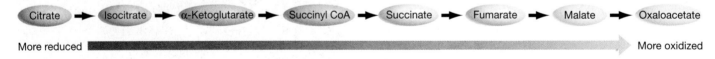

(b) The Krebs cycle hypothesis

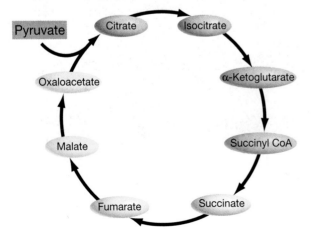

FIGURE 9.11 A Series of Carboxylic Acids Is Oxidized in the Krebs Cycle
(a) The eight molecules that are oxidized in actively respiring cells are pictured in order of their tendency to donate or accept electrons in redox reactions. **(b)** Hans Krebs proposed that if the pyruvate from glycolysis reacted with oxaloacetate to form citrate, then the eight reactions in part (a) would form a cycle.

reaction sequence was directly tied to the processing of pyruvate produced by glycolysis. To test these hypotheses, Krebs and a colleague set out to determine whether pyruvate—the endpoint of the glycolytic pathway—could react with oxaloacetate. Because oxaloacetate was the endpoint of the pathway in Figure 9.11a, the cycle hypothesis predicted that a reaction of oxaloacetate with pyruvate should produce the starting point of the pathway—the six-carbon carboxylic acid called citrate (**Figure 9.11b**).

Their result? As predicted, citrate formed when the biologists added oxaloacetate and pyruvate to cells. Based on this result, Krebs proposed that pyruvate is oxidized to carbon dioxide through a cycle of reactions. In honor of this insight, the pathway became known as the Krebs cycle.*

Converting Pyruvate to Acetyl CoA

When radioactive isotopes of carbon became available in the early 1940s, researchers in several labs used them to confirm the cyclical nature of the Krebs cycle. For example, by adding radioactively labeled citrate or pyruvate to cells and analyzing the radioactive compounds that resulted, it was possible to show that carbon atoms cycle through the sequence of reactions just as Krebs had proposed.

---

*The Krebs cycle is also known as the *citric acid cycle* because it starts with citrate, which becomes citric acid when protonated. Because citric acid has three carboxyl groups, the reaction sequence is also called the *tricarboxylic acid (TCA) cycle*.

Where do these reactions take place? In bacteria and archaea, the enzymes responsible for glycolysis and the Krebs cycle are located in the cytoplasm. In eukaryotes, though, the pyruvate produced by glycolysis is transported from the cytoplasm to mitochondria. As noted in Chapter 7, mitochondria are organelles found in virtually all eukaryotes. **Figure 9.12** shows a diagram and an electron micrograph of this organelle. Note that it has two membranes and that the interior is filled with layers of sac-like structures called **cristae**, which connect to the inner membrane. The region inside the inner membrane but outside the cristae is called the **mitochondrial matrix**. In eukaryotes, most of the enzymes responsible for the Krebs cycle are located in this matrix.

The pyruvate that links glycolysis and the Krebs cycle moves across the mitochondrion's outer membrane through small pores, then enters the mitochondrial matrix. Entry into the matrix occurs via active transport, through a membrane protein called the pyruvate carrier, located in the inner membrane. Because the transport of pyruvate into the mitochondrion is an active process, it represents an energy-consuming step in glucose oxidation.

What happens to pyruvate once it enters the mitochondrion? After Krebs's hypothesis was confirmed, researchers focused on understanding how pyruvate reacts with oxaloacetate to form citrate and start the cycle. This issue remained unresolved until it was discovered that an organic compound called **coenzyme A (CoA)** serves as a cofactor in a wide variety of reactions catalyzed by cellular enzymes. As we saw in Chapter 3, many enzymes

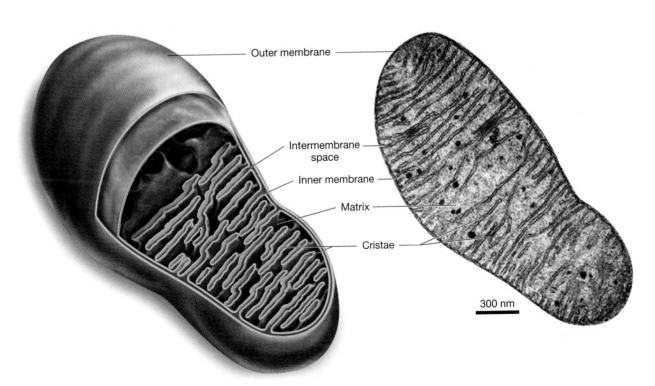

FIGURE 9.12 The Structure of the Mitochondrion
Mitochondria have inner and outer membranes. Saclike structures that contact the inner membrane are called cristae. The reactions of the Krebs cycle occur within the mitochondrial matrix.

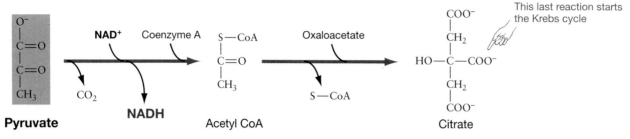

FIGURE 9.13 Pyruvate Is Oxidized to Acetyl CoA, Which Reacts with Oxaloacetate to Start the Krebs Cycle
Once it is inside the mitochondrion, the pyruvate produced by glycolysis is oxidized to acetyl CoA in a reaction catalyzed by pyruvate dehydrogenase. Acetyl CoA then reacts with oxaloacetate to form citrate in the first reaction of the Krebs cycle.

require a metal ion or a relatively small organic molecule called a coenzyme to function. Coenzymes often bind to the enzyme's active site and stabilize the reaction's transition state. Some coenzymes, for example, donate or accept electrons during the reaction; others transfer an acetyl ($-COCH_3$) group or carbon chains. CoA acts as a coenzyme by accepting and then transferring an acetyl group to a substrate. In fact, the "A" in "CoA" stands for *acetylation,* because an acetyl group is often bonded to a sulfur atom on one end of the molecule. When an acetyl group is bound to CoA, the acetyl group becomes activated and can be easily transferred to an acceptor molecule.

Soon after CoA's role as a coenzyme was revealed, other investigators found that pyruvate reacts with CoA to produce a molecule called **acetyl CoA.** Follow-up work showed that the conversion of pyruvate to acetyl CoA occurs in a series of steps inside an enormous and intricate enzyme complex called **pyruvate dehydrogenase.** The pyruvate dehydrogenase complex is located in the inner mitochondrial membrane and consists of three enzymes, which require a total of five cofactors. The coenzymes involved include thiamin, niacin, and pantothenic acid—molecules that you know as the B-complex vitamins.

As pyruvate is being processed, NAD^+ is reduced to NADH and one of the carbons in the pyruvate is oxidized to CO_2. The remaining two-carbon acetyl unit is transferred to CoA. The product, acetyl CoA, goes on to react with oxaloacetate to form citrate—the first compound in the Krebs cycle (**Figure 9.13**).

To summarize, the reactions catalyzed by pyruvate dehydrogenase act as a preparatory step for the Krebs cycle. The acetyl CoA that is produced goes on to react with oxaloacetate, forming citrate. When processing of pyruvate in the mitochondrion is complete, all three carbons in pyruvate are oxidized to carbon dioxide. One of the carbons is oxidized in the pyruvate dehydrogenase complex, and two are oxidized in the Krebs cycle itself. As **Figure 9.14** indicates, the energy released by the oxidation of one molecule of pyruvate is used to produce four molecules of NADH, one of $FADH_2$, and one of **guanosine triphosphate (GTP)** through substrate-level phosphorylation. GTP is then converted to ATP. Pyruvate and CoA

enter the sequence of reactions; CoA, carbon dioxide, ATP, NADH, and $FADH_2$ come out.

How Is the Krebs Cycle Regulated?

Like glycolysis, the Krebs cycle is carefully regulated. Reaction rates are high when ATP is scarce; reaction rates are low when ATP is abundant. The major control points are highlighted in **Figure 9.15**. Note that the pyruvate dehydrogenase complex is regulated, as are several enzymes in the cycle itself.

How does regulation occur? When supplies of ATP are abundant, the pyruvate dehydrogenase complex becomes phosphorylated. In this case, phosphorylation changes the shape of pyruvate dehydrogenase in a way that inhibits catalytic activity. High concentrations of other products—specifically, acetyl CoA and NADH—also increase the rate of phosphorylation of the enzyme complex. Similarly, the enzyme that converts acetyl CoA to citrate is shut down when ATP binds to it. Like the control of phosphofructokinase activity in glycolysis, these are examples of feedback inhibition. Reaction products feed back to stop or slow down the pathway.

As Figure 9.15 indicates, feedback inhibition also occurs at two points later in the cycle. At the first of these two points, NADH binds to the enzyme's active site. This is an example of competitive inhibition. At the second point, ATP binds to an allosteric regulatory site. In sum, the reaction pathway can be turned off at multiple points, via several different mechanisms of feedback inhibition.

In addition to these examples of inhibitory control, the presence of NAD^+, CoA, or adenosine monophosphate (AMP)—which indicates low supplies of ATP—speeds the conversion of pyruvate to acetyl CoA. Stated another way, the key steps in the processing of pyruvate are under both positive control and negative control. By speeding up the oxidation of pyruvate when ATP supplies are low and by slowing down the reactions when ATP is plentiful, cells carefully match the rate of cellular respiration to their energy requirements. Natural selection has favored the evolution of enzymes that allow cells to conserve pyruvate or acetyl CoA, just as it has favored glycolytic enzymes whose regulation allows cells to conserve glucose.

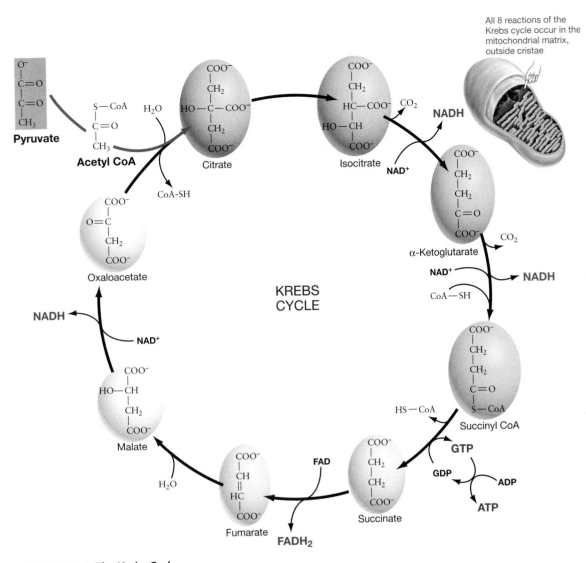

FIGURE 9.14 The Krebs Cycle
Acetyl CoA goes into the Krebs cycle, and carbon dioxide, NADH, FADH$_2$, and GTP come out. The GTP is produced by substrate-level phosphorylation and then converted to ATP. **EXERCISE** Mark one of the carbon atoms in acetyl CoA with a highlighter. Follow this atom from molecule to molecule around the cycle. What happens to it? Is it eventually given off as carbon dioxide?

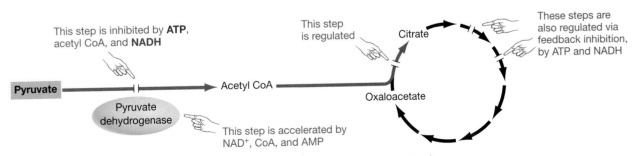

FIGURE 9.15 Pyruvate Dehydrogenase and the Krebs Cycle Are Carefully Regulated
The oxidation of pyruvate is under both negative and positive controls. Several steps in the Krebs cycle itself are also regulated. As a result, the reaction pathway speeds up when ATP supplies are low but slows down when ATP is plentiful.

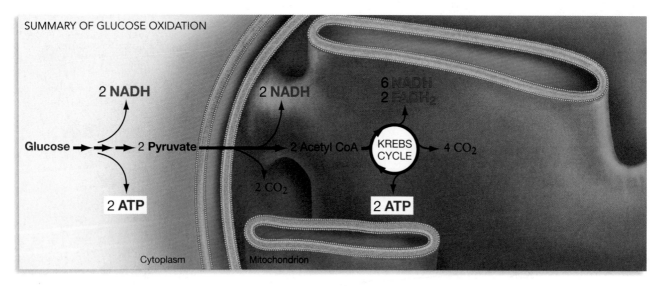

FIGURE 9.16 A Summary of Glucose Oxidation
Glucose is completely oxidized to carbon dioxide via glycolysis, the subsequent oxidation of pyruvate, and then the Krebs cycle. In eukaryotes, glycolysis occurs in the cytoplasm; pyruvate oxidation takes place in the inner mitchondrial membrane; the Krebs cycle occurs in the mitochondrial matrix.

What Happens to the NADH and FADH$_2$?

Figure 9.16 reviews the relationship between glycolysis and the Krebs cycle and identifies where each process takes place in eukaryotic cells. For each molecule of glucose that is fully oxidized to 6 carbon dioxide molecules, the cell produces 10 molecules of NADH, 2 of FADH$_2$, and 4 of ATP. The overall reaction for glycolysis and the Krebs cycle can be written as

$$C_6H_{12}O_6 + 10 \text{ NAD}^+ + 2 \text{ FAD} + 4 \text{ ADP} + 4P_i \rightarrow$$
$$6 \text{ CO}_2 + 10 \text{ NADH} + 2 \text{ FADH}_2 + 4 \text{ ATP}$$

The ATP molecules are produced by substrate-level phosphorylation and can be used to drive endergonic reactions, power movement, or run membrane pumps. The carbon dioxide molecules are a gas that is disposed of as waste—you exhale it.

What happens to the NADH and FADH$_2$ produced by glycolysis and the Krebs cycle? Recall that the overall reaction for glucose oxidation is

$$C_6H_{12}O_6 + 6 \text{ O}_2 \rightarrow 6 \text{ CO}_2 + 6 \text{ H}_2O + \text{energy}$$

Glycolysis and the Krebs cycle account for the glucose, the CO_2, and—because ATP is produced—some of the chemical energy that results from the overall reaction. But the O_2 and the H_2O that appear in the overall reaction for the oxidation of glucose are still unaccounted for. As it turns out, so is much of the chemical energy. The reaction that has yet to occur is

$$\text{NADH} + \text{FADH}_2 + \text{O}_2 + \text{ADP} + \text{P}_i \rightarrow$$
$$\text{NAD}^+ + \text{FAD} + \text{H}_2O + \text{ATP}$$

In this reaction, oxygen is reduced to form water. The electrons that drive the redox reaction come from NADH and FADH$_2$. These molecules are oxidized to NAD$^+$ + FAD.

In effect, glycolysis and the Krebs cycle transfer electrons from glucose to NAD$^+$ and FAD, creating NADH and FADH$_2$.

These molecules then carry the electrons to oxygen, which serves as the final electron acceptor in eukaryotic cells. When oxygen accepts electrons, water is produced. All the components of the overall reaction for glucose oxidation are accounted for.

How does this final part of the process occur? Specifically, how is ATP generated as electrons are transferred from NADH or FADH$_2$ to O$_2$? In the 1960s—decades after the details of glycolysis and the Krebs cycle had been worked out—a startling answer to these questions emerged.

✓ CHECK YOUR UNDERSTANDING

During glycolysis and the Krebs cycle, one molecule of glucose ($C_6H_{12}O_6$) is completely oxidized to six molecules of carbon dioxide (CO_2). You should be able to model these components of cellular respiration. To do this, pretend that a large piece of paper is a cell. Draw a large mitochondrion inside it. Cut out small squares of paper and label them as glucose, glycolytic reactions, Krebs cycle reactions, pyruvate dehydrogenase complex, pyruvate, acetyl CoA, CO$_2$, ATP, NADH, and FADH$_2$. Put each of the squares in the appropriate location in the cell. Using dimes for electrons, demonstrate how glucose is oxidized to CO$_2$ and where regulation occurs. Once your model is working, you'll be ready to consider what happens to the NADH and FADH$_2$ you've produced.

9.4   Electron Transport and Chemiosmosis

The answer to one fundamental question about the oxidation of NADH and FADH$_2$ turned out to be relatively straightforward. To determine where the oxidation reactions take place in eukaryotes, researchers isolated mitochondria by using differential centrifugation techniques (introduced in Chapter 7). Then they broke the organelles open and separated the inner and outer membranes from the mitochondrial matrix and the solution in the intermembrane space. The isolated membranes were capable of oxidizing NADH, but the matrix and intermembrane fluid were not. Follow-up work showed that the oxidation process takes place on the inner membrane of the mitochondria and the membranes of cristae. In prokaryotes, the oxidation of NADH occurs in the plasma membrane.

Biologists who analyzed the components of the mitochondrial inner membrane isolated molecules that switch between a reduced and an oxidized state during respiration. The molecules were hypothesized to be the key to processing NADH and $FADH_2$. What are these molecules, and how do they work?

Components of the Electron Transport Chain

Researchers made several fundamental observations about the molecules involved in the oxidation of NADH and $FADH_2$ in mitochondria. Collectively, the molecules involved are designated the **electron transport chain (ETC)**:

- Most of these molecules are proteins that contain distinctive chemical groups where the redox events take place. The active groups include ring-containing structures called flavins, iron-sulfur complexes, or iron-containing heme groups. Each subunit is readily reduced or oxidized.

- The inner membrane also contains a molecule called **ubiquinone**, which is not a protein. Ubiquinone got its name because it belongs to a family of compounds called quinones and because it is nearly ubiquitous in organisms. It is also called **coenzyme Q** or simply **Q**. Ubiquinone consists of a carbon-containing ring attached to a long tail made up of isoprene subunits. The structure of Q determines the molecule's function. The long, isoprene-rich tail is hydrophobic. As a result, Q is lipid soluble and can move throughout the mitochondrial membrane efficiently. In contrast, all but one of the proteins in the ETC are anchored in the membrane.

- The molecules involved in processing NADH and $FADH_2$ differ in electronegativity, or their tendency to hold electrons.

Because Q and the proteins can cycle between a reduced state and an oxidized state, and because they differ in electronegativity, investigators realized that it should be possible to arrange these molecules into a logical sequence. The idea was that electrons would pass from a molecule with lower electronegativity to one with higher electronegativity, via a redox reaction. As electrons moved through the chain and formed covalent bonds with each molecule in turn, they would be held more and more tightly. A small amount of energy would be released in each reaction, and the potential energy in each successive bond would lessen.

Using poisons, researchers were able to confirm that electron transport chains actually exist. Investigators found that cyanide, carbon monoxide, and other toxic compounds inhibit particular proteins in the inner membrane. If treatment with a particular toxin caused some proteins to sit in a reduced state while others

remained oxidized, by inference the poison had stopped the movement of electrons along the chain. The oxidized molecules had to be "downstream" from the reduced compounds. For example, when an electron transport chain is treated with the drug antimycin A, cytochrome *b* and Q remain reduced; cytochrome *c*, cytochrome *a*, and cytochrome *a₃* remain oxidized. This pattern makes sense if electrons flow from NADH and $FADH_2$ to cytochrome *b* and Q, but then stay there because the next element in the chain is poisoned. If they are downstream of the poisoned component in the chain, then cytochrome *c*, cytochrome *a*, and cytochrome *a₃* would stay oxidized. This too is logical, because they would pass their electrons on to O_2 but could not pick up any new electrons from the poisoned component. Data from poisoning experiments, along with measurements of the tendency for each protein and Q to gain or lose electrons, allowed researchers to place the molecules in the order shown in **Figure 9.17**.

Experiments with particular poisons also showed that NADH donates an electron to a flavin-containing protein at the top of

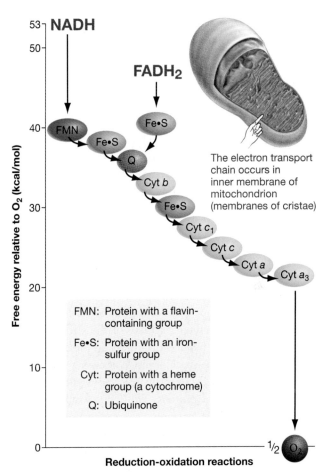

FIGURE 9.17 A Series of Reduction-Oxidation Reactions Occurs in an Electron Transport Chain

Electrons step down in potential energy from the electron carriers NADH and $FADH_2$ through an electron transport chain to a final electron acceptor. When oxygen is the final electron acceptor, water is formed. The overall free-energy change of 53 kcal/mol (from NADH to oxygen) is broken into small steps. EXERCISE Label the most and least electronegative molecules in the figure.

the chain, while $FADH_2$ donates electrons to an iron-sulfur-containing protein that then passes them directly to Q. After passing through each of the remaining components in the chain, the electrons are finally accepted by oxygen. Under standard conditions of temperature and pressure in the laboratory, the total potential energy difference from NADH to oxygen is 53 kilocalories/mole (abbreviated kcal/mol).

To summarize, electrons from NADH and $FADH_2$ pass through an electron transport chain consisting of a series of proteins and Q. Each successive molecule in the chain has slightly higher electronegativity than the previous one. As the electrons move from one link in the chain to the next, they are held tighter and tighter and their potential energy lessens. A molecule with particularly high electronegativity—oxygen, in plants, animals, and fungi—acts as the final electron acceptor. Electron transport takes place in the inner mitochondrial membrane.

Once the nature of the electron transport chain became clear, biologists understood the fate of the electrons carried by NADH and $FADH_2$ and how oxygen acts as the final electron acceptor. All of the electrons that were originally present in glucose were now accounted for. This is satisfying, except for one crucial question: How do these redox reactions generate ATP? Does substrate-level phosphorylation take place in the electron transport chain, just as it does in glycolysis and the Krebs cycle?

The Chemiosmotic Hypothesis

Throughout the 1950s most biologists interested in cellular respiration assumed that electron transport chains include enzymes that catalyze substrate-level phosphorylation. Despite intense efforts, however, no one was able to find a component of the ETC that phosphorylated ADP to produce ATP.

In 1961 Peter Mitchell made a radical break with prevailing ideas by proposing that the connection between electron transport and ATP production is indirect. Mitchell hypothesized that the real job of the electron transport chain is to pump protons from the matrix of the mitochondrion through the inner membrane and out to the intermembrane space or the interior of cristae. According to Mitchell, the pumping activity of the electron transport chain would lead to a buildup of protons in these areas. In this way, the intermembrane space or inside of cristae would become positively charged relative to the matrix and would have a much higher concentration of protons. The result would be a strong electrochemical gradient favoring the movement of protons back into the matrix. This **proton-motive force**, he hypothesized, is used by an enzyme in the inner membrane to synthesize ATP.

Mitchell called the production of ATP via a proton gradient **chemiosmosis**, and his proposal of an indirect link between electron transport and ATP production is the **chemiosmotic hypothesis**. Although proponents of a direct link between electron transport and substrate-level phosphorylation objected vigorously to Mitchell's idea, several key experiments supported it.

One of these experiments demonstrated the existence of a key element in Mitchell's scheme (**Figure 9.18**): A mitochondri-

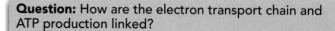

Question: How are the electron transport chain and ATP production linked?

Chemiosmotic hypothesis: The linkage is indirect. The ETC creates a proton-motive force that drives ATP synthesis by a mitochondrial protein.

Null hypothesis: The linkage is not indirect. The ETC does not produce a proton-motive force that drives ATP synthesis.

Experimental setup:

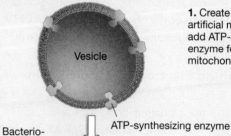

Vesicle

ATP-synthesizing enzyme

Bacterio-rhodopsin

1. Create vesicles from artificial membranes; add ATP-synthesizing enzyme found in mitochondria.

2. Add bacteriorhodopsin, a protein that acts as a light-activated proton pump.

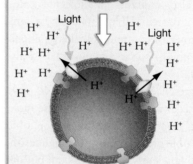

Light   Light

H^+

3. Illuminate vesicle so that bacteriorhodopsin pumps protons out of vesicle, creating a proton gradient.

Prediction of chemiosmotic hypothesis: ATP will be produced within the vesicle.

Prediction of null hypothesis: No ATP will be produced.

Results:

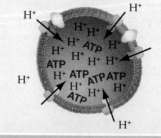

H^+ ATP

ATP is produced within the vesicle, in the absence of the electron transport chain.

Conclusion: The linkage between electron transport and ATP synthesis is indirect; the movement of protons drives the synthesis of ATP.

FIGURE 9.18 Evidence for the Chemiosmotic Hypothesis
QUESTION Do you regard this as a convincing test of the chemiosmotic hypothesis? Why or why not?

al enzyme can use a proton gradient to synthesize ATP. The biologists who did this experiment began by making vesicles from artificial membranes that contained an ATP-synthesizing enzyme found in mitochondria. Along with this enzyme, they inserted bacteriorhodopsin. Bacteriorhodopsin is a well-studied membrane protein that acts as a light-activated proton pump. When light strikes this protein, the protein absorbs some of the light energy and changes conformation in a way that results in the pumping of protons from the interior of a cell or membrane-bound vesicle to the outside. As a result, the experimental vesicles could establish a strong electrochemical gradient favoring proton movement to the interior. When the vesicles were illuminated to initiate proton pumping, ATP began to be produced from ADP inside the vesicles. This result provided strong support for the chemiosmosis hypothesis. Mitchell's prediction was correct: ATP production depended solely on the existence of a proton-motive force. Consequently, it could occur in the absence of an electron transport chain.

Based on this experiment and others, most biologists accepted the chemiosmotic hypothesis as valid. Instead of being produced by enzymes inside the mitochondrion, ATP is produced by a flow of protons. In combination, the electron transport chain, the mitochondrial membrane, and ATP synthase function much like a hydroelectric dam. The electron transport chain is analogous to a series of gigantic pumps that force water up and behind a dam. The inner mitochondrial membrane functions as the dam, and ATP synthase is like the turbines inside the dam. At a hydroelectric dam, the movement of water makes turbines spin and generate electricity. But in a mitochondrion, protons are pumped instead of water. When protons move through ATP synthase, the protein spins and generates ATP.

Electron transport chains and ATP synthase occur in organisms throughout the tree of life. They are humming away in your cells now. Let's look in more detail at how they function.

How Is the Electron Transport Chain Organized?

Once the predictions of the chemiosmotic hypothesis were verified, researchers focused on understanding the three-dimensional structure of the components of the electron transport chain and on determining how electron transport is coupled to proton pumping. This work has shown that the ETC components are organized into four large complexes of proteins and cofactors (**Figure 9.19**). Protons are pumped by three of the complexes. Q and the protein **cytochrome c** act as shuttles that transfer electrons between complexes.

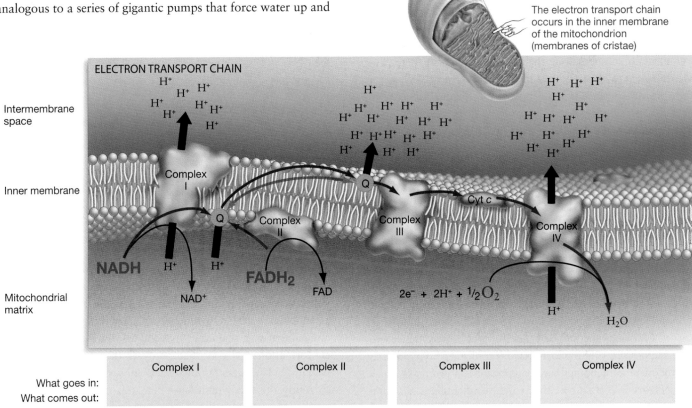

FIGURE 9.19 How Is the Electron Transport Chain Organized?
The individual components of the electron transport chain diagrammed in Figure 9.17 are grouped into large multiprotein complexes. Electrons are carried from one complex to another by Q and by cytochrome c. Complexes I, III, and IV use the potential energy released by the redox reactions to pump protons from the mitochondrial matrix to the intermembrane space. **EXERCISE** Label the solid blue arrows to indicate what they represent. Add an arrow across the membrane and label it "Proton gradient." In the boxes at the bottom, list "What goes in" and "What comes out" for each complex.

The structural studies completed to date confirm that in complexes I and IV, protons actually pass directly through a sequence of electron carriers. The exact route taken by the protons is still being worked out. It is also not clear how the redox reactions taking place inside each complex—as electrons step down in potential energy—make proton movement possible.

The best-understood interaction between electron transport and proton pumping takes place in complex III. Research has shown that when Q accepts electrons from complex I or complex II, it also gains protons. The reduced form of Q then diffuses to the outer side of the inner membrane, where its electrons are used to reduce a component of complex III near the intermembrane space. The protons held by Q are released to the intermembrane space. In this way, Q shuttles electrons and protons from one side of the membrane to the other. The electrons proceed down the transport chain, and the protons released to the intermembrane space contribute to the proton-motive force.

Now, how does this proton gradient make the production of ATP possible?

The Discovery of ATP Synthase

In 1960 Efraim Racker made several key observations about how ATP is synthesized in mitochondrial membranes. When he used these structures to make vesicles, Racker noticed that some happened to form with their membrane inside out. Electron microscopy revealed that the inside-out membranes had numerous large proteins studded along their surfaces. Each of these proteins appeared to have a base in the membrane and a stalk and a knob projecting out from the membrane (**Figure 9.20a**). But when Racker shook the vesicles or treated them with a compound called urea, the stalks and knobs fell off.

Racker seized on this technique to isolate the stalks and knobs and do experiments with them. For example, he found that isolated stalks and knobs could hydrolyze ATP, forming ADP and inorganic phosphate. The vesicles that contained just the base component, without the stalks and knobs, could not process ATP. The base components were, however, capable of transporting protons across the membrane.

Based on these observations, Racker proposed that the stalk-and-knob component of the protein was an enzyme that both hydrolyzes and synthesizes ATP. To test this idea, he added the stalk-and-knob components back to vesicles that had been stripped of them and confirmed that the vesicles were then capable of synthesizing ATP. Follow-up work confirmed his hypothesis that the membrane-bound base component is a proton channel.

As **Figure 9.20b** shows, the structure of this protein complex is now reasonably well understood. The ATPase "knob" component is called the F_1 unit; the membrane-bound, proton-transporting base component is the F_o unit; and the F_1 and F_o units are connected by the stalk. The entire complex is known as **ATP synthase**. According to the current model for the way this enzyme functions, a flow of protons through the F_o unit

(a) A vesicle formed from an "inside-out" mitochondrial membrane

Membrane proteins with a knob

(b) The F_o unit is the base; the F_1 unit is the knob.

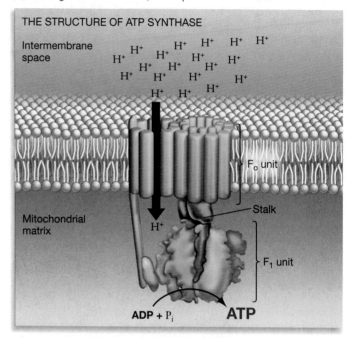

THE STRUCTURE OF ATP SYNTHASE

Intermembrane space

F_o unit

Stalk

Mitochondrial matrix

H^+

F_1 unit

ADP + P_i ATP

FIGURE 9.20 The Structure of ATP Synthase
(a) When patches of mitochondrial membrane turn inside out and form vesicles, proteins that have a stalk-and-knob structure face outward. Normally, the stalk and knob face inward, toward the mitochondrial matrix. **(b)** ATP synthase has two major components, designated F_o and F_1.

causes the stalk connecting the two subunits to spin. By attaching long actin filaments to the stalk and examining them with a videomicroscope, researchers have been able to see movement at an estimated 50 revolutions per second. As the F_1 unit rotates along with the stalk, its subunits are thought to change conformation in a way that catalyzes the phosphorylation of ADP to ATP. Understanding exactly how this reaction occurs is currently the focus of intense research. ATP synthase makes most of the ATP that keeps you alive.

Oxidative Phosphorylation

The formation of ATP through the combination of proton pumping by electron transport chains and the action of ATP synthase is called **oxidative phosphorylation**. This term is appropriate because the phosphorylation of ADP is based on the oxidation of NADH and FADH$_2$.

Figure 9.21 summarizes glucose oxidation and cellular respiration by tracing the fate of the carbon atoms and electrons in glucose. Notice that electrons from glucose are transferred to NADH and FADH$_2$, passed through the electron transport chain, and accepted by oxygen. The pumping of protons during electron transport creates the proton-motive force that drives ATP synthesis. The diagram also indicates the approximate yield of ATP from each component of the process. Recent research has shown that about 30 ATP molecules are produced from each molecule of glucose; of these, 26 ATP molecules are produced by ATP synthase. The fundamental message here is that the vast majority of the "payoff" from the oxidation of glucose occurs as a result of oxidative phosphorylation.

It is important to recognize, however, that cellular respiration can occur without oxygen. Oxygen is the electron acceptor used by all eukaryotes and a wide diversity of bacteria and archaea. Species that depend on oxygen as an electron acceptor are said to use **aerobic respiration** and are called aerobic organisms. (The Latin root *aero* means "air.") Many thousands of bacterial and archaeal species rely on electron acceptors other than oxygen, however, and electron donors other than glucose. As Chapter 27 will show, some bacteria and archaea use H$_2$, H$_2$S, CH$_4$, or other inorganic compounds as electron donors. For species in oxygen-poor environments, nitrate (NO$_3^-$) and sulphate (SO$_4^{2-}$) are particularly common electron acceptors. Cells that depend on electron acceptors other than oxygen are said to use **anaerobic** ("no air") **respiration.** Even though the starting and ending points of cellular respiration are different, these cells are still able to use electron transport chains to create a proton-motive force that drives the synthesis of ATP.

Oxygen is the most effective of all electron acceptors, however, because of its high electronegativity. Because oxygen holds electrons so tightly, the potential energy of electrons in a bond between an oxygen atom and a non-oxygen atom is low. As a result, there is a large difference between the potential energy of electrons in NADH and the potential energy of electrons bonded to an oxygen atom (see Figure 9.17). The large differential in potential energy means that the electron transport chain can generate a large proton-motive force. Cells that do not use oxygen as an electron acceptor cannot generate such a large potential energy difference. As a result, they cannot make as much ATP as do cells that use aerobic respiration and thus tend to grow much more slowly. If cells that use anaerobic respiration compete with cells using aerobic respiration, the cells that use oxygen as an electron acceptor almost always grow faster and reproduce more. As a result, species that rely on anaerobic respiration usually live in environments where oxygen is not present, where they do not have to compete with aerobic species.

What happens when oxygen or other electron acceptors are temporarily used up and are unavailable? Without oxygen or another electron acceptor in place, the electrons carried by NADH have no place to go. When this happens, the electron transport chain stops. All of the NAD$^+$ in the cell quickly becomes NADH. This situation is life threatening. When there is no longer any NAD$^+$ to supply the reactions of glycolysis, no ATP can be produced. If NAD$^+$ cannot be regenerated somehow, the cell will die. How do cells cope?

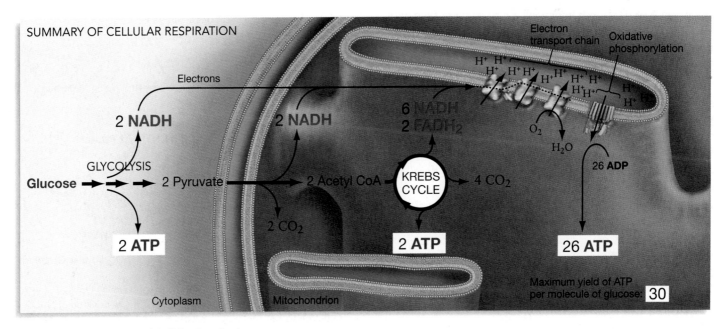

FIGURE 9.21 A Summary of Cellular Respiration

EXERCISE Across the top of the figure, write the overall reaction for the oxidation of glucose. Explain what happens to each reactant. Identify the source of each product.

✓CHECK YOUR UNDERSTANDING

As electrons from NADH and FADH$_2$ move through the electron transport chain, protons are pumped into the intermembrane space of mitochondria. The resulting electrochemical gradient drives protons through ATP synthase, resulting in the production of ATP from ADP. You should be able to add paper cutouts labeled ETC and ATP synthase to the model you made in Section 9.3, and then explain the steps in electron transport and chemiosmosis, using dimes to represent electrons and pennies to represent protons.

9.5   Fermentation

Fermentation is a metabolic pathway that allows continued production of ATP in the absence of an electron acceptor. It occurs when the pyruvate that is produced by glycolysis (or a molecule derived from pyruvate), rather than oxygen, accepts electrons from NADH. When NADH gets rid of electrons in this way, NAD$^+$ is produced. With NAD$^+$ present, glycolysis can continue to produce ATP via substrate-level phosphorylation and the cell can stay alive. In many cases, the molecule that is formed by the addition of an electron to pyruvate (or another electron acceptor) cannot be used by the cell. In some cases, this by-product is toxic and is excreted from the cell as waste (**Figure 9.22a**).

In organisms that usually use oxygen as an electron acceptor, fermentation is an alternative mode of energy production when oxygen supplies temporarily run out. For example, if you sprint a long distance, your muscles begin metabolizing glucose so fast that your lungs and circulatory system cannot supply oxygen rapidly enough to keep electron transport chains active. When oxygen is absent, the electron transport chains shut down and NADH cannot donate its electrons there. The pyruvate produced by glycolysis then begins to accepts electrons from NADH, and fermentation takes place (**Figure 9.22b**). The result of this process, called **lactic acid fermentation**, is the formation of lactate and regeneration of NAD$^+$.

Figure 9.22c illustrates a different type of fermentation pathway. These reactions, called **alcohol fermentation**, occur in the fungus *Saccharomyces cerevisiae*—baker's and brewer's yeast. When fungal cells are placed in an environment such as bread dough or a bottle of champagne and begin growing there, they quickly use up all the available oxygen. They continue to use glycolysis to metabolize sugar, however, by enzymatically converting pyruvate to the two-carbon compound acetaldehyde. This reaction gives off carbon dioxide, which causes bread to rise and creates the bubbles in champagne and beer. Acetaldehyde then accepts electrons from NADH, forming the NAD$^+$ required to keep glycolysis going. The addition of electrons to acetaldehyde results in the formation of ethanol as a waste product. Ethanol is the active ingredient in alcoholic beverages.

Species of bacteria and archaea exhibit a huge variety of other fermentative pathways. For example, bacteria and ar-

(a) Fermentation pathways allow cells to regenerate NAD$^+$ for glycolysis.

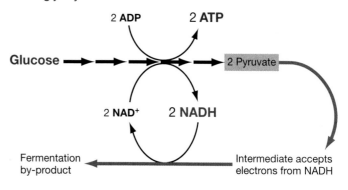

(b) Lactic acid fermentation occurs in humans.

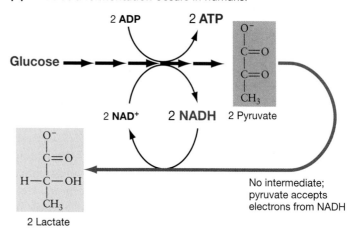

(c) Alcohol fermentation occurs in yeast.

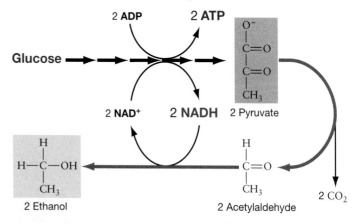

FIGURE 9.22 Fermentation Regenerates NAD$^+$ So that Glycolysis Can Continue
Among the bacteria, archaea, and eukaryotes, many types of fermentation occur. **(a)** All types result in the regeneration of NAD$^+$, so glycolysis can continue to produce ATP even when the electron acceptors required by electron transport chains are unavailable. **(b)** Lactic acid fermentation and **(c)** alcohol fermentation are two of the best-studied types.

chaea that exist exclusively through fermentation are present in phenomenal numbers in the oxygen-free environment of your small intestine and in the rumen of cows. The rumen is a specialized digestive organ that contains over 10^{10} (ten billion) bacterial and archaeal cells per *milliliter* of fluid. The fermentations that occur in these cells result in the production of an array of fatty acids. Cows use these fermentation by-products as a source of energy. Cells that employ other types of fermentations are used commercially in the production of soy sauce, tofu, yogurt, cheese, vinegar, and other products.

Even though fermentation is a widespread and commercially important type of metabolism, it is extremely inefficient compared with cellular respiration. Recall that fermentation produces just two molecules of ATP for each molecule of glucose metabolized, while cellular respiration produces about 30—approximately 15 times more energy per glucose molecule than fermentation. The reason for this disparity is that electron acceptors such as pyruvate and acetaldehyde have low electronegativity compared with an electron acceptor such as oxygen. As a result, the potential energy drop between the start and end of fermentation is a tiny fraction of the potential energy change that occurs during cellular respiration.

Based on these observations, it should not be surprising that organisms capable of both processes never use fermentation when an appropriate electron acceptor is available for cellular respiration. Organisms that can switch between fermentation and cellular respiration that uses oxygen as an electron acceptor are called **facultative aerobes**. The term *aerobe* refers to using oxygen, while *facultative* reflects the ability to use this pathway only when oxygen is present.

9.6   How Does Cellular Respiration Interact with Other Metabolic Pathways?

The enzymes, products, and intermediates involved in cellular respiration and fermentation do not exist in isolation. Instead, they are part of a huge and dynamic inventory of chemicals inside the cell. Because metabolism includes thousands of different chemical reactions, the amounts and identities of molecules inside cells are constantly in flux (**Figure 9.23**). Fermentation pathways, electron transport, and other aspects of carbohydrate metabolism may be crucial to the life of a cell, but they also have to be seen as parts of a whole.

To make sense of the chemical inventory inside cells and the full scope of metabolism, it is critical to recognize that cells have two fundamental requirements to stay alive, grow, and reproduce: energy and carbon. More formally, cells need a source of high-energy electrons for generating chemical energy in the form of ATP, and a source of carbon-containing molecules that can be used to synthesize DNA, RNA, proteins, fatty acids, and other molecules. Reactions that result in the breakdown of molecules and the production of ATP are called **catabolic pathways**; reac-

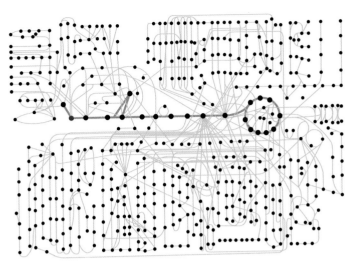

FIGURE 9.23 Metabolic Pathways Interact
A representation of a few of the thousands of chemical reactions that occur in cells. The dots represent molecules, and the lines represent enzyme-catalyzed reactions. **EXERCISE** Label the red dot, and circle the 10 reactions of glycolysis. Draw a box around the Krebs cycle.

tions that result in the synthesis of larger molecules from smaller components are called **anabolic pathways.**

This section introduces the ways in which glycolysis and the Krebs cycle interact with other catabolic pathways and with anabolic pathways. Let's consider how molecules other than carbohydrates are used as fuel in eukaryotes, and then examine how molecules involved in glycolysis and the Krebs cycle are sometimes used as building blocks in the synthesis of cell components.

Processing Proteins and Fats as Fuel

Most organisms ingest, synthesize, or absorb a wide variety of carbohydrates. These molecules range from sucrose, maltose, and other simple sugars to large polymers such as glycogen and starch. As noted in Chapter 5, glycogen is the major form of stored carbohydrate in animals, while starch is the major form of stored carbohydrate in plants. Recall that both glycogen and starch are polymers of glucose, but they differ in the way their long chains of glucose branch. Using enzyme-catalyzed reactions, cells can produce glucose from glycogen, starch, and most simple sugars. Glucose and fructose can then be processed by the enzymes of the glycolytic pathway.

Carbohydrates are not the only important source of carbon compounds used in the catabolic pathways, however. As pointed out in Chapter 6, fats are highly reduced macromolecules consisting of glycerol bonded to chains of fatty acids. In cells, fats are routinely broken down by enzymes to form glycerol and acetyl CoA. Glycerol enters the glycolytic pathway once it has been oxidized and phosphorylated to form glyceraldehyde-3-phosphate—one of the intermediates in the 10-reaction sequence. Acetyl CoA, in contrast, enters the Krebs cycle.

Proteins can also be catabolized. Once they are broken down to their constituent amino acids, the amino ($-NH_2$)

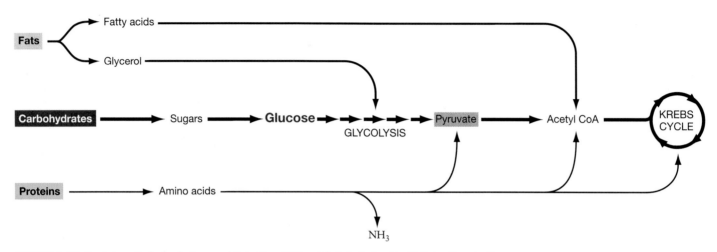

FIGURE 9.24 Proteins, Carbohydrates, and Fats Can All Furnish Substrates for Cellular Respiration
A variety of carbohydrates can be converted to glucose and processed by glycolysis. If carbohydrates are scarce, cells
can obtain high-energy compounds from fats or ultimately proteins for ATP production. These are catabolic reactions.

groups are removed in enzyme-catalyzed reactions. These
amino groups are excreted in urine as waste. The carbon compounds that remain after this catabolic step are converted to
pyruvate, acetyl CoA, and other intermediates in glycolysis and
the Krebs cycle.

Figure 9.24 summarizes the catabolic pathways of carbohydrates, fats, and proteins and shows how their breakdown
products feed an array of steps in glucose oxidation and cellular respiration. When all three types of molecules are available
in the cell to generate ATP, carbohydrates are used up first, then
fats, and finally proteins.

Anabolic Pathways Synthesize Key Molecules

Where do cells get the precursor molecules required to synthesize amino acids, RNA, DNA, phospholipids, and other cell
components? Not surprisingly, the answer often involves intermediates in carbohydrate metabolism. For example:

- In humans, about half of the 20 required amino acids can be
  synthesized from molecules siphoned off from the Krebs cycle.

- Acetyl CoA is the starting point for anabolic pathways that
  result in the synthesis of fatty acids.

- The molecule that is produced by the first reaction in glycolysis
  can be oxidized to start the synthesis of ribose-5-phosphate—
  a key intermediate in the production of ribonucleotides and
  deoxyribonucleotides. These nucleotides, in turn, are required
  for manufacturing RNA and DNA.

- If ATP is abundant, pyruvate and lactate (from fermentation) can be used as a substrate in the synthesis of glucose.
  Excess glucose is converted to glycogen and stored.

Figure 9.25 summarizes how intermediates in carbohydrate
metabolism are drawn off to synthesize macromolecules. The
same molecule can serve many different functions in the cell. As a
result, catabolic and anabolic pathways are closely intertwined.

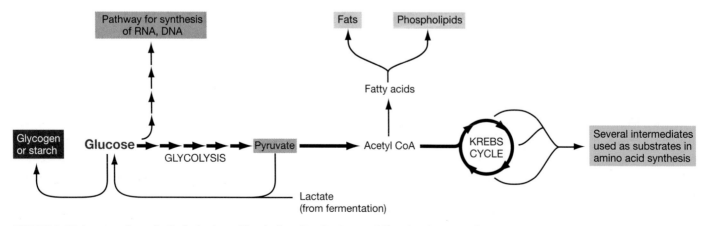

FIGURE 9.25 Intermediates in Carbohydrate Metabolism Can Be Drawn Off to Synthesize Cell Components
Several of the intermediates in carbohydrate metabolism act as precursor molecules in anabolic reactions
leading to the synthesis of RNA, DNA, glycogen or starch, amino acids, and fatty acids.

ESSAY ATP Production during Exercise

In an adult animal, the highest demands for ATP are made by exercising muscles. As Chapter 46 will describe, muscles contain a motor protein called myosin. Like the motor proteins kinesin and dynein, introduced in Chapter 7, myosin undergoes conformational changes in response to the addition of ATP and the subsequent release of a phosphate group. This movement by myosin is the physical basis of muscle contraction, and muscle contraction is the physical basis of movement.

On average, a muscle contains only enough ATP to sustain about 15 seconds of intense exercise. For muscle contractions to continue, massive amounts of ATP are required. Depending on the level and duration of activity, the muscles being exercised may produce the ATP they need either by cellular respiration or by fermentation. Sustained periods of moderate activity such as jogging are powered by aerobic respiration. In contrast, short periods of intense activity such as sprinting are powered by a combination of aerobic respiration and fermentation.

To understand the relationship between cellular respiration and fermentation in muscle cells, we'll focus on the constraints experienced by two types of runners. What aspects of cellular respiration affect the performance of a marathoner? What aspects of fermentation limit the duration of sprints?

A marathon course is 26 miles, 385 yards (42 km, 195 m) long. This race was inaugurated with the first of the modern Olympic games, which took place in 1899 in Athens, Greece. The event was designed to commemorate Pheidippides, a Greek soldier who reportedly ran from the plains of Marathon in southeast Greece to Athens immediately after the Greek army had triumphed over the invading Persians in 490 B.C. According to legend, Pheidippides shouted "Victory!" when he reached Athens, then collapsed and died of exhaustion.

Contemporary marathon runners rarely collapse and die, but virtually all experience a phenomenon known as "hitting the wall." What happens when a marathon runner hits the wall? Marathon running is an aerobic activity fueled by cellular respiration. The glucose and glycogen stores in the body are adequate to fuel only about 20 miles of running, however. During

What happens when a marathon runner hits the wall?

the last 6 miles of the race, marathoners rely exclusively on fatty-acid metabolism to keep cellular respiration going. Hitting the wall is a feeling of exhaustion that occurs as the last of the glycogen stores are used up. Because fatty-acid metabolism is markedly slower than glycogen metabolism, ATP delivery is reduced and running becomes much more difficult.

Sprinters do not hit the wall. Instead they "feel the burn." During intense activity, metabolism switches from aerobic to anaerobic and a burning sensation is felt in the muscles. This feeling is caused by a buildup of lactic acid due to sustained production of ATP by fermentation. As fermentation proceeds, lactic acid levels can build up, dropping the pH of muscle cells. The acidification may damage enzymes and structural proteins and contribute to pain and fatigue. Formally, muscle fatigue is defined as the inability of a muscle to continue contracting with maximum force. In addition, anaerobic metabolism produces a great deal of heat, which can damage muscle proteins if it is not controlled.

For both sprinters and marathoners, an extended cooldown period after exercise is important for muscle health. Several important events occur during recovery from exercise. Lactic acid is transported from muscles to the liver, where it is converted to pyruvate and then to glucose. Oxygen-rich blood is delivered to muscles, where cellular respiration restores normal levels of ATP. In addition, heat is transferred from muscles to blood. The blood carries the heat to the body surface, where it is radiated to the environment, cooling the body.

Recent research has shown that aerobic training can increase the number of mitochondria in muscle cells and thus increase ATP supplies during exercise. Similarly, sprint training can increase the number of glycolytic enzymes in muscle cells and thus the rate of ATP delivery during bursts of activity. Increasingly, an understanding of cellular respiration and fermentation is being used to design training programs for athletes.

CHAPTER REVIEW

Summary of Key Concepts

■ **Cellular respiration produces ATP from molecules with high potential energy—often glucose.**

ATP is the currency that cells use to pump ions, drive endergonic reactions, move cargo, and perform other types of work. When proteins bind ATP or are phosphorylated, they respond by changing conformation, or shape, in a way that allows them to act as enzymes, pumps, or motors.

Cells produce ATP from sugars or other compounds with high free energy by using one of two general pathways: (1) cellular respiration or (2) fermentation. Cellular respiration involves the transfer of electrons from a compound with high free energy, such as glucose, to a molecule with lower free energy, such as oxygen, through an electron transport chain. Fermentation involves the transfer of electrons from one organic compound to another without participation by an electron transport chain.

■ **Glucose processing has three components: (1) glycolysis, (2) the Krebs cycle, and (3) electron transport coupled with oxidative phosphorylation.**

In eukaryotes, glycolysis takes place in the cytoplasm. The Krebs cycle occurs in the mitochondrial matrix, and electron transport and oxidative phosphorylation proceed in the inner membranes of mitochondria.

Glycolysis is a 10-step reaction sequence in which glucose is broken down into two molecules of pyruvate. For each molecule of glucose processed during glycolysis, two molecules of ATP are produced by substrate-level phosphorylation and two molecules of NAD^+ are reduced to two molecules of NADH.

Prior to the Krebs cycle, each pyruvate is oxidized to acetyl CoA by a series of reactions that results in the production of one molecule of carbon dioxide and the synthesis of one of NADH. The Krebs cycle begins when acetyl CoA reacts with oxaloacetate to form citrate. The ensuing series of reactions results in the regeneration of oxaloacetate, the production of two molecules of carbon dioxide, the synthesis of one molecule of ATP by substrate-level phosphorylation, the reduction of one molecule of FAD to $FADH_2$, and the reduction of three molecules of NAD^+ to three of NADH.

In essence, glycolysis and the Krebs cycle strip electrons from glucose and use them to reduce NAD^+ and FAD. The resulting electron carriers—NADH and $FADH_2$—then donate electrons to an electron transport chain, which gradually steps the electrons down in potential energy until they are finally accepted by oxygen or by another final electron acceptor, resulting in the formation of water in aerobic organisms. The components of the electron transport chain use the energy released by the oxidation of NADH and $FADH_2$ to pump protons across the inner mitochondrial membrane. The pumping activity creates an electrochemical gradient, or proton-motive force, that ATP synthase uses to produce ATP. Oxidative phosphorylation is the production of ATP by electron transport and the generation of a proton-motive force.

Web Tutorial 9.1   Glucose Metabolism

■ **Fermentation pathways allow glycolysis to continue when the lack of an electron acceptor shuts down electron transport chains.**

If no electron acceptor such as oxygen is available, cellular respiration stops, because electron transport chains cannot continue. As a result, all NAD^+ is converted to NADH and glycolysis must stop. Fermentation pathways regenerate NAD^+ when an organic molecule such as pyruvate accepts electrons from NADH. Depending on the molecule that acts as an electron acceptor, fermentation pathways produce lactate, ethanol, or other reduced organic compounds as a by-product.

Web Tutorial 9.2   Chemiosmosis

■ **Cellular respiration and fermentation are carefully regulated.**

When supplies of ATP, NADH, and $FADH_2$ are high in the cell, feedback regulation occurs—that is, product molecules bind to and inhibit enzymes involved in ATP production. The glycolytic pathway slows when ATP binds to phosphofructokinase, and the pyruvate dehydrogenase complex is inhibited when it is phosphorylated by ATP. The enzyme that converts acetyl CoA to citrate slows when ATP binds to it, and certain enzymes in the Krebs cycle are inhibited when NADH or ATP bind to them. As a result, ATP is produced only when needed.

Questions

Content Review

1. When does feedback inhibition occur?
   a. when lack of an appropriate electron acceptor makes an electron transport chain stop
   b. when an enzyme that is active early in a metabolic pathway is inhibited by a product of the pathway
   c. when ATP synthase reverses and begins pumping protons out of the mitochondrial matrix
   d. when cellular respiration is inhibited and fermentation begins

2. Where does the Krebs cycle occur in eukaryotes?
   a. in the cytoplasm
   b. in the matrix of mitochondria
   c. in the inner membrane of mitochondria
   d. in the intermembrane space of mitochondria

3. What does the chemiosmotic hypothesis claim?
   a. Substrate-level phosphorylation occurs in the electron transport chain.
   b. Substrate-level phosophorylation occurs in glycolysis and the Krebs cycle.
   c. The electron transport chain is located in the inner membrane of mitochondria.
   d. Electron transport chains generate ATP indirectly, by the creation of a proton-motive force.

4. What is the function of the reactions in a fermentation pathway?
   a. to generate NADH from NAD^+, so electrons can be donated to the electron transport chain
   b. to synthesize pyruvate from lactate
   c. to generate NAD^+ from NADH, so glycolysis can continue
   d. to synthesize electron acceptors, so that cellular respiration can continue

5. When do cells switch from cellular respiration to fermentation?
   a. when electron acceptors are not available
   b. when the proton-motive force runs down
   c. when NADH and $FADH_2$ supplies are low
   d. when pyruvate is not available

6. Why are NADH and $FADH_2$ said to have "reducing power?"
   a. They are the reduced forms of NAD^+ and FAD.
   b. They donate electrons to components of the ETC, reducing those components.
   c. They travel between the cytoplasm and the mitochondrion.
   d. They have the power to reduce carbon dioxide to glucose.

Conceptual Review

1. Explain why NADH and FADH$_2$ are called electron carriers. Where do these molecules get electrons, and where do they deliver them? In eukaryotes, what molecule do these electrons reduce?

2. Compare and contrast substrate-level phosphorylation and oxidative phosphorylation.

3. What is the relationship between cellular respiration and fermentation? Why does cellular respiration produce so much more ATP than fermentation does?

4. Diagram the relationship among the three components of cellular respiration: glycolysis, the Krebs cycle, and electron transport. What molecules connect these three processes? Where does each process occur in a eukaryotic cell?

5. Explain the relationship between electron transport and oxidative phosphorylation. What does ATP synthase look like, and how does it work?

6. Describe the relationship among carbohydrate metabolism, the catabolism of proteins and fats, and anabolic pathways.

Group Discussion Problems

1. Cyanide (C≡N⁻) blocks complex IV of the electron transport chain. Suggest a hypothesis for what happens to the ETC when complex IV stops working. Your hypothesis should explain why cyanide poisoning is fatal.

2. The presence of many saclike cristae results in a large amount of membrane inside mitochondria. Suppose that some mitochondria had few cristae. How would their output of ATP compare with that of mitochondria with many cristae? Explain your answer.

3. When yeast cells are placed into low-oxygen environments, the mitochondria in the cells become reduced in size and number. Suggest an explanation for this observation.

4. Most agricultural societies have come up with ways to ferment the sugars in barley, wheat, rice, corn, or grapes to produce alcoholic beverages. Historians argue that this was an effective way for farmers to preserve the chemical energy in grains and fruits in a form that would not be eaten by rats or spoiled by bacteria or fungi. Why does a great deal of chemical energy remain in the products of fermentation pathways?

Answers to Multiple-Choice Questions   **1.** b; **2.** b; **3.** d; **4.** c; **5.** a; **6.** b

10 Photosynthesis

KEY CONCEPTS

- Photosynthesis consists of two distinct sets of reactions. In reactions driven by light, ATP and the electron carrier NADPH are produced. In subsequent reactions that do not depend directly on light, the ATP and NADPH are used to reduce carbon dioxide (CO_2) to carbohydrate $(CH_2O)_n$. In eukaryotic cells, both processes take place in chloroplasts.

- The light-dependent reactions transform the energy in sunlight to chemical energy in the form of electrons with high potential energy. Excited electrons either are used to produce NADPH or are donated to an electron transport chain, which results in the production of ATP.

- The light-independent reactions start with the enzyme rubisco, which catalyzes the addition of CO_2 to a five-carbon molecule. The compound that results undergoes a series of reactions that use ATP and NADPH and lead to the production of sugar.

Plants and other photosynthetic organisms convert the energy in sunlight to chemical energy in the bonds of sugar. The sugar produced by photosynthetic organisms fuels cellular respiration and growth. Photosynthetic organisms, in turn, are consumed by animals, fungi, and a host of other organisms. Directly or indirectly, most organisms on Earth get their energy from photosynthesis.

About three billion years ago, a novel combination of light-absorbing molecules and enzymes gave a bacterial cell the capacity to convert light energy into chemical energy in the carbon-hydrogen bonds of sugar. When sunlight is used to manufacture carbohydrate, **photosynthesis** is said to occur. The origin of photosynthesis ranks as one of the great events in the history of life. Since this process evolved, photosynthetic organisms have dominated the Earth in terms of abundance and mass.

The vast majority of organisms alive today rely on photosynthesis, either directly or indirectly, to stay alive. Photosynthetic organisms such as trees and mosses and ferns are termed **autotrophs** ("self-feeders"), because they make all of their own food from ions and simple molecules. Non-photosynthetic organisms such as humans and fungi and the bacterium *Escherichia coli* are called **heterotrophs** ("different-feeders") because they have to obtain the sugars and many of the other macromolecules they need from other organisms. Because there could be no heterotrophs without autotrophs, photosynthesis is fundamental to almost all life. Glycolysis may be the oldest set of energy-related chemical reactions in terms of evolutionary history; but ecologically, photosynthesis is easily the most important.

All organisms perform respiration or fermentation, but only selected groups are also capable of photosynthesis. This chapter presents a step-by-step analysis of how photosynthetic species manufacture sugar from sunlight, carbon dioxide, and water. After studying how plants do this remarkable chemistry, you should have a deeper appreciation for the food you eat and the oxygen you breathe.

10.1 An Overview of Photosynthesis

Research on photosynthesis began very early in the history of biological science. Starting in the 1770s, a series of experiments showed that photosynthesis takes place only in the green parts of plants; that sunlight, carbon dioxide (CO_2), and water (H_2O) are required; and that oxygen (O_2) is produced as a by-product. By the early 1840s enough was known about this process for biologists to propose that photosynthesis allows plants to convert sunlight into chemical energy in the bonds of carbohydrates. Eventually the overall reaction was understood to be

$$CO_2 + 2\,H_2O + \text{light energy} \rightarrow (CH_2O)_n + H_2O + O_2$$

The $(CH_2O)_n$ is a generic carbohydrate. (The "n" indicates that different carbohydrates have different multiples of CH_2O.) In essence, energy from light is transformed to chemical energy in the C–H bonds of carbohydrates. When glucose is the carbohydrate produced, the reaction can be written as

$$6\,CO_2 + 12\,H_2O + \text{light energy} \rightarrow$$
$$C_6H_{12}O_6 + 6\,O_2 + 6\,H_2O$$

How does it happen? Based on the overall reaction, early investigators assumed that CO_2 and H_2O react directly to form CH_2O, and that the oxygen atoms in carbon dioxide are given off as oxygen gas (O_2). Both hypotheses proved to be incorrect, however. Let's see why.

Photosynthesis: Two Distinct Sets of Reactions

During the 1930s two independent lines of research on photosynthesis converged, leading to a major advance. The first research program, led by Cornelius van Niel, focused on how photosynthesis occurs in purple sulfur bacteria. Van Niel and his group found that these cells can grow in the laboratory on a food source that lacks sugars. Based on this observation, he concluded that they must be autotrophs that manufacture their own carbohydrates. But to grow, the cells had to be exposed to sunlight and hydrogen sulfide (H_2S). Van Niel also showed that these cells did not produce oxygen as a by-product of photosynthesis. Instead, elemental sulfur (S) accumulated in their medium. In these organisms, the overall reaction for photosynthesis was

$$CO_2 + 2\,H_2S + \text{light energy} \rightarrow (CH_2O)_n + H_2O + 2\,S$$

Van Niel's work was crucial for two reasons. First, it showed that CO_2 and H_2O do *not* combine directly during photosynthesis. Instead of acting as a reactant in photosynthesis in these species, H_2O is a product of the process. Second, van Niel's data showed that the oxygen atoms in CO_2 are not released as oxygen gas (O_2). This conclusion was logical because no oxygen was produced by the purple sulfur bacteria, even though carbon dioxide participated in the reaction—just as it did in plants.

Based on these findings, biologists hypothesized that the oxygen atoms that are released during plant photosynthesis must come from water. This proposal was supported by experiments with isolated chloroplasts, which produced oxygen in the presence of sunlight even if no CO_2 was present. The hypothesis was confirmed when heavy isotopes of oxygen—^{18}O compared with the normal isotope, ^{16}O—became available to researchers. Biologists then exposed algae or plants to H_2O that contained ^{18}O, collected the oxygen gas that was given off as a by-product of photosynthesis, and confirmed that the released oxygen gas contained the heavy isotope. As predicted, the reaction that produced this oxygen occurred only in the presence of sunlight.

A second major line of research helped support these discoveries. When the radioactive isotope ^{14}C became available in the mid-1940s, Melvin Calvin and others began feeding labeled carbon dioxide ($^{14}CO_2$) to algae and identifying the molecules that subsequently became labeled with the radioisotope. The investigators also noticed that labeled carbon dioxide could become incorporated into complex organic compounds in algae in the dark. This was a remarkable observation. Photosynthesis obviously depended on light, but for some reason the reduction of CO_2 did not. Because Calvin played an important role in detailing the exact sequence of reactions that made this key finding possible, the light-independent component of photosynthesis came to be known as the **Calvin cycle**. The light-independent reactions of photosynthesis reduce carbon dioxide and result in the production of sugar.

To summarize, early research on photosynthesis showed that this process consists of two distinct sets of reactions. One set is dependent on light; the other set can occur in the absence of light. The **light-dependent reactions** result in the production of oxygen from water; the **light-independent reactions** result in the production of sugar from carbon dioxide.

In photosynthesizing cells, the light-dependent and light-independent reactions occur simultaneously. How are these two sets of reactions connected? The short answer is by electrons. More specifically, electrons are released when water is split to form oxygen gas. During the light-dependent reactions, these electrons are transferred to a phosphorylated version of NAD^+, called $NADP^+$, that is abundant in photosynthesizing cells. This reaction forms NADPH. Like NADH, NADPH is an electron

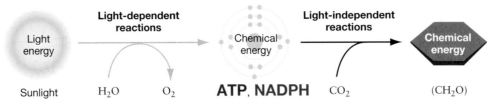

FIGURE 10.1 Photosynthesis Has Two Distinct Components
In the light-dependent reactions of photosynthesis, light energy is transformed to chemical energy. ATP is produced during these reactions. In addition, electrons are removed from water and are used to form NADPH, with oxygen (O_2) being released as a by-product. In the light-independent reactions, the ATP and NADPH produced in the light-dependent reactions are used to reduce carbon dioxide to carbohydrate.

carrier. ATP is also produced in the light-dependent reactions (see the middle portion of **Figure 10.1**). The light-independent reactions of the Calvin cycle use the electrons in NADPH and the potential energy in ATP to reduce CO_2 to carbohydrate (the right-hand side of Figure 10.1). The resulting sugars are used in cellular respiration to produce ATP for the cell. Plants oxidize sugars in their mitochondria and consume O_2 in the process, just as animals and other eukaryotes do.

Where does all this activity take place?

The Structure of the Chloroplast

Experiments with various plant tissues established that photosynthesis takes place only in the green portions of plants. Follow-up work with light microscopes suggested that the reactions occur inside the bright green organelles called **chloroplasts** ("green-formed"). Leaf cells typically contain from 40 to 50 chloroplasts, and a square millimeter of leaf averages about 500,000 chloroplasts (**Figure 10.2a**). When membranes derived from chloroplasts were found to release oxygen after

(a) Leaves contain millions of chloroplasts.

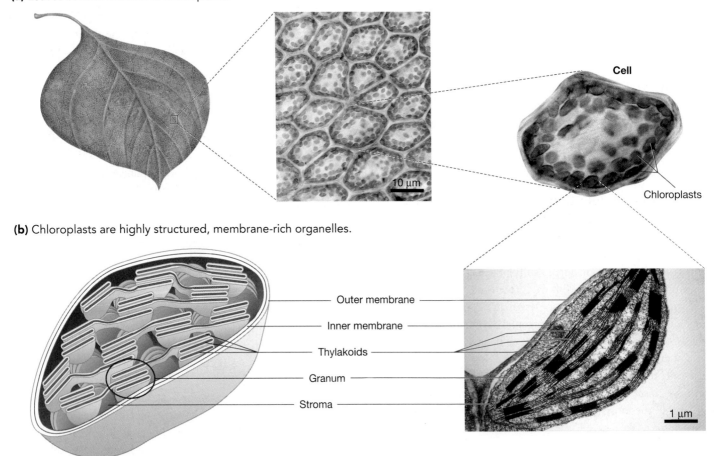

(b) Chloroplasts are highly structured, membrane-rich organelles.

FIGURE 10.2 Photosynthesis Takes Place in Chloroplasts
(a) In plants, photosynthesis takes place in organelles called chloroplasts. **(b)** The internal membranes of chloroplasts form flattened, vesicle-like structures called thylakoids, some of which form stacks called grana.

exposure to sunlight, the hypothesis that chloroplasts are the site of photosynthesis became widely accepted.

When electron microscopy became available in the 1950s, researchers observed that chloroplasts are extremely membrane rich. As **Figure 10.2b** shows, the organelle is enclosed by an outer membrane and an inner membrane. The interior is dominated by vesicle-like structures called **thylakoids**, which often occur in interconnected stacks called **grana** (singular: *granum*). The space inside a thylakoid is its **lumen**. (Recall that *lumen* is a general term for the interior of any sac-like structure. Your stomach and intestines have a lumen.) The fluid-filled space between the thylakoids and the inner membrane is the **stroma**.

When researchers analyzed the chemical composition of thylakoid membranes, they found huge quantities of pigments.

Pigments are molecules that absorb only certain wavelengths of light—other wavelengths are either transmitted or reflected. Pigments have colors because we see the wavelengths that pass through or bounce off them. The most abundant pigment found in the thylakoid membranes turned out to be chlorophyll. **Chlorophyll** ("green-leaf") absorbs blue light and red light and reflects or transmits green light. As a result, it is responsible for the green color of plants, some algae, and many photosynthetic bacteria.

Developmental studies showed that chloroplasts are derived from colorless organelles called **proplastids** ("before-plastids"). Proplastids are found in the cells of embryonic plants and in the rapidly dividing tissues of mature plants. As cells mature, proplastids develop into chloroplasts or other types of plastids specialized for that cell's particular task (see **Box 10.1**).

BOX 10.1 Types of Plastids

Plastids are a family of double-membrane-bound organelles found in plants. As **Figure 10.3** shows, they develop from small, unspecialized organelles called proplastids. As a cell matures and takes on a specialized function in the plant (that is, *differentiates*), its proplastids also differentiate. If a developing plant cell in a stem or leaf is exposed to light, for example, its

proplastids are usually stimulated to develop into chloroplasts.

There are three major types of plastids: (1) chloroplasts, (2) leucoplasts, and (3) chromoplasts. Figure 10.2 detailed the structure of chloroplasts. *Leucoplasts* ("white-formed") often function as energy storehouses. More specifically, leucoplasts may store chemical energy in the bonds of molecules with high potential

energy. Some leucoplasts store oils; others synthesize and sequester the carbohydrate called starch; still others store proteins. *Chromoplasts* ("color-formed") are brightly colored because they synthesize and hoard large amounts of orange, yellow, or red pigments in their vacuoles. High concentrations of chromoplasts are responsible for many of the bright colors of fruits and some flowers.

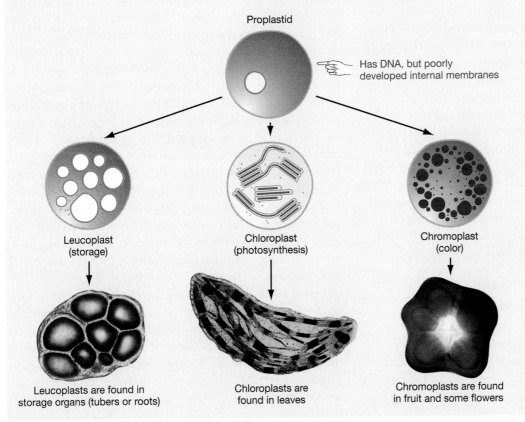

Proplastid

Has DNA, but poorly developed internal membranes

Leucoplast (storage)

Chloroplast (photosynthesis)

Chromoplast (color)

Leucoplasts are found in storage organs (tubers or roots)

Chloroplasts are found in leaves

Chromoplasts are found in fruit and some flowers

FIGURE 10.3 Chloroplasts Are a Specialized Type of Plastid Each type of plastid has a distinctive structure and function, but all plastids develop from organelles called proplastids.

Before plunging into the details of how photosynthesis occurs inside a chloroplast, let's consider just how astonishing the process is. Chemists have synthesized an amazing diversity of compounds from relatively simple starting materials, but their achievements pale in comparison to cells that synthesize sugar from just carbon dioxide, water, and sunlight. If it is not *the* most sophisticated chemistry on Earth, photosynthesis is certainly a contender. What's more, photosynthetic cells accomplish this feat in environments ranging from mountaintop snowfields to the open ocean, tropical rain forests, and polar ice caps. Photosynthesis can occur in virtually any habitat where light is available.

10.2 How Does Chlorophyll Capture Light Energy?

Photosynthesis begins with the light-dependent reactions, and these reactions begin with the simple act of sunlight striking chlorophyll. To understand the consequences of this event, it's helpful to review the nature of light. Light is a type of electromagnetic radiation, which is a form of energy. The essence of photosynthesis is converting electromagnetic energy in the form of sunlight into chemical energy in the C–H bonds of sugar.

Physicists describe light's behavior as both wavelike and particle-like. As is true of all waves, including waves of water or air, electromagnetic radiation is characterized by its **wavelength**, the distance between two successive wave crests (or wave troughs). The wavelength determines the type of electromagnetic radiation.

Figure 10.4 illustrates the **electromagnetic spectrum**—the range of wavelengths of electromagnetic radiation. Humans cannot see all these wavelengths, however. Electromagnetic radiation that humans can see is called **visible light**; it ranges in wavelength from about 400 to about 710 nanometers (nm, or 10^{-9} m). Shorter wavelengths of electromagnetic radiation contain more energy than longer wavelengths do, so blue light and ultraviolet light contain much more energy than red light and infrared light do.

To emphasize the particle-like nature of light, physicists point out that it exists in discrete packets called **photons**. In understanding photosynthesis, the important point is that each photon and each wavelength of light has a characteristic amount of energy. Pigment molecules absorb this energy. How?

Photosynthetic Pigments Absorb Light

When a photon strikes an object, the photon may be absorbed, transmitted, or reflected. A pigment absorbs particular wavelengths of light. Sunlight is white light, which consists of all wavelengths in the visible portion of the electromagnetic spectrum at once. If a pigment absorbs all of the

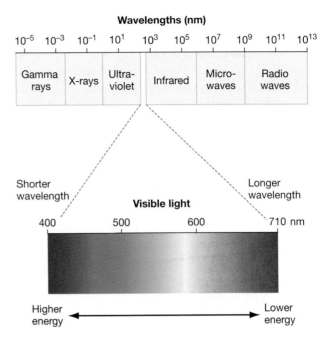

FIGURE 10.4 The Electromagnetic Spectrum
Electromagnetic energy radiates through space in the form of waves. Humans can see radiation at wavelengths between about 400 nm and about 710 nm. The shorter the wavelength of electromagnetic radiation, the higher its energy. **QUESTION** Does ultraviolet light contain more or less energy than infrared light?

visible wavelengths, no visible wavelength of light is reflected back to your eye, and the pigment appears black. If a pigment absorbs many or most of the wavelengths in the blue and green parts of the spectrum but transmits or reflects red wavelengths, it appears red.

What wavelengths do various plant pigments absorb? In one approach to answering this question, researchers grind up leaves and add a solvent to them. The solvent extracts pigment molecules from the leaf mixture. As **Figure 10.5a** shows, the pigments in the extract can then be separated from each other using a technique called **paper chromatography**. To begin, spots of a raw extract are placed near the bottom of a piece of filter paper. The filter paper is then placed in a solvent solution. As the solvent wicks upward along the paper, the pigment molecules in the mixture are carried along. Because the pigment molecules in the extract vary in size, solubility, or both, they are carried along with the solvent at different rates.

Figure 10.5b shows a chromatograph from a grass-leaf extract. Note that this leaf contains an array of pigments. To find out which wavelengths are absorbed by each of these molecules, researchers cut out a single region (color band) of the filter paper, extract the pigment, and use an instrument called a spectrophotometer to record the wavelengths absorbed (**Box 10.2**).

(a) ISOLATING PIGMENTS VIA PAPER CHROMATOGRAPHY

(b) A finished chromatograph

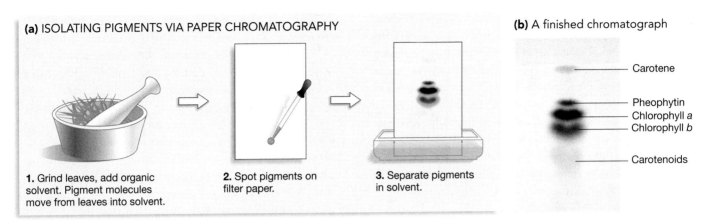

1. Grind leaves, add organic solvent. Pigment molecules move from leaves into solvent.

2. Spot pigments on filter paper.

3. Separate pigments in solvent.

Carotene

Pheophytin
Chlorophyll *a*
Chlorophyll *b*

Carotenoids

FIGURE 10.5 Pigments Can Be Isolated by Using Paper Chromatography
(a) Paper chromatography is an effective way to isolate the various pigments in photosynthetic tissue.
(b) Photosynthetic tissues, including those from the grass leaves used to make this chromatograph, typically contain several pigments. Different species of photosynthetic organisms may contain different types and quantities of pigments.

BOX 10.2   How Do Researchers Measure Absorption Spectra?

Researchers use an instrument called a **spectrophotometer** to measure the wavelengths of light absorbed by a particular pigment. As **Figure 10.6** shows, a solution of purified pigment molecules is exposed to a specific wavelength of light. If the pigment absorbs a great deal of this wavelength, then little light will be transmitted through the sample. But if the pigment absorbs the wavelength poorly, then most of the incoming light will be transmitted. The light that passes through the sample strikes a photoelectric tube, which converts it to an electric current. The amount of the electric current that is generated, as measured by a galvanometer, is proportional to the intensity of the incoming light. High current readings indicate low absorption at a particular wavelength, while low current readings signal high absorption. By testing one wavelength after another, investigators can measure the full absorption spectrum of a pigment.

HOW DOES A SPECTROPHOTOMETER WORK?

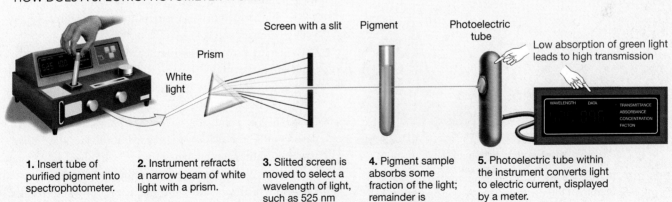

Screen with a slit   Pigment   Photoelectric tube

Prism

White light

Low absorption of green light leads to high transmission

WAVELENGTH   DATA   TRANSMITTANCE
ABSORBANCE
CONCENTRATION
FACTON

1. Insert tube of purified pigment into spectrophotometer.

2. Instrument refracts a narrow beam of white light with a prism.

3. Slitted screen is moved to select a wavelength of light, such as 525 nm (green), to shine through the sample.

4. Pigment sample absorbs some fraction of the light; remainder is transmitted.

5. Photoelectric tube within the instrument converts light to electric current, displayed by a meter.

FIGURE 10.6 Spectrophotometers Can Measure the Light Wavelengths Absorbed by Pigments
QUESTION Suppose that you were analyzing a sample of chlorophyll *b* and that you changed the wavelength selector in the spectrophotometer to 480 nm (blue light). How would the amount of absorption change?

Using this approach, biologists have produced data like those shown in **Figure 10.7a**. This graph is a plot of light absorbed versus wavelength and is called an **absorption spectrum**. Research based on these techniques has confirmed that there are two major classes of pigment in plant leaves: chlorophylls and carotenoids. The **chlorophylls**, designated chlorophyll *a* and chlorophyll *b*, absorb strongly in the blue and red regions of the visible spectrum and therefore reflect and transmit green light. Beta-carotene (*β*-carotene) and other **carotenoids** constitute a different family of pigments, which absorb in the blue and green parts of the visible spectrum. Thus carotenoids appear yellow, orange, or red.

(a) Different pigments absorb different wavelengths of light.

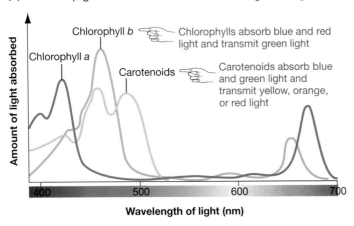

(b) Pigments that absorb blue and red photons are the most effective at triggering photosynthesis.

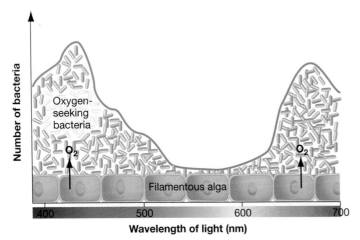

FIGURE 10.7 There Is a Strong Correlation between the Absorption Spectrum of Pigments and the Action Spectrum for Photosynthesis
(a) Each photosynthetic pigment has a distinct absorption spectrum.
(b) The "action spectrum" for photosynthesis—meaning the wavelengths of light that trigger the process—correlates with the absorption spectra of photosynthetic pigments. **EXERCISE** Water absorbs strongly in all but the blue and green parts of the spectrum. Add a curve to part (b) predicting the action spectrum for photosynthesis in plants and algae that live at the bottom of lake or ocean habitats.

Which of these wavelengths drive photosynthesis? T. W. Englemann answered this question by laying a filamentous alga across a glass slide that was illuminated with a spectrum of colors. The idea was that the alga would begin performing photosynthesis in response to the various wavelengths of light and produce oxygen as a by-product. To determine exactly where oxygen was being produced, Englemann added bacterial cells from a species that is attracted to oxygen. As **Figure 10.7b** shows, most of the bacteria congregated in the blue and red regions of the slide. Because wavelengths in these parts of the spectrum were associated with high oxygen concentrations, Englemann concluded that they defined the **action spectrum** for photosynthesis. The data suggested that blue and red photons are the most effective at driving photosynthesis. Because the chlorophylls absorb these wavelengths, the data also suggested that chlorophylls are the main photosynthetic pigments.

Before analyzing what happens during the absorption event itself, let's look quickly at the structure and function of the chlorophylls and carotenoids.

What Is the Role of Carotenoids and Other Accessory Pigments?

Carotenoids are called **accessory pigments**, because they absorb light and pass the energy on to chlorophyll. The carotenoids found in plants belong to two classes, called **carotenes** and **xanthophylls**. **Figure 10.8a** shows the structure of *β*-carotene (beta-carotene), which gives carrots their orange color. A xanthophyll called zeaxanthin, which gives corn kernels their bright yellow color, is nearly identical to *β*-carotene, except that the ring structures on either end of the molecule contain a hydroxyl (–OH) group.

Both xanthophylls and carotenes are found in chloroplasts. In autumn, when the leaves of deciduous trees begin to die and their chlorophyll degrades, the wavelengths scattered by carotenoids turn northern forests into spectacular displays of yellow, orange, and red.

What do carotenoids do? Because these pigments absorb wavelengths of light that are not absorbed by chlorophyll, they extend the range of wavelengths that can drive photosynthesis. But researchers discovered an even more important function by analyzing what happens to leaves when carotenoids are destroyed. Many herbicides, for example, work by inhibiting enzymes that are involved in carotenoid synthesis. Plants lacking carotenoids rapidly lose their chlorophyll, turn white, and die. Based on these results, researchers have concluded that carotenoids also serve a protective function.

To understand the molecular basis of carotenoid function, recall from Chapter 2 that energy from photons—especially the high-energy, short-wavelength photons in the ultraviolet part of the electromagnetic spectrum—contain enough energy to knock electrons out of atoms and create free radicals. Free radicals, in turn, trigger reactions that degrade molecules. Fortu-

(a) β-carotene

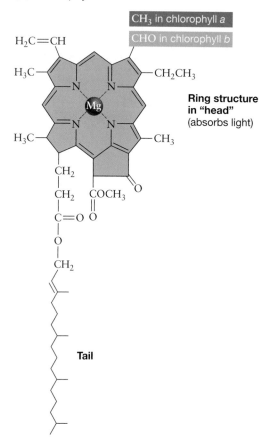

(b) Chlorophylls *a* and *b*

Tail

FIGURE 10.8 Photosynthetic Pigments Contain Ring Structures
(a) *β*-Carotene is an orange pigment found in carrot roots and other plant tissues. **(b)** Although chlorophylls *a* and *b* are very similar structurally, they have the distinct absorption spectra shown in Figure 10.7a.

nately, carotenoids can "quench" or stabilize free radicals and protect chlorophyll molecules from harm. When carotenoids are absent, chlorophyll molecules are destroyed. As a result, photosynthesis stops. Starvation and death follow.

Carotenoids are not the only molecules that protect plants from the damaging effects of sunlight, however. Researchers have recently analyzed individuals of the mustard plant *Arabidopsis thaliana* that are unable to synthesize pigments called flavonoids, which are normally stored in the vacuoles of leaf cells. Because **flavonoids** absorb ultraviolet radiation, individuals that lack flavonoids are subject to damage from UV light. In effect, these pigments function as a sunscreen for leaves and stems. Without them, chlorophyll molecules would be broken apart by high-energy radiation.

The general message here is that the energy in sunlight is a double-edged sword. It makes photosynthesis possible, but it can also lead to the formation of free radicals that damage cells. The role of carotenoids and flavonoids as protective pigments is crucial.

The Structure of Chlorophyll As **Figure 10.8b** shows, chlorophyll *a* and chlorophyll *b* are very similar in structure as well as in absorption spectra. They differ in just three atoms. In land plants, chloroplasts typically have about three molecules of chlorophyll *a* in their internal membranes for every molecule of chlorophyll *b*.

Notice from Figure 10.8b that chlorophyll molecules have two fundamental parts: a long tail made up of isoprene subunits (introduced in Chapter 6) and a "head" that consists of a large ring structure with a magnesium atom in the middle. The tail keeps the molecule embedded in the thylakoid membrane; the head is where light is absorbed.

But just what is "absorption?" Stated another way, what happens when a photon of a particular wavelength—say, red light with a wavelength of 680 nm—strikes a chlorophyll molecule?

When Light Is Absorbed, Electrons Enter an Excited State

When a photon strikes a chlorophyll molecule, the photon's energy can be transferred to an electron in the chlorophyll molecule's head region. In response, the electron is "excited," or raised to a higher electron shell—one with greater potential energy. As **Figure 10.9** shows, the excited electron states that are possible in a particular pigment are discrete—that is, incremental rather than continuous—and can be represented as lines on an energy scale. In the figure, the ground state, or unexcited state, is shown as 0 and the higher energy states are designated 1 and 2. If the difference between the possible energy states is

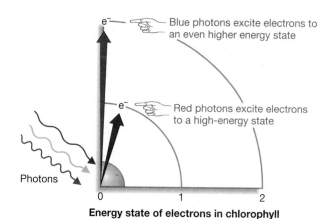

Energy state of electrons in chlorophyll

FIGURE 10.9 Electrons Are Promoted to High-Energy States when Photons Strike Chlorophyll
When a photon strikes chlorophyll, an electron can be promoted to a higher energy state, depending on the energy in the photon.

the same as the energy in the photon, then the photon can be absorbed and an electron is excited to that energy state.

In chlorophyll, for example, the energy difference between the ground state and state 1 is equal to the energy in a red photon, while the energy difference between state 0 and state 2 is equal to the energy in a blue photon. Thus chlorophyll can readily absorb red photons and blue photons. Chlorophyll does not absorb green light well because there is no discrete step—no difference in possible energy states for its electrons—that corresponds to the amount of energy in a green photon.

Wavelengths in the ultraviolet part of the spectrum have so much energy that they may actually eject electrons from a pigment molecule. In contrast, wavelengths in the infrared regions have so little energy that in most cases they merely increase the movement of atoms in the pigment, generating heat rather than exciting electrons.

If an electron is excited by a photon, it means that energy in the form of electromagnetic radiation is transferred to an electron, which now has high potential energy. If the excited electron falls back to its ground state, though, some of the absorbed energy is released as heat—meaning molecular movement—while the rest is released as electromagnetic radiation. This is the phenomenon known as **fluorescence**. Because some of the energy in the original photon is transformed to heat, the electromagnetic radiation that is given off during fluorescence has lower energy and a longer wavelength than the original photon does.

Figure 10.10 shows the fluorescence that occurs when a pure solution of chlorophyll is exposed to ultraviolet light. The isolated pigment absorbs the photons but simply fluoresces red in response. The amount of heat given off is equal to the energy difference between the higher-energy ultraviolet photons that are absorbed and the lower-energy red photons that are released.

FIGURE 10.10 Fluorescence
A pure solution of chlorophyll exposed to ultraviolet light. Electrons are excited to a high-energy state but immediately fall back to a low-energy state, emitting red photons and heat. **EXERCISE** Add to the diagram in Figure 10.9 to show what is happening here. Start with the electron promoted to a high-energy state by blue light; have it fall back to the ground state (0); and indicate the fate of the energy released.

How Do the Chlorophyll Molecules in Leaves Work?

When chlorophyll is in a chloroplast, only about 2 percent of the red and blue photons that it absorbs normally produce fluorescence. What happens to the other excited electrons? An answer to this question began to emerge as investigators struggled to interpret the experimental result described in **Figure 10.11**.

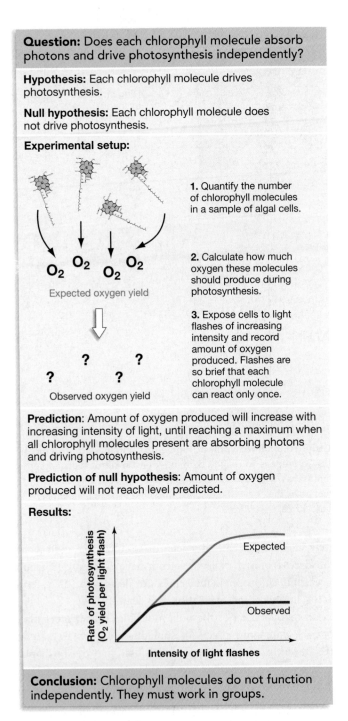

Question: Does each chlorophyll molecule absorb photons and drive photosynthesis independently?

Hypothesis: Each chlorophyll molecule drives photosynthesis.

Null hypothesis: Each chlorophyll molecule does not drive photosynthesis.

Experimental setup:

O_2 O_2 O_2 O_2
Expected oxygen yield

?   ?
?   ?
Observed oxygen yield

1. Quantify the number of chlorophyll molecules in a sample of algal cells.

2. Calculate how much oxygen these molecules should produce during photosynthesis.

3. Expose cells to light flashes of increasing intensity and record amount of oxygen produced. Flashes are so brief that each chlorophyll molecule can react only once.

Prediction: Amount of oxygen produced will increase with increasing intensity of light, until reaching a maximum when all chlorophyll molecules present are absorbing photons and driving photosynthesis.

Prediction of null hypothesis: Amount of oxygen produced will not reach level predicted.

Results:

[Graph: Rate of photosynthesis (O_2 yield per light flash) on y-axis versus Intensity of light flashes on x-axis, showing "Expected" curve rising higher and "Observed" curve plateauing lower]

Conclusion: Chlorophyll molecules do not function independently. They must work in groups.

FIGURE 10.11 Chlorophyll Molecules Work in Groups, Not Individually
The experimental results outlined here were a complete surprise to the investigators.

Robert Emerson and William Arnold designed this experiment to test the hypothesis that all of the chlorophyll molecules present in a photosynthetic cell absorb photons and drive photosynthesis. The key to the setup was that the researchers could estimate the number of chlorophyll molecules present in an average cell of their study organism—a green alga called *Chlorella*. These data allowed them to predict how much photosynthesis should occur in response to a flash of light. As predicted, the amount of photosynthesis—measured as the amount of oxygen produced—increased steadily as the researchers increased the intensity of the light flashes. But much to Emerson and Arnold's surprise, the total amount of photosynthesis leveled off far below the predicted value. Long before all or even most chlorophyll molecules were active, the total amount of photosynthesis reached a maximum.

What was happening? To make sense of the result, the researchers hypothesized that chlorophyll molecules do not work individually. Instead, chlorophylls must work in groups. Follow-up work has shown that in the thylakoid membrane, 200–300 chlorophyll molecules and accessory pigments such as carotenoids are grouped together in an array of proteins, forming a complex called a **photosystem**. Each photosystem, in turn, has two major elements: an antenna complex and a reaction center.

The Antenna Complex When a red or blue photon strikes a pigment molecule in the **antenna complex**, the energy is absorbed and an electron is excited in response. This energy—but not the electron itself—is passed along to a nearby chlorophyll molecule, where another electron is excited in response. As the energy is transmitted, the original excited electron falls back to its ground state. In this way, energy is transmitted from one chlorophyll molecule to the next inside the antenna complex (**Figure 10.12a**). The system acts like a radio antenna that receives a specific wavelength of electromagnetic radiation and transmits it to a receiver. In a photosystem, the receiver is called the reaction center.

The Reaction Center When energy from the antenna complex reaches the reaction center of a photosystem, an all-important energy transformation event occurs. At the **reaction center**, excited electrons are transferred to a molecule that acts as an electron acceptor (**Figure 10.12b**). When this molecule becomes reduced, the energy transformation event that started with the absorption of light becomes permanent: Electromagnetic energy is transformed to chemical energy. The redox reaction that occurs in the reaction center results in the production of chemical energy from sunlight.

The key to understanding the reaction center is that, in the absence of light, chlorophyll cannot reduce the electron acceptor because the reactions are endergonic. But when light excites electrons in chlorophyll to a high-energy state, the reactions become exergonic.

Now, what happens to these high-energy electrons? Specifically, how are they used to manufacture sugar?

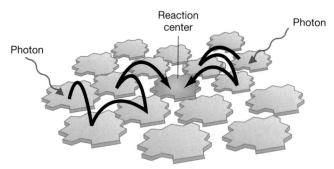

(a) Chlorophyll molecules transmit energy from excited electrons in the antenna complex to a reaction center.

Chlorophyll molecules in antenna complex

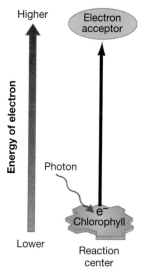

(b) At the reaction center, excited electrons are passed to an electron acceptor.

FIGURE 10.12 The Antenna Complex Captures Light Energy and Transmits It to the Reaction Center
(a) The antenna complex is aptly named. It captures specific wavelengths of light and transfers the energy to the reaction center.
(b) Redox reactions in the reaction center result in the addition of a high-energy electron to an electron acceptor.

10.3 The Discovery of Photosystems I and II

During the 1950s the fate of the high-energy electrons in photosystems was the central issue facing biologists interested in photosynthesis. Ironically, a central insight into this issue came from a simple experiment on how green algae responded to various wavelengths of light.

The experimental setup was based on the observation that photosynthetic cells respond to very specific wavelengths of light. In particular, the algal cells being studied responded strongly to wavelengths of 700 nm and 680 nm, which are in the far-red and red portions of the visible spectrum, respectively. In a key experiment, Robert Emerson found that if cells were illuminated with either far-red light or red light, the photosynthetic response was moderate (**Figure 10.13**). But if cells were exposed to a combination of far-red and red light, the rate of photosynthesis increased dramatically. When both wavelengths were present, the photosynthetic rate was much more than the sum of the rates produced by each wavelength independently. This phenomenon was called the *enhancement effect*. Why it occurred was a complete mystery at the time.

The puzzle posed by the enhancement effect was eventually solved by Robin Hill and Faye Bendall. These biologists synthesized data emerging from an array of labs and proposed that green algae and plants have two distinct types of reaction centers rather than just one. Hill and Bendall proposed that one reaction center, which came to be called **photosystem II**, interacts with a different reaction center, now referred to as **photosystem I**. According to the two-photosystem hypothesis, the enhancement effect occurs because photosynthesis is much more efficient when both photosystems are operating together.

Subsequent work has shown that the two-photosystem hypothesis is correct. In green algae and land plants, thylakoid membranes contain photosystems that differ in structure and function but complement each other.

To figure out how the two photosystems work, investigators chose not to study them together. Instead, they focused on species of photosynthetic bacteria that have photosystems similar to either photosystem I or II, but not both. Once each type of photosystem was understood in isolation, they turned to understanding how the two photosystems work in combination in green algae and land plants. Let's do the same—we'll analyze photosystem II, then photosystem I, and then how the two interact.

How Does Photosystem II Work?

To analyze photosystem II, researchers focused on studying species from the purple nonsulfur bacteria and the purple sulfur bacteria. These cells have a single photosystem that has many of the same components observed in photosystem II of cyanobacteria ("blue-green bacteria"), algae, and plants.

In photosystem II, the action begins when the antenna complex transmits energy to the reaction center and the molecule pheophytin comes into play. Structurally, **pheophytin** is very similar to chlorophyll. The two molecules are identical except that pheophytin lacks a magnesium atom in its head region. Functionally, though, they are different. Instead of acting as a pigment that promotes an electron when it absorbs a photon, pheophytin acts as an electron acceptor. When an electron in the reaction center chlorophyll is excited energetically, the electron binds to pheophytin and the reaction center chlorophyll is oxidized. When pheophytin is reduced in this way, the energy transformation step that started with the absorption of light is completed.

Question: Light at both the red and far-red wavelengths stimulates photosynthesis. How does a combination of these wavelengths affect the rate of photosynthesis?

Hypothesis: When red light and far-red light are combined, the rate of photosynthesis will double.

Null hypothesis: When red light and far-red light are combined, the rate of photosynthesis will not double.

Experimental setup:

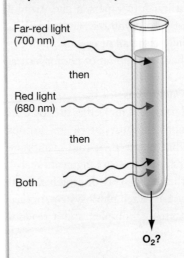

Far-red light (700 nm)

then

Red light (680 nm)

then

Both

O₂?

1. Expose algal cells to an intensity of far-red (700 nm) or red (680 nm) light that maximizes rate of photosynthesis at each wavelength. Then expose same algal cells to same intensities of each wavelength at same time.

2. Record the rate of photosynthesis as the amount of oxygen produced.

Prediction: When the wavelengths are combined, the rate of photosynthesis will be double the maximum rate observed for each wavelength independently.

Prediction of null hypothesis: When the wavelengths are combined, the rate of photosynthesis will not be double the maximum rate observed for each wavelength independently.

Results:

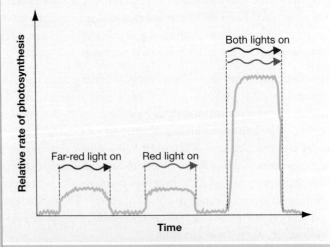

Conclusion: There is an enhancement effect for red and far-red light. The combination of 700 nm and 680 nm wavelengths more than doubles the rate of photosynthesis.

FIGURE 10.13 Discovery of the "Enhancement Effect" of Red and Far-Red Light

Electrons from Pheophytin Enter an Electron Transport Chain Electrons that reach pheophytin are passed to an electron transport chain in the thylakoid membrane. In both structure and function, this group of molecules is similar to the electron transport chain in the inner membrane of mitochondria (Chapter 9). For example, the electron transport chain associated with photosystem II contains several quinones and cytochromes. Electrons in both chains participate in a series of reduction-oxidation reactions and are gradually stepped down in potential energy. In mitochondria as well as chloroplasts, the redox reactions result in protons being pumped from one side of an internal membrane to the other. In both organelles, the resulting proton gradient drives ATP production via ATP synthase.

Figure 10.14a details the sequence of events in the electron transport chain of thylakoids. One of the key molecules involved is a quinone called **plastoquinone**, symbolized PQ. Recall from Chapter 9 that quinones are small hydrophobic molecules. Because plastoquinone is lipid soluble and not an-

chored to a protein, it is free to move from one side of the thylakoid membrane to the other. When it receives electrons from pheophytin, plastoquinone carries them to the other side of the membrane and delivers them to more electronegative molecules in the chain. These electron acceptors are found in a complex that contains a cytochrome similar to mitochondrial cytochromes.

In this way, plastoquinone shuttles electrons from pheophytin to the cytochrome complex. The electrons are then passed through a series of iron- and copper-containing proteins in the cytochrome complex. The potential energy released by these reactions allows protons to be added to other plastoquinone molecules, which carry them to the lumen side of the thylakoid membrane. As **Figure 10.14b** shows, the protons transported by plastoquinone result in a large concentration of protons in the thylakoid lumen. When photosystem II is active, the pH of the thylakoid interior reaches 5 while the pH of the stroma hovers around 8. Because the pH scale is logarithmic, the difference of 3 units means that the concentration of H^+ is $10 \times 10 \times 10 = 1000$ times higher in the lumen than in the stroma. In addition, the stroma becomes negatively charged relative to the thylakoid lumen. The net effect of electron transport, then, is to set up a large proton gradient that will drive H^+ out of the thylakoid lumen and into the stroma. Based on your reading of Chapter 9, it should come as no surprise that this proton-motive force drives the production of ATP.

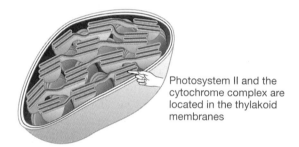

Photosystem II and the cytochrome complex are located in the thylakoid membranes

(a) In photosystem II, excited electrons feed an electron transport chain.

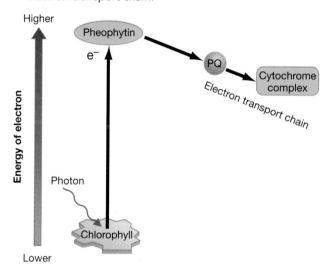

(b) Plastoquinone carries protons to the inside of thylakoids, creating a proton-motive force.

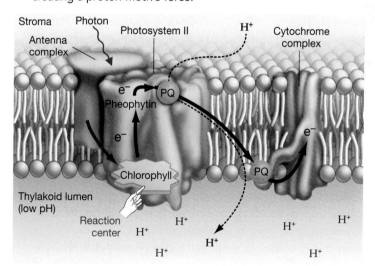

FIGURE 10.14 Photosystem II Feeds an Electron Transport Chain That Pumps Protons
(a) When an excited electron leaves the chlorophyll molecule in the reaction center of photosystem II, the electron is accepted by pheophytin, transferred to plastoquinone (PQ), and then stepped down in energy along an electron transport chain. **(b)** PQ carries electrons from photosystem II along with protons from the stroma. The electrons are passed to the cytochrome complex, and the protons are released in the thylakoid lumen. **EXERCISE** Add an arrow to part (b), indicating the direction of the proton-motive force.

ATP Synthase Uses the Proton-Motive Force to Phosphorylate ADP In mitochondria, NADH and $FADH_2$ donate electrons to an electron transport chain, and the redox reactions that occur in the chain result in protons being pumped out of the matrix and into the intermembrane space. In photosystem II, pheophytin donates electrons to an electron transport chain, and the redox reactions that occur result in protons being pumped out of the stroma and into the thylakoid lumen.

In both mitochondria and chloroplasts, protons diffuse down the resulting electrochemical gradient. This is an exergonic process that drives the endergonic synthesis of ATP. More specifically, the flow of protons through the enzyme ATP synthase causes conformational changes that drive the phosphorylation of ADP. In photosystem II, then, the light energy captured by chlorophyll is transformed to chemical energy stored in ATP. This process is called **photophosphorylation.** Substrate-level phosphorylation, oxidative phosphorylation, and photophosphorylation all result in the production of ATP.

To summarize, photosystem II starts with an electron being promoted to a high-energy state and ends with the production of ATP. The story is not complete, however, because we haven't accounted for the electrons that flow through the system. The electron that was transferred from chlorophyll to pheophytin in the photosystem II reaction center needs to be replaced. In addition, the electron transport chain of photosystem II needs to donate its electrons to some final electron acceptor.

Where do the electrons required by photosystem II come from, and where do they go? The parallel between the photosystem in purple sulfur bacteria and photosystem II of plants ends here. In purple sulfur bacteria, cytochrome donates an electron back to the reaction center, so the same electron can again be promoted to a high-energy state when a photon is absorbed. In this way, electrons cycle through the system. But in plants, algae, and cyanobacteria, electrons from photosystem II replenish the electrons that leave photosystem I in response to light. How are the electrons that leave photosystem II replaced?

Photosystem II Obtains Electrons by Oxidizing Water

To understand where the electrons that feed photosystem II come from, think back to the overall reaction for photosynthesis: $CO_2 + 2 H_2O + \text{light energy} \rightarrow (CH_2O) + H_2O + O_2$. In the presence of sunlight, carbon dioxide and water are used to produce carbohydrate, water, and oxygen gas. Recall that experiments with radioisotopes of oxygen showed that the oxygen atoms in O_2 come from water, not from carbon dioxide. It turns out that the electrons that enter photosystem II come from water. The oxygen-generating reaction can be written as

$$2 H_2O \rightarrow 4 H^+ + 4 e^- + O_2$$

Because electrons are removed from water, the molecule becomes oxidized. This reaction is referred to as "splitting" water. It supplies a steady stream of electrons for photosystem II and is catalyzed by enzymes that are physically integrated

into the photosystem II complex. When excited electrons leave photosystem II and enter the electron transport chain, the photosystem becomes so electronegative that enzymes can strip electrons away from water, leaving protons and oxygen.

Among all life-forms, photosystem II is the only protein complex that can catalyze the splitting of water molecules. Organisms such as cyanobacteria, algae, and plants that have this type of photosystem are said to perform **oxygenic** ("oxygen-producing") **photosynthesis,** because they generate oxygen as a by-product of the process. The purple sulfur and purple nonsulfur bacteria cannot oxidize water. They perform **anoxygenic** ("no oxygen-producing") **photosynthesis.**

It is difficult to overstate the importance of this unique reaction. The oxygen that is keeping you alive right now was produced by it. O_2 was, in fact, almost nonexistent on Earth prior to the evolution of enzymes that could catalyze the oxidation of water. According to the fossil record, oxygen levels in the atmosphere and oceans began to rise only about 2 billion years ago, as organisms that perform oxygenic photosynthesis increased in abundance. This change was a disaster for anaerobic organisms, because oxygen is toxic to them. But as oxygen became even more abundant, certain bacterial cells evolved the ability to use it as an electron acceptor during cellular respiration. This was a momentous development. O_2 is so electronegative that it creates a huge potential energy drop for the electron transport chains involved in cellular respiration. As a result, organisms that use O_2 as an electron acceptor in cellular respiration can produce much more ATP than can organisms that use other electron acceptors. Aerobic organisms grow so efficiently that they have long dominated our planet. Biologists rank the evolution of the oxygen-rich atmosphere as one of the most important events in the history of life.

Despite its fundamental importance, though, the mechanism responsible for the oxygen-generating reaction is not yet understood. Determining exactly how photosystem II splits water may be the greatest challenge currently facing researchers interested in photosynthesis. This issue has important practical applications as well, because if human chemists could replicate the reaction in an industrial setting, it might be possible to produce huge volumes of O_2 and hydrogen gas (H_2) from water. If this could be accomplished inexpensively, the H_2 produced could be used as a clean fuel for cars and trucks.

How Does Photosystem I Work?

Recall that researchers dissected photosystem II by studying similar, but simpler, photosystems in purple nonsulfur and purple sulfur bacteria. To understand the structure and function of photosystem I, they turned to heliobacteria ("sun-bacteria").

Like purple nonsulfur and purple sulfur bacteria, heliobacteria use the energy in sunlight to promote electrons to a high-energy state. But instead of being passed to an electron transport chain that pumps protons across a membrane, the

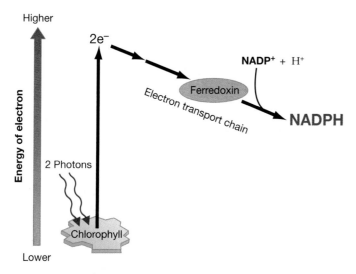

FIGURE 10.15 Photosystem I Produces NADPH
When excited electrons leave the chlorophyll molecule in the reaction center of photosystem I, they pass through a series of iron- and sulfur-containing proteins until they are accepted by ferredoxin. In an enzyme-catalyzed reaction, the reduced form of ferredoxin reacts with $NADP^+$ to produce NADPH.

high-energy electrons in heliobacteria are used to reduce NAD^+. When NAD^+ gains two electrons and a proton, NADH is produced. In the cyanobacteria and algae and land plants, a similar set of reactions reduces a phosphorylated version of NAD^+, symbolized $NADP^+$, yielding NADPH. Both NADH and NADPH function as electron carriers.

Figure 10.15 shows how the system works in photosystem I. When the reaction center in photosystem I absorbs a photon, excited electrons are passed through a series of iron- and sulfur-containing proteins inside the photosystem, and then to a molecule called **ferredoxin**. The electrons then move from ferredoxin

to the enzyme ferredoxin/$NADP^+$ oxidoreductase—also called $NADP^+$ reductase—which transfers two electrons and a proton to $NADP^+$. This reaction forms NADPH. The photosystem itself and $NADP^+$ reductase are anchored in the thylakoid membrane; ferredoxin is closely associated with the bilayer.

To summarize, photosystem I results in the production of NADPH, and photosystem II results in the production of a proton gradient that drives the synthesis of ATP. NADPH is similar in function to the NADH and $FADH_2$ produced by the Krebs cycle. It is an electron carrier that can donate electrons to other compounds and thus reduce them.

In combination, then, photosystems I and II produce both chemical energy stored in ATP and reducing power in the form of NADPH. Although several groups of bacteria have just one of the two photosystems, the cyanobacteria, algae, and plants have both. In these organisms, how do the two photosystems interact?

The Z Scheme: Photosystems I and II Work Together

When they realized that photosystems I and II have distinct but complementary functions, Robin Hill and Faye Bendall proposed that these systems interact as shown in **Figure 10.16**. The diagram illustrates a model, known as the **Z scheme**, that furnished a breakthrough in research on photosynthesis. The name was inspired by the shape of the proposed path of electrons through the two photosystems, when that path was plotted on a vertical axis representing the changes that occur in their potential energy.

Following the path of electrons through the Z scheme—by tracing the route of electrons through Figure 10.16 with your finger—will help drive home how photosynthesis works. The process starts when photons excite electrons in the chlorophyll molecules of photosystem II's antenna complex. When

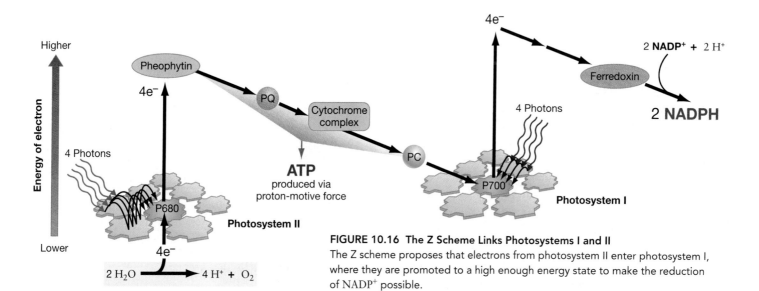

FIGURE 10.16 The Z Scheme Links Photosystems I and II
The Z scheme proposes that electrons from photosystem II enter photosystem I, where they are promoted to a high enough energy state to make the reduction of $NADP^+$ possible.

the energy in the excited electron is transmitted to the reaction center, a special pair of chlorophyll molecules named P680 passes excited electrons to pheophytin. From there the electron is gradually stepped down in potential energy through redox reactions among a series of quinones and cytochromes, which act as an electron transport chain. Using the energy released by the redox reactions, plastoquinone (PQ) carries protons across the thylakoid membrane. ATP synthase uses the resulting proton-motive force to phosphorylate ADP, creating ATP.

When electrons reach the end of photosystem II's electron transport chain, they are passed to a small diffusible protein called **plastocyanin** (symbolized PC in Figure 10.16). Plastocyanin picks up an electron from the cytochrome complex, diffuses through the lumen of the thylakoid, and donates the electron to photosystem I. A single plastocyanin molecule can shuttle over 1000 electrons per second between photosystems. In this way, plastocyanin forms a physical link between photosystem II and photosystem I.

The flow of electrons between photosystems, by means of plastocyanin, replaces electrons that are carried away from a chlorophyll molecule called P700 in the photosystem I reaction center. The electrons that emerge from P700 are eventually transferred to the protein ferredoxin, which then passes electrons to an enzyme that catalyzes the reduction of $NADP^+$ to NADPH. The electrons that initially left photosystem II are replaced by electrons that are stripped away from water, producing oxygen gas as a by-product.

The Z scheme helps explain the enhancement effect in photosynthesis documented in Figure 10.13. When algal cells are illuminated with wavelengths at 680 nm, in the red portion of the spectrum, only photosystem II can run at a maximum rate. The overall rate of electron flow through the Z scheme is moderate because photosystem I's efficiency is reduced. Similarly, when cells receive only wavelengths at 700 nm, in the far red, only photosystem I is capable of peak efficiency; photosystem II is working at a below-maximum rate, so the overall rate of electron flow is reduced. But when both wavelengths are available at the same time, both photosystem II and photosystem I are activated by light and work at a maximum rate, leading to enhanced efficiency.

Recent evidence indicates that a different electron path also occurs in green algae and plants. This pathway, called **cyclic photophosphorylation**, is illustrated in **Figure 10.17**. During cyclic photophosphorylation, photosystem I transfers electrons back to the electron transport chain, to augment ATP generation through photophosphorylation. This "extra" ATP is required for the chemical reactions that reduce carbon dioxide (CO_2) and produce sugars. In this way, cyclic photophosphorylation coexists with the Z scheme and produces additional ATP.

Although the Z-scheme model has held up well under experimental tests, several unresolved questions remain. The

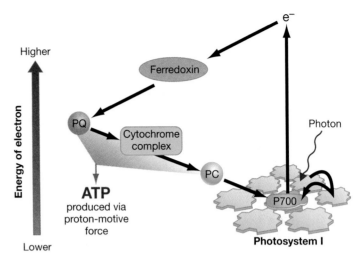

FIGURE 10.17 Cyclic Photophosphorylation Produces ATP
Cyclic electron transport is an alternative to the Z scheme. Instead of being donated to $NADP^+$, electrons cycle through the system and result in the production of additional ATP via photophosphorylation.

precise three-dimensional structure of each photosystem has been determined in bacteria but not yet in eukaryotes. Biologists are also trying to get a better understanding of how the two complexes are situated with respect to one another in the thylakoid membranes. As **Figure 10.18** shows, photosystems I and II are found in different parts of a single granum. Photosystem II is much more abundant in the interior, stacked membranes of grana, while photosystem I is much more common in the exterior, unstacked membranes. This physical separation between the photosystems is perplexing, given that their functions are so tightly integrated according to the Z scheme. Why they are found in different parts of the thylakoid is the focus of intense debate.

In contrast, the fate of the ATP and NADPH produced by photosystems I and II is well documented. Chloroplasts use ATP and NADPH to reduce carbon dioxide to sugar. Your life, and the life of most other organisms, depends on this process. How does it happen?

✓ CHECK YOUR UNDERSTANDING

Photosystem II contributes high-energy electrons to an electron transport chain that pumps protons, creating a proton-motive force that drives ATP synthase. Photosystem I makes NADPH. To check your understanding of the light-dependent reactions, you should be able to make a model of the Z scheme using paper cutouts. On pieces of paper, label the following: the antenna systems of photosystems II and I, pheophytin, plastoquinone and the electron transport chain, plastocyanin, ferredoxin, and the reaction that splits water. Using dimes to represent electrons, explain how they flow through the photosystems.

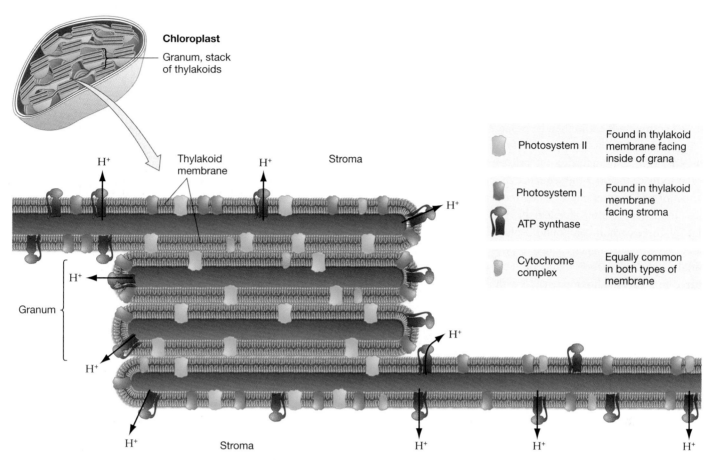

FIGURE 10.18 Photosystems I and II Occur in Separate Regions of Thylakoid Membranes within Grana
Virtually all of the active photosytem II is found in membranes facing the inside of the chloroplast's grana. In contrast, virtually all of the photosystem I and ATP synthase are found in membranes that face the stroma. The cytochrome complex, plastoquinone, and plastocyanin are equally common in both types of membranes. **EXERCISE** On the figure, draw the path of an electron that follows the Z scheme from photosystem II to photosystem I. Then draw the path of an electron that participates in cyclic photophosphorylation.

10.4 How Is Carbon Dioxide Reduced to Produce Glucose?

The reactions analyzed in Section 10.3 occur only in the presence of light. This is logical, because their entire function is focused on energy transformation—the conversion of electromagnetic energy in the form of sunlight to chemical energy in the phosphate bonds of ATP and the electrons of NADPH. The reactions that lead to the production of sugar from carbon dioxide, in contrast, do not depend directly on the presence of light. Although these reactions still require the ATP and NADPH produced by the light-dependent reactions of photosynthesis, it is possible for the reactions that lead to the reduction of CO_2 and production of sugars to occur in darkness.

The realization that the energy transformation and carbon dioxide reduction components of photosynthesis are two separate processes was a fundamental insight. Research on the exact sequence of light-independent reactions gained momentum just after World War II, when radioactive isotopes of carbon became available for research purposes. Between 1945 and 1955, a team led by Melvin Calvin carried out a groundbreaking series of experiments based on exposing green algae to radioactively labeled carbon dioxide ($^{14}CO_2$). By isolating and identifying product molecules that contained ^{14}C, the researchers gradually documented which intermediate compounds are produced as carbon dioxide is reduced to sugar.

The Calvin Cycle

To unravel the reaction sequence that reduces carbon dioxide, Calvin's group used the pulse-chase strategy introduced in Chapter 7. Recall that pulse-chase experiments introduce a pulse of labeled compound followed by a chase of unlabeled compound. The fate of the labeled compound is then followed through time. In this case, the researchers fed green algae a pulse of $^{14}CO_2$ (**Figure 10.19**). After waiting a specified amount of time, they ground the cells up to form a crude extract, separated individual molecules in the extract via paper chromatography, and laid X-ray film over the filter paper. If radioactively labeled molecules were present on the filter paper, the energy they emitted would expose the film and create a dark spot. The labeled compounds could then be isolated and identified.

By varying the amount of time between starting the pulse of labeled $^{14}CO_2$ and analyzing the cells, Calvin and co-workers began to piece together the sequence in which various intermediates formed. For example, when the team analyzed cells almost immediately after starting the $^{14}CO_2$ pulse, they found that the three-carbon compound 3-phosphoglycerate predominated. This result suggested that 3-phosphoglycerate was the initial product of carbon reduction. Stated another way, it appeared that carbon dioxide reacted with some unknown molecule to produce 3-phosphoglycerate.

This was an interesting result, since 3-phosphoglycerate is one of the ten intermediates in glycolysis. The finding that glycolysis and carbon reduction share intermediates was intriguing because of the relationship between the two pathways. The light-independent reactions lead to the manufacture of carbohydrate; glycolysis breaks it down. Because the two processes are related in this way, it was logical that at least some intermediates in glycolysis and CO_2 reduction are the same.

As Calvin's group pieced together the sequence of events in carbon dioxide reduction, an important question remained unanswered: What compound reacts with CO_2 to produce 3-phosphoglycerate? This was the key, initial step. The group searched in vain for a two-carbon compound that might serve as the initial carbon dioxide acceptor and yield 3-phosphoglycerate. Then, while Calvin was running errands one day, it occurred to him that the molecule reacting with carbon dioxide might contain five carbons, not two. The idea was that adding CO_2 to a five-carbon molecule would produce a six-carbon compound, which could then split in half to form two three-carbon molecules.

Experiments to test this hypothesis confirmed that the five-carbon compound **ribulose bisphosphate (RuBP)** is the initial reactant. Eventually the three phases of CO_2 reduction were worked out and became known as the Calvin cycle (**Figure 10.20**):

1. *Fixation phase.* The events begin when CO_2 reacts with RuBP. This phase "fixes" carbon dioxide by attaching it to a more complex molecule. It also leads to the production of two molecules of 3-phosphoglycerate. **Carbon fixation** is the addition of carbon dioxide to an organic compound, putting CO_2 into a biologically useful form.

2. *Reduction phase.* Next, 3-phosphoglycerate is phosphorylated by ATP and then reduced by electrons from NADPH. The product is the phosphorylated sugar **glyceraldehyde-3-phosphate (G3P)**. Some of the resulting G3P is drawn off to manufacture glucose and fructose, which are linked to form the disaccharide sucrose.

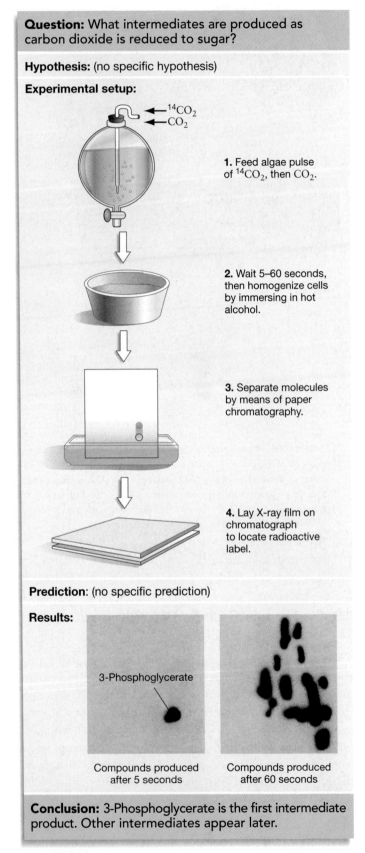

Question: What intermediates are produced as carbon dioxide is reduced to sugar?

Hypothesis: (no specific hypothesis)

Experimental setup:

$^{14}CO_2$
CO_2

1. Feed algae pulse of $^{14}CO_2$, then CO_2.

2. Wait 5–60 seconds, then homogenize cells by immersing in hot alcohol.

3. Separate molecules by means of paper chromatography.

4. Lay X-ray film on chromatograph to locate radioactive label.

Prediction: (no specific prediction)

Results:

3-Phosphoglycerate

Compounds produced after 5 seconds

Compounds produced after 60 seconds

Conclusion: 3-Phosphoglycerate is the first intermediate product. Other intermediates appear later.

FIGURE 10.19 Experiments Revealed the Reaction Pathway Leading to Reduction of CO_2

QUESTION Why wasn't this experiment based on a specific hypothesis and set of predictions?

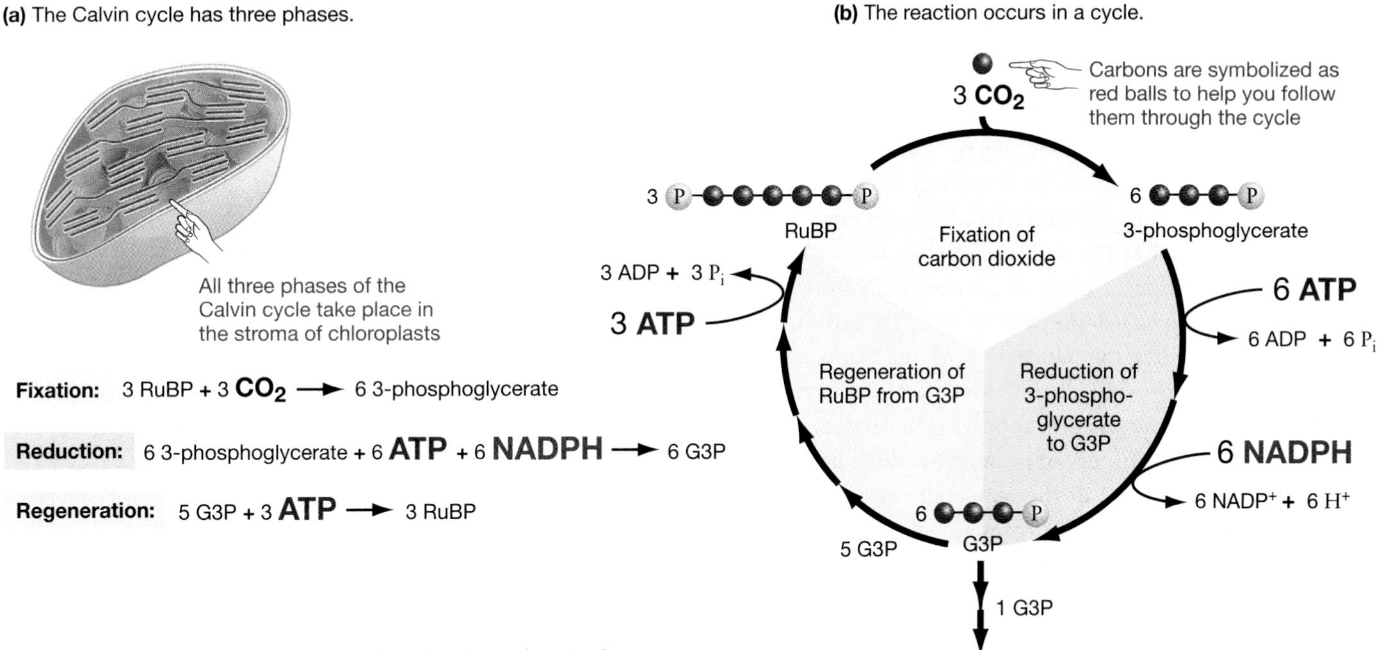

(a) The Calvin cycle has three phases.

All three phases of the Calvin cycle take place in the stroma of chloroplasts

Fixation: 3 RuBP + 3 CO_2 ⟶ 6 3-phosphoglycerate

Reduction: 6 3-phosphoglycerate + 6 **ATP** + 6 **NADPH** ⟶ 6 G3P

Regeneration: 5 G3P + 3 **ATP** ⟶ 3 RuBP

(b) The reaction occurs in a cycle.

3 CO_2

Carbons are symbolized as red balls to help you follow them through the cycle

3 (P)●●●●●(P)
RuBP

Fixation of carbon dioxide

6 ●●●(P)
3-phosphoglycerate

3 ADP + 3 P_i
3 ATP

6 ATP
6 ADP + 6 P_i

Regeneration of RuBP from G3P

Reduction of 3-phospho-glycerate to G3P

6 NADPH
6 $NADP^+$ + 6 H^+

6 ●●●(P)
G3P

5 G3P

1 G3P

Glucose

FIGURE 10.20 Carbon Dioxide Is Reduced in the Calvin Cycle
The reactions of the Calvin cycle do not depend directly on the presence of light.

3. *Regeneration phase.* The rest of the G3P keeps the cycle going by serving as the substrate for the third phase in the cycle: reactions that result in the regeneration of RuBP.

All three phases take place in the stroma of chloroplasts.

The discovery of the Calvin cycle clarified how the ATP and NADPH produced by light-dependent reactions allow cells to reduce CO_2 to carbohydrate $(CH_2O)_n$. Because sugars store a great deal of potential energy, producing them takes a great deal of chemical energy—transferred by ATP and NADPH.

Once the reaction sequence in the Calvin cycle was confirmed, attention focused on the initial phase—the reaction between RuBP and CO_2. It is one of only two reactions that are unique to the Calvin cycle. Most reactions involved in reducing CO_2 also occur during glycolysis or other metabolic pathways.

The reaction between CO_2 and RuBP starts the transformation of carbon dioxide gas from the atmosphere to sugars. Plants use sugars to fuel cellular respiration and build leaves, roots, flowers, seeds, tree trunks, and other structures. Millions of non-photosynthesizers organisms—including fish, insects, fungi, and mammals—also depend on this reaction to provide the sugars they need for cellular respiration. Ecologically, the addition of CO_2 to RuBP may be the most important chemical reaction on Earth. The enzyme that catalyzes it is fundamental to all life. What does this molecule look like, and how does it work?

The Discovery of Rubisco

To find the enzyme that fixes CO_2, Arthur Weissbach and colleagues ground up spinach leaves, purified a large series of proteins from the resulting cell extracts, and then tested each protein to see if it could catalyze the incorporation of $^{14}CO_2$

into RuBP to form 3-phosphoglycerate. Eventually they were able to isolate an enzyme that catalyzes the reaction. The enzyme turned out to be extremely abundant in leaf tissue. The researchers' data suggested that the enzyme constituted at least 10 percent of the total protein found in spinach leaves.

The CO_2-fixing enzyme was eventually purified and analyzed. Ribulose-1,5-bisphosphate carboxylase/oxygenase is its full name, but it is commonly referred to as **rubisco**. Rubisco is found in all photosynthetic organisms that use the Calvin cycle to fix carbon. It is thought to be the most abundant enzyme on Earth. Its three-dimensional structure has now been determined (**Figure 10.21**). The molecule is shaped like a cube and has a total of eight active sites where CO_2 is fixed.

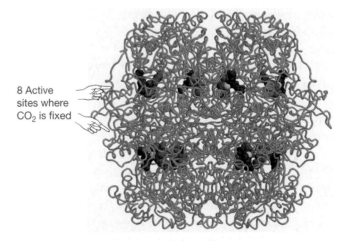

8 Active sites where CO_2 is fixed

FIGURE 10.21 Rubisco "Fixes" Carbon Dioxide
A three-dimensional model of rubisco. The red and blue molecules represent substrates at the eight active sites.

Even though it has a large number of active sites, rubisco is a very slow enzyme. Each active site catalyzes just three reactions per second; other enzymes typically catalyze thousands of reactions per second. Plants synthesize huge amounts of rubisco, possibly as an adaptation compensating for its lack of speed.

Besides being slow, rubisco is extremely inefficient. The inefficiency occurs because the enzyme catalyzes the addition of O_2 to RuBP as well as the addition of CO_2 to RuBP. Oxygen and carbon dioxide compete at the enzyme's active sites, and this competition slows the rate of CO_2 reduction. Why would an active site of rubisco accept both molecules? Given rubisco's importance in producing food for photosynthetic species, this detail is puzzling. It appears to be **maladaptive**—a trait that reduces the fitness of individuals. One hypothesis to explain the dual nature of the active site is based on the observation that rubisco was present in photosynthetic organisms long before the evolution of oxygenic photosynthesis. As a result, O_2 was extremely rare in the atmosphere when rubisco evolved. According to this hypothesis, rubisco's inefficiency is a historical artifact. The idea is that rubisco is adapted to an atmosphere that no longer exists—one that was extremely rich in CO_2 and poor in O_2.

Unfortunately, the reaction of O_2 with RuBP does more than simply compete with the reaction of CO_2 at the same active site. One of the molecules that results from the addition of oxygen to RuBP is processed in reactions that consume ATP and release CO_2. Part of this pathway occurs in chloroplasts, and part in peroxisomes and mitochondria. The reaction sequence resembles respiration, because it consumes oxygen and produces carbon dioxide. As a result, it is called **photorespiration**. Because photorespiration consumes energy and undoes carbon fixation, it can be considered a reverse photosynthesis (**Figure 10.22**). When photorespiration occurs, the overall rate of photosynthesis declines.

The oxygenation reaction that triggers photorespiration is favored when oxygen concentrations are high and CO_2 concentrations are low. But as long as carbon dioxide concentrations in leaves are high, the CO_2-fixation reaction is favored and photorespiration is relatively rare.

How Is Carbon Dioxide Delivered to Rubisco?

Carbon dioxide is present in the atmosphere and is continuously used as a reagent in photosynthesizing cells. It would seem straightforward, then, for CO_2 to diffuse directly into plants along a concentration gradient. But the situation is not this simple, because plants are covered with a waxy coating called a cuticle. This lipid layer prevents water from evaporating out of tissues, but it also prevents CO_2 from entering them.

How does CO_2 get into photosynthesizing tissues? A close look at a leaf surface, such as the one in **Figure 10.23a**, provides the answer. The leaf surface is dotted with openings bordered by two distinctively shaped cells. The paired cells are called **guard cells**, the opening is called a **pore**, and the entire structure is

Rubisco catalyzes competing reactions with very different outcomes.

Reaction with carbon dioxide during photosynthesis:

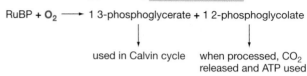

Reaction with oxygen during "photorespiration":

$RuBP + O_2 \longrightarrow$ 1 3-phosphoglycerate + 1 2-phosphoglycolate

used in Calvin cycle   when processed, CO_2 released and ATP used

FIGURE 10.22 Photorespiration Competes with Photosynthesis
QUESTION After studying these reactions, explain why biologists say that photorespiration "undoes" photosynthesis.

(a) Leaf surfaces contain stomata.

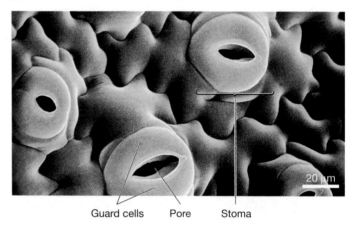

Guard cells   Pore   Stoma

(b) Carbon dioxide diffuses into leaves through stomata.

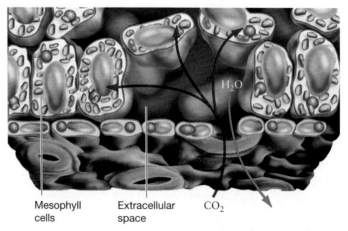

H_2O

Mesophyll cells   Extracellular space CO_2

FIGURE 10.23 Leaf Cells Obtain Carbon Dioxide through Stomata
(a) Stomata consist of two guard cells and a pore. **(b)** When a stoma is open, CO_2 diffuses into the leaf along a concentration gradient.

called a **stoma** (plural: **stomata**). If CO_2 concentrations inside the leaf are low as photosynthesis gets under way, chemical signals activate proton pumps in the membranes of guard cells. These pumps establish a charge gradient across the membrane. In response, potassium ions (K^+) move into the guard cells. When water follows along the newly created osmotic gradient, the cells swell and create a pore. As **Figure 10.23b** shows, an open stoma allows CO_2 from the atmosphere to diffuse into the air-filled spaces inside the leaf, and from there into the extracellular fluid surrounding photosynthesizing cells. Eventually the CO_2 diffuses along a concentration gradient into the chloroplasts of the cells. A strong concentration gradient favoring entry of CO_2 is maintained by the light-independent reactions, which constantly use up the CO_2 in chloroplasts.

Stomata are normally open during the day, when photosynthesis is occurring, and closed at night. But if the daytime is extremely hot and dry, leaf cells may begin losing a great deal of water to evaporation through their stomata. When this occurs, they must either close the openings and halt photosynthesis or risk death from dehydration. When conditions are hot and dry, then, photosynthesis and growth stop. How do plants that live in hot, dry environments cope? An answer emerged as biologists struggled to understand a surprising experimental result.

C₄ Photosynthesis After the Calvin cycle had been worked out in algae, researchers in a variety of labs used the same pulse-chase approach to investigate how carbon fixation occurs in other species. Just as Calvin had done, Hugo Kortschack and colleagues and Y. S. Karpilov and associates exposed leaves of sugarcane and maize (corn) to radioactive carbon dioxide ($^{14}CO_2$) and sunlight and then characterized the products. Both research teams expected to find the first of the radioactive carbon atoms in 3-phosphoglycerate—the normal product of carbon fixation by rubisco. Instead, they found that in some plant species the radioactive carbon atom ended up in four-carbon compounds such as malate and aspartate—not in three-carbon sugars.

The experiments revealed a twist on the usual pathway for carbon fixation. Instead of creating a three-carbon sugar, it appeared that in some species CO_2 fixation produced four-carbon sugars. The two pathways became known as C_3 and C_4 photosynthesis, respectively (**Figure 10.24**).

Researchers who followed up on the initial reports found that, in some plant species, carbon dioxide can be added to

RuBP by rubisco *or* to three-carbon compounds by an enzyme called **PEP carboxylase**. They also showed that the two enzymes are found in distinct cell types within the same leaf. PEP carboxylase is common in **mesophyll cells** near the surface of leaves, while rubisco is found in **bundle-sheath cells** that surround the vascular tissue in the interior of the leaf (**Figure 10.25a**). **Vascular tissue** conducts water and nutrients in plants.

Based on these observations, Hal Hatch and Roger Slack proposed a three-step model to explain how CO_2 that is fixed to a four-carbon sugar feeds the Calvin cycle (**Figure 10.25b**):

1. PEP carboxylase fixes CO_2 in mesophyll cells.

2. The four-carbon organic acids that result travel to bundle-sheath cells.

3. The four-carbon organic acids release a CO_2 molecule that rubisco uses as a substrate to form 3-phosphoglycerate. This step initiates the Calvin cycle.

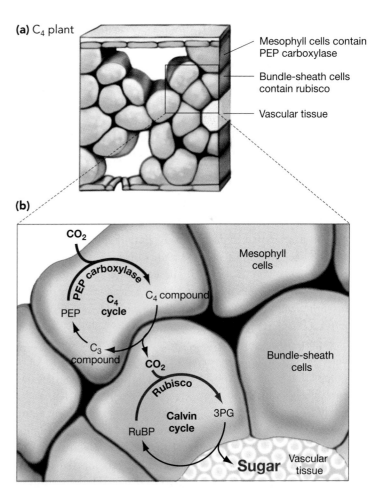

(a) C_4 plant

- Mesophyll cells contain PEP carboxylase
- Bundle-sheath cells contain rubisco
- Vascular tissue

(b)

FIGURE 10.25 In C₄ plants, Carbon Fixation Occurs Independently of the Calvin Cycle

(a) The carbon-fixing enzyme PEP carboxylase is located in mesophyll cells, while rubisco is in bundle-sheath cells. **(b)** CO_2 is fixed to the three-carbon compound PEP by PEP carboxylase, forming a four-carbon organic acid.

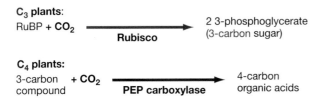

C₃ plants:
RuBP + CO_2 $\xrightarrow{\text{Rubisco}}$ 2 3-phosphoglycerate (3-carbon sugar)

C₄ plants:
3-carbon compound + CO_2 $\xrightarrow{\text{PEP carboxylase}}$ 4-carbon organic acids

FIGURE 10.24 Initial Carbon Fixation in C₄ Plants Is Different from That in C₃ Plants

In effect, then, the C_4 pathway acts as a CO_2 pump. The reactions that take place in mesophyll cells require energy in the form of ATP, but they increase CO_2 concentrations in cells where rubisco is active. Because it increases the ratio of carbon dioxide to oxygen in photosynthesizing cells, less O_2 binds to rubisco's active sites. Stated another way, CO_2 fixation is favored over O_2 fixation when carbon dioxide concentrations in leaves are high. As a result, the C_4 pathway limits the damaging effects of photorespiration. The pathway is an adaptation that keeps CO_2 concentrations in leaves high. Later experiments supported the Hatch and Slack model in almost every detail.

Logically enough, the C_4 pathway is found almost exclusively in plants that thrive in hot, dry habitats. Sugarcane, maize (corn), and crabgrass are some familiar C_4 plants, but the pathway is actually found in several thousand species in 19 distinct lineages of flowering plants. These observations suggest that the C_4 pathway has evolved independently several times. It is not the only mechanism that plants use to continue growth under hot, dry conditions, however.

CAM Plants   Some years after the discovery of C_4 photosynthesis, researchers studying a group of flowering plants called the Crassulaceae came across a second mechanism for limiting the effects of photorespiration. This photosynthetic pathway became known as **crassulacean acid metabolism**, or **CAM**. Like the C_4 pathway, CAM is a CO_2 pump that acts as an additional, preparatory step to the Calvin cycle. It also has the same effect: It increases the concentration of CO_2 inside photosynthesizing cells. But unlike the C_4 pathway, CAM occurs at a different time than the Calvin cycle does—not in a different place.

CAM occurs in cacti and other species that occupy environments that are so hot and dry that individuals routinely keep their stomata closed all day. When night falls and conditions become cooler and moister, CAM plants open their stomata and take in huge quantities of CO_2. These molecules are temporarily fixed to organic acids and stored in the central vacuoles of photosynthesizing cells. During the day, the molecules are processed in reactions that release the CO_2 and feed the Calvin cycle.

Figure 10.26 summarizes the similarities and differences between C_4 photosynthesis and CAM. Both function as CO_2 pumps that minimize the amount of photorespiration that occurs when stomata are closed and CO_2 cannot diffuse in directly from the atmosphere. Both are found in flowering plant species that live in hot, dry environments. But while C_4 plants stockpile CO_2 in cells where rubisco is not active, CAM plants store CO_2 at a time when rubisco is inactive. In C_4 plants, the reactions catalyzed by PEP-carboxylase and rubisco are separated in space; in CAM plants, the reactions are separated in time.

Obtaining and reducing CO_2 is fundamental to photosynthesis. In a larger sense, photosynthesis is fundamental to the

(a) C_4 plants sequester CO_2 in certain cells.

(b) CAM plants sequester CO_2 at night.

CO_2 stored in one cell ...

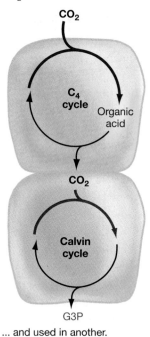

... and used in another.

CO_2 stored at night ...

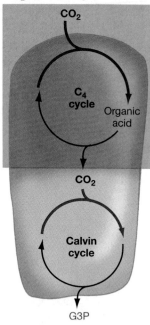

... and used during the day.

FIGURE 10.26 C_4 Photosynthesis and CAM Accomplish the Same Task in Different Ways
(a) In C_4 plants, CO_2 is fixed to organic acids in some cells and then released to other cells where the Calvin cycle enzymes are located. **(b)** CAM plants open their stomata at night and fix CO_2 to organic acids.

millions of species that depend on plants, algae, and cyanobacteria for food. What do photosynthetic organisms do with the sugar they synthesize? More specifically, what happens to the G3P that is drawn off from the Calvin cycle?

What Happens to the Sugar That Is Produced by Photosynthesis?

The G3P molecules that exit the Calvin cycle enter one of several reaction pathways. The most important of these pathways results in the production of the monosaccharides glucose and fructose, which in turn combine to form the disaccharide su-

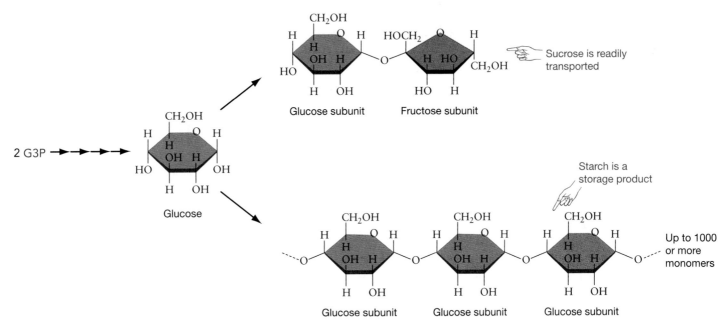

Sucrose is readily transported

Glucose subunit   Fructose subunit

2 G3P

Glucose

Starch is a storage product

Up to 1000 or more monomers

Glucose subunit   Glucose subunit   Glucose subunit

FIGURE 10.27 Sucrose and Starch Are the Main Photosynthetic Products
In plants, sugars are transported in the form of sucrose and stored in the form of starch.
EXERCISE Sucrose can be converted to starch, and starch to sucrose. Add an element to the diagram to indicate this.

crose (**Figure 10.27**). The reaction sequence starts with G3P, involves a series of other phosphorylated three-carbon sugars, includes the synthesis of the familiar six-carbon sugar glucose, and ends with the production of sucrose. An alternative pathway results in the production of glucose molecules that polymerize to form starch. Starch production occurs inside the chloroplast; sucrose synthesis takes place in the cytosol.

All of the intermediates involved in the production of glucose, as well as of G3P, also occur in glycolysis. The general observation here is that many of the enzymes and intermediates involved in glucose processing are common to both respiration and photosynthesis.

When photosynthesis is taking place slowly, almost all the glucose that is produced is used to make sucrose. As noted in Chapter 5, sucrose is a disaccharide ("two-sugar") that consists of a glucose molecule bonded to a fructose molecule. Sucrose is water soluble and is readily transported to other parts of the plant. If the sucrose is delivered to rapidly growing parts of the plant, it is broken down to fuel cellular respiration and growth. If it is transported to storage cells in roots, it is converted to starch and stored for later use.

When photosynthesis is proceeding rapidly and sucrose is abundant, glucose is used to synthesize starch in the chloroplasts of photosynthetic cells. Recall from Chapter 5 that starch is a polymer of glucose. In photosynthesizing cells, starch acts as a temporary sugar-storage product. Starch is not water soluble, so it cannot be transported from photosynthetic cells to other areas of the plant. At night, the starch that is temporarily stored in leaf cells is broken down and used to manufacture sucrose molecules that are then used by the photosynthetic cell in respiration or transported to other parts of the plant. In this way, chloroplasts provide sugars for cells throughout the plant by day and by night.

If a mouse eats the starch that is stored in a chloroplast or root cell, however, the chemical energy in the C–H bonds of the starch is used to fuel the mouse's growth and reproduction. If the mouse is then eaten by an owl, the chemical energy in the mouse's tissues fuels the predator's growth and reproduction. In this way, virtually all cell growth and reproduction can be traced back to the chemical energy that was originally captured by photosynthesis. Photosynthesis is the staff of life.

✔CHECK YOUR UNDERSTANDING

To make sure that you understand the light-independent reactions of photosynthesis, you should be able to summarize the three major phases of the Calvin cycle and explain the relationships among CO_2, G3P, RuBP, rubisco, glucose, and 3-phosphoglycerate. You should also be able to identify three potential sources of CO_2: organic acids in mesophyll cells, organic acids synthesized at night and stored in vacuoles, and direct diffusion through stomata.

ESSAY Are Rising CO_2 Levels in the Atmosphere Affecting the Rate of Photosynthesis?

The concentration of carbon dioxide in the atmosphere has increased dramatically over the past 100 years. In the late 1800s the atmospheric carbon dioxide concentration is thought to have been about 280 μL/L. Today, CO_2 is present in the atmosphere at 360 μL/L. Most of the increase is due to CO_2 that was released when natural gas, gasoline, coal, and other fossil fuels were burned for heat, manufacturing, transportation, and so on or when forests were burned to convert them to agricultural use. If present trends in fossil fuel use and deforestation continue, atmospheric CO_2 levels are expected to increase to 480 μL/L by the year 2050. If this prediction is correct, then CO_2 levels will have increased by 70 percent in just 150 years.

This increase in carbon dioxide concentration is causing dramatic increases in average temperatures around the globe. Global warming is occurring because carbon dioxide in the atmosphere absorbs electromagnetic radiation in the infrared part of the spectrum. These wavelengths radiate from Earth's surface after sunlight strikes it. If they are not absorbed by CO_2, they are lost to space. In this way, carbon dioxide traps heat in the atmosphere.

This process is called the greenhouse effect, because it mimics the effect of glass in a greenhouse. Increases in carbon dioxide concentrations are leading to abnormally large increases in the amount of heat retained in the atmosphere, leading to global warming. How are increases in CO_2 affecting plants?

According to the overall reaction for photosynthesis, increases in a reactant such as CO_2 could lead to increased rates of photosynthesis. Stated another way, plant productivity should rise. Biologists have confirmed this prediction experimentally by increasing CO_2 levels in controlled environmental chambers and in natural habitats. For example, a research team set up a series of experimental plots of land in a 13-year-old pine forest in North Carolina. Some plots were ringed with towers that emitted enough CO_2 to bring average levels inside the plot to 560 μL/L; other plots were left with normal air as a control treatment. As predicted, the growth rate of pine trees in the CO_2-augmented plots increased 25 percent relative to that of the controls.

Experiments like this suggest that rising CO_2 levels may lead to increased rates of photosynthesis in at least some habitats. This is important because increased growth of trees and shrubs removes carbon dioxide from the atmosphere and sequesters the carbon in wood. Similarly, increased growth of algae and photosynthetic bacteria in marine environments sequesters carbon in cellulose or calcium carbonate, which may then drop to the bottom of the ocean when the organisms die. As a result, increased growth by photosynthetic organisms should act as a feedback mechanism that helps offset rising CO_2 levels in the atmosphere. Based on this logic, biologists

have suggested that tree planting and forest restoration programs could play a role in a coordinated, worldwide effort to counteract global warming (**Figure 10.28**).

Because many plants respond strongly to augmented CO_2 it is clear that carbon dioxide can be an important limiting nutrient in plant growth. But at some point, plants should stop responding to increased carbon dioxide levels because water availability or some other nutrient—perhaps nitrogen or phosphorus—will become limiting instead. As a result, biologists caution that plant growth rates will eventually stop responding to increased CO_2 availability.

> *... plant growth rates will eventually stop responding to increased CO_2 availability.*

The other major prediction regarding plant responses to increased CO_2 concentrations focuses on desert-dwelling species. To understand this prediction, recall that plants lose water when their stomata are open to admit carbon dioxide for photosynthesis. If carbon dioxide levels rise, then plants will have to open their stomata less to obtain the CO_2 they need, thus losing less water to the atmosphere. As a result, they should be able to grow faster. Consistent with this prediction, a team of biologists found that when they artificially increased CO_2 levels in experimental plots in the Mojave Desert of southwestern North America, plant growth increased. The effect occurred only during a wet year, however—no increased growth was observed in a drought year. What will be the long-term effect of rising CO_2 on desert plants? The answer is not known. Research continues.

FIGURE 10.28 Trees Are a "Carbon Sink"—They Take in Carbon Dioxide and Store Carbon Atoms in Wood

CHAPTER REVIEW

Summary of Key Concepts

Photosynthesis is the conversion of light energy to chemical energy, stored in the bonds of carbohydrates. The sucrose generated by photosynthesis fuels cellular respiration and supplies a substrate for the synthesis of complex carbohydrates, amino acids, fatty acids, and other cell components. As the primary food source for a diverse array of heterotrophs, photosynthetic organisms provide the energy that sustains most life on Earth.

Web Tutorial 10.1 Photosynthesis

■ **Photosynthesis consists of two distinct sets of reactions. In reactions driven by light, ATP and the electron carrier NADPH are produced. In subsequent reactions that do not depend directly on light, the ATP and NADPH are used to reduce carbon dioxide (CO_2) to carbohydrate $(CH_2O)_n$. In eukaryotic cells, both processes take place in chloroplasts.**

The light-dependent reactions occur in internal membranes of the chloroplast that are organized into structures called thylakoids in stacks known as grana. The light-independent reactions, known as the Calvin cycle, take place in a fluid portion of the chloroplast called the stroma.

■ **The light-dependent reactions transform the energy in sunlight to chemical energy in the form of electrons with high potential energy. Excited electrons either are used to produce NADPH or are donated to an electron transport chain, which results in the production of ATP.**

The energy transformation step of photosynthesis begins when a pigment molecule in an antenna complex absorbs a photon in the blue or red part of the visible spectrum. When absorption occurs, the energy in the photon is transferred to an electron in the pigment molecule. The electron is raised to an excited state equivalent to the energy in the photon. If the electron falls back to the normal, or ground, state, energy is given off as light (fluorescence) and heat. But in photosynthetic organisms, the energy in the excited electron is eventually transferred to chlorophyll molecules that act as reaction centers. There the high-energy electron is transferred to an electron acceptor, which becomes reduced. In this way, light energy is transformed to chemical energy.

Plants and algae have two types of reaction centers, which are part of larger complexes called photosystem I and photosystem II. Each photosystem consists of an antenna complex with 200–300 chlorophyll and carotenoid molecules, a reaction center, and an electron acceptor that completes energy transformation.

In photosystem II, high-energy electrons are accepted by the electron acceptor pheophytin. Electrons are then passed along an electron transport chain. As electrons move through this chain, they are gradually stepped down in potential energy. The energy released by these reduction-oxidation (redox) reactions is used to pump protons across the thylakoid membrane. The resulting proton gradient drives the synthesis of ATP by ATP synthase. This method of producing ATP is called photophosphorylation. Electrons donated to the electron transport chain by photosystem II are replaced by electrons taken from water, resulting in the production of oxygen as a by-product.

In photosystem I, high-energy electrons are accepted by iron- and sulfur-containing proteins and passed to ferredoxin. In an enzyme-catalyzed reaction, the reduced form of ferredoxin passes electrons to $NADP^+$ to form NADPH. NADPH carries electrons required for the redox reactions that result in the synthesis of sugars and other cell materials. Photosystem I produces the electron carrier NADPH.

The Z scheme describes how photosystems I and II are thought to interact. The scheme begins with the movement of an electron from photosystem II to the electron transport chain. At the end of the chain, the protein plastocyanin carries electrons to photosystem I. There the electrons are promoted to a very high energy state in response to the absorption of a photon, and they are subsequently used to reduce $NADP^+$. Electrons from photosystem I may occasionally be passed to the electron transport chain instead of being used to reduce $NADP^+$, resulting in a cyclic flow of electrons between the two photosystems to produce the additional ATP needed to reduce carbon dioxide.

■ **The light-independent reactions start with the enzyme rubisco, which catalyzes the addition of CO_2 to a five-carbon molecule. The compound that results undergoes a series of reactions that use ATP and NADPH and lead to the production of sugar.**

The light-independent reactions of photosynthesis depend on the products of the light-dependent reactions and are called the Calvin cycle. The process of reducing carbon dioxide to sugar begins when CO_2 is attached to a five-carbon compound called ribulose bisphosphate (RuBP). This reaction is catalyzed by the enzyme rubisco. The six-carbon compound that results immediately splits in half to form two molecules of 3-phosphoglycerate. Subsequently, 3-phosphoglycerate is reduced to a sugar called glyceraldehyde-3-phosphate (G3P). Some G3P is used to synthesize glucose and fructose, which combine to form sucrose; the rest participates in reactions that regenerate RuBP so the cycle can continue.

Rubisco catalyzes the addition of oxygen as well as carbon dioxide to RuBP. The reaction with oxygen leads to a loss of fixed CO_2 and ATP and is called photorespiration. Photorespiration is particularly important in hot, dry conditions, when stomata close to prevent excessive water loss. Because the closure of stomata reduces CO_2 levels in photosynthesizing cells, the reaction of O_2 with RuBP is favored. C_4 and CAM plants have distinct but functionally similiar mechanisms for augmenting CO_2 concentrations in photosynthesizing cells and thus for limiting photorespiration.

Web Tutorial 10.2 Strategies for Carbon Fixation

Questions

Content Review

1. What is the stroma of a chloroplast?
   a. the inner membrane
   b. the pieces of membrane that connect grana
   c. the interior of a thylakoid
   d. the fluid inside the chloroplast but outside the thylakoids

2. Why is chlorophyll green?
   a. It absorbs all wavelengths in the visible spectrum, transmitting ultraviolet and infrared light.
   b. It absorbs wavelengths only in the red and far-red portions of the spectrum (680 nm, 700 nm).
   c. It absorbs wavelengths in the blue and red parts of the visible spectrum and transmits wavelengths in the green part.
   d. It absorbs wavelengths only in the blue part of the visible spectrum and transmits all other wavelengths.

3. What does it mean to say that CO_2 becomes fixed?
   a. It becomes bonded to an organic compound.
   b. It is released during cellular respiration.
   c. It acts as an electron acceptor.
   d. It acts as an electron donor.

4. What do the light-dependent reactions of photosynthesis produce?
   a. G3P
   b. RuBP
   c. ATP and NADPH
   d. plastoquinone

5. Why do the absorption spectrum for chlorophyll and the action spectrum for photosynthesis coincide?
   a. Photosystems I and II are activated by different wavelengths of light.
   b. Wavelengths of light that are absorbed by chlorophyll trigger the light-dependent reactions.
   c. Energy from wavelengths absorbed by carotenoids is passed on to chlorophyll.
   d. The rate of photosynthesis depends on the amount of light received.

6. What happens when an excited electron is passed to an electron acceptor in a photosystem?
   a. It drops back down to its ground state, resulting in the phenomenon known as fluorescence.
   b. The chemical energy in the excited electron is released as heat.
   c. The electron acceptor is oxidized.
   d. Energy in sunlight is transformed to chemical energy.

Conceptual Review

1. Explain how the energy transformation step of photosynthesis occurs. How is light energy converted to chemical energy in the form of ATP and NADPH?

2. Explain how the carbon reduction step of photosynthesis occurs. How is carbon dioxide fixed? Why are both ATP and NADPH required to produce sugar?

3. Sketch the Z scheme. Explain how photosystem I and photosystem II interact by tracing the path of an electron through the Z scheme. What molecule connects the two photosystems?

4. In what sense does photorespiration "undo" photosynthesis?

5. Make a sketch showing how C_4 photosynthesis and CAM separate CO_2 acquisition from the Calvin cycle in space and time, respectively.

6. Why do plants need both chloroplasts and mitochondria?

Group Discussion Problems

1. Compare and contrast mitochondria and chloroplasts. In what ways are their structures similar and different? What molecules or systems function in both types of organelles? Which enzymes or processes are unique to each organelle?

2. The Calvin cycle and rubisco are found in lineages of bacteria and archaea that evolved long before the origin of oxygenic photosynthesis. Based on this observation, biologists infer that rubisco evolved in an environment that contained little, if any, oxygen. Some biologists propose that this inference explains why photorespiration occurs today. Do you agree with the hypothesis that photorespiration is an evolutionary "holdover?" Why or why not?

3. In addition to providing their protective function, carotenoids absorb certain wavelengths of light and pass the energy on to the reaction centers of photosystem I and II. Based on their function, predict exactly where carotenoids are located in the chloroplast. Explain your rationale. How would you test your hypothesis?

4. Consider plants that occupy the top, middle, or ground layer of a forest, and algae that live near the surface of the ocean or in deeper water. Would you expect the same photosynthetic pigments to be found in species that live in these different habitats? Why or why not? How would you test your hypothesis?

Answers to Multiple-Choice Questions   1. d; 2. c; 3. a; 4. c; 5. b; 6. d

The Cell Cycle

11

KEY CONCEPTS

▦ After chromosomes are copied, mitosis distributes one chromosome copy to each of two daughter cells. Mitosis and cytokinesis produce two cells that are identical to the parent cell.

▦ Over their life span, cells go through a life cycle that consists of four carefully controlled phases.

▦ Uncontrolled cell growth leads to cancer. Different types of cancer result from different types of defects in control over the cell cycle.

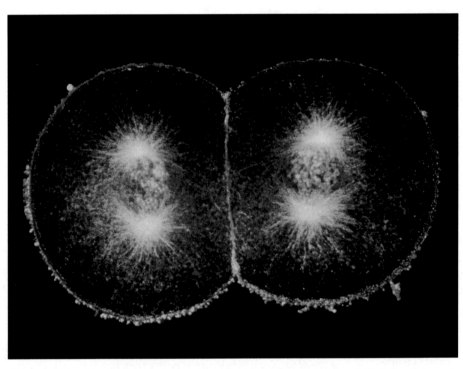

This sea urchin embryo consists of two cells, each of which is beginning to divide. Many thousands more cell divisions will occur as the individual develops into an adult.

I n Chapter 1 we considered the cell theory, which maintains that all organisms are made of cells and that all cells arise from preexisting cells. Although the cell theory was widely accepted among biologists by the 1860s, a great deal of confusion remained about how cells reproduced. Most proponents of the cell theory believed that new cells arose within preexisting cells by a process that resembled crystallization. But Rudolf Virchow proposed that new cells arise through the division of preexisting cells—that is, **cell division**.

In the late 1800s, careful microscopic observations of newly developing individuals, or **embryos**, confirmed Virchow's hypothesis. Research documented that multicellular individuals start life as single-celled embryos and grow through a series of cell divisions. As better microscopes be-

came available, biologists were able to describe the division process in more detail. In particular, researchers focused their attention on the nuclei of dividing cells and the fate of the chromosomes. Chromosomes are the carriers of hereditary material—the instructions for building and operating the cell. These early studies revealed two fundamentally different ways that nuclei divide prior to cell division. In animals, one type of nuclear division leads to the production of sperm and eggs, and the other type of nuclear division leads to the production of all other cell types. Sperm and eggs are male and female reproductive cells, termed **gametes**; all other cell types are referred to as **somatic** ("body-belonging") **cells**. In both kinds of cell division, a so-called *parent cell* is said to give rise to *daughter cells*.

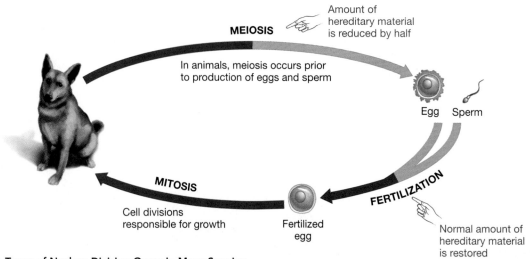

FIGURE 11.1 Two Types of Nuclear Division Occur in Many Species
In animals, meiosis leads to the production of eggs and sperm. Mitosis is responsible for producing somatic cells.

During the type of nuclear division that leads to the production of sperm and eggs, the amount of hereditary material found in the parent cell nucleus is reduced by half. As a result, the daughter cells that become sperm or eggs do not contain the same genetic material as the parent cell. This type of nuclear division is called **meiosis,** and it is involved only in the production of reproductive cells (as in the example shown in **Figure 11.1**). Meiosis is the basis of sexual reproduction and genetic inheritance. Chapter 12 explores the mechanism and consequences of meiosis in detail.

When nuclei divide prior to the formation of new somatic cells, the amount of hereditary material in the original cell and the daughter cells remains constant. **Mitosis** is a division of the genetic material that produces daughter cells that are genetically identical to their parent cell. Mitosis is usually accompanied by **cytokinesis** ("cell movement")—the division of the cytoplasm into the two daughter cells. Mitosis, followed by cytokinesis, supplies the cells required for several key activities in eukaryotes—wound repair, reproduction, and growth:

- New cells are required to repair damaged tissues in multicellular organisms. When you suffer a scrape or cut, the cells that repair your skin and heal the wound are generated via mitosis and cytokinesis.

- When yeast cells greatly increase in number in a pile of bread dough or in a vat of beer, they are reproducing by mitosis and cytokinesis. In both unicellular and multicellular species, mitosis followed by cytokinesis is the basis of asexual reproduction. **Asexual reproduction** results in the production of offspring that are genetically identical to the parent.

- Mitosis and cytokinesis are also responsible for growth in multicellular organisms. For example, the ancestry of the

trillions of genetically identical cells that make up your body can be traced back through a series of mitotic divisions to a single fertilized egg—the result of the union of a sperm and an egg from your parents.

For this chapter, our goals are to explore how mitosis occurs and how cell division is regulated. The first section introduces the relationship between mitosis and other major stages in a cell's life cycle. The next two sections provide an in-depth look at each event in mitosis and explore how mitosis and other events in the cell's life cycle are regulated. The chapter concludes by examining why uncontrolled cell division and cancer can result when the systems that regulate cell division break down.

11.1   Mitosis and the Cell Cycle

In the course of studying cell division, nineteenth-century biologists found that certain chemical dyes made threadlike structures visible in the nuclei of dividing cells. In 1879 Walther Flemming followed up on this discovery by documenting how the threadlike structures in salamander embryos change as the cells divide. As **Figure 11.2a** shows, the threads that Flemming observed were paired when they first appeared, just before cell division. Prior to cell division, each pair of threads split to produce single, unpaired threads. Flemming introduced the term *mitosis,* from the Greek *mitos* ("thread"), to describe this division process.

Soon after Flemming made that discovery, similar observations were reported for the roundworm *Ascaris.* In addition to confirming that each pair of threads split in this species, investigators reported that the total number of threads in a cell remained constant during subsequent divisions. Thus, all of the cells in a roundworm's body had the same number and types of threads.

(a)   **(b)**

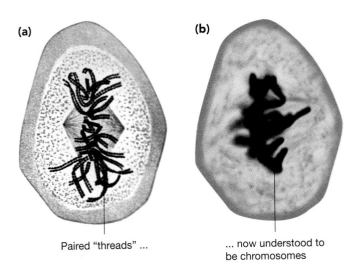

Paired "threads" ...   ... now understood to
be chromosomes

FIGURE 11.2 Chromosomes Move during Mitosis
(a) Walther Flemming's 1879 drawing of mitosis in the salamander
larva. The black threads are chromosomes. (b) Chromosomes can be
stained with dyes and observed in the light microscope.

In 1888 Wilhelm Waldeyer coined the term **chromosome**
("colored-body") to refer to the threadlike structures observed
in dividing cells (**Figure 11.2b**). As Chapters 14 and 15 will
show, chromosomes are made up in part of deoxyribonucleic
acid (DNA). More specifically, a chromosome consists of a single, long DNA double helix that is wrapped around proteins in
a highly organized manner. DNA encodes the cell's hereditary
information, or genetic material. By observing how chromosomes moved during mitosis, biologists realized that the purpose of mitosis was to distribute the parent cell's genetic
material to daughter cells during cell division. Prior to mitosis,
each chromosome is copied. During mitosis, one of the copies is
distributed to each of two daughter cells.

Before delving into the details of how mitosis occurs,
though, let's examine how it fits into the other events in the life
of a cell.

The Cell Cycle

As early workers studied the fate of chromosomes during cell
division, they realized that even rapidly growing plant and animal cells do not divide continuously. Instead, growing cells
cycle between a dividing phase called the **mitotic** (or **M**) **phase**
and a nondividing phase called **interphase** ("between-phase").
Chromosomes can be stained and observed with a light microscope only during M phase, when they are condensed into compact structures. Cells actually spend most of their time in
interphase, however. No dramatic changes are observed in the
nucleus during interphase, when chromosomes are uncoiled
into extremely long, thin structures. Even when stains are used,
individual interphase chromosomes are not normally visible in
the light microscope.

To describe the regular alternation between M phase and interphase, biologists began referring to the cell cycle. The **cell
cycle** is the orderly sequence of events that occurs from the formation of a eukaryotic cell, through the duplication of its chromosomes, to the time it undergoes division itself. During the
cycle, two key events are (1) the **replication**, or copying, of the
hereditary material in chromosomes and (2) the partitioning of
copied chromosomes to the two daughter cells. The hereditary
material is duplicated, with one copy going to each daughter
cell during mitosis. As a result, daughter cells contain genetic
information that is identical to that of the parent cell. Let's take
a closer look.

When Does Chromosome Replication Occur?

Using the light microscope, nineteenth-century biologists could
see chromosomes move to daughter cells during mitosis. Because each daughter cell ended up with the same number of
chromosomes as the parent cell, it was logical to infer that the
chromosomes were duplicated in the parent cell at some point
in the cell cycle. Did the paired threads observed in salamander
cells represent the replicated chromosomes? If so, when did this
replication step occur? Because no dramatic changes to chromosomes are visible during interphase, some biologists hypothesized that replication must occur early in M phase as
chromosomes condense and become visible. But because chromosomes already appear as doublets early in M phase, it was
possible that replication occurred during interphase.

Which hypothesis is correct? This question was not answered until the 1950s, when two technical innovations provided the experimental tools needed to answer it. One advance
was the ability to grow eukaryotic cells outside the source organism, in culture. Cultured cells are powerful experimental
tools because they can be manipulated much more easily than
cells in an intact organism can (see **Box 11.1**, page 231). A second advance was the availability of radioactively labeled
deoxyribonucleotides—the building blocks of DNA. If radioactive deoxyribonucleotides are present as DNA is being synthesized, they will be incorporated into the new DNA molecule
and label it.

To determine when chromosome replication takes place
during the cell cycle, Alma Howard and Stephen Pelc exposed
cultured cells to radioactive thymidine and nonradioactive
adenosine, guanosine, and cytidine. These deoxyribonucleotides
are components of DNA and are incorporated into DNA molecules only during replication. In the cultured cells, M phase lasted about 30 minutes, and the entire cell cycle lasted about a day.
To test the hypothesis that chromosome replication occurs during M phase, Howard and Pelc let the radioactive thymidine
stay in the cultured cells for 30 minutes. If the M-phase hypothesis is correct, then any cells undergoing mitosis during the labeling period would incorporate radioactive thymidine into
their replicated chromosomes.

After washing the radioactive thymidine out of the culture, the biologists removed a sample of cells, spread them out, and laid a sheet of X-ray film over them. As the photograph in **Figure 11.3** shows, small black dots appeared where radioactive thymidine molecules had exposed the film.

The key observation in this experiment was that radioactive labels were not found in M-phase nuclei—only in interphase nuclei. Thus, only interphase cells had incorporated radioactive thymidine. Because only replicating DNA would incorporate radioactive thymidine into newly synthesized DNA molecules, the researchers concluded that chromosome replication occurs during interphase. Howard and Pelc had identified a new stage in the cell cycle called **synthesis** (or **S**) **phase**, for DNA synthesis. Their data showed that duplication of the genetic material occurs independently of mitosis—the process that distributes chromosome copies to daughter cells.

Discovery of the Gap Phases

To determine how long it takes a cell to complete S phase, other investigators repeated Howard and Pelc's experiment but waited various lengths of time to examine the cells exposed to a pulse of radioactive thymidine. Recall that these cells take about a day to complete the cell cycle. As **Figure 11.4** indicates, experiments showed that no labeled M-phase cells appeared until 4 to 5 hours after the pulse of radioactive thymidine had ended. Between 5 and 12 hours after the labeling period, though, all of the labeled cells had exhibited the chromosomal changes associated with mitosis.

Cells in these cultures are dividing continuously. Thus, the key to interpreting Figure 11.4 is to realize that the first labeled cells to enter mitosis must represent cells that were just completing chromosome replication when they were labeled. If so, then the 4- to 5-hour time lag between the end of the pulse and the appearance of the first labeled mitotic nuclei corresponds to a time lag that occurs between the end of S phase

---

Question: Does chromosome replication occur during M phase or interphase?

Hypothesis: Chromosome replication occurs during the 30-minute M phase.

Alternate hypothesis: Chromosome replication occurs during interphase.

Experimental setup:

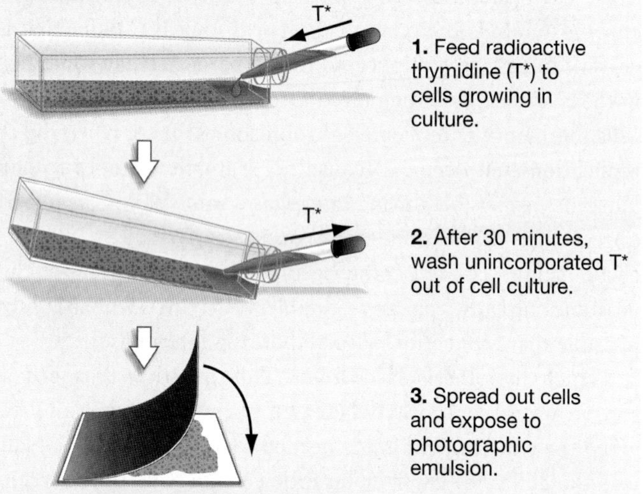

1. Feed radioactive thymidine (T*) to cells growing in culture.

2. After 30 minutes, wash unincorporated T* out of cell culture.

3. Spread out cells and expose to photographic emulsion.

Prediction: The radioactive label indicating the replication of chromosomes will appear in M-phase cells.

Prediction of alternate hypothesis: The radioactive label indicating the replication of chromosomes will appear in interphase cells.

Results: Only interphase cells are labeled with T*; label is localized in nucleus

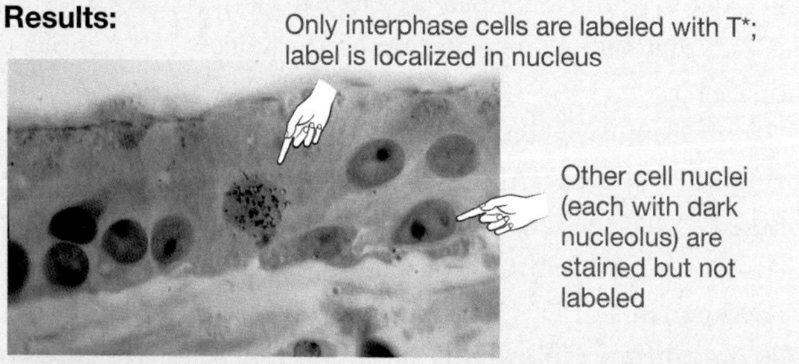

Other cell nuclei (each with dark nucleolus) are stained but not labeled

Conclusion: Chromosomes are replicated during interphase.

FIGURE 11.3 When Are Chromosomes Replicated?
Because thymidine is incorporated into newly synthesized DNA, the experimental protocol shown established that the chromosomes are replicated during interphase (in particular, the synthesis [S] phase), rather than M phase, of the cell cycle.

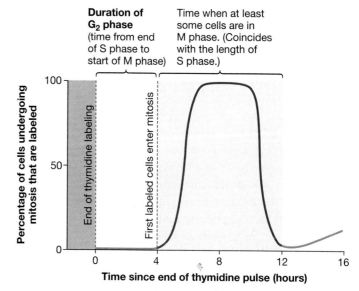

FIGURE 11.4 A Gap Occurs between Chromosome Replication and Mitosis
A plot of the percentage of cells undergoing mitosis that are labeled versus the amount of time that has elapsed since the cells were fed a pulse of radioactive thymidine. If at least some cells are in M phase for 8 hours, it means that S phase lasts 8 hours.

BOX 11.1 Cell-Culture Methods

For researchers, there are important advantages to growing plant and animal cells outside the organism itself. Cell cultures provide homogenous populations of a single type of cell and the opportunity to control experimental conditions precisely.

The first successful attempt to culture animal cells occurred in 1907, when a researcher cultivated amphibian nerve cells in a drop of fluid from the spinal cord. But it was not until the 1950s and 1960s that biologists could routinely culture plant and animal cells in the laboratory. The long lag time was due to the difficulty of recreating conditions that exist in the intact organism.

To grow in culture, animal cells must be provided with a liquid mixture that includes nutrients, vitamins, and hormones that stimulate growth. Initially, this mixture was provided through the use of *serum*, which is the liquid portion of blood; now serum-free media are available for certain cell types. Serum-free media are preferred because they are much more precisely defined chemically than serum. In addition, many types of animal cells will not grow in culture unless they are provided with a solid surface that mimics the types of surfaces to which cells in the intact organisms adhere. As a result, cells are typically cultured in flasks like the one shown in Figure 11.3.

Even under optimal conditions, though, normal cells display a finite life span in culture. In contrast, many cultured cancerous cells grow indefinitely.

They also do not adhere tightly to the surface of the culture flask and do not need growth factors in the media.

Because of their immortality and relative ease of growth, cultured cancer cells are commonly used in research. For example, the first human cell type to be grown in culture was isolated in 1951 from a malignant tumor of the uterine cervix. These cells are called HeLa cells in honor of their source, Henrietta Lacks, who died soon thereafter from the cervical cancer. HeLa cells continue to grow in laboratories around the world. They have been used in numerous studies of human cell function, including the experiments that documented chromosome replication occurring independently of mitosis.

---

and the beginning of M phase. Stated another way, there is a gap in the cycle. The gap represents the period when chromosome replication is complete but mitosis has not yet begun.

This "lag" in the cell cycle came to be called **G$_2$ phase**, for second gap. It was considered the second gap because data indicate that another gap exists. To follow the logic, note in Figure 11.4 that labeled nuclei undergoing mitosis are observed over a period of about 6 to 8 hours. Because all of these cells had to be in S phase when radioactive thymidine was available, it is logical to conclude that S phase lasts 6 to 8 hours. When the times to get through the S, G$_2$, and M phases are added up and compared with the 24 hours it takes these cells to complete one cell cycle, though, there is a discrepancy of 7 to 9 hours. This discrepancy represents the gap called **G$_1$ phase**, for first gap. As **Figure 11.5** shows, G$_1$ phase occurs after M phase but before S phase. In these cells, the G$_1$ phase is about twice as long as G$_2$.

Why do the gap phases exist? In addition to copying their chromosomes during S phase, dividing cells must replicate organelles and manufacture additional cytoplasm. Before mitosis can take place, the parent cell must grow large enough and synthesize enough organelles that its daughter cells will be normal in size and function. The two gap phases provide the time required to accomplish these tasks. They allow the cell to complete all the requirements for cell division other than chromosome replication. In addition, the long G$_1$ phase is the time when cells perform their normal functions. In your body, for example, most cells have stopped dividing and will spend the rest of your life in G$_1$.

Given this overview of the major events in the life of a cell, let's turn now to M phase and delve into the process of mitosis. Once the genetic material has been copied, how do cells divide it up between daughter cells? Recall that a **gene** is a segment of DNA that contains the information for synthesizing a particular polypeptide or RNA molecule. How do cells ensure that

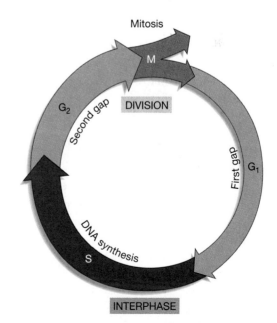

FIGURE 11.5 The Cell Cycle Has Four Phases
A representative cell cycle. The time required for the G$_1$ and G$_2$ phases varies dramatically among cells and organisms.

BOX 11.2 How Do Bacteria Divide?

To reproduce, bacterial cells divide into two genetically identical daughters. This process is called **binary fission**. Although the structure of bacterial cells is very different from that of eukaryotic cells, bacteria face challenges similar to those faced by eukaryotes in replicating and partitioning their hereditary material during cell division.

Figure 11.6 sketches the major events in bacterial cell division. Recall that most bacteria contain a single, circular chromosome composed of DNA and that this chromosome is coiled upon itself. Because bacteria lack a nucleus, the bacterial chromosome is located in the cytoplasm. After the chromosome is replicated, the two daughter chromosomes become attached to different sites on the plasma membrane. A contractile ring composed of FtsZ fibers then forms between the two chromosomes. Recall from Chapter 7 that FtsZ fibers are the major component of the bacterial cytoskeleton and are similar in structure to microtubules. As the FtsZ ring closes, the bacterial cytoplasm is divided in two, completing cell division.

STEPS IN BACTERIAL CELL DIVISION

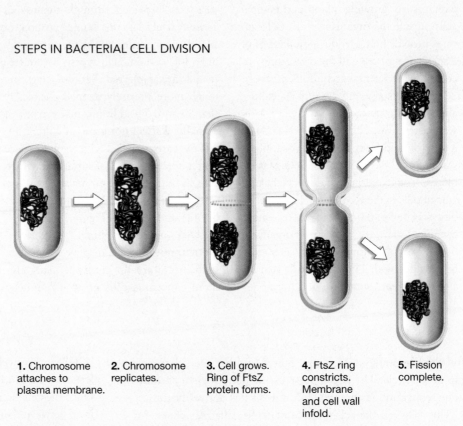

1. Chromosome attaches to plasma membrane.

2. Chromosome replicates.

3. Cell grows. Ring of FtsZ protein forms.

4. FtsZ ring constricts. Membrane and cell wall infold.

5. Fission complete.

FIGURE 11.6 Steps in Bacterial Cell Division
QUESTION Are the daughter cells of bacterial cell division identical to each other in chromosomal makeup, or are they different?

each daughter cell receives an identical complement of chromosomes and thus an identical complement of genes? (The discussion that follows describes mitosis in plant cells and in animal cells; **Box 11.2** describes cell division in bacteria.)

11.2 How Does Mitosis Take Place?

Early observations of cell division focused on the fate of the parent cell's chromosomes. Because chromosomes that are undergoing mitosis are visible under the light microscope when they are stained, investigators could watch mitosis occur. As a result, the major events in mitosis were well understood long before the cell cycle was fully described.

Recall that mitosis results in the division of chromosomes and the formation of two daughter nuclei. Mitosis is usually accompanied by cytokinesis—cytoplasmic division and the formation of two daughter cells. Let's take a closer look at the events in mitosis, beginning with observations about the nature of a eukaryotic cell's chromosomes.

Events in Mitosis

Figure 11.7a shows the chromosomes found in a hypothetical plant cell or animal cell. The number of chromosomes in each cell varies widely among species. Both humans and potato plants have a total of 46 chromosomes in each cell; a maize (corn) plant has 20, dogs have 66, and fruit flies have 8. In Figure 11.7a there are a total of four chromosomes per cell. (They are shown partially condensed simply to make them visible.) Recall from Section 11.1 that eukaryotic chromosomes normally exist as extremely long, threadlike strands consisting of DNA associated with globular proteins called **histones**. In eukaryotes the DNA-protein complex is called **chromatin**.

Prior to mitosis, the DNA in each chromosome is copied. As mitosis begins, chromatin condenses to form a much more compact structure. **Figure 11.7b** shows what chromosomes look like when they have been replicated and condensed. Replicated, condensed chromosomes correspond to the paired threads that were observed in salamander cells by early biologists (see Figure 11.2a).

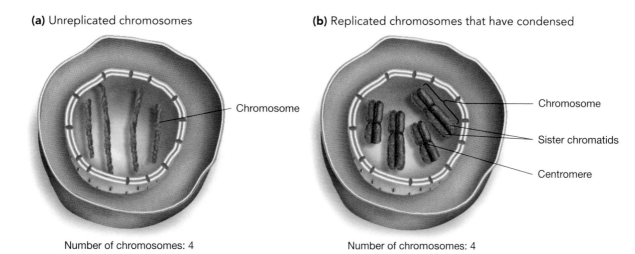

(a) Unreplicated chromosomes

Chromosome

Number of chromosomes: 4

(b) Replicated chromosomes that have condensed

Chromosome

Sister chromatids

Centromere

Number of chromosomes: 4

FIGURE 11.7 A Closer Look at Chromosomes
(a) Chromosomes consist of long strands of DNA that are associated with proteins. **(b)** When chromosomes have replicated, the two strands are called sister chromatids and are joined along their entire length as well as at a structure called the centromere. (Recall that chromosomes are visible in the light microscope only when they have condensed and are undergoing mitosis.)

Each of the DNA copies in a replicated chromosome is called a **chromatid**. The two strands are joined together along their entire length as well as at a specialized region of the chromosome called the **centromere**. Chromatids from the same chromosome are referred to as **sister chromatids**. Once replication is complete, each chromosome consists of two sister chromatids. Sister chromatids represent exact copies of the same genetic information. Each chromatid contains one long DNA double helix.

Recall from Section 11.1 that chromosomes are replicated before mitosis begins. At the start of M phase, then, chromosomes consist of two sister chromatids that are attached to one another at the centromere (**Figure 11.8**). During mitosis, the two sister chromatids separate to form independent chromosomes, and one copy of each chromosome goes to each of the two daughter cells. As a result, each daughter cell receives a copy of the genetic information that is contained in each chromosome. Every daughter cell ends up with exactly the same complement of chromosomes as the parent cell had prior to replication, and thus every daughter cell receives the same genetic information.

Although mitosis is a continuous process, biologists routinely identify several subphases within M phase on the basis of distinctive events that occur. These subphases of mitosis are designated prophase, prometaphase, metaphase, anaphase, and telophase. Some students use the mnemonic device IPPMAT to remind themselves that *i*nterphase is followed by the mitotic subphases *p*rophase, *p*rometaphase, *m*etaphase, *a*naphase, and *t*elophase.

To understand how mitosis proceeds, let's look at each subphase in turn.

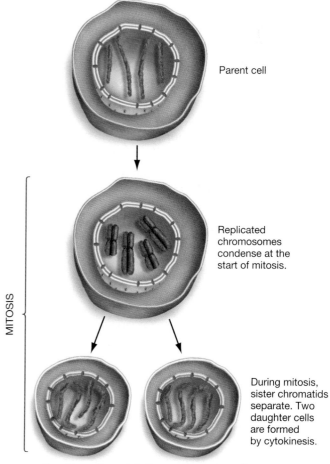

Parent cell

MITOSIS

Replicated chromosomes condense at the start of mitosis.

During mitosis, sister chromatids separate. Two daughter cells are formed by cytokinesis.

Daughter cells contain the same complement of chromosomes as the parent cell.

FIGURE 11.8 An Overview of Mitosis
Chromosomes are replicated prior to mitosis. During mitosis, the replicated chromosomes are partitioned to the two daughter nuclei. In most cases, mitosis is followed by cytokinesis.

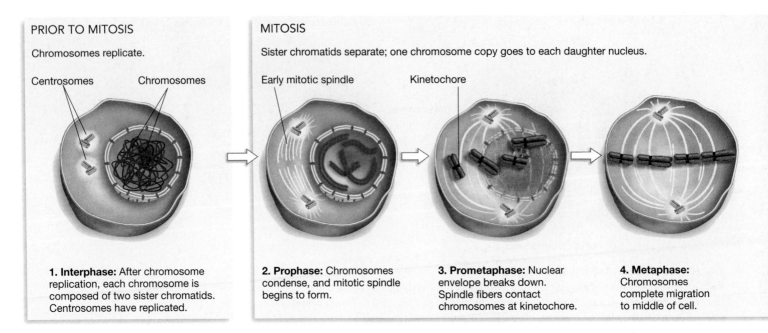

FIGURE 11.9 Mitosis and Cytokinesis

Prophase Mitosis begins with the events of **prophase** ("before-phase"), as shown in **Figure 11.9**. The chromosomes have already replicated during interphase (Figure 11.9, step 1); during prophase (step 2), they condense into compact structures. Chromosomes first become visible in the light microscope during prophase.

In the cytoplasm, prophase is marked by the formation of the mitotic spindle. The **mitotic spindle** is a structure that produces mechanical forces that pull chromosomes into the daughter cells during mitosis. The mitotic spindle consists of an array of microtubules—components of the cytoskeleton that were introduced in Chapter 7. Groups of microtubules attach to the chromosomes and are called **spindle fibers**. In fungi, spindle fibers radiate from a microtubule-organizing center called the *spindle pole body*, while in plants the mitotic spindle forms from microtubule-organizing centers associated with the nuclear envelope. In animals, the microtubule-organizing centers responsible for mitotic spindle formation are called **centrosomes**. Recall from Chapter 7 that animal centrosomes contain structures called **centrioles**. Centrioles are not required for mitotic spindle formation, however, and their function is currently a topic of intense research.

The main point is that some sort of microtubule-organizing center is responsible for the formation of the mitotic spindle. Although the nature of this structure varies among groups, the function of the spindle itself is the same. During prophase, the mitotic spindles begin moving to opposite sides of the cell, or they form on opposite sides.

Prometaphase Once chromosomes have condensed, the nucleolus disappears and the nuclear envelope breaks down. After the nuclear envelope has disintegrated, spindle fibers from each mitotic spindle attach to one of the two sister chromatids of each chromosome. These events occur during **prometaphase** ("before middle-phase"); see Figure 11.9, step 3.

The attachment between the spindle fibers and each chromatid is made at a structure called the **kinetochore**. Kinetochores are located at the centromere region of the chromosome, where sister chromatids are attached to each other.

During prometaphase in animals, the centrosomes continue their movement to opposite poles of the cell. In all groups, the microtubules that are attached to the kinetochores begin moving the chromosomes to the middle of the cell.

Metaphase During **metaphase** ("middle-phase"), animal centrosomes complete their migration to the opposite poles of the cell (Figure 11.9, step 4). In all groups, the kinetochore microtubules finish moving the chromosomes to the middle of the cell. When metaphase is over, the chromosomes are lined up along an imaginary plane called the **metaphase plate**. At this point, the formation of the mitotic spindle is complete. Each chromatid is attached to spindle fibers that run from its kinetochore to one of the two poles of the cell. Each chromosome is held by kinetochore spindle fibers reaching to opposite poles and exerting the same amount of tension or pull. A tug of war is occurring, with kinetochore spindle fibers pulling each chromosome in opposite directions.

Anaphase At the start of **anaphase** ("throughout-phase"), the centromeres that are holding sister chromatids together split (Figure 11.9, step 5). Because they are under tension, sister chromatids are pulled apart equally—with the same amount of force—to create independent chromosomes. The kinetochore spindle fibers then begin to shorten. As they do, motor proteins pull the chromosomes to opposite poles of the

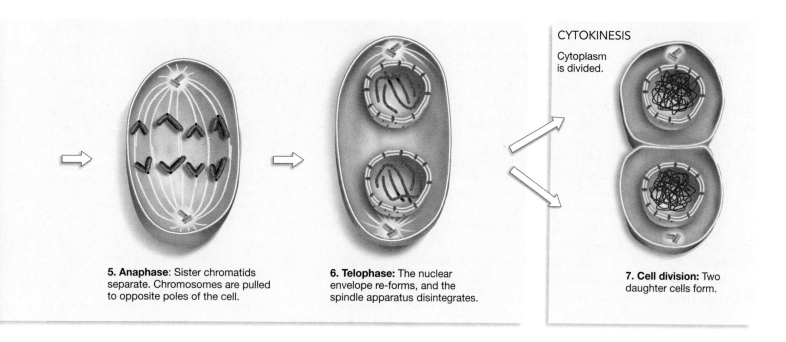

5. Anaphase: Sister chromatids separate. Chromosomes are pulled to opposite poles of the cell.

6. Telophase: The nuclear envelope re-forms, and the spindle apparatus disintegrates.

CYTOKINESIS

Cytoplasm is divided.

7. Cell division: Two daughter cells form.

cell. The two poles of the cell are also pushed away from each other by motor proteins associated with microtubules that are not attached to chromosomes. During anaphase, replicated chromosomes split into two identical sets of unreplicated chromosomes.

The separation of sister chromatids to opposite poles is a critical step in mitosis, because this is the step that ensures that each daughter cell receives the same complement of chromosomes. When anaphase is complete, each pole of the cell has an equivalent and complete collection of chromosomes that are identical to those present in the parent cell prior to chromosome replication.

Telophase   During **telophase** ("end-phase"), a nuclear envelope begins to form around each set of chromosomes (Figure 11.9, step 6). The mitotic spindle disintegrates, and the chromosomes begin to de-condense. Once two independent nuclei have formed, mitosis is complete.

Cytokinesis

Prior to the onset of M phase, all of the cell's organelles besides the nucleus have replicated, and the rest of the cell contents have grown. During cytokinesis (Figure 11.9, step 7), the cytoplasm divides to form two daughter cells, each with its own nucleus and complete set of organelles. Although cytokinesis normally occurs immediately following mitosis, in some cases it does not. As the muscle cells in your upper arm grew and developed, for example, repeated mitoses that were not followed by cytokinesis resulted in the formation of cells containing multiple nuclei. Such cells are referred to as *multinucleate*.

In animals, fungi, and slime molds, cytokinesis begins with the formation of a **cleavage furrow** (**Figure 11.10a**). The furrow appears because a ring of actin and myosin filaments forms just inside the plasma membrane, in a plane that bisects the cell. Recall from Chapter 7 that myosin is a motor protein that binds to actin filaments. When myosin binds to ATP or ADP, part of the protein moves in a way

(a) Cytokinesis in animals

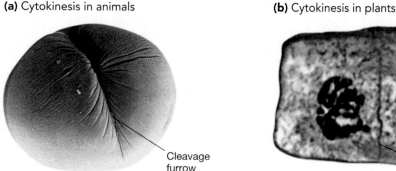

Cleavage furrow

(b) Cytokinesis in plants

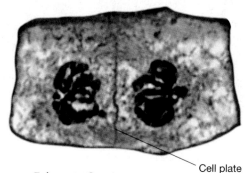

Cell plate

FIGURE 11.10 The Mechanism of Cytokinesis Varies among Eukaryote Groups
(a) In animals, the cytoplasm is divided by a cleavage furrow that pinches the parent cell in two. **(b)** In plants, the cytoplasm is divided by a cell plate that forms in the middle of the parent cell.   [(a) ©Dr. Richard Kessel/Visuals Unlimited]

that causes actin filaments to slide. As myosin moves the ring of actin filaments on the inside of the plasma membrane, the ring tightens. Because the actin-myosin ring is attached to the plasma membrane, the tightening ring pulls the membrane with it. As a result, the plasma membrane pinches inward. The actin and myosin filaments continue to slide past each other, tightening the ring further, until the original membrane is pinched in two and cell division is complete.

In plants, the mechanism of cytokinesis is different. A series of microtubules and other proteins define and organize the region where the new plasma membranes and cell walls will form. Vesicles from the Golgi apparatus are then transported to the middle of the dividing cell, where they form a structure called the **cell plate** (**Figure 11.10b,** page 235). The vesicles carry components of the cell wall and plasma membrane that gradually build up, forming the cell plate and dividing the two daughter cells.

To help you review the major events in cell division, **Table 11.1** summarizes the key structures involved, and **Figure 11.11** shows photographs of cells in interphase and undergoing mitosis and cytokinesis.

TABLE 11.1 Some Structures Involved in Mitosis

Structure	Definition
Chromosome	A structure composed of a DNA molecule and associated proteins.
Chromatin	The material that makes up eukaryotic chromosomes. Consists of a DNA molecule complexed with histone proteins.
Chromatid	One strand of a replicated chromosome.
Sister chromatids	The two strands of a replicated chromosome. When chromosomes are replicated, they consist of two sister chromatids. The genetic material in sister chromatids is identical. When sister chromatids separate during mitosis, they become independent chromosomes.
Centromere	The structure that joins sister chromatids.
Kinetochore	The structure on sister chromatids where spindle fibers attach.
Microtubule organizing center	Any structure that organizes microtubules.
Centrosome	The microtubule organizing center in animals.
Centriole	A paired structure inside animal centrosomes.

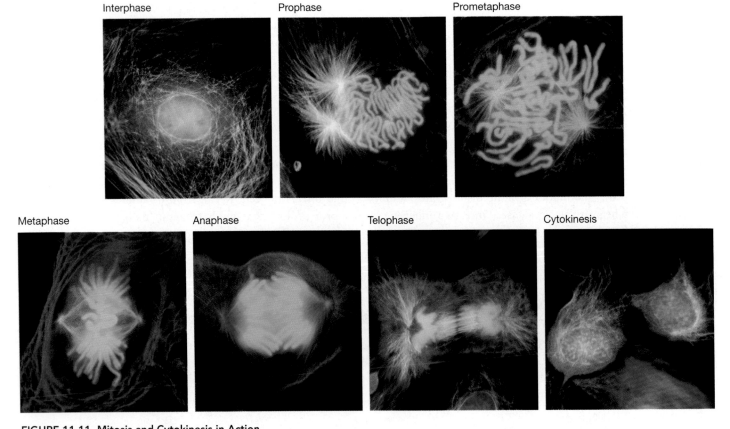

FIGURE 11.11 Mitosis and Cytokinesis in Action
Micrographs showing newt cells in interphase and undergoing mitosis and cytokinesis. Chromosomes are stained blue; microtubules are green, and actin filaments are red. **EXERCISE** Next to each image, describe what is happening. In at least two of the photographs, label the chromosomes, centrosomes, and mitotic spindle. In the metaphase cell, label the metaphase plate.

Once these structures and processes had been described in detail, biologists turned their attention to understanding the molecular mechanisms that are responsible for them. In particular, biologists wanted to know two things about mitosis: (1) How do sister chromatids separate to become independent chromosomes, and (2) how do those chromosomes move to daughter cells? The exact and equal partitioning of genetic material to the two daughter cells is the most fundamental aspect of cell division. How does this process occur?

How Do Chromosomes Move during Mitosis?

To understand how sister chromatids separate and move to daughter cells, biologists have tried to understand how the mitotic spindle functions. Do spindle microtubules act as railroad tracks, the way they do in vesicle transport? Is some sort of motor protein involved? And what is the nature of the kinetochore, where the chromosome and microtubules are joined?

Mitotic Spindle Forces Spindle fibers are composed of microtubules. Recall from Chapter 7 that microtubules are composed of α-tubulin and β-tubulin dimers, that the length of a microtubule is determined by the number of tubulin dimers it contains, and that microtubules are asymmetric—they have a plus end and a minus end. These observations suggest a mechanism for the movement of chromosomes during anaphase: Is the spindle microtubule shortening due to a loss of tubulin dimers from one end? Alternatively, it is possible that microtubules slide past each other, much as actin and myosin filaments do during cytokinesis in animal cells, and that this sliding action pulls chromosomes toward the poles of the cell.

To test these hypotheses, biologists introduced fluorescently labeled tubulin subunits into prophase or metaphase cells. This treatment made the entire mitotic spindle visible. (See step 1 in the "Experimental setup" section of **Figure 11.12.**) Then, once anaphase had begun, the researchers marked a region of the spindle by irradiating it with a bar-shaped beam of laser light. The laser quenched the fluorescence in the exposed region, making that section of the spindle dark, or "photobleached," although it was still functional (see step 2 of the "Experimental setup"). The "Results" section of Figure 11.12 shows that as anaphase progressed, the photobleached region remained stationary while the chromosomes moved toward this region.

To explain this result, the biologists concluded that the kinetochore microtubules must remain stationary during anaphase. The hypothesis suggested that, instead of sliding past each other as actin fibers do, microtubules shorten at the kinetochore because tubulin subunits are lost from the kineto-

Question: How do microtubules shorten to pull sister chromatids apart at anaphase?

Hypothesis: Microtubules shorten at the kinetochore.

Alternate hypothesis: Microtubules slide past each other like actin filaments.

Experimental setup:

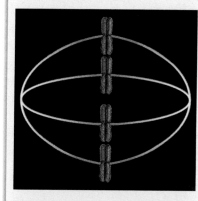

1. Use fluorescent labels to make the metaphase chromosomes fluoresce blue and the microtubules fluoresce yellow.

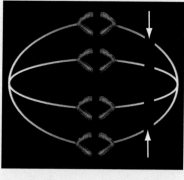

2. At the start of anaphase, photobleach a section of microtubules to mark them without changing their function.

Prediction: The photobleached section will not move. Instead, microtubules on one side of the photobleached section will shorten.

Prediction of alternate hypothesis: The photobleached section will move.

Results:

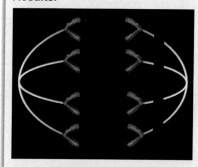

By late anaphase, the distance between chromosomes and photobleached section lessened.

Conclusion: Microtubules shorten near the chromosome.

FIGURE 11.12 During Anaphase, Microtubules Shorten at the Kinetochore

EXERCISE Add drawings to show the outcomes you would expect if microtubules shortened at the end opposite the chromosome or if microtubules slide past each other.

chore ends (**Figure 11.13a**). As kinetochore microtubules shorten at their minus end, the chromosomes are pulled along. How does this happen?

A Kinetochore Motor   Recall that kinetochores are structures located at the centromere of replicated chromosomes and that microtubules attach to each sister chromatid at its kinetochore. If microtubules shorten at the kinetochore end, how does the chromosome remain attached to the microtubule? Although the answer is not yet clear, the structure and function of the kinetochore is gradually becoming better understood.

Recent research has shown that the kinetochore contains dyneins and other motor proteins that "walk" chromosomes down microtubules—from their plus ends near the kinetochore toward their minus ends at the spindle (**Figure 11.13b**). This observation suggests that the mechanism of chromosome movement is reminiscent of the way kinesin walks down microtubules during vesicle transport. As explained in Chapter 7, motor proteins such as kinesin and dynein change shape when they are phosphorylated or bind to ATP. These shape changes can produce movement. Biologists hypothesize that, during mitosis, dyneins and other kinetochore motor proteins detach near the chromosome and then reattach to the kinetochore microtubule farther along its length after they move. As this detach-move-reattach cycle repeats, the microtubule shortens and the chromosome is pulled to one end of the mitotic spindle.

(a) What is happening at the kinetochore?

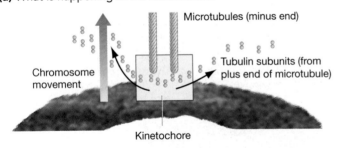

(b) Kinetochores contain motor proteins.

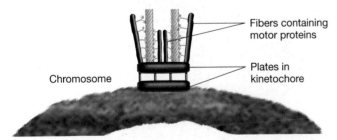

FIGURE 11.13 How Do Microtubules Move Chromosomes during Mitosis?
(a) According to one hypothesis, microtubules shorten as tubulin subunits split off at the kinetochore. **(b)** The kinetochore consists of an inner plate, an outer plate, and associated fibers that contain motor proteins. The fibers are thought to act as "legs" that walk the chromosomes down the length of the kinetochore microtubules.

Efforts to understand spindle structure and movement bring us to the frontier of research on mitosis. Having explored how the process occurs, let's focus on how it is controlled. When does a cell divide, and when does it stop dividing? How is cell division regulated? These questions are fundamental. When cell division occurs in an uncontrolled manner, cancerous tumors can form.

✓ CHECK YOUR UNDERSTANDING

After chromosomes replicate, mitosis distributes one copy of each chromosome to each daughter cell. As a result, mitosis and cytokinesis lead to the production of cells with the same genetic material as that of the parent cell. You should be able to (1) draw a replicated chromosome and an unreplicated chromosome and label the sister chromatids and the centromere, and (2) diagram what happens to chromosomes and the nuclear envelope during prophase, prometaphase, metaphase, anaphase, and telophase.

11.3   Control of the Cell Cycle

Although the events of mitosis are virtually identical in all eukaryotes, other aspects of the cell cycle can be extremely variable. For example, the length of the cell cycle can vary enormously among different cell types, even in the same individual. In humans, intestinal cells routinely divide more than twice a day to renew tissue that is lost during digestion; mature human nerve and muscle cells do not divide at all. Most of these differences are due to variation in the length of the G_1 phase. In rapidly dividing cells, G_1 is essentially eliminated. Most nondividing cells, in contrast, are permanently stuck in G_1. Researchers refer to this arrested stage as the G_0 state, or simply "G zero."

A cell's division rate can also vary in response to changes in conditions. For example, human liver cells normally divide about once per year. But if part of the liver is damaged or lost, the remaining cells divide every one or two days until repair is accomplished. Cells of unicellular organisms such as yeasts, bacteria, or archaea divide rapidly only if the environment is rich in nutrients; otherwise, they enter a quiescent (inactive) state.

To explain the existence of so much variability, biologists hypothesized that the cell cycle must be regulated in some way and that regulation varies among cells and organisms. Understanding how the cell cycle is controlled is now the most prominent issue in research on cell division—partly because defects in control can lead to uncontrolled, cancerous growth. What evidence first suggested that regulatory molecules control the cell cycle?

The Discovery of Cell-Cycle Regulatory Molecules
The first solid evidence for cell-cycle control molecules came to light in 1970, when researchers published the results of

experiments on fusing pairs of cultured mammalian cells. In the presence of certain chemicals, viruses, or an electric shock, the membranes of two cells can be made to fuse. The hybrid cell that results has two nuclei.

How did cell fusion experiments point to the existence of cell-cycle control molecules? When investigators fused cells that were in different stages of the cell cycle, certain nuclei changed phases. For example, when a cell in M phase was fused with one in interphase, the nucleus of the interphase cell initiated M phase (**Figure 11.14a**). The biologists hypothesized that the cytoplasm of M-phase cells contains a regulatory molecule that induces interphase cells to enter M phase.

This hypothesis was confirmed by experiments on the South African claw-toed frog, *Xenopus laevis*. As the eggs of these frogs mature, they change from a cell called an **oocyte**, which is arrested in a phase similar to G_2, to a mature egg that has entered M phase. The eggs are attractive to study, partly because they are more than 1 mm in diameter. Their large size makes it

relatively easy to purify large amounts of cytoplasm and to use microsyringes to inject the eggs with cytoplasm from eggs in different stages of development. When biologists purified cytoplasm from M-phase frog eggs and injected it into the cytoplasm of frog oocytes arrested in the G_2-like phase, the immature oocytes entered M phase (**Figure 11.14b**). But when cytoplasm from interphase cells was injected into G_2 oocytes, the cells remained in the G_2-like phase. The researchers concluded that the cytoplasm of M-phase cells—but not the cytoplasm of interphase cells—contains a factor that drives immature oocytes into M phase to complete their maturation.

This factor was eventually purified and is now called **mitosis-promoting factor**, or **MPF**. Subsequent experiments showed that MPF induces mitosis in all eukaryotes. For example, injecting M-phase cytoplasm from mammalian cells into immature frog eggs results in egg maturation. Human MPF can trigger mitosis in yeast cells. The molecule appears to be a general signal that says "Start mitosis." How does it work?

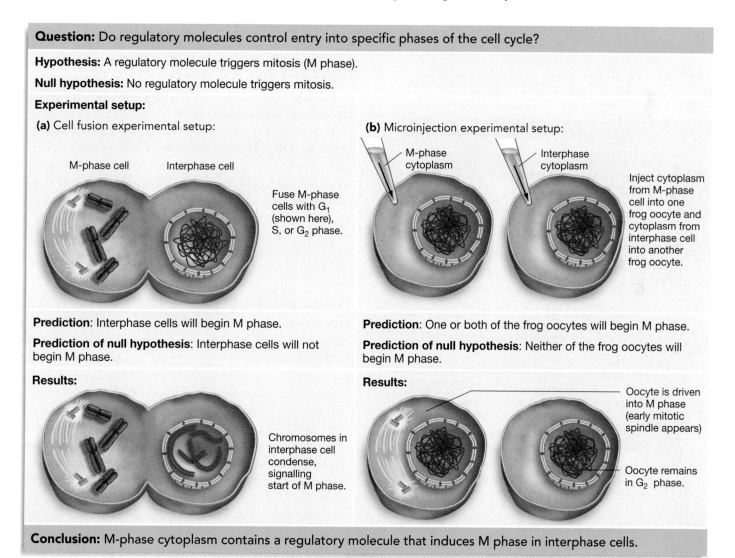

Question: Do regulatory molecules control entry into specific phases of the cell cycle?

Hypothesis: A regulatory molecule triggers mitosis (M phase).

Null hypothesis: No regulatory molecule triggers mitosis.

Experimental setup:

(a) Cell fusion experimental setup:

M-phase cell   Interphase cell

Fuse M-phase cells with G_1 (shown here), S, or G_2 phase.

(b) Microinjection experimental setup:

M-phase cytoplasm   Interphase cytoplasm

Inject cytoplasm from M-phase cell into one frog oocyte and cytoplasm from interphase cell into another frog oocyte.

Prediction: Interphase cells will begin M phase.

Prediction of null hypothesis: Interphase cells will not begin M phase.

Results:

Chromosomes in interphase cell condense, signalling start of M phase.

Prediction: One or both of the frog oocytes will begin M phase.

Prediction of null hypothesis: Neither of the frog oocytes will begin M phase.

Results:

Oocyte is driven into M phase (early mitotic spindle appears)

Oocyte remains in G_2 phase.

Conclusion: M-phase cytoplasm contains a regulatory molecule that induces M phase in interphase cells.

FIGURE 11.14 Experimental Evidence for Cell-Cycle Control Molecules
(a) When M-phase cells are fused with cells in G_1, S, or G_2 phase, the interphase chromosomes condense and begin M phase. **(b)** Microinjection experiments supported the hypothesis that a regulatory molecule induces M phase.

MPF Contains a Protein Kinase and a Cyclin

Once MPF had been isolated and purified, researchers found that it is made up of two distinct polypeptide subunits. One of the components is a **protein kinase**—an enzyme that catalyzes the transfer of a phosphate group from ATP to a target protein. Recall from Chapter 3 that proteins can be activated or inactivated by phosphorylation. Because the addition of a phosphate group changes the target protein's shape and activity, protein kinases frequently act as regulatory elements in the cell.

These observations suggested that MPF acts by phosphorylating a protein that triggers the onset of mitosis. But research showed that the concentration of MPF protein kinase is more or less constant throughout the cell cycle. How can MPF trigger mitosis if the protein kinase subunit is always present?

The answer to this question lies in the second MPF subunit, which belongs to a family of proteins called the **cyclins**. Cyclins got their name because their concentrations fluctuate throughout the cell cycle. As **Figure 11.15a** shows, the cyclin associated with MPF builds up in concentration during interphase and peaks during M phase. This increase in concentration is important because the protein kinase subunit in MPF can be active only when it is bound to the cyclin subunit. As a result, the protein kinase subunit of MPF is called a **cyclin-dependent kinase**, or **Cdk**. MPF is a dimer that consists of a cyclin subunit and a cyclin-dependent kinase subunit.

According to Figure 11.15a, the number of complete MPF dimers builds up steadily during interphase. Why doesn't this increasing concentration of MPF trigger the onset of M phase? The answer is that MPF's cyclin unit is phosphorylated when it is initially synthesized. Phosphorylation of the protein changes its conformation in a way that renders the protein inactive. Late in G_2 phase, however, enzymes catalyze the dephosphorylation of cyclin. These reactions activate MPF.

Once MPF is activated, it phosphorylates the proteins listed in **Figure 11.15b**. Chromosomal proteins activated by MPF cause chromosomes to condense into the threads visible during M phase. Microtubule-associated proteins phosphorylated by MPF may be involved in assembling the mitotic spindle apparatus. In this way, MPF triggers the onset of M phase.

MPF also activates an enzyme complex that promotes the degradation of MPF's own cyclin subunit, however. MPF causes its own destruction. This is an example of negative feedback—similar to the mechanisms of feedback inhibition introduced in Chapter 9. In response to MPF activity, then, the concentration of cyclin declines rapidly. Slowly, it builds up again during interphase. In this way, an oscillation in cyclin concentration is set up.

The dramatic oscillation in cyclin concentration and activation acts as a clock that drives the ordered events of the cell cycle. These events are happening millions of times each day in locations throughout your body. Over a 24-hour period, you swallow millions of cheek cells. Millions of additional cells are lost from your intestinal lining each day and leave your body as waste. To replace them, cells in your cheek and intestinal tissue are constantly making and degrading cyclin and pushing themselves through the cell cycle.

(a) Cyclin concentration regulates MPF concentration.

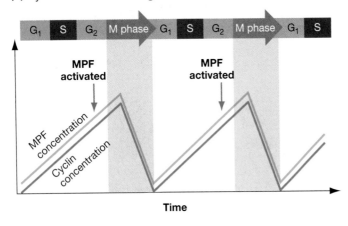

(b) Activated MPF (with unphosphorylated cyclin subunit) has an array of effects.

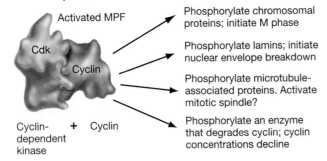

Activated MPF

Cdk

Cyclin

Cyclin-dependent kinase + Cyclin

→ Phosphorylate chromosomal proteins; initiate M phase

→ Phosphorylate lamins; initiate nuclear envelope breakdown

→ Phosphorylate microtubule-associated proteins. Activate mitotic spindle?

→ Phosphorylate an enzyme that degrades cyclin; cyclin concentrations decline

FIGURE 11.15 M-Phase Promoting Factor Is Created When a Cyclin Binds to a Protein Kinase
(a) Cyclin concentrations cycle in dividing cells, reaching a peak in M phase. Cyclin binds to a protein kinase, creating MPF.
(b) MPF activates proteins that initiate M phase, as well as proteins that degrade cyclin.

Cell-Cycle Checkpoints

MPF is only one of many protein complexes involved in regulating the cell cycle. For example, a different cyclin and protein kinase are involved in triggering the passage from G_1 phase into S phase, and several regulatory proteins are involved in maintaining the G_0 state of quiescent cells. A complex array of regulatory molecules is involved in either holding cells in particular stages or in stimulating passage to the next phase.

To make sense of these observations, Leland Hartwell and Ted Weinert introduced the concept of a cell-cycle checkpoint. A **cell-cycle checkpoint** is a critical point in the cell cycle that is regulated. Hartwell and Weinert identified checkpoints by analyzing yeast cells with defects in the cell cycle. The defective cells lacked a specific checkpoint and, as a result, kept dividing under culture conditions when normal cells stopped growing. In the body, cells that keep growing in this way form a mass of cells called a **tumor**.

As **Figure 11.16** indicates, biologists have since obtained evidence of three distinct checkpoints. In effect, interactions

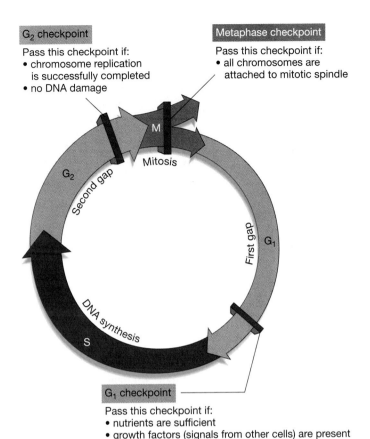

G₂ checkpoint

Pass this checkpoint if:
• chromosome replication is successfully completed
• no DNA damage

Metaphase checkpoint

Pass this checkpoint if:
• all chromosomes are attached to mitotic spindle

G₁ checkpoint

Pass this checkpoint if:
• nutrients are sufficient
• growth factors (signals from other cells) are present
• cell size is adequate
• DNA is undamaged
Note: Mature cells do not pass this checkpoint (they enter G₀ state).

FIGURE 11.16 The Three Cell-Cycle Checkpoints

QUESTION Why is it advantageous for a cell to arrest at the M-phase checkpoint if not all chromosomes are attached to the mitotic spindle?

among regulatory molecules at each checkpoint allow a cell to "decide" whether to proceed with division. If these regulatory molecules are defective, the checkpoint may fail. As a result, cells may start growing in an uncontrolled fashion.

The first cell-cycle checkpoint occurs late in G₁. For most cells, this checkpoint is the most important in establishing whether the cell will continue through the cycle and divide. What determines whether a cell passes the G₁ checkpoint?

- Because a cell must reach a certain size before its daughter cells will be large enough to function normally, biologists hypothesize that some mechanism exists to arrest the cell cycle if the cell is too small.

- Unicellular organisms arrest at the G₁ checkpoint if nutrient conditions are poor.

- Cells in multicellular organisms pass through the G₁ checkpoint in response to signaling molecules from other cells, or what are termed *social signals.*

- If DNA is physically damaged, the protein **p53** activates genes that either stop the cell cycle until the damage can be repaired or lead to the cell's programmed, controlled de-

struction—a phenomenon known as **apoptosis**. In this way, p53 acts as a brake on the cell cycle. Because cancer can develop if "brake" molecules such as p53 are defective, these regulatory proteins are called **tumor suppressors.**

The general message here is that the components of the G₁ checkpoint have the same function: ensuring that the cell is healthy and should proceed to replicating its DNA and dividing.

The second checkpoint occurs after S phase, at the boundary between the G₂ and M phases. Specifically, cells appear to arrest at the G₂ checkpoint if chromosome replication has not been completed properly or if DNA is damaged. Because MPF is the key signal that triggers the onset of M phase, investigators were not surprised to find that it is involved in the G₂ checkpoint. Although much remains to be learned, data suggest that if DNA is damaged or if chromosomes are not replicated correctly, the dephosphorylation and activation of MPF are blocked. When MPF is not activated, cells remain in G₂ phase. Some data indicate that cells at this checkpoint may also respond to signals from other cells and to internal signals relating to their size.

The final checkpoint occurs during mitosis. If not all chromosomes are properly attached to the mitotic spindle, M phase arrests at metaphase. Specifically, anaphase is delayed until all kinetochores are properly attached to mitotic spindle fibers. If this checkpoint did not exist, some chromosomes might not separate properly, and daughter cells would receive an incorrect number of chromosomes during anaphase. Because they would receive too much or too little genetic material, the effect on the daughter cells could be disastrous.

To summarize, the three cell-cycle checkpoints have the same purpose: They prevent the division of cells that are damaged or that have other problems, and they prevent the growth of mature cells that should not grow any more. If one of the checkpoints fails, the affected cells may begin growing in an uncontrolled fashion. For the organism as a whole, the consequences of uncontrolled cell division are dire: cancer.

✔ CHECK YOUR UNDERSTANDING

The cell cycle consists of four carefully controlled phases. You should be able to diagram the cell cycle and indicate (1) the locations of cell-cycle checkpoints, (2) the level of Cdk at each stage, (3) the level of cyclin at each stage, and (4) the levels of MPF activity during each stage.

11.4 Cancer: Out-of-Control Cell Division

Few diseases inspire more fear than cancer. The fear springs from the difficulty of treating many forms of cancer, the potentially fatal nature of many cancers, and their frequency. Most of us know someone who has had some form of cancer, and most of us know someone who has died from the disease. According to the American Cancer Society, 50 percent of American men

(a) Cancer rates in males have changed over time.

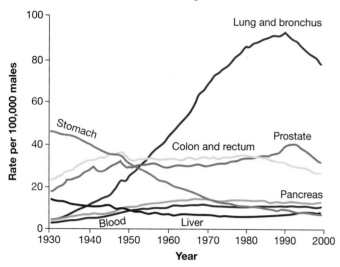

(b) Cancer rates in females have changed over time.

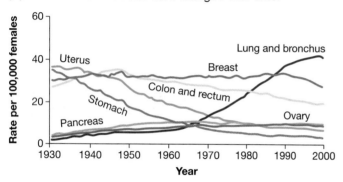

FIGURE 11.17 Changes in Cancer Rates over Time
Changes in the frequencies of various types of cancer in **(a)** men and **(b)** women in the United States.

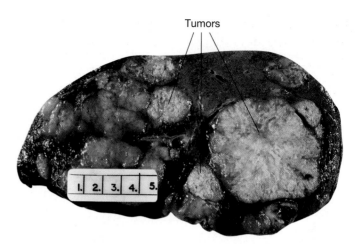

FIGURE 11.18 Cancers May Spread to New Locations in the Body
The liver of a human who died of cancer. A large number of tumors have formed due to the migration of cancerous cells from the original tumor, which occurred in the patient's colon. The numbers provide a scale in centimeters.

and 33 percent of American women will develop cancer during their lifetime (**Figure 11.17**). In the United States, one in four of all deaths are from cancer. It is the second leading cause of death, exceeded only by heart disease.

Humans suffer from at least 200 types of cancer. Stated another way, cancer is not a single illness but a complex family of diseases that affect an array of organs, including the breast, colon, brain, lung, and skin. In addition, several types of cancer can affect the same organ. Skin cancers, for example, come in multiple forms. Some are relatively easy to treat; others are often fatal. Although cancers vary in time of onset, growth rate, seriousness, and cause, all cancers have a unifying feature: They arise from cells in which cell-cycle checkpoints have failed.

To get an understanding of why cancer occurs, let's review general characteristics of the disease and then delve into the details of why regulatory mechanisms become defective.

Properties of Cancer Cells

When even a single cell in a multicellular organism begins to divide in an uncontrolled fashion, a mass of cells called a

tumor results. For example, most cells in the adult human brain do not divide. But if a single abnormal brain cell begins unrestrained division, the growing tumor that results may disrupt the brain's function. What can be done? If the tumor can be removed without damaging the affected organ, a cure might be achieved. This is why surgical removal of the tumor is usually the first step in the treatment of a cancer. Often, though, surgery does not cure cancer. Why?

In addition to growing quickly, cancer cells are *invasive*—that is, able to spread throughout the body via the bloodstream or the lymphatic vessels (introduced in Chapter 49). Invasiveness is a defining feature of a **malignant tumor**—one that is cancerous. Masses of noninvasive cells are noncancerous and form **benign tumors**, such as a wart. Noncancerous cells may become malignant, however, if they gain the ability to detach from the original tumor and invade other tissues. By spreading from the primary tumor site where uncontrolled growth originated, cancer cells can establish secondary tumors elsewhere in the body (**Figure 11.18**). This process is called **metastasis**. If metastasis has occurred by the time the original tumor is detected, then secondary tumors have begun to form and surgical removal of the primary tumor will not lead to a cure. As a result, the disease can be very difficult to treat. This is why early detection as the key to treating cancer most effectively.

Cancer Involves Loss of Cell-Cycle Control

If cancer is caused by uncontrolled cell growth, what is the molecular nature of the disease? Recall that when many cells mature, they enter the G_0 phase—meaning their cell cycle is arrested at the G_1 checkpoint. In contrast, cells that do pass through the G_1 checkpoint are irreversibly committed to replicating their DNA and entering G_2. Based on this observation, biologists hypothesized that many or even most types of cancer involve defects in the G_1 checkpoint. To understand the molecular nature of the disease, then, researchers focused on understanding the normal mechanisms that operate at that checkpoint. In this way, cancer research and research on the normal cell cycle have become two sides of the same coin.

Social Control In unicellular organisms, passage through the G_1 checkpoint is thought to depend primarily on cell size and the availability of nutrients. If nutrients are plentiful, cells pass through the checkpoint and grow rapidly. In multicellular organisms, however, cells receive a constant supply of adequate nutrients via the bloodstream. Because nutrients generally are not limiting, most cells in multicellular organisms divide in response to some other type of signal. Because these signals arrive from other cells, biologists refer to *social control* over cell division. The general idea is that individual cells should be allowed to divide only when their growth is in the best interests of the organism as a whole.

The most important signals involved in social control of the cell cycle are called growth factors. **Growth factors** are polypeptides or small proteins discovered in the course of working out techniques for growing cells in culture. When researchers isolated mammalian cells in culture and provided them with adequate nutrients, the cells arrested in G_1 phase. They began to grow only when biologists added **serum**—the liquid that remains after blood clots and the blood cells have been removed. Some component of serum allowed cells to pass through the G_1 checkpoint. What was it?

In 1974 biologists succeeded in identifying one of the serum components responsible for stimulating cell division. The component was a protein called **platelet-derived growth factor (PDGF)**. As its name implies, PDGF is released by blood components called **platelets**, which promote blood clotting at wound sites. PDGF that is secreted by platelets binds to receptor tyrosine kinases in the plasma membranes of cells in the area. When these cells receive the growth signal, they are stimulated to divide. The increased cell numbers facilitate wound healing.

Researchers subsequently found that PDGF is produced by an array of cell types. Investigators also succeeded in isolating and identifying a diverse array of other growth factors. For different types of cells to grow in culture, different combinations of growth factors must be supplied. Based on this result, biologists infer that different types of cells in an intact multicellular organism are controlled by different combinations of growth factors. Cancer cells are another story, however. Cancerous cells can often be cultured successfully without externally supplied growth factors. This observation suggests that the normal social controls on the G_1 checkpoint have broken down in cancer cells.

Social Controls and Cell-Cycle Checkpoints How do growth factors stimulate division in normal cells? Recall from Section 11.3 that for a cell to begin mitosis or to pass through the G_1 checkpoint, a specific cyclin-Cdk complex must be activated. Because activation depends on the presence of a high concentration of the cyclin involved, investigators hypothesized that growth factors trigger cyclin synthesis. This hypothesis turned out to be correct. As **Figure 11.19** shows, the arrival of growth factors initiates cell division by stimulating the production of cyclins.

The link between growth factors and cyclin synthesis also turned out to be important in cancer biology. In some human cancers, the cyclin that is active at the G_1 checkpoint is overproduced. Cyclin overproduction can result from the presence of excessive amounts of growth factors or from cyclin production in the absence of growth signals.

When the G_1 cyclin is overproduced, the Cdk to which it binds phosphorylates target proteins continuously. In many cancers, the key target protein is the retinoblastoma (Rb) protein. This protein was discovered by researchers interested in a childhood cancer called retinoblastoma, which occurs in about 1 in 20,000 children. The disease is characterized by the appearance of malignant tumors in the light-sensing tissue, or retina, of the eye.

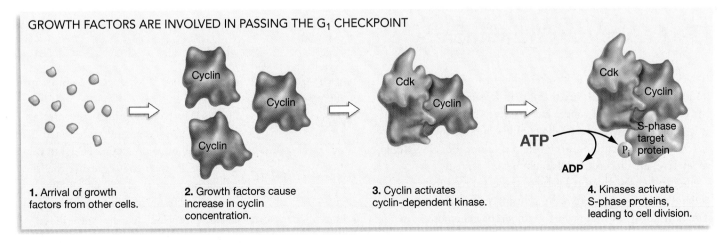

GROWTH FACTORS ARE INVOLVED IN PASSING THE G_1 CHECKPOINT

1. Arrival of growth factors from other cells.

2. Growth factors cause increase in cyclin concentration.

3. Cyclin activates cyclin-dependent kinase.

4. Kinases activate S-phase proteins, leading to cell division.

FIGURE 11.19 How Is the G_1 Checkpoint Controlled?

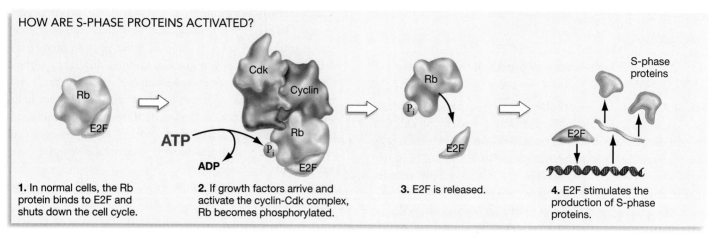

HOW ARE S-PHASE PROTEINS ACTIVATED?

1. In normal cells, the Rb protein binds to E2F and shuts down the cell cycle.

2. If growth factors arrive and activate the cyclin-Cdk complex, Rb becomes phosphorylated.

3. E2F is released.

4. E2F stimulates the production of S-phase proteins.

FIGURE 11.20 The Role of Rb Protein in Cell Division
QUESTION What would happen if the Rb protein were defective or not produced at all? What would happen if cyclin concentrations were high all the time, irrespective of the arrival of growth factors?

The role of Rb protein in controlling normal cell division is diagrammed in **Figure 11.20**. In normal, nondividing cells, Rb binds to a protein called E2F and inhibits it—shutting down the cell cycle. In this way, Rb acts as a stop signal for cell growth. But if a growth factor stimulates the production of cyclin and activates the cyclin-Cdk complex, the Rb protein becomes phosphorylated. In response, it no longer binds to E2F. Once it is free of Rb, E2F begins stimulating the production of molecules needed in S phase.

In effect, a cell's commitment to dividing depends on whether Rb stays bound to the E2F protein. As long as Rb binds to E2F, the cell remains in G_1 phase. Rb acts as a brake on the cell cycle, much as p53 does. Rb is also classified as a tumor suppressor, because its normal function in the cell is to shut down the cell cycle. But if Rb is defective, the E2F protein is always active and cell division is uncontrolled. This is exactly what happens in several types of cancer. In retinoblastoma and some breast, bladder, and lung cancers, the Rb protein is nonfunctional in the cancerous cells or not produced at all.

To summarize, overproduction of cyclins or defects in Rb itself can lead to the inactivation of Rb as a tumor suppressor.

The result is failure of the G_1 checkpoint, uncontrolled cell growth, and cancer.

Cancer Is a Family of Diseases   One of the broadest messages to come out of research on the Rb protein is that a wide variety of defects can lead to the failure of the G_1 checkpoint and the onset of cancer. In addition, researchers now realize that cancer is seldom due to a single defect. Most cancers develop only after a number of genetic errors combine to break cell-cycle control and induce uncontrolled growth and metastasis. Each type of cancer is due to a unique combination of errors. Stated another way, cancer can be caused by hundreds if not thousands of different defects.

Because cancer is actually a family of diseases with a complex and highly variable molecular basis, there will be no "magic bullet," or single therapy that cures all forms of the illness. Still, recent progress in understanding the cell cycle and the molecular basis of cancer has been dramatic, and cancer prevention and early detection programs are increasingly effective. The prognosis for many cancer patients is remarkably better now than it was even a few years ago.

ESSAY Cancer Chemotherapy

If a malignant tumor is detected before metastasis occurs, doctors may be able to cure the patient by surgically removing the growth. But if the cancer has begun to spread by the time it is detected, stray cancer cells could be hiding anywhere, regardless of the location of the original tumor. For the patient to survive, all cancerous cells must be eradicated.

In these cases, doctors often propose surgery followed by **chemotherapy**—treatment with anticancer chemicals or drugs. You may know someone who has undergone chemotherapy for cancer. If so, you've witnessed the side effects: weakness, acute nausea, weight loss, and hair loss. Why does chemotherapy make people so sick?

Why does chemotherapy make people so sick?

Most of the drugs currently approved for chemotherapy have the same mode of action: They kill rapidly dividing cells or cause them to arrest in the G_1 phase of the cell cycle. To heighten the probability of killing all stray cancer cells, physicians usually prescribe high doses of these drugs. They are applied to the entire body, often through injection into the bloodstream. Besides killing cancer cells, though, anticancer drugs also kill normal cells that divide rapidly. These include blood cells, intestinal cells, and hair follicles. As a result, chemotherapy leads to low blood-cell counts, loss of appetite, and loss of hair, among other symptoms.

Physicians who prescribe chemotherapy must balance the probability of eradicating all cancer cells with the patient's ability to tolerate the side effects of treatment. One approach to achieving this balance is to mitigate the most dangerous side effects, such as low counts of the white blood cells that battle infections. Patients with low blood-cell counts may be given growth factors that stimulate the production of white blood cells. These patients may also be given antibiotics to help them fight infections.

Another approach in chemotherapy is to prevent side effects altogether by targeting the drugs more precisely. If the cancer has not spread far from the original tumor site, anticancer drugs may be applied to just a limited region of the body. This is rarely possible, however, so many researchers are focusing on an alternative strategy—attaching chemotherapy drugs to proteins that bind to specific membrane proteins on the surfaces of cancer cells. One idea is to package drugs inside liposomes that are coated with a protein that binds to cancer cells. The hope is that the liposomes will fuse with the plasma membrane of the cancer cells and deliver the drugs directly to the malignant cells. If this research is successful, chemotherapy may eventually become largely free of side effects and patient suffering.

CHAPTER REVIEW

Summary of Key Concepts

In the late nineteenth century, observations of growing embryos revealed that cells reproduce by division of preexisting cells. Mitosis is the series of events that results in a complete copy of the chromosomes being distributed to daughter nuclei. Cytokinesis is the cell division that usually follows mitosis.

▪ **After chromosomes are copied, mitosis distributes one chromosome copy to each of two daughter cells. Mitosis and cytokinesis produce two cells that are identical to the parent cell.**

Most unicellular eukaryotes and some multicellular eukaryotes reproduce by mitosis followed by cytokinesis. Mitotic divisions are also responsible for building the bodies of multicellular organisms from a single-celled embryo.

Mitosis can be described as a sequence of five phases:

Prophase   Chromosomes condense, and the mitotic spindle apparatus begins to form.

Prometaphase   The nuclear envelope disintegrates, and spindle fibers make contact with chromosomes.

Metaphase   Spindle fibers move chromosomes to the metaphase plate, an imaginary plane.

Anaphase   Spindle fibers pull sister chromatids apart.

Telophase   Chromosomes are pulled to opposite poles of the cell, and a nuclear envelope forms around each set.

In most cells, mitosis is followed immediately by division of all cell contents, or cytokinesis.

Web Tutorial 11.1   The Phases of Mitosis

▪ **Over their life span, cells go through a life cycle that consists of four carefully controlled phases.**

Dividing cells alternate between the dividing phase, called M phase, and a nondividing phase known as interphase. This alternation describes the cell cycle—the ordered sequence of events that occur as a cell duplicates its chromosomes and divides.

Chromosome synthesis and replication, or S phase, occurs during interphase. S phase and M phase are separated by gap phases called G_1 and G_2. Cell growth and replication of nonnuclear cell components occur during the gap phases. The order of cell-cycle phases is $G_1 \rightarrow S \rightarrow G_2 \rightarrow M \rightarrow G_1$ and so forth.

Progression through the cell cycle is regulated at three checkpoints. Passage through the G_1 checkpoint represents an irreversible commitment to divide and is contingent on cell size, nutrient availability, lack of DNA damage, and/or growth signals from other cells. At the G_2 checkpoint, progression through the cycle is delayed until chromosome replication has been successfully completed. At the metaphase checkpoint during M phase, anaphase is delayed until all chromosomes are correctly attached to the mitotic spindle.

An array of different molecules regulates progression through the cell-cycle checkpoints. The most important of these regulatory molecules are cyclin-dependent kinases (Cdks) and cyclins, which are found in all eukaryotic cells. Active Cdks are enzymes that trigger progress through a checkpoint by phosphorylating important target proteins. Cyclins are proteins whose concentrations oscillate during the cell cycle, regulating the activity of Cdks. In multicellular organisms, cyclin concentrations are partially controlled by growth factors from other cells. As a result, the G_1 checkpoint is said to be under social control.

Web Tutorial 11.2   The Cell Cycle

▪ **Uncontrolled cell growth leads to cancer. Different types of cancer result from different types of defects in control over the cell cycle.**

Cancer is a common disease that is characterized by loss of control at the G_1 checkpoint, resulting in cells that divide in an uncontrolled fashion. Some cancer cells also have the ability to spread throughout the body, which makes treatment difficult. In normal cells, growth-factor signals are required to activate G_1 cyclin-Cdk complexes and trigger division. An important target of activated G_1 cyclin-Cdk complexes is the retinoblastoma (Rb) protein, which regulates the production of enzymes and other proteins required for S phase. Defects in G_1 cyclin and Rb protein are common in human cancers. Cancerous cells usually have defects in several genes involved in cell-cycle control, however.

Questions

Content Review

1. Which statement about the daughter cells of mitosis is correct?
   a. They differ genetically from one another and from the parent cell.
   b. They are genetically identical to one another and to the parent cell.
   c. They are genetically identical to one another but are different from the parent cell.
   d. Only one of the two daughter cells is genetically identical to the parent cell.

2. Progression through the cell cycle is regulated by oscillations in the concentration of which type of molecule?
   a. actin
   b. tubulin
   c. cyclin-dependent kinases
   d. cyclins

3. According to the data in Figure 11.4, the first labeled mitotic cells appear about 4 hours after the labeling period ends. From these data, researchers concluded that G_2 lasted about 4 hours. Why?
   a. The lengths of the total cell cycle and the G_1, S, and M phases were known, so the length of G_2 could be found by subtraction.
   b. It confirmed other data indicating that G_2 lasts about 4 hours in cultured cells of this type.
   c. Cyclins were labeled, so their concentration had to build up for 4 hours.
   d. Labeled cells are in S phase, so 4 hours passed between the end of S phase and the onset of M phase.

4. What major events occur during anaphase of mitosis?
   a. Chromosomes replicate, so each chromosome consists of two identical sister chromatids.
   b. Chromosomes condense and the nuclear envelope disappears.
   c. The chromosomes end up at opposite ends of the cell and two nuclear envelopes form.
   d. Sister chromatids separate, forming independent chromosomes.

5. What evidence suggests that during anaphase, spindle fibers shorten at the kinetochore and not at the base of the mitotic spindle?
   a. Motor proteins are located at the kinetochore.
   b. Motor proteins are located at the kinetochore *and* at the base of the mitotic spindle.
   c. When fluorescing microtubules are bleached in the middle, the bleached segment stays stationary as the fibers shorten.
   d. When fluorescing microtubules are bleached in the middle, the bleached segment moves toward the base of the mitotic spindle as the fibers shorten.

6. The normal function of the Rb protein is to bind to E2F, inactivating it. What are the consequences of defects in Rb?
   a. It binds more tightly to E2F, so E2F cannot stimulate the production of molecules needed in S phase. The cell cannot progress through the cell cycle.
   b. It no longer responds to phosphorylation—it cannot be regulated.
   c. It does not bind to E2F properly, so E2F continuously stimulates the production of molecules needed in S phase.
   d. Rb acts as a tumor suppressor, similar in function to p53.

Conceptual Review

1. Sketch the phases of mitosis, listing the major events of each phase. Identify at least two events that must be completed successfully for daughter cells to share an identical complement of chromosomes.

2. What are the consequences for the cell if the G_1 checkpoint fails? Answer the same question for the G_2 and metaphase checkpoints.

3. Explain how cell fusion and microinjection experiments supported the hypothesis that specific molecules are involved in the transition from interphase to M phase.

4. Why are most protein kinases considered regulatory proteins?

5. Why are cyclins called cyclins? Explain their relationship to cyclin-dependent kinases and to growth factors.

6. Early detection is the key to surviving most cancers. Why?

Group Discussion Problems

1. In multicellular organisms, nondividing cells stay in G_1 phase. For the cell, why is it better to be held in G_1 rather than S, G_2, or M phase?

2. The Rb protein helps regulate the cell cycle in many types of cells. Children with hereditary retinoblastoma have a defective version of the Rb protein but get tumors only in their eyes—not elsewhere. Suggest a hypothesis to explain why tumors start only in the retinas of children. How could you test your hypothesis?

3. Predict the outcome of an experiment involving the fusion of a cell in G_1 phase with a cell in G_2 phase. What would happen to the G_1-phase nucleus? To the G_2-phase nucleus? Why?

4. Cancer is primarily a disease of older people. Further, a group of individuals may share a genetic predisposition to developing certain types of cancer yet vary a great deal in time of onset—or not get the disease at all. Discuss these observations in light of the claim that several defects usually have to occur for cancer to develop.

Answers to Multiple-Choice Questions 1. b; 2. d; 3. d; 4. d; 5. c; 6. c

Gene Structure and Expression

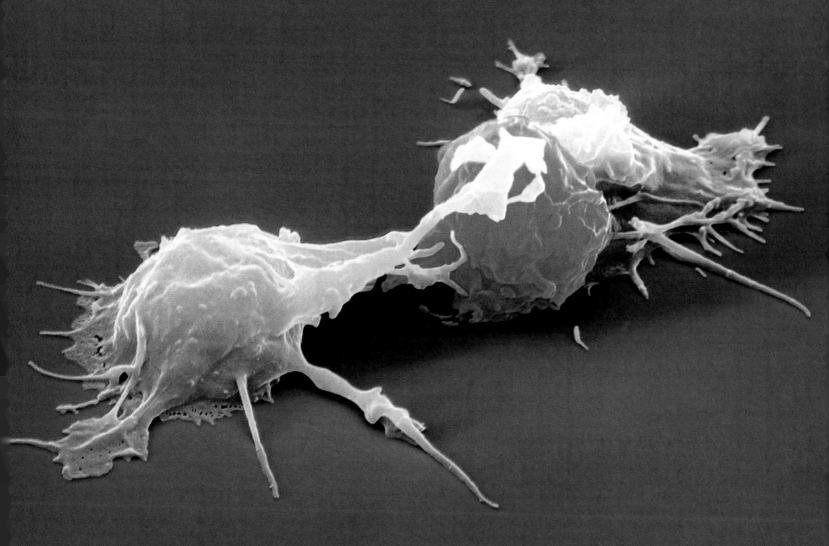

Two immune system cells attack a cancerous cell (colored red) in this micrograph. The three cells contain the same genes. Their shapes and behaviors differ, however, because they are expressing different genes.

12 Meiosis

▥ Meiosis is a type of nuclear division. It results in cells that have half as many chromosomes as the parent cell, and in animals it occurs prior to the formation of eggs and sperm. When an egg and a sperm cell combine to form an offspring, the original number of chromosomes is restored.

▥ Each cell produced by meiosis receives a different combination of chromosomes. Because genes are located on chromosomes, each cell produced by meiosis receives a different complement of genes. Meiosis leads to offspring that are genetically distinct from each other and from their parents. If offspring are genetically variable, some might thrive if the environment changes.

▥ If mistakes occur during meiosis, the resulting egg and sperm cells may contain the wrong number of chromosomes. It is rare for offspring with an incorrect number of chromosomes to develop normally.

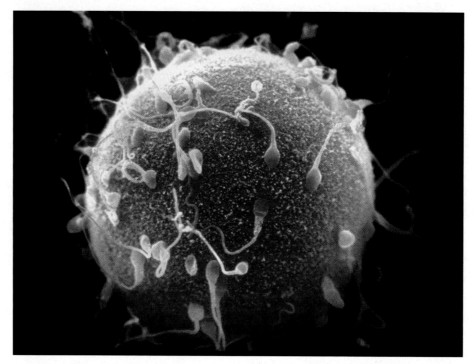

Scanning electron micrograph showing human sperm attempting to enter an egg. This chapter introduces the type of nuclear division called meiosis, which occurs prior to gamete formation in animals.

Why sex?

Simple questions—such as why sexual reproduction exists—are sometimes the best. This chapter asks what sexual reproduction is and why some organisms employ it. Understanding sex has been one of the great mysteries of classical and contemporary biology. After decades of effort, clear answers to the question of why sex exists are finally beginning to emerge.

Before we plunge ahead, it is important to realize that there are two basic levels of explanation in biology: ultimate explanation and proximate explanation. An **ultimate explanation** is evolutionary in nature. It explains *why* something happens. In terms of sex, biologists want to know why it evolved. This issue is important because most organisms do not reproduce sexual-

ly or do so only occasionally. Presumably, sexual reproduction increases the ability of certain organisms to produce offspring successfully in particular environments. Why is sex advantageous for some species? This question is the focus of Sections 12.2 and 12.3.

A **proximate explanation** is mechanistic in nature. It explains *how* something happens. In terms of sex, biologists want to know how the process of sexual reproduction occurs. They want to understand the molecular and cellular structures and events that are involved. For example, it has long been understood that during sexual reproduction, a male reproductive cell—a **sperm**—and a female reproductive cell—an **egg**—unite to form a new individual. The process of uniting sperm and egg is called **fertilization**. The first biologists to observe fertilization studied the large, translucent eggs of sea urchins. Due to the semitransparency of the sea urchin egg cell, researchers were able to see the nuclei of a sperm and an egg fuse.

When these results were published in 1876, they raised an important question at the proximate level, concerning the number of chromosomes found in parents, sperm, eggs, and offspring. Cell biologists had already established that the number of chromosomes is constant from cell to cell within a multicellular organism. It was also accepted that chromosome number is the same in the parent and daughter cells of mitosis. Biologists confirmed that all of the cells in a newly growing offspring, or **embryo**, are the products of mitotic divisions, and that all cells in the body are the direct descendants of the nucleus that formed at fertilization. The question is, How can the chromosomes from a sperm cell and an egg cell combine, but form an offspring that has the same chromosome number as its mother and its father?

A hint at the answer came in 1883, when a researcher noted that cells in the body of roundworms of the genus *Ascaris* have four chromosomes, while their sperm and egg nuclei have only two chromosomes apiece. Four years later, August Weismann formally proposed a hypothesis to explain the riddle: During the formation of **gametes**—reproductive cells such as sperm and eggs—there must be a distinctive type of cell division that leads to a reduction in chromosome number. Specifically, if the sperm and egg contribute an equal number of chromosomes to the fertilized egg, Weismann reasoned, they must each contain half of the usual number of chromosomes. Then, when sperm and egg combine, the resulting cell has the same chromosome number as its mother's cells and its father's cells have.

In the decades that followed, biologists confirmed this hypothesis by observing gamete formation in a wide variety of plant and animal species. Eventually this form of cell division came to be called meiosis ("lessening-act"). **Meiosis** is nuclear division that leads to a halving of chromosome number. It precedes the formation of eggs and sperm, and provides a satisfying proximate explanation for how sexual reproduction occurs. Let's begin with a closer look at how meiosis occurs, then shift to the ultimate level of explanation and ask why it exists.

12.1 How Does Meiosis Occur?

When cell biologists began to study the cell divisions that lead to gamete formation, they made an important observation: Each organism has a characteristic number of chromosomes. Consider the drawing in **Figure 12.1**, based on a paper published by Walter Sutton in 1902. It shows the chromosomes of the lubber grasshopper during the cell divisions leading up to the formation of a sperm. There is a total of 24 chromosomes in the cell. Sutton realized, however, that there are just 12 dis-

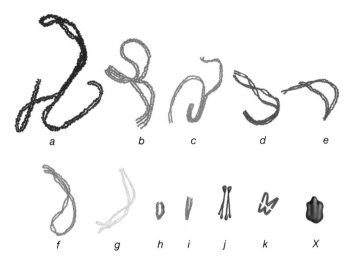

FIGURE 12.1 Cells Contain Different Types of Chromosomes, and Chromosomes Come in Pairs
Letters designate each of the 12 distinct types of chromosomes found in lubber grasshopper cells. There are two of each type of chromosome. The two members of a chromosome pair are called homologs. **EXERCISE** Draw a line over one of the two homologs in each of the 12 pairs shown, to distinguish the two homologs from each other.

tinct types of chromosomes based on size and shape. In this species, there were two chromosomes of each type.

Sutton designated 11 of the chromosomes by the letters *a* through *k* and the twelfth by the letter *X*. He referred to the pairs of chromosomes as **homologous chromosomes**. (They can also be called **homologs**.) The two chromosomes labeled *c*, for example, have the same size and shape and are homologous.

In Figure 12.1, most of the homologous chromosome pairs happen to be joined at several places. The two chromosomes designated X are joined so tightly that they look like a single chromosome. The two X chromosomes are associated with the sex of the individual, and each is designated a **sex chromosome**. Two types of sex chromosomes, known as X and Y, exist in lubber grasshoppers. Males have one X and one Y chromosome, while females have two X chromosomes. Non-sex chromosomes, such as *a–k* in a grasshopper cell, are known as **autosomes**.

Later work showed that homologous chromosomes are similar not only in size and shape but also in content. Homologous chromosomes carry the same genes. A **gene** is a section of DNA that influences one or more hereditary traits in an individual. For example, each copy of chromosome *c* found in lubber grasshoppers might carry genes that influence eye formation, body size, singing behavior, or jumping ability. The versions of a gene found on homologous chromosomes may differ, however. Biologists use the term **allele** to denote different versions of a particular gene. For example, the alleles on each copy of chromosome *c* in a lubber grasshopper might contribute to rounder eyes versus narrower eyes, larger body size versus smaller body size, faster songs

BOX 12.1   Karyotyping Techniques

Although chromosomes maintain their individuality and physical integrity throughout the cell cycle, they are readily visible only during mitosis or meiosis. Early in cell division, chromosomes condense into highly compact structures that can be observed with the light microscope. To describe an individual's karyotype, then, biologists must study cells undergoing cell division.

The first step in generating a karyotype is to obtain a sample of cells from the individual being studied. Cancer researchers might collect cells from a tumor. Physicians who are concerned about the possibility of birth defects might obtain a few cells from the developing embryo inside the mother. The next step is to grow the cells in culture, using techniques introduced in Chapter 11. When the cells are dividing rapidly, they are treated with a compound called colchicine. Colchicine stops mitosis at metaphase by disrupting the formation of the mitotic spindle. At this stage the chromosomes are easy to study, because they are condensed and consist of sister chromatids. The chromosomes of colchicine-treated cells are then stained and examined with the light microscope.

Researchers can distinguish stained chromosomes by size; by the position of the *centromere*, which holds sister chromatids together; and by striping or banding patterns. Subtler differences among chromosomes are apparent when a higher-resolution technique for karyotyping called **spectral karyotyping (SKY)**, or **chromosome painting**, is used. The "painting" is done with fluorescent dyes that are attached to short DNA molecules. The dyed pieces of DNA bind to particular regions of particular chromosomes. By using a combination of dyes, technicians can give each pair of homologous chromosomes a distinctive suite of colors (**Figure 12.2a**). The high-resolution image produced by this technique allows clinicians to diagnose an array of chromosomal abnormalities. Consider some examples:

- **Figure 12.2b** shows chromosomal changes associated with a cancer called *chronic myelogenous leukemia*. The defect involved is a **translocation**—a swapping of chromosome segments. In this case, a small piece of chromosome 9 and a large piece of chromosome 22 have changed places. The translocation causes a genetic change that leads to uncontrolled cell growth.

- **Figure 12.2c** shows the karyotype of an individual with **Klinefelter syndrome**, which develops in people who have two X chromosomes and a Y chromosome instead of a single X and a single Y. People with this syndrome have male sex organs but are sterile. They may also develop some female-like characteristics, such as enlarged breasts.

Karyotyping is an important diagnostic tool. As techniques for generating karyotypes improve, our ability to detect and interpret chromosomal defects also improves.

---

versus slower songs, and so on. Homologous chromosomes carry the same genes, but each homolog may contain different alleles.

At this point in his study, Sutton had succeeded in determining the lubber grasshopper's **karyotype**—meaning the number and types of chromosomes present (see **Box 12.1**). As karyotyping studies expanded, cell biologists realized that, like lubber grasshoppers, the vast majority of plants and animals have more than one of each type of chromosome. These investigators designated terms to identify the number of chromosome copies they observed. Organisms such as lubber grasshoppers, humans, and cedar trees are called **diploid** (literally, "double-form"), because they have two versions of each type of chromosome. Diploid organisms have two alleles of each gene—one on each of the homologous pairs of chromosomes. Organisms such as bacteria, archaea, and many algae are called **haploid** ("single-form"), because their cells contain just one of each type of chromosome. Haploid organisms do not contain homologous chromosomes. They have just one allele of each gene.

Researchers also invented a compact notation to indicate the number of chromosome sets in a particular organism or type of cell. By convention, the letter n stands for the number of distinct types of chromosomes in a given cell and is called the **haploid number**. If sex chromosomes are present, they are counted as a single type in the haploid number. To indicate the number of complete chromosome sets observed, a number is placed before the n. Thus, a cell can be n, or $2n$, or $3n$, and so on. The combination of a number and n is termed the cell's **ploidy**. Diploid cells or species are designated $2n$, because two chromosomes of each type are present—one from each parent. Haploid cells or species are labeled simply n, because they have just one set of chromosomes—no homologs are observed. In haploid cells, the number 1 in front of n is implied and is not written out.

Later work revealed that it is common for species in some lineages—particularly land plants such as ferns—to contain more than two of each type of chromosome. Instead of having two homologous chromosomes per cell, as many organisms do, **polyploid** ("many-form") species may have three or more of each type of chromosome in each cell. Depending on the number of homologs present, such species are called *triploid* ($3n$), *tetraploid* ($4n$), *hexaploid* ($6n$), *octoploid* ($8n$), and so on.

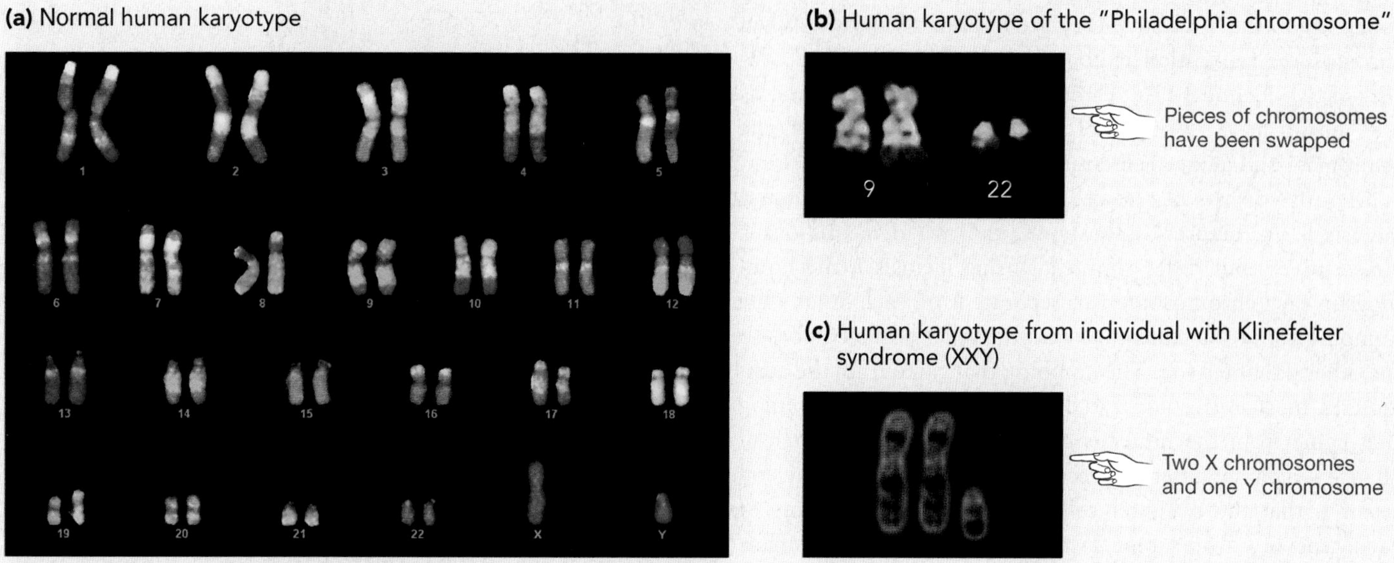

(a) Normal human karyotype

(b) Human karyotype of the "Philadelphia chromosome"

Pieces of chromosomes have been swapped

(c) Human karyotype from individual with Klinefelter syndrome (XXY)

Two X chromosomes and one Y chromosome

FIGURE 12.2 Human Chromosomes
(a) Condensed chromosomes that are undergoing mitosis are arranged randomly when first observed with the microscope. To determine a karyotype, a technician uses a computer to separate the image of each chromosome. The images of homologous chromosomes are then placed side by side, and the homologous pairs are arranged by number. **(b)** In the individual with this karyotype, the purple sections of chromosome 9 and chromosome 22 have been swapped via translocation. **(c)** Individuals with Klinefelter syndrome have two X chromosomes and a Y chromosome. **QUESTION** Is the karyotype in part (a) normal, or does it reveal the presence of an extra chromosome or the absence of a chromosome?

Why some species are haploid versus diploid or tetraploid is currently the subject of debate and research.

To summarize, the haploid number n indicates the number of distinct types of chromosomes present. Different species have different haploid numbers (see **Table 12.1**). Human cells contain 23 distinct types of chromosomes, so $n = 23$. In the grasshopper cells that Sutton studied, $n = 12$. In contrast, a cell's ploidy (n, $2n$, $3n$, etc.) indicates the number of each type of chromosome present. Stating a cell's ploidy is the same as stating the number of haploid chromosome sets present. Because most human and grasshopper cells contain two of each type of chromosome, they are diploid. In humans $2n = 46$; in Sutton's grasshoppers $2n = 24$. Ploidy refers to the number of each type of chromosome present; haploid number identifies how many different types of chromosomes occur.

Table 12.2, page 253, summarizes the vocabulary that biologists use to describe the number and types of chromosomes found in a cell.

Sutton and the other early cell biologists did more than describe the karyotypes observed in their study organisms, however. Through careful examination, they were able to track how chromosome numbers change during meiosis. These studies confirmed Weismann's hypothesis that a special type of cell division occurs during gamete formation. This result was a major advance in understanding sex at the proximate level.

TABLE 12.1 The Number of Chromosomes Found in Some Familiar Organisms

Organism	Number of Different Types of Chromosomes (haploid number n)	Diploid Chromosome Number ($2n$)
Humans	23	46
Domestic dog	36	72
Fruit fly	4	8
Chimpanzee	24	48
Bulldog ant	1	2
Garden pea	7	14
Corn (maize)	10	20

An Overview of Meiosis

Recall that cells replicate each of their chromosomes before undergoing meiosis. At the start of the process, then, chromosomes are in the same state they are in prior to mitosis. When chromosome replication is complete, each chromosome consists of two identical **sister chromatids**. These sister chromatids are joined at a portion of the chromosome called the **centromere** and along their entire length (**Figure 12.3a**).

Meiosis consists of two cell divisions, called **meiosis I** and **meiosis II**. As **Figure 12.3b** shows, the two divisions occur consecutively but differ sharply. During meiosis I, the homologs in each chromosome pair separate from each other. One homolog goes to one daughter cell; the other homolog goes to the other daughter cell. The homolog that came from the individual's mother is colored red in Figure 12.3; the homolog that came from the father is colored blue. It is a matter of chance which daughter cell receives which homolog. The end result is that the daughter cells from meiosis I have one of each type of chromosome instead of two, and thus half as many chromosomes as the parent cell has. During meiosis I, the diploid ($2n$) parent cell produces two haploid (n) daughter cells. Each chromosome still consists of two identical sister chromatids, however.

During meiosis II, sister chromatids from each chromosome separate. One sister chromatid goes to one daughter cell; the other sister chromatid goes to the other daughter cell. The cell that starts meiosis II has one of each type of chromosome, but each chromosome has been replicated (meaning it still consists of two sister chromatids). The cells produced by meiosis II also have one of each type of chromosome, but now the chromosomes are unreplicated. Recall from Chapter 11 that a chromosome is considered one chromosome whether it is replicated or unreplicated. This point is crucial. A replicated chromosome consists of two sister chromatids but still represents just one chromosome.

To reiterate, sister chromatids separate during meiosis II, just as they do during mitosis. Meiosis II is actually equivalent to mitosis occurring in a haploid cell. As in mitosis, chromosome movements during meiosis I and II are caused by spindle fibers that attach at the centromere of each chromosome.

Sutton and a host of other early cell biologists worked out this sequence of events through careful observation of cells with the light microscope. Based on these studies, they came to a key realization: The outcome of meiosis is a reduction in chromosome number. For this reason, meiosis is known as a *reduction division*. In most plants and animals, the original cell is diploid and the four daughter cells are haploid. These four haploid daughter cells, each containing one of each homologous chromosome, eventually go on to form egg cells or sperm cells via a process called **gametogenesis** ("gamete-origin"), which is described in Chapter 21. When two gametes fuse during fertilization, a full complement of chromosomes is restored (**Figure 12.3c**). The cell that results from fertilization is diploid

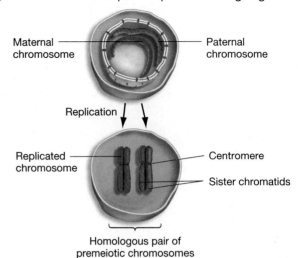

(a) Each chromosome replicates prior to undergoing meiosis.

Maternal chromosome — Paternal chromosome

Replication

Replicated chromosome — Centromere — Sister chromatids

Homologous pair of premeiotic chromosomes

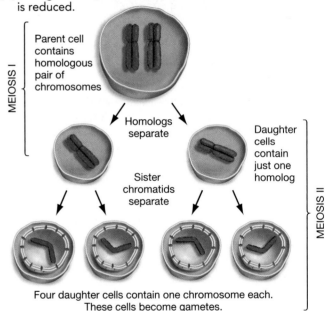

(b) During meiosis, chromosome number in each cell is reduced.

MEIOSIS I

Parent cell contains homologous pair of chromosomes

Homologs separate

Daughter cells contain just one homolog

Sister chromatids separate

MEIOSIS II

Four daughter cells contain one chromosome each. These cells become gametes.

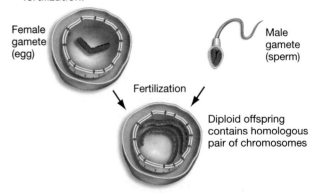

(c) A full complement of chromosomes is restored during fertilization.

Female gamete (egg)

Male gamete (sperm)

Fertilization

Diploid offspring contains homologous pair of chromosomes

FIGURE 12.3 The Major Events in Meiosis
Meiosis reduces chromosome number by half. In diploid organisms, the products of meiosis are haploid. **EXERCISE** In parts (b) and (c), write n or $2n$ next to each cell to indicate its ploidy.

TABLE 12.2 Vocabulary Used to Indicate the Chromosomal Makeup of a Cell

Term	Definition	Example or Comment
Chromosome	Structure made up of DNA and proteins; carries the cell's hereditary information (genes)	Eukaryotes have linear, threadlike chromosomes; bacteria and archaea often have just one circular chromosome
Sex chromosome	Chromosome associated with an individual's sex	X and Y chromosomes of humans: Males are XY, females XX. Also, Z and W chromosomes of birds: Males are ZZ, females ZW.
Autosome	A non-sex chromosome	Chromosomes 1–22 in humans.
Unreplicated chromosome	A chromosome that consists of a single copy (in eukaryotes, a single "thread")	
Replicated chromosome	A chromosome that has been copied; consists of two linear structures joined at the centromere	Centromere
Sister chromatids	The chromosome copies in a replicated chromosome	Sister chromatids
Homologous chromosomes (homologs)	In a diploid organism, chromosomes that are similar in size, shape, and gene content	
Non-sister chromatids	The chromosome copies in homologous chromosomes	Non-sister chromatids
Tetrad	Synapsed homologous chromosomes	Tetrad
Haploid number	The number of different types of chromosomes in a cell; symbolized n	Humans have 23 different types of chromosomes ($n = 23$)
Ploidy	The number of each type of chromosome present; equivalent to the number of haploid chromosome sets present	
Haploid	Having one of each type of chromosome (n)	Bacteria and archaea are haploid, as are many algae; gametes are haploid
Diploid	Having two of each type of chromosome ($2n$)	Most familiar plants and animals are diploid
Polyploid	Having more than two of each type of chromosome; cells may be triploid ($3n$), tetraploid ($4n$), hexaploid ($6n$), and so on	Seedless bananas are triploid; many ferns are tetraploid; bread wheat is hexaploid

and is called a **zygote**. In this way, each diploid individual receives both a haploid chromosome set from its mother and a haploid set from its father. Homologous chromosomes are therefore referred to as either maternal or paternal in origin. **Maternal chromosomes** come from the mother; **paternal chromosomes** come from the father.

When these studies were published, the mystery of fertilization was finally solved. To appreciate the consequences of meiosis fully, though, we need to analyze the events in more detail and will do so next.

The Phases of Meiosis I

Meiosis begins after chromosomes have been replicated during S phase. Prior to the start of meiosis, chromosomes are extremely long structures, just as they are during interphase of the normal cell cycle. The major steps that occur once meiosis

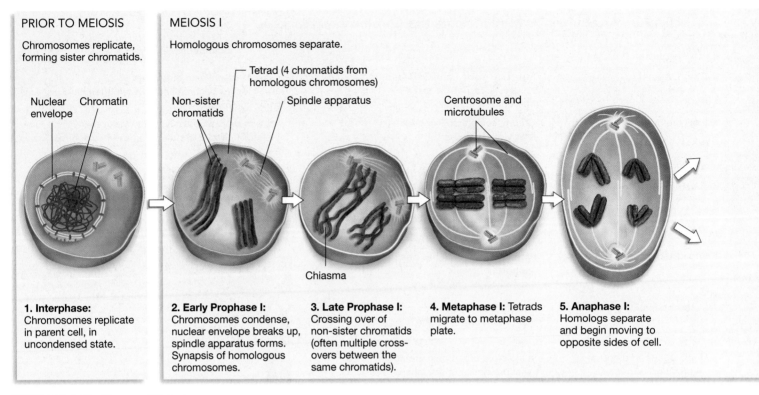

FIGURE 12.4 The Phases of Meiosis
EXERCISE In step 3, circle all of the other chiasmata.

begins are shown in **Figure 12.4**, using a diploid species with a haploid number of 2 as an example ($n = 2$; $2n = 4$). As in Figure 12.3, maternal chromosomes are red and paternal chromosomes are blue.

During **early prophase I** the chromosomes condense, the spindle apparatus forms, and the nuclear envelope begins to disappear. The next event illustrated, still during early prophase of meiosis I, is crucial: Homologous chromosome pairs come together. This pairing process is called **synapsis** and is illustrated in step 2 of Figure 12.4. Synapsis is possible because regions of homologous chromosomes that are similar at the molecular level attract one another, via mechanisms that are currently the subject of intense research. The structure that results from synapsis is called a **tetrad** (*tetra* means "four" in Greek). A tetrad consists of two homologous chromosomes, and each homolog consists of two sister chromatids. The chromatids from the homologs are referred to as **non-sister chromatids**. In the figure, non-sister chromatids are colored red versus blue.

During **late prophase I**, the non-sister chromatids begin to separate at many points along their length. They stay joined at certain locations, however, and look as if they cross over one another. Each crossover forms an X-shaped structure called a **chiasma** (plural: *chiasmata*). (In the Greek alphabet, the letter X is "chi.") Normally, at least one chiasma forms in every pair of homologous chromosomes; usually there are several chiasmata. As step 3 of Figure 12.4 shows, the chromatids involved in chiasma formation are homologous but not sisters. Consistent with this observation, Thomas Hunt Morgan proposed that a

physical exchange of paternal and maternal chromosomes occurs at chiasmata. According to this hypothesis, paternal and maternal chromatids break and rejoin at each chiasma, producing chromatids that have both paternal and maternal segments. Morgan called this process of chromosome exchange **crossing over**. In step 4 of Figure 12.4, the result of crossing over is illustrated by chromosomes with a combination of red and blue segments. When crossing over occurs, the chromosomes that result have a mixture of maternal and paternal alleles.

The next major stage in meiosis I occurs during **metaphase I**, when pairs of homologous chromosomes (tetrads) are moved to a region called the **metaphase plate** by spindle fibers from the centrosomes (step 4). Two points are key here: Each tetrad moves to the metaphase plate independently of the other tetrads, and the alignment of maternal and paternal homologs from each chromosome is random. During **anaphase I**, homologs separate and begin moving to opposite sides of the cell (step 5). Meiosis I concludes with **telophase I**, when the homologs finish moving to opposite sides of the cell (step 6). When meiosis I is complete, the cell divides.

The end result of meiosis I is that one chromosome of each homologous pair is distributed to a different daughter cell. As a result of crossing over and the random distribution of maternal and paternal homologues during metaphase, each daughter cell from meiosis I gets a random and unique assortment of maternal and paternal chromosomes.

Study steps 1 through 6 of Figure 12.4 to see that the daughter cells of meiosis I are haploid. In each chromosome, however,

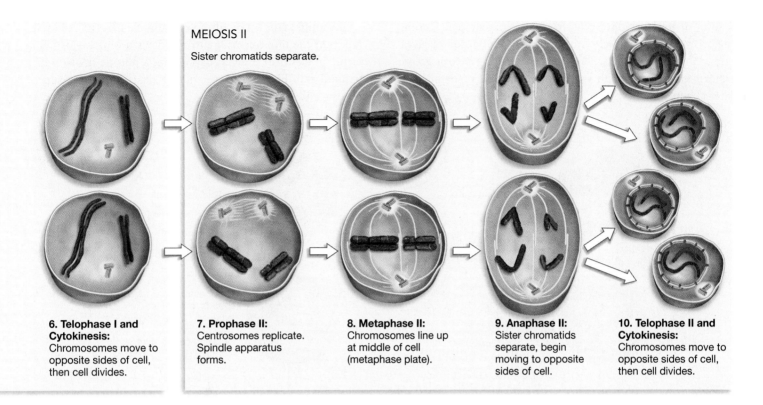

MEIOSIS II

Sister chromatids separate.

6. Telophase I and Cytokinesis: Chromosomes move to opposite sides of cell, then cell divides.

7. Prophase II: Centrosomes replicate. Spindle apparatus forms.

8. Metaphase II: Chromosomes line up at middle of cell (metaphase plate).

9. Anaphase II: Sister chromatids separate, begin moving to opposite sides of cell.

10. Telophase II and Cytokinesis: Chromosomes move to opposite sides of cell, then cell divides.

the sister chromatids remain attached. At the end of meiosis I, each daughter cell still contains replicated chromosomes.

Although meiosis I is a continuous process, biologists summarize the events by identifying these distinct phases:

- Early Prophase I—Replicated chromosomes condense, the spindle apparatus forms, and the nuclear envelope disappears. Synapsis of homologs forms pairs of homologous chromosomes (tetrads). Spindle fibers attach to chromosome centromeres.

- Late Prophase I—Crossing over results in a mixing of chromosome segments from maternal and paternal chromosomes.

- Metaphase I—Pairs of homologous chromosomes migrate to the metaphase plate, and homologs line up.

- Anaphase I—Homologs separate and begin moving to opposite ends of the cell.

- Telophase I—Homologs finish moving to opposite sides of the cell. In some species, a nuclear envelope re-forms around each set of chromosomes.

When meiosis I is complete, **cytokinesis** (division of cytoplasm) occurs and two haploid daughter cells form.

The Phases of Meiosis II

Recall that chromosome replication occurred prior to meiosis I. Throughout meiosis I, sister chromatids remain attached. That is, replicated chromosomes remain intact during meiosis I. But no chromosome replication occurs between meiosis I and meiosis II. At the start of meiosis II, each chromosome still consists

of two identical sister chromatids. Because only one member of each homologous pair is present, the cell is haploid.

Next, during **prophase II**, a spindle apparatus forms in both daughter cells. Spindle fibers attach to each side of the replicated chromosomes and move toward the middle of each cell (step 7 of Figure 12.4). In **metaphase II**, the replicated chromosomes are lined up at the metaphase plate (step 8). The sister chromatids of each chromosome separate during **anaphase II** (step 9) and move to different daughter cells during **telophase II** (step 10). Once they are separated, each chromatid is considered an independent chromosome. Meiosis II results in four haploid cells, each with one chromosome of each type.

As in meiosis I, biologists routinely designate distinct phases in meiosis II:

- Prophase II—The spindle apparatus forms. If a nuclear envelope formed at the end of meiosis I, it breaks apart.

- Metaphase II—Replicated chromosomes, consisting of two sister chromatids, are lined up at the metaphase plate.

- Anaphase II—Sister chromatids separate. The unreplicated chromosomes that result begin moving to opposite sides of the cell.

- Telophase II—Chromosomes finish moving to opposite sides of the cell. A nuclear envelope forms around each haploid set of chromosomes.

When meiosis II is complete, each cell divides to form two daughter cells. Because meiosis II occurs in exactly the same way in both daughter cells of meiosis I, the process results in a total of four daughter cells from each parent cell. During meiosis, one diploid cell with replicated chromosomes gives rise to four haploid cells. The cells produced by meiosis go on to form gametes, through a series of events that are detailed in later chapters.

It should make sense, after you examine Figure 12.4 in detail, that the movement of chromosomes during meiosis II is virtually identical to what happens in a mitotic division in a haploid cell. **Figure 12.5** provides a detailed comparison of mitosis and meiosis.

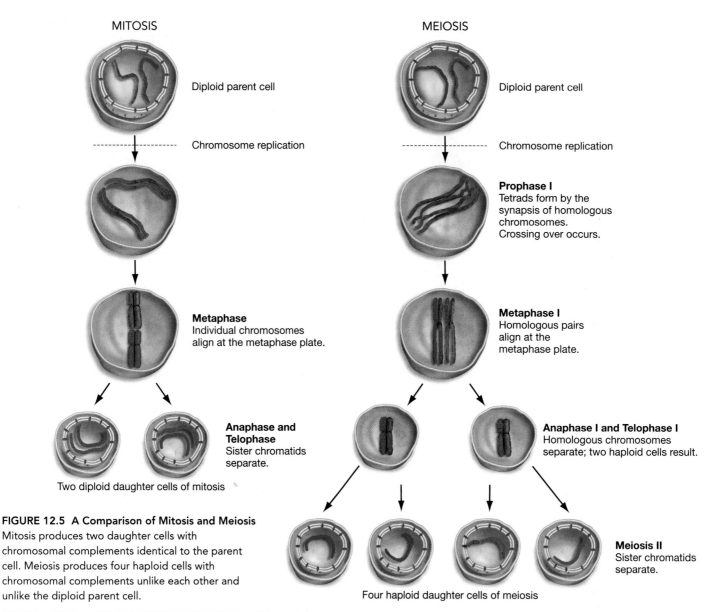

FIGURE 12.5 A Comparison of Mitosis and Meiosis
Mitosis produces two daughter cells with chromosomal complements identical to the parent cell. Meiosis produces four haploid cells with chromosomal complements unlike each other and unlike the diploid parent cell.

Feature	Mitosis	Meiosis
Number of cell divisions	One	Two
Number of chromosomes in daughter cells, compared with parent cell	Same	Half
Synapsis of homologs	No	Yes
Number of crossing-over events	None	One or more per pair of homologous chromosomes
Makeup of chromosomes in daughter cells	Identical	Different—only one of each chromosome type present, paternal and maternal segments mixed within chromosomes
Role in life cycle	Asexual reproduction in eukaryotes; cell division for growth of multicellular organisms	Precedes production of gametes in sexually reproducing animals

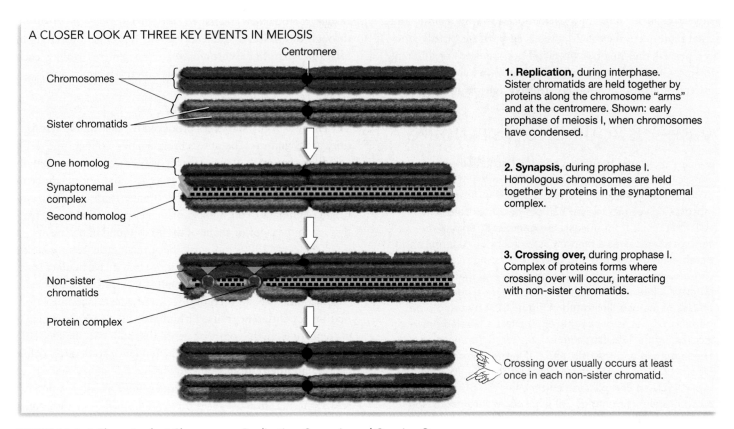

A CLOSER LOOK AT THREE KEY EVENTS IN MEIOSIS

Centromere

Chromosomes

Sister chromatids

1. Replication, during interphase. Sister chromatids are held together by proteins along the chromosome "arms" and at the centromere. Shown: early prophase of meiosis I, when chromosomes have condensed.

One homolog

Synaptonemal complex

Second homolog

2. Synapsis, during prophase I. Homologous chromosomes are held together by proteins in the synaptonemal complex.

Non-sister chromatids

Protein complex

3. Crossing over, during prophase I. Complex of proteins forms where crossing over will occur, interacting with non-sister chromatids.

Crossing over usually occurs at least once in each non-sister chromatid.

FIGURE 12.6 A Closer Look at Chromosome Replication, Synapsis, and Crossing Over

A Closer Look at Key Events in Prophase of Meiosis I

Figure 12.6 provides more detail on how several important events in prophase of meiosis I occur. Step 1 of Figure 12.6 shows that after chromosome replication is complete, sister chromatids stay very tightly joined along their entire length. When homologs synapse, two pairs of non-sister chromatids are brought together and are held there by a network of proteins called the **synaptonemal complex** (step 2). Crossing over can occur at many

locations along the length of this paired structure. Crossing over typically occurs at least once in each pair of homologs and two or three times in some species. It occurs when a complex of proteins cuts the chromosomes and then reattaches segments from homologs (step 3).

This more-detailed look at replication, synapsis, and crossing over should help you interpret the micrographs in **Figure 12.7,** which show grasshopper cells undergoing meiosis. Once you have a solid understanding of how meiosis occurs, you will be

(a) Prophase of meiosis I

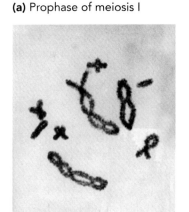

(b) Anaphase of meiosis I

(c) Metaphase of meiosis II

(d) Anaphase of meiosis II

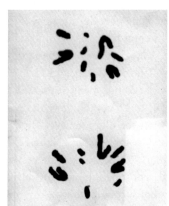

FIGURE 12.7 Meiotic Chromosomes in Grasshoppers
EXERCISE (a) Label several chiasmata, sister chromatids, tetrads, and non-sister chromatids; (b) add centrosomes; (c) label sister chromatids and the metaphase plate, and add a spindle apparatus; (d) label the haploid number and ploidy of each nucleus.

ready to consider ultimate explanations for why meiosis and sexual reproduction exist. Meiosis is an intricate, tightly regulated process that involves dozens if not hundreds of different proteins. Given this complexity, it is logical to hypothesize that this type of nuclear division is extremely important. Why?

✓ CHECK YOUR UNDERSTANDING

Meiosis is called a reduction division because the total number of chromosomes present is cut in half. During meiosis, a single diploid parent cell with replicated chromosomes gives rise to four haploid daughter cells. The cells that result from meiosis are genetically unique, because each receives a random assortment of maternal and paternal chromosomes and because crossing over leads to chromosomes with random combinations of maternal and paternal alleles. You should be able to (1) demonstrate the phases of meiosis illustrated in Figure 12.4 by using pipe cleaners or pieces of spaghetti; (2) make a labeled drawing like Figure 12.5 that summarizes the differences between mitosis and meiosis; and, (3) in the micrographs of Figure 12.7, (a) identify each phase of meiosis, (b) identify replicated and unreplicated chromosomes, (c) label chromosomes and chromatids, (d) label sister chromatids and non-sister chromatids, (e) label chiasmata, (f) add colors to indicate segments of maternal and paternal chromosomes that exchanged as a result of crossing over, and (g) determine whether each cell is haploid or diploid.

12.2   The Consequences of Meiosis

The cell biologists who worked out the details of meiosis in the late 1800s and early 1900s realized that the process solved the riddle of fertilization. Weissman's hypothesis—that a reduction division precedes gamete formation—was confirmed. This was an important advance in providing a proximate explanation for sexual reproduction. But researchers also understood that meiosis has another important outcome: Thanks to the independent shuffling of maternal and paternal chromosomes and crossing over during meiosis I, the chromosomes in gametes are different from the chromosomes in parental cells. Subsequently, fertilization brings haploid sets of chromosomes from a mother and father together to form a diploid offspring. The chromosome complement of this offspring is unlike that of either parent. It is a combination of contributions from each parent.

At the ultimate level of causation, this change in chromosomal complement is crucial. The critical observation is that changes in chromosome configuration occur only during sexual reproduction and not during asexual reproduction. **Asexual reproduction** refers to any mechanism of producing offspring that does not involve the fusion of gametes. Asexual reproduction in eukaryotes is usually based on mitosis, and the chromosomes in the daughter cells of mitosis are identical to the chromosomes in the parental cell (see Chapter 11). In contrast,

sexual reproduction refers to the production of offspring through the fusion of gametes. Sexual reproduction results in offspring that have chromosome complements unlike each other and their parents. Why is this difference important?

Chromosomes and Heredity

The changes in chromosomes produced by meiosis and fertilization are significant because chromosomes contain the cell's hereditary material. Stated another way, chromosomes contain the instructions for specifying what a particular trait might be in an individual. These inherited traits range from eye color and height in humans to the number or shape of the bristles on a fruit fly's leg to the color or shape of the seeds found in pea plants.

In the early 1900s biologists began using the term **gene** to refer to the inherited instructions for a particular trait. Chapter 13 explores the experiments that confirmed that each chromosome is composed of a series of genes encoding information for different traits. In humans, for example, a single chromosome might comprise genes that influence height, hair color, the spacing of teeth, and the tendency to develop colon cancer; another chromosome might include genes that affect eye color, susceptibility to allergies, and predisposition to schizophrenia. In most cases there are hundreds or thousands of genes on each chromosome. Recall from Section 12.1 that *allele* refers to a particular version of a gene and that homologous chromosomes may carry different alleles. You might have a chromosome containing alleles that tend to produce medium height, black hair color, and a low predisposition to colon cancer, as well as a homologous chromosome containing alleles that contribute to short height, blond hair color, and a predisposition to colon cancer. Other individuals you know might have alleles associated with extreme height, red hair color, and a moderate predisposition to colon cancer.

Because chromosomes are composed of genes, the offspring produced during asexual reproduction are genetically identical to one another as well as to their parent. The offspring of asexual reproduction are **clones**—or exact copies—of their parent. But the offspring produced by sexual reproduction are genetically different from one another and unlike either their mother or their father. Let's analyze three aspects of meiosis that create variation among chromosomes—and hence the genetic makeup—of sexually produced offspring: (1) separation and distribution of homologous chromosomes, (2) crossing over, and (3) fertilization.

How Does the Separation and Distribution of Homologous Chromosomes Produce Genetic Variation?

Each cell in your body contains 23 homologous pairs of chromosomes and 46 chromosomes in total. Half of these chromosomes came from your mother, and half came from your father. Each chromosome is composed of genes that influence particular traits. For example, one gene that affects your eye color might be located on one chromosome, while one of the genes that affects your hair color might be located on a different chromosome (**Figure 12.8a**).

Suppose that the chromosomes you inherited from your mother contains alleles that tend to produce brown eyes and black hair, but the chromosomes you inherited from your father includes alleles that tend to specify green eyes and red hair. (In reality, many genes with various alleles interact in complex ways to produce human eye color and hair color.) Will any particular gamete you produce contain the genetic instructions inherited from your mother or the instructions inherited from your father?

To answer this question, study the diagram of meiosis in **Figure 12.8b**. It shows that when pairs of homologous chromosomes line up during meiosis I and the homologs separate, a variety of combinations of maternal and paternal chromosomes can result. Each daughter cell gets a random assortment of maternal and paternal chromosomes. In the example given here, meiosis results in gametes with alleles for brown eyes and black hair, like your mother, and green eyes and red hair, like your father. But two additional combinations also occur: brown eyes and red

hair, or green eyes and black hair. Four different combinations of paternal and maternal chromosomes are possible when two chromosomes are distributed to daughter cells during meiosis I.

How many different combinations of maternal and paternal homologs are possible when more chromosomes are involved? In an organism with three chromosomes per haploid set ($n = 3$), eight types of gametes can be generated by randomly grouping maternal and paternal chromosomes. In general, a diploid organism can produce 2^n combinations of maternal and paternal chromosomes, where n is the haploid chromosome number. This means a human ($n = 23$) can produce 2^{23}, or about 8.4 million, gametes that differ in their combination of maternal and paternal chromosome sets. Clearly, the random assortment of whole chromosomes generates an impressive amount of genetic variation among gametes.

The Role of Crossing Over

Recall from Section 12.1 that segments of paternal and maternal chromatids exchange at each chiasma that forms during meiosis I. Thus, crossing over produces new combinations of alleles on the same chromosome—combinations that did not exist in either parent. This phenomenon is recombination. **Genetic recombination** is any change in the combination of alleles on a given chromosome. In species that reproduce sexually, recombination occurs via crossing over during meiosis. But genetic recombination also occurs in haploid organisms such as bacteria, which cannot undergo meiosis. Processes that produce genetic recombination in bacteria are described in later chapters.

Crossing over and recombination are important because they dramatically increase the genetic variability of gametes produced by meiosis. Recall that the separation and distribution of homologous chromosomes during meiosis varies the combination of chromosomes present. But in addition, crossing over varies the combinations of alleles within each chromosome. Recent data on humans, for example, indicate that an average of about 50 chiasmata occur in each cell undergoing meiosis I. As a result, the number of genetically different gametes that an individual can produce is much more than the 8.4 million produced by the separation and distribution of homologs. When crossing over occurs a total of 50 or more times in random locations throughout the entire suite of tetrads, the number of genetically distinct gametes that you can produce is virtually limitless.

Biologists say that meiosis "shuffles" alleles, because each daughter cell gets a random assortment of maternal and paternal chromosomes and because most chromosomes contain both maternal and paternal alleles. As Section 12.3 will show, the genetic variation produced by meiosis has enormous consequences for the ability of offspring to survive and reproduce.

How Does Fertilization Affect Genetic Variation?

Crossing over and the random mixing of maternal and paternal chromosomes ensure that each gamete is genetically unique. Even if two gametes produced by the same individual fuse to

(a) Example: individual who is heterozygous at two genes

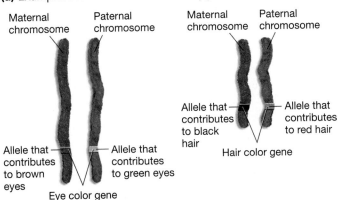

(b) During meiosis I, tetrads can line up two different ways before the homologs separate.

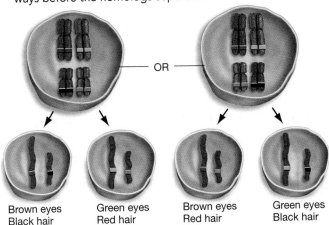

Brown eyes
Black hair

Green eyes
Red hair

Brown eyes
Red hair

Green eyes
Black hair

FIGURE 12.8 Separation of Homologous Chromosomes Leads to New Combinations of Genes
(a) A hypothetical example: Genes that influence eye color and hair color in humans are on different chromosomes. **(b)** How gametes with different combinations of genes result from separation of homologous chromosomes during meiosis I.

EVEN SELF-FERTILIZATION LEADS TO GENETICALLY VARIABLE OFFSPRING

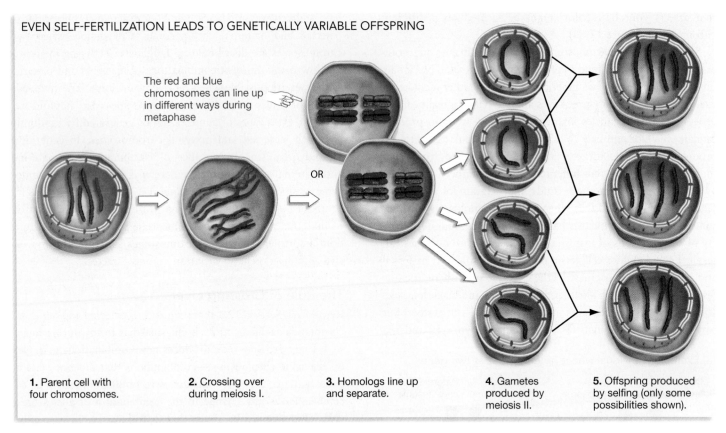

The red and blue chromosomes can line up in different ways during metaphase

OR

1. Parent cell with four chromosomes.

2. Crossing over during meiosis I.

3. Homologs line up and separate.

4. Gametes produced by meiosis II.

5. Offspring produced by selfing (only some possibilities shown).

FIGURE 12.9 Even If Self-Fertilization Takes Place, Offspring Are Genetically Variable
Some possible results of self-fertilization in an organism with four chromosomes ($2n = 4$). **EXERCISE** In step 2, label where two chiasmata are causing a double crossover event. Add sketches showing the chromosome complements of offspring produced by other pairings of gametes indicated in step 4.

form a diploid offspring—meaning **self-fertilization**, or "selfing," takes place—the offspring are very likely to be genetically different from the parent (**Figure 12.9**). Selfing is common in some plant species. It also occurs in the many animal species in which individuals contain both male and female sex organs.

Self-fertilization is rare or nonexistent in many sexually reproducing species, however. Instead, gametes from different individuals combine to form offspring. This is called **outcrossing**. Outcrossing increases the genetic diversity of offspring because it combines chromosomes from different individuals, which are likely to contain different alleles.

How many genetically distinct offspring can be produced when outcrossing occurs? Let's answer this question using humans as an example. Recall that a single human can produce about 8.4 million different gametes—even in the absence of crossing over. When a person mates with a member of the opposite sex, the number of different genetic combinations that can result is equal to the product of the number of different gametes produced by each parent. In humans this means that potentially 8.4 million × 8.4 million = 70.6×10^{12} genetically distinct offspring can result from any one mating. This number is far greater than the total number of people who have ever lived—and the calculation does not even take into

account variation generated by crossing over, which occurs at least once along each chromosome. Sexual reproduction results in genetically diverse offspring. Why is this important?

✓ CHECK YOUR UNDERSTANDING

Meiosis results in the production of haploid cells from a diploid parent cell. The daughter cells produced by meiosis are genetically different from the parent cell because maternal and paternal homologs align randomly at metaphase of meiosis I and because crossing over leads to recombination within chromosomes. You should be able to (1) draw a parent cell with $n = 3$ (three types of chromosomes) and six of the many genetically distinct types of daughter cells that may result when this parent cell undergoes meiosis; (2) compare and contrast the degree of genetic variation that results from asexual reproduction, selfing, and outcrossing.

12.3 Why Does Meiosis Exist? Why Sex?

Meiosis and sexual reproduction occur in only a small fraction of the lineages on the tree of life. Bacteria and archaea undergo only asexual reproduction; most algae, fungi, and some land

New shoots sprout from underground stems.

A genetically identical clone of trees results.

FIGURE 12.10 Asexual Reproduction in Quaking Aspen
Quaking aspen trees send out underground stems, from which new individuals sprout. Both the underground stems and the new shoots are produced by mitosis. As this process continues, a large group of genetically identical individuals forms. The photograph shows such a group in autumn, when quaking aspen leaves turn bright yellow.

plants reproduce asexually as well as sexually. Recall that asexual reproduction in eukaryotes occurs via mitosis. Quaking aspen trees, for example, can produce new individuals by sending up shoots from underground stems (**Figure 12.10**). Asexual reproduction is found even among the vertebrates. For example, several species of guppy in the genus *Poeciliopsis* reproduce exclusively via mitosis.

Sexual reproduction is common among multicellular ("many-celled") organisms, however. It is the major mode of reproduction in lineages such as the insects, with millions of species.

Even though sex plays an important role in the life of many organisms, until recently it was not clear why it occurs. On the basis of theory, biologists had good reason to think that sexual reproduction should not exist. Let's take a closer look.

The Paradox of Sex

In 1978 John Maynard Smith pointed out that the existence of sexual reproduction presents a paradox. Maynard Smith developed a mathematical model showing that because asexually reproducing individuals do not have to produce male offspring, their progeny can produce twice as many grand-offspring as can individuals that reproduce sexually. **Figure 12.11** diagrams this result by showing the number of females (♀) and males (♂) produced over several generations by asexual versus sexual reproduction. In this example, each individual produces four offspring over the course of his or her lifetime. In the asexual population, each individual is a female that produces four offspring. But in the sexual population, it takes two individuals—one male and one female—to produce four offspring.

Based on Maynard Smith's analysis, what will happen when asexual and sexual individuals exist in the same population and compete with one another? If all other things are equal, individuals that reproduce asexually should increase in frequency while individuals that reproduce sexually should decline in frequency. In fact, Maynard Smith's model predicts that sexual reproduction is so inefficient that it should be completely eliminated. At the ultimate level of explanation, the existence of sex is a paradox.

To resolve the paradox of sex, biologists began examining the assumption "If all other things are equal." Stated another way, biologists began looking for ways that meiosis and outcrossing could lead to the production of offspring that reproduce more than asexually produced individuals do.

The Changing-Environment Hypothesis   One hypothesis to explain sexual reproduction focuses on the benefits of producing genetically diverse offspring. Here's the key idea: If the

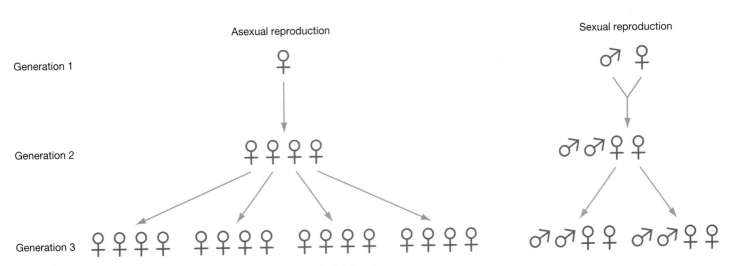

FIGURE 12.11 Asexual Reproduction Confers a Large Numerical Advantage
Each female symbol (♀) and male symbol (♂) represents an individual. In this hypothetical example, every individual produces four offspring over the course of a lifetime, sexually reproducing individuals produce half males and half females, and all offspring survive to breed.   **QUESTION** How many asexually produced offspring would be present in generation 4? How many sexually produced offspring?

environment changes from one generation to the next, then off-spring that are genetically different from their parents and from each other are more likely to survive and produce offspring of their own. Conversely, offspring that are genetic clones of their parents are less likely to thrive if the environment changes.

What type of environmental change would favor genetically diverse offspring? The possibilities include changes in temperature regimes, moisture, predators, competitors, and food sources. But recently researchers have focused on one particular component of environmental change: the emergence of new strains of disease-causing agents such as bacteria, parasitic worms, the eukaryote that causes malaria, and viruses.

The genetic characteristics of disease-causing organisms and viruses tend to change very quickly over time. In your own lifetime, for example, several new disease-causing agents have emerged that afflict humans; they include the SARS virus and new strains of HIV, the parasite that causes malaria, and the tuberculosis bacterium. In addition to employing drugs to combat these types of agents, humans have hundreds of genes that are involved in defense. At many of these genes, certain alleles help host individuals fight off particular strains of bacteria, parasitic worms, or viruses. As you might predict, the presence of certain disease-fighting alleles is crucial for plants and animals that cannot rely on drug therapies for help.

What happens if all of the offspring produced by an individual are genetically identical? If a new strain of disease-causing agent evolves, then all of the asexually produced offspring are likely to be susceptible to that new strain. But if the offspring are genetically variable, then it is likely that at least some offspring will have combinations of alleles that enable them to fight off the new disease and produce offspring of their own.

The logic of the hypothesis is sound. Do any data support it?

Testing the Changing-Environment Hypothesis   Curtis Lively and colleagues tested the changing-environment hypothesis by studying a species of snail that is native to New Zealand. This type of snail lives in ponds and other freshwater habitats and is parasitized by over a dozen species of trematode worms. Snails that become infected cannot reproduce—the worms eat their reproductive organs. In this way, the parasites cause illness. The parasites are rare in some habitats and common in others.

The biologists were interested in working on this snail species because some individuals reproduce only sexually while others reproduce only asexually. If the changing-environment hypothesis for the advantage of sex is correct, then the frequency of sexually reproducing individuals should be much higher in habitats where parasites are common than it is in habitats where parasites are rare (**Figure 12.12**). The logic here is that asexually reproducing individuals should have high fitness in environments where parasites are rare. In contrast, sexually reproducing individuals should have high fitness in habitats where parasites are common.

Question: Why does sexual reproduction occur?

Hypothesis: Sexually produced offspring have higher fitness than do asexually produced individuals in habitats where parasitism is common.

Null hypothesis: There is no relationship between the presence of parasites and method of reproduction.

Study design:

1. Collect snails from a wide array of habitats.

2. Document percentage of males in each population, as an index of frequency of sexual reproduction.

3. Note two types of populations: In one, males are common; in the other, males are almost nonexistent. Infer that sexual reproduction is either common or almost nonexistent.

4. Document percentage of individuals infected with parasites in sexually versus asexually reproducing populations.

Prediction: In populations where sexual reproduction is common, parasitism rates are high. In populations with only asexual reproduction, infection rates are low.

Prediction of null hypothesis: No difference in parasitism rate between populations that reproduce sexually versus asexually.

Results:

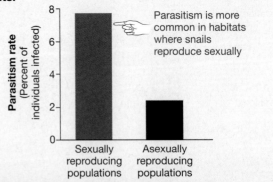

Parasitism is more common in habitats where snails reproduce sexually

Conclusion: Sexual reproduction is common in habitats where parasitism is common. Asexual reproduction is common in habitats where parasitism is rare.

FIGURE 12.12   Is Sexual Reproduction Favored when Disease or Parasitism Rates Are High?

To test these predictions, the researchers collected a large number of individuals from different habitats. Lively and co-workers examined snails in habitats where parasites were more or less common and calculated the frequency of individuals that reproduce sexually versus those that reproduce asexually. The results are plotted in Figure 12.12. The data show that habitats where parasite infection rates are high have a relatively large number of sexually reproducing individuals compared with habitats that have low parasite incidence.

This result and a variety of other studies support the changing-environment hypothesis. Although the paradox of sex remains an active area of research, more biologists are becoming convinced that sexual reproduction is an adaptation that increases the fitness of individuals in environments where disease-causing organisms are common.

When Do Meiosis and Fertilization Occur during the Life of an Organism?

At the proximate level, all sexually reproducing organisms are thought to undergo meiosis in virtually the same way. At the ultimate level, the changing-environment hypothesis may turn out to be a valid and universal explanation for why sex can be advantageous. But on the question of when meiosis occurs during the course of an organism's lifetime, biologists are faced with a striking amount of variation that remains unexplained.

To bring this question into focus, consider the life cycles observed in various species. A **life cycle** is the sequence of events that occurs during the life span of an individual, from fertilization to the production of offspring. When sexually reproducing species are compared, biologists find that many of the same events occur: meiosis, production of haploid gametes, fertilization and formation of a diploid zygote, and mitosis. What varies among species is the timing of these events and the extent of the haploid and diploid stages. Some of this variation is highlighted in **Figure 12.13**.

In humans and most other animals, individuals are diploid for most of their lives (**Figure 12.13a**). Meiosis occurs in the reproductive organs of an adult, and gametes represent the only haploid phase of the individual's life. In contrast, many species of eukaryotes are haploid for most of their lives. The green alga shown in **Figure 12.13b** illustrates this pattern. In these organisms, the diploid stage consists of a single-celled structure that can resist heat and drying. This cell then undergoes meiosis. The haploid cells that result divide by mitosis to generate a haploid multicellular organism.

Land plants and several groups of algae have yet another type of life cycle, known as **alternation of generations**. This life cycle features a multicellular haploid stage and a multicellular diploid stage (**Figure 12.13c**). In some land plants, such as angiosperms (flowering plants) and ferns, the haploid stage is tiny and relatively short lived. In other species, such as mosses, the haploid stage is larger and longer lived than the diploid plant.

(a) Diploid dominant

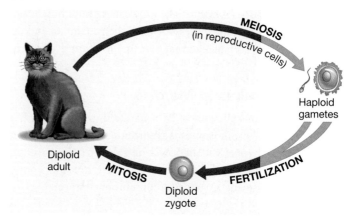

(b) Haploid dominant

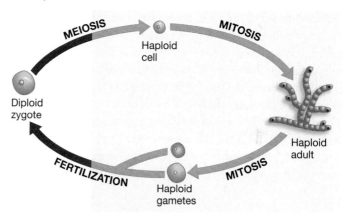

(c) Alternation of generations

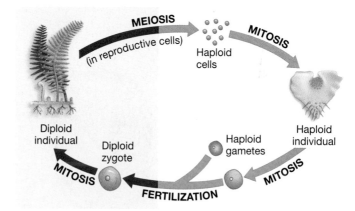

FIGURE 12.13 Life Cycles Differ in the Timing of Meiosis and Fertilization and in the Extent of Haploid versus Diploid Stages
(a) In most animals, gametes are the only haploid cells. Meiosis occurs in special reproductive tissues of adults. **(b)** In many algae, the fertilized egg is the only diploid cell. When this cell undergoes meiosis, the haploid cells that are produced form a multicellular adult. **(c)** In land plants and some algae, there is a multicellular diploid stage and a multicellular haploid stage. Typically, one of these two stages is larger and longer lived than the other. In ferns, the diploid stage is more prominent. QUESTION The green alga illustrated in part (b) is haploid during virtually all of its life. Does it undergo sexual reproduction?

What mechanisms are responsible for all this variation in life cycles, and why does it occur? The answers are not known. Explaining variation in life cycles—at both the proximate and ultimate levels—is a challenge that remains for future research.

12.4 Mistakes in Meiosis

When homologous chromosomes separate during meiosis I, a complete set of chromosomes is transmitted to each daughter cell. But what happens if there is a mistake and the chromosomes are not properly distributed? What are the consequences for offspring if gametes contain an abnormal set of chromosomes?

In 1866 Langdon Down described a distinctive suite of concurrent conditions observed in some humans. The syndrome was characterized by mental retardation, a high risk for heart problems and leukemia, and a degenerative brain disorder similar to Alzheimer's disease. **Down syndrome**, as the disorder came to be called, is observed in about 0.15 percent of live births (3 infants in every 2000). For over 80 years the cause of the syndrome was unknown. Then in the late 1950s a researcher published observations on the chromosome sets of nine Down syndrome children. The data suggested that the condition is associated with the presence of an extra copy of chromosome 21. This situation is called a **trisomy** ("three-bodies")—in this case, trisomy-21—because each cell has three copies of the chromosome. To explain the anomaly, the biologist proposed that the extra chromosome resulted from a mistake during meiosis in one of the parents.

How Do Mistakes Occur?

For a gamete to get one complete set of chromosomes, two steps in meiosis must be perfectly executed. During the first meiotic division in humans, for instance, 23 pairs of homologous chromosomes must separate, or *disjoin*, from each other so that only one homolog ends up in each daughter cell. If both homologs move to the same pole of the parent cell, however, the products of meiosis will be abnormal. This sort of meiotic error, illustrated in **Figure 12.14**, is referred to as **nondisjunction**, because the homologs do not separate or disjoin. Note that two daughter cells have two copies of the same chromosome (blue in Figure 12.14), while the other two lack that chromosome entirely. Gametes that contain an extra chromosome are symbolized as $n + 1$; gametes that lack one chromosome are symbolized as $n - 1$. If an $n + 1$ gamete is fertilized by a normal n gamete, the resulting zygote will be $2n + 1$. This situation is a trisomy. If the $n - 1$ gamete is fertilized by a normal n gamete, the resulting zygote will be $2n - 1$. This situation is called **monosomy**. Cells that have too many or too few chromosomes are said to be **aneuploid** ("without-form").

More rarely, abnormal $n + 1$ and $n - 1$ gametes are produced during nondisjunction in the second meiotic division. If

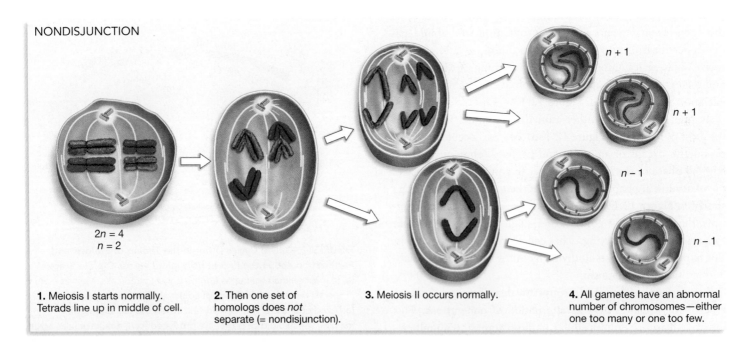

NONDISJUNCTION

$2n = 4$
$n = 2$

$n + 1$

$n + 1$

$n - 1$

$n - 1$

1. Meiosis I starts normally. Tetrads line up in middle of cell.

2. Then one set of homologs does *not* separate (= nondisjunction).

3. Meiosis II occurs normally.

4. All gametes have an abnormal number of chromosomes—either one too many or one too few.

FIGURE 12.14 Nondisjunction Leads to Gametes with Abnormal Chromosome Numbers
If homologous chromosomes fail to separate during meiosis I, the gametes that result will have an extra chromosome or will lack a chromosome. EXERCISE Nondisjunction also results when meiosis I occurs normally but two of the sister chromatids illustrated in step 3 fail to separate. Redraw steps 2–4 to show how this mechanism of nondisjunction occurs.

sister chromatids fail to separate and move to opposite poles of the dividing cell during meiosis II, then the resulting daughter cells will be $n + 1$ and $n - 1$.

Meiotic mistakes occur at a relatively high frequency. In humans, for example, researchers estimate that nondisjunction events occur in as many as 10 percent of meiotic divisions. The types of mistakes vary, but the consequences are almost always severe when defective gametes participate in fertilization. In a recent study of human pregnancies that ended in early embryonic or fetal death, 38 percent of the 119 cases involved atypical chromosome complements that resulted from mistakes in meiosis. Trisomy accounted for 36 percent of the abnormal karyotypes found; the incorrect number of complete chromosome sets, called *triploidy* ($3n$), for 30 percent; aberrantly sized or shaped chromosomes for 4 percent; and monosomy ($2n - 1$) for 2 percent. Mistakes in meiosis are a major cause of spontaneous abortion in humans.

Why Do Mistakes Occur?

The leading hypothesis to explain the incidence of trisomy and other meiotic mistakes is that they are accidents—random errors that occur during meiosis. Consistent with this proposal, there does not seem to be any genetic or inherited predisposition to trisomy or other types of dysfunction. Most cases of Down syndrome, for example, occur in families with no history of the condition.

Even though meiotic errors may be random, there are still strong patterns in their occurrence:

- With the exception of trisomy-21, most of the trisomies and monosomies observed in humans involve the sex chromosomes. Klinefelter syndrome, which develops in XXY individuals—as described in Box 12.1—occurs in about 1 in 2000 live births. Trisomy X (karyotype XXX) occurs in about 1 in 1000 live births and results in females who are healthy and fertile. **Turner syndrome** develops in XO individuals and occurs in about 1 in 5000 live births. Individuals with this syndrome are female but are sterile.

- **Table 12.3** shows data collected on trisomies in the autosomes of human fetuses and infants. Three observations deserve mention: (1) Trisomy is much more common in the smaller chromosomes (numbers 13–22) than it is in the larger chromosomes (numbers 1–12); (2) trisomy-21 is far and away the most common type of trisomy observed; and (3) maternal errors account for most incidences of trisomy. For example, over 90 percent of cases of Down syndrome are due to chromosomal defects in eggs.

- Maternal age is an important factor in the occurrence of trisomy. As **Figure 12.15** shows, the incidence of Down syndrome increases dramatically in mothers over 35 years old.

Why do these patterns occur? Biologists still do not have a good explanation for why most cases of aneuploidy in humans are due to problems with meiosis in females. In addition, it is still a mystery why there is such a strong correlation between maternal age and frequency of trisomy-21. But to explain why most instances of aneuploidy involve chromosome 21 or the sex chromosomes, biologists offer two hypotheses: (1) Individuals with other types of aneuploidy do not develop normally and are spontaneously aborted, long before birth; and (2) the sex chromosomes and chromosome 21 are more susceptible to aneuploidy than are other types of chromosomes. These hypotheses are not mutually exclusive, so both may be correct.

Studies on the mechanism of aneuploidy continue. In the meantime, one overall message is clear: Undergoing meiosis correctly is critical to the health and welfare of offspring.

TABLE 12.3 The Incidence of Trisomy in Humans: Effects of Chromosome Number and Paternal versus Maternal Origin

Trisomy (chromosome number)	Total Number of Cases	Due to Error in Sperm	Due to Error in Egg	Maternal Errors (%)
2–12	16	3	13	81
13	7	2	5	71
14	8	2	6	75
15	11	3	8	73
16	62	0	62	100
18	73	3	70	96
21	436	29	407	93
22	11	0	11	100

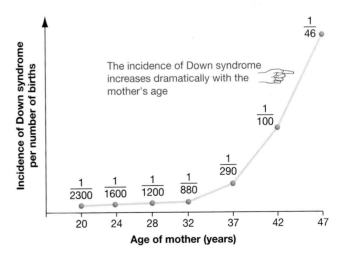

FIGURE 12.15 The Frequency of Down Syndrome Increases as a Function of a Mother's Age

QUESTION Suppose that you are an obstetrician. Based on these data, at what age would you recommend that pregnant mothers undergo procedures to check the karyotype of the embryos they carry?

ESSAY Seedless Fruits

Have you ever sat outside on a hot summer night eating watermelon and having a seed-spitting contest? It is likely that your children will never have the experience, because seedless varieties of watermelon have recently begun to dominate the market (**Figure 12.16**). New varieties of watermelon aren't the only example of a seedless fruit, either. Imagine biting into a banana and crunching down on hard, inedible seeds. The virtually seedless and sterile bananas we eat today are the result of selective breeding and artificial selection. As pointed out in Chapter 1, artificial selection occurs when humans allow only certain offspring of domesticated plants and animals to breed. Bananas have undergone artificial selection for fruit size, fruit sweetness, and seedlessness. The little black specks inside a banana are aborted seeds.

How did these seedless fruits come to be? One set of chromosomes makes the difference between the presence and absence of seeds in bananas and watermelons. In both of these species, diploid ($2n$) or tetraploid ($4n$) varieties make seeds while triploid ($3n$) strains are sterile and seedless.

Many cultivated plant species are polyploid, meaning that they have more than two sets of chromosomes. As long as there is a homolog for each chromosome during synapsis, meiosis I proceeds as it would in a diploid species and results in a halving of the chromosome number. For normal seed production, the key is that there must be an even number of chromosome sets. When the ploidy number is even, homologous chromosomes are able to synapse and later separate correctly. The wheat used in bread making, for example, is hexaploid ($6n$). It generates $3n$ gametes by meiosis. These fuse to form normal $6n$ offspring.

What happens when an individual has an odd number of chromosome sets? Consider triploid bananas and watermel-

ons. Early in meiosis I, each group of three homologous chromosomes tries to pair. Spindle fibers move the three chromosomes to the metaphase plate. When the homologs separate, however, one daughter cell typically receives two copies of a given chromosome while the other daughter cell receives one copy. This asymmetrical distribution of chromosomes occurs randomly for each chromosome in the cell. As a result, virtually all of the gametes that are produced have an unbalanced complement of chromosomes.

What happens when an individual has an odd number of chromosome sets?

In plants, as in animals, additions or deletions of specific chromosomes almost always reduce an individual's ability to produce offspring, and thus its fitness. In bananas and watermelons, gametes with unbalanced chromosome sets cannot produce viable embryos. Because they lack growing embryos, their seeds die and a seedless fruit results. Triploid bananas and watermelons are seedless because they can't complete meiosis correctly.

Where do these triploid, seedless strains come from in the first place? Triploid individuals originate in several ways. For example, massive errors can occur during meiosis I in a diploid parent and lead to all of the chromosomes being pulled to a single daughter cell. If the $2n$ gamete that results fuses with a normal gamete, a triploid offspring results. In other cases, plant breeders cross tetraploid and diploid parents from closely related species to form triploid offspring. Because the triploid is sterile, it has to be propagated through cuttings or some other mode of asexual reproduction.

In tropical regions where bananas are an important export crop, it is routine for thousands of acres to be planted with asexually propagated, genetically identical individuals. Because virtually all of these individuals have the same alleles for genes that fight disease, banana plantations are extremely susceptible to new strains of disease-causing organisms—particularly fungal epidemics. For decades banana farmers have been keeping disease-causing fungi at bay with massive doses of fungicides. But even though these chemicals kill most fungi, some fungal individuals happen to have alleles that make them less susceptible to fungicides. These fungicide-resistant individuals have increased in frequency as fungicide use has continued. In many parts of the world, banana plantations are now being threatened with outbreaks of disease from fungi that have evolved resistance to all of the currently available fungicides. To solve the problem, researchers may have to find diploid and tetraploid species of wild bananas and cross them to make entirely new triploid strains.

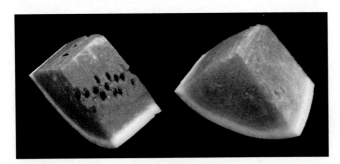

FIGURE 12.16 Seeded Watermelons Are Diploid; Seedless Watermelons Are Triploid
Diploid watermelons undergo meiosis normally, and they produce viable seeds. Triploid watermelons cannot undergo meiosis normally, and they produce aborted seeds.

CHAPTER REVIEW

Summary of Key Concepts

■ Meiosis is a type of nuclear division. It results in cells that have half as many chromosomes as the parent cell, and in animals it occurs prior to the formation of eggs and sperm. When an egg and a sperm cell combine to form an offspring, the original number of chromosomes is restored.

When biologists confirmed that sperm and egg nuclei fuse during fertilization, it led to the hypothesis that a special type of cell division must precede gamete formation. Specifically, the proposal was that a sperm and egg must each have half of the normal number of chromosomes found in other cells. This hypothesis was confirmed when researchers observed meiosis and established that it results in gametes with half the normal chromosome number.

As the details of meiosis were being worked out, biologists realized that chromosomes exist in sets. In diploid organisms, individuals have two versions of each type of chromosome. One of the versions is inherited from the mother, and one from the father. The similar, paired chromosomes are called homologs. Haploid organisms, in contrast, have just one of each type of chromosome.

Each chromosome is replicated before meiosis begins. At the start of meiosis I, each chromosome consists of a pair of sister chromatids joined along their length and at a centromere. Homologous pairs of chromosomes synapse early in meiosis I, forming a tetrad—a group of two homologous chromosomes. After non-sister chromatids from the homologous chromosomes undergo crossing over, the pair of homologous chromosomes migrates to the metaphase plate. At the end of meiosis I, the homologous chromosomes separate and are distributed to two daughter cells. During meiosis II, sister chromatids separate and are distributed to two daughter cells.

Web Tutorial 12.1   Meiosis

■ Each cell produced by meiosis receives a different combination of chromosomes. Because genes are located on chromosomes, each cell produced by meiosis receives a different complement of genes. Meiosis leads to offspring that are genetically distinct from each other and from their parents. If offspring are genetically variable, some might thrive if the environment changes.

When meiosis and outcrossing occur, the chromosome complements of offspring differ from one another and from their parents, for three reasons: (1) Maternal and paternal homologs are distributed randomly when chromosomes separate at the end of meiosis I; (2) maternal and paternal homologs exchange segments during crossing over; and (3) outcrossing results in a combination of chromosome sets from different individuals. The consequences of these differences became clear when biologists realized that chromosomes contain the hereditary material. Meiosis leads to genetic differences among offspring and between parents and offspring.

According to the changing-environment hypothesis, meiosis exists primarily because genetically diverse offspring are better able to resist parasites than are genetically uniform offspring. This hypothesis has been tested by studying populations of snails that contain both sexually reproducing and asexually reproducing individuals. The data from this research support the prediction that sexual reproduction is more common in habitats where parasite infection is frequent.

■ If mistakes occur during meiosis, the resulting egg and sperm cells may contain the wrong number of chromosomes. It is rare for offspring with an incorrect number of chromosomes to develop normally.

Mistakes during meiosis lead to gametes and offspring with an unbalanced set of chromosomes. Children with Down syndrome, for example, have an extra copy of chromosome 21. The leading hypothesis to explain these mistakes is that they are random accidents that result in a failure of homologous chromosomes or sister chromatids to separate properly during meiosis.

Web Tutorial 12.2   Mistakes in Meiosis

Questions

Content Review

1. In the roundworm *Ascaris*, eggs and sperm have two chromosomes, but all other cells have four. Observations such as this inspired which important hypothesis?
   a. Before gamete formation, a special type of cell division leads to a quartering of chromosome number.
   b. Before gamete formation, a special type of cell division leads to a halving of chromosome number.
   c. After gamete formation, half of the chromosomes are destroyed.
   d. After gamete formation, either the maternal or the paternal set of chromosomes disintegrates.

2. What are homologous chromosomes?
   a. chromosomes that are similar in their size, shape, and gene content
   b. similar chromosomes that are found in different individuals of the same species
   c. the two "threads" in a replicated chromosome (they are identical copies)
   d. the products of crossing over, which contain a combination of segments from maternal chromosomes and segments from paternal chromosomes

3. What is a tetrad?
   a. the "X" that forms when chromatids from homologous chromosomes cross over
   b. a group of four chromatids produced when homologs synapse
   c. the four points where homologous chromosomes touch as they synapse
   d. the group of four genetically identical daughter cells produced by mitosis

4. What is genetic recombination?
   a. the new combination of maternal and paternal homologs that results when chromosomes separate at meiosis I
   b. the new combination of maternal and paternal chromosome segments that results when homologs cross over
   c. the new combinations of chromosome segments that result when outcrossing occurs
   d. the combination of a prominent haploid phase *and* a prominent diploid phase in a life cycle

5. What is meant by a paternal chromosome?
   a. the largest chromosome in a set
   b. a chromosome that does not separate correctly during meiosis I
   c. the member of a homologous pair that was inherited from the mother
   d. the member of a homologous pair that was inherited from the father

6. Meiosis II is similar to which process?
   a. mitosis in haploid cells
   b. nondisjunction
   c. outcrossing
   d. meiosis I

Conceptual Review

1. Triploid ($3n$) watermelons are produced by crossing a tetraploid ($4n$) strain with a diploid ($2n$) plant. Briefly explain why this mating produces a triploid individual. Why can mitosis proceed normally in triploid cells, but meiosis cannot?

2. Meiosis is called a reduction division, but all of the reduction occurs during meiosis I—no reduction occurs during meiosis II. Explain why meiosis I is a reduction division but meiosis II is not.

3. Some plant breeders are concerned about the resistance of asexually cultivated plants, such as seedless bananas, to new strains of disease-causing bacteria, viruses, or fungi. Briefly explain their concern by discussing the differences in the genetic "outcomes" of asexual and sexual reproduction.

4. Explain why nondisjunction leads to trisomy and other types of abnormal chromosome complements. In what sense are these chromosome complements "unbalanced?"

5. Examine Figure 12.1, which shows the karyotype of the lubber grasshopper. The researcher drew these chromosomes as they were undergoing meiosis. What event is occurring in the drawing?

6. Lay two pens and two pencils on a tabletop, and imagine that they represent chromosomes in a diploid cell with $n = 2$. Explain the phases of meiosis by moving the pens and pencils around. (If you don't have enough pens and pencils, use strips of paper or fabric.)

Group Discussion Problems

1. The gibbon has 44 chromosomes per diploid set, and the siamang has 50 chromosomes per diploid set. In the 1970s a chance mating between a male gibbon and a female siamang produced an offspring. Predict how many chromosomes were observed in the somatic cells of the offspring. Do you predict that this individual would be able to form viable gametes? Why or why not?

2. Meiosis results in a reassortment of maternal and paternal chromosomes. If $n = 3$ for a given organism, there are eight different combinations of paternal and maternal chromosomes. If no crossing over occurs, what is the probability that a gamete will receive *only* paternal chromosomes?

3. Suppose that while studying coral reefs in Indonesia, you discovered a new species of algae. When you examine the cells of different individuals in the laboratory, you find that some individuals have cells with 20 chromosomes while other, identical-looking individuals have cells with 10 chromosomes. What is going on?

4. The data on snail populations that were used to test the changing-environment hypothesis have been criticized because they are observational and not experimental in nature. As a result, they do not control for factors other than parasites that might affect the frequency of sexually reproducing individuals.
   a. Design an experimental study that would provide stronger evidence that the frequency of parasite infection causes differences in the frequency of sexually versus asexually reproduced individuals in this species of snail.
   b. In defense of the existing data, comment on the value of observing patterns like this in nature, versus under controlled conditions in the laboratory.

Answers to Multiple-Choice Questions: 1. b; 2. a; 3. b; 4. b; 5. d; 6. a

Mendel and the Gene

13

Gregor Mendel performed some of the most brilliant experiments in the history of biology in this garden.

KEY CONCEPTS

▓ Mendel discovered that in garden peas, individuals have two alleles, or versions, of each gene. Prior to the formation of eggs and sperm, the two alleles of each gene separate so that one allele is transmitted to each egg or sperm cell.

▓ Genes are located on chromosomes. The separation of homologous chromosomes during meiosis I explains why alleles of the same gene segregate to different gametes.

▓ If genes are located on different chromosomes, then the alleles of each gene are transmitted to egg cells and sperm cells independently of each other.

▓ Important exceptions exist to the rules that individuals have two alleles of each gene and that alleles of different genes are transmitted independently. Genes on the same chromosome are not transmitted independently of each other. If a gene is on a sex chromosome, then not every gamete will receive an allele of that gene.

The science of biology is built on a series of great ideas. Two of these—the cell theory and the theory of evolution—were introduced in Chapter 1. The cell theory describes the basic structure of organisms; the theory of evolution by natural selection clarifies why species change through time. These theories explain fundamental features of the natural world and answer some of our most profound questions about the nature of life: What are organisms made of? Where did species come from?

A third great idea in biology addresses an equally important question: Why do offspring resemble their parents? An Austrian monk named Gregor Mendel provided part of the answer in 1865 when he announced that he had worked out

the rules of inheritance through a series of experiments on garden peas. The other part of the answer was provided by the biologists who described the details of meiosis during the final decades of the nineteenth century. In 1903 Walter Sutton and Theodor Boveri linked these two parts by formulating the *chromosome theory of inheritance*. This theory contends that meiosis, introduced in Chapter 12, causes the patterns of inheritance that Mendel observed. It also asserts that the hereditary factors called *genes* are located on chromosomes.

This chapter focuses on the evidence for the chromosome theory of inheritance. Let's begin with a basic question: What are the rules of inheritance that Mendel discovered?

13.1 Mendel's Experiments with a Single Trait

Gregor Mendel was a monk who lived and worked in the city of Brünn, located 70 miles north of Vienna. (Brünn was then part of Austria. Today the city is called Brno and is part of the Czech Republic.) Mendel was educated in the natural sciences at the University of Vienna and also studied physics and mathematics under Christian Doppler, who discovered the Doppler effect for sound and light waves.

In Mendel's day, questions about **heredity**—meaning inheritance, or the transmission of traits from parents to offspring—were primarily the concern of animal breeders and horticulturists. A **trait** is any characteristic of an individual, ranging from overall height to the primary structure of a particular membrane protein. In Brünn, for example, there was a particular interest in how selective breeding could result in hardier and more productive varieties of sheep, fruit trees, and vines. To that end, an Agricultural Society had been formed. Its members emphasized the importance of research that would help breeding programs become more efficient. Mendel was an active member of this society; the monastery he belonged to was also devoted to scientific teaching and research.

What Questions Was Mendel Trying to Answer?

Mendel set out to address the most fundamental of all issues concerning heredity: What are the basic patterns in the transmission of traits from parents to offspring? At the time, two hypotheses had been formulated to answer this question. The first, called *blending inheritance*, claimed that the traits observed in a mother and father blend together to form the traits observed in their offspring. For example, blending inheritance contended that black sheep have hereditary determinants for black wool and that white sheep have hereditary determinants for white wool. When these individuals mate, their hereditary determinants blend to form a new hereditary determinant for gray wool—their offspring should be gray. Mendel's scientific mentor, the widely respected Carl Nägeli, was a proponent of the blending-inheritance hypothesis.

The second hypothesis was called the *inheritance of acquired characters*, which claimed that traits present in parents are modified, through use, and passed on to their offspring in the modified form. The classical prediction of this hypothesis is that adult giraffes acquire longer necks by straining to reach leaves high in the tops of trees and that they subsequently produce longer-necked offspring. The idea here is that the genetic determinants present in an individual are modified through use. Jean-Baptiste Lamarck originally formulated this hypothesis in the eighteenth century; in Mendel's day it was championed by Charles Darwin.

These hypotheses were being promoted by the greatest scientists of Mendel's time. But are they correct? What *are* the basic patterns of inheritance?

Garden Peas Serve as the First Model Organism in Genetics

Mendel was certainly not the first scientist interested in studying the basic mechanisms of heredity. Why was he successful where others failed? Several factors came into play. One of the most important was that Mendel chose an appropriate model organism to study. A species that serves as a **model organism** consists of individuals that are usually small, short lived, inexpensive to care for, prolific in producing offspring, and easy to manipulate experimentally. Such species are called models because the conclusions drawn from studying them turn out to apply to many other species as well.

Which model organism did Mendel choose? After investigating and discarding several candidates, he chose the pea plant *Pisum sativum*. His reasons were practical: Peas are inexpensive, are easy to propagate, and have a relatively short reproductive cycle. These features made it possible for Mendel to continue experiments over several generations and to collect data from a large number of individuals. Because of his choice, garden peas became the first model organism in genetics. **Genetics** is the branch of biology that focuses on the inheritance of traits.

Two additional features of the pea made it possible for Mendel to design his experiments: He could control which parents were involved in a mating, and he could arrange matings between individuals that differed in easily recognizable traits such as flower color or seed shape. Why was this important?

How Did Mendel Arrange Matings? **Figure 13.1a** shows a garden pea flower, including its male and female reproductive organs. Sperm cells are produced in **pollen grains**, which are small sacs that mature in the male reproductive structure of the plant. Eggs are produced in the female reproductive structure. Fertilization begins when pollen grains are deposited on a tubelike section of the female reproductive organs. Sperm cells travel down this tube to the egg cells, where fertilization takes place.

Under normal conditions, garden peas pollinate themselves rather than requiring pollen from other pea plants for fertilization to occur. **Self-fertilization** (or *selfing*) takes place when pollen from one flower falls on the female reproductive organ of that same flower. Selfing is common because pollen from other plants rarely reaches the flower—its petals form a compartment that encloses the male and female reproductive organs and tends to exclude bees and other types of pollinating insects.

As **Figure 13.1b** shows, however, Mendel could circumvent this arrangement by removing the male reproductive organs from a flower before any pollen formed. Later he could put pollen from another flower on that flower's female reproductive organ. This type of mating is referred to as a **cross-pollination**, or simply a *cross*. Using this technique, Mendel could control the matings of his model organism.

What Traits Did Mendel Study? Mendel conducted his experiments on varieties of peas that differed in seven traits: seed

(a) Self-pollination

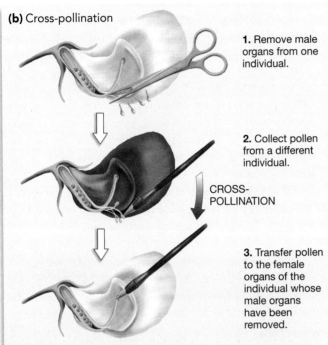

Female organ (receives pollen)

SELF-POLLINATION

Male organs (produce pollen grains, which produce male gametes)

Eggs

(b) Cross-pollination

1. Remove male organs from one individual.

2. Collect pollen from a different individual.

CROSS-POLLINATION

3. Transfer pollen to the female organs of the individual whose male organs have been removed.

FIGURE 13.1 Peas Can Be Self-Pollinated or Cross-Pollinated
(a) The petals of a pea form an enclosed compartment. As a result, most fertilization takes place when pollen grains from the male reproductive organ of one flower fall on the female reproductive organ of the same flower. **(b)** Mendel arranged matings between individuals by removing the male organs from one flower and then dusting its female organ with pollen collected from a different flower.

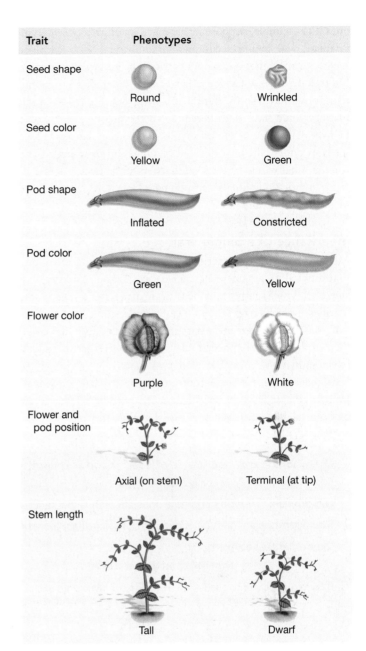

Trait	Phenotypes	
Seed shape	Round	Wrinkled
Seed color	Yellow	Green
Pod shape	Inflated	Constricted
Pod color	Green	Yellow
Flower color	Purple	White
Flower and pod position	Axial (on stem)	Terminal (at tip)
Stem length	Tall	Dwarf

FIGURE 13.2 Mendel Studied Seven Traits That Were Variable in Garden Peas
Two distinct phenotypes existed for each of the seven traits that Mendel studied in garden peas.

shape, seed color, pod shape, pod color, flower color, flower and pod position, and stem length. As **Figure 13.2** shows, each trait exhibited one of two forms. Biologists refer to the observable features of an individual, such as the shape of a pea seed or the eye color of a human, as its **phenotype** (literally, "show-type"). In the pea populations that Mendel studied, two distinct phenotypes existed for all seven traits.

Mendel began his work by obtaining individuals from what breeders called pure lines or true-breeding lines. A **pure line** consists of individuals that produce offspring identical to themselves when they are self-pollinated or crossed to another member of the same population. For example, earlier breeders had developed pure lines for wrinkled seeds and round seeds. During two years of trial experiments, Mendel confirmed that individuals that germinated from his wrinkled seeds produced only

wrinkled-seeded offspring when they were mated to themselves or to another pure-line individual that germinated from a wrinkled seed; individuals from his round seeds produced only round-seeded offspring when they were mated to themselves or to another pure-line individual from a round seed.

Why is this result important? Remember that Mendel wanted to find out how traits are transmitted from parents to offspring. Once he had confirmed that he was working with pure lines, he could predict how matings within each line would turn out; in other words, he knew what the offspring from these matings would look like. He could then compare these results with the outcomes of crosses between individuals from different pure lines. For example, suppose he arranged matings between an individual with round seeds and an individual with wrinkled seeds. He knew that one parent carried a hereditary determinant for round seeds, while the other carried a hereditary determinant for wrinkled seeds. But the offspring that resulted from this mating would have both hereditary determinants. They would be **hybrids**—a mix of the two types. Would they have wrinkled seeds, round seeds, or a blended combination of wrinkled and round? What would be the seed shape in subsequent generations when hybrid individuals self-pollinated or were crossed with members of the pure lines?

Inheritance of a Single Trait

Mendel's first set of experiments involved crossing pure lines that differed in just one trait, such as seed shape. Working with single traits was important because it made the results of the matings more interpretable. Once he understood how a single trait was transmitted from parents to offspring, Mendel could then explore what happened when crosses were performed between individuals that differed in two traits.

Mendel began his single-trait crosses by crossing individuals from round-seeded and wrinkled-seeded pure lines. The adults used in an initial experimental cross such as this represent the **parental generation**. Their progenies (that is, offspring) are called the **F$_1$ generation**. (F$_1$ stands for "first filial"; *filial* refers to the relationship between parent and offspring.) Subsequent generations are symbolized as the F$_2$ generation, F$_3$ generation, and so on.

Certain Traits "Recede"   In his first set of crosses, Mendel took pollen from round-seeded plants and placed it on the female reproductive organs of plants from the wrinkled-seeded line. As **Figure 13.3** shows, all of the progeny seeds resulting from this cross were round. This was remarkable, for two reasons. First, the traits did not blend together to form an intermediate phenotype. Instead, the round-seeded form appeared intact. This result was in stark contrast to the predictions of the blending-inheritance hypothesis. Second, the genetic determinant for wrinkled seeds seemed to have disappeared. Did this disappearance occur because the determinant for wrinkled seeds was located in the egg (produced by the female flower) instead of in the pollen (produced by the male flower)? In general, did it matter which parent had a particular genetic determinant?

To answer these questions, Mendel performed a second set of crosses—this time with pollen taken from an individual germinated from a wrinkled-seeded pea (see Figure 13.3). These

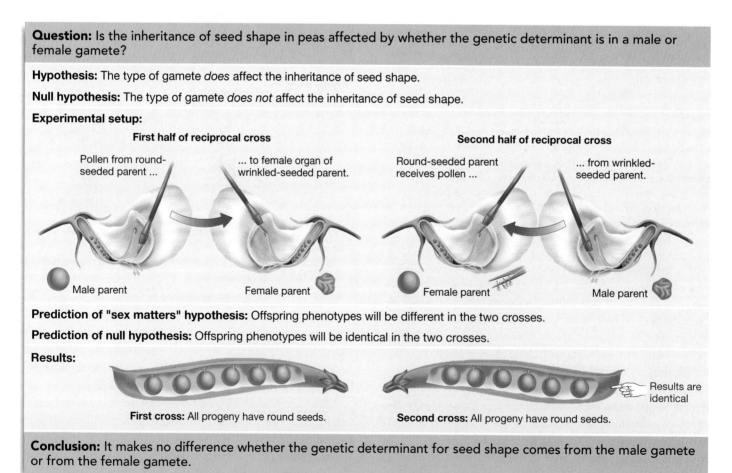

Question: Is the inheritance of seed shape in peas affected by whether the genetic determinant is in a male or female gamete?

Hypothesis: The type of gamete *does* affect the inheritance of seed shape.

Null hypothesis: The type of gamete *does not* affect the inheritance of seed shape.

Experimental setup:

First half of reciprocal cross

Pollen from round-seeded parent ...   ... to female organ of wrinkled-seeded parent.

Male parent   Female parent

Second half of reciprocal cross

Round-seeded parent receives pollen ...   ... from wrinkled-seeded parent.

Female parent   Male parent

Prediction of "sex matters" hypothesis: Offspring phenotypes will be different in the two crosses.

Prediction of null hypothesis: Offspring phenotypes will be identical in the two crosses.

Results:

First cross: All progeny have round seeds.   Second cross: All progeny have round seeds.

Results are identical

Conclusion: It makes no difference whether the genetic determinant for seed shape comes from the male gamete or from the female gamete.

FIGURE 13.3 A Reciprocal Cross

QUESTION What is the purpose of a reciprocal cross?

crosses completed a **reciprocal cross**—a set of matings where the mother's phenotype in the first cross is the father's phenotype in the second cross, and the father's phenotype in the first cross is the mother's phenotype in the second cross.

In this case the results of the reciprocal crosses were identical: All of the F_1 progeny in the second cross had round seeds, just as in the first cross. This second cross established that it does not matter whether the genetic determinants for seed shape are located in the male or female parent. But what had happened to the genetic determinant for wrinkled seeds?

Dominant and Recessive Traits Mendel planted the F_1 seeds and allowed the individuals to self-pollinate when they matured. He collected the seeds that were produced by many plants in the F_2 generation and observed that 5474 were round and 1850 were wrinkled. This observation was striking. The wrinkled seed shape reappeared in the F_2 generation after disappearing completely in the F_1 generation! No one had observed this phenomenon before because it had been customary for biologists to stop their breeding experiments with F_1 offspring. Mendel was successful where others failed, in part because he extended his experiments to a second generation.

Mendel invented some important terms to describe this result. He designated the genetic determinant for the wrinkled shape as **recessive**. This was an appropriate term because none of the F_1 individuals had wrinkled seeds—meaning the determinant for wrinkled seeds appeared to recede or temporarily become latent. In contrast, Mendel referred to the genetic determinant for round seeds as **dominant**. This term was apt because the round-seed determinant appeared to dominate over the wrinkled-seed determinant when both were present. It's important to note, though, that in genetics the term *dominant* has nothing to do with the everyday English usage as powerful or superior. Subsequent research has shown that individuals with the dominant phenotype do not necessarily have higher fitness than do individuals with the recessive phenotype. In fact, there are many examples of dominant genetic determinants that lead to the death of the individual carrying them. Nor are dominant genetic determinants necessarily more common than recessive ones. In genetics, the terms *dominance* and *recessiveness* refer *only* to which phenotype is observed in individuals carrying two different genetic determinants.

What is the relationship between these two types of determinants? That is, how do they interact? Mendel made an important start in answering these questions when he noticed that the round and wrinkled seeds of the F_2 generation were present in a ratio of 2.96:1, or essentially 3:1. The 3:1 ratio means that for every four individuals, on average three had the dominant phenotype and one had the recessive phenotype. In other words, about 3/4 of the F_2 seeds were round and 1/4 were wrinkled.

Before trying to interpret this pattern, however, it was important for Mendel to establish that the results were not restricted to inheritance of seed shape. So he repeated the experiments with each of the six other traits of pea plants. In each case, he obtained similar results: The products of reciprocal crosses were the same; one form of the trait was always dominant regardless of the parent it came from; the F_1 progeny showed only the dominant trait and did not exhibit an intermediate phenotype; and in the F_2 generation, the ratio of individuals with dominant and recessive phenotypes was 3 to 1.

How could these patterns be explained? Mendel answered this question with a series of propositions about the nature and behavior of the hereditary determinants. These hypotheses rank as some of the most brilliant insights in the history of biology.

The Nature and Behavior of the Hereditary Determinants

Mendel's results were clearly inconsistent with the hypothesis of blending inheritance. To explain the patterns that he observed, Mendel proposed a competing hypothesis called *particulate inheritance*. He maintained that the hereditary determinants for traits do not blend together or acquire new or modified characteristics through use. In fact, hereditary determinants maintain their integrity from generation to generation. Instead of blending together, they act like discrete entities or particles.

Mendel's hypothesis was the only way to explain the observation that phenotypes disappeared in one generation and reappeared intact in the next. It also represented a fundamental break with ideas that had prevailed for hundreds of years.

What Are Genes, Alleles, and Genotypes? Today geneticists use the word **gene** to indicate the hereditary determinant for a trait. For example, the hereditary factor that determines the difference between round and wrinkled seeds in garden peas is referred to as the gene for seed shape.

Mendel's insights were even more penetrating, however. He also proposed that each individual has two versions of each gene. Today different versions of the same gene are called **alleles**. Different alleles are responsible for the variation in the traits that Mendel studied. In the case of the gene for seed shape, one allele of this gene is responsible for the round form of the seed while another allele is responsible for the wrinkled form. The alleles that are found in a particular individual are called its **genotype**. An individual's genotype has a profound effect on its phenotype—its physical traits.

The hypothesis that alleles exist in pairs was important because it gave Mendel a framework for explaining dominance and recessiveness. He proposed that some alleles are dominant and others are recessive. Dominance and recessiveness identify which phenotype actually appears in an individual when both alleles are present. In garden peas, the allele for round seeds is dominant; the allele for wrinkled seeds is recessive. Therefore, as long as one allele for round seeds is present, seeds are round. When both alleles present are for wrinkled seeds (thus no allele for round seeds is present), seeds are wrinkled.

These hypotheses explain why the phenotype for wrinkled seeds disappeared in the F_1 generation and reappeared in the F_2 generation. But why did round- and wrinkled-seeded plants exist in a 3:1 ratio in the F_2 generation?

The Principle of Segregation To explain the 3:1 ratio of phenotypes in F_2 individuals, Mendel reasoned that the two alleles of each gene must *segregate*—that is, separate—into different gamete cells during the formation of eggs and sperm in the parents. As a result, each gamete contains one allele of each gene. This idea is called the **principle of segregation**.

To show how this principle works, Mendel used a letter to indicate the gene for a particular trait. For example, *R* represents the gene for seed shape. He used uppercase (*R*) to symbolize a dominant allele and lowercase (*r*) to symbolize a recessive allele. (Note that the symbols for genes are always italicized.)

Using this notation, Mendel could describe the genotype of the individuals in the pure line with round seeds (dominant) as *RR*. The genotype of the pure line with wrinkled seeds (recessive) is *rr*. Because *RR* and *rr* individuals have two copies of the same allele, they are said to be **homozygous** for the seed-shape gene (*homo* is the Greek root for "same," while *zygo* means "yoked together"). Pure-line individuals always produce offspring with the same phenotype because they are homozygous—no other allele is present.

Figure 13.4a diagrams what happened to these alleles when Mendel crossed the *RR* and *rr* pure lines. According to his analysis, *RR* parents produce eggs and sperm that carry the *R* allele, while *rr* parents produce gametes with the *r* allele. When two gametes—one from each parent—are fused together, they create offspring with the *Rr* genotype. Such individuals, with two different alleles for the same gene, are said to be **heterozygous** (*hetero* is the Greek root for "different"). Because the *R* allele is dominant, all of these F_1 offspring produced round seeds.

Why do the two phenotypes appear in a 3:1 ratio in the F_2 generation? A mating between parents that are both heterozygous at the gene in question is called a **monohybrid cross**. Mendel proposed that during gamete formation in the F_1 (heterozygous) individuals, the paired *Rr* alleles separate into different gamete cells. As a result, about half of the gametes carry the *R* allele and half carry the *r* allele (**Figure 13.4b**). During self-fertilization, a given sperm has an equal chance of fertilizing either an *R*-bearing egg or an *r*-bearing egg. R. C. Punnett invented a straightforward technique for predicting the genotypes and phenotypes that should appear in the resulting offspring. In a **Punnett square**, each gamete genotype produced by one parent is shown as a row, and each gamete genotype produced by the other parent is shown as a column. Then the boxes of the square or rectangle composed of the rows and columns are filled in with the offspring genotypes that result from fusion of these gametes. Finally, the proportions of each offspring genotype and phenotype can be calculated by tallying

the offspring genotypes and phenotypes present in the boxes. (**Box 13.1** explains the logic behind these calculations.) For example, the Punnett square in Figure 13.4b predicts that 1/4 of the F_2 offspring will be *RR*, 1/2 will be *Rr*, and 1/4 will be *rr*. Because the *R* allele is dominant to the *r* allele, 3/4 of the offspring should be round seeded and 1/4 should be wrinkled seeded.

(a) A cross between two homozygotes

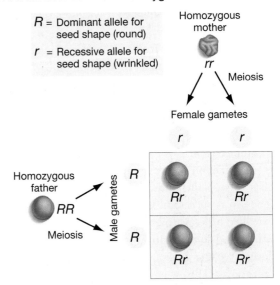

Offspring genotypes: All *Rr* (heterozygous)
Offspring phenotypes: All round seeds

(b) A cross between two heterozygotes

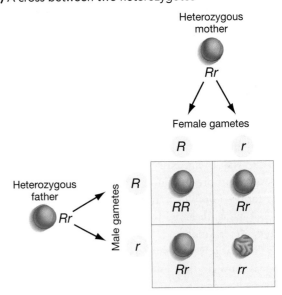

Offspring genotypes: 1/4 *RR* : 1/2 *Rr* : 1/4 *rr*
Offspring phenotypes: 3/4 round : 1/4 wrinkled

FIGURE 13.4 Mendel Analyzed the F_1 and F_2 Offspring of a Cross Between Pure Lines

QUESTION In constructing a Punnett square, does it matter whether the male or female gametes go on the left or across the top? Why or why not?

BOX 13.1 Why Do Punnett Squares Work?

Punnett squares are powerful tools because they allow biologists to predict the outcome of a particular mating. The calculations involved in setting up and analyzing Punnett squares have a strong theoretical basis. Punnett squares work because they are based on two fundamental rules of probability, called the "both-and rule" and the "either-or rule." Each rule pertains to a distinct situation.

The *both-and rule* applies when you want to know the probability that two or more independent events occur together. Let's use the rolling of two dice as an example. What is the probability of rolling two sixes? These two events are independent, because the probability of rolling a six on one die has no effect on the probably of rolling a six on the other die. (In the same way, the probability of getting a gamete with allele R from one parent has no effect on the probability of getting a gamete with allele R from the other parent. Gametes fuse randomly.) The probability of rolling a six on the first die is 1/6. The probability of rolling a six on the second die is also 1/6. The probability of rolling a six on both die, then, is $1/6 \times 1/6 = 1/36$. In other words, if you rolled two dice 36 times, on average you would expect to roll two sixes once. It should make sense that the both-and rule is also called the *multiplication rule* or *product rule*. In the case of the cross diagrammed in Figure 13.4b, the probability of getting allele R from the father is 1/2 and the probability of getting R from the mother is 1/2. Thus, the probability of getting both alleles and creating an offspring with genotype RR is $1/2 \times 1/2 = 1/4$.

The *either-or rule*, in contrast, applies when you want to know the probability of an event happening when there are several ways for the event to occur. In this case, the probability that the event will occur is the sum of the probabilities of each way that it can occur. For example, suppose you wanted to know the probability of rolling either a one or a six when you toss a die. The probability of tossing each is 1/6, so the probability of getting one or the other is $1/6 + 1/6 = 1/3$. (The either-or rule is also called the *addition rule* or *sum rule*.) If you rolled a die three times, on average you'd expect to get a one or a six once. In Figure 13.4b, the probability of getting an R allele from the father and an r allele from the mother is $1/2 \times 1/2 = 1/4$. Similarly, the probability of getting an r allele from the father and an R allele from the mother is $1/2 \times 1/2 = 1/4$. Thus, the combined probability of getting the Rr genotype in either of the two ways is $1/4 + 1/4 = 1/2$. Because a Punnett square shows every possible way to generate a particular genotype, you can calculate the total probability of getting a particular genotype or phenotype simply by calculating how often it occurs in the square.

These results are *exactly* what Mendel found in his experiments with peas. In the simplest and most elegant fashion possible, his interpretation explains the 3:1 ratio of round to wrinkled seeds observed in the F_2 offspring and the mysterious reappearance of the wrinkled seeds.

The term **genetic model** refers to a set of hypotheses that explains how a particular trait is inherited. **Figure 13.5** summarizes Mendel's model for explaining the basic patterns in the transmission of traits from parents to offspring. These hypotheses are sometimes referred to as *Mendel's rules*. His genetic model was a radical break from the hypotheses of blending inheritance and inheritance of acquired characters that previously dominated scientific thinking about heredity.

Testing the Model

Mendel's model explained his results in a logical way. But is it correct? To answer this question, Mendel conducted a series of experiments with the F_2 progeny described in Figure 13.4b. These experiments tested two important predictions:

1. Plants with wrinkled seeds are *rr*. Thus, they should produce only *rr* offspring when they are self-pollinated or crossed with another individual with wrinkled seeds.

MENDEL'S MODEL

1. **Peas have two versions, or alleles, of each gene.** This also turns out to be true for many other organisms.

2. **Alleles do not blend together.** The hereditary determinants maintain their integrity from generation to generation.

3. **Each gamete contains one allele of each gene.** The alleles of each gene segregate during the formation of gametes.

4. **Males and females contribute equally to the genotype of their offspring.** When gametes fuse, offspring acquire a total of two alleles for each gene—one from each parent.

5. **Some alleles are dominant to other alleles.** When a dominant allele and a recessive allele for the same gene are found in the same individual, that individual has the dominant phenotype.

FIGURE 13.5 Mendel Created a Model to Explain the Results of a Cross between Pure Lines

QUESTION What is the difference between genes and alleles, a genotype and a phenotype, a homozygous individual and a heterozygous individual, and dominant alleles and recessive alleles?

2. Plants with the dominant phenotype may be either *Rr* or *RR*. These two genotypes should be present in the ratio 2:1. (That is, there should be twice as many heterozygotes as homozygotes among individuals with round seeds.) Individuals with the *RR* genotype should produce only *RR* offspring when they are self-pollinated. In contrast, *Rr* individuals should produce offspring with the same 3:1 ratio of round:wrinkled phenotypes observed in the cross diagrammed in Figure 13.4b.

Mendel planted the F_2 seeds and allowed the plants to self-pollinate when they matured. He then examined the phenotypes of the F_3 seeds. He quickly confirmed the first prediction: F_2 plants with wrinkled seeds always produced offspring with wrinkled seeds. This result was consistent with the claim that these F_2 plants have an *rr* genotype.

What about the offspring of parents with the dominant phenotype? Mendel let 565 round-seeded plants self-pollinate. Of these, 193 plants produced only round-seeded offspring. Mendel inferred that these parents had the *RR* genotype. In contrast, 372 of the round-seeded parents produced seeds that were either round or wrinkled. Mendel inferred that in these individuals, the round-seeded parental genotype was *Rr*. In this experiment, the ratio of *Rr* to *RR* in the parents—based on the number of round and wrinkled seeds they produced—was 1.93:1. This is extremely close to the prediction of 2:1. Mendel observed the same patterns when he let F_2 individuals from the other six crosses (involving seed color and the other traits) self-fertilize. These results were a ringing confirmation of his model.

13.2   Mendel's Experiments with Two Traits

Working with one trait at a time allowed Mendel to establish that blending inheritance does not occur. It also allowed him to infer that each gene he was working with had two alleles and to recognize the principle of segregation. His next step was to extend these results. The most important question he addressed was whether the principle of segregation holds true if parental lines differ with respect to two traits.

To explore this issue, Mendel crossed a pure-line parent that produced round, yellow seeds with a pure-line parent that produced wrinkled, green seeds. The F_1 offspring of this cross should be heterozygous at both genes. A mating between parents that are both heterozygous for two traits is called a **dihybrid cross**. **Figure 13.6** shows how Mendel set up one of his dihybrid crosses.

Mendel's earlier experiments had established that the allele for yellow seeds was dominant to the allele for green seeds; these alleles were designated *Y* for yellow and *y* for green. As Figure 13.6 indicates, two distinct possibilities existed for how the alleles of these two different genes—the gene for seed

shape and the gene for seed color—would be transmitted to offspring. The first possibility was that the allele for seed shape and the allele for seed color present in each parent would separate from one another and be transmitted independently. This hypothesis is called **independent assortment,** because the two alleles would separate and sort themselves independently of each other (Figure 13.6a). The second possibility was that the allele for seed shape and the allele for seed color would be transmitted to gametes together. This hypothesis can be called *dependent assortment*, because the transmission of one allele would depend on the transmission of another (Figure 13.6b).

As Figure 13.6 shows, the F_1 offspring of Mendel's mating are expected to have the dominant round and yellow phenotypes whether the alleles were transmitted together or independently. When Mendel did the cross and observed the F_1 individuals, this is exactly what he found. All of the F_1 offspring had round, yellow seeds. All of these individuals were heterozygous at both genes.

In contrast to the situation in the F_1 generation, however, the two hypotheses make radically different predictions about what Mendel should have observed when the F_1 individuals were allowed to self-fertilize and produce an F_2 generation. If the alleles assort independently and combine randomly to form gametes, then each heterozygous parent should produce four different gamete genotypes, as illustrated in Figure 13.6a. A 4-row-by-4-column Punnett square results, and it predicts that there should be 9 different offspring genotypes and 4 phenotypes. Further, the yellow-round, green-round, yellow-wrinkled, and green-wrinkled phenotypes should be present in the frequencies 9/16, 3/16, 3/16, and 1/16, respectively. This is equivalent to a ratio of 9:3:3:1. But if the alleles from each parent stay together, then a 2-row-by-2-column Punnett square would predict only three possible offspring genotypes and just two phenotypes, as Figure 13.6b shows. The hypothesis of dependent assortment predicts that F_2 offspring should be yellow-round or green-wrinkled, present in a ratio of 3:1. Note that the Punnett squares are making explicit predictions about the outcome of an experiment, based on a specific hypothesis about which alleles are present in each parent and how they are transmitted.

When Mendel examined the phenotypes of the F_2 offspring, he found that they conformed to the predictions of the hypothesis of independent assortment. Four phenotypes were present in frequencies that closely approximated the predicted frequencies of 9/16, 3/16, 3/16, and 1/16 and the predicted ratio of 9:3:3:1 (Figure 13.6c). Based on these data, he accepted the hypothesis that alleles of different genes are transmitted independently of one another. This result became known as the **principle of independent assortment.**

As an aside, it's interesting to analyze how these data relate to the 3:1 ratio of phenotypes observed for single traits. According to the data in Figure 13.6c, the ratio of round seeds to

(a) Hypothesis of independent assortment:
Alleles of different genes don't stay together when gametes form.

(b) Hypothesis of dependent assortment:
Alleles of different genes stay together when gametes form.

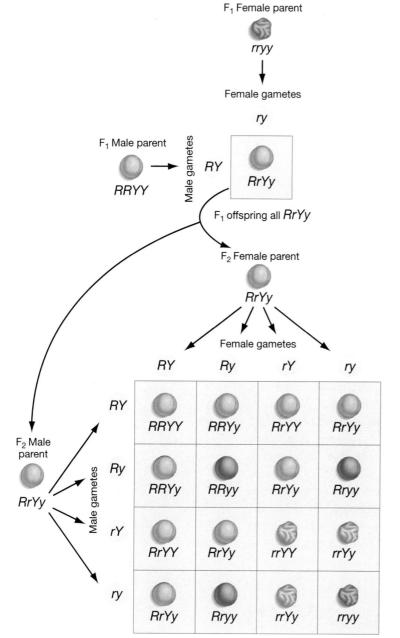

F₂ offspring genotypes: 9/16 *R–Y–* : 3/16 *R–yy* : 3/16 *rrY–* : 1/16 *rryy*
F₂ offspring phenotypes: 9/16 ⬤ : 3/16 ⬤ : 3/16 ◉ : 1/16 ◉

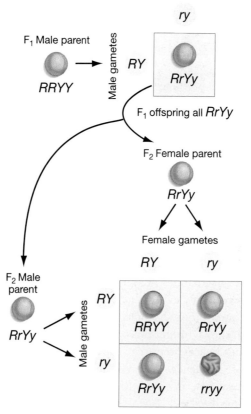

F₂ offspring genotypes: 1/4 *RRYY* : 1/2 *RrYy* : 1/4 *rryy*
F₂ offspring phenotypes: 3/4 ⬤ : 1/4 ◉

R = Dominant allele for seed shape (round)
r = Recessive allele for seed shape (wrinkled)
Y = Dominant allele for seed color (yellow)
y = Recessive allele for seed color (green)

(c) Mendel's results

F₂ generation phenotype	⬤	◉	⬤	◉	556 total
Number	315	101	108	32	556 total
Fraction of offspring	9/16	3/16	3/16	1/16	☞ These data are consistent with the predictions of independent assortment

FIGURE 13.6 Mendel Analyzed the F₁ and F₂ Offspring of a Cross between Pure Lines for Two Traits
Either of two events could occur when alleles of different genes are transmitted to offspring: The alleles
could sort themselves into gametes independently of each other, or alleles from the same parent could be
transmitted together, generation after generation.

wrinkled seeds in the F_2 offspring was 423 to 133, or 3.18 to 1. (**Box 13.2** explains why Mendel used such large sample sizes in his experiments.) Likewise, the ratio of yellow to green seeds was 416 to 140—almost exactly 3 to 1. When seed shape and seed color are considered separately, then, the F_2 individuals exhibited the same 3:1 ratio of dominant to recessive phenotypes observed in the first set of experiments. In each case, the dominant phenotype showed up in 3/4 of the F_2 offspring, while the recessive phenotype was observed in 1/4. Because the probability of observing each of the dominant phenotypes is 3/4, the probability of observing both is $3/4 \times 3/4 = 9/16$.

Using a Testcross to Confirm Predictions

Mendel did experiments with combinations of traits other than seed shape and color and obtained results similar to those in Figure 13.6c. Each paired set of traits produced a 9:3:3:1 ratio of progeny phenotypes in the F_2 generation. He even did a limited set of crosses examining three traits at a time. Although all of these data were consistent with the principle of independent assortment, his most powerful support for the hypothesis came from a different type of experiment.

In designing this study, Mendel's goal was to test the prediction that an *RrYy* plant produces four different types of ga-

---

BOX 13.2   Sample Size and Chance Fluctuations

Each time Mendel designed a cross to answer a question about the inheritance of a trait, he made sure to analyze the results in a large number of offspring. Collecting data from a large sample is a goal of scientific studies. Large sample sizes reduce chance fluctuations in the outcome of an experiment and make it easier to recognize patterns in the data.

Chance fluctuations are inevitable in the outcome of any experiment or observational study. For example, consider the data shown in the accompanying table. They indicate the seed-shape phenotypes that Mendel observed in offspring from heterozygous (*Rr*) parents that self-fertilized.

From parent to parent, the ratios of round and wrinkled seeds fluctuated

from almost 1:1 to over 4:1. This is because gametes combine at random and each parent's offspring represent a small sample. For example, even though only half the gametes available when parent plant 4 self-fertilized contained an *r* allele, there happened to be a run of fertilizations between *r*- and *r*-containing gametes, so the ratio of round to wrinkled phenotypes for this parent is lower than usual. Small samples frequently have skewed results like this, just due to chance. But when Mendel pooled the data into a large sample, it became clear that the overall ratio of round to wrinkled phenotypes was about 3:1 (**Figure 13.7**). If Mendel had been able to obtain a larger number of offspring from parent plant 4, it

is very likely that the observed ratio of phenotypes from that parent would have been closer to 3:1.

Rr × *Rr*			
Plant Number	Round (*RR* or *Rr*)	Wrinkled (*rr*)	Ratio
1	45	12	3.75:1
2	27	8	3.37:1
3	24	7	3.42:1
4	19	16	1.19:1
5	32	11	2.91:1
6	26	6	4.33:1
7	88	24	3.66:1
8	22	10	2.20:1
9	28	6	4.66:1
10	25	7	3.57:1
Total	**336**	**107**	**3.14:1**

Pea pod contains 9 smooth seeds and 3 wrinkled seeds, a 3:1 ratio

FIGURE 13.7   When an *Rr* Individual Self-Fertilizes, Offspring Phenotypes Are in a Ratio of 3:1
Garden pea seed pod showing the offspring of an *Rr* individual that was allowed to self-fertilize. In this pod, the offspring phenotypes are in exact 3:1 proportions. QUESTION Do you predict that seed pods with exact 3:1 ratios are observed rarely or frequently in this cross? Explain your answer.

metes in equal proportions. To accomplish this, Mendel invented a technique called a testcross. A **testcross** uses a parent that contributes only recessive alleles to its offspring to help determine the unknown genotype of the second parent. Testcrosses are useful because the genetic contribution of the homozygous recessive parent is easy to predict and analyze. As a result, a testcross allows experimenters to test the genetic contribution of the other parent. If the other parent has the dominant phenotype but an unknown genotype, the results of the testcross allow researchers to infer whether that parent is homozygous or heterozygous for the dominant allele.

In this case, Mendel performed a testcross between parents that were *RrYy* and *rryy*. The types and proportions of offspring that should result can be predicted with the Punnett square shown in **Figure 13.8**. If the principle of independent assortment is valid, there should be four types of offspring in equal proportions.

What were the actual proportions observed? Mendel did this experiment and examined the seeds produced by the progeny. He found that 31 were round and yellow, 26 were round and green, 27 were wrinkled and yellow, and 26 were wrinkled and green. As predicted, these numbers are nearly identical to the 27.5 individuals expected with each genotype, given the total of 110 individuals. The predicted ratio of phenotypes was 1:1:1:1, which matched the observed ratio. The testcross had confirmed the principle of independent assortment.

Mendel's Contributions to the Study of Heredity

Mendel introduced an approach to studying heredity that is still in use today. This approach includes the following considerations:

- The choice of an appropriate model organism;

- The use of pure lines with discrete differences in traits to explore how traits are transmitted to offspring;

- The examination of large numbers of progeny for each type of cross;

- The application of the rules of probability to predict the numbers and types of progeny produced from crosses;

- The use of reciprocal crosses to test the hypothesis that an allele's transmission depends on the sex of the parent; and

- The use of testcrosses to establish the genotype of a parent.

In short, Mendel's work provided a powerful conceptual framework for thinking about heredity. He was the first individual to describe correctly the basic rules of **transmission genetics**—the patterns that occur as alleles pass from one generation to the next. All of the types and proportions of progeny that Mendel observed in his F_1 and F_2 generations could be explained as the consequence of two processes: (1) the segregation of discrete, paired alleles into separate gametes and (2) the independent assortment of alleles that affect different traits.

Mendel's experiments were brilliant in design, execution, and interpretation. Unfortunately, they were ignored for 34 years.

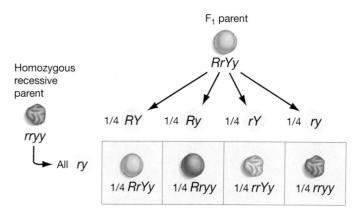

FIGURE 13.8 The Predictions Made by the Principle of Independent Assortment Can Be Evaluated in a Testcross
If the principle of independent assortment is correct and *RrYy* parents produce four types of gametes in equal proportions, then a mating between *RrYy* and *rryy* parents should produce four types of offspring in equal proportions, as this Punnett square shows.

✓CHECK YOUR UNDERSTANDING

Mendel discovered that individuals have two alleles of each gene, that each gamete receives one of the two alleles present in a parent, and that alleles from different genes are transmitted to gametes independently of each other. The alleles he analyzed were either dominant or recessive, meaning heterozygous individuals had the dominant phenotype. You should be able to (1) create and analyze Punnett squares to predict the genotypes and phenotypes that will occur in the F_1 and F_2 offspring of a cross, then calculate the expected frequency of each genotype and phenotype. Given the outcome of a cross, you should also be able to (2) infer the genotypes and phenotypes of the parents. (Use the questions at the end of this chapter to practice these skills.)

13.3 The Chromosome Theory of Inheritance

Historians of science frequently debate why Mendel's work was overlooked for so long. It is almost undoubtedly true that his use of probability theory in explaining his results and his quantitative treatment of data were difficult for biologists of that time to understand and absorb. It may also be true that the theory of blending inheritance was so well entrenched that there was a tendency to dismiss his results as peculiar or unbelievable. Whatever the reason, Mendel's work was not appreciated until other biologists, working with a variety of plants and animals, independently reproduced his results in the early 1900s.

The rediscovery of Mendel's work, 16 years after his death, ignited the young field of genetics. Mendel's experiments established the basic rules that govern how traits are passed from parents to offspring. They described the pattern of inheritance. But what process is responsible for these patterns? Two biologists, working independently, came up with the answer. Walter Sutton and Theodor Boveri each realized that meiosis could be responsible for

Mendel's rules. When this hypothesis was published in 1903, research in genetics exploded.

Recall from Chapter 12 that meiosis is the type of cell division that precedes gamete formation. The details of the process were worked out in the final decades of the nineteenth century. What Sutton and Boveri grasped is that meiosis not only reduces chromosome number by half, but also explains the principle of segregation and the principle of independent assortment.

The cell nucleus at the top of **Figure 13.9a** illustrates Sutton and Boveri's central insight—the hypothesis that chromosomes are composed of Mendel's hereditary determinants, or genes. In this example, the gene for seed shape is shown at a particular position along a chromosome. This location is known as a **locus** ("place"; plural: *loci*). A genetic locus is the physical location of a gene. The paternal and maternal chromosomes shown in Figure 13.9a happen to possess different alleles of the gene for seed shape: One allele specifies round seeds (*R*), while the other specifies wrinkled seeds (*r*).

The subsequent steps in Figure 13.9a show how these alleles segregate into different daughter cells during meiosis I, when homologous chromosomes separate. This physical separation of alleles produces Mendel's principle of segregation.

Figure 13.9b follows the fate of the alleles for two different genes—in this case, for seed shape and seed color—as meiosis proceeds. Because these genes are located on different nonhomologous chromosomes, they assort independently of one another at meiosis I. Four types of gametes, produced in equal proportions, result. This is the physical basis of Mendel's principle of independent assortment. Most of the genes that Mendel analyzed assort independently from one another because they are each located on different chromosomes.

Sutton and Boveri formalized these observations in the **chromosome theory of inheritance**. Like other theories in biology, the chromosome theory consists of a pattern—a set of observations about the natural world—and a process that explains the pattern. The chromosome theory states that Mendel's rules can be explained by the independent alignment and separation of homologous chromosomes at meiosis I.

(a) Principle of segregation

Rr parent

Dominant allele for seed shape — Recessive allele for seed shape

Chromosomes replicate

Meiosis I
Alleles segregate

Meiosis II

Gametes

Each gamete carries only one allele for seed shape, because the alleles have segregated during meiosis.

(b) Principle of independent assortment

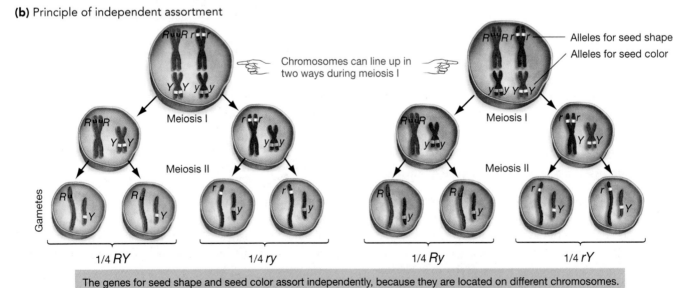

Chromosomes can line up in two ways during meiosis I

Alleles for seed shape
Alleles for seed color

Meiosis I

Meiosis II

Gametes

1/4 *RY*   1/4 *ry*   1/4 *Ry*   1/4 *rY*

The genes for seed shape and seed color assort independently, because they are located on different chromosomes.

FIGURE 13.9 Meiosis Is Responsible for the Principles of Segregation and Independent Assortment
(a) A parent's two alleles segregate into different gametes, as Mendel hypothesized, because homologous chromosomes separate during meiosis I. **(b)** The alleles for different traits assort independently, again as Mendel hypothesized, because nonhomologous chromosomes assort independently during meiosis I. Maternal and paternal chromosomes are shown in different colors for clarity.

When Sutton and Boveri published their findings, however, the hypothesis that chromosomes consist of genes was untested. What experiments confirmed that chromosomes contain genes?

13.4 Testing and Extending the Chromosome Theory

During the first decade of the twentieth century, an unassuming insect rose to prominence as a model organism for testing the chromosome theory of inheritance. This organism—the fruit fly *Drosophila melanogaster*—has been at the center of genetic studies ever since. *Drosophila melanogaster* has all the attributes of a useful model organism for experimental studies in genetics: small size, ease of rearing in the lab, a short reproductive cycle (about 10 days), and abundant offspring (up to a few hundred per mating). The elaborate external anatomy of this insect also makes it possible to identify interesting phenotypic variation among individuals (**Figure 13.10a**).

The Discovery of Sex-Linked Traits

Drosophila research was pioneered by Thomas Hunt Morgan and his students. But because *Drosophila* is not a domesticated species like the garden pea, Morgan had no readily available phenotypic variants such as Mendel's round and wrinkled seeds.

(a) The fruit fly *Drosophila melanogaster*

1 mm

(b) Eye color is a variable trait.

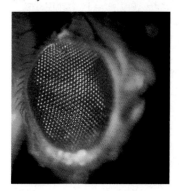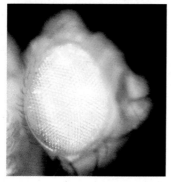

FIGURE 13.10 The Fruit Fly *Drosophila melanogaster* Is an Important Model Organism in Genetics
EXERCISE In part (b), label the phenotype that is considered wild type. Label the phenotype that is a rare mutant.

Consequently, an early goal of Morgan's research was simply to find and characterize individuals with different phenotypes.

Morgan's group referred to individuals with the most common phenotype as **wild type**. But while examining his cultures, Morgan discovered a male fly that had white eyes rather than the wild-type red eyes (**Figure 13.10b**). This individual had a discrete and easy-to-recognize phenotype different from the normal phenotype. Morgan inferred that the white-eyed phenotype resulted from a **mutation**—a change in a gene (in this case, a gene that affects eye color). Individuals with white eyes (or other traits attributable to mutation) are referred to as **mutants**.

To explore how the white-eye trait is inherited in fruit flies, Morgan mated a red-eyed female fly with the mutant white-eyed male fly. Because all of the F_1 progeny had red eyes, Morgan tentatively concluded that the white-eyed phenotype was due to a single allele that was recessive to the wild-type red-eyed allele. To test this hypothesis, Morgan allowed the F_1 males and females to breed with one another. As expected, he observed a 3-to-1 ratio of red-eyed to white-eyed phenotypes in the F_2 offspring. But Morgan noted something peculiar: All of the white-eyed progeny were male. Half of the males in the F_2 generation had white eyes, and half had red. But all of the F_2 females had red eyes. This outcome was remarkable. There appeared to be some sort of association between eye color and sex.

To test the sex-association hypothesis, Morgan crossed F_1 red-eyed females with white-eyed males. Some of the female offspring had white eyes. This told Morgan that females could have white eyes. Obtaining white-eyed females also allowed Morgan to perform the second part of a reciprocal cross. (Recall that Mendel used reciprocal crosses to show that the inheritance of traits in peas is not affected by the transmission of alleles by males versus females.) When Morgan mated white-eyed females to red-eyed males from a pure line, the results were striking: All F_1 females had red eyes, but all F_1 males had white eyes.

Mendel's reciprocal crosses had always given results that were similar to each other. But Morgan's reciprocal crosses did not. The experiment suggested a definite relationship between the sex of the progeny and the inheritance of eye color. Even though the white-eyed phenotype appeared to be recessive in the first set of crosses, it appeared in the F_1 males when the reciprocal cross was performed. How could these observations be reconciled with Mendel's rules of inheritance?

The Discovery of Sex Chromosomes

Nettie Stevens began studying the karyotypes of insects about the time that Morgan began his work with *Drosophila*. One of her outstanding observations was of a striking difference in the chromosome complements of males and females in the beetle *Tenebrio molitor*. In females of this species, diploid cells contain 20 large chromosomes. But diploid cells in males contain 19 large chromosomes and 1 small one. Stevens called the small chromosome the Y chromosome. This Y chromosome paired

(a) Sex chromosomes from humans

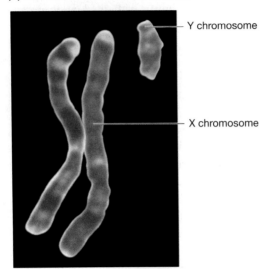

(b) Sex chromosomes pair at meiosis I.

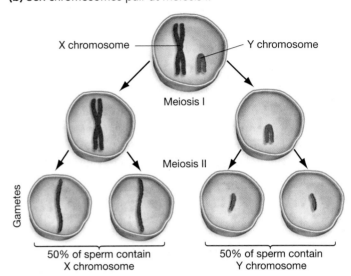

FIGURE 13.11 Sex Chromosomes Pair during Meiosis I, Then Segregate to Form X-Bearing and Y-Bearing Gametes
Sex chromosomes synapse at meiosis I in male fruit flies, even though the X and Y chromosomes differ in size and shape. (No crossing over occurs in male fruit flies.) Thus, half the sperm cells that result from meiosis bear an X chromosome; half have a Y chromosome.

with one of the large chromosomes at meiosis I, which Stevens called the X chromosome. The X and Y were different in size and shape, but they acted like homologs during meiosis.

In addition to discovering the X and Y chromosomes, Stevens observed that all eggs in this species had 10 large chromosomes. Sperm, however, could be divided into two categories. Because the X chromosome and Y chromosome paired up during meiosis and separated into different gamete cells, about 50 percent of the sperm contained 10 large chromosomes, including the X chromosome; the other 50 percent had 9 large chromosomes plus one small (Y) chromosome. Stevens observed similar patterns in other species of insects.

Based on these descriptive studies, Stevens developed a hypothesis to explain how sex determination occurs in these species. She proposed that a male is produced when an egg is fertilized by a sperm carrying a Y chromosome but that a female is produced when an egg is fertilized by a sperm carrying an X chromosome. We now call this pair of chromosomes **sex chromosomes (Figure 13.11a)**. Sex chromosomes do not carry the same genes. As **Figure 13.11b** shows, the equal ratio of X-bearing and Y-bearing sperm explains why the sexes are produced in nearly equal proportions.

X-Linked Inheritance and the Chromosome Theory

To explain the results of his crosses with white-eyed flies, Morgan put his genetic data together with Stevens's observations on sex chromosomes. *Drosophila* females, like *Tenebrio* females, have two X chromosomes; male fruit flies carry an X and a Y. As a result, half the gametes produced by a male fruit fly should carry an X chromosome; the other half, a Y chromosome.

Morgan realized that the transmission pattern of the X chromosome in males and females could account for the results of his

reciprocal crosses. Specifically, he proposed that the gene for white eye color in fruit flies is located on the X chromosome and that the Y chromosome does not carry an allele of this gene. This hypothesis is called **X-linked inheritance** or **X-linkage**. (Proposing that a gene resides on the Y chromosome is called **Y-linked inheritance** or **Y-linkage**. Proposing that a gene is on either sex chromosome is termed **sex-linked inheritance** or **sex-linkage**.) The key observation is that even though the X and Y chromosomes synapse during prophase of meiosis I, they differ in size, shape, and gene content.

According to the hypothesis of X-linkage, a female has two copies of the gene that specifies eye color because she has two X chromosomes. One of these chromosomes came from her female parent, and the other from her male parent. A male, in contrast, has only one copy of the eye-color gene because he has only one X chromosome, which comes from his mother.

The crosses and the Punnett squares in **Figure 13.12** show that Morgan's hypothesis of X-linkage explains his experimental results. In this figure, the allele for red eyes is denoted w^+, while the allele for white eyes is denoted w. (In fruit-fly genetics, the + symbol always indicates the wild-type trait.) When reciprocal crosses give different results, such as those illustrated in Figure 13.12, it is likely that the gene in question is located on a sex chromosome. Recall from Chapter 12 that non-sex chromosomes are called **autosomes**. Genes on non-sex chromosomes are said to show **autosomal inheritance**.

Morgan's discovery of X-linked inheritance carried an even more important message, however. In *Drosophila*, the gene for white eye color is clearly correlated with inheritance of the X chromosome. This correlation was important evidence in support of the hypothesis that chromosomes contain genes. The discovery of X-linked inheritance convinced most biologists that the chromosome theory of inheritance was correct.

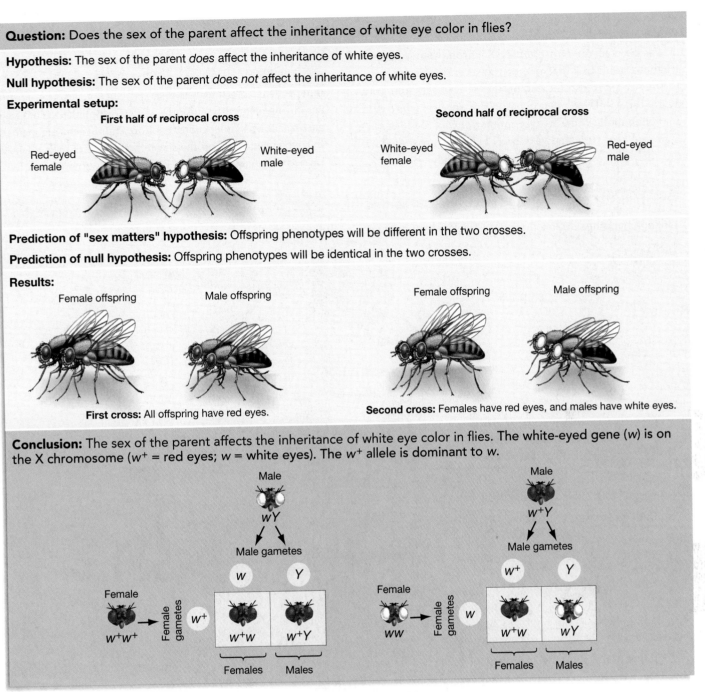

Question: Does the sex of the parent affect the inheritance of white eye color in flies?

Hypothesis: The sex of the parent *does* affect the inheritance of white eyes.

Null hypothesis: The sex of the parent *does not* affect the inheritance of white eyes.

Experimental setup:

First half of reciprocal cross

Red-eyed female — White-eyed male

Second half of reciprocal cross

White-eyed female — Red-eyed male

Prediction of "sex matters" hypothesis: Offspring phenotypes will be different in the two crosses.

Prediction of null hypothesis: Offspring phenotypes will be identical in the two crosses.

Results:

Female offspring   Male offspring

Female offspring   Male offspring

First cross: All offspring have red eyes.

Second cross: Females have red eyes, and males have white eyes.

Conclusion: The sex of the parent affects the inheritance of white eye color in flies. The white-eyed gene (w) is on the X chromosome (w^+ = red eyes; w = white eyes). The w^+ allele is dominant to w.

Male wY → Male gametes: w, Y

Female w^+w^+ → Female gametes: w^+

	w	Y
w^+	w^+w	w^+Y

Females — Males

Male w^+Y → Male gametes: w^+, Y

Female ww → Female gametes: w

	w^+	Y
w	w^+w	wY

Females — Males

FIGURE 13.12 Reciprocal Crosses Confirm that Eye Color in *Drosophila* Is an X-Linked Trait
When Morgan crossed red-eyed females with white-eyed males and then crossed white-eyed females with red-eyed males, he observed strikingly different results. This was consistent with his hypothesis that eye color is an X-linked trait in fruit flies. **EXERCISE** Morgan also crossed red-eyed w^+w females with red-eyed males, w^+Y. Create a Punnett square to predict the types and proportions of offspring genotypes and phenotypes that would result.

What Happens When Genes Are Located on the Same Chromosome?

When later experiments confirmed that genes are indeed the physical components of chromosomes, the result prompted Morgan and other geneticists to reevaluate Mendel's principle of independent assortment. The key issue was that genes could not undergo independent assortment if they were lo-

cated on the same chromosome. The physical association of genes that are found on the same chromosome is called **linkage**. (Notice that the terms *linkage* and *sex-linkage* are different in meaning. If two or more genes are linked, it means that they are located on the same chromosome. If a single gene is sex-linked, it means that it makes up part of a sex chromosome.)

The first examples of linked genes involved the X chromosome of fruit flies. After Morgan established that the white-eye gene was located on *Drosophila*'s X chromosome, he and colleagues established that one of the several genes that affects body color is also located on the X. Red eyes and gray body are the wild-type phenotypes in this species; white eyes and a yellow body occur as rare mutant phenotypes. The alleles for red eyes (w^+) and gray body (y^+) also are dominant to the alleles for white eyes (w) and yellow body (y).

Thus, it seemed logical to predict that the linked genes would always be transmitted together during gamete formation. Stated another way, linked genes should violate the principle of independent assortment. Recall from Section 13.2 that independent assortment is observed when genes are on different chromosomes, because the alleles of unlinked genes segregate to gametes independently of one another during meiosis I. But when genes are on the same chromosome, their alleles are carried to gametes together. **Figure 13.13** shows that a female

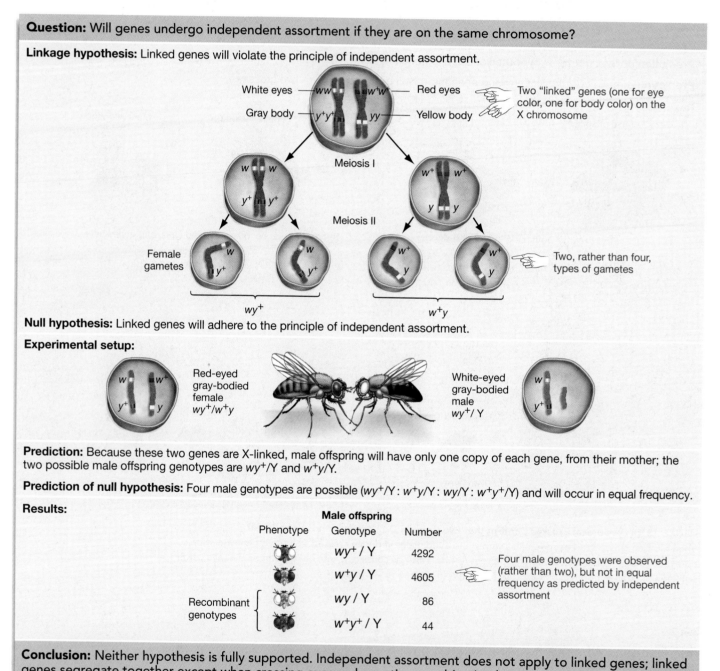

Question: Will genes undergo independent assortment if they are on the same chromosome?

Linkage hypothesis: Linked genes will violate the principle of independent assortment.

Null hypothesis: Linked genes will adhere to the principle of independent assortment.

Experimental setup:

Red-eyed gray-bodied female wy^+/w^+y

White-eyed gray-bodied male wy^+/Y

Prediction: Because these two genes are X-linked, male offspring will have only one copy of each gene, from their mother; the two possible male offspring genotypes are wy^+/Y and w^+y/Y.

Prediction of null hypothesis: Four male genotypes are possible ($wy^+/Y : w^+y/Y : wy/Y : w^+y^+/Y$) and will occur in equal frequency.

Results:

	Male offspring		
Phenotype	Genotype	Number	
	wy^+/Y	4292	
	w^+y/Y	4605	Four male genotypes were observed (rather than two), but not in equal frequency as predicted by independent assortment
Recombinant genotypes	wy/Y	86	
	w^+y^+/Y	44	

Conclusion: Neither hypothesis is fully supported. Independent assortment does not apply to linked genes; linked genes segregate together except when crossing over and genetic recombination have occurred.

FIGURE 13.13 Linked Genes Are Inherited Together unless Recombination Occurs
Independent assortment does not occur when genes are linked. Morgan crossed pure-line white-eyed, gray-bodied *Drosophila* females with red-eyed, yellow-bodied pure-line males. When he let the F_1 offspring breed, he observed four phenotypes among the male F_2 progeny. Two of the phenotypes were not predicted by complete linkage as in the "Linkage hypothesis" above.

fruit fly with one X chromosome carrying the w and y^+ alleles and with a second X chromosome carrying the w^+ and y alleles should generate just two classes of gametes in equal numbers during meiosis, instead of the four classes that are predicted under the principle of independent assortment. Is this what actually occurs?

The First Studies of Linked Genes To determine whether linked traits behave as predicted, Morgan performed the cross described in the "Experimental setup" of Figure 13.13. This figure introduces some new notation. When you express a genotype in writing, you are simply listing the relevant alleles that are present. When referring to linked genes, biologists use a slash symbol (/) to separate the alleles that are found on homologous chromosomes. For example, the female fruit flies in Morgan's experiment had wy^+ alleles on one X chromosome and w^+y alleles on the other X chromosome. This genotype is written as wy^+/w^+y.

The interesting result in Figure 13.13 is contained in the "Results" table, which summarizes the phenotypes and genotypes observed in the male offspring of this experimental cross. Most of these males carried an X chromosome with one of the two combinations of alleles found in their mothers: wy^+ or w^+y. Thus, *white* and *yellow* alleles do not segregate independently of each other most of the time. But a small percentage of males had novel phenotypes and genotypes: wy and w^+y^+. Morgan referred to these individuals as **recombinant**, because the combination of alleles on their X chromosome was different from the combinations of alleles present in the parental generation.

To explain this result, Morgan proposed that gametes with new, recombinant genotypes were generated when crossing over occurred during prophase of meiosis I in the females. Recall from Chapter 12 that *crossing over* involves a physical exchange of segments from homologous chromosomes. Crossing over occurs at least once in every synapsed pair of homologous chromosomes, and usually multiple times. (Male fruits flies are an exception to this rule. No crossing over occurs in male fruit flies.) As **Figure 13.14** shows, a crossing-over event occurred somewhere between the w and y genes in the wy^+/w^+y females. The chromosomes that resulted would have the genotypes wy and w^+y^+ (see the middle of the bottom row of Figure 13.14). If these chromosomes ended up in a male offspring, they would produce individuals with yellow bodies and white eyes, along with individuals with gray bodies and red eyes. This is exactly what Morgan observed.

The take-home message of Morgan's experiments was that linked genes are inherited together unless crossing over occurs. When crossing over takes place, genetic recombination occurs. These results cemented the connection between the events of meiosis I and Mendel's laws. Linked alleles segregate together unless there is a physical crossover between homologous chromosomes.

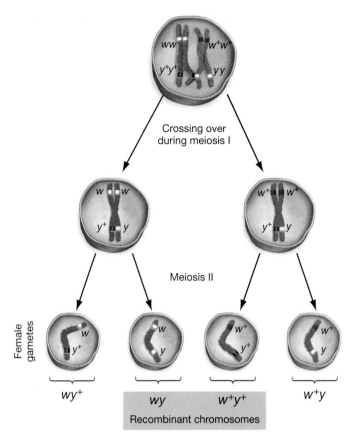

FIGURE 13.14 Genetic Recombination Results from Crossing Over
To explain the results in Figure 13.13, Morgan proposed that crossing over occurred between the w genes and the y genes in a small percentage of the F₁ females during meiosis I. The recombinant chromosomes that resulted would produce the recombinant phenotypes observed in F₂ males.

Linkage Mapping In the experiment that is summarized in Figure 13.13, about 1.4 percent of the male offspring were recombinant genotypes. But when Morgan and his co-workers performed the same types of crosses with different pairs of X-linked traits, the team found that the fraction of recombinant gametes varied. For example, when crosses involved fruit flies with X-linked genes for a yellow body and a mutant phenotype called singed bristles (*sn*), males with recombinant chromosomes for these two genes were produced about 21 percent of the time.

Morgan explained these observations by making a conceptual breakthrough: He proposed that genes are arranged in a linear array along a chromosome. According to Morgan's hypothesis, the physical distance between genes determines the frequency at which crossing over occurs between them. Recall from Chapter 12 that crossing over occurs at random and can take place anywhere along the length of the chromosome. Based on these observations, it is logical to predict that the shorter the distance between any two genes

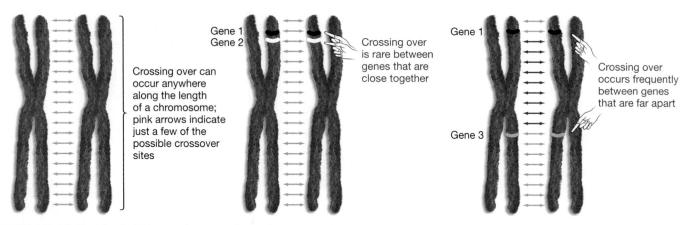

FIGURE 13.15 The Physical Distance between Genes Determines the Frequency of Crossing Over
Crossing over occurs at least once in every meiosis in each homologous pair of chromosomes, and often multiple times.

on a chromosome, the lower the probability that crossing over will take place somewhere in between. The fundamental idea here is that greater physical distance between genes increases the chance that crossing over will take place between them (**Figure 13.15**).

In 1911 A. H. Sturtevant, an undergraduate who was studying with Morgan, realized that variation in recombination frequency had an important implication: If genes are lined up along the chromosome and if the frequency of crossing over is a function of the physical distance between genes, then it should be possible to figure out where genes are in relation to each other based on the frequency of recombinants between various pairs. That is, it should be possible to create a **genetic map**.

Sturtevant proposed that in constructing a genetic map, the unit of distance along a chromosome should simply be the percentage of offspring that have recombinant phenotypes with respect to two genes. Sturtevant called this unit the *centiMorgan* (cM), in honor of his mentor. One map unit, or 1 centiMorgan, represents the physical distance that produces 1 percent recombinant offspring. For example, he proposed that the eye-color and body-color genes of fruit flies are 1.4 map units apart on the X chromosome, because recombination between these loci results in 1.4 percent recombinant offspring (**Figure 13.16a**). The *y* gene for body color and the gene for singed bristles, in contrast, are 21 map units apart. Where is the *white* gene relative to the *singed-bristles* gene? Because recombinants occurred in 19.6 percent of the gametes produced by females that are w^+sn/wsn^+, Sturtevant inferred that the gene for white eyes must be located between the genes for yellow body and singed bristles, as shown in Figure 13.16a.

Figure 13.16b provides a partial genetic map of the X chromosome in *Drosophila melanogaster* and the data on which the map positions are based. Using this logic and these data, Sturtevant assembled the first genetic map.

Do *All* Linked Genes Violate the Principle of Independent Assortment? When crossing over occurs, the new genotypes that are created match the ones that would be produced if the genes involved were located on different chromosomes. Stated another way, crossing over breaks up linkage and makes it appear as if independent assortment is occurring. To convince yourself that this is so, look back at the four gamete genotypes listed at the bottom of Figure 13.14. Recall that these were produced by a female with genotype wy^+/w^+y. The four gamete genotypes listed are wy^+, wy, w^+y^+, and w^+y. It should be clear that these are the same gamete genotypes that meiosis would produce if the female's genotype were w^+wy^+y, meaning the *white* and *yellow* genes were not linked.

If crossing over occurs frequently enough that all four of these gamete genotypes are produced in equal proportions, then the results of a cross involving linked genes would be indistinguishable from the predictions of independent assortment. When genes are 50 or more map units apart on the same chromosome, that is exactly what happens. Genes behave as if they assort independently when 50 percent of the gametes are recombinant with respect to the genes. As it turns out, this situation was occurring in Mendel's experiments. The genes for two of the traits he worked with are located on chromosome number 1 in peas, and at least two of the genes are located on chromosome 4. In each case, however, the genes he analyzed are located far apart. As a result, crossing over was frequent enough to produce the 9:3:3:1 ratios of F_2 phenotypes predicted by the hypothesis of independent assortment.

Before analyzing some extensions to Mendel's rules, let's step back and consider the events we've just surveyed—from the rediscovery of Mendel's work, around 1900, to Sturtevant's chromosome map. In 1899 geneticists did not understand the basic rules of heredity. But by 1911 they could map the locations of genes on chromosomes. A remarkable knowledge explosion had occurred.

(a) Frequency of recombinant offspring can be used to map genetic distance.

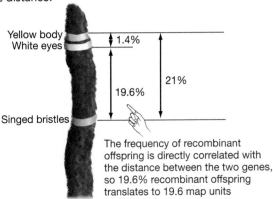

The frequency of recombinant offspring is directly correlated with the distance between the two genes, so 19.6% recombinant offspring translates to 19.6 map units

(b) Constructing a genetic map

Frequency of crossing over (%) between some genes on the X chromosome of fruit flies			
	Miniature Wings	Crossveinless Wings	Ruby Eyes
Yellow body	36.1	13.7	7.5
White eyes	34.7	12.3	6.1
Singed bristles	15.1	7.3	13.5
Miniature wings	—	22.4	28.6
Crossveinless wings	—	—	6.2

FIGURE 13.16 The Locations of Genes Can Be Mapped by Analyzing the Frequency of Recombination
(a) Recombinants between the yellow-body locus and the white-eye locus occur 1.4 percent of the time. Therefore, these genes are 1.4 map units apart on the chromosome. Recombinants between the yellow-body locus and the singed-bristles locus of another X-linked gene occur 21 percent of the time. But recombinants between the white-eye locus and the singed-bristles locus occur just 19.6 percent of the time. Therefore, the loci must be arranged as shown. **(b)** A partial genetic map of the X chromosome in fruit flies. **EXERCISE** On the chromosome in part (b), label the orange, blue, and beige genes.

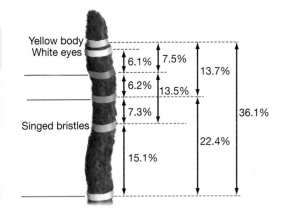

13.5   Extending Mendel's Rules

Biologists point out that Mendel analyzed the simplest possible genetic system. To understand why it's fair to label the traits and alleles that Mendel studied as simple, consider the following:

- The phenotypes that he observed are influenced by the alleles at a single gene, even though many traits in peas and other organisms are shaped by the effects of multiple genes.

- In the pure lines that he studied, a total of two alleles were present at each gene. Most populations in most species harbor dozens or even hundreds of different alleles at each gene.

- The alleles he studied were completely dominant or recessive to each other. In many cases, alleles don't show complete dominance or recessiveness, as the next section explains.

- Mendel worked with **discrete traits**—characteristics that are qualitatively different. His pea flowers were white or purple; the plants were tall or short; the seeds were wrinkled or round. But it's more common for traits to show what biologists call **quantitative variation**—differences in degree of variation. In humans, for example, characteristics such as height, skin color, and body shape do not show discrete variation. Instead of being either 1.5 m tall or 1.75 m tall or 2.0 m tall, humans may be any height within a wide range of variation. Traits that exhibit quantitative variation are harder to measure and study than discrete traits are.

- Peas do not have sex chromosomes, so sex-linked inheritance did not occur.

- The experimental results were not complicated by linkage, because the linked genes that Mendel analyzed were so far apart that crossing over between them occurred frequently. As a result, alleles from different genes appeared to assort independently.

The key point is that by studying a simple situation, Mendel was able to discover the most fundamental rules of inheritance. This is an extremely important research strategy in biological science. Researchers almost always choose to analyze the simplest situation possible before going on to explore more complex systems. Mendel probably would have failed, as so many others had done before him, had he worked with any of the complications just listed.

Once Mendel's work was rediscovered, though, researchers began to analyze traits and alleles whose inheritance was complex. If experimental crosses produced F_2 progeny that did not conform to the expected 3:1 or 9:3:3:1 ratios in phenotypes, researchers had a strong hint that something interesting was going on. In many cases it turned out that unraveling the cause of surprising results led to new insights into how genes and heredity worked. How can traits that don't appear to follow Mendel's rules contribute to a more complete understanding of heredity?

Incomplete Dominance and Codominance

The terms *dominant* and *recessive* describe which phenotype is observed when two different alleles of a gene occur in the same individual. In all seven traits that Mendel studied, only the phenotype associated with one allele—the "dominant" one—appeared in heterozygous individuals. But consider the flowers called four-o'clocks, pictured in **Figure 13.17a.** In this species, biologists have developed a pure line that has purple flowers and a pure line that has white flowers. When individuals from these strains are mated, all of their offspring are lavender (**Figure 13.17b**). In Mendel's peas, crosses between purple- and white-flowered parents produced all purple-flowered offspring. Why the difference?

Biologists answered this question by examining the phenotypes of F_2 offspring. These are the progeny of self-fertilization in lavender-flowered F_1 individuals. Of the F_2 plants, 1/4 have purple flowers, 1/2 have lavender flowers, and 1/4 have white flowers. This 1:2:1 ratio of *phenotypes* is unlike any we have seen to date, but it exactly matches the 1:2:1 ratio of *genotypes* that is produced when flower color is controlled by one gene with two alleles.

To convince yourself that this explanation is sound, study the genetic model shown in Figure 13.17b. According to the hypothesis shown in the diagram, the inheritance of flower color in four-o'clocks and peas is identical, except that the four-o'-clock alleles show incomplete dominance rather than complete dominance. When **incomplete dominance** occurs, heterozygotes have an intermediate phenotype. In the case of four-o'clocks, neither purple nor white alleles dominate. Instead, the F_1 progeny—all heterozygous—show a phenotype intermediate between the two parental strains.

Codominance Incomplete dominance illustrates an important general point: Dominance is not necessarily an all-or-none phenomenon. In fact, many alleles show a relationship called codominance. When **codominance** occurs, heterozygotes have the phenotype associated with *both* of the alleles present. Whereas incomplete dominance results in a phenotype that is intermediate between the phenotypes associated with both alleles present, codominance results in a phenotype that expresses both alleles.

As an example of codominance, consider the *MN* gene in humans. This gene codes for a membrane protein that is found in the plasma membrane of red blood cells. As a result, the phenotype associated with the *MN* gene is the type of membrane protein present. Different alleles of this gene lead to the appearance of membrane proteins with different amino acid sequences. In most human populations, only two alleles are present. These alleles are designated *M* and *N*. Thus, three genotypes are found: *MM, MN,* and *NN*. In *MM* individuals, all of the MN-type membrane proteins present in red blood cells have the primary sequence associated with the *M* allele. In *NN* individuals, all of the MN-type proteins have the *N* allele's

primary sequence. But in *MN* individuals, red blood cells have some membrane proteins with the M phenotype and some with the N phenotype. Because both alleles are represented in the phenotype, the alleles are considered codominant.

(a) Flower color is variable in four-o'clocks.

(b) Incomplete dominance in flower color

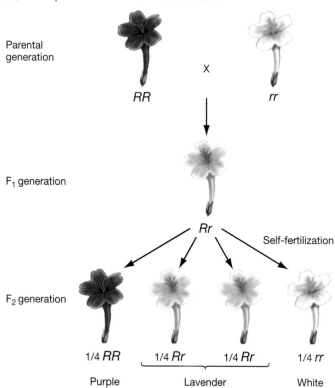

Parental generation

RR   X   rr

F_1 generation

Rr

Self-fertilization

F_2 generation

1/4 *RR*   1/4 *Rr*   1/4 *Rr*   1/4 *rr*

Purple   Lavender   White

FIGURE 13.17 When Incomplete Dominance Occurs, Heterozygotes Have Intermediate Phenotypes
(a) Four-o'clocks got their name because their flowers open in the late afternoon. Flower color is highly variable. **(b)** The hypothesis illustrated is that flower color is controlled by a single gene with two alleles, symbolized *R* and *r*. These alleles exhibit incomplete dominance.

Why Do Dominance Relationships Vary among Alleles?

Once you understand what genes do on the molecular level, it is straightforward to see why some alleles are dominant, recessive, codominant, or intermediate in dominance. Genes contain the information required to manufacture the molecules that are found in cells. Some of these molecules are proteins that form structures in the cell or body, such as the MN proteins in the plasma membranes of red blood cells. Because the *M* and *N* alleles code for proteins with different amino acid sequences, both are present in the cells of heterozygous individuals. Both gene products are detected in the phenotype, so the alleles are considered codominant. Codominance is fairly common in alleles that are associated with membrane proteins such as the MN protein.

Complete and incomplete dominance tend to occur in alleles that code for enzymes. The plant-height alleles Mendel studied, for example, code for enzymes involved in the synthesis of a key growth hormone. A recessive allele has a defect that keeps the enzyme from functioning. Homozygous recessive individuals lack the growth hormone and are dwarfed. The dominant allele, in contrast, codes for an enzyme that functions normally. In heterozygotes, the one normal allele that is present produces enough enzyme to make a normal amount of growth hormone. Normal height results, and the allele is considered dominant.

The situation in four-o'clock flowers is different. The *R* gene analyzed in Figure 13.17b codes for an enzyme involved in producing a purple pigment in flower petals. The recessive allele codes for a dysfunctional enzyme and leads to a complete lack of pigment. Recessive homozygotes are white as a result. In heterozygotes, though, the one normal copy of the gene can produce enough enzyme to result in a small amount of purple pigment. The lavender phenotype results. Homozygous dominant individuals, in contrast, produce large amounts of normal enzyme and thus enough pigment to produce purple petals.

The message here is that alleles can interact in many different ways, depending on what their products do and how they work. In addition, recall that dominance relationships have nothing to do with fitness—the ability of an individual to produce offspring in a particular environment. Dominant alleles do not necessarily confer high fitness, and they are not necessarily more common than recessive alleles. For instance, the allele that causes Huntington's disease is fatal, dominant, and rare.

Multiple Alleles and Polymorphic Traits

Mendel worked with a total of seven traits and just 14 alleles—two for each trait. In most populations, however, dozens of alleles can be identified at each genetic locus. The existence of more than two alleles of the same gene is known as **multiple allelism**. The gene for the hemoglobin protein in humans is among the most dramatic examples of multiple allelism studied to date. Hemoglobin carries oxygen from the lungs to tissues. As pointed out in Chapter 3, the protein has a quarternary structure made up of four distinct polypeptide chains. The instructions for making one of these polypeptide chains are en-

TABLE 13.1 The ABO Blood Types in Humans

In humans, the four different ABO blood types are produced by the alleles present at a single locus. Three alleles are common in most populations: *i*, *I^A*, and *I^B*.

Phenotype (blood type)	Genotype
O	*ii*
A	*I^A I^A* or *I^A i*
B	*I^B I^B* or *I^B i*
AB	*I^A I^B*

i = recessive
I^A and *I^B* = codominant

coded by the *β-globin* gene. Over the past few decades, biologists have identified and named over 500 different alleles of this gene. Many of these alleles are associated with distinctive phenotypes. Some alleles produce polypeptides with normal oxygen-carrying capacity, while others lead to reduced oxygen-carrying capacity and various types of anemia. Still other *β-globin* alleles are associated with adaptation to living at high altitudes, decreased stability at high temperatures, or resistance to the parasites that infect red blood cells and cause malaria.

When different combinations of alleles produce more than two distinct phenotypes, the trait is **polymorphic** ("many-formed"). Oxygen-carrying capacity in humans is a highly polymorphic trait. But the first polymorphic trait ever described involves the ABO blood group in humans. Red blood cells contain many membrane proteins in addition to those encoded by the MN gene; the ABO gene is responsible for one of these. Your blood has almost certainly been analyzed and assigned an O, A, B, or AB phenotype. These phenotypes were discovered in 1900; much later it was established that they are caused by carbohydrates located on the membranes of red blood cells.

As **Table 13.1** shows, the ABO phenotypes result from three alleles called *i*, *I^A*, and *I^B*. The *i* allele is recessive and produces the O phenotype when it is homozygous. The *A* and *B* alleles, in contrast, are dominant with respect to the *i* allele but codominant with respect to each other.

The ABO blood group involves three alleles, six genotypes, and four phenotypes. This gene is multiallelic, the phenotypes are polymorphic, and the alleles are completely dominant or codominant.

Pleiotropy

As far as is known, the alleles that Mendel analyzed affect just a single trait. The gene for seed color in garden peas, for example, does not appear to affect other aspects of the individual's phenotype. In contrast, many cases have been documented in which a single allele affects a wide variety of traits. A gene that influences many traits, rather than just one trait, is said to be **pleiotropic** ("more-turning"). The gene responsible for

Marfan syndrome in humans is a good example. Although current research suggests that just a single gene is involved, individuals with Marfan syndrome exhibit a wide array of phenotypic affects: increased height, disproportionately long limbs and fingers, an abnormally shaped chest, and potentially severe heart problems. A large percentage of these individuals also suffer from problems with their backbone. The gene associated with Marfan syndrome is pleiotropic.

Genes Are Affected by the Physical Environment and Genetic Environment

When Mendel analyzed height in his experiments, he ensured that each plant received a similar amount of sunlight and grew in similar soil. This was important because individuals with alleles for tallness are stunted if they are deprived of nutrients, sunlight, or water—so much so that they look similar to individuals with alleles for dwarfing. For Mendel to analyze the hereditary determinants of height, he had to control the environmental determinants of height. Let's consider how two aspects of the environment affect phenotypes: (1) the individual's physical surroundings and (2) the alleles present at other genes.

The Physical Environment Has a Profound Effect on Phenotypes

The phenotypes produced by most genes and alleles are strongly affected by the individual's physical environment. As a result, it is often the case that an individual's phenotype is as much a product of its physical environment as it is a product of its genotype. To drive this point home, consider two examples in humans:

1. If you have alleles for normal height and normal eyesight but are raised in an environment where the diet is poor in protein and vitamin A, then you are very likely to be short and to have poor eyesight. If one person could be raised in two different environments, it is highly likely that two different phenotypes would result.

2. People with the genetic disease called **phenylketonuria** (PKU) lack an enzyme that helps convert the amino acid phenylalanine to the amino acid tyrosine. As a result, phenylalanine and a related molecule, phenylpyruvic acid accumulate in these people's bodies. The molecules interfere with the development of the nervous system and produce profound mental retardation. But if PKU individuals are identified at birth and placed on a low-phenylalanine diet, then they develop normally. In many countries, all newborns are tested for the defect. PKU is a genetic disease, but it is neither inevitable nor invariant. Because of a simple change in their environment (their diet), individuals with a PKU genotype can have a normal phenotype.

Genetic traits are influenced by more than the physical environment that an individual experiences, however. Phenotypes are also influenced by the action of other genes.

Interactions with Other Genes Have a Profound Effect on Phenotypes

In Mendel's pea plants, a single locus influenced seed shape. Further, Mendel's data showed that the seed-shape phenotype does not appear to be affected by the action of genes for seed color, seed-pod color, seed-pod shape, or other traits. The pea seeds he analyzed were round or wrinkled regardless of the types of alleles present at other loci.

In many cases, however, genes are not as independent as the gene for seed shape in peas. As an example, take the inheritance of fruit color in bell peppers. As **Figure 13.18a** shows, bell peppers come in several colors. Are the color variations due to multiple alleles of a single gene, or are several genes involved? To answer this question, consider the data presented in **Figure 13.18b** on the F_1 and F_2 offspring of pure-line parental strains with yellow and brown fruit.

The results are surprising. Like the experimental cross with four-o'clocks, the F_1 phenotype is different from either parental strain. But a fourth phenotype (green), not seen in either parent or in the F_1, appears in the F_2 generation. Further, the progeny ratios in the F_2 generation are in sixteenths, not quarters. This combination of observations is completely different from anything that Mendel encountered.

A moment's thought should convince you that the F_2 data can't be explained by incomplete dominance, because a novel phenotype shows up. The parental generation consisted of pure lines, so multiple allelism can't be involved. The 9:3:3:1 pattern in F_2 phenotypes is interesting, though, because it matches the ratio that Mendel observed when he studied the inheritance of two different traits. These data suggest that two genes interact to produce pepper color. Further, we can infer that the red phenotype, present 9/16 of the time in the F_2, is due to a combination of the dominant alleles from each gene. Green might result from a combination of the recessive alleles for each gene, because this phenotype is present in just 1/16 of the progeny.

Pepper fruits start out with a green color due to the presence of the green pigment called chlorophyll. This observation suggests that one gene affects whether chlorophyll production stops or continues as a pepper develops. If chlorophyll production continues, a green or greenish fruit would result. The second gene might determine the nature of a second pigment in the fruit, such as red or yellow. **Figure 13.18c** shows how several phenotypes could result from the two alleles of each of these genes.

The key point is that the alleles of different genes affect each other in a way that influences the phenotype observed in these individuals. When these types of gene interactions occur, the phenotype produced by an allele depends on the action of alleles of other genes. This phenomenon is known as **epistasis** ("stopping" or "diminishing"). Epistasis occurs when one gene affects the action of another gene. In the case of peppers, the phenotype of an individual with the R allele depends on the al-

(a) Fruit color is highly variable in bell peppers.

(b) Crosses between pure lines produce novel colors.

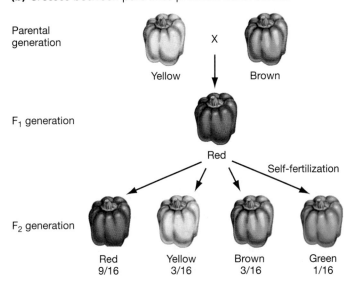

(c) Hypothesis to explain 9 : 3 : 3 : 1 pattern observed above: Two genes interact to produce pepper color.

Gene 1	Gene 2
R = Red	*Y* = Absence of green (no chlorophyll)
r = Yellow	*y* = Presence of green (chlorophyll)
(−) = *R* or *r*	(−) = *Y* or *y*

Genotype	Color	Explanation of color
R–Y–	Red	Red pigment + no chlorophyll
rrY–	Yellow	Yellow pigment + no chlorophyll
R–yy	Brown	Red pigment + chlorophyll
rryy	Green	Yellow pigment + chlorophyll

FIGURE 13.18 Fruit Color in Bell Peppers Results from Interactions among Genes

leles present at the *Y* gene. The term *epistasis* was chosen because, in many cases, the alleles at one gene mask or reduce the effects of alleles at a different gene.

Organelle Genomes

Chapter 7 pointed out that both mitochondria and chloroplasts have small, circular chromosomes that consist in part of DNA. Recall that mitochondria are the ATP-synthesizing centers in eukaryotic cells and that chloroplasts are the sugar-producing centers in photosynthetic eukaryotes. Mitochondria and chloroplasts have their own sets of genes, or genomes. The chromosomes that reside in mitochondria and chloroplasts contain genes that code for molecules used inside each of those organelles.

In most species examined to date, the genes that are located in mitochondria and chloroplasts are transmitted to offspring by only one parent—usually the mother. All of the mitochondrial genes in your cells, for example, came from your mother. In most individual trees and shrubs on your campus, both the mitochondrial genes and the chloroplast genes came from the individual's mother. In these species, none of the organelle genes found in a sperm are passed on to offspring; only the organelle genes found in an egg are imparted. As a result, individuals that are diploid for nuclear genes are haploid for all organelle genes. There is little to no genetic recombination in mitochondrial and chloroplast chromosomes, so all of the genes are inherited as if they were a single allele. And there is no segregation of alleles and no independent assortment. Mitochondrial genes and chloroplast genes are not inherited in a Mendelian fashion.

Quantitative Traits

Recall that Mendel worked with discrete traits. In garden peas, seed color is either yellow or green—no intermediate phenotypes exist. But many traits in peas and other organisms don't fall into discrete categories. In humans, for example, height, weight, and skin color fall anywhere on a continuous scale of measurement. People are not only 160 cm tall or 180 cm tall, with no other heights possible. Height and many other characteristics exhibit quantitative variation—meaning that individuals differ by degree, as described earlier in this section—and are called **quantitative traits**. Like discrete traits, quantitative traits are highly influenced by the physical environment. For example, the effects of nutrition on human height and intelligence have been well documented.

Quantitative traits share a common characteristic: When the frequencies of different values observed in a population are plotted on a histogram, they usually form a bell-shaped curve. This distribution is observed so frequently that it is often called a *normal distribution*. In a normal distribution, high and low values occur at low frequency; intermediate values

(a) A "living histogram"—distribution of height in a college class

(b) Normal distribution = bell-shaped curve

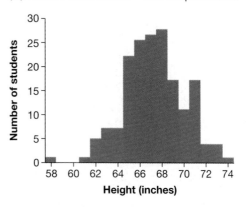

FIGURE 13.19 Quantitative Traits Are Normally Distributed
(a) Male undergraduates, sorted by height, at Connecticut Agricultural College in 1914. **(b)** A histogram plotting the heights of the students in part (a). The distribution of height in human populations forms a bell-shaped curve, called a normal distribution. QUESTION The shortest student in the photo is 4'10" (147 cm), and the tallest is 6'2" (188 cm). How would a histogram of men in your class compare with the distribution in part (b)?

occur at high frequency. **Figure 13.19** shows a classical example. In this case, the trait is human height and the population is a college class.

In the early 1900s Herman Nilsson-Ehle showed that if many genes each contribute a small amount to the value of a quantitative trait, then a continuous, bell-shaped (normal) distribution results. Nilsson-Ehle established this finding using strains of wheat that differed in kernel color. **Figure 13.20a** shows the results of a cross he performed between pure lines of white wheat and dark-red wheat. Note that the frequency of colors in F_2 progeny forms a bell-shaped curve. To explain these results, Nilsson-Ehle proposed the set of hypotheses illustrated in **Figure 13.20b**:

- The parental strains differ with respect to three genes that control kernel color: *AABBCC* produces dark-red kernels, and *aabbcc* produces white kernels.

- The three genes assort independently. When the *AaBbCc* F_1 individuals self-fertilize, white F_2 individuals would occur at a frequency of 1/4 (*aa*) × 1/4 (*bb*) × 1/4 (*cc*) = 1/64 *aabbcc*.

- The *a*, *b*, and *c* alleles do not contribute to pigment production, but the *A*, *B*, and *C* alleles contribute to pigment production in an equal and additive way. As a result, the degree of red pigmentation is determined by the number of *A*, *B*, or *C* alleles present. Each uppercase allele that is present makes a wheat kernel slightly darker red.

Later work showed that Nilsson-Ehle's model hypotheses were correct in virtually every detail. Quantitative traits are produced by the independent actions of many genes, al-

though it is now clear that some genes have much greater effects on the trait in question than other genes do. As a result, the transmission of quantitative traits is said to result from polygenic ("many-genes") inheritance. In **polygenic inheritance**, each gene adds a small amount to the value of the phenotype.

In the decades immediately after the rediscovery of Mendel's work, analyses of phenomena such as sex-linkage, incomplete dominance, multiple allelism, environmental effects, gene interactions, and polygenic inheritance provided a fairly comprehensive answer to the question of why offspring resemble their parents. **Table 13.2** summarizes some of the key exceptions and extensions to Mendel's rules and gives you a chance to compare and contrast their effects on patterns of inheritance.

✓ CHECK YOUR UNDERSTANDING

Meiosis is the process responsible for Mendel's principle of segregation and principle of independent assortment. X-linked genes violate the principle of segregation, because males can have only one allele for a given trait. Genes on the same chromosome violate the principle of independent assortment, because they are not transmitted to gametes independently of each other, unless crossing over occurs between them. If you understand incomplete dominance, codominance, multiple allelism, pleiotropy, environmental effects, epistasis, and polygenic inheritance, you should be able to (1) explain each of these phenomena and give an example of each, and (2) explain why each phenomenon does or does not violate the principle of segregation and the principle of independent assortment.

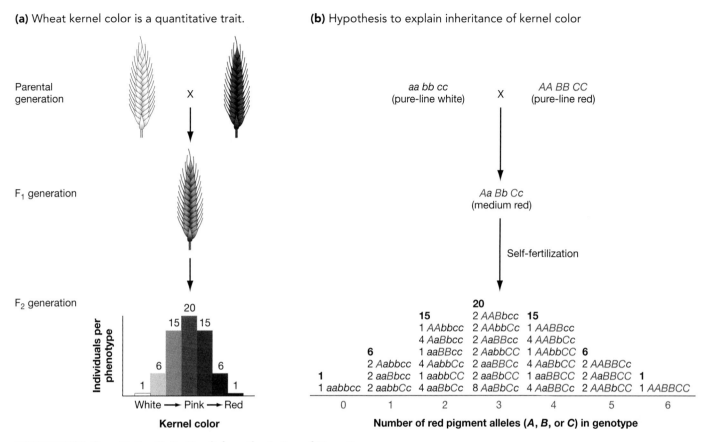

(a) Wheat kernel color is a quantitative trait.

(b) Hypothesis to explain inheritance of kernel color

FIGURE 13.20 Quantitative Traits Result from the Action of Many Genes
(a) When wheat plants with white kernels were crossed with wheat plants with red kernels, the F$_2$ offspring showed a range of kernel colors. The frequency of these phenotypes formed a normal distribution. **(b)** This model attempts to explain the results of part (a). **EXERCISE** Confirm that the distribution of genotypes shown in part (b) is correct by drawing a Punnett square based on the gametes produced by the F$_1$ parents and then filling in the F$_2$ offspring genotypes that result. (The square will be 8 × 8, with 64 boxes!)

TABLE 13.2 Some Extensions to Mendel's Rules: A Summary

Type of Inheritance	Definition	Consequences or Comments
Sex-linkage	Genes located on sex chromosomes.	Patterns of inheritance in males and females differ.
Linkage	Two genes found on same chromosome.	Linked genes violate principle of independent assortment.
Incomplete dominance	Heterozygotes have intermediate phenotype.	Polymorphism—heterozygotes have unique phenotype.
Codominance	Heterozygotes have phenotype of both alleles.	Polymorphism is possible—heterozygotes have unique phenotype.
Multiple allelism	In a population, more than two alleles present at a locus.	Polymorphism is possible.
Polymorphism	In a population, more than two phenotypes present.	Can result from actions of multiple alleles, incomplete dominance, and/or codominance.
Pleiotropy	A single allele affects many traits.	This is common.
Epistasis: Variation in the genetic environment	In discrete traits, alleles of other genes affect phenotype.	Same genotypes at a particular locus can be associated with different phenotypes.
Variation in the physical environment	Phenotype influenced by environment experienced by individual.	Same genotypes can be associated with different phenotypes.
Polygenic inheritance of quantitative traits	Many genes are involved in specifying traits that exhibit continuous variation.	Unlike alleles that determine discrete traits, each allele adds a small amount to phenotype.

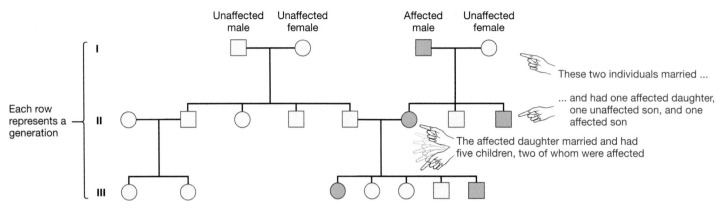

FIGURE 13.21   Reading a Pedigree
Pedigrees use special notation, shown here, as a compact way of communicating a great deal of information about which individuals in a family exhibit a certain phenotype.

13.6   Applying Mendel's Rules to Humans

When researchers set out to study how a particular gene is transmitted in wheat or fruit flies or garden peas, they start out by making a series of controlled experimental crosses. For obvious reasons, this research strategy is not possible with humans. But in many cases it is extremely valuable for biologists and physicians to know the answers to basic questions about the inheritance of certain human genes—particularly those associated with inherited diseases. Do the alleles present in the population have a simple recessive and dominant relationship? Is the gene involved autosomal or sex-linked? Is the trait influenced primarily by alleles at a single gene, or is inheritance polygenic? To answer these questions, investigators have to analyze human crosses that already exist. They do so by constructing a **pedigree**, or family tree, of affected individuals.

A pedigree, such as the one shown in **Figure 13.21**, records the genetic relationships among the individuals in a family along with each person's sex and phenotype with respect to the trait in question. If the trait is due to a single gene, then analyzing the pedigree may reveal whether the trait is due to a dominant or recessive allele and whether the gene responsible is located on a sex chromosome or on an autosome. Let's look at a series of specific case histories to see how this work is done.

Are Alleles Recessive or Dominant?

To analyze the inheritance of a trait that shows discrete variation, biologists begin by assuming that a single autosomal gene is involved and that the alleles present in the population have a simple dominant-recessive relationship. This is the simplest possible situation. If the pattern of inheritance fits this model, then the assumptions—of inheritance via a single gene and simple dominance—are supported. Let's first analyze the pattern of inheritance that is typical of autosomal recessive traits and then examine patterns that emerge in pedigrees for autosomal dominant traits.

Patterns of Inheritance: Autosomal Recessive Traits   In analyzing the inheritance of traits, it's helpful to distinguish conditions that *must* be met when a particular pattern of inheritance occurs versus conditions that are *likely* to be met. For example, if a phenotype is due to an autosomal recessive allele, then individuals with the trait must be homozygous. If the recessive allele is rare in the population, then the parents of individuals with the trait are heterozygous for the trait. Heterozygous individuals who carry a recessive allele for an inherited disease are referred to as **carriers** of the disease. These individuals carry the allele and transmit it even though they do not exhibit signs of the disease. When two carriers mate, they should produce offspring with the recessive phenotype about 25 percent of the time.

Figure 13.22a is the pedigree from a family in which an autosomal recessive disease, such as cystic fibrosis, occurs. The key feature to notice in this pedigree is that some boys and girls exhibit the trait even though their parents do not. This is the pattern you would expect to observe under the model of an autosomal recessive trait if the parents of individuals with the trait are heterozygous. In addition, it is logical to observe that when an affected individual has children, those children do not necessarily have the trait. This pattern is predicted if affected people marry individuals who are homozygous for the normal allele, which is likely to be true if the recessive allele is rare in the population.

The recessive phenotype should show up in offspring only when both parents carry that recessive allele and pass it on to their offspring. By definition, a recessive allele produces a given phenotype only when the individual is homozygous for that allele.

Patterns of Inheritance: Autosomal Dominant Traits   By definition, individuals who are homozygous or heterozygous for an autosomal dominant trait will have the dominant phenotype. For autosomal dominant traits, individuals with a single copy of the allele must have the dominant phenotype. Even if

(a) Pedigree of a family with an autosomal recessive disease

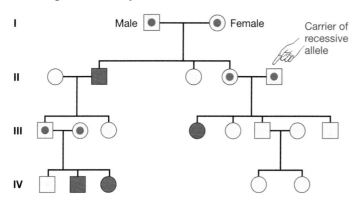

(b) Pedigree of a family with Huntington's disease

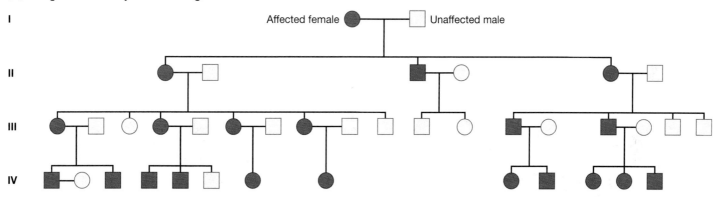

FIGURE 13.22 Pedigrees of Diseases Inherited on Autosomes
(a) Diseases, such as cystic fibrosis, that are inherited as autosomal recessives appear in both males and females. For an individual to be affected, both parents must carry the allele responsible. If the recessive allele is rare, affected individuals may not appear in every generation. **(b)** Diseases that are inherited as autosomal dominants may appear in both males and females and tend to appear in every generation. **QUESTION** In a pedigree where an autosomal recessive disease appears, an affected parent frequently has unaffected children. But unaffected parents can have affected offspring. Why?

one parent is heterozygous and the other is homozygous recessive, on average half of their children should show the dominant phenotype. And unless a new mutation has occurred, any child with the trait must have a parent with this trait. The latter observation is in strong contrast to the pattern seen in autosomal recessive traits.

Figure 13.22b shows the consequences of autosomal dominant inheritance, in the pedigree of a family affected by a degenerative brain disorder called **Huntington's disease**. The pedigree has two features that indicate this disease is passed to the next generation as an autosomal dominant allele. First, if a child shows the trait, then one of its parents shows the trait as well. Second, if families have a large number of children, the trait usually shows up in every generation—due to the high probability of heterozygous parents having affected children. Other examples of autosomal dominant traits in humans include a widow's peak, the presence of extra fingers or toes (polydactyly), and achondroplastic dwarfism, which is due to a defect in cartilage structure.

Is the Trait Autosomal or Sex-Linked?

When it is not possible to arrange reciprocal crosses, can data in a pedigree indicate whether a trait is autosomal or sex-

linked? The answer is based on a simple premise: If a trait appears about equally often in males and females, then it is likely to be autosomal. For example, the data in the Huntington's disease pedigree indicates that the disease appears in both males and females at about equal rates. But if males are much more likely to have the trait in question than females are, then the gene responsible may be found on the X chromosome. Because so few genes occur on the Y chromosome, Y-linked inheritance is rare.

To understand why a sex bias in phenotypes implicates sex-linked inheritance, recall from Section 13.4 that sex-linked genes are located on one of the sex chromosomes. Because human males have one X chromosome and one Y chromosome, they have just one copy of each X-linked gene. But because human females have two X chromosomes, they have two copies of each X-linked gene. These simple observations are critical. In humans—just as in fruit flies and in every other species that has sex chromosomes—the pattern of inheritance in sex-linked traits is different in men and women, because the complement of sex chromosomes differs in males and females.

What does the pedigree of an X-linked trait look like? To answer this question, let's consider the pedigree of a classic

X-linked trait—the occurrence of hemophilia in the descendants of Queen Victoria and Prince Albert, who were monarchs of England in the mid-to-late 1800s (**Figure 13.23a**). **Hemophilia** is caused by a defect in an important blood-clotting factor. Hemophiliacs are at a high risk of bleeding to death, because even minor injuries result in prolonged bleeding. The key point to notice in the pedigree of Queen Victoria's descendants is that only males developed hemophilia (**Figure 13.23b**). In addition, note the affected male in generation II (the second square from the right in row II). His two sons were unaffected, but the trait reappeared in a grandson. Stated another way, the occurrence of hemophilia skipped a generation.

This pattern, is logical because hemophilia is due to an X-linked *recessive* allele. Because males have only one X chromosome, the phenotype associated with an X-linked recessive allele appears in every male that carries it. Females express the recessive phenotype only if they are homozygous for that allele. Because the defective allele is rare in this and other populations, it is extremely unlikely to appear in the homozygous state.

Further, the appearance of an X-linked recessive trait skips a generation in a pedigree because the affected male passes his only X chromosome on to his daughters (his sons receive his Y chromosome). But because his daughters have two X chromosomes and are most likely to have received a normal allele from their mother, the daughters don't show the trait. They, however, will pass the defective allele on to about half of their sons—the probability is that about half will inherit the X chromosome with the defective allele, and half will inherit the other X chromosome. Thus about half of their sons will develop the trait. Females who can be inferred to be carriers, because one or more of their sons have hemophilia, are highlighted in Figure 13.23b.

In contrast, if an X-linked trait is dominant, it will appear in every individual who has the defective allele. A good indicator

(a) Queen Victoria and Prince Albert

FIGURE 13.23 A Pedigree of an X-Linked Recessive Disease
(a) Queen Victoria, who was a carrier for hemophilia, and Prince Albert. **(b)** A pedigree showing the occurrence of hemophilia in the descendants of Victoria and Albert. In this family, the condition shows up only in males. **EXERCISE** Redraw this pedigree, showing the pattern of inheritance if the allele for hemophilia were X-linked dominant.

(b) Occurrence of hemophilia in royal families of Europe

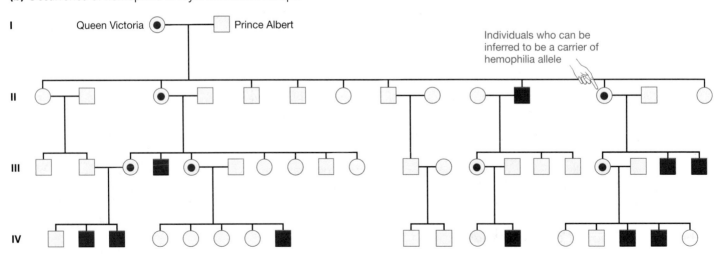

of an X-linked dominant trait is a pedigree in which an affected male has all affected daughters but no affected sons. This pattern is logical because every female offspring of an affected father gets an X chromosome from him and will herself be affected, while his sons get their only X chromosome from their unaffected mother. Besides the inherited form of a bone disease called **rickets**, however, very few diseases are known to be due to X-linked dominant alleles.

By analyzing pedigrees in this way, biomedical researchers have been able to characterize how most of the common genetic diseases in humans are inherited. By the 1940s, though, the burning question in genetics was no longer the nature of inheritance, but the nature of the gene itself. What are genes made of, and how are they copied so that parents pass their alleles on to their offspring? These are the questions we turn to in Chapter 14.

ESSAY Does "Genetic Determinism" Exist?

When people say that a trait is genetically determined, they usually mean that an individual with a particular genotype will have a particular phenotype, no matter what environment the individual experiences as it grows and matures. Is seed shape in garden peas genetically determined? Naïvely, the answer would appear to be yes. In Mendel's experiments, individuals with *RR* or *Rr* genotypes had round seeds, while individuals with *rr* genotypes had wrinkled seeds. At the seed-shape gene, then, there is a strong association between genotype and phenotype. There is also a strong resemblance between the phenotypes of parents and offspring. Is the same true for traits in humans, where data confirm a strong resemblance between parents and offspring? Are traits such as your height, weight, intelligence, and personality genetically determined?

The answer is that few traits are "determined" in the sense of being fixed or invariant based on a certain genotype. Among pea seeds, for example, the degree of wrinkling varies widely. There are two reasons for this. First, traits such as seed shape are affected by many aspects of the physical environment. For instance, the availability of water and nutrients and the presence of disease-causing fungi or viruses influence seed development and thus the phenotype. In garden peas, well-watered individuals with the *rr* genotype tend to have more wrinkled seeds than do individuals with the same genotype that experience prolonged water stress. Second, no gene acts alone. Mendel's seven traits also vary depending on the actions of alleles at other genes. Interaction among alleles, or epistasis, is the rule. In garden peas, researchers have identified at least three other genes that may affect the degree of wrinkling in seeds.

Mendel was able to analyze the effect of a single gene on the traits he studied only because he controlled for these effects carefully. He grew individuals in similar environments and analyzed strains with similar genetic backgrounds. (In the process of creating pure lines through breeding, Mendel ensured that individuals in his populations would have similar or identical genotypes for all traits, because they shared common ancestors.)

What does all this have to do with human traits? Each of your characteristics, like seed shape in pea plants, is a combination of your genotype at loci that have a major effect on the trait, environmental influences, and the effect of alleles at dozens or hundreds of other loci. Environmental effects happen to be particularly strong in the case of human height and intelligence.

Are traits such as your height, weight, intelligence, and personality genetically determined?

Over the past 75 years, the aspect of intelligence measured by IQ tests has increased dramatically in the industrialized nations. Yet no biologists support the hypothesis that these changes are a result of changes in the genetic makeup of these populations. Instead, most attribute the increase in IQ scores as a consequence of changes in the environment—specifically, better education.

Where does all this leave us? Few traits are genetically determined in the sense of being fixed or invariant. Similarly, it would be equally careless to refer to a trait as being environmentally determined. All traits have a genetic basis, and all have an environmental component. A genotype defines the range of phenotypes that are possible for an individual. This range is called a **norm of reaction**. Sometimes the range of possibilities is very small, as in the MN proteins of red blood cells or in the development of Huntington's disease. Sometimes the range of possibilities is very large, as in traits such as human height or intelligence. When the range of possibilities is wide, the environment plays a prominent role in determining the eventual phenotype. The relative importance of genetics and environmental effects varies from trait to trait.

In general, then, it is accurate to think of genotypes as contributing to phenotypic tendencies and predispositions or to the potential of an individual. But most genes can "determine" a trait only in the context of a carefully controlled environment and a precisely defined set of alleles at other loci. In the sense that the phrase is used in casual conversation, genetic determinism in complex traits such as intelligence and personality is rare or nonexistent.

CHAPTER REVIEW

Summary of Key Concepts

▪ **Mendel discovered that in garden peas, individuals have two alleles, or versions, of each gene. Prior to the formation of eggs and sperm, the two alleles of each gene separate so that one allele is transmitted to each egg or sperm cell.**

Gregor Mendel was the first individual to apply a modern scientific approach to the study of heredity. Mendel proposed two basic principles of transmission genetics. The first was the principle of segregation, which claimed that each trait was specified by paired hereditary determinants (alleles of genes) that separate from each other during gamete formation.

▪ **Genes are located on chromosomes. The separation of homologous chromosomes during meiosis I explains why alleles of the same gene segregate to different gametes.**

The chromosome theory of inheritance claimed that genes are located on chromosomes and that the movements of chromosomes during meiosis provide a physical basis for Mendel's observations. The chromosome theory provided a physical basis for the principle of segregation. Alleles of the same gene segregate from each other and end up in different gametes because homologous chromosomes segregate at anaphase of meiosis I.

▪ **If genes are located on different chromosomes, then the alleles of each gene are transmitted to egg cells and sperm cells independently of each other.**

Mendel's second basic conclusion was the principle of independent assortment, which stated that the segregation of one pair of genes—controlling a given trait—was not influenced by the segregation of other gene pairs. The chromosome theo-

ry provided a physical basis for the principle of independent assortment. Genes located on different chromosomes move to gametes independently of each other during meiosis. The principle of independent assortment holds because chromosomes line up independently of each other at the metaphase plate during meiosis I. Thus, gametes receive random combinations of chromosomes.

Web Tutorial 13.1   Mendel's Experiments

Web Tutorial 13.2   The Principle of Independent Assortment

▪ **Important exceptions exist to the rules that individuals have two alleles of each gene and that alleles of different genes are transmitted independently. Genes on the same chromosome are not transmitted independently of each other. If a gene is on a sex chromosome, then not every gamete will receive an allele of that gene.**

Thomas Hunt Morgan and colleagues extended Mendel's work by describing X-linked inheritance and by showing that genes located on the same chromosome do not exhibit independent assortment. Studies of X-linked traits helped confirm that genes are found on chromosomes, while studies of linked traits led to the first genetic maps showing the locations of genes on chromosomes. Later studies confirmed that many traits are influenced by the interaction of several genes and that phenotypes are influenced by the environment an individual has experienced as well as by its genotype. Analysis of pedigrees allowed researchers to work out the basic modes of inheritance for most of the common genetic diseases in humans.

Genetics Questions

1. In studies of how traits are inherited, what makes certain species candidates for model organisms?
   a. They are the first organisms to be used in a particular type of experiment, so they are a historical "model" of what researchers expect to find.
   b. They are easy to study because a great deal is already known about them.
   c. They are the best or most fit of their type.
   d. They are easy to maintain, have a short life cycle, produce many offspring, and yield data that are relevant to many other organisms.

2. Why is the allele for wrinkled seed shape in garden peas considered recessive?
   a. It "recedes" in the F_2 generation when homozygous parents are crossed.
   b. The trait associated with the allele is not expressed in heterozygotes.
   c. Individuals with the allele have lower fitness than that of individuals with the dominant allele.
   d. The allele is less common than the dominant allele. (The wrinkled allele is a rare mutant.)

3. The alleles found in haploid organisms cannot be dominant or recessive. Why?
   a. Dominance and recessiveness describe interactions between two alleles of the same gene in the same individual.
   b. Because only one allele is present, alleles in haploid organisms are always dominant.
   c. Alleles in haploid individuals are transmitted like mitochondrial DNA or chloroplast DNA.
   d. Most haploid individuals are bacteria, and bacterial genetics is completely different from eukaryotic genetics.

4. Biologists no longer use the term *pure line* except in a historical context. Instead, they simply refer to "pure-line" individuals as homozygotes. Why?
   a. They are highly inbred and are homozygous at every gene.
   b. Only two alleles are present at each gene in the populations to which these individuals belong.
   c. Calling them "pure" implies some sort of value judgment.
   d. All of the individuals in pure lines are homozygous at the gene in question.

5. The genes for the traits that Mendel worked with either are located on different chromosomes or are so far apart on the same chromosome that crossing over almost always occurs between them. How did this circumstance help Mendel recognize the principle of independent assortment?
   a. Otherwise, his dihybrid crosses would not have produced a 9:3:3:1 ratio of F_2 phenotypes.
   b. The occurrence of individuals with unexpected phenotypes led him to the discovery of recombination.
   c. It led him to the realization that the behavior of chromosomes during meiosis explained his results.
   d. It meant that the alleles involved were either dominant or recessive, which gave 3:1 ratios in the F_1 generation.

6. The text claims that Mendel worked with the simplest possible genetic system. Is this claim legitimate?
   a. Yes—discrete traits, two alleles, simple dominance and recessiveness, no sex chromosomes, and unlinked genes are the simplest situation known.
   b. Yes—the ability to self-fertilize or cross-pollinate made it simple for Mendel to set up controlled crosses.
   c. No—Mendel was unaware of meiosis and the chromosome theory of inheritance, so it was not easy to reach the conclusions he did.
   d. No—Mendel's experimental designs and his rules of inheritance are actually quite complex and sophisticated.

7. Mendel's rules do not correctly predict patterns of inheritance for tightly linked genes or the inheritance of alleles that show incomplete dominance or epistasis. Does this mean that his hypotheses are incorrect?
   a. Yes, because they are relevant to only a small number of organisms and traits.
   b. Yes, because not all data support his hypotheses.
   c. No, because he was not aware of meiosis or the chromosome theory of inheritance.
   d. No, it just means that his hypotheses are limited to certain conditions.

8. The artificial sweetener NutraSweet consists of a phenylalanine molecule linked to aspartic acid. The labels of diet sodas that contain NutraSweet include a warning to people with PKU. Why?
   a. Even though it is an artificial sweetener, NutraSweet stimulates the same taste receptors that natural sugars do.
   b. If an individual cannot metabolize phenylalanine, ingestion of NutraSweet will increase blood levels of phenylalanine.
   c. In people with PKU, phenylalanine reacts with aspartic acid to form a toxic compound.
   d. People with PKU cannot lead normal lives, even if their environment is carefully controlled.

9. When Sutton and Boveri published the chromosome theory of inheritance, research on meiosis had not yet established that paternal and maternal homologs assort independently of each other. Then, in 1913, Elinor Carothers published a paper about a grasshopper species with an unusual karyotype: One chromosome had no homolog (meaning no pairing partner at meiosis I); another chromosome had homologs that could be distinguished under the light microscope. If chromosomes assort independently, how often should Carothers have observed each of the four products of meiosis shown in the following figure? Explain.

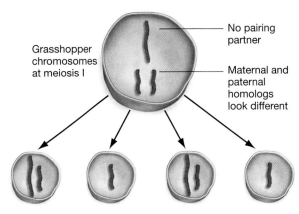

Four types of gametes possible
(each meiotic division can produce only two of the four)

   a. Only the gametes with one of each type of chromosome would occur.
   b. The four types of gametes should be observed to occur at equal frequencies.
   c. The chromosome with no pairing partner would disintegrate, so only gametes with one copy of the other chromosome would be observed.
   d. Gametes with one of each type of chromosome would occur twice as often as gametes with just one chromosome.

10. Consider the alleles involved in quantitative traits such as human height. Is the concept of dominance relevant to these alleles? Why or why not?
   a. Yes—the dominant alleles confer the highest values for the trait and the highest fitness.
   b. Yes—the dominant alleles are the most common.
   c. No—the alleles in quantitative traits are always codominant.
   d. No—dominance can occur only in discrete traits.

Genetics Problems

The best way to test and extend your knowledge of transmission genetics is to work problems. Most genetics problems are set up as follows: You are given some information about the genotypes or phenotypes of one or both parents, along with data on the phenotypes of F_1 or F_2 offspring. Your task is to generate a set of hypotheses—or what biologists call a genetic model—to explain the results. More specifically, you will need to generate a hypothesis to address each of the following questions:

- Is the trait involved discrete or quantitative?
- Is the phenotype a product of one gene or many genes?
- For each gene involved, how many alleles are present—one, two, or many?
- Do the alleles involved show complete dominance, incomplete dominance, or codominance?
- Are the genes involved sex-linked or autosomal?
- If more than one gene is involved, are they linked or unlinked? If they are linked, does crossing over occur frequently?

It's also helpful to ask yourself whether epistasis or pleiotropy might be occurring and whether it is safe to assume that the experimental design carefully controlled for effects of variation in other genes or the environment.

In working the problem, be sure to start with the simplest possible explanation. For example, if you are dealing with a discrete trait, you might hypothesize that the cross involves a single autosomal gene with two alleles that show complete dominance. Your next step is to infer what the parental genotypes are (if they are not already given), and then do a Punnett square to predict the offspring phenotypes and their frequencies based on your hypotheses. Next, check whether these predictions match the observed results given in the problem. If the answer is yes, you have a valid solution. But if the answer is no, you need to go back and change one of your hypotheses, redo the Punnett square, and check to see if the predictions and observations match. Keep redoing these steps until you have a model that fits the data.

11. In the plant species *Plectritis congesta*, known as rosy plectris, individuals produce fruits that either have or do not have prominent structures called wings. The alleles involved are W^+ = winged fruit; W^- = wingless fruit. Researchers collected an array of individuals from the field and performed a series of crosses. The results are given in the table below. Complete the table by writing down the genotype of the parent or parents involved in each cross. (Write down one genotype if selfing occurs and two genotypes if two different individuals were involved.)

Parental Phenotype(s)	Number of Offspring with Winged Fruits	Number of Offspring with Wingless Fruits	Parental Genotype(s)
Wingless (self-fertilized)	0	80	
Winged (self-fertilized)	90	30	
Winged × wingless	46	0	
Winged × winged	44	0	

12. *Tay-Sachs disease* causes nerve cells to malfunction and results in death by age 4. Two healthy parents know from blood tests that each parent carries a recessive allele responsible for Tay-Sachs. If their first three children have the disease, what is the probability that their fourth child will not? Assuming that they have not yet had a child, what is the probability that, if they have four children, all four will have the disease? If their first three children are male, what is the probability that their fourth child will be male?

13. Two black female mice are crossed with a brown male. In several litters, female I produced 9 blacks and 7 browns; female II produced 57 blacks. What deductions can you make concerning the inheritance of black and brown coat color in mice? What are the genotypes of the parents in this case?

14. Suppose that in garden peas the genes for seed color and seed-pod shape are linked, and that Mendel crossed *YYII* parents (which produce yellow seeds in inflated pods) with *yyii* parents (which produce green seeds in constricted pods).

- Draw the F_1 Punnett square and predict the expected F_1 phenotype(s).
- List the genotype(s) of gametes produced by F_1 individuals if no crossing over occurs.
- Draw the F_2 Punnett square if no crossing over occurs. Based on this Punnett square, predict the expected phenotype(s) in the F_2 generation and the expected frequency of each phenotype.
- If crossing over occurs during gamete formation in F_1 individuals, give the genotype of the recombinant gamete(s) that result.
- Add the recombinant gametes to the F_2 Punnett square above. Will any additional phenotypes be observed at low frequency in the F_2 generation? If so, what are they?

15. In Jimson weed, the allele that results in violet flowers (*V*) is dominant to one that results in white flowers (*v*); at another locus, the allele that makes prickly seed capsules (*P*) is dominant to one that results in smooth capsules (*p*). A plant with white flowers and prickly capsules is crossed with one that has violet flowers and smooth

capsules. The F_1 consisted of 47 plants with white flowers and prickly capsules, 45 plants with white flowers and smooth capsules, 50 plants with violet flowers and prickly capsules, and 46 plants with violet flowers and smooth capsules. What are the genotypes of the parents? Are these two genes linked?

16. In cats, the *Manx* allele (*M*) causes a short or absent tail, while a recessive allele *m* confers a normal, long tail. Cats of genotype *MM* die as embryos. If two Manx cats mate, what is the probability that each *surviving* kitten has a long tail?

17. A plant with orange-spotted flowers was grown in the greenhouse from a seed collected in the wild. The plant was self-pollinated and gave rise to the following progeny: 88 orange with spots, 34 yellow with spots, 32 orange with no spots, and 8 yellow with no spots. What can you conclude about the dominance relationships of the alleles responsible for the spotted and unspotted phenotypes? Of orange and yellow phenotypes? What can you conclude about the genotype of the original plant that had orange, spotted flowers?

18. As a genetic counselor, you routinely advise couples about the possibility of genetic disease in their offspring based on their family histories. This morning you met with an engaged couple, both of whom are phenotypically normal. The man, however, has a brother who died of Duchenne-type muscular dystrophy, an X-linked condition that results in death before the age of 20. The allele responsible for this disease is recessive. His prospective bride, whose family has no history of the disease, is worried that the couple's sons or daughters might be afflicted.

- How would you advise this couple?
- The sister of this man is planning to marry his fiancé's brother. How would you advise this second couple?

19. Study the experimental result illustrated in Figure 13.6a. To further test his hypothesis of independent assortment, Mendel let the F_2 progeny from this experiment self-fertilize. Then he examined the phenotypes of the F_3 generation.

- Consider the F_2 individuals that have one dominant and one recessive trait (yellow and wrinkled seeds or green and round seeds). When they self-fertilize, what fraction of the progeny will be homozygous for the traits that appeared in the parents?
- Consider the F_2 individuals that have both dominant traits. When they self-fertilize, what fraction of the progeny will be homozygous for the dominant traits?

20. Suppose that you are heterozygous for two genes that are located on different chromosomes. You carry alleles *A* and *a* for one gene and alleles *B* and *b* for the other. Draw a diagram that illustrates what happens to these genes and alleles when meiosis occurs in your reproductive tissues. Label the stages of meiosis, the homologous chromosomes, sister chromatids, nonhomologous chromosomes, genes, and alleles. Be sure to list all of the genetically different gametes that could form and how frequently each type should be observed. On the diagram, identify the events responsible for the principle of segregation and the principle of independent assortment.

21. In humans, the ABO blood type is a polymorphic trait. Describe the alleles and genotypes responsible for this polymorphism. Suppose a woman with blood type O married a man with blood type AB. What phenotypes and genotypes would you expect to observe in their offspring, and in what proportions?

22. Mr. Spock's father came from the planet Vulcan, and his mother was from planet Earth. A Vulcan has pointed ears and a heart located on the right side of the chest. Mr. Spock has both of these traits, which are known to be determined by two different genes, each with two alleles. Suppose that Mr. Spock married an Earth woman and that they had many children. About half of their children have pointed

ears and a right-sided heart, like Spock, and about half have round-ed ears and a left-sided heart, like their mother.

- What would Mendel predict the progeny phenotypes and ratios to be? Explain your answer using formal genetic terminology.
- How do you explain the actual results?

23. In Klingons, one gene determines hair texture; another gene determines whether the individual will have a sagittal crest (a protrusion on the forehead). The two genes are not linked.

K = curly Klingon hair (dominant)

k = silky earthling-like hair (recessive)

S = large sagittal crest (dominant)

s = smooth, flat, earthling-like forehead (recessive)

Kayless is a half-human, half-Klingon with the genotype *KkSs*. He mates with an individual who is also heterozygous for both genes.

- Set up a Punnett square for this dihybrid cross.
- What are the four possible phenotypes that may result from this mating? Include a description of both hair and forehead for each phenotype.
- What is the expected phenotypic ratio from the dihybrid cross?
- What fraction of the progeny are expected to be heterozygous for both genes?
- What fraction are expected to be homozygous for both genes?
- Are Kayless and his mate more likely to see an actual ratio close to the predicted values if they have 16 children or 160? Why? Explain why sample size does or does not affect observed phenotypic ratios.

24. The blending-inheritance hypothesis proposed that the genetic material from parents is irreversibly mixed in the offspring. As a result, offspring should always appear intermediate in phenotype to the parents. Mendel, in contrast, proposed that genes are discrete and that their integrity is maintained in the offspring and in subsequent generations. Suppose the year is 1890. You are a horse breeder and have just read Mendel's paper. You don't believe his results, however, because you often work with cremello (very light-colored) and chestnut (reddish-brown) horses. You know that if you cross a cremello individual from a pure-breeding line with a chestnut individual from a pure-breeding line, the offspring will be palomino—meaning they have an intermediate (golden-yellow) body color. What additional crosses would you do to test whether Mendel's model is valid in the case of genes for horse color? List the crosses and the offspring genotypes and phenotypes you'd expect to obtain. Explain why these experimental crosses would provide a test of Mendel's model.

25. Two mothers give birth to sons at the same time in a busy hospital. The son of couple 1 is afflicted with hemophilia A, which is a recessive X-linked disease. Neither parent has the disease. Couple 2 has a normal son even though the father has hemophilia A. The two couples sue the hospital in court, claiming that a careless staff member swapped their babies at birth. You appear in court as an expert witness. What do you tell the jury? Make a diagram that you can submit to the jury.

26. You have crossed two *Drosophila melanogaster* individuals that have long wings and red eyes—the wild-type phenotype. In the progeny, the mutant phenotypes, called curved wings and lozenge eyes, appear as follows:

Females
600 long wings, red eyes
200 curved wings, red eyes

Males
300 long wings, red eyes
300 long wings, lozenge eyes
100 curved wings, red eyes
100 curved wings, lozenge eyes

- According to these data, is the curved-wing allele autosomal recessive, autosomal dominant, sex-linked recessive, or sex-linked dominant?
- Is the lozenge-eye allele autosomal recessive, autosomal dominant, sex-linked recessive, or sex-linked dominant?
- What is the genotype of the female parent?
- What is the genotype of the male parent?

27. In parakeets, two autosomal genes that are located on different chromosomes control the production of feather pigment. Gene *B* codes for an enzyme that is required for the synthesis of a blue pigment, and gene *Y* codes for an enzyme required for the synthesis of a yellow pigment. Recessive, loss-of-function mutations are known for both genes. Suppose that a bird breeder has two green parakeets and mates them. The offspring are green, blue, yellow, and albino (unpigmented).

- Based on this observation, what are the genotypes of the green parents? What is the genotype of each type of offspring? What fraction of the total progeny should exhibit each type of color?
- Suppose that the parents were the progeny of a cross between two true-breeding strains. What two types of crosses between true-breeding strains could have produced the green parents? Indicate the genotypes and phenotypes for each cross.

28. The pedigree that follows is for the human trait called osteopetrosis, which is characterized by bone fragility and dental abscesses. Is the gene that affects bone and tooth structure autosomal or sex-linked? Is the allele for osteopetrosis dominant or recessive?

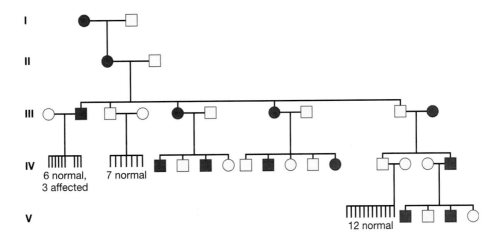

Answers to Genetics Questions 1. d; 2. b; 3. a; 4. d; 5. a; 6. a; 7. d; 8. b; 9. b; 10. d

Answers to Genetics Problems 11. First cross: W^-W^-; second cross: W^+W^-; third cross: $W^+W^+ \times W^-W^-$; fourth cross: $W^+W^+ \times W^+W^+$ or $W^+W^+ \times W^+W^-$. 12. 3/4; 1/256; 1/2. 13. The simplest explanation for the data is that these coat-color phenotypes are caused by a single gene where two alleles are present and where B = black and b = brown. Female I is Bb, female II is BB, and the male is bb. 14. Your answer should conform to the cross diagrammed in Figure 13.6b, except that different alleles and traits are being analyzed. 15. Parental genotypes are $vvPp \times Vvpp$. Based on these data, we can't tell if they are linked or not. Both hypotheses make the same prediction. 16. 1/3. 17. This is a dihybrid cross that yields progeny phenotypes in a 9:3:3:1 ratio. Let O stand for the allele for orange petals and o the allele for yellow petals; let S stand for the allele for spotted petals and s the allele for unspotted petals. Start with the hypothesis that O is dominant to o, that S is dominant to s, that the two genes are found on different chromosomes so they assort independently, and that the parent individual's genotype is $OoSs$. If you do a Punnett square for the $OoSs \times OoSs$ mating, you'll find that progeny phenotypes should be in the observed 9:3:3:1 proportions. 18. Let D stand for the normal allele and d stand for the allele responsible for Duchenne-type muscular dystrophy. The woman's family has no history of the disease, so her genotype is almost certainly DD. The man is not afflicted, so he must be DY. (The trait is X-linked, so he has only one allele; the "Y" stands for the Y chromosome.) Their children are not at risk. The man's sister could be a carrier, however—meaning she has the genotype Dd. If so, then half of the second couple's male children are likely to be affected. 19. 1/2; 1/4. 20. See website. 21. Half of their offspring should have the genotype iI^A and the type A blood phenotype. Half of their offspring should have the genotype iI^B and the type B blood phenotype. 22. Presumably, Vulcans and earthlings are homozygous for both traits. Spock, then, should be heterozygous at both loci. Because he has the Vulcan phenotype, though, we can infer that pointed ears and a right-sided heart are dominant to rounded ears and a left-sided heart. Let P be the allele for pointed ears and p be the allele for rounded ears; let R be the allele for a right-sided heart and r the allele for a left-sided heart. Spock is $PpRr$; his earthling wife is $pprr$. Mendel would predict that their offspring would have the genotypes $PpRr$, $Pprr$, $ppRr$, and $pprr$ in equal proportions. Thus a quarter of their children should have pointed ears and left-sided hearts or rounded ears and right-sided hearts. The only logical explanation for the actual results is that the two genes are closely linked. If so, then Spock's gametes are either PR or pr, and his wife's gametes are all pr. 23. Four phenotypes should be observed in the ratio 9:3:3:1, as follows: curly and large: curly and smooth: silky and large: silky and smooth. One-quarter of the progeny should be heterozygous for both traits. One-quarter should be homozygous for both traits. They are more likely to see the predicted ratios with a large family, because there would be fewer fluctuations due to chance. 24. According to Mendel's model, palomino individuals should be heterozygous at the locus for coat color. If you mated palomino individuals, you would expect to see a combination of chestnut, palomino, and cremello offspring. If blending inheritance occurred, however, all of the offspring should be palomino. 25. See website. 26. The curved-wing allele is autosomal recessive; the lozenge-eye allele is sex-linked (specifically, X-linked) recessive. Let w^+ be the allele for long wings and w be the allele for curved wings; let r^+ be the allele for red eyes and r the allele for lozenge eyes. The female parent is w^+wr^+r; the male parent is w^+wr^+Y. 27. Albinism indicates the absence of pigment, so let b stand for an allele that gives the absence of blue and y for an allele that gives the absence of yellow pigment. If blue and yellow pigment blend to give green, then both green parents are $BbYy$. The green phenotype is found in $BBYY$, $BBYy$, $BbYY$, and $BbYy$ offspring. The blue phenotype is found in $BByy$ or $Bbyy$ offspring. The yellow phenotype is observed in $bbYY$ or $bbYy$ offspring. Albino offspring are $bbyy$. The phenotypes of the offspring should be in the ratio 9:3:3:1 as green:blue:yellow:albino. Two types of crosses yield $BbYy$ F_1 offspring: $BByy \times bbYY$ (blue × yellow) and $BBYY \times bbyy$ (green × albino). 28. Autosomal dominant.

www.prenhall.com/freeman is your resource for the following: Web Tutorials; Online Quizzes and other Online Study Guide materials; Answers to Conceptual Review Questions; Solutions to Group Discussion Problems; Answers to Figure Caption Questions and Exercises; and Additional Readings and Research.

DNA Synthesis

14

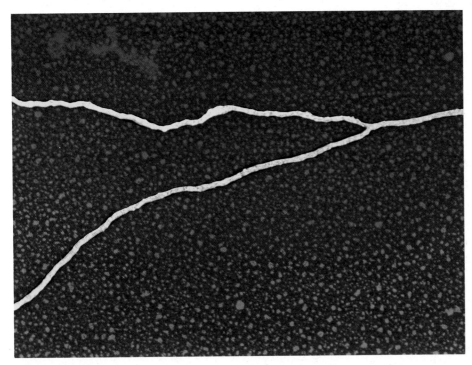

Electron micrograph showing DNA in the process of replication. The "Y" shape results from a structure called the replication fork, where DNA synthesis is taking place. The DNA double helix on the right is being replicated into two DNA double helices on the left.

KEY CONCEPTS

▥ Genes are made of DNA. When DNA is copied, each strand of a DNA double helix serves as the template for the synthesis of a complementary strand.

▥ DNA is synthesized only in the $5' \rightarrow 3'$ direction. When a DNA molecule is being copied, a large suite of specialized enzymes is involved in unwinding the double helix, continuously synthesizing the "leading strand" in the $5' \rightarrow 3'$ direction and synthesizing the "lagging strand" as a series of fragments that are then linked together.

▥ Most mistakes that occur in DNA synthesis are repaired by specialized enzymes. If these repair enzymes are defective, the mutation rate increases. If mutations occur in genes that control the cell cycle, cancer may develop.

hat are genes made of, and how are they copied so that they are faithfully passed on to offspring? These questions dominated biology during the middle of the twentieth century. After the chromosome theory of inheritance was published and confirmed in the early 1900s, understanding the molecular nature of the gene became *the* burning question in biological science. But for decades the answer remained a mystery.

During this time, the predominant research strategy in genetics was to conduct a series of experimental crosses, create a genetic model to explain the types and proportions of phenotypes that resulted, and then test the model's predictions through reciprocal crosses, testcrosses, or other techniques. This strategy was extremely productive. It led to virtually all of the discoveries analyzed in Chapter 13, including Mendel's rules, sex-linkage, linkage, genetic mapping, incomplete dominance and codominance, multiple allelism and polymorphic traits, pleiotropy, epistasis, and quantitative inheritance.

Entirely new research strategies came to the fore, however, when the molecular basis of inheritance was finally understood. How was the chemical nature of the gene eventually discovered? How are genes copied, so they can be transmitted to the next generation? Do mistakes ever occur when genes are being replicated prior to mitosis or meiosis?

The goal of this chapter is to answer these questions. We begin with studies that identified deoxyribonucleic acid (DNA) as the genetic material, and we then explore how this molecule is copied during the synthesis phase of the cell cycle.

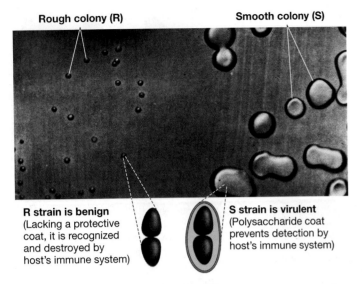

Rough colony (R)   **Smooth colony (S)**

R strain is benign
(Lacking a protective
coat, it is recognized
and destroyed by
host's immune system)

S strain is virulent
(Polysaccharide coat
prevents detection by
host's immune system)

FIGURE 14.1 There Are Two Strains of *Streptococcus pneumoniae*
Nonvirulent and virulent strains of *Streptococcus* differ in colony
appearance and at the cellular level.

14.1   DNA as the Hereditary Material

The first hint of what genes might be made of was published in
1928, when Frederick Griffith reported the discovery of a mysterious phenomenon involving hereditary traits. Griffith referred to
this phenomenon as **transformation**. His transformation experiments appeared to isolate the hereditary material.

In the 1920s Griffith had been doing experiments in an attempt to develop a vaccine against the *Streptococcus pneumoniae* bacterium, which is a leading cause of pneumonia,
earaches, sinusitis, and meningitis in humans. For his experiments, however, he worked with strains of *S. pneumoniae* that
infect mice. A **strain** is a population of genetically identical individuals. As is the case with the strains that affect humans, the
strains that affect mice vary in their **virulence**—their ability to
cause disease and death. Virulent strains cause disease; nonvirulent (or benign) strains do not.

As **Figure 14.1** shows, the strains that Griffith happened
to work with can be identified by eye when grown on a nutrient medium in a petri dish. Cells from the nonvirulent
strain form colonies that look rough; cells from the virulent
strain form colonies that look smooth. Logically enough,
Griffith called the nonvirulent strain R for rough and the virulent strain S for smooth.

To understand how the strains interact, Griffith carried out
the experiment illustrated in **Figure 14.2**. This figure shows
four experimental treatments that he designed. In the first treatment, he injected mice with cells of the nonvirulent R strain. As
he expected, these mice lived. In the second treatment, he injected mice with cells of the virulent S strain. Not surprisingly,
these mice died of pneumonia. So far, so good: These first two
treatments were controls showing the effect that each strain of
S. pneumoniae has on mice. In the third treatment, Griffith
killed cells of the virulent S strain by heating them and then in-

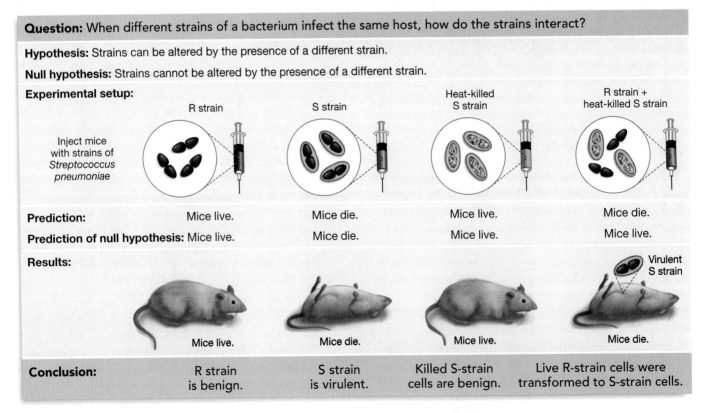

Question: When different strains of a bacterium infect the same host, how do the strains interact?

Hypothesis: Strains can be altered by the presence of a different strain.

Null hypothesis: Strains cannot be altered by the presence of a different strain.

Experimental setup:

	R strain	S strain	Heat-killed S strain	R strain + heat-killed S strain
Inject mice with strains of *Streptococcus pneumoniae*				
Prediction:	Mice live.	Mice die.	Mice live.	Mice die.
Prediction of null hypothesis:	Mice live.	Mice die.	Mice live.	Mice live.
Results:	Mice live.	Mice die.	Mice live.	Virulent S strain — Mice die.
Conclusion:	R strain is benign.	S strain is virulent.	Killed S-strain cells are benign.	Live R-strain cells were transformed to S-strain cells.

FIGURE 14.2 The Discovery of "Transformation"
Mice died after being injected with a combination of live R cells and heat-killed S cells.

jected them into mice. These mice lived. This experimental treatment was interesting because it showed that dead S cells do not cause disease. In the final treatment, Griffith injected mice with heat-killed S cells *and* live nonvirulent R cells. Unexpectedly, these mice died. Autopsies confirmed pneumonia as the cause of death. When Griffith isolated and grew *Streptococcus* from these dead mice, he found virulent S cells, *not* nonvirulent R cells.

What was going on? Griffith proposed that something from the heat-killed S cells had "transformed" the nonvirulent R cells. Stated another way, something had changed the appearance and behavior of the R cells from R-like to S-like. Because this "something" appeared in the growing population of cells that Griffith isolated from the dead mice, it had been passed on to the offspring of the transformed cells. It was clearly some sort of hereditary, or genetic, factor. Next the question became, What is it?

Is DNA the Genetic Material?

The chromosome theory of inheritance, introduced in Chapter 13, proposed that chromosomes are composed of genes. It had been known since the late 1800s that chromosomes are a complex of DNA and proteins. Because the chromosome theory had been confirmed around 1920, it was clear that Griffith's transforming factor had to consist of either protein or DNA.

Initially, most biologists backed the hypothesis that genes are made of proteins. The arguments in favor of this hypothesis were compelling. Hundreds, if not thousands, of complex and highly regulated chemical reactions occur in even the simplest living cells. The amount of information required to specify and coordinate these reactions is almost mind-boggling. With their almost limitless variation in structure and function, proteins are complex enough to contain this much information.

DNA, in contrast, was known to be composed of just four types of deoxyribonucleotides. It was also thought to be a simple molecule with some sort of repetitive and uninteresting structure. So when researchers published the first experimental evidence that DNA was the hereditary material, most biologists had the same reaction: They didn't believe it.

The Avery et al. Experiment In the early 1940s Oswald Avery, Colin M. MacLeod, and Maclyn McCarty set out to understand the molecular basis of Griffith's result. They used an elegant experimental strategy to isolate the transforming factor. To determine whether protein, RNA, or DNA was responsible for transformation, they grew quantities of virulent S cells in culture. A **culture** is a collection of cells that grows under controlled conditions—usually in a liquid suspension or on the surface of a dish on solid growth medium. Avery's group killed the cultured cells with heat, broke them open to create a cell extract, and then used chemical treatments to remove the lipids and carbohydrates from the extracts. These steps left a mixture

containing protein, RNA, and DNA from the virulent S cells. The researchers divided the sample into three treatments and used different enzymes to destroy a specific macromolecule in each. One sample was treated with proteases, enzymes that destroy proteins. Another sample was treated with ribonuclease, an enzyme that breaks apart RNA. A third sample was treated with DNAase, an enzyme that cuts up DNA (**Figure 14.3**). When small quantities of the three resulting solutions were added to cultures containing nonvirulent R cells, virulent S cells appeared in all of the cultures that still contained S-cell DNA; no S cells appeared in the sample that lacked DNA. The biologists concluded that DNA—not protein or RNA—must be the transforming factor.

This experiment provided strong evidence that DNA is the hereditary material. But the result was not widely accepted. Researchers who advocated the protein hypothesis claimed that the enzymatic treatments were not sufficient to remove all of the proteins present and that enough protein could remain to transform R cells.

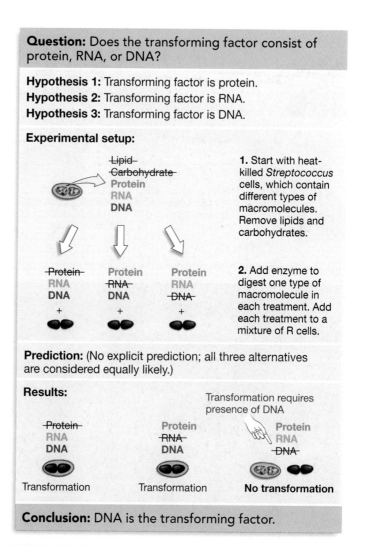

Question: Does the transforming factor consist of protein, RNA, or DNA?

Hypothesis 1: Transforming factor is protein.
Hypothesis 2: Transforming factor is RNA.
Hypothesis 3: Transforming factor is DNA.

Experimental setup:

1. Start with heat-killed *Streptococcus* cells, which contain different types of macromolecules. Remove lipids and carbohydrates.

2. Add enzyme to digest one type of macromolecule in each treatment. Add each treatment to a mixture of R cells.

Prediction: (No explicit prediction; all three alternatives are considered equally likely.)

Results: Transformation requires presence of DNA

Transformation — Transformation — **No transformation**

Conclusion: DNA is the transforming factor.

FIGURE 14.3 Experimental Evidence that DNA Is the Transforming Factor QUESTION If genes were made of protein, how would the results of this experiment change?

The Hershey-Chase Experiment Alfred Hershey and Martha Chase took up the question of whether genes were made of protein or DNA by studying how a virus called T2 infects the bacterium *Escherichia coli*. Hershey and Chase knew that T2 infections begin when the virus attaches to the cell wall of *E. coli* and injects its genes into the cell's interior (**Figure 14.4a**). These genes then direct the production of a new generation of virus particles inside the infected cell, which acts as a **host** for the parasitic virus. During the infection, the protein coat of the original, parent virus is left behind, still attached to the exterior of the host cell as a "ghost" (**Figure 14.4b**). Hershey and Chase also knew that T2 is made up almost exclusively of protein and DNA. But was it protein or DNA that entered the host cell and directed the production of new viruses?

Hershey and Chase's strategy for determining which part of the virus enters the cell and acts as the hereditary material was based on two facts: (1) Proteins present in T2 contain sulfur but not phosphorus, and (2) DNA contains phosphorus

but not sulfur. As **Figure 14.5** shows, the researchers began their work by growing viruses in the presence of either the radioactive isotope of sulfur (^{35}S) or the radioactive isotope of phosphorus (^{32}P). Because these molecules were incorporated

(a) ONLY VIRAL GENES ENTER A CELL THAT IS BEING INFECTED.

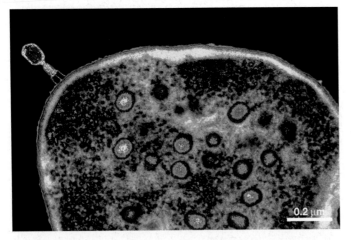

1. Start of infection. Virus genes enter host cell. Protein coat does not.

2. Virus genes direct the production of new virus particles.

3. End of infection. New generation of virus particles bursts from host cell.

(b) The virus's protein coat stays on the outside of the host cell.

FIGURE 14.4 Viruses Inject Genes into Bacterial Cells and Leave a Protein Coat Behind
(a) Viruses that infect a bacterial cell start by injecting their genes into the cell. **(b)** The protein coat of the original virus is left behind, attached to the bacterial cell wall.

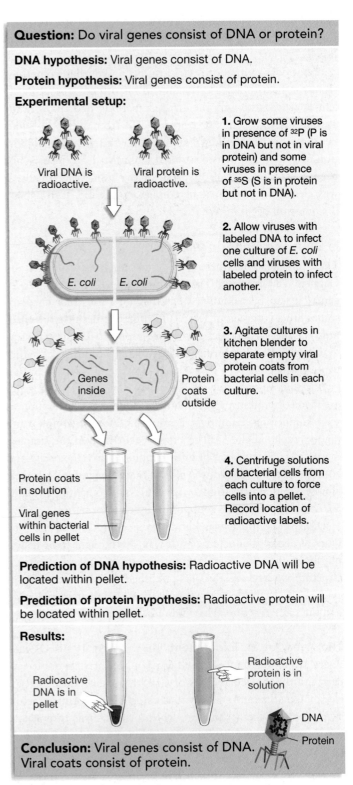

Question: Do viral genes consist of DNA or protein?

DNA hypothesis: Viral genes consist of DNA.

Protein hypothesis: Viral genes consist of protein.

Experimental setup:

Viral DNA is radioactive.

Viral protein is radioactive.

1. Grow some viruses in presence of ^{32}P (P is in DNA but not in viral protein) and some viruses in presence of ^{35}S (S is in protein but not in DNA).

E. coli   *E. coli*

2. Allow viruses with labeled DNA to infect one culture of *E. coli* cells and viruses with labeled protein to infect another.

Genes inside

Protein coats outside

3. Agitate cultures in kitchen blender to separate empty viral protein coats from bacterial cells in each culture.

Protein coats in solution

Viral genes within bacterial cells in pellet

4. Centrifuge solutions of bacterial cells from each culture to force cells into a pellet. Record location of radioactive labels.

Prediction of DNA hypothesis: Radioactive DNA will be located within pellet.

Prediction of protein hypothesis: Radioactive protein will be located within pellet.

Results:

Radioactive DNA is in pellet

Radioactive protein is in solution

DNA

Protein

Conclusion: Viral genes consist of DNA. Viral coats consist of protein.

FIGURE 14.5 More Experimental Evidence that DNA Is the Hereditary Material
The experiment shown here convinced most biologists that DNA is indeed the hereditary material.

into newly synthesized proteins and DNA, this step produced a population of viruses with radioactive proteins and a population with radioactive DNA. Then Hershey and Chase allowed each set of radioactive viruses to infect *E. coli* cells. If genes consist of DNA, then the radioactive protein should be found in the empty, or "ghost," protein coats outside the infected host cells, while the radioactive DNA should be located inside the cells. But if genes consist of proteins, the opposite should be true.

To test these predictions, Hershey and Chase sheared the ghosts off the cells by agitating each of the cultures in kitchen blenders. When the researchers spun the samples in a centrifuge, the ghosts stayed in the solution while the cells formed a pellet at the bottom of the centrifuge tube. As predicted by the DNA hypothesis, the biologists found that virtually all of the radioactive protein was in the ghosts, while virtually all of the radioactive DNA was inside the host cells. Because the injected component of the virus directs the production of a new generation of virus particles, it is this component that represents the virus's genes.

After these results were published, proponents of the protein hypothesis had to admit that DNA, not protein, must be the hereditary material. In combination, the evidence from the bacterial transformation experiments and the virus-labeling experiments was convincing. The claim that a seemingly simple molecule contained all the information for life's complexity was finally accepted.

The realization that genes are made of DNA was one of the great advances of twentieth-century biology. The insight raised two crucial questions, however: (1) How did the simple primary structure and secondary structure of DNA hold the information required to make life possible? (2) How is DNA copied, so that genetic information is faithfully passed from one cell to another during growth and from parents to offspring during reproduction? Understanding how DNA contains information is the focus of Chapter 15. The remainder of this chapter concentrates on the question of how genes are replicated prior to mitosis and meiosis.

14.2 Testing Early Hypotheses about DNA Replication

The DNA inside each cell is like an ancient text that has been painstakingly copied and handed down, generation after generation. But while the most ancient of all human texts contain messages that are thousands of years old, the DNA in living cells has been copied and passed down for *billions* of years. And instead of being copied by monks or clerks, DNA is replicated by molecular scribes. What molecules are responsible for copying DNA, and how do they work?

Chapter 4 introduced Watson and Crick's model for the secondary structure of DNA, which was proposed in 1953. Recall that DNA is a polymer made up of monomers called

nucleotides, which consist of a deoxyribose molecule, a phosphate group, and a nitrogenous base. The completed DNA molecule is a long, linear polymer that has two major components: (1) a "backbone" made up of the sugar and phosphate groups of nucleotides and (2) a series of nitrogen-containing bases that project from the backbone. Watson and Crick realized that if two of these long strands twist around each other, then certain of the nitrogen-containing bases fit together in pairs inside the spiral. The double-stranded molecule that results is called a **double helix** (**Figure 14.6**). The structure is stabilized by hydrogen bonds that form between the bases adenine (A) and thymine (T) and between the bases guanine (G) and cytosine (C). The specific pairing rules for hydrogen bonding of nitrogen-containing bases is called **complementary base pairing.**

Watson and Crick realized that the A-T and G-C pairing rules suggested a way for DNA to be copied prior to mitosis and meiosis. They suggested that the existing strands of DNA served as a template (pattern) for the production of new strands, with bases being added to the new strands according to complementary base pairing. For example, if the template strand contained a T, then an A would be added to the new strand to pair with that T. Similarly, a G on the template strand would dictate the addition of a C on the new strand.

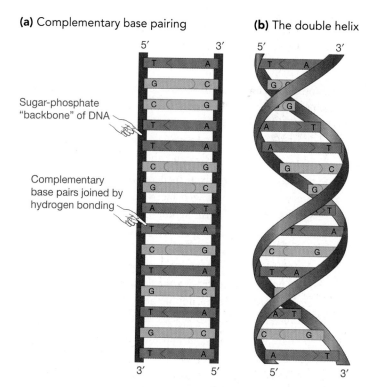

(a) Complementary base pairing   **(b)** The double helix

Sugar-phosphate "backbone" of DNA

Complementary base pairs joined by hydrogen bonding

FIGURE 14.6 DNA's Secondary Structure: The Double Helix
(a) Deoxyribonucleic acid normally has a secondary structure consisting of two strands, each with a sugar-phosphate backbone. Nitrogen-containing bases project from each strand and form hydrogen bonds. Only A-T and G-C pairs fit together in a way that allows hydrogen bonding to occur. **(b)** That bonding twists the molecule into a double helix.

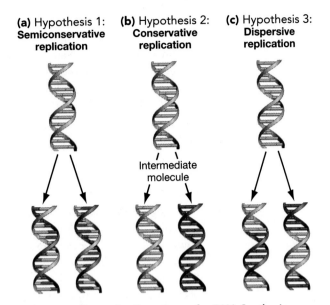

(a) Hypothesis 1: Semiconservative replication

(b) Hypothesis 2: Conservative replication

(c) Hypothesis 3: Dispersive replication

Intermediate molecule

FIGURE 14.7 Alternative Hypotheses for DNA Synthesis
The old strand of DNA is shown in gray, and the newly synthesized strand in red. **(a)** If replication were semiconservative, each strand in a DNA double helix would serve as the template for the synthesis of a daughter strand. **(b)** If replication were conservative, the original molecule would remain intact after synthesis, so the daughter molecule would consist of two newly synthesized strands. **(c)** If dispersive replication occurred, each strand in the daughter DNA would consist of a mixture of old and new DNA.

As **Figure 14.7** shows, however, biologists had three hypotheses about how the old and new strands might interact during replication:

1. If the old strands of DNA separated, each could then be used as a template for the synthesis of a new, daughter strand (Figure 14.7a). This hypothesis is called **semiconservative replication,** because each new daughter DNA molecule would consist of one old strand and one new strand.

2. If the bases temporarily turned outward so that complementary strands no longer faced each other, they could serve as a template for the synthesis of an entirely new double helix all at once. The new strands would bond to each other and be released from the template molecules. This hypothesis is called **conservative replication.** As Figure 14.7b shows, conservative replication results in two parental strands paired to form a double helix and two new daughter strands paired to form a second double helix.

3. If the parent helix was cut and unwound in short sections before being copied and put back together, then new and old strands would intermingle as shown in Figure 14.7c. This possibility is called **dispersive replication.**

Which of these three hypotheses is correct?

The Meselson-Stahl Experiment

Matthew Meselson and Frank Stahl realized that if they could tag parental and daughter strands of DNA in a way that would make them distinguishable from each other, they could determine whether replication was conservative, semiconservative, or dispersive. Soon after their results were published in 1958, the Meselson-Stahl work became recognized as a classic experiment in biological science.

Before they could do any tagging or publishing, however, they needed to choose an organism to study. They decided to work with a common inhabitant of the human gastrointestinal tract, the bacterium *Escherichia coli*. Because *E. coli* is small and grows quickly and readily in the laboratory, it had become a favored model organism in studies of biochemistry and molecular genetics.

Like all organisms, bacterial cells copy their entire complement of DNA, or their **genome**, before every cell division. To distinguish parental strands of DNA from daughter strands when *E. coli* replicates, Meselson and Stahl grew the cells for many generations in the presence of one of two isotopes of nitrogen: either ^{15}N or ^{14}N. Because ^{15}N contains an extra neutron, it is heavier than the normal isotope, ^{14}N.

This difference in mass, which creates a difference in density of ^{14}N-containing and ^{15}N-containing DNA, was the key to the experiment. The biologists reasoned that if different nitrogen isotopes were available in the growth medium when parental and daughter strands of DNA were produced, then the two types of strands should behave differently during centrifugation. More specifically, when intact, double-stranded DNA molecules are added to a solution that forms a gradient from low to high density during centrifugation, DNA strands that contain ^{14}N should form a band in the lower-density part of the centrifuge tube. In contrast, DNA strands that contain ^{15}N should form a band in the higher-density part of the centrifuge tube. Because the highest-density solution is at the bottom of the tube, DNA that contains ^{15}N should be found lower in the tube than DNA containing ^{14}N. In this way, DNA strands containing ^{15}N or ^{14}N should form separate bands. By exposing the centrifuge tube to X-ray film, the biologists could visualize the positions of the bands. How could this tagging system be manipulated to test whether replication is semiconservative, conservative, or dispersive?

Figure 14.8 summarizes Meselson and Stahl's experimental strategy. They began by growing *E. coli* cells in the presence of ^{15}N as the only nitrogen isotope in nitrogen-containing nutrient molecules. They purified DNA from a sample of these cells and transferred the rest of the culture to a growth medium containing only the ^{14}N isotope. After enough time had elapsed for these experimental cells to divide once, Meselson and Stahl removed a sample and isolated the DNA. After the remainder of the culture had divided again, they removed another sample and purified the DNA.

As Figure 14.8 shows, the conservative, semiconservative, and dispersive models make distinct predictions about the makeup of the DNA molecules after replication occurred in the first and second generation. For example, if replication

Question: Is replication semiconservative, conservative, or dispersive?

Hypothesis 1: Replication is semiconservative.

Hypothesis 2: Replication is conservative.

Hypothesis 3: Replication is dispersive.

Experimental setup:

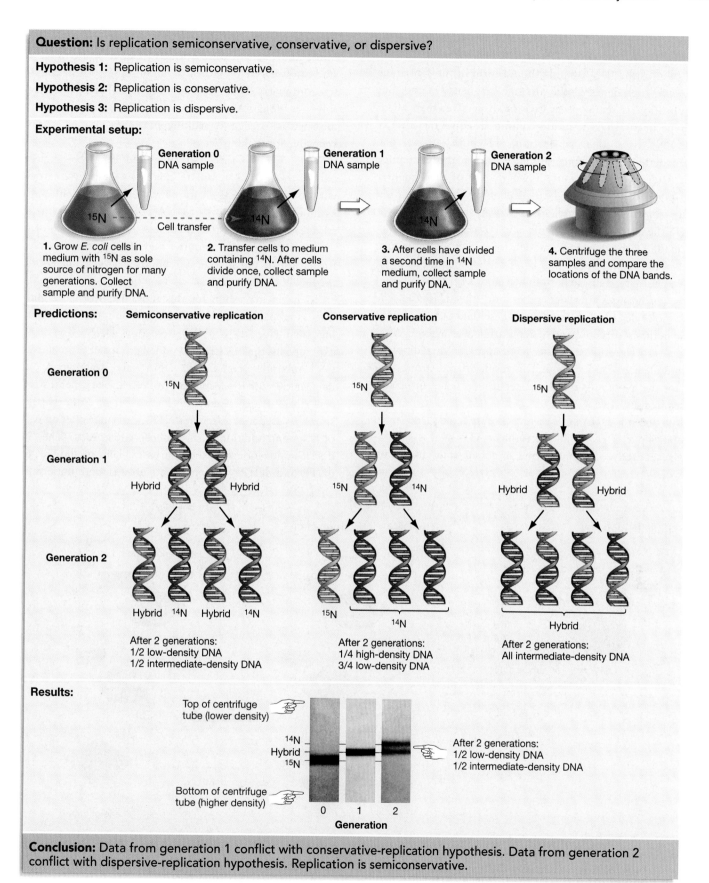

Conclusion: Data from generation 1 conflict with conservative-replication hypothesis. Data from generation 2 conflict with dispersive-replication hypothesis. Replication is semiconservative.

FIGURE 14.8 The Meselson-Stahl Experiment

EXERCISE Meselson and Stahl actually let their experiment run for four generations, with cultures continuing to grow in the presence of ^{14}N. Sketch what the data from third- and fourth-generation DNA should look like—that is, where the DNA band(s) should be.

is conservative, then the daughter cells should have double-stranded DNA with either ^{14}N or ^{15}N, but not both. As a result, two distinct DNA bands should form in the centrifuge tube—one high-density band and one low-density band. But if replication is semiconservative or dispersive, then all of the experimental DNA should contain an equal mix of ^{14}N and ^{15}N after one generation. In both cases, one intermediate-density band should form in the centrifuge tube. After two generations, however, half of the daughter cells should contain only ^{14}N if semiconservative replication occurs—meaning a second, lower-density band should appear in the centrifuge tube. This is in contrast to the prediction made by the dispersive model, for which the centrifuge tube should contain only one band at intermediate density.

The photograph at the bottom of Figure 14.8 shows the experiment's results. After one generation, the density of the DNA molecules was intermediate. These data suggested that the hypothesis of conservative replication was wrong. After two generations, a lower-density band appeared in addition to the intermediate-density band. This result offered strong support for the hypothesis that DNA replication is not dispersive but semiconservative.

How does the DNA synthesis reaction proceed? Does it require an input of energy in the form of ATP, or it is spontaneous? Is it catalyzed by an enzyme, or does it occur quickly on its own?

The Discovery of DNA Polymerase

In the mid-1950s Arthur Kornberg and co-workers set out to study how DNA replicates by attempting to generate a cell-free, or in vitro, DNA replication system. Their goal was to resolve a long-standing paradox: If genes contain the information needed for making an organism, then what is responsible for making genes?

Watson and Crick had suggested that the nucleotides paired to a DNA template might be "zippered" together without assistance from an enzyme. Kornberg, in contrast, assumed that some sort of DNA-polymerizing enzyme existed. He hypothesized that a cell-free system capable of replicating DNA would require an enzyme to catalyze the formation of phosphodiester bonds between nucleotides in the newly formed strands.

To begin his search for this molecule, Kornberg and colleagues prepared proteins from *E. coli* cultures and began testing them for their ability to copy DNA. To do these experiments, the investigators added groups of proteins to reaction mixtures containing DNA and monomers called **deoxynucleoside triphosphates**, or **dNTPs**. The general structure of a dNTP is diagrammed in **Figure 14.9a**. Note that the researchers used four different dNTPs, carrying either an A, T, G, or C base in addition to three phosphate groups. (The "N" in dNTP stands for any of the four bases. They used dATP, dTTP, dGTP, and dCTP.) The phosphate groups were important be-

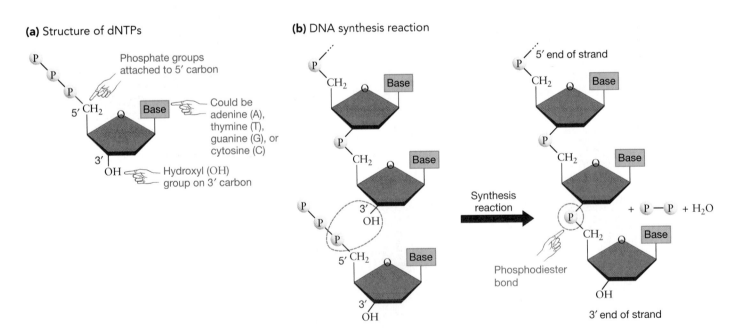

FIGURE 14.9 The DNA Synthesis Reaction

(a) A dNTP (deoxynucleoside triphosphate) monomer (b) is added to a DNA polymer when a phosphodiester bond forms between the 3′ carbon on the end of a DNA strand and the 5′ carbon on the dNTP in a condensation reaction.   QUESTION According to this reaction, does DNA synthesis proceed from the 5′ end of the new molecule toward the 3′ end, or in the 3′ to 5′ direction?   EXERCISE The "P–P" group (pyrophosphate) that is released reacts with water to form two phosphate ions (HPO_4^-). This reaction is highly exergonic. Add this reaction to the figure.

cause the potential energy stored in the bonds between the phosphates could be used to drive the energy-demanding synthesis reaction (**Figure 14.9b**). The biologists also used radioactive versions of these monomers, which provided an assay to detect whether dNTPs had been incorporated into newly synthesized DNA. If Kornberg and co-workers could detect radioactivity in a high-molecular-weight polymer instead of in low-molecular-weight monomers, it meant that DNA synthesis had occurred.

Kornberg and colleagues tested a variety of proteins with this in vitro assay. After many failed attempts, they finally succeeded in identifying an enzyme that allowed DNA synthesis to proceed. They called this molecule **DNA polymerase**. Its name was later changed to DNA polymerase I when researchers realized that there was more than one type of DNA polymerase in *E. coli* and other organisms. Each of these enzymes is specialized to perform a particular function in the cell. The enzyme that is primarily responsible for copying *E. coli*'s chromosome prior to cell division is called DNA polymerase III. When the double helix of bacterial DNA separates, DNA polymerase III adds the complementary bases and catalyzes the formation of phosphodiester bonds in the newly synthesized strand.

Follow-up experiments confirmed that DNA polymerase I and the other DNA polymerases do not catalyze the random addition of monomers to create a DNA molecule. Instead, they synthesize a growing strand of DNA by adding bases that are complementary to a template strand. To capture this point, biologists say that DNA synthesis is template directed.

The discovery of the DNA polymerases ended one chapter in research on DNA synthesis and started another. The pioneering phase of work on DNA replication was over. Researchers had established that DNA replication is semiconservative in nature, that it is enzyme dependent and template directed, and that several DNA polymerases exist. The next challenge was to understand the mechanics of the process in detail.

14.3 A Comprehensive Model for DNA Synthesis

How does DNA polymerase III accomplish DNA synthesis? Three results from the pioneering phase of work on DNA replication helped focus this question and point the way to an answer.

The first result was an observation about how the enzyme works. DNA polymerases catalyze the addition of a dNTP monomer to only the 3′ end of the DNA chain—never to the 5′ end (see Figure 14.9). As a result, DNA synthesis always proceeds in the 5′ → 3′ direction. Nucleotides are added only to the 3′ end of an existing molecule.

The second result, from experiments with in vitro systems, was that DNA polymerase III cannot work if the DNA

(a) Completely single stranded

Polymerase does not work; no DNA synthesis

(b) Completely double stranded

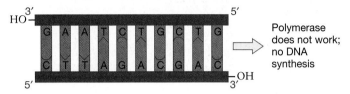

Polymerase does not work; no DNA synthesis

(c) Single strand as a template plus 3′ end to start DNA synthesis

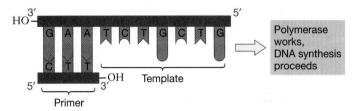

Polymerase works, DNA synthesis proceeds

FIGURE 14.10 DNA Polymerase Requires a Primer and a Template
When DNA templates are **(a)** completely single stranded or **(b)** double stranded, no synthesis occurs. **(c)** DNA polymerases work only when both a single-stranded template *and* a 3′ hydroxyl group are available.

that serves as a template is completely single stranded or completely double stranded (**Figure 14.10a, b**). The enzyme begins to work only when a 3′ end *and* a single-stranded template are available (**Figure 14.10c**). In effect, DNA polymerases only add nucleotides that complete a double helix. The existing strand acts as a template that dictates which nucleotide should be added next, while the strand to which nucleotides are being added provides a free 3′ hydroxyl ($-OH$) group that combines with the incoming dNTP to form a phosphodiester bond.

It's important to note, however, that DNA polymerase III can get started once even a few nucleotides are in place to form a double-stranded section on the parent DNA template. These few nucleotides are called a **primer**. Once a primer is added to a single-stranded template, DNA polymerase III begins working in the 5′ → 3′ direction and adds nucleotides to complete the complementary strand.

The third result that proved to be a major insight into the mechanism of DNA synthesis emerged when electron micrographs caught chromosome replication in action. A "bubble" forms in a chromosome when DNA is actively being synthesized,

(a) A chromosome being replicated

(b) Bacterial chromosomes have a single point of origin.

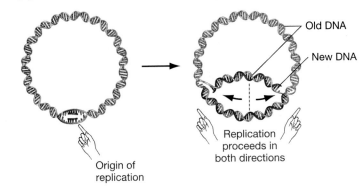

Old DNA

New DNA

Replication proceeds in both directions

Origin of replication

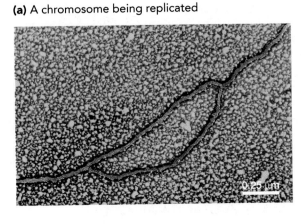

0.25 µm

(c) Eukaryotic chromosomes have multiple points of origin.

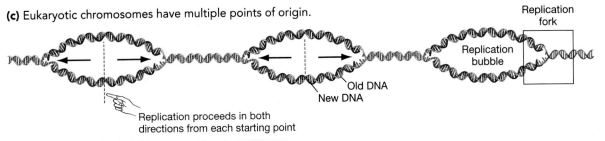

Replication fork

Replication bubble

Old DNA
New DNA

Replication proceeds in both directions from each starting point

FIGURE 14.11 DNA Synthesis Proceeds in Two Directions from a Point of Origin
(a) A micrograph of a "replication bubble." **(b)** In most bacteria, chromosomes are circular and there is a single point of origin during DNA replication. **(c)** Most eukaryotes have linear chromosomes; each contains several to many points of origin for DNA synthesis. EXERCISE Circle the two replication forks in part (a). Add arrows showing the direction of DNA synthesis in each fork. Note that because replication is bidirectional, each of the two forks should have arrows pointing in opposite directions.

as **Figure 14.11a** shows. Bacterial chromosomes have a single location where the replication process begins, and thus a single bubble forms. Initially, the replication bubble forms a point called the **origin of replication** (**Figure 14.11b**). The bubble grows as DNA replication proceeds, because synthesis is *bidirectional*—that is, it occurs in both directions (though always 5′ → 3′, because the strands are antiparallel) at the same time. Eukaryotes also have bidirectional replication; but they have multiple sites along each chromosome where DNA synthesis begins, and thus multiple replication bubbles (**Figure 14.11c**).

To put these observations into a comprehensive picture of how DNA synthesis occurs, let's focus on what happens in the corners of each replication bubble—at the structure called the replication fork. A **replication fork** is a Y-shaped region where the parent DNA double helix is split into two single strands, which are then copied. How does the splitting event occur, and how are each of the resulting strands replicated?

Opening the Helix

Several key events occur at the point where the double helix opens. An enzyme called a **helicase** catalyzes the breaking of hydrogen bonds between nucleotides. This reaction causes the two strands of DNA to separate. Proteins called **single-strand DNA-binding proteins** attach to the separated strands and prevent them from snapping back into a double helix. In combination, then, the helicase and single-strand DNA-binding proteins open up the double helix and make both strands available for copying (**Figure 14.12a**).

The "unzipping" process that occurs at the replication fork creates tension farther down the helix, however. To understand why, imagine what would happen if you started to pull apart the twisted strands of a rope. The untwisting movements at one end would force the intact section to rotate in response. If the intact end of the rope were fixed in place, though, it would eventually begin to coil on itself and kink in response to the twisting forces. This does not happen in DNA, because the twisting stress induced by helicase is relieved by proteins called topoisomerases. A **topoisomerase** (literally, "place-same-part") is an enzyme that cuts and rejoins the DNA downstream of the replication fork. Topoisomerases do this cutting and pasting in a way that undoes twists and knots.

Now, what happens once the DNA helix is open and has stabilized?

(a) Opening, unwinding, and priming the DNA

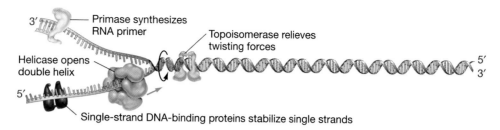

(b) Synthesis of leading strand

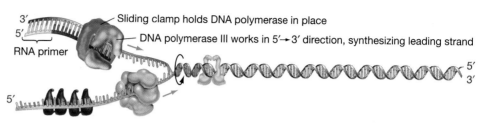

FIGURE 14.12 Synthesizing the Leading Strand during DNA Replication
(a) Once the double helix is opened and stabilized, **(b)** DNA polymerase synthesizes the leading strand in the 5′ to 3′ direction.

Synthesis of the Leading Strand

To understand what happens at the start of DNA synthesis, recall that DNA polymerase III works only in the 5′ → 3′ direction and that it requires a primer to get started. Before DNA synthesis can get under way, an enzyme called **primase** has to synthesize a short stretch of ribonucleic acid (RNA) that acts as a primer for DNA polymerase. Primase is a type of **RNA polymerase**—an enzyme that catalyzes the polymerization of ribonucleotides to RNA. Unlike DNA polymerases, primase and other RNA polymerases do not require a primer. These enzymes can simply match ribonucleotides directly by complementary base pairing on single-stranded DNA. In this way, primase creates a primer for DNA synthesis.

Once the primer is in place, DNA polymerase III begins adding deoxyribonucleotides to the 3′ end of the new strand, in a sequence that is complementary to the template strand. As **Figure 14.12b** shows, the shape of DNA polymerase III grips the DNA strand during synthesis, a little like your hand clasping a rope. Catalysis takes place in the groove inside the enzyme, at an active site between the enzyme's "thumb" and "fingers." As DNA polymerase moves along the DNA molecule, a doughnut-shaped structure behind it, called the **sliding clamp**, holds it in place. Its product is called the **leading strand**, or **continuous strand**, because it leads into the replication fork and is synthesized continuously.

Synthesis of the Lagging Strand

Synthesis of the leading strand is straightforward after an RNA primer is in place—DNA polymerase III chugs along, adding bases to the 3′ end of that strand. The enzyme moves into the replication fork, which "unzips" ahead of it. By comparison, events on the opposite strand are much more in-

volved. To understand why, recall from Chapter 4 that the two strands in the DNA double helix are *antiparallel*—meaning they are parallel to one another but oriented in opposite directions. DNA polymerase works in only one direction, however, so if the DNA polymerase that is synthesizing the leading strand works into the replication fork, then a DNA polymerase must work away from the replication fork to synthesize the other strand in the 5′ → 3′ direction. The strand that is synthesized in the opposite direction of the replication fork is called the **lagging strand**, because it lags behind the fork.

The synthesis of the lagging strand starts when primase synthesizes a short stretch of RNA that acts as a primer. DNA polymerase III then adds bases to the 3′ end of the lagging strand. The key observation is that the enzyme moves *away* from the replication fork. But behind it, helicase continues to open the replication fork and expose new single-stranded DNA on the lagging strand.

These events create a paradox: How can lagging-strand synthesis be completed if new single-stranded template DNA is constantly appearing behind DNA polymerase III—in the direction *opposite* the direction of synthesis? Consider the state of the lagging strand at time 1 and time 2, here:

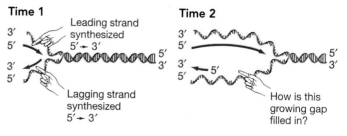

The puzzle posed by lagging-strand synthesis was resolved when Reiji Okazaki and colleagues tested a hypothesis called discontinuous replication. This hypothesis stated that, once primase synthesizes an RNA primer on the lagging strand (step 1 in **Figure 14.13**), DNA polymerase III might synthesize short fragments of DNA along the lagging strand and that these fragments would later be linked together to form a continuous whole. The idea was that primase would add a primer to the newly exposed single-stranded DNA at intervals and that DNA polymerase III would then synthesize the lagging strand until it reached the fragment produced earlier.

To test this hypothesis, Okazaki's group set out to test a key prediction: Could they document the existence of short DNA fragments produced during replication? Their critical experiment was based on the pulse-chase strategy introduced in Chapter 7. Specifically, they added a short "pulse" of radioactive thymidy-late to *E. coli* cells, followed by a large "chase" of nonradioactive thymidylate. Thymidylate is a nucleotide that contains a deoxyribose subunit attached to thymine. It serves as a building block of DNA. According to the discontinuous replication model, some of this radioactive thymidylate should end up in short, single-stranded fragments of DNA.

As predicted, the researchers succeeded in finding these fragments when they purified DNA from the experimental cells and separated the molecules by centrifugation. A small number of labeled pieces of DNA, about 1000 base pairs long, were present. These short sections came to be known as **Okazaki fragments** (steps 2 and 3 of Figure 14.13). Subsequent work showed that Okazaki fragments in eukaryotes are just 100 to 200 base pairs long.

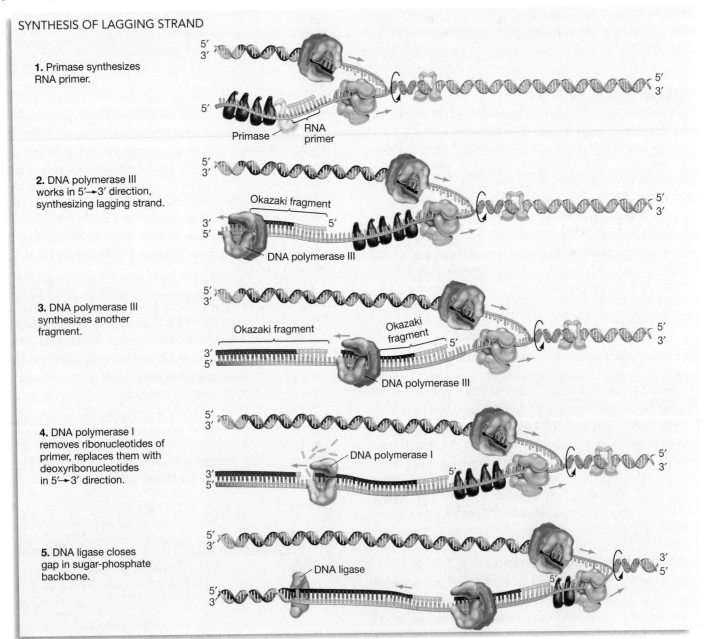

SYNTHESIS OF LAGGING STRAND

1. Primase synthesizes RNA primer.

2. DNA polymerase III works in 5'→3' direction, synthesizing lagging strand.

3. DNA polymerase III synthesizes another fragment.

4. DNA polymerase I removes ribonucleotides of primer, replaces them with deoxyribonucleotides in 5'→3' direction.

5. DNA ligase closes gap in sugar-phosphate backbone.

FIGURE 14.13 The Completion of DNA Replication
Synthesizing the lagging strand.

How are Okazaki fragments connected into a continuous whole? As step 4 of Figure 14.13 shows, DNA polymerase I removes the RNA primer at the start of each fragment and fills in the appropriate deoxyribonucleotides. Finally, an enzyme called **DNA ligase** catalyzes the formation of a phosphodiester bond between the bases synthesized by DNA polymerase I and DNA polymerase III (Figure 14.13, step 5). Because Okazaki fragments are synthesized independently and joined together later, the lagging strand is also called the **discontinuous strand.**

In combination, then, the enzymes that open the replication fork and manage the synthesis of the leading and lagging strands succeed in producing a faithful copy of the original DNA molecule prior to mitosis or meiosis. Given the number of enzymes and proteins involved in the synthesis machinery, it is not surprising that it took 25 years for biologists to assemble the results summarized in **Table 14.1**.

It is also worth mentioning the following points:

- Although Figures 14.12 and 14.13 show the enzymes involved in DNA synthesis at different locations around the replication fork, in reality most are joined into one large multi-enzyme machine at the replication fork. This complex of proteins opens the double helix and synthesizes both the leading strand and the Okazaki fragments on the lagging strand.

- The basic elements of DNA synthesis were worked out in experiments on *E. coli*. Follow-up research has shown that the fundamental elements of this process are almost identical in eukaryotes. The exceptions are that (1) some of the enzymes at the replication fork in eukaryotes are larger and more complex than similar enzymes in bacteria; and (2) on the lagging strand, two different DNA polymerases do the work done by DNA polymerase III in bacteria. In eukaryotes, enzymes called DNA polymerase δ and DNA polymerase α synthesize the lagging strand. (DNA polymerase δ also synthesizes the leading strand.) In almost every other detail, however, the critical components of DNA synthesis have been highly conserved over the course of evolution.

Most aspects of the model summarized in Figures 14.12 and 14.13 had emerged in the early 1980s, and subsequent decades saw major advances in our understanding of the detailed molecular structures of the major enzymes involved. But in addition to studying the replication process itself, biologists have been exploring other questions about DNA synthesis. This chapter closes with a look at three of these research areas: How are the ends of linear chromosomes replicated? What happens when a DNA polymerase inserts the wrong base into a sequence? And can DNA be repaired if it is damaged? As it turns out, all three questions have practical implications—particularly for the origin of certain types of cancer.

✓ CHECK YOUR UNDERSTANDING

DNA synthesis begins at specific points of origin on the chromosome, then proceeds in both directions. Once DNA replication is under way, synthesis can be analyzed as a three-step process that occurs at the replication fork. (1) Helicase, single-stranded DNA-binding proteins, and topoisomerase open the double helix to make single strands available for copying; (2) a DNA polymerase synthesizes the leading strand after primase has added an RNA primer; and (3) a series of enzymes synthesizes the lagging strand. Lagging-strand synthesis cannot be continuous, because it moves away from the replication fork. In bacteria, enzymes called primase, DNA polymerase III, DNA polymerase I, and ligase work in sequence to synthesize Okazaki fragments and link them into a continuous whole. You should be able to draw a replication fork, label the main components, and explain the function of each structure involved.

TABLE 14.1 Proteins Involved in DNA Synthesis

Process	Name	Function
Opening the helix	Helicase	Catalyzes the breaking of hydrogen bonds between base pairs and the opening of the double helix
	Single-strand DNA-binding proteins	Stabilizes single-stranded DNA
	Topoisomerase	Breaks and rejoins the DNA double helix to relieve twisting forces caused by the opening of the helix
Synthesizing the leading strand	Primase	Catalyzes the synthesis of the RNA primer
	DNA polymerase III	Extends the leading strand
	Sliding clamp	Holds DNA polymerase in place during strand extension
Synthesizing the lagging strand	Primase	Catalyzes the synthesis of the RNA primer on an Okazaki fragment
	DNA polymerase III	Extends an Okazaki fragment
	Sliding clamp	Holds DNA polymerase in place during strand extension
	DNA polymerase I	Removes the RNA primer and replaces it with DNA
	DNA ligase	Catalyzes the joining of Okazaki fragments into a continuous strand

14.4   Replicating the Ends of Linear Chromosomes

The circular DNA molecules in bacteria and archaea can be synthesized via the sequence of events diagrammed in Figures 14.12 and 14.13. Similarly, the leading and lagging strands of the linear DNA molecules found in eukaryotes are copied efficiently by the enzymes illustrated in those figures. But how can the *ends* of the linear chromosomes found in garden peas, fruit flies, humans, and other eukaryotes be copied faithfully?

The region at the end of a linear chromosome is called a **telomere** ("end-part"). **Figure 14.14** illustrates the problem that arises during the replication of telomeres. Steps 1 and 2 show that as the replication fork nears the end of a linear chromosome, DNA polymerase synthesizes the leading strand all the way to the end of the parent DNA template. As a result, leading-strand synthesis results in a normal copy of the DNA molecule. But the situation is different on the lagging strand. As

step 2 of Figure 14.14 illustrates, primase adds an RNA primer close to the tip of the chromosome. But after this primer is removed and DNA polymerase synthesizes the final Okazaki fragment on the lagging strand (step 3), the enzyme is unable to replace the RNA primer or to add DNA near the tip of the chromosome because DNA polymerase can add bases only to the 3′ end of an existing strand. As step 4 shows, there is no 3′ end available at the end of the lagging strand. As a result, the single stranded DNA that is left once the primer has been removed must stay single stranded.

The single-stranded DNA that remains after telomeres are copied is eventually degraded, meaning the chromosome shortens. If this process were to continue, every chromosome would shorten by 50–100 nucleotides on average each time that DNA replication occurred prior to mitosis or meiosis. Over time, linear chromosomes would be expected to disappear completely.

Bacteria and archaea do not have this end-replication problem, because virtually all of these cells have a single, circular

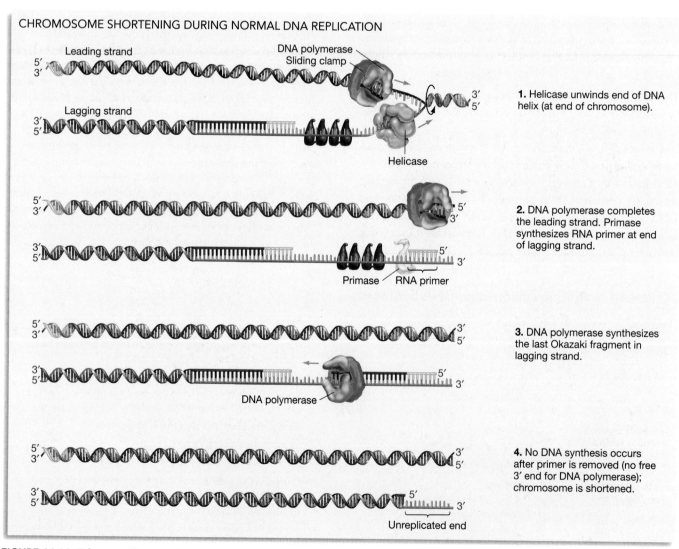

CHROMOSOME SHORTENING DURING NORMAL DNA REPLICATION

1. Helicase unwinds end of DNA helix (at end of chromosome).

2. DNA polymerase completes the leading strand. Primase synthesizes RNA primer at end of lagging strand.

3. DNA polymerase synthesizes the last Okazaki fragment in lagging strand.

4. No DNA synthesis occurs after primer is removed (no free 3′ end for DNA polymerase); chromosome is shortened.

FIGURE 14.14 Telomeres Shorten during Normal DNA Replication
An RNA primer is added to the lagging strand near the end of the chromosome. Once the primer is removed, it cannot be replaced with DNA. As a result, the chromosome shortens.

chromosome. How do eukaryotes maintain the integrity of their linear chromosomes? An answer emerged after two important discoveries were made:

1. Telomeres do not contain genes. Instead, they consist of short stretches of bases that are repeated over and over. In humans, for example, the base sequence TTAGGG is repeated thousands of times. (Because DNA and RNA sequences are always written $5' \rightarrow 3'$, the base sequence on the complementary strand is CCCTAA rather than AATCCC.) Human telomeres consist of a total of about 10,000 nucleotides with the sequence TTAGGGT-TAGGGTTAGGG....

2. An interesting enzyme called telomerase is involved in replicating telomeres. **Telomerase** is remarkable because it catalyzes the synthesis of DNA from an RNA template. The enzyme carries an RNA molecule with it that acts as a built-in template. The sequence of this RNA is complementary to the repeated sequence in telomere DNA. As a result, telomerase can add repeats onto the end of a chromosome.

Figure 14.15 shows how this process works. Step 1 shows the unreplicated segment at the $3'$ end of the lagging strand at the telomere. It forms a single-strand "overhang." In step 2, telomerase binds to this overhanging section of single-stranded parent DNA and adds repeats to it. This step lengthens the overhang. The normal machinery of DNA synthesis—primase, DNA polymerase, and ligase—then synthesizes the lagging strand in the $5' \rightarrow 3'$ direction. The result is that the lagging strand becomes slightly longer than it was originally.

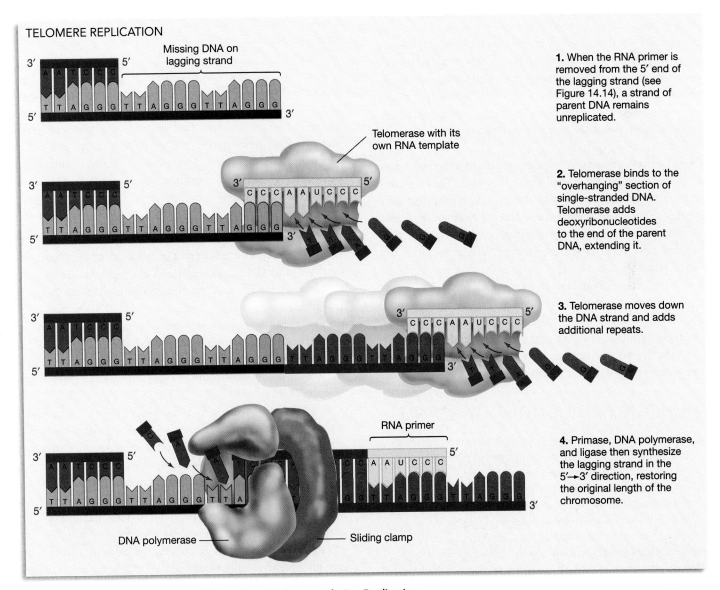

FIGURE 14.15 Telomerase Prevents Shortening of Telomeres during Replication
By extending the number of repeated sequences in the $5' \rightarrow 3'$ direction, telomerase provides room for enzymes to add an RNA primer to the lagging strand. DNA polymerase can then fill in the missing section of the lagging strand.

It is important to recognize, though, that telomerase is not active in most types of cells. In humans, for example, active telomerase is found primarily in the cells of reproductive organs—specifically, in the cells that eventually undergo meiosis and produce gametes. Cells that are not involved in gamete formation, or what biologists call **somatic cells**, normally lack telomerase. As predicted, the chromosomes of somatic cells gradually shorten with each mitotic division, getting progressively smaller as an individual grows and ages.

These observations inspired a pair of important hypotheses. The first hypothesis was that telomere shortening might eventually cause cells to stop dividing altogether. The idea here was that telomere loss would signal cells to enter the nondividing state typical of adult cells. The second hypothesis was that if telomerase were mistakenly activated in a somatic cell, the maintenance of telomeres with normal length might induce the cell to keep dividing and possibly contribute to uncontrolled growth and cancer.

The hypothesis that telomere shortening signals cells to stop growing has been supported by recent experiments with human cells grown in culture. Biologists found that if human cells growing in vitro are experimentally provided with functioning copies of telomerase, they continue dividing long past the age when otherwise identical cells stop growing. These results have convinced most biologists that telomere shortening is involved in the normal aging process in cells.

A link between continued telomerase activity and cancer formation has been harder to nail down, however. The most important work to date involves mice that have a mutation in a gene that contains information for making telomerase. These mice fail to produce telomerase, yet many develop tumors. This result suggests that telomerase activity is not required for tumor formation. On the other hand, many cancerous cells in humans and other organisms do have functioning telomerase or some other mechanism for maintaining telomere length. Noncancerous cells in these organisms do not have functioning telomerase. Would developing drugs that knock out telomerase be an effective way to fight cancer? To date, the data on this question are equivocal. Research continues.

✓CHECK YOUR UNDERSTANDING

During normal chromosome replication, linear chromosomes in most somatic cells shorten because the end of the lagging strand lacks a primer and cannot be synthesized. In certain cells—particularly those found in reproductive organs—telomerase adds short, repeated DNA sequences to the template strand after normal replication is complete. In this way, telomerase extends the length of the chromosome, so primase has room to add an RNA primer to the lagging strand. DNA polymerase can then fill in missing sections of the lagging strand and restore the chromosome's original length. You should be able to diagram the sequence of events involved in telomere replication, and add labels to indicate the enzymes involved and the polarity of each DNA strand.

14.5   Repairing Mistakes in DNA Synthesis

DNA polymerases work fast. In yeast, for instance, the replication fork is estimated to move at a rate of about 50 bases per second. But the replication process is also astonishingly accurate. In organisms ranging from *E. coli* to animals, the error rate during DNA replication averages less than one mistake per *billion* nucleotides. This level of accuracy is critical. Humans, for example, develop from a fertilized egg that has DNA containing over 6 billion nucleotides. The DNA inside the fertilized egg is replicated over and over to create the trillions of cells that will eventually make up the adult body. If more than one or two mutations occurred during each cell division cycle as a human grew, genes would be riddled with errors by the time the individual reached maturity. Genes that contain errors are often defective. Based on these observations, it is no exaggeration to claim that the accurate replication of DNA is a matter of life and death.

How can the enzymes involved in replicating DNA be as precise as they are? The answer to this question has several parts. DNA polymerase is highly selective in matching complementary bases correctly. This enzyme can also catch mismatched base pairs during the synthesis process and correct them, meaning that it can "proofread." Finally, if mistakes remain after synthesis is complete, repair enzymes can correct mismatched base pairs.

How Is DNA Polymerase Proofread?

DNA polymerases are selective because the correct base pairing (A-T and G-C) is energetically the most favorable of all possibilities for the pairing of nitrogen-containing bases. As DNA polymerase marches along a parent DNA template, hydrogen bonding occurs between incoming nucleotides and the nucleotides on the template strand. Due to the precise nature of these interactions, the enzyme inserts the incorrect nucleotide only about once every thousand bases added (**Figure 14.16a**).

What happens when a G-T, an A-C, or another type of mismatch occurs? An answer to this question emerged when researchers found mutants in *E. coli* with error rates that were 100 times greater than normal. Recall from Chapter 13 that mutants are individuals with traits caused by mutation and that mutation is a change in the gene responsible for those traits. Many mutations change the individual's phenotype. The change may result in a trait such as white eyes in fruit flies or an elevated mutation rate in *E. coli*. At the molecular level, a mutant phenotype usually results from a change in an enzyme or other type of protein.

In the case of *E. coli* cells with high mutation rates, biologists found that the mutation was localized to a particular portion of the DNA polymerase III enzyme, called the ε (epsilon)

(a) Mismatched bases

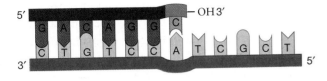

(b) DNA polymerase can repair mismatches.

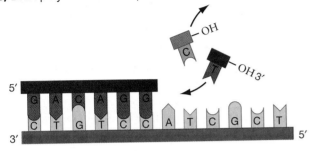

FIGURE 14.16 DNA Polymerase Can Proofread
(a) In bacteria, DNA polymerase adds an incorrect base to a growing strand of DNA about once in every thousand bases added. The result is a mismatch such as the pairing of A with C. **(b)** DNA polymerase can act as a 3′ → 5′ exonuclease, meaning that it can remove bases in that direction. The enzyme then adds the correct base.

subunit. Further analyses showed that this subunit of the enzyme acts as an **exonuclease**—meaning an enzyme that removes nucleotides from DNA (**Figure 14.16b**). The DNA polymerase III exonuclease activity removes nucleotides only from the 3′ end of DNA, and only if they are not hydrogen bonded to a base on the complementary strand. These results led to the conclusion that DNA polymerase III can proofread. If the wrong base is added during DNA synthesis, the enzyme pauses, removes the mismatched base that was just added, and then proceeds with synthesis.

Eukaryotic DNA polymerases have the same type of proofreading ability. Typically, proofreading reduces a DNA polymerase's error rate to about 1×10^{-7} (one mistake per 10 million bases). Is this accurate enough? No. If DNA replication were this sloppy, then at least 600 mistakes would occur every time a human cell replicated its DNA. Given that about a million-billion cell divisions take place in the course of a human's lifetime, an error rate this high would be a disaster. Once replication is complete, other repair mechanisms come into play.

Mismatch Repair

When DNA polymerase leaves a mismatched pair behind in the DNA sequence by mistake, a battery of enzymes springs into action. The proteins responsible for **mismatch repair** were discovered in the same way that the proofreading capability of DNA polymerase III was uncovered—by analyzing *E. coli* mu-

tants. In this case, the mutants had normal DNA polymerase III but abnormally high mutation rates. The first mutant gene that caused a deficiency in mismatch repair was identified in the late 1960s and was called *mutS*. (The *mut* is short for "mutator.") By the late 1980s researchers had identified 10 proteins involved in the identification and repair of base-pair mismatches in *E. coli*.

The observation that repair enzymes fix mismatched bases raised a fundamental question: How do the enzymes know which of the two bases in a mismatched pair is right and which is wrong? A hypothesis to answer this question was inspired by the observation that several minutes after DNA polymerase III completes the synthesis of bacterial DNA, another enzyme adds methyl ($—CH_3$) groups to several of the adenosines on the new strand. The addition of methyl groups is called **methylation**. But until this reaction takes place, there is a marked difference between the parent and daughter strands. The parent DNA template is methylated, but the newly synthesized DNA is unmethylated. When a mismatched pair occurs, the wrong base must be found on the newly synthesized, unmethylated strand. Can the repair enzymes recognize which strand is methylated and use it as a template to fix the mistake?

To test the methyl-directed repair hypothesis, Paul Modrich and co-workers began experimenting with small, circular, double-stranded DNA molecules that they constructed. As **Figure 14.17** indicates, the experimental DNAs had two interesting features: a single base-pair mismatch and an adenine base that had a methyl group attached. The presence of one methylated strand and one unmethylated strand simulated the situation that occurs in *E. coli* immediately after DNA synthesis is complete. When the team added the other enzymes and molecules required for mismatch repair, they observed that repair proceeded only when a methyl group was present. Further, the repair proteins always removed the mismatched base on the unmethylated strand.

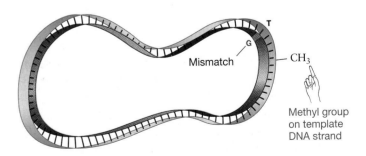

FIGURE 14.17 Methylated DNA Loop
A methyl group (–CH₃) on a loop of double-stranded DNA. In bacteria, an enzyme attaches a methyl group to adenosine bases a few minutes after replication is complete.

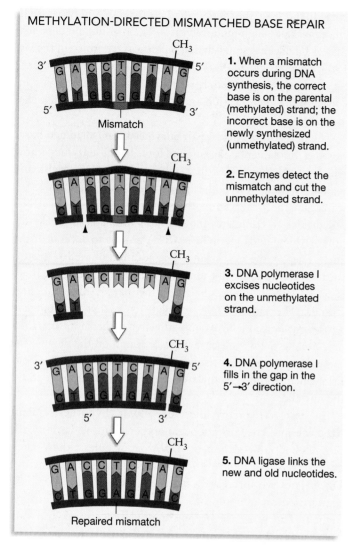

FIGURE 14.18 The Role of Methylation in Mismatch Repair

QUESTION Why is it valid to say that the methylated parent strand serves as a template for repair?

Based on results like these, researchers have been able to piece together the picture of mismatch repair that is shown in **Figure 14.18**. This model is called *methylation-directed mismatched base repair*. Although the model works well when applied to bacteria, researchers have realized that mismatch repair in eukaryotes is not based on a methylation signal. Many of the enzymes involved in mismatch repair in bacteria and eukaryotes are similar, but researchers have not been able to determine how the eukaryotic enzymes recognize the newly synthesized strand versus the original strand. Research on the eukaryotic repair pathway continues, and it has recently acquired some added urgency. As the essay at the end of this chapter explains, cancer can develop when the genes involved in the repair pathway are disabled by mutation.

14.6   Repairing Damaged DNA

The DNA that makes up your genome is under constant assault. The chemical bonds that hold DNA together can break spontaneously, sending nucleotides flying or snapping DNA's sugar-phosphate backbone in two. When even small amounts of X-ray, gamma-ray, or ultraviolet radiation bombard cells, one or both strands of DNA can break. In addition, certain molecules found in cells can actively attack and degrade nucleotides. These molecules include the hydroxyl (OH) radicals produced during aerobic metabolism, the aflatoxin B1 found in moldy peanuts and corn, and the benzo[α]pyrene in cigarette smoke.

Under normal conditions, biologists estimate that thousands of nucleotides are altered or lost from the DNA in every human cell every day due to spontaneous, radiation-induced, and chemical-caused damage. If these mistakes are not repaired, they would lead to permanent changes in genes. The vast majority of alterations would be harmful.

To cope with these insults, cells have a sophisticated battery of repair enzymes that act like maintenance crews. These proteins patrol DNA, identify damaged bases or sections of the sequence, and repair them. Collectively, these coordinated groups of molecules are called **excision repair systems**.

DNA Nucleotide Excision Repair: An Overview

Excision repair systems share a common mechanism: They excise a stretch of single-stranded DNA around a damaged site, then resynthesize a new strand based on the information in the intact, complementary strand. The mismatch repair system introduced in Section 14.5, which corrects mistakes made during DNA replication, is an example of an excision repair system.

The symmetry and regularity of DNA's secondary structure makes it possible for the proteins involved in repair to recognize mismatched base pairs or damaged sections after synthesis is complete. As Figure 14.18 showed, kinks in the double helix occur at the site of mismatched pairs, and damaged bases often produce a bulge or other irregularity in the molecule. Once a damaged or error-containing region is recognized, the presence of a DNA strand complementary to the damaged strand provides a template for resynthesis of the defective sequences. In this way, DNA's structure makes accurate repair possible. Because the molecule is repairable, DNA's structure supports its function as the cell's information-processing center.

Many of the proteins involved in the excision repair systems first came to light when researchers isolated strains of *E. coli* or yeast with abnormally high mutation rates. Follow-up studies were able to trace the cause of the defect. Much of this early work focused on how *E. coli* repairs damage induced by exposure to ultraviolet (UV) radiation. Rapid progress in this area inspired studies on DNA repair in a much more unusual model organism. In contrast to *E. coli* and fruit flies and garden peas, this model organism reproduces slowly and is extremely expensive to maintain. The organism? Human beings.

Xeroderma Pigmentosum: A Case Study

Xeroderma pigmentosum (XP) is a rare autosomal recessive disease in humans. Individuals with this condition are extremely sensitive to ultraviolet (UV) light. Their skin develops lesions after even slight exposure to sunlight. In normal individuals, these kinds of lesions develop only after extensive exposure to UV light, X-rays, or other forms of high-energy radiation.

In 1968 James Cleaver proposed a connection between XP and DNA excision repair systems. He knew that in *E. coli*, mutations in certain genes cause DNA excision repair proteins to fail. Cells with these mutations have an increased sensitivity to radiation. Cleaver's hypothesis was that people with XP have similar mutations. He claimed that they are extremely sensitive to sunlight because they are unable to repair the damage that occurs when the nitrogen-containing bases of DNA absorb UV light.

What is the nature of this damage? The most common defect caused by UV radiation is the formation of a covalent bond between adjacent pyrimidine bases. The thymine-thymine pair illustrated in **Figure 14.19a** is an example. This defect is called a *thymine dimer*, and it creates a kink in the secondary structure of DNA. This structural flaw keeps the gene from functioning normally and stalls the movement of the replication fork during DNA replication. If the error is not repaired, the cell may die.

Cleaver's hypothesized connection between DNA damage, faulty error repair, and XP turned out to be correct. Much of the work that he and other investigators did relied on the use of cell cultures. In this case, the researchers collected skin cells from people with XP and from people with a normal phenotype for excision repair. When these cell populations were grown in culture and exposed to increasing amounts of ultraviolet radiation, a striking difference emerged: Cells from XP individuals died off much more rapidly than did normal cells (**Figure 14.19b**).

The connection to excision repair systems was confirmed when Cleaver exposed cells from normal individuals and cells from XP individuals to various amounts of UV light, then fed the cells radioactive thymidine to label DNA synthesized during the

(a) UV-induced thymine dimers cause DNA to kink.

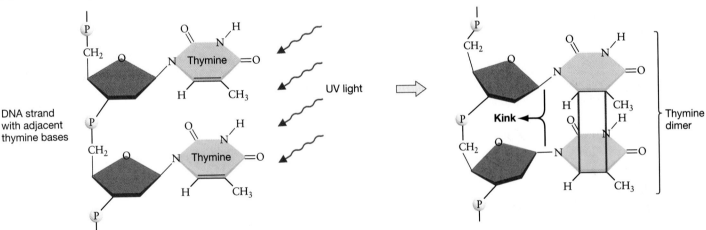

(b) Vulnerability of cells to UV light damage

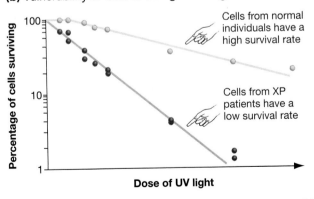

(c) Ability of cells to repair damage

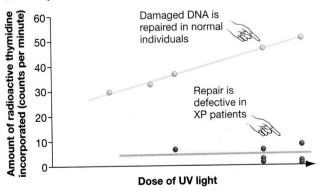

FIGURE 14.19 DNA Damage from UV Light Is Not Repaired Properly in Individuals with XP
(a) When UV light strikes a section of DNA with adjacent thymines, the energy can break bonds within each base and result in the formation of bonds *between* them. The thymine dimer that is produced causes a kink in DNA. **(b)** When cell cultures from normal individuals and from XP patients are irradiated with various doses of UV light, the percentage of cells that survive is strikingly different. **(c)** When cell cultures from normal individuals and from XP patients are irradiated with various doses of UV light and then fed radioactive thymidine, only normal individuals incorporate the labeled base.

repair period. If repair is defective in the XP individuals, then their cells should incorporate virtually no radioactive thymidine into their DNA. Cells from normal individuals, in contrast, should incorporate large amounts of labeled thymidine into their DNA. As **Figure 14.19c** (page 321) shows, this is exactly what happens. These data are consistent with the hypothesis that repair synthesis is virtually nonexistent in XP individuals.

More recently, genetic analyses of XP patients have shown that the condition can result from mutations in any one of seven genes. This result is not surprising in light of the large number of enzymes involved in excision repair even in bacteria.

Finally, as the essay at the end of this chapter points out, defects in the genes required for excision repair are frequently associated with cancer. Individuals with xeroderma pigmentosum, for example, are 1000 to 2000 times more likely to get skin cancer than are individuals with normal excision repair systems. To explain this pattern, biologists suggest that if mutations in the genes involved in the cell cycle (see Chapter 7) go unrepaired, the cell may begin to grow in an uncontrolled manner. Tumor formation could result. Stated another way, if the overall mutation rate in a cell is elevated because of defects in DNA repair genes, then the mutations that trigger cancer become more likely. In this case, research on fundamental aspects of DNA replication and repair, using *Escherichia coli* as a model system, led directly to a major advance in understanding a form of cancer in humans.

ESSAY DNA Mismatch Repair and Cancer

Cancer geneticists have known for decades that families can have predispositions to certain types of cancers, including breast cancer and the skin cancers associated with XP. A common type of colon cancer called **hereditary nonpolyposis colorectal cancer (HNPCC)** also runs in families. Affected individuals frequently develop tumors of the colon, ovary, and other organs before age 50 years (**Figure 14.20**). A major breakthrough in understanding HNPCC occurred in the early 1990s, when techniques introduced in later chapters allowed researchers to determine that a gene associated with susceptibility to HNPCC mapped to a specific region of chromosome 2.

Meanwhile, a different group of investigators was trying to determine whether humans have mismatch repair genes similar to those found in *E. coli*. The pace of this work accelerated when the DNA sequence of the *mutS* gene was determined and when a mismatch repair gene identified in the yeast *Saccharomyces cerevisiae* turned out to be extremely similar to *mutS*. The bacterial and yeast genes were so similar that they were considered **homologous**, meaning that they trace their ancestry to a gene in a common ancestor.

Using DNA sequence information from these two homologous genes, researchers were able to locate a similar sequence in the human genome. This gene came to be called *hMSH* (for human *mutS* homolog). Sparks really began to fly when biologists discovered that the *hMSH* gene mapped to the same region of chromosome 2 as the HNPCC susceptibility gene.

> *... because humans, yeast, and bacteria share a common evolutionary history, ... early experiments made a major advance in cancer biology possible.*

The link between a defect in mismatch repair and HNPCC was cemented when it was confirmed that individuals with this type of cancer have mutated forms of *hMSH*. As predicted, cells from these patients have mutation rates 100 times that of normal cells when both cell types are grown in culture. Follow-up studies confirmed that this increased mutation rate is indeed due to defective mismatch repair in cells derived from HNPCC patients. People who inherit a nonfunctional copy of the *hMSH* gene have a genetic predisposition for developing HNPCC.

This breakthrough validates the use of model organisms to study basic questions in biology. When experiments on mutated strains of *E. coli* began, they appeared to be pure, or basic, research—without practical application. But because humans, yeast, and bacteria share a common evolutionary history, and because DNA synthesis and repair are so fundamental to the functioning of all cells, the results of those early experiments made a major advance in cancer biology

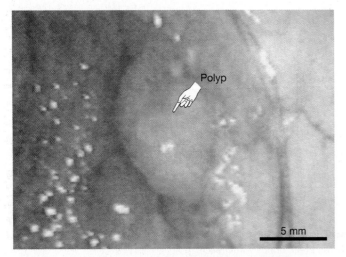

FIGURE 14.20 The Presence of Polyps Indicates HNPCC
This knob of tissue is a polyp that resulted from HNPCC. To diagnose this cancer, physicians try to detect polyps at the earliest stage of their formation.

possible. If individuals with mutant forms of *hMSH* can be identified early in life, dietary changes and therapy could significantly reduce their risk of developing cancer.

The HNPCC story also highlights one of the major questions in cancer biology today. Tissues in the colon and certain other organs appear to be extremely sensitive to mutations in *hMSH*. Stated another way, it appears that if cells in those tissues acquire mutations in *hMSH*, they are likely to become cancerous. But if cells in the brain, lung, breast, skin, or many other organs acquire mutations in *hMSH*, they are not likely to become cancerous. Chapter 8 introduced the same phenomenon in the cancer called retinoblastoma, which results from mutations in the *Rb* gene but tends to produce cancer cells only in the retina—not in other tissues. Why are certain types of genetic defects associated with the development of certain types of cancers, but only in certain types of cells? This question brings us to a cutting edge in cancer research. In most cases, the answer is still unknown.

CHAPTER REVIEW

Summary of Key Concepts

▓ **Genes are made of DNA. When DNA is copied, each strand of a DNA double helix serves as the template for the synthesis of a complementary strand.**

By labeling DNA with ^{15}N or ^{14}N, Meselson and Stahl were able to validate the hypothesis that DNA replication is semiconservative. Each strand of a parent DNA molecule provides a template for the synthesis of a daughter strand, resulting in two complete DNA double helices.

▓ **DNA is synthesized only in the $5' \rightarrow 3'$ direction. When a DNA molecule is being copied, a large suite of specialized enzymes is involved in unwinding the double helix, continuously synthesizing the "leading strand" in the $5' \rightarrow 3'$ direction and synthesizing the "lagging strand" as a series of fragments that are then linked together.**

Kornberg showed that DNA synthesis is an enzyme-catalyzed reaction by purifying DNA polymerase I from *E. coli* and using this enzyme to synthesize DNA in vitro. Follow-up work over several decades showed that DNA synthesis takes place only in the $5' \rightarrow 3'$ direction, requires both a template and a primer sequence, and takes place at the replication fork where the double helix is being opened.

Synthesis of the leading strand in the $5' \rightarrow 3'$ direction is straightforward, but synthesis of the lagging strand is more complex because DNA polymerase has to work away from the replication fork. By feeding *E. coli* cells a short pulse of radioactive thymidine, Okazaki and co-workers confirmed that short DNA fragments form on the lagging strand. These Okazaki fragments are primed by a short strand of RNA and are linked together after synthesis.

At the ends of linear chromosomes in eukaryotes, the enzyme telomerase adds short, repeated sections of DNA so that the lagging strand can be synthesized without shortening the chromosome. Telomerase is active in reproductive cells that eventually undergo meiosis. As a result, gametes contain chromosomes of normal length.

Web Tutorial 14.1   DNA Synthesis

▓ **Most mistakes that occur in DNA synthesis are repaired by specialized enzymes. If these repair enzymes are defective, the mutation rate increases. If mutations occur in genes that control the cell cycle, cancer may develop.**

DNA replication is remarkably accurate, because DNA polymerase proofreads and because mismatch repair enzymes excise incorrect bases once synthesis is complete and replace them with the correct sequence. In addition, DNA repair occurs after bases have been damaged by spontaneous breakage of bonds or by chemicals or radiation. Various excision repair systems cut out damaged portions of genes and replace them with correct sequences. Several types of human cancers are associated with defects in the genes responsible for DNA repair.

Questions

Content Review

1. What constitutes an individual's genome?
   a. all of its proteins   b. all of its mRNAs
   c. all of its DNA   d. all of its organelles

2. The experiment that removed proteins, RNA, or DNA from *Streptococcus* extracts purported to show that DNA is the hereditary material. Why did it fail to convince many skeptics?
   a. It had no control treatments, so it was poorly designed.
   b. Genes had already been shown to be made of protein.
   c. The trait that was studied—virulence—is not genetic.
   d. Critics argued that the chemical treatments had not removed all of the protein present.

3. How did researchers first isolate DNA polymerase?
   a. They used transmission electron microscopy to see molecules attached to DNA during synthesis.
   b. They isolated mutants that could not synthesize DNA.
   c. They added cell extracts to an in vitro synthesis system and found one that worked.
   d. They used certain recombinant DNA techniques (that is, genetic engineering).

4. Why did the lagging strand get its name?
   a. Replication of the lagging strand is discontinuous.
   b. The lagging strand consists of Okazaki fragments that are ligated late in the replication sequence.
   c. DNA ligase is active on this strand and is the "lagging" enzyme—it is the last enzyme in the replication sequence.
   d. Synthesis of this strand lags behind synthesis of the leading strand.

5. Where and how are Okazaki fragments synthesized?
   a. at the leading strand, oriented in a $5' \rightarrow 3'$ direction
   b. at the leading strand, oriented in a $3' \rightarrow 5'$ direction
   c. at the lagging strand, oriented in a $5' \rightarrow 3'$ direction
   d. at the lagging strand, oriented in a $3' \rightarrow 5'$ direction

6. What does telomerase do?
   a. It adds an RNA primer to the ends of linear chromosomes.
   b. It adds double-stranded DNA to the "blunt end" of a linear chromosome.
   c. It adds double-stranded DNA to the lagging strand at the end of a linear chromosome.
   d. It adds single-stranded DNA to the lagging strand at the end of a linear chromosome.

Conceptual Review

1. Researchers try to design experiments so that an experimental treatment shows the effect that one, and only one, condition or agent has on the phenomenon being studied. Examine Figure 14.2, and decide which treatments in the transformation study acted as experimental treatments and which acted as controls. Write a general statement explaining the role of control treatments in experimental design.

2. Why is the activity of helicase, single-strand DNA-binding proteins, and topoisomerase required during DNA synthesis?

3. What does it mean to say that DNA replication is "bidirectional?"

4. Why is the synthesis of the lagging strand of DNA discontinuous? How is it possible for the synthesis of the leading strand to be continuous?

5. List in chronological order the events that increase the accuracy of DNA replication. Indicate which events involve DNA polymerase III and which involve specialized repair enzymes.

6. Explain how the structure of DNA makes it relatively easy for proteins to recognize base-pair mismatches or damaged bases. How does DNA's secondary structure make it possible for damaged sections or incorrect bases to be removed and repaired?

Group Discussion Problems

1. If DNA polymerase III did not require a primer, which steps in DNA synthesis would differ from what is observed? Would any special enzymes be required to replicate telomeres? Explain your answers.

2. In the late 1950s Herbert Taylor grew bean root-tip cells in a solution of radioactive thymidine and allowed them to undergo one round of DNA replication. He then transferred the cells to a solution without the radioactive nucleotide, allowed them to replicate again, and examined their chromosomes for the presence of radioactivity. His results are shown in the following figure.

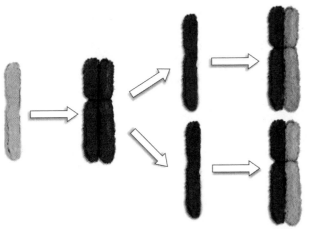

1. One round of DNA replication in radioactive solution   **2.** Mitosis   **3.** One round of DNA replication in non-radioactive solution

   a. Draw diagrams explaining the pattern of radioactivity observed in the sister chromatids after the first and second rounds of replication.
   b. What would the results of Taylor's experiment be if eukaryotes used a conservative mode of DNA replication?

3. The graph that follows shows the survival of four different *E. coli* strains after exposure to increasing doses of ultraviolet light. The wild-type strain is normal, but the other strains have a mutation in genes called *uvrA*, *recA*, or both *uvrA* and *recA*.

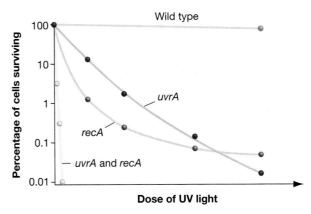

   a. Which strains are most sensitive to UV light? Which strains are least sensitive?
   b. What are the relative contributions of these genes to the repair of UV damage?

4. One widely used test to identify whether certain chemicals, such as pesticides or herbicides, might be carcinogenic (cancer causing) involves exposing bacterial cells to a chemical and recording whether the exposure leads to an increased mutation rate. In effect, this test equates cancer-causing chemicals with mutation-causing chemicals. Why is this an informative test?

Answers to Multiple-Choice Questions   **1.** c; **2.** d; **3.** c; **4.** d; **5.** c; **6.** d

www.prenhall.com/freeman is your resource for the following: Web Tutorials; Online Quizzes and other Online Study Guide materials; Answers to Conceptual Review Questions; Solutions to Group Discussion Problems; Answers to Figure Caption Questions and Exercises; and Additional Readings and Research.

How Genes Work

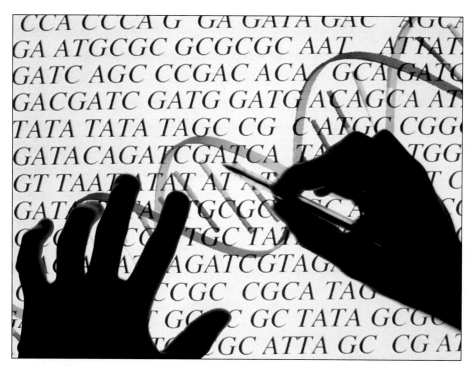

15

Each A, T, G, and C here represents a nitrogen-containing base in DNA. This chapter explores how biologists determined that genes are made of DNA and how a base sequence in DNA is translated into an amino acid sequence in a protein.

KEY CONCEPTS

▨ Most genes code for enzymes that catalyze specific chemical reactions in the cell.

▨ In cells, information flows from DNA to RNA to proteins. DNA is transcribed to messenger RNA by RNA polymerase, and then messenger RNA is translated to proteins in ribosomes. Each amino acid in a protein is specified by a group of three bases in RNA.

NA has been called the blueprint of life. If an organism's DNA is like a set of blueprints, then its cells are like a construction site. How does the DNA inside each cell specify the types and amounts of lumber, nails, and cement that are needed as the cell grows and as its structure and function change? If the enzymes inside a cell are like construction workers, how does DNA organize them into a team that can build and maintain the cell, and perhaps remodel it when conditions change?

Mendel provided insights that made it possible to ask these questions. He discovered that particular alleles are associated with certain phenotypes and that alleles are transmitted faithfully from parent to offspring. The chromosome theory of inheritance established that genes are found in chromosomes, and the theory detailed how the movement of

chromosomes during meiosis explains Mendel's results. Based on these early advances, it was clear that genes carry information—the instructions for making and maintaining an individual. But biologists still did not know how this information is translated into action. How does an organism's genotype specify its phenotype?

This chapter introduces a series of classic results that answered this question and revealed how genes work at the molecular level. Understanding how genes work, in turn, triggered a major transition in the history of biological science. Instead of thinking about genes solely in relation to their effects on eye color in fruit flies or on seed shape in garden peas, biologists could begin analyzing the molecular composition of both genes and their products. The molecular revolution in biology was under way.

15.1 What Do Genes Do?

In 1902 a physician named Archibald Garrod provided the first clue in the mystery of how genes work. A formal hypothesis addressing this mystery was not published until almost 40 years later, however—about the time evidence began to suggest that genes are made of DNA. Let's consider Garrod's results first and then explore a classical experiment that turned his hint into a solid understanding of what genes do.

The Molecular Basis of Hereditary Diseases

Chapter 13 examined how pedigrees can be used to determine patterns of inheritance in human traits. By analyzing the pedigrees of families that were afflicted with particular illnesses, Garrod became convinced that certain human diseases can be inherited. He was particularly interested in a condition known as **alkaptonuria**, which is characterized by arthritis and other symptoms. This condition is easy to diagnose, because an affected person's urine turns black. Garrod observed that people suffering from alkaptonuria excrete huge quantities of a molecule called *homogentisic acid* in their urine. To explain why alkaptonurics accumulate homogentisic acid, Garrod suggested they lack an enzyme that catalyzes a reaction involving this molecule. His idea was that if the enzyme did not function normally, then homogentisic acid would not be broken down or used as a reactant as it normally would. As a result, homogentisic acid would accumulate in the body.

In forming this hypothesis, Garrod was inspired by a fundamentally important concept. Biologists who were studying glycolysis and other processes had established that in cells, chemical reactions frequently take place in a series of steps called a **metabolic pathway**. As **Figure 15.1a** shows, molecules are often formed or broken apart in a distinct sequence of steps, from product to reactant by way of intermediates in a specific order. The glycolytic pathway, Krebs cycle, fermentation pathways, and Calvin cycle introduced in Chapters 9 and 10 are examples of particularly important metabolic pathways.

In most cases, every step in a metabolic pathway is catalyzed by a different enzyme. If the enzyme is not available, the reaction that it catalyzes does not occur, and the substrate for that particular reaction accumulates. For example, **Figure 15.1b** shows part of the metabolic pathway that involves homogentisic acid. If the enzyme that catalyzes the reaction marked with a red "X" in the figure does not function correctly, homogentisic acid cannot be transformed into the next compound in the sequence. As a result, it accumulates in the body and causes the symptoms of alkaptonuria. Using the same logic, Garrod identified the biochemical basis of other genetic diseases.

Garrod had correctly identified the cause of an inherited defect, or what he called an inborn error of metabolism. In doing so, he became the first researcher to draw a connection between genes and the chemical reactions that occur inside cells, or what biologists call **metabolism**.

Genes and Enzymes

Despite Garrod's brilliant work, an explicit hypothesis explaining what genes do did not appear until 1941. That year George Beadle and Edward Tatum published a series of breakthrough experiments on a bread mold called *Neurospora crassa*. The

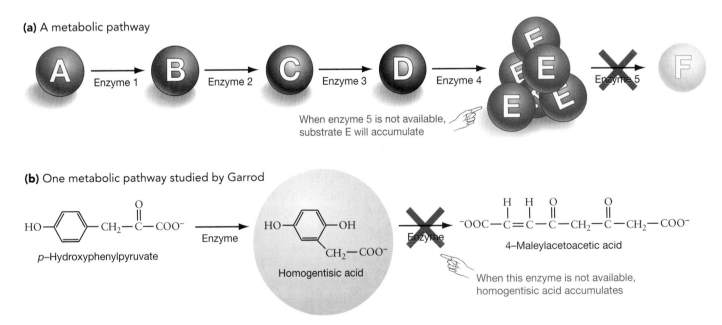

(a) A metabolic pathway

When enzyme 5 is not available, substrate E will accumulate

(b) One metabolic pathway studied by Garrod

p–Hydroxyphenylpyruvate

Homogentisic acid

4–Maleylacetoacetic acid

When this enzyme is not available, homogentisic acid accumulates

FIGURE 15.1 Enzymes Catalyze the Reactions in Metabolic Pathways
(a) A generalized metabolic pathway. Each reaction is catalyzed by a different enzyme. **(b)** Garrod hypothesized that alkaptonurics lack the enzyme that uses homogentisic acid as a substrate, thus causing the acid to accumulate. Excess homogentisic acid is the cause of alkaptonuria.

work inspired their **one-gene, one-enzyme hypothesis**. This was the idea that genes contain the information needed to make proteins, most of which function as enzymes.

Three years later, Adrian Srb and Norman Horowitz published a rigorous test of the one-gene, one-enzyme hypothesis that focused on the ability of *N. crassa* individuals to synthesize the amino acid arginine. In the lab, normal cells of this bread mold grow well on a laboratory culture medium that lacks arginine. This is possible because *Neurospora* cells are able to synthesize their own arginine.

Earlier, Hans Krebs had established that cells from mammals synthesize arginine via a series of steps. Krebs determined that compounds called ornithine and citrulline act as intermediates in the metabolic pathway that leads to arginine, and that enzymes are required to convert ornithine to citrulline and citrulline to arginine. Srb and Horowitz hypothesized that *N. crassa* cells synthesize arginine via the same pathway, using similar enzymes. More importantly, the researchers hypothesized that specific genes in *N. crassa* cells are responsible for producing each of the enzymes involved. Krebs had established the biochemical aspects of arginine synthesis; Srb and Horowitz set out to link the biochemistry to genetics.

This research was inspired by an idea that was brilliant in its simplicity and expressed concisely by Beadle: "One ought to be able to discover what genes do by making them defective." The logic ran as follows: If *N. crassa* synthesizes arginine via a metabolic pathway, and if the one-gene, one-enzyme hypothesis is correct, then Srb and Horowitz should be able to find individuals with a defective gene for each of the enzymes in the pathway. Their goal was to find out how each gene in the pathway acts by observing what happens when each gene, individually, does *not* act.

In analyzing the effects of defective genes, the early *Neurospora* researchers pioneered a research strategy that is still widely used. Contemporary biologists are often able to infer a great deal about what a gene does by comparing the phenotypes of individuals with and without a functioning copy of that gene. Today, alleles that do not function at all are called **knock-out mutants, null mutants,** or **loss-of-function mutants**. In effect, Srb and Horowitz set out to create knock-out mutants for each of the enzymes involved in arginine synthesis. If this approach were successful, the result would provide strong support for the hypothesis that genes make enzymes.

To pursue that strategy, Srb and Horowitz had to create a large number of mutant individuals. They did this with an approach that Beadle and Tatum had developed: exposing *N. crassa* cultures to high-energy radiation such as X-rays or ultraviolet light. As Chapter 14 indicated, high-energy radiation damages the double-helical structure of DNA—often in a way that makes the affected section nonfunctional. As a result, exposing organisms to radiation is an effective way to create knock-out mutants.

It is important to recognize that most treatments that create mutants are random with respect to the gene involved. High-energy radiation is equally likely to damage DNA in any part of the organism's genome. Most organisms have thousands of genes. Of the tens of thousands of mutants that the biologists created, then, it was likely that only a handful contained knock-out mutations in the pathway for arginine synthesis.

To find cells that contained a knock-out mutation in the pathway for arginine synthesis, the researchers performed what is now known as a genetic screen. A **genetic screen** is a technique for picking certain types of mutants out of many thousands of randomly generated mutants. In this case, Srb and Horowitz began their screen by raising the irradiated cells on a medium that included arginine, the final product in the pathway. Then they grew a sample of each type of cell on a medium that lacked amino acids, and they were *not* able to grow on that medium. This test identified cells in the original cultures that were potential knock-out mutants in the pathway for arginine synthesis. If a new sample of cells was added to a medium containing arginine and the cells were subsequently able to grow, then it was likely that they had a defect somewhere in the metabolic pathway for synthesizing arginine. The biologists followed up by confirming that the offspring of these cells also had this defect. Based on these data, they were confident that they had isolated individuals with mutations in the gene or genes for arginine synthesis.

Did some of these cells have knock-out mutations affecting different enzymes in the metabolic pathway? To answer this question, the biologists grew the mutants on media that contained ornithine, or citrulline, or arginine. The results were dramatic. Some of the mutant cells were able to grow in the presence of some of these compounds but not in the presence of others. More specifically, the mutants fell into three distinct classes, which the researchers called *arg1*, *arg2*, and *arg3*:

Supplement

Mutant Type	no ornithine no citrulline no arginine	Ornithine no citrulline no arginine	no ornithine Citrulline no arginine	no ornithine no citrulline Arginine
arg1	no growth	GROWTH	GROWTH	GROWTH
arg2	no growth	no growth	GROWTH	GROWTH
arg3	no growth	no growth	no growth	GROWTH

To interpret these data, the biologists drew the following conclusions:

- *arg1* cells lack the enzyme that converts a precursor molecule to ornithine;

- *arg2* cells lack the enzyme that converts ornithine to citrulline; and

- *arg3* cells lack the enzyme that converts citrulline to arginine.

Metabolic pathway for arginine synthesis:

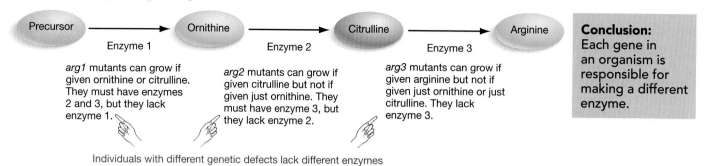

arg1 mutants can grow if given ornithine or citrulline. They must have enzymes 2 and 3, but they lack enzyme 1.

arg2 mutants can grow if given citrulline but not if given just ornithine. They must have enzyme 3, but they lack enzyme 2.

arg3 mutants can grow if given arginine but not if given just ornithine or just citrulline. They lack enzyme 3.

Conclusion: Each gene in an organism is responsible for making a different enzyme.

Individuals with different genetic defects lack different enzymes

FIGURE 15.2 There Are Several Steps in the Metabolic Pathway for Arginine Synthesis
Different *Neurospora crassa* mutants have defects in different parts of the arginine pathway. This result supports the hypothesis that each gene codes for a different enzyme.

As **Figure 15.2** shows, these are exactly the results predicted by Krebs's finding that ornithine is converted to citrulline and that citrulline is then converted to arginine. The experiments had shown that each type of mutant lacked a different, specific step in a metabolic pathway because of a defect in a particular gene. These results convinced most investigators that the one-gene, one-enzyme hypothesis was correct. Follow-up work showed that genes are responsible for all of the different types of proteins produced by cells—not just enzymes. Biologists finally understood what most genes do: They contain the instructions for making proteins.

15.2 The Genetic Code

How does a gene specify the production of a protein? After Beadle and Tatum's hypothesis had been supported in a variety of organisms, this question became a central one. Part of the answer lay in understanding the molecular nature of the gene. Biochemists knew that the primary components of DNA were four nitrogen-containing bases: the pyrimidines thymine (abbreviated T) and cytosine (C), and the purines adenine (A) and guanine (G). They also knew that these bases were connected in a linear sequence by a sugar-phosphate backbone. Watson and Crick's model for the secondary structure of the DNA molecule, introduced in Chapter 4, revealed that two strands of DNA are wound into a double helix, held together by hydrogen bonds between the complementary base pairs A-T and G-C.

Given DNA's structure, it appeared extremely unlikely that DNA directly catalyzed the reactions that produce proteins. Its shape was too regular to suggest that it could bind a wide variety of substrate molecules and lower the activation energy for chemical reactions. Instead, Crick proposed that the sequence of bases in DNA might act as a code. The idea was that DNA was *only* an information-storage molecule. The instructions it contained would have to be read and then translated into proteins.

Crick offered an analogy with Morse code. Morse code is a message-transmission system using dots and dashes to represent the letters of the alphabet. Crick's idea was that different combinations of bases could specify the 20 amino acids, just as different combinations of dots and dashes specify the 26 letters of the alphabet.

A particular stretch of DNA, then, could contain the information needed to specify the amino acid sequence of a particular enzyme. In code form, the tremendous quantity of information needed to build and run a cell could be stored compactly. This information could also be copied through complementary base pairing and transmitted efficiently from one generation to the next.

DNA's structure correlates closely with this proposed function. As **Figure 15.3** shows, the bases that compose DNA are exposed if the two strands in the double helix are separated. These bases are then able to pair with complementary nucleotides to form two newly synthesized strands of DNA or a new strand of RNA. When this pairing occurs, biologists say that the original strands form a template and are "read."

It soon became apparent, however, that the information encoded in the base sequence of DNA is not translated into the amino acid sequence of proteins directly. Instead, the link between DNA as information repository and proteins as cellular machines is indirect.

DNA sequences can be read when strands in the double helix are separated

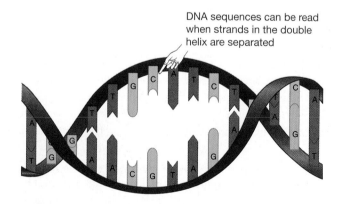

FIGURE 15.3 The Sequence of Bases in DNA Provides Encoded Information
The information encoded in the sequences of bases in DNA can be read when strands are separated.

RNA as the Intermediary between Genes and Proteins

The first clue that biological information does not flow directly from DNA to proteins came from data on the structure of cells. In eukaryotic cells, DNA is enclosed within a membrane-bound organelle called the nucleus (see Chapter 7). But the organelles called ribosomes, where protein synthesis takes place, are outside the nucleus, in the cytoplasm. This observation began to make sense after François Jacob and Jacques Monod suggested that RNA molecules act as a link between genes and the protein-manufacturing centers. Jacob and Monod's hypothesis is illustrated in **Figure 15.4**. They predicted that short-lived molecules of RNA called **messenger RNA**, or **mRNA**, carry information from DNA to the site of protein synthesis. Messenger RNA is one of several distinct types of RNA in cells.

Follow-up research confirmed that the messenger RNA hypothesis is correct. One particularly important piece of evidence was the discovery of an enzyme that catalyzes the synthesis of RNA. This protein is called **RNA polymerase**, because it polymerizes ribonucleotides into strands of RNA. The key observation was that RNA polymerase synthesizes RNA molecules according to the information provided by the sequence of bases in a particular stretch of DNA.

Figure 15.5a shows how the transfer of information from DNA to mRNA works. To determine this connection, researchers created a reaction mix containing three critical elements: (1) the enzyme RNA polymerase; (2) ribonucleotides containing the bases adenine (A), uracil (U), guanine (G), and cytosine (C); and (3) copies of a strand of synthetic DNA that contained deoxyribonucleotides in which the only base was thymine (T). When they allowed the polymerization reaction to proceed, the only base in the RNA molecules that resulted was adenine. This result strongly supported the hypothesis that RNA polymerase synthesizes RNA according to the rules of complementary base pairing introduced in Chapter 4: Thymine

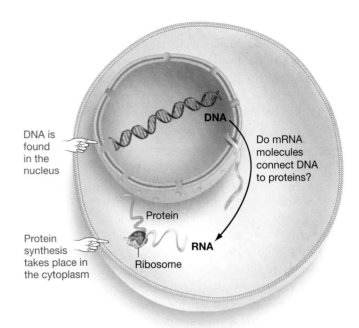

FIGURE 15.4 The Messenger RNA Hypothesis
In the cells of plants, animals, fungi, and other eukaryotes, most DNA is found only in the nucleus, but proteins are manufactured outside the nucleus, at ribosomes. Biologists proposed that the information coded in DNA is carried from inside the nucleus to the ribosomes by messenger RNA (mRNA). **QUESTION** Why was the choice of the term *messenger* appropriate?

pairs with adenine. A sequence of DNA containing thymine as the only base directed the polymerization of a complementary RNA sequence containing adenine as the only base. Synthetic DNAs containing no bases other than cytosine result in the production of RNA molecules containing no bases other than guanine. **Figure 15.5b** summarizes how the sequence of bases in DNA maps to the sequence of bases in RNA. These experiments were possible because RNA polymerase does not require a primer.

(a) Experimental evidence that RNA is synthesized by complementary base pairing

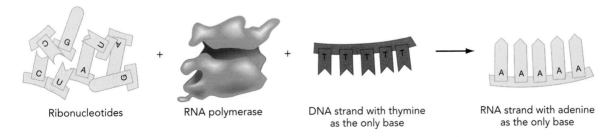

| Ribonucleotides | RNA polymerase | DNA strand with thymine as the only base | RNA strand with adenine as the only base |

(b) RNA sequences are complementary to DNA sequences.

FIGURE 15.5 Complementary Base Pairing Transfers Information from DNA to RNA
(a) This experiment supported the hypothesis that RNA molecules are synthesized with a base sequence that is complementary to the corresponding DNA sequence. **(b)** RNA sequences are complementary to the DNA sequences of genes.

The next challenge was to understand the final link between DNA and proteins: Exactly how does the sequence of bases in a strand of mRNA code for the sequence of amino acids in a protein? If this question could be answered, biologists would have cracked the **genetic code**—the rules that specify the relationship between a sequence of nucleotides in DNA or RNA and the sequence of amino acids in a protein. Researchers from all over the world took up the challenge. The race was on.

How Long Is a Word in the Genetic Code?

The first step in cracking the genetic code was to determine how many bases make up a "word." In a sequence of mRNA, how long is a message that specifies one amino acid? Based on some simple logic, George Gamow suggested that each code word contains three bases. His reasoning derived from the observation that there are 20 amino acids, and the hypothesis that each amino acid must be specified by a particular sequence of mRNA. Because there are only four different bases in ribonucleotides (A, U, G, and C), a one-base code could specify only four amino acids. Similarly, a two-base code could represent just 4 × 4, or 16, amino acids. A three-base code, though, could specify 4 × 4 × 4, or 64, different amino acids. As a result, a three-base code would provide more than enough messages to code for all 20 amino acids (**Figure 15.6**). This three-base code is known as a **triplet code**.

These simple calculations highlight an important point: Gamow's hypothesis predicted that the genetic code is **redundant**. That is, more than one triplet of bases might specify the same amino acid. The idea was that different three-base se-

quences in an mRNA—say, AAA and AAG—might code for the same amino acid—say, lysine. The group of three bases that specifies a particular amino acid is called a **codon**. According to the triplet code hypothesis, many of the 64 codons that are possible actually specify the same amino acids.

Work by Francis Crick and colleagues confirmed that codons are three bases long. These researchers used chemical dyes called *acridines* to induce a mutation that adds or deletes a single base in the DNA of a virus called T4, which infects *E. coli* cells. As predicted for a triplet code, a 1-base addition or deletion in the base sequence led to a loss of function in the gene being studied. This is because a single addition or deletion mutation throws the sequence of codons, or the **reading frame**, out of register. To understand how this works, consider the sentence "*The fat cat ate the rat.*" The reading frame of this sentence is a three-letter word and a space. If the fourth letter in this sentence—the "f" in "fat"—were deleted but the reading frame stayed intact, the sentence would be transformed into "*The atc ata tet her at.*" This is gibberish.

Figure 15.7a illustrates how the addition or deletion of a single base disrupts the reading frame in a sequence of codons, assuming that the genetic code is read in triplets. The parent DNA sequence (top line) shows the reading frame for a series of AAT codons. The next two lines in Figure 15.7a illustrate how the addition or deletion of a single base causes a different group of triplets to be read. If each codon specifies a distinct amino acid, then the protein produced from the altered sequence will have an entirely different sequence of amino acids than normal. Enzymes with such radically altered primary structure are almost guaranteed to be completely nonfunctional. In this way, the addition or deletion of a single base leads to a knock-out

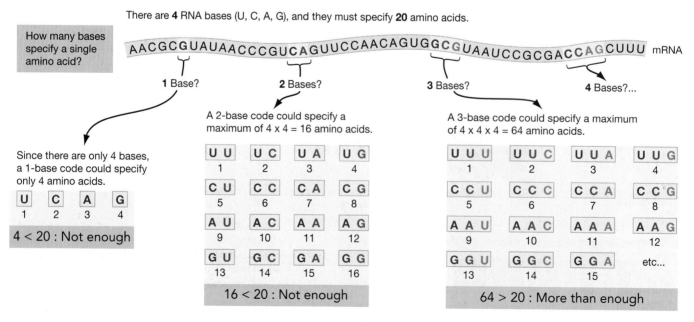

FIGURE 15.6 In the Genetic Code, How Many Bases Form a "Word"?

(a) Single deletion or addition of base leads to nonfunctional protein.

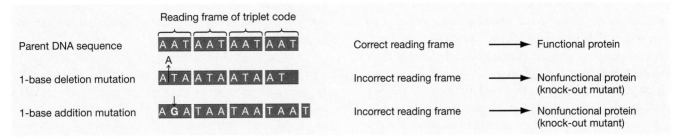

(b) Through recombination, chromosomes can obtain more than one addition or deletion.

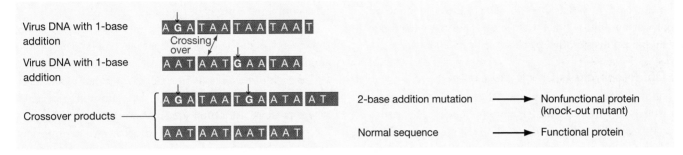

(c) Two deletions or additions of bases lead to nonfunctional protein.

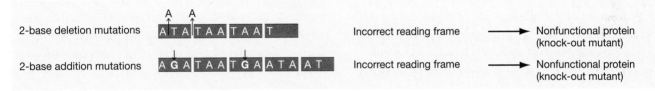

(d) Three deletions or additions of bases lead to functional protein.

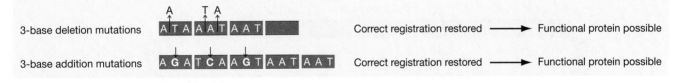

Conclusion: The code is read in groups of three bases, not one or two.

FIGURE 15.7 The Experiments That Confirmed the Triplet Code
(a) If a deletion or addition mutation (indicated by an arrow) occurs, the reading frame of a DNA sequence goes out of register. **(b)** When crossing over occurs between different addition or deletion mutants, the recombinant DNA that results may have multiple addition or deletion mutations or none at all. **(c)** DNA sequences with two addition or deletion mutations are out of register and do not produce a functional protein. **(d)** Three deletion mutations, or three addition mutations, bring the reading frame back into register.

mutation—a complete loss of function. Crick and his colleagues could collect these types of mutants by identifying viruses that were unable to infect *E. coli* cells successfully.

It's important to recognize, though, that the addition or deletion of a single base would also disrupt a code with a reading frame that was two bases long or four bases long. To test the three-base hypothesis rigorously, Crick and co-workers allowed T4 viruses with different single-base-pair additions or deletions to undergo crossing over and recombination. Some of the crossover products contained two 1-base addition mutations

(thus, a 2-base addition, **Figure 15.7b**), or two 1-base deletion mutations. These viruses were still unable to infect *E. coli* cells as well as normal viruses could, however (**Figure 15.7c**). This result suggested that the reading frame of the gene involved was still disrupted, leading to a knock-out mutation and nonfunctional protein. From this result, the biologists concluded that the reading frame of the genetic code was not two bases long.

In contrast, viruses carrying two deletion mutations could be allowed to recombine with viruses carrying a single deletion, resulting in a virus chromosome with three deletion mutations. In the sentence "The fat cat ate the rat," the combination of removing one letter from each of the first three words might result in "Tha tca ate the rat." Just as the altered sentence still conveys some meaning, viruses with three deletion mutations were able to produce a functional protein and successfully infect *E. coli* cells. The biologists interpreted these results as strong evidence in favor of the triplet code hypothesis. As **Figure 15.7d** (page 331) shows, the combination of three deletion mutations, or three addition mutations, would restore the reading frame to normal, possibly resulting in a protein with a functional amino acid sequence.

These experiments provided persuasive evidence that the genetic code is read in triplets of bases. The results also launched a long, laborious, and ultimately successful effort to determine which amino acid is specified by each of the 64 codons.

How Did Researchers Crack the Code?

The initial advance in deciphering the genetic code came when Marshall Nirenberg and Heinrich Matthaei created a method for synthesizing RNAs of known sequence. Their method was based on an enzyme called *polynucleotide phosphorylase*, which randomly catalyzes the formation of phosphodiester bonds between ribonucleotides. By providing ribonucleotides whose only base was uracil (U) to a reaction mix containing this enzyme, Nirenberg and Matthaei were able to create a long polymer of uracil-containing ribonucleotides. These synthetic RNAs were added to an in vitro system for synthesizing proteins. The researchers analyzed the resulting amino acid chain and determined that it was *polyphenylalanine*—a polymer consisting of the amino acid phenylalanine.

This result could mean only one thing: The RNA triplet UUU codes for the amino acid phenylalanine. By complementary base pairing, it was clear that the corresponding DNA sequence would be AAA. This initial observation was followed by experiments with synthetic RNAs consisting of only A or C. RNAs consisting of only AAAAA... produced polypeptides consisting of only lysine; poly-C RNAs produced polypeptides composed entirely of proline.

Nirenberg and Philip Leder later devised a system for synthesizing specific codons. Once they had copies of specific codons, they performed a series of experiments in which they added each of the codons to a cell extract that included the 20 different amino acids, ribosomes (the multimolecular machines where proteins are synthesized), and other molecules required for protein synthesis. Then the researchers determined which amino acid was bound to the ribosomes when a particular codon was present. For example, when the codon CAC was in the reaction mix, the amino acid histidine would bind to the ribosomes. This result confirmed that CAC codes for histidine. These ribosome-binding experiments allowed Nirenberg and Leder to determine which of the 64 codons coded for each of the 20 amino acids (**Figure 15.8**).

First base	Second base: U	Second base: C	Second base: A	Second base: G	Third base
U	UUU, UUC — Phenyl-alanine UUA, UUG — Leucine	UCU, UCC, UCA, UCG — Serine	UAU, UAC — Tyrosine UAA Stop codon UAG Stop codon	UGU, UGC — Cysteine UGA Stop codon UGG Tryptophan	U C A G
C	CUU, CUC, CUA, CUG — Leucine	CCU, CCC, CCA, CCG — Proline	CAU, CAC — Histidine CAA, CAG — Glutamine	CGU, CGC, CGA, CGG — Arginine	U C A G
A	AUU, AUC, AUA — Isoleucine AUG Methionine (start codon)	ACU, ACC, ACA, ACG — Threonine	AAU, AAC — Asparagine AAA, AAG — Lysine	AGU, AGC — Serine AGA, AGG — Arginine	U C A G
G	GUU, GUC, GUA, GUG — Valine	GCU, GCC, GCA, GCG — Alanine	GAU, GAC — Aspartic acid GAA, GAG — Glutamic acid	GGU, GGC, GGA, GGG — Glycine	U C A G

FIGURE 15.8 The Genetic Code
To read the code, match the first base in an mRNA codon, in the red band on the left side, with the second base in a codon, in the blue band along the top, and the third base in a codon, in the green band on the right side. The 64 codons, along with the amino acid or stop (termination) signal that they specify, are given in the boxes. By convention, codons are always written in the 5′ → 3′ direction. EXERCISE Pick four codons at random. Next to each of them, write the three bases in DNA that specify the codon. (Remember to write the DNA sequence in the 5′ → 3′ direction.)

In addition to matching codons to amino acids, researchers discovered that certain codons are punctuation marks that signal "start of message" or "end of message." These codons relay information that the protein chain is complete or that protein synthesis should start at a given codon. There is one **start codon** (AUG), which codes for the amino acid methionine, to signal that protein synthesis should start at that point on the mRNA molecule. There are three **stop codons** (UAA, UAG, and UGA), also known as *termination codons*, which signal that the protein is complete.

The deciphering of the full genetic code, presented in Figure 15.8 is a tremendous achievement. It represents more than five years of work by several teams of researchers. As predicted, the genetic code is redundant. All amino acids except methionine and tryptophan are coded by more than one codon. The code is also unambiguous, however, because one codon never codes for more than one amino acid. Later work showed that the genetic code is also nearly universal: With a few minor exceptions, all codons specify the same amino acids in all organisms. **Box 15.1** explores the evolutionary implications of the universal code.

BOX 15.1   The Evolution of the Genetic Code

Figure 15.9 is a simplified version of the tree of life that was introduced in Chapter 1. Recall that the three largest branches on this tree represent what biologists call the three domains of life: the Bacteria, the Archaea, and the Eukarya. Smaller branches on the tree represent groups of species within each of the three domains, such as the cyanobacteria, α-proteobacteria, land plants, fungi, and animals.

Virtually all of the organisms on this tree use the genetic code shown in Figure 15.8. Exceptions are few. For example, in the single-celled eukaryotes *Tetrahymena* and *Paramecium*, UAA and UAG code for glutamine instead of stop; and in the yeast *Candida cylindracea*, CUG codes for serine instead of leucine. A few species initiate translation at other codons as well as AUG.

To explain why the genetic code is nearly universal, biologists hypothesize that the common ancestor of all species living today used this same code. This common ancestor is indicated at the base of the tree in Figure 15.9. The logic here is that if all species living today use the same genetic code, then it is reasonable to infer that their common ancestor did also.

Given that the genetic code has been in existence for a long time, the question arises, Is the code arbitrary? That is, is the genetic code random with respect to the ability of organisms to survive and produce offspring? A brief examination of Figure 15.8 suggests that the answer is no. Note that when several codons specify the same amino acid, the first two bases in those codons are almost always identical. This observation suggests that the code does not represent a random assemblage of bases, like letters drawn from a hat. In fact, extensive analysis has shown that, compared with randomly generated codes, the existing genetic code is structured in a way that efficiently minimizes the phenotypic effects of small changes in DNA and errors during translation. This result suggests that the genetic code has been honed by natural selection. Stated another way, it is likely that other genetic codes may have existed early in the history of life. But because the code in Figure 15.8 is particularly efficient, it emerged as the most successful version.

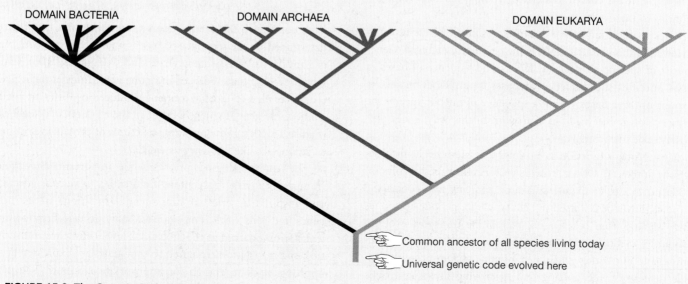

FIGURE 15.9 The Genetic Code Is Universal

✓ CHECK YOUR UNDERSTANDING

Genes code for proteins, but they do so indirectly. The sequence of bases in DNA is used to produce messenger RNA (mRNA). The sequence of bases in an mRNA molecule is complementary to the DNA sequence of a gene. The sequence of bases in mRNA forms a code in which particular combinations of three bases lead to the addition of a specific amino acid to the protein encoded by the gene. This code is redundant because 64 combinations of three bases occur, but only 20 amino acids and a stop "punctuation mark"—indicating the end of the coding sequence—need to be specified. You should be able to (1) draw a DNA double helix, (2) choose one of the two strands of the double helixes the coding strand, (3) write the sequence of the mRNA molecule that would be synthesized (in the 5′ → 3′ direction!) from that strand, and (4) write the amino acid sequence of the protein that would be produced by that mRNA.

15.3 The Central Dogma of Molecular Biology

Even before the genetic code was fully revealed, Francis Crick articulated what became known as the central dogma of molecular biology. The **central dogma** summarizes the flow of information in cells. It simply states that DNA codes for RNA, which codes for proteins:

$$DNA \rightarrow RNA \rightarrow proteins$$

Besides clarifying information flow in the cell, the central dogma points out that information flow occurs in only one direction—from DNA to RNA to proteins. RNA does not code for the production of other RNA or DNA, and proteins do not code for the production of RNA or DNA or other proteins. (The essay at the end of this chapter points out important exceptions to these rules, however.)

Crick's simple statement encapsulates much of the research reviewed in this chapter. DNA is the hereditary material. Genes consist of specific stretches of DNA. The sequence of bases in DNA specifies the sequence of bases in an RNA molecule. Groups of three bases within a sequence of mRNA specify the sequence of amino acids in a protein. In this way, genes ultimately code for proteins.

Most proteins function as enzymes that catalyze chemical reactions in the cell. Other proteins perform the types of roles introduced in earlier chapters: Motor proteins and contractile proteins move the cell itself or cellular cargo, peptide hormones carry signals from cell to cell, membrane transport proteins conduct specific ions or molecules across the plasma membrane, and antibodies and other immune system proteins provide defense by recognizing and destroying invading viruses and bacteria.

In addition, some genes code for RNA molecules that do not function as mRNAs. The various types of RNA molecules that exist have an array of functions in the cell. Most of these functions are related to managing the flow of information from DNA to mRNA to proteins.

Biologists use specialized vocabulary to summarize this sequence of events encapsulated in the central dogma. For example, biologists say that DNA is *transcribed* to RNA. In everyday English, the word *transcription* simply means making a copy of information. The scientific use of the term is appropriate because DNA acts as a permanent record—an archive or blueprint containing the information needed to build and run the cell. This permanent record is copied, via **transcription,** to the short-lived form called mRNA. The information is then transferred to a new molecular form—a sequence of amino acids. In everyday English, the word *translation* refers to the transferring of information from one language to another. In biology, the synthesis of protein from mRNA is called **translation.** Translation is the transferring of information from one type of molecule to another—from the "language" of nucleic acids to the "language" of proteins. **Figure 15.10a** summarizes the relationship between transcription and translation, as well as the relationship among DNA, RNA, and proteins. In diagrammatic form,

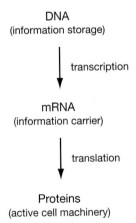

According to the central dogma, an organism's genotype is determined by the sequence of bases in its DNA; its phenotype is a product of the proteins it produces (**Figure 15.10b**). Later work revealed that alleles of the same gene differ in their DNA sequence. As a result, the proteins produced by different alleles of the same gene frequently differ in their amino acid sequence (**Figure 15.10c**). If the primary structures of proteins vary, their functions are likely to vary as well.

The central dogma provided an important conceptual framework for the burgeoning field called *molecular genetics,* and inspired a series of fundamental questions about how genes and cells work. For example, how do transcription and translation proceed at the molecular level? How are transcription and translation regulated so that genes are expressed at an appropriate time and in an appropriate amount? The next three chapters are devoted to exploring these questions.

(a) Information flows from DNA to RNA to proteins.

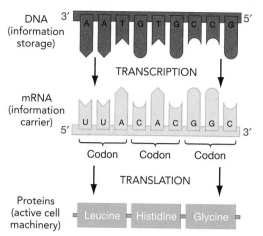

DNA (information storage) — 3′ A A T G T G C C G 5′

TRANSCRIPTION

mRNA (information carrier) — 5′ U U A C A C G G C 3′

Codon   Codon   Codon

TRANSLATION

Proteins (active cell machinery) — Leucine — Histidine — Glycine

(b) DNA sequences define the genotype; proteins create the phenotype.

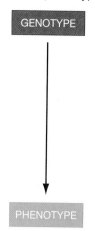

GENOTYPE

PHENOTYPE

(c) Changes in the genotype may lead to changes in the phenotype.

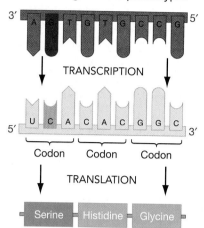

3′ A G T G T G C C G 5′

TRANSCRIPTION

5′ U C A C A C G G C 3′

Codon   Codon   Codon

TRANSLATION

Serine — Histidine — Glycine

FIGURE 15.10 The Flow of Information in the Cell
(a) DNA functions as the information archive in the cell, analogous to a set of blueprints. mRNA is a short-lived copy of the information in DNA, used to manufacture proteins. Most of the machinery and many of the structures in cells consist of proteins. **(b)** The central dogma revealed the connection between genotype and phenotype. **(c)** Alleles of genes differ in their DNA base sequence. As a result, they may produce proteins with different primary structures and thus different functions. In this way, different alleles may be associated with different phenotypes.

ESSAY Exceptions to the Central Dogma

The old adage "Rules are made to be broken" applies to biology. In 1955, just three years after DNA had been confirmed as the hereditary material, Heinz Fraenkel-Conrat discovered that the virus he was studying (the *tobacco mosaic virus*, or TMV; **Figure 15.11**) does not contain DNA. The genes of this

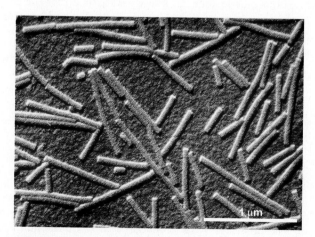

FIGURE 15.11 Tobacco Mosaic Virus
The genes inside this virus consist of RNA.

virus are made of RNA instead. Fraenkel-Conrat showed that when TMV infects a cell of a tobacco plant, the viral genes are translated directly into viral proteins. Thus, in RNA viruses such as TMV, the flow of information is simply RNA → proteins. Over the years since Fraenkel-Conrat's work was published, researchers have identified hundreds of viruses with RNA genomes. These include the viruses that cause polio, hepatitis, flu, and the common cold in humans.

With respect to breaking rules, though, an even bigger bombshell exploded in 1970.

With respect to breaking rules, though, an even bigger bombshell exploded in 1970. That year David Baltimore, Howard Temin, and Satoshi Mizutani announced the discovery of RNA viruses that violate the central dogma's tenet of one-way information flow. The genomes of these viruses include sequences that code for an enzyme called **reverse transcriptase**. This protein catalyzes the formation of DNA from an RNA molecule that acts as a template—with deoxyribonucleotides being matched to the complementary

(Continued on next page)

(Essay continued)

bases in RNA. When one of these viruses infects a host cell, reverse transcriptase "reverse transcribes" the viral RNA genome to DNA. This viral DNA usually incorporates itself into the host's chromosome. Messenger RNAs are then transcribed from the viral DNA and translated into proteins. In these viruses, the direction of information flow is RNA → DNA → RNA → proteins. As a result, they came to be called **retroviruses** ("backward-viruses"). The human immunodeficiency virus (HIV)—the virus that causes AIDS—is a retrovirus.

The experiments leading to this discovery were elegant. To confirm the existence of reverse transcriptase, Temin and Mizutani purified large quantities of the *Rous sarcoma virus* (a virus associated with cancer formation in chickens). Then they disrupted the lipid-containing envelope that surrounds the virus by treating the particles with a detergent. This step released the contents of the virus particles into solution. The researchers added

deoxyribonucleotides labeled with a radioactive isotope and then incubated the mixture. Later they observed the formation of radioactive DNA molecules in the mixture. This observation was a smoking gun—strong evidence that the virus contains a molecule capable of synthesizing DNA from an RNA template.

Temin and Mizutani did a follow-up experiment. It was the same experiment just described, with one change: Along with the labeled deoxyribonucleotides, the researchers added ribonuclease, which breaks RNA molecules apart. In this treatment, no DNA formed. This result was consistent with the prediction that the enzyme requires an RNA template.

With the discovery of reverse transcriptase and the retroviruses, biologists could no longer be quite so dogmatic about the central dogma. Although all *cells* conform to the central dogma, some viruses do not.

CHAPTER REVIEW

Summary of Key Concepts

▦ **Most genes code for enzymes that catalyze specific chemical reactions in the cell.**

The question "What do genes do?" was solved in a series of experiments on the bread mold *Neurospora crassa*. Researchers isolated mutants that cannot grow unless they are supplied with a specific amino acid and then showed that different mutants were unable to make different chemical precursors of the amino acid. The one-gene, one-enzyme hypothesis explains these results by proposing that most genes code for proteins and that most proteins act as enzymes that catalyze specific reactions in metabolic pathways.

Web Tutorial 15.1   The One-Gene, One-Enzyme Hypothesis

▦ **In cells, information flows from DNA to RNA to proteins. DNA is transcribed to messenger RNA by RNA polymerase,**

and then messenger RNA is translated to proteins in ribosomes. Each amino acid in a protein is specified by a group of three bases in RNA.

Experiments confirmed that DNA does not code for proteins directly. Instead, mRNA molecules are transcribed from DNA and then translated into proteins. One-way flow of information, from DNA to RNA to proteins, is called the central dogma of molecular biology. By synthesizing RNAs of known base composition and then observing how they functioned during the process of translation, researchers were able to unravel the genetic code. It is now established that the code is read in triplets and that the code is "redundant"—meaning most of the 20 amino acids are specified by more than one codon.

Web Tutorial 15.2   The Triplet Nature of the Genetic Code

Questions

Content Review

1. What is metabolism?
   a. the chemical reactions that occur in organisms
   b. the synthesis of macromolecules through spark discharges and other reactions during chemical evolution
   c. the in vitro synthesis of macromolecules in a test tube
   d. carefully controlled chemical reactions that occur in an industrial setting

2. What does the one-gene, one-enzyme hypothesis state?
   a. Genes are composed of stretches of DNA.
   b. Genes are made of protein.
   c. Genes code for ribozymes.
   d. A single gene codes for a single protein.

3. DNA's primary structure is made up of just four different bases, and its secondary structure is regular and highly stable. How can a molecule with these characteristics hold all of the information required to build and maintain a cell?
   a. The information is first transcribed, then translated.
   b. The messenger RNA produced from DNA has much more complex primary and secondary structures, and thus holds much information.
   c. A protein produced (indirectly) from DNA has much more complex primary and secondary structures, and thus holds much information.
   d. The information in DNA is in code form.

4. Why did researchers suspect that DNA does not code for proteins directly?
   a. In eukaryotes, DNA is found inside the nucleus, but proteins are produced outside the nucleus.
   b. In prokaryotes, DNA and proteins are never found together.
   c. When DNA was damaged by ultraviolet radiation or other sources of energy, the proteins in the cell did not change accordingly.
   d. There are several distinct types of RNA, of which only one functions as messenger RNA.

5. Which of the following describes an important experimental strategy in deciphering the genetic code?
   a. comparing the amino acid sequences of proteins with the base sequence of their genes
   b. analyzing the sequence of RNAs produced from known genes
   c. analyzing mutants that changed the code
   d. examining the proteins produced when RNAs of known sequence were translated

6. Recessive alleles often contain a knock-out mutation. When a normal allele and a knock-out allele are paired in a heterozygous individual, the normal copy of the gene is frequently able to produce enough functional protein to give a normal phenotype. Why are these facts important?
   a. They explain why some mutants are still able to infect bacterial cells efficiently.
   b. They explain the phenomenon of dominance and recessiveness discovered by Mendel.
   c. They offer a serious challenge to the one-gene, one-enzyme hypothesis.
   d. They illustrate why the central dogma is considered central.

Conceptual Review

1. DNA is referred to as an "information-storage molecule." How can the base sequence of DNA store information?

2. Draw a hypothetical metabolic pathway involving a sequence of five substrates, five enzymes, and a product called BiolSciazine. Number the substrates 1–5, and label the enzymes A–E, in order. (For instance, enzyme A applies to the reaction between substrates 1 and 2.)
   - Suppose a mutation made the gene for enzyme C nonfunctional. What molecule would accumulate in the affected cells?
   - Suppose some individuals with a mutation affecting this metabolic pathway can survive if given substrate 5 in the diet. But they die even if given substrates 1, 2, 3, and 4. State a hypothesis for which enzyme in the pathway is affected by this mutation.

3. Why did experiments with *Neurospora crassa* mutants support the one-gene, one-enzyme hypothesis?

4. When researchers discovered that a combination of three deletion mutations or three addition mutations would restore the function of a gene, most biologists were convinced that the genetic code was read in triplets. Explain the logic behind this conclusion.

5. Why is the genetic code redundant?

6. Explain how a single-base deletion disrupts the reading frame of a gene.

Group Discussion Problems

1. Examine the genetic code as illustrated in Figure 15.8. Note that when several codons specify the same amino acid, the first two bases in those codons are almost always identical. Does this observation support the hypothesis that the genetic code is arbitrary? That is, does the code represent a random assemblage of bases, like letters drawn from a hat, or does it have distinct patterns? (If patterns exist, you could hypothesize that the code is structured in a way that helps transcription or translation occur more efficiently.) How would you test the hypothesis that the code is indeed random?

2. Recall that DNA and RNA are synthesized only in the $5' \rightarrow 3'$ direction and that DNA and RNA sequences are always written in the $5' \rightarrow 3'$ direction. Consider the following DNA sequence:

5′ T T G A A A T G C C C G T T T G G A G A T C G G G G T T A C A G C T A G T C A A A G 3′
3′ A A C T T T A C G G G C A A A C C A C T A G C C C A A T G T C G A T C A G T T T C 5′

   - Identify the bases in the bottom strand that code for start and stop codons. Write the mRNA sequence that would be transcribed between them if the bottom strand served as the template.
   - Write the amino acid sequence that would be translated from the mRNA sequence you just wrote.

3. Scientists say that a phenomenon is a "black box" if they can describe it and study its effects but don't yet know the underlying mechanism that causes it. In what sense was genetics—meaning the transmission of heritable traits—a black box before the central dogma of molecular biology was understood?

4. One of the possibilities that researchers interested in the genetic code had to consider was that the code was overlapping, meaning that a single base could be part of more than one codon. Make a diagram showing how an overlapping code would work, assuming that each codon is three bases long.

Answers to Multiple-Choice Questions   1. a; 2. d; 3. d; 4. a; 5. d; 6. b

16

Transcription and Translation

KEY CONCEPTS

▨ After the enzyme RNA polymerase binds to a specific site in DNA with the help of other proteins, it catalyzes the production of an RNA molecule. The base sequence of the RNA produced is complementary to the base sequence of the DNA template strand.

▨ In eukaryotes, genes consist of coding regions called exons interspersed with noncoding regions called introns. After a eukaryotic mRNA is produced, introns must be spliced out. The ends of the molecule receive a cap and tail.

▨ Inside ribosomes, mRNAs are translated to proteins via intermediary molecules called transfer RNAs. Transfer RNAs carry an amino acid and have a three-base-pair anticodon, which binds to an mRNA codon. The amino acid carried by the transfer RNA is then added to the growing protein via formation of a peptide bond.

▨ Many mutations produce changes in the phenotype.

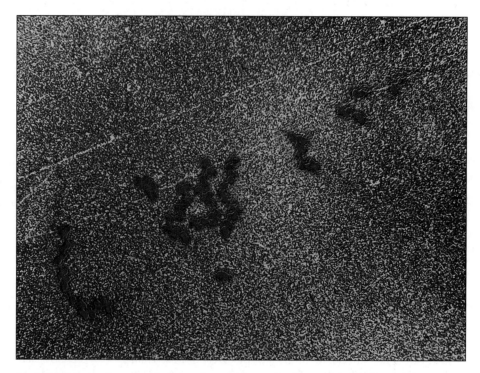

A micrograph showing transcription and translation occurring simultaneously in a bacterial cell. The yellow lines are strands of DNA, and the red structures are ribosomes.

Proteins are the stuff of life. They give cells their shape and control the chemical reactions that go on inside them. No one knows exactly how many different proteins can be made in each of your body's cells, but 100,000 is a reasonable guess. While some of those 100,000 different proteins may not be manufactured at all in certain types of cells, other proteins may be present in quantities ranging from millions of copies to fewer than a dozen. If even one of those types of protein is defective, the cell may sicken or even die.

A cell builds the proteins it needs from instructions encoded in its genome. Chapter 15 explored the basic strategy that cells use to retrieve these instructions and convert them into action in the form of proteins. The first step in the process is the transcription of a gene and the production of a messenger RNA, or mRNA. The RNA message is a short-lived copy of the archived instructions in DNA. The sequence of ribonucleotides in the mRNA is then translated into a sequence of amino acids. This series of events defines the flow of information in the cell:

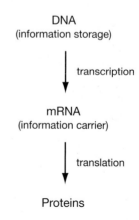

DNA
(information storage)

↓ transcription

mRNA
(information carrier)

↓ translation

Proteins

The discovery of the relationship between DNA and proteins was one of the great scientific advances of the twentieth century. But once the relationship between transcription and translation had been firmly established, biologists puzzled over how cells actually accomplish the feat of converting genetic information into protein. How do transcription and translation work? For example, how does RNA polymerase know where to start transcribing a gene and where to end? Once an RNA message is produced, how is the linear sequence of ribonucleotides translated into the linear sequence of amino acids in a protein?

This chapter explores the experiments that answered these questions. In addition to providing a deeper understanding of how cells work, the results summarized here have had a significant impact on biomedicine and drug development. Many of the antibiotics prescribed today work by disrupting translation in bacteria that cause disease. In addition, research on transcription and translation set the stage for an explosion of work, introduced in Chapters 17 and 18, on how the expression of specific genes is regulated. Most genes are transcribed and translated only at certain times or in certain types of cells.

Let's begin by examining research on the mechanisms of transcription in bacteria and eukaryotes, then delve into the mechanics of translation. The chapter concludes by examining how changes in DNA can lead to changes in protein structure and function.

16.1 Transcription in Bacteria

The first step in converting genetic information into proteins is to synthesize a messenger RNA version of the instructions archived in DNA. The enzyme RNA polymerase, introduced in Chapters 14 and 15, is responsible for synthesizing mRNA. Like the DNA polymerases introduced in Chapter 14, RNA polymerase performs a template-directed synthesis in the 5′ → 3′ direction. (Recall that 5′ and 3′ refer to carbons on the sugar subunits of ribonucleotides.) But unlike DNA polymerases, RNA polymerase does not require a primer to begin transcription.

Transcription occurs when RNA polymerase matches the base in a ribonucleotide triphosphate with the complementary base in a gene—a section of DNA. Once a matching ribonucleotide is in place, RNA polymerase catalyzes the formation of a phosphodiester bond between the 3′ end of the growing mRNA chain and the new ribonucleotide. As this matching-and-catalysis process continues, an RNA that is complementary to the gene is synthesized (**Figure 16.1**). Note that only one of the two DNA strands is used as a template and transcribed, or "read," by RNA polymerase. The strand that is read by the enzyme is called the **template strand**; the other strand is called the **non-template strand**. Because RNA has uracil (U) rather than thymine (T), an adenine (A) in the DNA template strand specifies a U in the complementary RNA strand.

Biologists reached these conclusions by studying RNA synthesis in cell-free, or in vitro, systems. Once the basic chemical

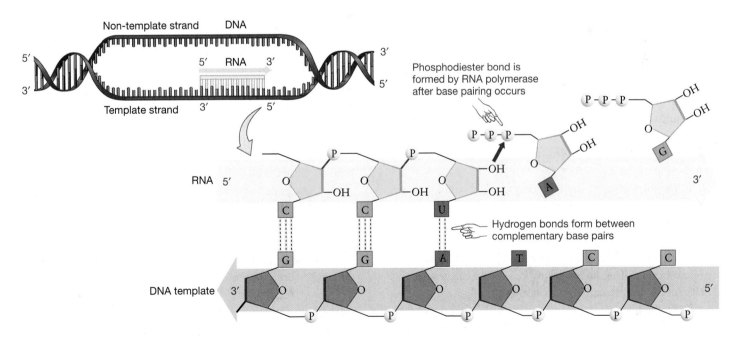

FIGURE 16.1 Transcription Is the Synthesis of RNA from a DNA Template
The reaction catalyzed by RNA polymerase results in the formation of a phosphodiester bond between ribonucleotides. RNA polymerase produces an RNA strand whose sequence is complementary to the bases in the DNA template. **QUESTION** In which direction is RNA synthesized, 5′ → 3′ or 3′ → 5′? In which direction is the DNA template "read"?

reaction was understood, an entirely new set of questions arose:

- What does RNA polymerase look like?
- Are any other proteins or factors involved in transcription, or does RNA polymerase act alone?
- How does the enzyme know where to start transcription on the DNA template?

RNA Polymerase Structure and Function

To understand the structure of RNA polymerase, researchers have employed the technique called X-ray crystallography introduced in Chapter 4. Recall that this procedure allows biologists to obtain information about the three-dimensional structure of large, complex molecules.

The most recent models of bacterial RNA polymerase indicate that the enzyme is large and globular and has several prominent channels that run through the interior. The enzyme's active site, where phosphodiester bonds form, is located where several of these channels intersect. The active site contains a strategically placed magnesium atom (Mg^{2+}), which plays a critical role in stabilizing the transition state as the polymerization reaction takes place.

How does this structure correlate with the enzyme's function? It is logical to predict that DNA fits into one of the enzyme's channels and that the two strands in the double helix separate inside the enzyme to expose a single-stranded template at the active site. If so, how do the enzyme and DNA come together?

Initiation: How Does Transcription Begin in Bacteria?

Soon after the discovery of RNA polymerase, researchers realized that the enzyme cannot initiate transcription on its own. Instead, a detachable protein subunit called **sigma** must bind to RNA polymerase before transcription can begin. RNA polymerase and sigma form what biologists call a **holoenzyme** ("whole enzyme"). A holoenzyme consists of a **core enzyme**, which contains the active site for catalysis, and other required proteins (**Figure 16.2**).

If RNA polymerase is the core enzyme of this holoenzyme, what does sigma do? When researchers mixed RNA polymerase, sigma, and DNA together, they found that the holoenzyme bound tightly to specific sections of DNA. These binding sites were named **promoters**, because they are sections of DNA where transcription begins. The discovery of promoters suggested that sigma's function is regulatory in nature. Sigma appeared to be responsible for guiding RNA polymerase to specific locations where transcription should begin. What is the nature of these specific locations? What do promoters look like, and what do they do?

David Pribnow offered an initial answer to these questions in the mid-1970s. When Pribnow analyzed the base sequence of promoters from a variety of bacteria and from viruses that infect bacteria, he found that the promoters are all 40–50 base

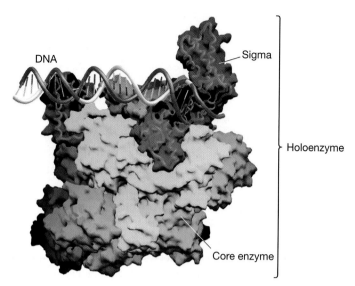

FIGURE 16.2 The Three-Dimensional Structure of RNA Polymerase Bacterial RNA polymerase is a globular enzyme. Sigma is a regulatory component, and the core enzyme contains the active site.

pairs long and that a particular section in each promoter looked similar. This similar segment of DNA had a series of bases identical or similar to TATAAT. This six-base-pair sequence is now known as the **−10 box**, because it occurs about 10 bases from the point where RNA polymerase starts transcription. (DNA that is located in the direction that RNA polymerase moves during transcription is said to be **downstream**; DNA in the opposite direction is said to be **upstream**. Thus, the −10 box is 10 bases upstream from the transcription start site.) Soon after, researchers recognized that the sequence TTGACA occurred in these same promoters about 35 bases upstream from the start of the mRNA. This second key sequence is called the **−35 box**.

Follow-up work showed that transcription begins when sigma binds to the −35 and −10 boxes (**Figure 16.3**, step 1). This is a key point: Sigma, and not RNA polymerase, makes the initial contact with DNA that starts transcription in bacteria. This finding supports the hypothesis that sigma is a regulatory protein. Sigma tells RNA polymerase where and when to start synthesizing RNA.

Once sigma binds to a promoter, the DNA helix opens and creates two strands of single-stranded DNA. As shown in step 2 of Figure 16.3, the template strand is threaded through a channel that leads to the active site inside RNA polymerase. Monomers known as *ribonucleoside triphosphates*, or NTPs, enter a channel at the bottom of the enzyme and diffuse to the active site. When an incoming NTP pairs with a complementary base on the template strand of DNA, RNA polymerization begins. The reaction catalyzed by RNA polymerase is exergonic and spontaneous because NTPs have so much potential energy, owing to their three phosphate groups. As step 3 of Figure 16.3

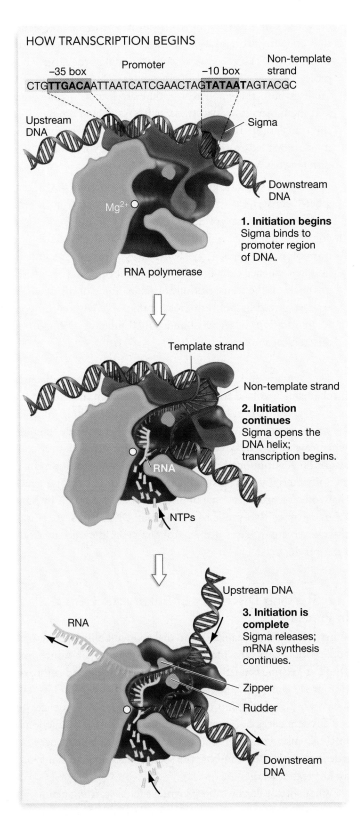

HOW TRANSCRIPTION BEGINS

-35 box   Promoter   -10 box   Non-template strand

CTG**TTGACA**ATTAATCATCGAACTAG**TATAAT**AGTACGC

Upstream DNA

Sigma

Mg^{2+}

Downstream DNA

RNA polymerase

1. Initiation begins
Sigma binds to promoter region of DNA.

Template strand

Non-template strand

2. Initiation continues
Sigma opens the DNA helix; transcription begins.

RNA

NTPs

Upstream DNA

RNA

3. Initiation is complete
Sigma releases; mRNA synthesis continues.

Zipper

Rudder

Downstream DNA

FIGURE 16.3 In Bacteria, Sigma Plays a Key Role in Initiating Transcription
When sigma attaches to RNA polymerase, the holoenzyme is able to bind to the -35 box and the -10 box of a promoter. Once it has bound to a promoter, sigma opens the DNA helix and threads the template strand through the enzyme's active site. When transcription is under way, sigma disengages from RNA polymerase. EXERCISE Circle and label RNA polymerase's active site.

shows, sigma is released once RNA synthesis is under way. The **initiation phase** of transcription is complete.

Recent work has shown that most bacteria have several types of sigma proteins, each with a distinctive structure and function. *Escherichia coli* has seven different sigma proteins, for example, while *Streptomyces coelicolour* has more than 60. Each of these proteins binds to promoters with slightly different DNA base sequences. Although all promoters have a -10 box and a -35 box, the sequences within each box vary somewhat. These variations in DNA sequences affect the ability of different sigma proteins to bind. As a result, each type of sigma protein allows RNA polymerase to bind to a different type of gene. The identity of the sigma protein in the RNA polymerase holoenzyme determines which types of genes will be transcribed. Thus by controlling which sigma proteins are active, bacterial cells control which genes are expressed.

Elongation and Termination

During the **elongation phase** of transcription, RNA polymerase moves along the DNA template in the $3' \rightarrow 5'$ direction of the template strand, synthesizing RNA in the $5' \rightarrow 3'$ direction. In the interior of the enzyme, a group of projecting amino acids called the enzyme's *zipper* helps open the double helix at the upstream end, and a nearby group of amino acids called the *rudder* helps steer the template and non-template strands through channels inside the enzyme. Meanwhile, the enzyme's active site catalyzes the addition of nucleotides to the 3' end of the growing RNA molecule at the rate of about 50 nucleotides per second.

Note that during the elongation phase of transcription, three prominent channels or grooves in the enzyme are filled. Double-stranded DNA goes into and out of one groove; ribonucleoside triphosphates enter another; and the growing RNA strand exits to the rear. In this way, the enzyme's structure correlates closely with its function.

Transcription ends with a **termination phase**. In most cases, transcription stops when RNA polymerase reaches a stretch of DNA sequence that functions as a **transcription termination signal**. The bases that make up the termination signal code for a stretch of RNA with an unusual property: As soon as it is synthesized, the RNA sequence folds back on itself and forms a short double helix that is held together by complementary base pairing. The secondary structure that results is called a *hairpin* (see Chapter 4). The formation of the hairpin structure is thought to disrupt the interaction between RNA polymerase and the RNA transcript, resulting in the physical separation of the enzyme and its product.

To summarize, transcription begins when sigma binds to the promoter at the start of a gene. Once binding occurs, RNA polymerase begins to synthesize mRNA. Transcription ends when a termination signal at the end of the gene leads to the formation of a hairpin in the mRNA, disrupting the transcription complex.

16.2 Transcription in Eukaryotes

The data reviewed in Section 16.1 established how transcription occurs in bacteria. How do these results compare with research on the mechanism of transcription in eukaryotes? What similarities and differences have come to light?

One of the first conclusions about transcription in eukaryotes is that it is similar to bacterial transcription in an important respect: RNA polymerase does not bind directly to promoter sequences by itself. Instead, proteins that came to be called **basal transcription factors** initiate eukaryotic transcription by matching the enzyme with the appropriate promoter region in DNA. The function of basal transcription factors observed in eukaryotes is analogous to the function of the sigma proteins in bacteria.

The discovery of basal transcription factors emerged from studies with cell-free, or in vitro, systems. The basic approach was to purify RNA polymerase from human cells, add template DNA, and analyze the RNAs that were produced. But in early experiments, researchers found that RNA polymerase began copying at random locations on the template DNA instead of specifically at promoter regions. In addition, when RNA polymerase was alone, both strands of the DNA were transcribed instead of just one. This observation inspired the hypothesis that eukaryotic cells must contain sigma-like proteins that are required for normal transcription. To test this hypothesis, biologists added proteins isolated from human cells to the cell-free reaction system. By adding one or a few proteins at a time and recording which ones enabled RNA polymerase to bind to promoters and correctly transcribe the template strand, researchers were gradually able to characterize the suite of proteins that came to be called basal transcription factors.

Follow-up work confirmed several important distinctions about how transcription works in bacteria and eukaryotes:

- In bacteria a single sigma protein binds to a promoter and initiates transcription, but in eukaryotes many basal transcription factors are required to initiate transcription.
- Eukaryotes have three distinct types of RNA polymerase (abbreviated RNA pol) instead of just one. As **Table 16.1** shows, RNA pol I, pol II, and pol III each transcribe a discrete class of RNA. Only RNA pol II transcribes the genes that code for proteins. Stated another way, RNA pol II produces mRNA. RNA pol I, in contrast, makes the large RNA molecules that are found in ribosomes. RNA pol III manufactures both the small RNAs that are found in ribosomes and the molecules called transfer RNAs, which are required for translation. **Box 16.1** explains how research on toxic mushrooms led to the discovery that eukaryotes have three different RNA polymerases.

- Although eukaryotic genomes contain promoters that signal where transcription should begin, just as bacteria do, the promoters in eukaryotic DNA are much more diverse and complex than bacterial promoters. Many of the eukaryotic promoters recognized by RNA polymerase II include a unique sequence called the **TATA box**, located 30 base pairs upstream of the transcription start site. Some of the promoters recognized by pol II do not contain a TATA box, however. In addition, RNA pol I and pol III interact with entirely different promoters.

The overall message of this research is that the molecular mechanisms involved in transcription are much more complex in eukaryotes than in bacteria. An even more striking contrast between bacterial and eukaryotic gene expression emerged when researchers discovered that eukaryotic genes do not consist of one continuous DNA sequence that codes for a product, as do bacterial genes. Instead, the regions of eukaryotic genes that code for proteins come in pieces that are separated by hundreds or many thousands of intervening DNA bases. Although these intervening bases are part of the gene, they do not code for a product.

The Startling Discovery of Eukaryotic Genes in Pieces

When researchers realized that the protein-coding regions of eukaryotic genes are interrupted by stretches of noncoding DNA, it became clear that the nature of the gene and information processing are different in bacteria and eukaryotes. In eukaryotes, the conversion of information in DNA sequences to mRNA sequences does not occur directly. To make a functional mRNA, eukaryotic cells must dispose of certain sequences inside genes and then combine the separated coding sections into an integrated whole.

What sort of data would provoke such a startling claim? To answer this question, consider work that Phillip Sharp and colleagues carried out in the late 1970s to determine how DNA templates are transcribed. They began one of their experiments by heating DNA molecules enough to break the hydrogen bonds between complementary bases. This treatment separated the two strands. The single-stranded DNA was then incubated with the mRNA encoded by the sequence. The team's idea was to promote base pairing between the mRNA and the single-

TABLE 16.1 Eukaryotic RNA Polymerases

Name of Enzyme	Type of Gene Transcribed
RNA polymerase I (RNA pol I)	Genes that code for most of the large RNA molecules (rRNAs) found in ribosomes (see Section 16.3)
RNA polymerase II (RNA pol II)	Protein-coding genes (produce mRNAs)
RNA polymerase III (RNA pol III)	Genes that code for transfer RNAs (see Section 16.4), and genes that code for one of the small RNA molecules (rRNAs) found in ribosomes
RNA pol II and RNA pol III	RNA molecules found in snRNPs (see Section 16.2)

BOX 16.1 Toxins and Transcription

Although wild mushrooms can be delicious, dining on certain species may be lethal. For example, in late summer and fall the reproductive structures of the death cap mushroom (*Amanita phalloides*) appear in forests, fields, and backyards in North America and Europe (**Figure 16.4**). The death cap is among

FIGURE 16.4 The Death Cap Mushroom These mushrooms contain the toxin α-amanitin, which at low concentrations poisons the eukaryotic enzyme RNA polymerase II. QUESTION State a hypothesis to explain why the presence of α-amanitin increases the fitness of death cap mushrooms.

the most poisonous of all mushrooms. Worldwide, it is thought to account for 90 percent of mushroom poisoning deaths. Eating a single death cap can kill a healthy adult.

The active agent in death caps is a toxin called α-amanitin. By happenstance, researchers who were investigating α-amanitin's mode of action made a crucial discovery about transcription in eukaryotes. The researchers observed that mRNA synthesis stopped when eukaryotic cells were treated with low concentrations of α-amanitin. The production of other types of RNAs—specifically, transfer RNAs (tRNAs) and ribosomal RNAs (rRNAs)—continued, however. At very high concentrations of the toxin, tRNA production also stopped but transcription of rRNA genes continued. These observations eventually led to the discovery that eukaryotes have three RNA polymerases, each with a distinct role in the transcription of mRNA, tRNA, and rRNA. Low concentrations of α-amanitin inhibit

RNA polymerase II, which synthesizes mRNAs. As a result, low concentrations of the toxin halt protein synthesis. High concentrations of this poison block RNA polymerase III, which synthesizes tRNAs. RNA polymerase I is not affected by α-amanitin at any concentration.

Toxins such as α-amanitin are useful in research on transcription and translation for the same reason that knock-out mutations, which completely disable genes, are useful in genetics. Toxins allow researchers to understand what happens when a molecule or a particular stage in a process does *not* work. This is similar, in concept, to learning how a car works by removing a part or a system and carefully recording the machine's response; or to learning how the brain works by studying the behavior of people whose brains have been damaged in specific, quantifiable ways by accidents. Toxins and knock-out mutations provide a molecular dissecting kit. They let biologists explore the intricacies of cell function by inactivating key components.

stranded DNA. Under these conditions, the research team expected the mRNA to form base pairs with the DNA sequences that act as the template for its synthesis. When the researchers examined the DNA-RNA hybrid molecules with the electron microscope, however, they observed the structure shown in

Figure 16.5a. Instead of matching up exactly, parts of the DNA formed loops. As **Figure 16.5b** shows, Sharp and his co-workers interpreted these loops as stretches of nucleotides that are present in the DNA template strand but are *not* in the corresponding mRNA.

(a) Micrograph of DNA-RNA hybrid

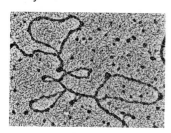

(b) Interpretation of micrograph

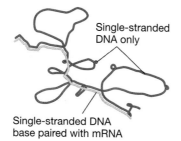

Single-stranded DNA only

Single-stranded DNA base paired with mRNA

(c) Genes and RNA transcripts differ in length.

Intron Exon

Size of gene (DNA)

Size of RNA transcript

FIGURE 16.5 The Discovery of Noncoding Regions of DNA **(a)** Electron micrograph of single-stranded DNA that is bonded via complementary base pairing to the mRNA it encodes. **(b)** Researchers' interpretation of the photograph in part (a). The loops represent regions of DNA that do not have an equivalent sequence in the mRNA. These intervening regions are "extra" DNA compared with the sequences in the mRNA. **(c)** Because eukaryotic genes contain both introns and exons, these genes are much larger than the corresponding mature mRNA.

To make sense of these and other findings, Sharp's group and a team headed by Richard Roberts proposed that there is *not* a one-to-one correspondence between the nucleotide sequence of a eukaryotic gene and its mRNA. By analogy, eukaryotic genes do not carry messages such as "Biology is my favorite course of all time." Instead, eukaryotic genes carry messages that read something like "Biolτηεπροτεινχοδινγρεγιονσοφγενεσογy is my favor αρειντερρυπτεδβψνονχοδινγΔite course ofανδηαϖετοβεσπλιχεδτογετηερall time." The sections of noncoding sequence must be removed from the mRNA before it can carry an intelligible message to the translation machinery.

When it became clear that the genes-in-pieces hypothesis was correct, Walter Gilbert suggested that regions of eukaryotic genes that are part of the final mRNA be referred to as **exons** (because they are *ex*pressed) and the untranslated stretches as **introns** (because they are *int*ervening). Introns are sections of genes that are not represented in the final mRNA product. As a result, eukaryotic genes are much larger than their corresponding mature RNA transcripts (**Figure 16.5c,** page 343).

Exons, Introns, and RNA Splicing

The transcription of eukaryotic genes by RNA polymerase generates a **primary RNA transcript** that contains both the exon and intron regions (**Figure 16.6a**). As transcription proceeds, the introns are removed from the growing mRNA strand by a process known as **splicing**. In this phase of information processing, pieces of the primary transcript are removed and the remaining segments are joined together. Splicing occurs while transcription is still under way and results in an mRNA that contains an uninterrupted genetic message.

Figure 16.6b provides more detail about how introns are removed from genes. In the case illustrated here, splicing is catalyzed by a complex of proteins and small RNAs known as small nuclear ribonucleoproteins, or **snRNPs** (pronounced "snurps"). The process begins when snRNPs bind to specific sequences on the primary RNA transcript that define the exon-intron boundaries. Once the initial snRNPs are in place, other snRNPs arrive to form a multipart complex called a **spliceosome**. Recent data have shown that the spliceosomes found in human cells contain about 145 different proteins and RNAs, making them the most complex molecular machines known. Once the spliceosome forms, the intron forms a loop with an adenine ribonucleotide at its base (step 1, Figure 16.6b). The loop breaks when the 2′ hydroxyl group on the adenine attacks a phosphodiester bond at the 5′ end of the intron.

Once the initial cut in the RNA is made, the free 5′ end of the intron becomes attached to the adenine nucleotide and forms a "lariat" of RNA (step 2, Figure 16.6b). Then the free 3′ end of the first exon reacts with the 5′ end of the second exon. This reaction

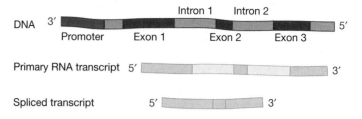

(a) Noncoding regions must be removed from RNA transcripts.

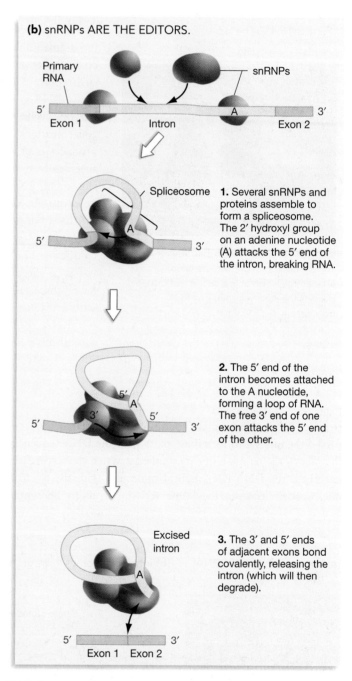

(b) snRNPs ARE THE EDITORS.

1. Several snRNPs and proteins assemble to form a spliceosome. The 2′ hydroxyl group on an adenine nucleotide (A) attacks the 5′ end of the intron, breaking RNA.

2. The 5′ end of the intron becomes attached to the A nucleotide, forming a loop of RNA. The free 3′ end of one exon attacks the 5′ end of the other.

3. The 3′ and 5′ ends of adjacent exons bond covalently, releasing the intron (which will then degrade).

FIGURE 16.6 Introns Are Spliced Out of the Original mRNA
(a) RNA polymerase produces a primary RNA transcript that contains exons and introns. Subsequent processing generates a mature transcript. **(b)** snRNPs are responsible for the splicing reactions that take place in the nucleus of eukaryotes.

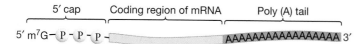

FIGURE 16.7 In Eukaryotes, mRNAs Are Given a Cap and a Tail
Eukaryotic mRNAs have a cap consisting of a molecule called 7-methylguanylate (symbolized as m^7G) bonded to three phosphate groups; the tail is made up of a long series of adenine residues.

breaks the 3' end of the intron and covalently joins the two exons into a contiguous coding sequence (step 3, Figure 16.6b). Splicing is now complete. In most cases, the excised intron is degraded to ribonucleotide monophosphates.

Current data suggest that both the cutting and rejoining reactions that occur during splicing are catalyzed by RNA molecules in the spliceosome. Recall from Chapter 4 that RNA molecules with catalytic ability are called ribozymes. In eukaryotes, ribozymes play a key role in the production of proteins.

Other Aspects of Transcript Processing: Caps and Tails

Intron splicing is not the only way that primary mRNA transcripts are processed in eukaryotes. As soon as the 5' end of a eukaryotic mRNA emerges from RNA polymerase, enzymes add a structure called the **5' cap** (**Figure 16.7**). The cap consists of the molecule 7-methylguanylate and three phosphate groups. When transcription is complete, an enzyme cleaves the 3' end of most mRNAs, and other enzymes add a long tract of 100–250 adenine nucleotides. This sequence is known as the **poly (A) tail.**

Not long after the caps and tails on eukaryotic mRNAs were described, evidence began to accumulate that they protect mRNAs from degradation by ribonucleases and that they enhance the efficiency of translation. For example, experimental mRNAs that have a cap and a tail last longer when they are introduced into cells than do experimental mRNAs that lack a cap, a tail, or both a cap and a tail. The mRNAs with caps and tails also produce more proteins than do mRNAs without caps and tails.

Recent work has shown why the cap and tail are so critical to mRNAs. The 5' cap serves as a recognition signal for the translation machinery, and the poly (A) tail extends the life span of an mRNA by protecting the message from degradation by ribonucleases in the cytoplasm.

✓ CHECK YOUR UNDERSTANDING

Transcription occurs when RNA polymerase catalyzes the $5' \rightarrow 3'$ synthesis of an RNA molecule. RNA synthesis is driven by the potential energy stored in nucleotide triphosphates and is based on matching complementary base pairs to the sequence in a template DNA strand. **Table 16.2** summarizes some of the similarities and differences observed in transcription in bacteria versus eukaryotes. You should be able to diagram (1) how transcription initiation, elongation, and termination occur in bacteria and eukaryotes and (2) how mRNAs are processed in eukaryotes.

16.3 An Introduction to Translation

In translation, the sequence of bases in a messenger RNA molecule is converted into a sequence of amino acids in a polypeptide. The genetic code specifies the relationship between the bases of a triplet codon in mRNA and the amino acid it codes for (see Chapter 15). But how are the amino acids assembled into a polypeptide according to the information in messenger RNA?

Studying how translation occurred in cell-free systems proved to be an extremely productive approach to answering this question. Once in vitro translation systems had been developed from human cells, *E. coli*, and a variety of other organisms, it became clear that the basic mechanisms of translation are fundamentally the same throughout the tree of life. The sequence of events that occurs during protein synthesis is similar in bacteria and eukaryotes.

TABLE 16.2 Comparing Transcription in Bacteria and Eukaryotes

Aspect	Bacteria	Eukaryotes
RNA polymerase	One	Three; each produces a different class of RNA
Promoter structure	Typically contains a −35 box and a −10 box	Complex and variable; often includes a TATA box −30 from start of gene
Protein(s) involved in contacting promoter	Sigma; different versions of sigma bind to different promoters	Many basal transcription factors
RNA processing steps	None; translation occurs while transcription is still under way	Extensive; several processing steps occur in nucleus before RNA is exported to cytoplasm for translation:
		1. Enzyme-catalyzed addition of 5' cap
		2. Splicing (intron removal) by spliceosome
		3. Enzyme-catalyzed addition of 3' poly (A) tail

(a) In bacteria, transcription and translation are tightly coupled.

Ribosome translates mRNA as it is being synthesized by RNA polymerase

5′ end of mRNA

Protein

Ribosome

RNA polymerase

Start of gene (3′ end of template strand)

End of gene (5′ end of template strand)

(b) In eukaryotes, transcription and translation are separated in space and time.

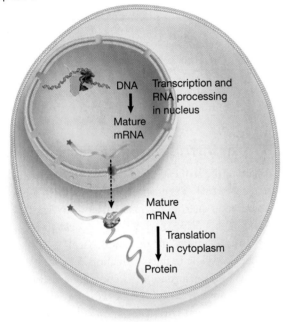

DNA   Transcription and RNA processing in nucleus

Mature mRNA

Mature mRNA

Translation in cytoplasm

Protein

FIGURE 16.8 Transcription and Translation Occur Simultaneously in Bacteria but Not in Eukaryotes
(a) In bacteria, ribosomes attach to mRNA transcripts and begin translation while RNA polymerase is still transcribing the DNA template strand. **(b)** In eukaryotes, mRNA transcripts are processed inside the nucleus before they are exported to the cytoplasm. There, they are translated by ribosomes. **EXERCISE** In bacteria, many ribosomes attach to an mRNA transcript at the same time—so many copies of the protein product are made simultaneously. Add two ribosomes to the longest mRNA transcript in part (a), along with the protein each is producing.

Ribosomes Are the Site of Protein Synthesis

The first question that biologists answered about translation concerned where it occurs. The answer was inspired by a simple observation: There is a strong, positive correlation between the number of small cellular structures known as **ribosomes** in a given type of cell and the rate at which that cell synthesizes proteins. For example, immature human red blood cells divide rapidly, synthesize millions of copies of the protein hemoglobin, and contain large numbers of ribosomes. In contrast, the same cells at maturity have low rates of protein synthesis and very few ribosomes. Based on this correlation, investigators proposed that ribosomes are the site of protein synthesis in the cell.

To test this hypothesis, Roy Britten and collaborators did a pulse-chase experiment similar in design to experiments introduced in Chapter 7. Recall that the goal of a pulse-chase experiment is to label a population of molecules as they are being produced. The location of the tagged molecules is then followed through time. The tagging is done by supplying a pulse

of radioactive atoms that are incorporated into the molecule as it is being synthesized, followed by a chase of unlabeled atoms. Britten's group wanted to label amino acids with radioactive atoms and then track the molecules as translation occurred.

The biologists began by feeding a pulse of radioactive sulfate ($^{35}SO_4^{-2}$) to growing cultures of *E. coli*. They expected the cells to incorporate the radioactive sulfur into the amino acids methionine and cysteine—which contain sulfur (see Chapter 3)—and then into newly synthesized proteins. Fifteen seconds after adding the radioactive sulfate, the researchers "chased" the label by adding a large excess of nonradioactive sulfate to the culture medium. The chase vastly reduced the incorporation of radioactive sulfur into proteins. If the ribosome hypothesis was correct, then the radioactive signal should be associated with ribosomes for a short period of time. Later, all of the radioactivity should be found in proteins independently of the ribosomes.

This is exactly what the researchers found. Soon after the pulse of labeled sulfate ended, the radioactive atoms were found in free amino acids or in ribosomes. Later, all of the radioactive atoms were found on completed proteins. Based on these data, biologists concluded that proteins are synthesized at ribosomes and then released.

About a decade after the ribosome hypothesis was confirmed, electron micrographs like this chapter's opening photograph showed bacterial ribosomes in action. The images confirmed that, in bacteria, ribosomes attach to mRNAs and begin synthesizing proteins even before transcription is complete (**Figure 16.8a**). Transcription and translation can occur concurrently in bacteria because there is no nuclear envelope to separate the two processes. In bacteria, transcription and translation are physically connected. But in eukaryotes, mRNAs are spliced in the nucleus and then exported to the cytoplasm. Once RNA messages are outside the nucleus, ribosomes attach to them and begin translation (**Figure 16.8b**). In eukaryotes, transcription and translation are separated in time and space.

How Does an mRNA Triplet Specify an Amino Acid?

During translation, genetic information is converted from nucleic acids to proteins. In effect, the hereditary instructions are translated from one distinctive chemical language into another. The discovery of the genetic code revealed that each triplet codon in mRNA specifies a particular amino acid in a protein. How does this conversion happen?

One early hypothesis was that mRNA codons and amino acids interact directly. The proposal was that the bases in a particular codon were complementary in shape or charge to the side group of a particular amino acid (**Figure 16.9a**). But Francis Crick pointed out that the chemistry involved didn't make

sense. For example, how could the nucleic acid bases interact with a hydrophobic amino acid side group, which does not form hydrogen bonds?

Crick proposed an alternative hypothesis. As **Figure 16.9b** shows, he suggested that some sort of adapter molecule holds amino acids in place while interacting directly and specifically with a codon in mRNA via hydrogen bonding. In essence, Crick predicted the existence of a chemical go-between that produced a physical connection between the two types of molecules. Crick, as it turns out, was right.

16.4   The Role of Transfer RNA

Crick's adapter molecule was discovered by accident. Biologists were trying to work out a cell-free protein synthesis system derived from mammalian liver cells and had discovered that ribosomes, mRNA, amino acids, ATP, and a molecule called guanosine triphosphate, or GTP, had to be present for translation to occur. (GTP is similar to ATP but contains guanosine instead of adenosine.) These results were logical—ribosomes provide the catalytic machinery, mRNAs contribute the message to be translated, amino acids are the building blocks of proteins, and ATP and GTP supply potential energy to drive the endergonic polymerization reactions responsible for forming proteins. But in addition, a cellular fraction that contained a previously unknown type of RNA turned out to be indispensable. If this type of RNA is missing, protein synthesis does not occur. What is this mysterious RNA, and why is it essential to translation?

The novel class of RNAs eventually became known as **transfer RNA (tRNA)**. The role of tRNA in translation was a mystery until some researchers happened to add a radioactive amino acid—leucine—to an in vitro protein synthesis system. The treatment was actually done as a control for an unrelated experiment. To the researchers' amazement, some of the radioactive leucine attached to tRNA molecules. Follow-up experiments showed that the attachment of an amino acid to a tRNA requires an input of energy in the form of ATP. A tRNA molecule that becomes covalently linked to an amino acid, such as any of the molecules observed in the lucky experiment, is called an **aminoacyl tRNA**. More recent research has shown that enzymes called **aminoacyl tRNA synthetases** are responsible for catalyzing the addition of amino acids to tRNAs. For each of the 20 major amino acids, there is a different aminoacyl synthetase and one or more tRNAs. **Figure 16.10**, page 348, summarizes how amino acids are transferred from aminoacyl tRNA synthetases to tRNAs.

What happens to the amino acid that is carried by an aminoacyl tRNA? Biologists who tracked the fate, over time, of radioactive leucine molecules that were attached to tRNAs discovered that the amino acids are transferred from aminoacyl tRNAs to proteins. The data that support this conclusion are

(a) Hypothesis 1: Amino acids interact directly with mRNA codons.

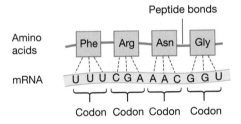

(b) Hypothesis 2: Adapter molecules hold amino acids and interact with mRNA codons.

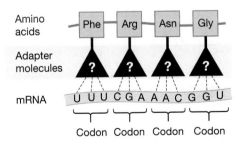

FIGURE 16.9 Two Hypotheses for How mRNA Codons Interact with Amino Acids

(a) An early hypothesis suggested that amino acids and mRNA codons interact directly. **(b)** The adapter-molecule hypothesis predicts that mRNA codons do not interact directly with the amino acids being added to a growing polypeptide. Instead, an intermediary adapter molecule carries an amino acid and aligns it with the appropriate mRNA codon.

HOW AMINO ACIDS ARE LOADED ONTO tRNAs

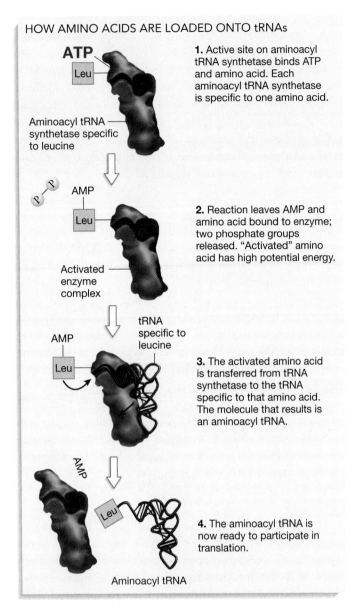

1. Active site on aminoacyl tRNA synthetase binds ATP and amino acid. Each aminoacyl tRNA synthetase is specific to one amino acid.

2. Reaction leaves AMP and amino acid bound to enzyme; two phosphate groups released. "Activated" amino acid has high potential energy.

3. The activated amino acid is transferred from tRNA synthetase to the tRNA specific to that amino acid. The molecule that results is an aminoacyl tRNA.

4. The aminoacyl tRNA is now ready to participate in translation.

FIGURE 16.10 Aminoacyl tRNA Synthetases Load Amino Acids onto the Appropriate tRNA
An aminoacyl tRNA synthetase—in this case, for leucine—becomes activated in a reaction that requires ATP. Activated aminoacyl tRNA synthetases transfer an amino acid to a tRNA, producing an aminoacyl tRNA.

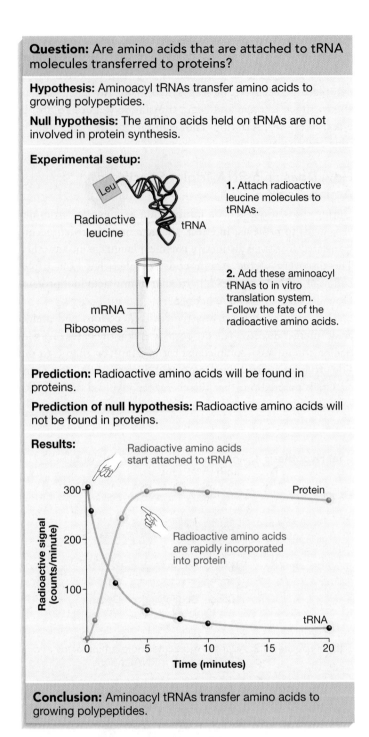

Question: Are amino acids that are attached to tRNA molecules transferred to proteins?

Hypothesis: Aminoacyl tRNAs transfer amino acids to growing polypeptides.

Null hypothesis: The amino acids held on tRNAs are not involved in protein synthesis.

Experimental setup:

1. Attach radioactive leucine molecules to tRNAs.

2. Add these aminoacyl tRNAs to in vitro translation system. Follow the fate of the radioactive amino acids.

Prediction: Radioactive amino acids will be found in proteins.

Prediction of null hypothesis: Radioactive amino acids will not be found in proteins.

Results:

Conclusion: Aminoacyl tRNAs transfer amino acids to growing polypeptides.

FIGURE 16.11 Experimental Evidence That Amino Acids Are Transferred from tRNAs to Proteins

shown in **Figure 16.11**. The graph's message is that radioactive amino acids are lost from tRNAs and incorporated into polypeptides synthesized in ribosomes. These results inspired the use of *transfer* in tRNA's name. The experiment also confirmed that aminoacyl tRNAs act as the interpreter in the translation process. tRNAs are Crick's adapter molecules.

Connecting Structure with Function

Transfer RNAs serve as the chemical go-betweens that allow amino acids to interact with an mRNA template. But precisely how does this connection occur?

This question was answered by research on tRNA's molecular structure. The initial studies established the sequence of nucleotides in various tRNAs, or what is termed their primary structure. Transfer RNAs are relatively short, ranging from 75 to 85 nucleotides in length. When biologists studied the primary sequence closely, however, they noticed that certain parts of the molecules can form secondary structures. Specifically, some of the bases in the tRNA molecule can form hydrogen bonds with complementary bases in a different region of the

(a) Secondary structure of tRNA

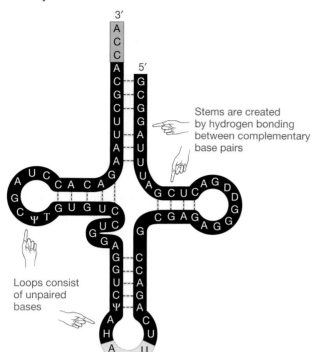

(b) Early model of tRNA function

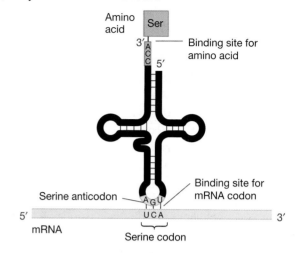

(c) Revised model incorporating tertiary structure of tRNA

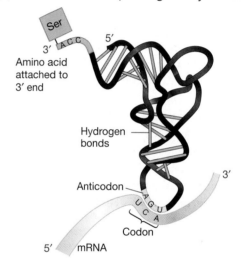

FIGURE 16.12 The Structure of Transfer RNA
(a) The secondary structure of tRNA resembles a cloverleaf.
(b) If amino acids are attached to the 3′ end of a tRNA with an anticodon appropriate for that amino acid, then the tRNA anticodon will form a complementary base pair with the mRNA codon. **(c)** Recent data indicate that tRNAs have an L-shaped tertiary structure.

same molecule. As a result, the entire tRNA molecule could assume the cloverleaf shape illustrated in **Figure 16.12a**. The stems in the cloverleaf are produced by complementary base pairing between different portions of the molecule, while the loops consist of single-stranded RNA.

Two aspects of this secondary structure proved especially interesting. The CCA sequence at the 3′ end of the molecule offered a binding site for amino acids, while the triplet on a loop on the opposite end of the cloverleaf could serve as an anticodon. An **anticodon** is a set of three nucleotides that forms base pairs with the mRNA codon. **Figure 16.12b** depicts an early model of how the anticodon of a tRNA connects an mRNA codon with the appropriate amino acid.

This model had to be modified, however, when X-ray crystallography studies revealed the tertiary structure of tRNAs. Recall from Chapter 3 that the tertiary structure of a molecule is defined by the three-dimensional arrangement of its atoms and is usually a product of folding. According to the X-ray crystallography data, the cloverleaf structure folds over to produce an L-shaped molecule (**Figure 16.12c**). All of

the tRNAs in a cell have the same L-shaped structure, but each has a distinct anticodon and attached amino acid. The tertiary structure of tRNAs is important because it results in a precise separation between the anticodon and the attached amino acid. As we'll see, this separation is key to the positioning of the amino acid and the anticodon in the ribosome.

How Many tRNAs Are There?

When research succeeded in characterizing all of the different types of tRNAs available in cells, a paradox arose. According to the genetic code introduced in Chapter 15, the 20 amino acids found in proteins are specified by 61 different mRNA codons. Instead of containing 61 different tRNAs with 61 different anticodons, though, most cells contain only about 40. How can all 61 mRNA codons in the genome be translated with only two-thirds of the tRNAs required?

To resolve this paradox, Francis Crick proposed what is known as the **wobble hypothesis**. To understand his logic, recall from Chapter 15 that many amino acids are specified by more than one codon. Further, recollect that codons for the same amino acid tend to have the same nucleotides at the first and second positions but a different nucleotide at the third position. For example, both of the codons CAA and CAG code for the amino acid glutamine. (Codons are always written in the $5' \rightarrow 3'$ direction.) Surprisingly, experimental data have shown that a tRNA with an anticodon of GUU can base pair with both CAA and CAG in mRNA. (Anticodons are always written in the $3' \rightarrow 5'$ direction.) The GUU anticodon matches the first two bases (C and A) in both codons, but the U in the third position forms a nonstandard base pair with a G in the CAG codon.

Crick proposed that inside the ribosome, tRNAs with non-standard base pairing at the third position can still bind successfully to an anticodon. If so, it would allow a limited flexibility, or "wobbling," in the base pairing. According to the wobble hypothesis, a nonstandard base pair in the third position—such as G-U—is acceptable as long as it does not change the amino acid for which the codon codes.

Follow-up research has shown that certain tRNAs are especially prone to pairing with more than one mRNA codon. These tRNAs have a nitrogen-containing base called *inosine* in the third position of their anticodon. (Inosine is derived from adenine by a chemical modification that occurs after the tRNA is produced.) During protein synthesis, an inosine in the anticodon can base pair with a A, U, or C in an mRNA codon. For example, the tRNA with the anticodon GAI carries the amino acid leucine and binds to four of the six codons that specify leucine: CUA, CUC, CUU, and UUA. (The UUA is possible because the G in the anticodon sometimes pairs with U in the mRNA. This interaction between

anticodons and codons is one of the only known exceptions to the standard base-pairing rules.) In this way, "wobble" in the third position of a codon allows just 40 or so tRNAs to bind to all 61 mRNA codons.

16.5   Ribosomes and the Mechanism of Translation

During translation, the sequence of bases in an RNA message is converted to a sequence of amino acids in a polypeptide. The conversion begins when the anticodon of a tRNA that carries an amino acid binds to a codon in mRNA. The conversion is complete when a peptide bond forms between that amino acid and the growing polypeptide chain.

Both of these events occur inside a ribosome. Biologists have known since the 1930s that ribosomes contain a considerable amount of protein along with a great deal of **ribosomal RNA (rRNA)**. When high-speed centrifuges became available, researchers found that ribosomes could also be separated into two major substructures, called the *large subunit* and *small subunit*. Each subunit consists of a complex of RNA molecules and proteins. For example, the large subunit in *E. coli* contains two different rRNA molecules and 34 proteins. More recent work has shown that the small subunit holds the mRNA in place during translation and that peptide bond formation takes place in the large subunit.

The three-dimensional model of the ribosome in **Figure 16.13a** shows how all of the molecules and reactions required for translation fit together. Note that during protein synthesis,

(a) Ribbon model of ribosome during translation

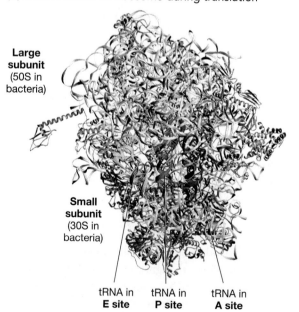

Large subunit (50S in bacteria)

Small subunit (30S in bacteria)

tRNA in **E site**   tRNA in **P site**   tRNA in **A site**

(b) Diagram of ribosome during translation

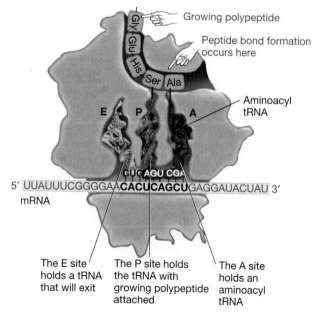

Growing polypeptide

Peptide bond formation occurs here

Aminoacyl tRNA

5′ UUAUUUCGGGGAA**CACUCAGCU**GAGGAUACUAU 3′
mRNA

The E site holds a tRNA that will exit

The P site holds the tRNA with growing polypeptide attached

The A site holds an aminoacyl tRNA

FIGURE 16.13 The Structure of the Ribosome
(a) During translation, three tRNAs line up side by side inside the ribosome. **(b)** Different events occur at each of the three sites inside the ribosome where tRNAs are found. **EXERCISE** Circle and label the active site in the ribosome.

three distinct tRNAs are lined up inside the ribosome. All three are bound to their corresponding mRNA codon at the base of the structure. The tRNA that is on the right (colored red) in the figure carries an amino acid. This tRNA's position in the ribosome is called the **A site**, for acceptor or aminoacyl. The tRNA that is in the middle (green) holds the growing polypeptide chain and occupies the **P site**, for peptidyl, inside the ribosome. (Think of "P" for "peptide-bond formation.") The left-hand (yellow) tRNA no longer has an amino acid attached and is about to leave the ribosome. It occupies the ribosome's **E site**, for exit. **Figure 16.13b** summarizes what happens to each tRNA inside the ribosome. Because all tRNAs have the same secondary and tertiary structure, they all fit equally well in the A, P, and E sites.

The ribosome is a molecular machine that synthesizes proteins in a three-step sequence:

1. An aminoacyl tRNA diffuses into the A site; its anticodon binds to a codon in mRNA.

2. A peptide bond forms between the amino acid held by the aminoacyl tRNA in the A site and the existing polypeptide, which is held by a tRNA in the P site.

3. The ribosome moves ahead, and all three tRNAs move one position down the line. The tRNA in the E site exits; the tRNA in the P site moves to the E site; and the tRNA in the A site switches to the P site.

The protein that is being synthesized grows by one amino acid each time this three-step sequence repeats. The process occurs up to 20 times per second in bacterial ribosomes and about 2 times per second in eukaryotic ribosomes.

This introduction to how tRNAs, mRNAs, and ribosomes interact during protein synthesis leaves a number of key questions unanswered, however. How do mRNAs and ribosomes get together to start the process? Once protein synthesis is under way, how is peptide bond formation catalyzed inside the ribosome? And how does protein synthesis conclude when the ribosome reaches the end of the message? Let's consider each question in turn.

Initiation

To translate an mRNA, a ribosome must begin at a specific point in the message, translate the mRNA up to the message's termination codon, and then stop. Biologists call these three phases of protein synthesis **initiation**, **elongation**, and **termination**, respectively. One key to understanding initiation is to recall, from Chapter 15, that the codon AUG is found near the 5′ end of all mRNAs and codes for the amino acid methionine. The presence of this start codon is an aspect of initiation that is common to both bacteria and eukaryotes.

Figure 16.14 shows how translation gets under way in bacteria. The process begins when a section of rRNA in the small subunit of a ribosome binds to a complementary sequence on an mRNA. The mRNA region is called the **ribosome binding site** or **Shine-Dalgarno sequence**, after the biologists who discovered it. The site is about six nucleotides upstream from the AUG start codon. It consists of all or part of the bases 5′ AGGAGGU 3′. The complementary sequence in the rRNA of the small subunit reads 3′ UCCUCCA 5′. The interaction between the small subunit and the message is mediated by proteins called **initiation factors**. In eukaryotes, initiation factors bind to the 5′ cap on mRNAs and guide it to the ribosome.

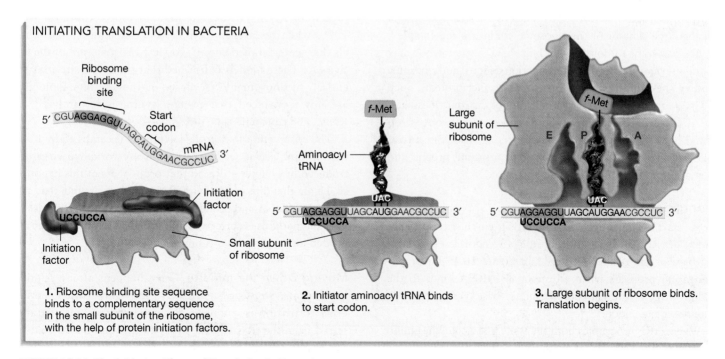

INITIATING TRANSLATION IN BACTERIA

1. Ribosome binding site sequence binds to a complementary sequence in the small subunit of the ribosome, with the help of protein initiation factors.

2. Initiator aminoacyl tRNA binds to start codon.

3. Large subunit of ribosome binds. Translation begins.

FIGURE 16.14 The Initiation Phase of Translation in Bacteria

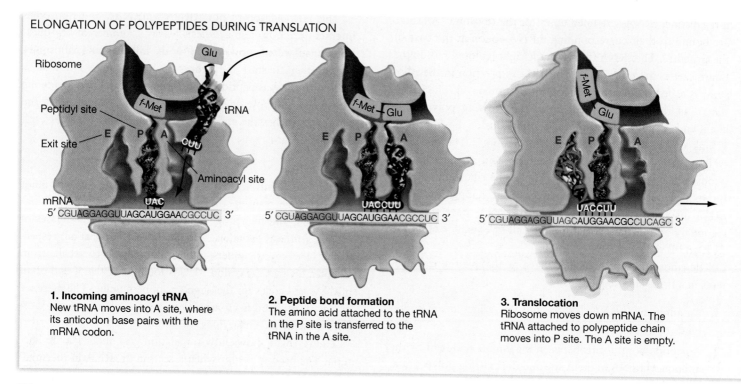

ELONGATION OF POLYPEPTIDES DURING TRANSLATION

1. Incoming aminoacyl tRNA
New tRNA moves into A site, where its anticodon base pairs with the mRNA codon.

2. Peptide bond formation
The amino acid attached to the tRNA in the P site is transferred to the tRNA in the A site.

3. Translocation
Ribosome moves down mRNA. The tRNA attached to polypeptide chain moves into P site. The A site is empty.

FIGURE 16.15 The Elongation Phase of Translation

Once the Shine-Dalgarno sequence has attached to the small ribosomal subunit, an aminoacyl tRNA bearing a modified form of methionine called *N*-formylmethionine (abbreviated *f*-met) binds to the AUG start codon. Initiation is complete when the large subunit joins the complex. When the ribosome is completely assembled, the tRNA bearing *f*-met occupies the P site.

To summarize, initiation is a three-step process in bacteria: (1) The mRNA binds to the small subunit of the ribosome, (2) the initiator aminoacyl tRNA binds to the start codon, and (3) the large subunit of the ribosome completes the complex.

It's important to note that in bacteria a single mRNA may contain several start codons, termination codons, and ribosome binding sites and thus code for several distinct proteins. Such mRNAs are said to be **polycistronic**. In polycistronic messages, an AUG start sequence near the 5′ end of the mRNA is followed by a termination signal farther along, which in turn is followed by an AUG start sequence for a second protein and then a second termination signal, and so on.

Elongation

At the start of elongation, the E and A sites in the ribosome are empty of tRNAs. As a result, an mRNA codon is exposed at the base of the A site. As step 1 in **Figure 16.15** illustrates, elongation proceeds when an aminoacyl tRNA binds to the codon in the A site via complementary base pairing between anticodon and codon.

When a tRNA occupies both the P site and A site, the amino acids they hold are placed in the ribosome's active site. This is where peptide bond formation—the essence of protein

synthesis—occurs. Peptide bond formation is considered one of the most important reactions that takes place in cells, because manufacturing proteins is among the most fundamental of all cell processes. The question is, how does it happen?

Is the Ribosome an Enzyme or a Ribozyme?   Because ribosomes contain both proteins and RNA, researchers had argued for decades over whether the active site consisted of protein or RNA. The debate was not resolved until the year 2000, when researchers completed three-dimensional models that were detailed enough to view the structure of the active site. These models confirmed that the active site consists entirely of ribosomal RNA. Based on these results, biologists are now convinced that protein synthesis is catalyzed by RNA. The ribosome is a ribozyme—not an enzyme.

The observation that protein synthesis is catalyzed by RNA is important, because it supports the RNA world hypothesis introduced in Chapter 4. Recall that proponents of this hypothesis claim that life began with RNA molecules and that the presence of DNA and proteins in cells evolved later. If the RNA world hypothesis is correct, then it is logical to observe that the production of proteins is catalyzed by RNA.

Moving Down the mRNA   What happens after a peptide bond forms? Step 2 in Figure 16.15 shows that when peptide bond formation is complete, the polypeptide chain is transferred from the tRNA in the P site to the amino acid held by the tRNA in the A site. Step 3 shows the process called **translocation**, which occurs when the ribosome moves down

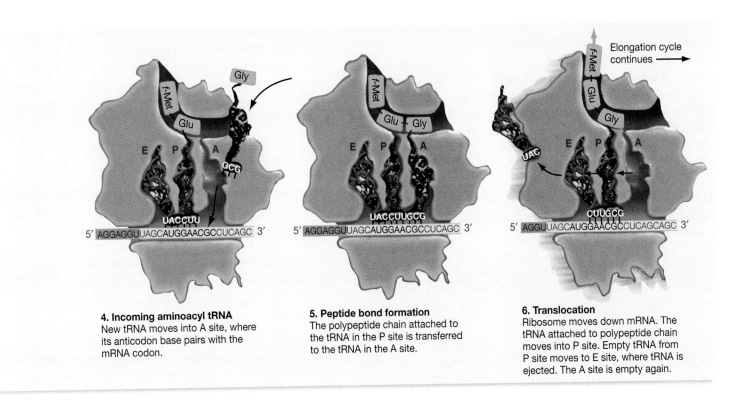

4. Incoming aminoacyl tRNA
New tRNA moves into A site, where its anticodon base pairs with the mRNA codon.

5. Peptide bond formation
The polypeptide chain attached to the tRNA in the P site is transferred to the tRNA in the A site.

6. Translocation
Ribosome moves down mRNA. The tRNA attached to polypeptide chain moves into P site. Empty tRNA from P site moves to E site, where tRNA is ejected. The A site is empty again.

the mRNA in the $5' \rightarrow 3'$ direction. Translocation does several things: It moves the empty tRNA into the E site; it moves the tRNA containing the growing polypeptide into the P site; and it opens the A site and exposes a new mRNA codon. If the E site is occupied when translocation occurs, the tRNA there is ejected into the cytoplasm.

The three steps in elongation—(1) arrival of aminoacyl tRNA, (2) peptide bond formation, and (3) translocation—repeat down the length of the mRNA. By adding specific molecules to in vitro translation systems, researchers have confirmed that each elongation cycle depends on an input of energy from several GTP molecules as well as assistance from proteins called **elongation factors.** Further, recent three-dimensional models of ribosomes in various stages of the translation sequence show that the machine as a whole is highly dynamic during the process. The ribosome constantly changes shape as tRNAs come and go and catalysis and translocation occur. Its structure changes in conjunction with its multipart function: coordinating interactions between tRNAs and an mRNA, and catalyzing peptide bond formation.

Electron microscopy has also confirmed that once elongation is under way, additional ribosomes bind to the start site in the same message and initiate translation. As this process continues, strings of ribosomes called **polyribosomes** assemble along an mRNA. Each ribosome synthesizes a copy of the protein (**Figure 16.16**). Even though it may take a minute or more for a single ribosome to produce a large polypeptide, the presence of polyribosomes can quicken the overall pace of protein production. Polyribosomes are observed routinely in both bacteria and eukaryotes.

Termination

How does protein synthesis end? To answer this question, recall from Chapter 15 that the genetic code includes three stop codons: UAA, UAG, and UGA. In most cells, no aminoacyl tRNA has an anticodon that binds to these sequences. When

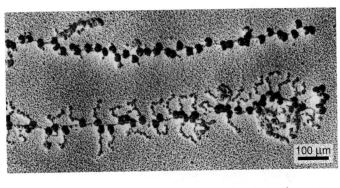

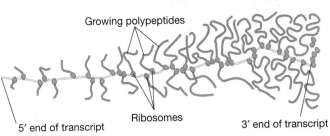

FIGURE 16.16 Polyribosomes

In both bacteria and eukaryotes, many ribosomes synthesize proteins from the same mRNA at the same time. The resulting structures are called polyribosomes.

translocation opens the A site and exposes one of the stop codons, a protein called a **release factor** fills the A site (**Figure 16.17**). Release factors are proteins that fit tightly into the A site because their size, shape, and electrical charge are extremely similar to the same properties in a tRNA. Release factors do not carry an amino acid, however. When a release factor occupies the A site, the active site catalyzes the hydrolysis of the bond linking the tRNA in the P site with the polypeptide chain. This reaction frees the polypeptide, and it is released from the ribosome. The ribosome then separates from the mRNA, and the two ribosomal subunits dissociate. The subunits are ready to attach to the start codon of another message and start translation anew.

Post-Translational Modifications Proteins are not fully formed and functional when termination occurs. From earlier chapters, it should be clear that most proteins go through an extensive series of processing steps, collectively called **post-translational modification**, before they are ready to go to work in a cell:

- Recall from Chapter 3 that a protein's function depends on its shape and that a protein's shape depends on how it folds. Folding actually begins during elongation, long before termination occurs and ribosomes disassemble. Although it occurs spontaneously, in the sense that no energy input is required, folding is frequently speeded up by proteins called **molecular chaperones**. Recent data have shown that, in some bacteria, chaperone proteins actually bind to the ribosome near the "tunnel" where the growing polypeptide emerges from the ribosome. This result suggests that folding occurs as the polypeptide is emerging from the ribosome.

- Chapter 7 pointed out that many eukaryotic proteins are extensively modified after they are synthesized. For example, small chemical groups may be added to proteins in the organelles called the rough endoplasmic reticulum and the Golgi apparatus. Some proteins receive a sorting signal that serves as an address and ensures that the molecule will be carried to the correct location in the cell. Certain proteins are also augmented with sugar or lipid groups that are critical for normal functioning.

- Many proteins are altered by enzymes that add or remove a phosphate group. Phosphorylation (addition of phosphate) and dephosphorylation (removal of phosphate) of proteins were introduced in Chapter 3 and Chapter 8. Recall that because a phosphate group has two negative charges, adding or removing a phosphate group causes major changes in the shape and chemical reactivity of proteins. These changes have a dramatic effect on the protein's activity—often switching it from an inactive state to an active state or vice versa.

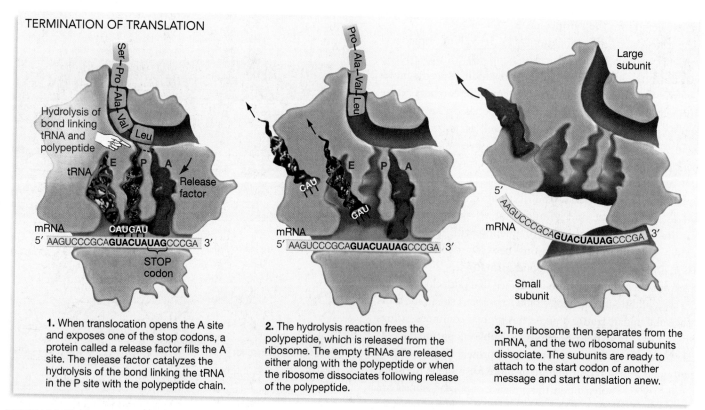

TERMINATION OF TRANSLATION

1. When translocation opens the A site and exposes one of the stop codons, a protein called a release factor fills the A site. The release factor catalyzes the hydrolysis of the bond linking the tRNA in the P site with the polypeptide chain.

2. The hydrolysis reaction frees the polypeptide, which is released from the ribosome. The empty tRNAs are released either along with the polypeptide or when the ribosome dissociates following release of the polypeptide.

3. The ribosome then separates from the mRNA, and the two ribosomal subunits dissociate. The subunits are ready to attach to the start codon of another message and start translation anew.

FIGURE 16.17 The Termination Phase of Translation

The overall message is that the manufacture of a completely functional protein depends not only on translation in the ribosome but also on a wide array of other molecules and events that take place throughout the cell.

✓ CHECK YOUR UNDERSTANDING

In bacteria, translation begins when the ribosome binding site on an mRNA binds to an rRNA sequence in the small subunit of the ribosome. The initiator aminoacyl tRNA binds to the start codon in the mRNA, and the large subunit of the ribosome attaches to the small subunit to complete the ribosome.

In both bacteria and eukaryotes, polypeptides elongate when the appropriate aminoacyl tRNA enters the A site. A peptide bond forms between the amino acid held by that tRNA in the A site and the polypeptide held by the tRNA in the P site. The ribosome then moves down the mRNA one codon, and the process occurs again. Translation ends when the ribosome reaches a stop codon. You should be able to (1) diagram a ribosome during translation initiation, elongation, and termination; (2) label each major structure in the diagrams; and (3) describe the function of each major structure.

16.6   The Molecular Basis of Mutation

This chapter has focused on how the information in DNA is transcribed to RNA and then translated into a functioning protein.

Although biologists continue to work on issues such as how basal transcription factors in eukaryotes interact with RNA polymerase and how ribosomes change shape during translation, the molecular mechanisms responsible for information flow in the cell, from DNA to RNA to proteins, are now considered fairly well understood.

Before we go on to consider how transcription and translation are regulated in bacteria and eukaryotes in the next two chapters, it will be helpful to tie together some major points covered in the last three chapters by analyzing the molecular basis of mutation. A **mutation** is any change in an organism's DNA. It is a change in a cell's information archive—a change in its genotype. What are the consequences of mutation?

The molecular machinery introduced in this chapter faithfully transcribes and translates mutations in DNA sequences. The result is the production of novel types of proteins. In this way, changes in DNA lead to changes in molecules that affect the organism's phenotype.

Figure 16.18 shows how a common type of mutation occurs. If DNA polymerase mistakenly inserts the wrong base as it synthesizes a new strand of DNA, and if proofreading by DNA polymerase and the mismatch repair system fail to correct the mismatched base, a change in the sequence of bases in DNA results. A single base change such as this is called a **point mutation**.

What happens when point mutations are transcribed and translated? To answer this question, consider the first point mutation ever described. This mutation occurs in the human gene for the protein hemoglobin. Recall that hemoglobin is abundant in red blood cells and that it carries oxygen to tissues.

Figure 16.19a, page 356, shows a small section of DNA sequence from the normal gene for hemoglobin, along with the same sequence from the mutant allele found in individuals who suffer from sickle-cell disease. The mutant form has a thymine in place of an adenine at the second position in the sixth codon specified by this gene. A glance back at the genetic code in Chapter 15 confirms why this point mutation is significant: During protein synthesis, the mutant codon specifies valine instead of glutamic acid in the amino acid chain of hemoglobin. Point mutations that cause changes in the amino acid sequence of proteins are called **missense mutations** or **replacement mutations**. In the case of hemoglobin, the single change in primary structure causes the protein to crystallize when oxygen levels in the blood are low. When hemoglobin crystallizes, it causes red blood cells to become sickle shaped (**Figure 16.19b**). The misshapen cells get stuck in blood vessels, so nearby cells are starved for oxygen. The result is intense pain and anemia.

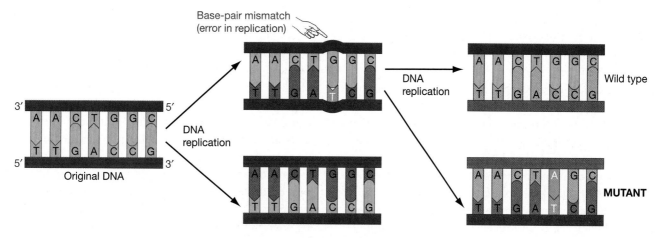

FIGURE 16.18 Unrepaired Mistakes in DNA Synthesis Lead to Point Mutations
QUESTION Why is it logical that the type of mutation illustrated here is termed a point mutation?

(a) DNA point mutation can lead to a different amino acid sequence.

(b) Phenotype

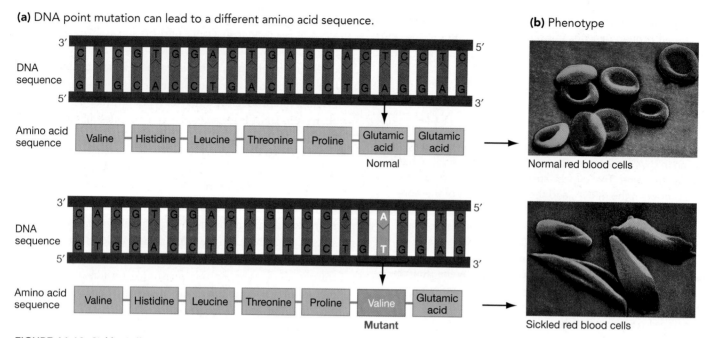

FIGURE 16.19 Sickle-Cell Disease Results from a Point Mutation in the Gene for Hemoglobin
(a) These normal and mutant genes that code for hemoglobin differ by a single nucleotide. The sequences shown are just a small portion of the gene. **(b)** The mutation in part (a) changes the primary sequence of the gene product. The change in amino acid sequence causes hemoglobin molecules to crystallize when oxygen levels in the blood are low. As a result, red blood cells sickle and get stuck in small blood vessels. **EXERCISE** The mRNA codon GUG specifies the valine at the start (left) of this protein. Label the template strand and the non-template strand of DNA, and indicate their 5′ and 3′ ends.

Missense mutations such as that found in the sickling allele can be **deleterious**, meaning they reduce an individual's fitness—the ability to survive and reproduce. But missense mutations may also be beneficial. For example, the sickling allele confers a fitness advantage in regions of the world where the life-threatening disease malaria is common. The advantageous effect is possible because the parasite that causes malaria infects red blood cells. If an individual has one normal hemoglobin allele and one sickling allele, red blood cells that become infected with the parasite tend to sickle while others function normally. The deformed cells are quickly destroyed by the body, eliminating the parasitic cells inside the blood cells. As a result, individuals who have one copy of the mutant gene have lower parasite loads and tend to be much healthier than people who have two normal copies of the gene.

In contrast to having a negative or positive effect on the organism, many point mutations have virtually no effect at all. To understand why, suppose that at the next site in the hemoglobin gene of Figure 16.19 (the third position in the same sixth codon), a thymine had been substituted for the cytosine. This point mutation would have no consequence. Why? Both GAA and GAG specify glutamic acid. In this case, the amino acid sequence of the gene product does not change even though the DNA sequence is altered due to mu-

tation. This type of alteration in the base sequence is called a **silent mutation**. Silent mutations are said to be **neutral** in their effect on an individual's fitness. It is possible for point mutations to be silent and neutral with respect to fitness because of the redundancy in the genetic code.

The take-home message from this analysis of hemoglobin, sickle-cell disease, and malaria is simple: It is possible for point mutations to be deleterious, beneficial, or neutral. **Figure 16.20a** summarizes the types of point mutations that have been documented and reviews their consequences for the amino acid sequences of proteins.

In addition to documenting various types of point mutations, biologists study mutations that involve larger-scale changes in the composition of chromosomes. Chapter 9 introduced the phenomena called polyploidy and aneuploidy, which involve changes in the number of chromosome complements present and the addition or deletion of a chromosome, respectively. Just as point mutations are caused by random errors made by DNA polymerase, chromosome changes result from chance mistakes in the partitioning of chromosomes during meiosis or mitosis. In addition, errors that occur in crossing over during prophase of meiosis I (Chapter 12) can lead to gene duplications or deletions, as illustrated in **Figure 16.20b**. Finally, when accidental breaks in

(a) Types of point mutations

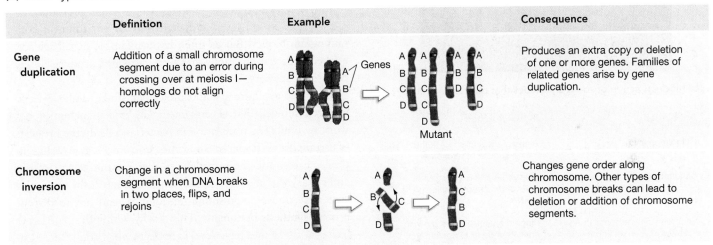

	Definition	Example	Consequence
		Original DNA sequence — TAT TGG CTA GTA CAT Original polypeptide — Tyr—Trp—Leu—Val—His	
Silent	Change in nucleotide that does not change amino acid specified by codon	TAC TGG CTA GTA CAT Tyr—Trp—Leu—Val—His	Change in genotype but no change in phenotype
Missense (Replacement)	Change in nucleotide that changes amino acid specified by codon	TAT TGT CTA GTA CAT Tyr—**Cys**—Leu—Val—His	Change in primary structure of protein
Nonsense	Change in nucleotide that results in early stop codon	TAT TGA CTA GTA CAT Tyr **STOP**	Premature termination—polypeptide is truncated
Frameshift	Addition or deletion of a nucleotide	TAT TCG GCT AGT ACA T Tyr—**Ser**—**Ala**—**Ser**—**Thr**	Reading frame is shifted—massive missense

(b) Other types of mutations

	Definition	Example	Consequence
Gene duplication	Addition of a small chromosome segment due to an error during crossing over at meiosis I—homologs do not align correctly		Produces an extra copy or deletion of one or more genes. Families of related genes arise by gene duplication.
Chromosome inversion	Change in a chromosome segment when DNA breaks in two places, flips, and rejoins		Changes gene order along chromosome. Other types of chromosome breaks can lead to deletion or addition of chromosome segments.

FIGURE 16.20 A Mutation Is a Change in an Organism's DNA
(a) Point mutations in DNA produce an array of effects on the amino acid sequences of proteins, ranging from no change to extensive change. **(b)** Many mutations involve large-scale changes in the composition of chromosomes. In addition to gene duplication and chromosome inversions, Chapter 9 introduced polyploidy (multiple chromosome complements), trisomy (additional chromosomes), and aneuploidy (loss or gain of a chromosome).

chromosomes occur, the chromosome segments may become flipped and rejoin—a phenomenon known as a **chromosome inversion** (Figure 16.20b).

When changes in an organism's genotype are transcribed and translated, they may result in changes in its phenotype. In this way, mutations produce new alleles and new traits. At the level of populations, mutations furnish the heritable variation that Mendel and Morgan analyzed and that makes evolution possible. At the level of individuals, mutations can cause disease or death or lead to increases in fitness.

ESSAY Antibiotics That Poison the Ribosome

Antibiotics are molecules that kill bacteria or slow their growth. Since their initial discovery by biomedical researchers in the 1930s, their use has saved tens of millions of human lives that were threatened by bacterial infections. It is no exaggeration to claim that the era of modern medicine began with the widespread availability of inexpensive antibiotics.

(a) Erythromycin binds to ribosomes and plugs the exit tunnel for the polypeptide.

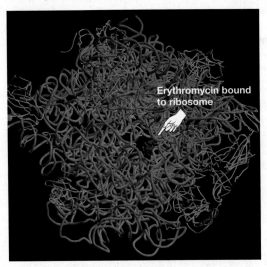

Erythromycin bound to ribosome

(b) Mode of action of some antibiotics that poison translation

Antibiotic	Mode of Action
Chloramphenicol	Binds at the active site in the large subunit of ribosome
Clindamycin	Binds to A and P sites and prevents correct positioning of tRNAs
Erythromycin, carbomycin A, clarithromycin, roxithromycin, spiramycin, tylosin, azithromycin	Plugs the tunnel where growing polypeptide exits ribosome
Puromycin	Enters A site and bonds to growing polypeptide, but blocks addition of more amino acids
Spectinomycin	Inhibits translocation of ribosome
Streptomycin, parmomycin	Binds to the small subunit of ribosome and prevents accurate reading of the mRNA
Tetracycline	Prevents aminoacyl tRNAs from entering A site in ribosome

FIGURE 16.21 Many Antibiotics Work by Disrupting Translation in Bacteria
[(a) "Structural basis for the interaction of antibiotics with peptidyl transferase centre in eubacteria: by Schluenzen et al." Figure 5 in *Nature* (2001) 413: 814–821 by Schluenzen et al.]

Many of the antibiotics that physicians prescribe today are produced naturally by soil-dwelling fungi. These organisms compete with bacterial cells for space and resources. For a fungus, the ability to synthesize antibiotics is an adaptation that helps individuals destroy competitors and increase their access to nutrients. Antibiotics are weapons in chemical warfare that has gone on among competing soil-dwelling organisms for millions of years. Pharmaceutical companies still search for new antibiotics by studying naturally occurring fungi in the soil.

> *Recent research on antibiotics has focused on two seemingly distinct topics: the evolution of bacteria that resist them and the molecular mechanism of drug action.*

Recent research on antibiotics has focused on two seemingly distinct topics: the evolution of bacteria that resist them and the molecular mechanism of drug action. **Antibiotic resistance** is the ability to grow and reproduce efficiently in an environment containing antibiotics. But as often happens in biological science, efforts to answer questions that seem unrelated produce closely connected answers.

How are evolutionary and molecular studies linked? Early research showed that several medically important antibiotics work by inhibiting translation in bacterial cells. In the presence of certain drugs, bacterial ribosomes stop working, protein synthesis ceases, and the targeted cell dies. Fortunately, bacterial and eukaryotic ribosomes are different enough at the molecular level that ribosome-poisoning antibiotics tend not to disrupt protein synthesis in humans. This is why antibiotics that target bacterial ribosomes have few side effects when administered to humans as a drug.

To see how evolutionary and molecular studies of ribosomes intersect, consider the latest three-dimensional models of bacterial ribosomes from bacteria treated with the antibiotic erythromycin. As **Figure 16.21a** shows, the drug binds to the tunnel where growing polypeptide chains leave the ribosome. When erythromycin binds to this site, it inhibits translation and kills the cell. A number of bacterial strains have evolved resistance to this antibiotic, however. When researchers analyzed the structure of ribosomes from drug-resistant populations of bacteria, they found that the ribosomes contained novel amino acids and rRNA sequences at the site where erythromycin binds. The altered amino acids and bases make erythromycin much less likely to bind to the ribosome. As a result, the drug no longer inhibits translation effectively.

Similar studies have now been done on a wide array of antibiotics. The results have documented the binding site and mode of action of many commonly prescribed drugs (**Figure 16.21b**). In many cases, investigators have also been able to confirm that drug-resistant strains of bacteria have mutations precisely at these binding sites.

To make sense of these findings, biomedical researchers acknowledge that mutations are constantly creating variations in the makeup of ribosomes within bacterial populations. The researchers hypothesize that extensive use of antibiotics has led to intensive natural selection for cells with mutations that make them resistant to one or more

drugs. In areas where antibiotic use is widespread, the mutant bacteria have higher fitness and come to dominate their population.

Antibiotic resistance is now so pervasive among disease-causing bacteria that some physicians claim that we have entered a post-antibiotic era. Researchers and physicians have already spent several decades in a race to develop new antibiotics that are effective against drug-resistant strains of bacteria. Research on ribosome structure and the mechanisms of translation has taken on an urgent tone. The more we know about the molecular mechanisms of protein synthesis, the better we can run the race against drug-resistant bacteria.

CHAPTER REVIEW

Summary of Key Concepts

The instructions for making and operating a cell are archived in DNA, transcribed into messenger RNA, and then translated into protein.

After the enzyme RNA polymerase binds to a specific site in DNA with the help of other proteins, it catalyzes the production of an RNA molecule. The base sequence of the RNA produced is complementary to the base sequence of the DNA template strand.

Early experiments established that RNA polymerase begins transcription by binding to promoter sequences in DNA. In bacteria, this binding occurs in conjunction with a regulatory protein called sigma. Sigma recognizes particular sequences within promoters that are located 10 bases and 35 bases upstream from the start of the actual genetic message. These binding sites ensure that the correct DNA strand is transcribed in the correct direction. Eukaryotic promoters are more complex and variable than bacterial promoters, and interact with many more transcription factors.

Web Tutorial 16.1   Transcription

In eukaryotes, genes consist of coding regions called exons interspersed with noncoding regions called introns. After a eukaryotic mRNA is produced, introns must be spliced out. The ends of the molecule receive a cap and tail.

After transcription of a eukaryotic mRNA is complete, several key events occur. Stretches of noncoding RNA called introns are spliced out by complex molecular machines called spliceosomes, a "cap" signal is added to the 5′ end, and a poly (A) tail is added to the 3′ end. Experiments with modified mRNAs established that both the 5′ cap and the poly (A) tail serve as recognition signals for the translation machinery and protect the message from degradation by ribonucleases.

Inside ribosomes, mRNAs are translated to proteins via intermediary molecules called transfer RNAs. Transfer RNAs

carry an amino acid and have a three-base-pair anticodon, which binds to an mRNA codon. The amino acid carried by the transfer RNA is then added to the growing protein via formation of a peptide bond.

Experiments with radioactively labeled amino acids confirmed that ribosomes are the site of protein synthesis, and that transfer RNAs (tRNAs) serve as the chemical bridge between the RNA message and the polypeptide product. tRNAs have a predicted secondary structure that resembles a cloverleaf and an L-shaped tertiary structure. One leg of the L contains the anticodon, which forms a base pair with the mRNA codon, while the other leg holds the amino acid appropriate for that codon. Because imprecise pairing—or "wobbling"—is allowed in the third positions of anticodons, only about 40 different tRNAs are required to translate the 61 codons that code for amino acids.

Through intensive study of ribosome structure and function, researchers established that the process of elongating a polypeptide involves three steps: (1) an incoming aminoacyl tRNA occupies a position in the ribosome called the A site; (2) the growing polypeptide chain is transferred from a peptidyl tRNA in the ribosome's P site to the amino acid bound to the tRNA in the A site, and a peptide bond is formed; and (3) the ribosome is translocated to the next codon on the mRNA, accompanied by ejection of the empty tRNA from the E site. Recent data on the three-dimensional structure of the ribosome has confirmed that peptide bond formation is catalyzed by a ribozyme, not an enzyme.

While translation is in progress, proteins fold into their three-dimensional conformation (tertiary structure), sometimes with the aid of chaperone proteins. Some proteins are targeted to specific locations in the cell by the presence of signal sequences, while others remain inactive until modified by phosphorylation or the removal of certain amino acids.

Web Tutorial 16.2   Translation

■ **Many mutations produce changes in the phenotype.**

When mutations occur in DNA, the altered sequences are faithfully transcribed and translated into gene products that are part of the organism's phenotype. Depending on the location and type of alteration in DNA and its impact on the resulting RNA or protein product, a mutation can be beneficial, deleterious, or neutral with respect to fitness. Mutations produce novel proteins and RNAs. They are the source of the heritable variation that makes evolution possible.

Questions

Content Review

1. How did the A site of the ribosome get its name?
   a. It is where amino acids are affixed to tRNAs, producing aminoacyl tRNAs.
   b. It is where the amino group on the growing polypeptide chain is available for peptide bond formation.
   c. It is the site occupied by incoming aminoacyl tRNAs.
   d. It is surrounded by α-helices of ribosomal proteins.

2. How did the P site of the ribosome get its name?
   a. It is where the promoter resides.
   b. It is the site where peptide bond formation takes place.
   c. It is where peptidyl tRNAs reside.
   d. It is the site where the growing polypeptide chain is phosphorylated.

3. What is a molecular chaperone?
   a. a protein that recognizes the promoter and guides the binding of RNA polymerase
   b. a protein that activates or deactivates another protein by adding or removing a phosphate group
   c. a protein that is a component of the large ribosomal subunit and that assists with peptide bond formation
   d. a protein that helps newly translated proteins fold into their proper three-dimensional configuration

4. The three types of RNA polymerase found in eukaryotic cells transcribe different types of genes. What does RNA polymerase II produce?
   a. rRNAs
   b. tRNAs
   c. mRNAs
   d. spliceosomes

5. What is an anticodon?
   a. the part of an RNA message that signals the termination of translation
   b. the part of an RNA message that signals the start of translation
   c. the part of a tRNA that binds to a complementary codon in mRNA
   d. the part of a tRNA that accepts an amino acid, via a reaction catalyzed by tRNA synthetase

6. Which of the following questions about transcription was *not* answered by the discovery of the -35 and -10 sequences in bacterial promoters?
   a. Does RNA polymerase act alone to initiate transcription, or are other proteins involved?
   b. How do sigma proteins bind to RNA polymerase?
   c. How does RNA polymerase "know" where to initiate transcription?
   d. How does RNA polymerase "know" which strand of DNA acts as the template?

Conceptual Review

1. Explain the relationship among eukaryotic promoter sequences, basal transcription factors, and RNA polymerase. Explain the relationship among bacterial promoter sequences, sigma, and RNA polymerase.

2. According to the wobble rules, the correct amino acid can be added to a growing polypeptide chain even if the third base in an mRNA codon does not correctly match the corresponding base in a tRNA anticodon. How do the wobble rules relate to the redundancy of the genetic code?

3. Why does splicing occur in eukaryotic mRNAs? Where does it occur, and how are snRNPs involved?

4. Sketch the structure of a tRNA molecule. Label the anticodon and the CCA sequence where amino acids bind. Explain how a tRNA interacts with tRNA synthetase. Sketch the structure of an aminoacyl tRNA. Explain how an aminoacyl tRNA interacts with an mRNA in the ribosome.

5. Explain the sequence of events that occurs during translation as a protein elongates by one amino acid. At each step, specify what is happening in the ribosome's A site, P site, and E site.

6. What evidence supports the hypothesis that peptide bond formation is catalyzed by a ribozyme?

Group Discussion Problems

1. The 5' cap and poly (A) tail that are attached to eukaryotic mRNAs appear to help the message last longer by protecting it from degradation by ribonucleases. But why is an enzyme such as a ribonuclease in the cell in the first place? What function would an enzyme that destroys messages serve?

2. Look back at Figure 16.1, which shows the formation of a phosphodiester bond during RNA polymerization. Then study the structure of the nucleotide shown below, called cordycepin. If cordycepin triphosphate, which has three phosphate groups bonded to the 5' hydroxyl group in the figure, is added to a cell-free transcription reaction, the nucleotide is added onto the growing RNA chain. This observation confirms that synthesis occurs by the addition of monomers in the form of triphosphates to the 3' end of the growing chain. Briefly explain why. Be sure to describe the expected result if synthesis occurred at the 5' end of the polymer.

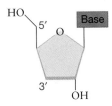

3. Carl Woese has determined the nucleotide sequence of rRNA molecules from a diverse array of organisms and compared them with each other. According to his data, certain portions of the rRNAs in the large subunit are very similar in all organisms. To make sense of this result, Woese suggests that the conserved sequences have an important functional role. His logic is that these conserved sequences are so important to cell function that any changes in the sequence cause death. Woese also claims that the existence of the conserved sequences supports other data indicating that peptide bond formation is catalyzed by the large rRNA, not by ribosomal proteins. Explain the thinking behind his claim.

4. Recent work has revealed why α-amanitin inhibits transcription by RNA pol II. Structural models show that α-amanitin binds to a site inside the enzyme, but not to the active site itself. Based on the model of bacterial RNA polymerase in Figure 16.2, predict where α-amanitin binds and why it inhibits transcription.

Answers to Multiple-Choice Questions   1. c; **2.** b; **3.** d; **4.** c; **5.** c; **6.** d

17

Control of Gene Expression in Bacteria

KEY CONCEPTS

▦ Changes in gene expression allow bacterial cells to respond to environmental changes.

▦ Gene expression can be controlled at three levels: transcription, translation, or post-translation (protein activation).

▦ Transcriptional control can be positive or negative. In the case of the genes for lactose metabolism in *E. coli*, positive control occurs when a regulatory protein binds to DNA and increases the transcription rate. Negative control occurs because another regulatory protein must be released before transcription can occur.

▦ Many regulatory proteins bind to specific sites in DNA. Because different regulatory proteins have different amino acid sequences, they bind to different DNA sequences.

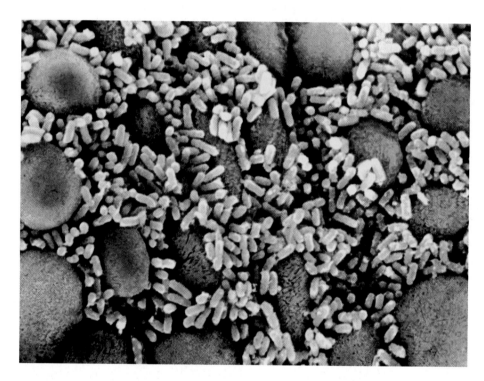

The red mounds in this micrograph are human intestinal cells; the yellow structures are bacteria. In the intestine, the nutrients available to bacteria constantly change. This chapter explores how changes in gene expression help bacteria respond to environmental changes.

I magine waiting anxiously to hear the opening lines of a wonderfully melodic symphony, played by a renowned orchestra. The crowd applauds as the celebrated conductor comes onstage, then hushes as he takes the podium. He cocks the baton; the musicians raise their instruments. As the baton comes down, every instrument begins blaring a different tune at full volume. A tuba plays "Dixie"; a violinist renders "In-A-Gadda-Da-Vida": and a cellist begins Mexico's national anthem. A snare drum lays down beats for "Baby Got Back" while the bass drum simulates the cannons in the "1812 Overture." Instead of music, there is pandemonium. The conductor staggers off stage, clutching his heart.

A cacophony like this would result if a bacterial cell "played" all its genes at full volume all the time. The *Escherichia coli* cells living in your gut right now have over 4300 genes. If all of those genes were expressed at the fastest possible

rate at all times, the *E. coli* cells would stagger off the stage, too. But this does not happen. Cells are extremely selective about which genes are expressed, in what amounts, and when.

This chapter explores how bacterial cells control the activity of their genes. **Gene expression** is said to occur when a protein or other gene product is synthesized and is active in the cell.

Understanding how cells regulate gene expression is a fundamental issue in biological science, for two reasons. First, bacteria are the most abundant organisms on Earth, and they occupy virtually every habitat known. In many cases, bacterial cells are able to grow and reproduce successfully because they respond rapidly to dramatic changes in temperature, pH, light, competitors, and nutrients. Changes in gene expression give bacteria the ability to cope with these types of environmental change. Second, questions about bacterial gene expression have enormous practical significance. For example, biologists are trying to understand which genes are expressed when disease-causing bacteria colonize a person and start an infection in the hope that drugs can be developed that alter the expression of key genes. Bacterial cells are also used commercially to manufacture important products such as human insulin and growth hormone. Efficient use of bacteria depends on a solid understanding of how transcription and translation are regulated.

Understanding gene expression is fundamental to our understanding of life. This chapter's goal is to introduce key concepts in the regulation of gene expression. Let's begin by reviewing some of the environmental challenges that bacterial cells face and then explore how these organisms meet them.

17.1 Gene Regulation and Information Flow

Bacteria are found in virtually every habitat on Earth, from boiling hot springs to alpine snowfields and from the open ocean to crevices miles underground. The millions of bacterial species existing today have evolved a bewildering variety of ways to solve a central problem of living: obtaining the materials and energy required for growth and reproduction. Although some bacteria specialize by using just one type of food, the vast majority of species are able to switch among several different sources of carbon and energy, according to which nutrients are available in the environment at the time. The fundamental question addressed in this chapter is *how* this switching occurs. Each type of nutrient requires a different membrane transport protein to bring the nutrient molecule into the cell and a different suite of enzymes to process it. How does a bacterial cell turn some genes on and others off so that it can take advantage of alternative food sources? More generally, how do bacteria regulate gene expression so that a cell makes only the products that it needs?

As a case study, we explore the strategies of *Escherichia coli*. These cells can use a wide array of carbohydrates to supply the carbon and energy they need. The carbohydrates are in your food. As your diet changes from day to day, the availability of different sugars in your intestines varies. Precise control of gene expression gives *E. coli* the ability to respond to these changes in its environment and use the different sugars.

To appreciate why precise control over gene expression is so important to these cells, it is critical to realize that bacteria can be packed an inch thick along your intestinal walls. The organisms represent many different species that are competing for space and nutrients. For a cell to survive and reproduce in this environment, it must use resources efficiently—particularly resources that provide energy and carbon compounds. For example, recall from Chapter 4 that carbohydrates are polymers made up of simple sugars and from Chapter 9 that carbohydrates provide cells with energy and chemical building blocks. *Escherichia coli* is capable of metabolizing a wide variety of carbohydrates. But it would be energetically wasteful for *E. coli* to produce all of the enzymes required to process all of the various carbohydrates all of the time. Instead, it is logical to predict that the enzymes being produced match the sugars that are available at a given time. Efficient use of resources, via tight control over gene expression, is critical if cells are going to be able to compete successfully for space and nutrients.

On the basis of this reasoning, transcription and translation of individual genes in bacteria should be triggered by specific signals from the environment, such as the presence of specific sugars. Did you drink milk at your last meal, or eat french fries and a candy bar? Each type of food contains different sugars. Each sugar should induce a different response by the *E. coli* cells in your intestine. Just as a conductor needs to regulate the orchestra's musicians, cells need to regulate which proteins they produce at any given time.

Mechanisms of Regulation—an Overview

Gene expression can be controlled at any step between the synthesis of RNA and the activation of the final gene product. Three steps occur as information flows from DNA to proteins, represented by arrows in the following expression:

$$DNA \rightarrow mRNA \rightarrow protein \rightarrow activated\ protein$$

The arrow from DNA to RNA represents transcription—the making of messenger RNA (mRNA). The arrow from RNA to protein represents translation, in which ribosomes read the information in mRNA and use that information to synthesize a protein. The arrow from protein to activated protein represents post-translational modifications—including folding, addition of carbohydrate or lipid groups, or perhaps phosphorylation.

How can a bacterial cell conserve ATP and amino acids by avoiding the production of proteins that are not needed at a particular time? A look at the flow of information from DNA to protein suggests three possible mechanisms:

1. The cell could avoid making the mRNAs for particular enzymes. If there is no mRNA, then ribosomes cannot make the gene product. For example, various regulatory proteins affect the ability of RNA polymerase to bind to a promoter and initiate transcription. Genes that are controlled in this way are said to undergo **transcriptional control**:

$$DNA \xrightarrow{\;\;\times\;\;} mRNA \rightarrow protein \rightarrow activated\ protein$$

2. If the mRNA for an enzyme has been transcribed, the cell might have a way to prevent the mRNA from being translated into protein. Mechanisms that alter the length of time an mRNA survives before it is degraded by ribonucleases, that affect translation initiation, or that affect elongation factors and other proteins during the translation process are forms of **translational control**:

$$DNA \rightarrow mRNA \xrightarrow{\;\;\times\;\;} protein \rightarrow activated\ protein$$

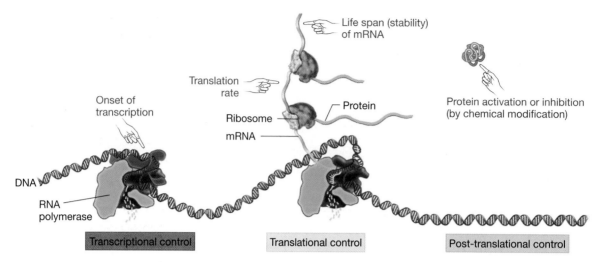

FIGURE 17.1 Gene Expression Can Be Regulated at Three Levels
Although this chapter focuses on how regulatory molecules affect the ability of RNA polymerase to initiate transcription, translational and post-translational controls also occur in bacteria.

3. Chapter 16 pointed out that some proteins are manufactured in an inactive form and have to be activated by chemical modification, such as the addition of a phosphate group. This type of regulation is **post-translational control:**

DNA → mRNA → protein → activated protein

Which of these three forms of control occur in bacteria? The short answer to this question is "all of the above." As **Figure 17.1** shows, many factors affect how much active protein is produced from a particular gene. Transcriptional control is particularly important due to its efficiency—it saves the most energy for the cell, because it stops the process at the earliest possible point. Translational control is advantageous because it allows a cell to change quickly which proteins are produced. Post-translational control is significant as well, and it provides the most rapid response of all three mechanisms. Among these mechanisms of gene regulation, there is a clear trade-off between the speed of response and the conservation of ATP, amino acids, and other resources. Transcriptional control is slow but efficient in resource use. Post-translational control is fast but energetically expensive.

Although this chapter focuses almost exclusively on mechanisms of transcriptional control in bacteria, it is important to keep in mind that both translational and post-translational controls occur in these organisms. It's also important to recognize that some genes—such as those that code for the enzymes required for glycolysis—are transcribed all the time, or **constitutively.** But the expression of other genes is regulated, meaning that they may be induced or repressed. Finally, it is critical to realize that gene expression is not an all-or-none proposition. Genes are not just "on" or "off"— instead, the level of expression is highly variable. What factors determine how much a particular gene is expressed at a particular time?

Metabolizing Lactose—A Model System

As Chapters 13 through 16 have shown, many of the fundamental advances in genetics have been achieved through the analysis of a model system. Studying the inheritance of seed shape in garden peas revealed the fundamental patterns of gene transmission. Exploring transcription in viruses and *E. coli* led to the discovery of RNA polymerase, transcription factors (such as the sigma proteins), and promoters. In studies of gene regulation, the key model system has been the metabolism of the sugar lactose in *E. coli*.

Jacques Monod, François Jacob, and many colleagues introduced lactose metabolism in *E. coli* as a model system during the 1950s and 1960s. Although they worked with a single species of bacterium, their results had a profound effect on thinking about gene regulation in all organisms. Some details turned out to be specific to the *E. coli* genes responsible for lactose metabolism, but many of Monod and Jacob's results are universal.

Escherichia coli can use a wide variety of sugars for ATP production, via glycolysis and the electron transport chain or via fermentation. These sugars also serve as a raw material in the synthesis of amino acids, vitamins, and other complex compounds. Glucose is *E. coli*'s preferred carbon source, however, meaning that it is the source of energy and carbon atoms that the organism uses most efficiently. This observation is logical, because glycolysis begins with glucose and is the main pathway for the production of ATP. Lactose, the sugar found in milk, is also used by *E. coli*, but only when glucose supplies are depleted. Recall from Chapter 5 that lactose is a disaccharide made up of one molecule of glucose and one molecule of galactose.

To use lactose, *E. coli* must first transport the sugar into the cell. Once lactose is inside the cell, *E. coli* can break the sugar down into glucose and galactose by means of the enzyme β-galactosidase. The glucose released by this reaction is used directly via glycolysis; other enzymes convert the galactose to an intermediate in the glycolytic pathway.

In the early 1950s, biologists discovered that *E. coli* produces high levels of β-galactosidase only when lactose is present in the environment. If lactose is absent, little β-galactosidase is synthesized. Because lactose appears to induce the production of the enzyme that breaks itself down, researchers proposed that lactose itself regulates the gene for β-galactosidase. More formally, the hypothesis was that lactose acts as an inducer. An **inducer** is a molecule that stimulates the expression of a specific gene.

In the late 1950s Jacques Monod investigated how the presence of glucose affects the regulation of the β-galactosidase gene. Would *E. coli* produce high levels of β-galactosidase when both glucose and lactose were present in the surrounding environment? **Figure 17.2** summarizes the experimental approach by

which Monod found that the answer was no. This meant that β-galactosidase production is induced by the availability of lactose, but only when glucose is unavailable. The key observations can be summarized as follows:

- Cells grown on glucose only: no β-galactosidase production
- Cells grown on glucose + lactose: no β-galactosidase production
- Cells grown on lactose only: β-galactosidase production

Monod teamed up with François Jacob to investigate exactly how lactose and glucose regulate the genes responsible for lactose metabolism. Research on how these genes are regulated opened a window on how genes in all organisms are controlled. Research on this system is still continuing, some 50 years later.

17.2 Identifying the Genes Involved in Lactose Metabolism

To understand how *E. coli* controls production of β-galactosidase and the membrane transport protein that brings lactose into the cell, Monod and Jacob first had to find the genes that code for these proteins. To do this, they employed the same tactic used in the pioneering studies of DNA replication, transcription, and translation reviewed in earlier chapters: They isolated and analyzed mutant individuals. In this case, their goal was to find *E. coli* cells that were *not* capable of metabolizing lactose. Cells that can't use lactose must lack either β-galactosidase or the lactose-transporter protein.

Finding mutants with respect to a particular trait was a two-step process. The researchers' first step was to generate a large number of individuals with mutations at random locations in their genomes. To produce mutant cells, Monod and others exposed *E. coli* populations to X-rays, UV light, or chemicals that damage DNA and increase mutation rates. The second step is to screen the mutants to find individuals with defects in the process or biochemical pathway in question—in this case, defects in lactose metabolism. A technique that allows researchers to identify individuals with a particular type of mutation is called a **screen**.

Monod and colleagues were looking for cells that cannot grow in an environment that contains only lactose as an energy source. Normal cells grow well in this environment. How could the researchers select cells on the basis of *lack* of growth?

Screening Mutants—Replica Plating and Indicator Plates

Replica plating and growth on indicator plates were key techniques in the search for mutants with defects in lactose metabolism. **Replica plating** begins with spreading mutagenized

Question: *E. coli* produces β-galactosidase when lactose is present. Does *E. coli* produce β-galactosidase when both glucose and lactose are present?

Hypothesis: *E. coli* does not produce β-galactosidase when glucose is present, even if lactose is present. (Glucose is the preferred food source.)

Alternative hypothesis: *E. coli* produces β-galactosidase whenever lactose is present, regardless of the presence or absence of glucose.

Experimental setup:

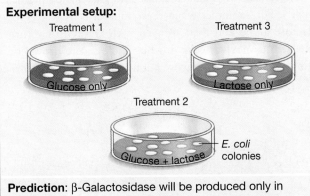

Prediction: β-Galactosidase will be produced only in treatment 3.

Prediction of alternative hypothesis: β-Galactosidase will be produced in treatments 2 and 3.

Results:

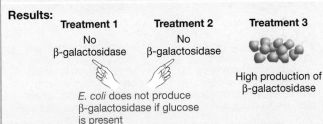

Conclusion: Glucose prevents the expression of the gene for β-galactosidase. The presence of lactose without glucose stimulates the expression of that gene.

FIGURE 17.2 Glucose Affects the Regulation of the β-Galactosidase Gene

bacteria on a plate that is filled with gelatinous agar containing glucose, as illustrated in **Figure 17.3**. This plate is known as the **master plate**. The bacteria are then allowed to grow, so that each cell produces a single colony. (A bacterial colony consists of a large number of identical cells that are descended from a single cell.) Next, a block covered with a piece of sterilized velvet is pressed onto the master plate. Because of the contact, cells from each colony on the master plate are transferred to the velvet. Then the velvet is pressed onto a plate containing a medium that differs from that of the master plate by a single component. For instance, in Figure 17.3, the second medium has only lactose as a source of carbon and energy. Cells from the velvet stick to the plate's surface, producing an exact copy of the colonies on the master plate. This copy is called the **replica plate**. After these cells grow, an investigator can compare the colonies that thrive on the replica plate's medium with those on the master plate. In this case, colonies that grow on the master plate but that are missing on the replica plate represent mutants deficient in lactose metabolism. By picking these particular colonies from the master plate, researchers build a collection of lactose mutants.

Monod also used an alternative strategy based on **indicator plates**, where mutants with metabolic deficiencies are observed directly. Consider what happens when mutagenized bacteria are grown on agar plates containing lactose and are then sprayed with a solution containing the colorless compound ONPG (o-nitrophenyl-β-D-galactoside). ONPG consists of two subunits—a galactose molecule and a molecule called o-nitrophenol—that are joined by the same β-1,4 glycosidic linkage that joins the galactose and glucose subunits of lactose. Consequently, ONPG acts as an alternative substrate for β-galactosidase. When the enzyme cleaves ONPG, galactose and o-nitrophenol are released. This is significant, because o-nitrophenol is intensely yellow. When colonies exposed to ONPG turn yellow, it means that they have β-galactosidase activity. Colonies that stay white are unable to cleave ONPG, meaning they have a defect in the enzyme or its production.

Different Classes of Lactose Metabolism Mutants

The initial mutant screen yielded the three types of mutants summarized in **Table 17.1**. In one class, the mutant cells were unable to cleave ONPG even if lactose was present inside the cells to induce production of the β-galactosidase protein. The investigators concluded that these mutants must lack a functioning version of the β-galactosidase protein—meaning the gene that encodes β-galactosidase is defective. This gene was designated *lacZ*, and the mutant allele *lacZ⁻*.

In the second class of mutants, the cells failed to accumulate lactose inside the cell. In contrast, the concentration of lactose in wild-type cells is about 100 times that of lactose in the surrounding environment. To interpret this result, Jacob and Monod hypothesized that the mutant cells had defective copies of the membrane protein responsible for transporting lactose into the cell. This protein was identified and named **galactoside permease**; the gene that encodes it was designated *lacY*.

The third and most surprising class of mutants did not regulate expression of β-galactosidase and galactoside permease normally. For example, when these mutant cells were grown on lactose alone and then sprayed with ONPG, they turned yellow just as normal cells do. But if they were grown on a medium that contained glucose but no lactose, they still turned

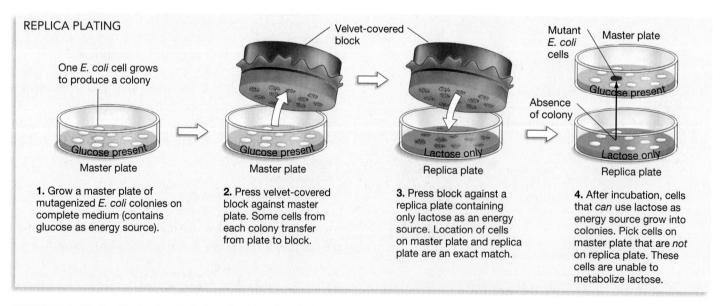

FIGURE 17.3 Replica Plating Is a Technique for Identifying Mutant Cells
Replica plating, used here to isolate mutant *E. coli* cells with a deficiency in lactose metabolism. There are two keys to this technique: (1) The location of colonies on the master plate and replica plate must match exactly, and (2) the media in the two plates must differ by just one type of component. **QUESTION** How would you alter this protocol to isolate mutant cells with a deficiency in the enzymes required to synthesize tryptophan?

TABLE 17.1 Three Distinct Types of Mutants in Lactose Metabolism of *E. coli*

Mutant Phenotype	Interpretation	Inferred Genotype
Cells cannot cleave ONPG even if lactose is present as an inducer.	No β-galactosidase; gene for β-galactosidase is defective. Call this gene *lacZ*.	*lacZ⁻*
Cells cannot accumulate lactose inside the cell.	No membrane protein (galactoside permease) required for import of lactose; gene for galactoside permease is defective. Call this gene *lacY*.	*lacY⁻*
ONPG is cleaved even if lactose is absent (no inducer).	Constitutive expression of *lacZ* and *lacY*; gene for regulatory protein that shuts down *lacZ* and *lacY* is defective—it does not need to be induced by lactose. Call this gene *lacI*.	*lacI⁻*

yellow after being sprayed with ONPG. Normal cells, in contrast, do not produce β-galactosidase under these conditions. Normal cells remain white when grown on glucose and then sprayed with ONPG.

Cells that are abnormal because they produce a product at all times are called **constitutive mutants**. Unlike normal cells, their product is always part of the cell's constitution. The gene that mutated to produce constitutive β-galactosidase expression was named *lacI*. The use of "I" was appropriate because these mutants did not need an inducer to express β-galactosidase or galactoside permease. Recall that, in normal cells, the expression of these genes is induced by the presence of lactose. But in cells with a mutant form of *lacI* (*lacI⁻* mutants), gene expression occurred with or without lactose. This meant that *lacI⁻* mutants have a defect in gene regulation. In these mutants, the gene remains "on" when it should be turned off.

Based on these observations, it is logical to infer that the normal product of the *lacI* gene prevents the transcription of *lacZ* and *lacY* when lactose is absent. Because lactose acts as an inducer for the production of β-galactosidase, it is reasonable to expect that the *lacI* gene or gene product interacts with lactose in some way. (Later work showed that the inducer is a de-

rivative of lactose called *allolactose*. For the sake of historical accuracy and simplicity, however, this discussion refers to lactose itself as the inducer.)

Several Genes Are Involved in Metabolizing Lactose

Jacob and Monod had succeeded in identifying three genes involved in lactose metabolism: *lacZ*, *lacY*, and *lacI*. They had concluded that *lacZ* and *lacY* code for proteins involved in the metabolism and import of lactose, while *lacI* is responsible for some sort of regulatory function. The *lacI* gene or gene product repressed the expression of *lacZ* and *lacY* in the absence of lactose, but when lactose was present, the opposite occurred—transcription of *lacZ* and *lacY* was induced.

When Jacob and Monod followed up on these experiments by mapping the physical location of the three genes on *E. coli*'s circular chromosome, they found that the genes are close together (**Figure 17.4**). This was a crucial result, because it suggested that both *lacZ* and *lacY* might be controlled by *lacI*. Could one regulatory gene manage more than one protein-encoding gene? If so, how does *lacI* actually work? And why do lactose and glucose have opposite effects on it?

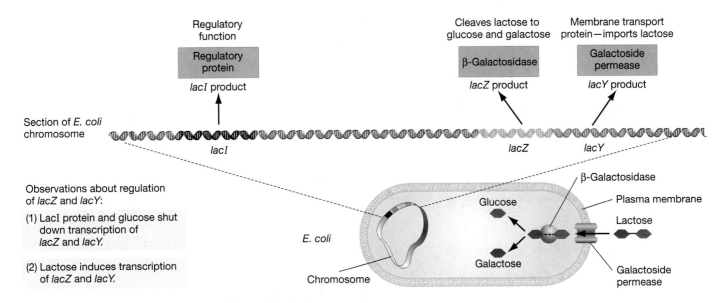

FIGURE 17.4 The *lac* Genes Are in Close Physical Proximity
Early experiments on lactose use in *E. coli* identified the *lacI*, *lacZ*, and *lacY* genes, documented the function of their protein products, and mapped their locations on the *E. coli* chromosome.

17.3 The Discovery of the Repressor

In principle, there are two ways that transcription can be regulated: via positive control or negative control. **Positive control** is in effect when something must be added for transcription to occur; **negative control** is in effect when something must be taken away for transcription to occur. In starting up a car, positive control occurs when you step on the gas pedal, meaning that you add fuel; negative control occurs when the parking brake is set. It turned out that the *lacZ* and *lacY* genes in *E. coli* are controlled by both a gas pedal and a parking brake—that is, they are under positive control *and* negative control.

The hypothesis that the *lacZ* and *lacY* genes might be under negative control originated with Leo Szilard in the late 1950s. Szilard suggested to Jacques Monod that the *lacI* gene codes for a product that represses transcription of the *lacZ* and *lacY* genes. Recall from Section 17.2 that, in *lacI⁻* mutants, the production

of β-galactosidase and galactoside permease is constitutive, even if lactose is not available. Szilard proposed that regulation fails because the LacI⁺ protein—which normally prevents enzyme synthesis—is inactive. Stated another way, the *lacI* gene produces an inhibitor that exerts negative control over the *lacZ* and *lacY* genes. This transcription inhibitor was called a **repressor** and was thought to bind directly to DNA near or on the promoter for the *lacZ* and *lacY* genes (**Figure 17.5a**). To explain how the presence of lactose triggers transcription in normal cells, Szilard and Monod proposed that lactose interacts with the repressor in a way that makes the repressor release from its binding site (**Figure 17.5b**). The idea was that lactose induces transcription by removing negative control. Lactose releases the parking brake. As **Figure 17.5c** shows, it was logical to observe that constitutive transcription occurs in *lacI⁻* mutants because a functional repressor is absent—the parking brake is broken.

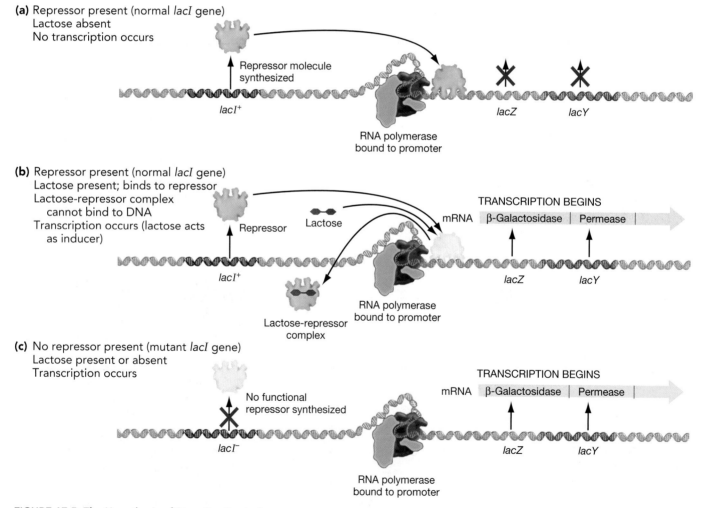

FIGURE 17.5 The Hypothesis of Negative Control
The negative-control hypothesis maintains **(a)** that transcription of genes involved in lactose use is normally blocked by a repressor molecule that binds to DNA on or near the promoter for *lacZ* and *lacY*, **(b)** that lactose induces transcription of *lacZ* and *lacY* by interacting with the repressor, and **(c)** that, when a functional repressor is absent, transcription proceeds.

How could researchers assess these hypotheses? This was no trivial task. But in a set of brilliant experiments, each of the predictions of the negative-control hypothesis was tested and confirmed. The key studies were based on creating *E. coli* cells that were diploid for the genes involved in lactose metabolism. For example, consider an experiment that begins with cells that are haploid and have the genotype $I^- lacZ^+ lacY^+$. These cells lack the hypothesized repressor but have normal copies of the genes for β-galactosidase or galactoside permease. As predicted by the hypothesis of negative control, they produce β-galactosidase constitutively. But what happens when these cells are made diploid by the addition of DNA with the genotype $I^+ lacZ^- lacY^-$? The second set of genes codes for a functioning copy of the repressor, but not of β-galactosidase or galactoside permease. Once the cells are made diploid, β-galactosidase production declines and then stops. This result supports the hypothesis that the repressor codes for a protein that shuts down transcription. As predicted, the repressor is the "parking brake" on transcription. But if an inducer such as lactose is added to these diploid cells, β-galactosidase activity continues. This result supports the hypothesis that lactose removes the repressor—it releases the parking brake. Stated another way, lactose acts as an inducer by removing negative control.

The *lac* Operon

Jacob and Monod summarized the results of their experiments with a comprehensive model of negative control that was published in 1961. In essence, their experimental data had confirmed the hypotheses illustrated in Figure 17.5. One of their key conclusions was that the genes for β-galactosidase and galactoside permease are controlled together. To capture this point, they coined the term **operon** for a set of coordinately regulated bacterial genes that are transcribed together into one mRNA. Logically enough, the group of genes involved in lactose metabolism was termed the ***lac* operon.**

Three hypotheses are central to the Jacob-Monod model of *lac* operon regulation:

1. The *lacZ* and *lacY* genes are adjacent and are transcribed into one mRNA initiated from the single promoter of the *lac* operon. As a result, the expression of these two genes is coordinated.

2. The repressor is a protein encoded by *lacI* that binds to DNA and prevents transcription of *lacZ* and *lacY*. The binding site for the repressor is a section of DNA in the *lac* operon called the **operator.** Jacob and Monod proposed that *lacI* is expressed constitutively and that its product (the repressor) prevents transcription by blocking RNA polymerase from making contact with the promoter.

3. The inducer (lactose) interacts directly with the repressor by binding to it. As a result, the repressor changes shape in a way that causes it to drop off the DNA strand. Transcription is then able to proceed because RNA polymerase can access the promoter. Recall from Chapter 3 that this form of control over protein function is called *allosteric regulation*. When allosteric regulation occurs, a small molecule binds directly to a protein and causes it to change its shape and activity. When the inducer binds to the repressor, negative control ends.

Testing the *lac* Operon Model

Did Jacob and Monod "get it right"? After their model of negative control was published, Jacob and Monod confirmed the existence of the operator. They did this by finding *E. coli* mutants that expressed *lacZ* constitutively yet had normal forms of the repressor. In these new mutants, the repressor protein was unable to function because the nucleotide sequence of the operator was altered. In 1967 Walter Gilbert and Benno Müller-Hill were able to tag copies of the repressor protein with a radioactive atom. Their experiments showed that the repressor physically binds to the DNA sequences of the operator. This result confirmed that the operator is not a protein or an RNA product, but instead is part of the DNA sequence of the *lac* operon.

The only major modification of the original *lac* operon model is based on the recent recognition that the repressor does not physically block RNA polymerase from contacting the promoter, as Jacob and Monod had proposed. Instead, the repressor prevents the RNA polymerase holoenzyme from opening the DNA helix once that enzyme has bound to the promoter. As a result, the template strand cannot be threaded through the enzyme's active site so that transcription can begin.

In addition, a third gene of the *lac* operon, *lacA*, was discovered after the original model of negative control was published. The *lacA* gene is tightly linked to *lacY* and *lacZ* and is transcribed as part of the same mRNA strand that carries *lacZ* and *lacY*. The gene codes for the enzyme transacetylase, which transfers acetyl (CH_3COO^-) groups to β-galactoside sugars. The protein's function is protective in nature: The acetylation step allows sugars that cannot be metabolized to be exported from the cell instead of accumulating to dangerous levels within the cell.

An mRNA like the one produced by the *lac* operon is known as a polycistronic mRNA. **Polycistronic mRNAs** contain more than one protein-encoding segment. In most cases, bacterial genes that are transcribed as a polycistronic mRNA code for products devoted to a common end, such as the *lacZ* and *lacY* products required for lactose metabolism. Similarly, the polycistronic *trp* (pronounced "trip") operon in *E. coli* consists of five genes, each of which is required for different steps in the synthesis of the amino acid tryptophan. The *trp* operon is also under negative control—but with a twist explained in **Box 17.1**, page 370. Biologists interpret the polycistronic structure of many bacterial genes as an adaptation

BOX 17.1   Almost Parallel Worlds: Similarities and Contrasts in Control of the *trp* and *lac* Operons

The genes in the *lac* operon code for proteins that are responsible for breaking down a compound in the cell, lactose. Logically enough, the expression of these genes depends on the presence of lactose, the compound that is to be broken down. When the inducer, a lactose derivative, binds to the repressor, the repressor comes off the operator and transcription begins.

Now imagine how a cell would regulate an operon that is responsible for synthesizing a molecule instead of breaking it down. In this case, the cell should stimulate gene expression when the molecule is absent but should reduce transcription when the molecule is present. To see how cells accomplish this task, let's explore how *E. coli* regulates the **trp operon**—a set of five cotranscribed genes, each of which is required for the synthesis of the amino acid tryptophan.

If RNA polymerase binds to the promoter for the *trp* operon and begins transcription, all five *trp* operon genes are transcribed and translated together. Once translation is complete, the synthesis of tryptophan commences.

The *trp* operon is under negative control. It has an operator sequence that overlaps the promoter, and a repressor protein that binds to the operator to prevent transcription—as in the *lac* operon.

In the *trp* operon, tryptophan binds to the repressor to regulate transcription, just as allolactose binds to the *lac* operon's repressor to regulate transcription. There is a crucial difference in the two systems, however. If tryptophan binds to the repressor, it means that tryptophan is present in the cell and transcription should not proceed.

This is exactly what biologists observe: As **Figure 17.6a** shows, the *trp* repressor binds operator DNA and exerts negative control only when the repressor is complexed with tryptophan. When tryptophan is absent, the repressor does not bind to the operator and the operon is expressed constitutively (**Figure 17.6b**). This is exactly the opposite of how the presence of lactose affects the *lac* repressor. And it makes perfect sense for the *trp* operon: Bacterial cells will have higher fitness if the operon is transcribed when tryptophan levels are low but shut down when tryptophan is abundant.

In addition to offering a new twist on negative control, the *trp* operon has provided other fascinating insights into gene regulation. Experiments have revealed that negative control in the *trp* operon is supplemented by a regulatory mechanism called **attenuation**. Attenuation works like a dimmer switch, because it superimposes fine-tuned control on what would otherwise be a simple on-off system. In the case of the *trp* operon, the dimmer is sensitive to tryptophan levels. When tryptophan levels are high, attenuation causes transcription to terminate early in the operon. When tryptophan levels are low, transcription proceeds through the entire operon. Attenuation that is based on amino acid concentration is seen in many other operons devoted to amino acid synthesis.

After considering how the *lac* and *trp* operons are regulated in *E. coli*, it's difficult not to have new respect for "simple" forms of life. Bacteria may be small and unicellular, but they perform fantastically complex, finely controlled chemistry. Even the simplest cells are marvelous machines.

(a) When tryptophan is present, transcription is blocked.

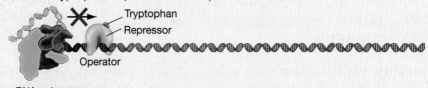

RNA polymerase bound to promoter

(b) When tryptophan is absent, transcription occurs.

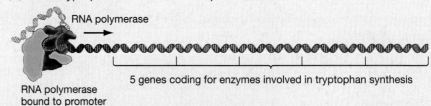

RNA polymerase bound to promoter

5 genes coding for enzymes involved in tryptophan synthesis

FIGURE 17.6 The *trp* Operon Is under Negative Control
(a) When tryptophan is present, it binds to a repressor protein. The tryptophan-repressor complex binds to the operator and shuts down transcription, which ends the synthesis of tryptophan. **(b)** The *trp* operon contains five coding genes.

that increases the efficiency of gene expression. By binding to one promoter, regulatory proteins can control the expression of several genes that are required for the same process. Polycistronic mRNAs are common in bacteria but rare or nonexistent in most eukaryotes.

The Impact of the *lac* Operon Model

Regulation of the *lac* operon provided an immensely important model system in genetics, partly because aspects of the model turned out to be important in many other species, genes, and operons. For example, follow-up work has shown that numerous bacterial genes are under negative control via repressor proteins. Even more important, the *lac* operon model introduced the idea that gene expression is regulated by physical contact between specific regions in regulatory proteins and specific regulatory sites in DNA. Stated another way, the model introduced the idea that gene expression could be modified by proteins that bind directly to regulatory sequences in DNA. Publication of the *lac* operon model was a watershed event in the history of biological science. Its impact on thinking about gene regulation is hard to overestimate.

As elegant as the model presented in Figure 17.5 is, however, something is missing. Where does glucose fit in? This is the final element of the regulatory system for lactose metabolism. As it turns out, it is also the element that is based on positive control of transcription.

✔ CHECK YOUR UNDERSTANDING

Negative control occurs when something must be taken away for transcription to occur. The *lac* operon repressor exerts negative control over three protein-coding genes by binding to the operator site in DNA near the promoter. For transcription to occur, an inducer molecule (a derivative of lactose) must bind to the repressor, causing it to release from the operator. You should be able to (1) explain why it is logical to observe that a molecule derived from lactose induces transcription of the *lac* operon; (2) diagram the *lac* operon, showing the relative positions of the operator, the promoter, and the three protein-coding genes; and (3) show what that operon looks like in the presence and absence of lactose.

17.4 Catabolite Repression and Positive Control

When glucose is present in the environment, transcription of the *lac* operon is reduced. This is true even when lactose is available to induce β-galactosidase expression. Given that glucose is the preferred carbon source, this aspect of control over the *lac* operon is logical. Glucose is produced when β-galactosidase cleaves lactose. When glucose is present, it is not necessary for the cell to cleave lactose and produce still more glucose.

Recall from Chapter 9 that biologists use the term *catabolism* to refer to reactions that break larger molecules into simpler subunits. The hydrolysis of lactose into its glucose and galactose subunits is an example of catabolism (**Figure 17.7a**). In many cases, operons that encode catabolic enzymes are inhibited when the end product of the reaction, the **catabolite**, is abundant. This form of end-product inhibition is **catabolite repression**. In the case of the *lac* operon, glucose is the catabolite. When glucose is abundant in the cell, transcription of the *lac* operon is decreased by catabolite repression (**Figure 17.7b**).

How does glucose prevent expression of the *lac* operon? An answer to this question began to emerge when researchers discovered a second major control element in the *lac* operon. This regulatory switch consists of a DNA sequence known as the **CAP binding site**, which is located just upstream of the *lac* promoter, and a regulatory protein called **catabolite activator protein (CAP)**, which binds to this DNA sequence.

To understand how CAP works, notice that not all promoters are created equal. Strong promoters allow efficient initiation of transcription by RNA polymerase; weak ones support much less efficient initiation of transcription. The *lac* promoter is weak. But when the CAP regulatory protein is bound to the

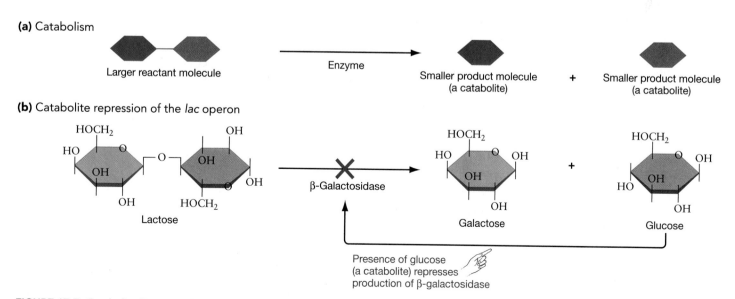

(a) Catabolism

(b) Catabolite repression of the *lac* operon

FIGURE 17.7 Catabolite Repression Is a Mechanism of Gene Regulation
(a) A generalized example of catabolism. **(b)** Catabolite repression occurs when one of the small product molecules represses the production of the enzyme responsible for the reaction. In the case of lactose metabolism, the production of β-galactosidase is suppressed when glucose is present.

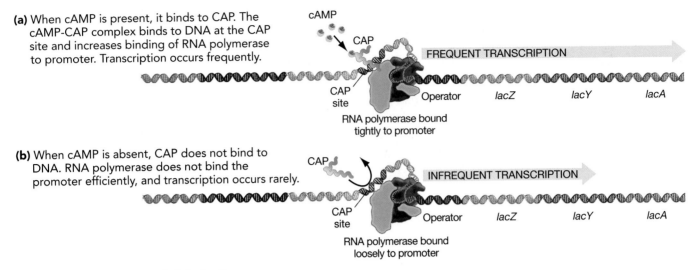

(a) When cAMP is present, it binds to CAP. The cAMP-CAP complex binds to DNA at the CAP site and increases binding of RNA polymerase to promoter. Transcription occurs frequently.

(b) When cAMP is absent, CAP does not bind to DNA. RNA polymerase does not bind the promoter efficiently, and transcription occurs rarely.

FIGURE 17.8 Positive Control of the lac Operon
(a) When glucose levels in an *E. coli* cell are low, cAMP is produced. cAMP then interacts with CAP to increase transcription of the *lac* operon. **(b)** When glucose is abundant, cAMP is rare in the cell and positive control does not occur.

CAP site just upstream of the *lac* promoter, the protein interacts with RNA polymerase in a way that allows transcription to begin much more frequently. CAP binding greatly strengthens the *lac* promoter. In this way, CAP exerts positive control of the *lac* operon. When CAP is active, transcription increases.

Researchers also discovered that CAP, like the repressor protein, is allosterically regulated. CAP changes shape when the regulatory molecule **cyclic AMP (cAMP)** binds to it. Only when cAMP is attached to it can CAP bind to DNA (**Figure 17.8a**). The same type of allosteric regulation goes on with the repressor and allolactose during negative control, except that the repressor binds to DNA only in the *absence* of the inducer. During positive control, the CAP-cAMP complex binds to the CAP binding site. As long as a repressor isn't bound to the operator, the complex increases the efficiency of transcription. If no cAMP is bound to CAP, then CAP has a conformation that does not allow binding to the CAP site (**Figure 17.8b**).

How Does Glucose Influence Formation of the CAP-cAMP Complex?

Where does the elusive molecule glucose fit into this scheme? Glucose's role in positive control of the *lac* operon is indirect. More specifically, it is mediated by the influence that glucose has on the concentration of cAMP. Glucose levels outside the cell and cAMP levels inside the cell are inversely related. When extracellular glucose concentrations are high, intracellular cAMP concentrations are low; when extracellular glucose concentrations are low, intracellular cAMP concentrations are high. This seesaw is driven by the enzyme **adenylyl cyclase**, which produces cAMP from ATP (**Figure 17.9a**). Adenylyl cyclase's activity is inhibited by extracellular glucose. To see the consequences of this fact, imagine a situation in which glucose is abundant outside the cell (**Figure 17.9b**). In this state, adeny-

lyl cyclase activity is low. Therefore, cAMP levels inside the cell are low. CAP is not in a CAP-cAMP complex, so it does not have the conformation that allows it to bind to the CAP site and stimulate *lac* operon transcription. Conversely, when the extracellular concentration of glucose is low, there is an increase in the intracellular level of cAMP. In this case, the CAP-cAMP complex forms, binds to the CAP site, and allows RNA polymerase to initiate transcription efficiently. As long as the repressor is not bound to the operator, transcription of the *lac* operon is initiated frequently and lactose can be employed as an alternative energy source.

It is important to note that the CAP-cAMP system influences many genes in addition to the *lac* operon. CAP sites are found adjacent to the promoters for several operons that are required for the metabolism of sugars other than glucose. When glucose levels fall and cAMP concentrations rise, the effect on gene regulation is similar to ringing an alarm bell. In response to the alarm, genes that encode enzymes required for the use of lactose, maltose, glycerol, and other food sources may be turned on—but only if the metabolite of a particular operon is available to the cell. Conversely, cAMP levels are low when glucose supplies are adequate. In this case, the genes for these enzymes are expressed only at low levels even when the appropriate metabolite is present.

Figure 17.10 summarizes how positive control and negative control combine to regulate the *lac* operon. The general message of this figure is that interactions among regulatory elements produce finely tuned control over gene expression. Because positive and negative control elements are superimposed, *E. coli* fully activates the genes for lactose metabolism only when lactose is available and glucose is scarce or absent. In this way, control over gene expression increases the ability of these cells to compete, grow, and reproduce efficiently.

(a) Glucose inhibits the activity of the enzyme adenylyl cyclase, which produces cAMP from ATP.

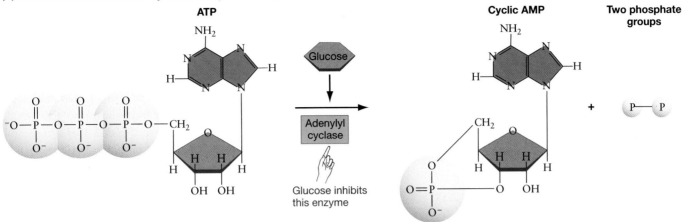

(b) The amount of cAMP and the rate of transcription of *lac* operon are inversely related to the concentration of glucose.

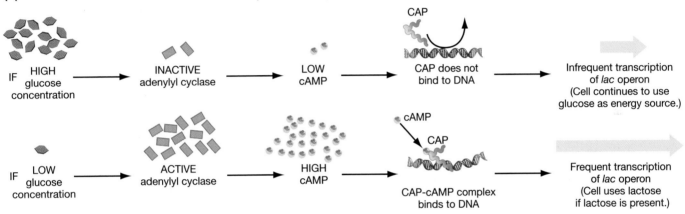

FIGURE 17.9 Cyclic AMP (cAMP) Is Synthesized When Glucose Levels Are Low

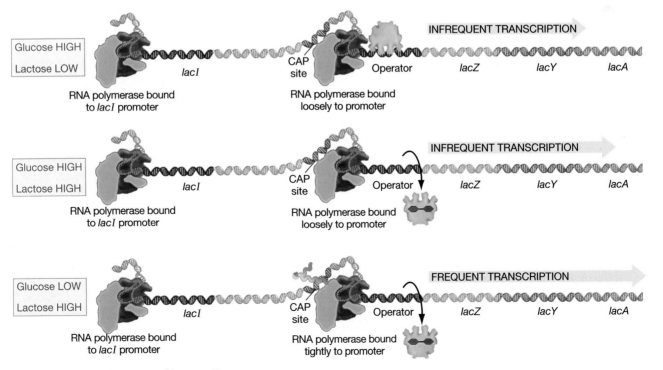

FIGURE 17.10 An Overview of Lactose Use

✔CHECK YOUR UNDERSTANDING

Positive control occurs when something must be added for transcription to occur. CAP exerts positive control over the *lac* operon by binding to the CAP site and increasing the transcription rate. CAP can bind only when it is complexed with cAMP—a signaling molecule indicating that glucose levels are low. You should be able to (1) diagram what the operon looks like when cAMP levels are low and high and (2) explain how positive control and negative control interact, based on the presence or absence of glucose and lactose.

17.5 The Operator and the Repressor— an Introduction to DNA-Binding Proteins

The insights that biologists gained from studying the *lac* operon proved to be relevant to many other genes and species. For example, transcription of bacterial and eukaryotic genes is either constitutive or controlled. When regulation occurs, it is negative, positive, or based on a combination of negative and positive factors. Negative control is based on repressor proteins, such as the *lacI* product, whereas positive control depends on transcription activator proteins, such as CAP. In both bacteria and eukaryotes, as in the *lac* operon, it is particularly common for genes to be regulated by a combination of several control systems. Catabolite repression is also extremely common.

Currently, biologists who are interested in gene regulation are trying to understand how DNA-binding proteins such as CAP and the repressor can control transcription. Precisely how do these proteins interact with DNA? How do small regulatory molecules such as cAMP and lactose induce changes in the conformation of these large DNA-binding proteins? Once again,

investigators are using the *lac* operon as a model system that offers insights applicable to all cells.

Before taking a detailed look at the repressor and operator, let's step back and think about what is required for a DNA-binding protein to control transcription. First, the protein must bind to a specific DNA sequence associated with a specific gene or operon. Often an investigator's first task is to identify what that sequence is. Second, the protein's structure must allow it to bind DNA, interact with a regulatory molecule such as cAMP or lactose, and affect RNA polymerase. What structural features make each of these functions possible? As it turns out, the *lac* operon's operator DNA sequence and repressor protein structure have features that are common to many regulatory elements in bacteria and eukaryotes.

Finding the Operator

To understand how the *lac* repressor controls transcription, a logical first step is to identify the DNA sequences it targets: the operator. By mapping mutants with defects in the operator, researchers were able to confirm that the target site is just downstream from the promoter. But what is the actual DNA sequence?

Investigators use a technique called DNA footprinting to find and characterize the sequences targeted by DNA-binding proteins. DNA footprinting was used to characterize the first promoter ever discovered (see Chapter 16); the technique is explained in detail in **Box 17.2**.

Footprinting experiments with the *lac* repressor yielded two new discoveries. First, the *lac* operon actually contains three sites where the repressor protein may bind. As **Figure 17.12a** shows, these sites are called O_1, O_2, and O_3. When the repressor protein is bound to DNA, the protein simultaneously binds O_1 and one of the two other operators. This finding required a modification of the original Jacob-Monod model, which proposed the existence of a single operator. Second, all three

BOX 17.2   DNA Footprinting

DNA footprinting is used to characterize DNA sequences that are bound by regulatory proteins. Researchers begin a footprinting study by obtaining DNA from a region of interest—the *lac* operon of *E. coli*, for example. Then they generate many copies of the sequence, using techniques introduced in Chapter 19. The identical DNA fragments are then radioactively labeled at one end (see step 1 of **Figure 17.11**). Once labeling is complete, half of the fragments are mixed with the DNA-binding protein being studied—in this case, the repressor—and

half are left without the DNA-binding protein (step 2). Both sets of fragments are then treated with a nuclease—an enzyme that cleaves DNA in random locations. The concentration of nuclease is so low and the exposure time is so short, however, that most fragments are never cut; those that are cut are cut just once. As step 3 shows, the nuclease treatment produces a population of DNA fragments of different lengths. In the sample containing the DNA-binding protein, the binding site is protected from the nuclease.

The DNA fragments produced by the nuclease treatment are separated by gel electrophoresis and visualized via autoradiography (see Chapter 3) or other techniques. As step 4 shows, the sample with the repressor lacks any cuts and thus any fragments in the segment that was protected by that protein. The gap in the series of fragments is the "footprint" of a DNA-binding protein. By aligning the footprinted region with the gene's DNA sequence, researchers can identify the nucleotide sequence of the binding site.

DNA FOOTPRINTING

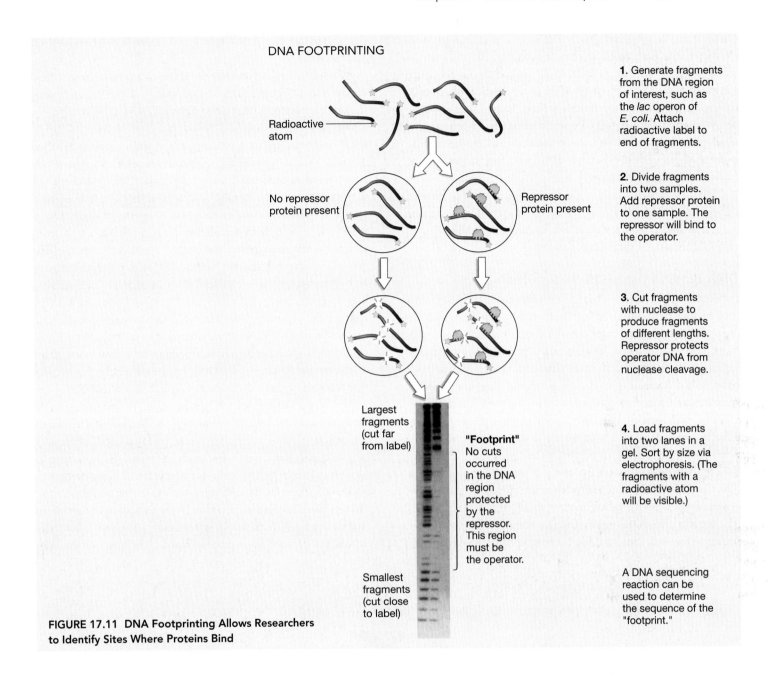

Radioactive atom

No repressor protein present

Repressor protein present

Largest fragments (cut far from label)

"Footprint"
No cuts occurred in the DNA region protected by the repressor. This region must be the operator.

Smallest fragments (cut close to label)

1. Generate fragments from the DNA region of interest, such as the *lac* operon of *E. coli*. Attach radioactive label to end of fragments.

2. Divide fragments into two samples. Add repressor protein to one sample. The repressor will bind to the operator.

3. Cut fragments with nuclease to produce fragments of different lengths. Repressor protects operator DNA from nuclease cleavage.

4. Load fragments into two lanes in a gel. Sort by size via electrophoresis. (The fragments with a radioactive atom will be visible.)

A DNA sequencing reaction can be used to determine the sequence of the "footprint."

FIGURE 17.11 DNA Footprinting Allows Researchers to Identify Sites Where Proteins Bind

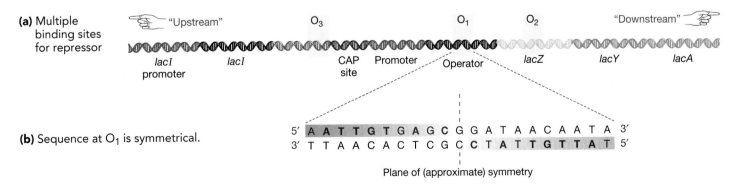

(a) Multiple binding sites for repressor

"Upstream" — O_3 — O_1 — O_2 — "Downstream"

lacI promoter — *lacI* — CAP site — Promoter — Operator — *lacZ* — *lacY* — *lacA*

(b) Sequence at O_1 is symmetrical.

5′ **A A T T G T** G A G **C** G G A T A A C A A T A 3′
3′ T T A A C A C T C G C **C T A** T **T G T T A T** 5′

Plane of (approximate) symmetry

FIGURE 17.12 Operator Sequences
(a) The lac operon contains three operator sequences, called O_1, O_2, and O_3, where the repressor can bind. A repressor simultaneously binds O_1 and either O_2 or O_3. **(b)** Many sequences targeted by DNA-binding proteins are symmetrical. In this example, all but three bases are identical on the two sides of the plane of symmetry.

operators have a similar DNA sequence that has an unusual characteristic: It contains an axis of symmetry. To understand what an axis of symmetry in DNA is, look at the O_1 sequence shown in **Figure 17.12b** (page 375). Note that if the sequence were rotated 180°, it would be little changed. This type of symmetry is known as **two-fold rotational symmetry** or **dyad symmetry**. Follow-up work showed that the nucleotide sequences recognized by many DNA-binding proteins have two-fold rotational symmetry.

Now that we've examined the operator, we need to consider the repressor. How does this DNA-binding protein recognize specific sequences along the double helix? In particular, why is two-fold rotational symmetry important?

DNA Binding via the Helix-Turn-Helix Motif

Like other proteins, the *lac* operon's repressor resembles a hand tool in one important respect: Both have distinct regions or parts. Consider a screwdriver, which consists of a handle, a shaft, and a blade. Each of these parts has a distinctive three-dimensional structure and function. Proteins also have regions or parts, called **domains**, that have a distinctive three-dimensional structure and function. Just as screwdrivers of various shapes and sizes all have a handle, shaft, and blade, it is common to observe the same domain in proteins of various shapes and sizes. A domain that is observed in many different proteins is known as a **motif**. Each discrete screwdriver has a handle domain, a shaft domain, and a blade domain. As a group, screwdrivers are characterized by a handle motif, a shaft motif, and a blade motif. The handle and shaft motifs are also found in hammers and socket wrenches, but the blade motif is unique to screw-drivers. Hundreds of different motifs have been characterized in proteins.

When investigators analyzed repressor proteins from different bacteria, they discovered that many of the bacteria contained the domain illustrated in **Figure 17.13a**. This structure is a **helix-turn-helix motif**. As its name indicates, it consists of two α-helices connected by a short stretch of amino acids that forms a turn. Physical models that were constructed during the 1980s suggested that the helix-turn-helix region is the section of the repressor that binds to DNA. This hypothesis was later supported by structural studies based on X-ray crystallography.

To see how a helix-turn-helix domain works, recall from Chapter 4 that the exterior of the DNA double helix has a minor groove and a major groove (**Figure 17.13b**). A protein that has a helix-turn-helix domain can bind to DNA because one of the helices interacts with the sugar-phosphate backbone of a DNA strand while the other helix binds to the base pairs in the major groove (**Figure 17.13c**). DNA does not have to become single stranded for the helix-turn-helix domain of the repressor to bind to the operator. This observation holds for nearly all proteins that recognize specific sequences of double-stranded DNA.

The section of the helix-turn-helix domain that binds inside the major groove is called the **recognition sequence**. The amino acid sequence of this section is crucial, because specific sequences of amino acids bind to specific sequences of DNA. Each type of regulatory protein with a helix-turn-helix motif has a unique sequence of amino acids in its recognition sequence. As a result, each of these regulatory proteins binds to a unique regulatory sequence in DNA. In many cases, precise control over gene expression is based on a precise chemical interactions between

(a) Helix-turn-helix motif of DNA-binding protein

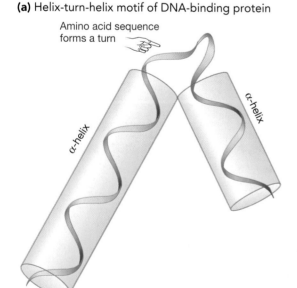

Amino acid sequence forms a turn

α-helix

α-helix

(b) Grooves of DNA

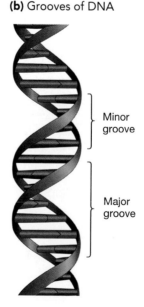

Minor groove

Major groove

(c) Recognition sequence of helix-turn-helix motif binds to DNA sequences in major groove.

FIGURE 17.13 Helix-Turn-Helix Motifs in DNA-Binding Proteins

QUESTION The part of a helix-turn-helix motif that interacts directly with DNA is called the recognition sequence. Why?

the amino acid sequence in a recognition sequence and the DNA sequence in the major groove of a regulatory site.

Biologists used X-ray crystallography to document the precise way that the *lac* repressor binds to the DNA of the operator. Consider the structure of the repressor alone, shown in **Figure 17.14a**. Note that the protein consists of four identical polypeptide chains, and that four helix-turn-helix domains project from the main body of the molecule. When researchers analyzed the structure of the repressor while it was bound to the operator sequence, they confirmed that each of these helix-turn-helix motifs binds operator DNA inside the major groove.

Now recall that the operator essentially is composed of two symmetrical halves. Each half of the operator sequence is bound by one of the repressor's helix-turn-helix motifs. The major grooves in the operator and the recognition helices of the repressor can pair up in this way because the operator is symmetrical and because the repressor consists of identical poly-peptides that are flipped 180° with respect to each other. In addition, because the repressor contains four helix-turn-helix domains and two of the four bind per operator, a total of two operators can be bound simultaneously by one repressor. When this occurs, a loop of DNA is created between O_1 and whichever other operator (O_2 or O_3) is bound (**Figure 17.14b**). This type of DNA looping is observed at many other bacterial promoters and at virtually all eukaryotic promoters.

(a) Structure of repressor protein

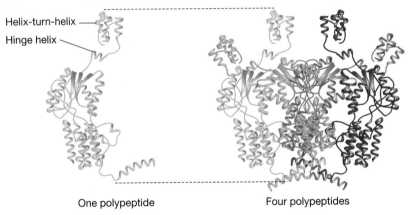

Helix-turn-helix

Hinge helix

One polypeptide   Four polypeptides

(b) DNA binding of repressor

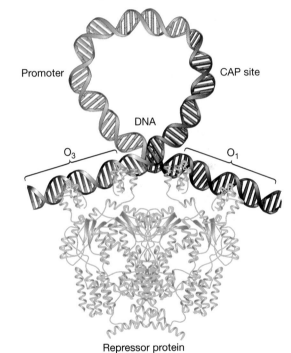

Promoter   CAP site

DNA

O_3 O_1

Repressor protein

(c) Interaction of inducer and repressor

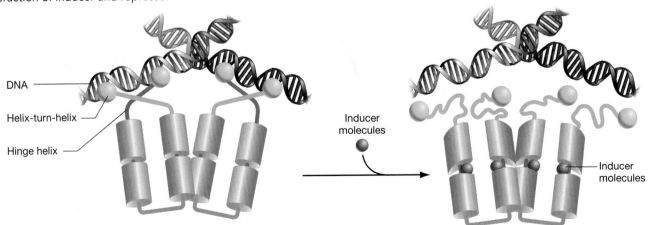

DNA

Helix-turn-helix

Hinge helix

Inducer molecules

Inducer molecules

FIGURE 17.14 How Does the Repressor Interact with the Operator and the Inducer?
(a) The repressor is a complex of four polypeptides, each a product of the *lacI* locus. **(b)** The repressor protein has four helix-turn-helix motifs to bind DNA at O_1 and either O_2 or O_3. **(c)** "Hinge helices" in the repressor exist only when the inducer is not present (left). When the inducer binds to the repressor, the hinge helices relax and cause the protein to fall off the operator DNA (right).

How Does the Inducer Change the Repressor's Affinity for DNA?

Negative control of the *lac* operon occurs when the repressor binds to the operator. In this state, the helix-turn-helix motifs of the repressor are attached to DNA. Even though RNA polymerase can bind to the promoter nearby, the enzyme cannot open the double helix and begin transcription. This physical interference is typical of how negative regulators work. Positive regulators, in contrast, work by stabilizing the interaction between RNA polymerase and the promoter.

Negative control continues as long as the repressor is bound to DNA. But when lactose is present, the repressor dissociates from the operator. How does the inducer affect the repressor in a way that disrupts DNA binding?

To answer this question, researchers analyzed the three-dimensional structure of the repressor when it was bound to a molecule called IPTG, which acts as an inducer in the *lac* system. Their data showed that IPTG binds at locations in the middle of the repressor and that it produces an important change in the shape of the protein near the helix-turn-helix domains. To see this change in shape, compare the left and right sides of **Figure 17.14c** (page 377). The diagram on the left summarizes the repressor's shape without the inducer. Pay attention to the structures labeled **hinge helices** near the helix-turn-helix motifs. The hinge helices lock the DNA-binding motifs into place when they contact the operator. As the diagram on the right shows, however, the hinge helices uncoil when IPTG or the inducer binds to the repressor. As a result, the two DNA-binding regions can't bind to the operator at the same time, and the protein releases from the site.

The "unhinging" of the hinge helices that occurs in the *lac* repressor turns out to be a unique mechanism for changing the activity of a regulatory protein. Still, several general lessons emerge from the study of this repressor. First, the shape and activity of a protein may change when a small molecule binds to it. Allosteric shifts like that observed in the *lac* repressor occur in thousands of different proteins. Second, key elements in transcriptional control are regulated at the **post-translational level**. For example, the repressor protein is always present because it is transcribed and translated constitutively at low levels. When a rapid change in *lac* operon activity is required, it does not occur via changes in the transcription or translation of the repressor. Rather, the activity of the already translated protein is altered. In virtually all cases, the activity of key regulatory proteins is controlled by post-translational modifications.

Future Directions

Now that biologists understand the physical basis of negative control and induction, what does the future hold? One promising line of research is focused on how the activity of many genes and operons is coordinated. For example, suppose that an *E. coli* cell in your gut is faced with starvation. Does it adjust the activity of hundreds or even thousands of its genes in order to survive? If so, how are changes in gene expression coordinated throughout the genome? Because biologists have a fairly solid understanding of how individual genes and operons are regulated, the current challenge is to explore how multiple genes and operons are controlled in a way that increases the fitness of bacterial cells. Answering these questions may be important to our understanding of how certain bacteria are able to sustain infections that threaten human health and how some bacterial species are able to detoxify pollutants by using them as food. Chapter 19 introduces some techniques that biologists use to study coordinated changes in gene expression.

ESSAY Control of Gene Expression in Bacterial Pathogens

A staggering number of bacteria are living in and on your body right now. To give just one example, the number of bacterial cells in your mouth at this moment far exceeds the number of people who have ever lived, including the world's current population of more than 6.4 billion. Virtually all of these bacteria are **commensal**—meaning they live in association with their host without causing harm.

As you know, however, a few bacterial species can cause disease. How are certain bacteria able to invade and harm us? How are disease-causing bacteria, or **pathogens**, transmitted from person to person? And how can bacterial infections be controlled or prevented? In many cases, the answers depend on an understanding of how the expression of key genes is controlled. To drive this point home, consider *Vibrio cholerae*, the causative agent of cholera (**Figure 17.15**).

Cholera is a disease characterized by watery diarrhea that, if untreated, can lead to dehydration and death. The disease is currently most prevalent in Africa and Asia, where 5 million cases occur each year. Because *V. cholerae* is transmitted from person to person when diarrhea from an infected individual contaminates drinking water, cholera occurs primarily in regions with poor sanitation.

A staggering number of bacteria are living in and on your body right now.

Vibrio cholerae is extraordinarily flexible regarding the habitats it occupies. For example, near the confluence of the Ganges and Brahmaputra Rivers in the Bengal region of India and Bangladesh, *V. cholerae* can be found as free-swimming cells, inside two different eukaryotes, in dense layers coating rocks and other underwater surfaces, and inside humans. In the Bengal, most incidences of human infection occur during an outbreak that coincides with the end of

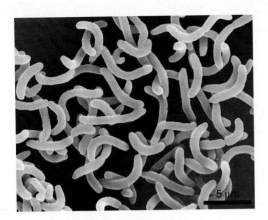

FIGURE 17.15 *Vibrio cholerae*, the Bacterium That Causes Cholera
[©Dr. Dennis Kunkel/Visuals Unlimited].

the monsoon rains in October and November each year. This observation suggests that environmental signals trigger *V. cholerae*'s yearly emergence as a human pathogen.

What genes allow *V. cholerae* to act as a human pathogen, and how are these genes regulated? Once it is ingested by a person, *V. cholerae* confronts an environment that is markedly different from its other habitats. In fact, most of the ingested bacteria are killed in the harsh, acidic environment of the stomach. The few that survive enter the small intestine, where they encounter a thick layer of mucus that protects the intestinal cells. Then, in response to a still-unknown signal in the environment, major changes in *V. cholerae* gene expression occur. The new proteins that are produced allow *V. cholerae* cells to move through the mucus layer and make contact with the intestinal cells. Each invading cell then produces a short, hairlike projection called a pilus, which allows the *V. cholerae* cell to attach to an intestinal cell. The proteins required to create this structure result from the coordinately regulated transcription of pilus-specific genes.

Once the bacterial cells have attached to intestinal cells, *V. cholerae* begins to manufacture and secrete a toxin composed of two different polypeptide chains. When intestinal cells take up the toxin, a devastating chain of events occurs. The toxin activates a host protein that stimulates human adenylyl cyclase. Massive amounts of cAMP are produced in response, resulting in a flow of Cl^- ions from the affected intestinal cells into the center (lumen) of the small intestine. Water follows by osmosis. The result is the watery diarrhea symptomatic of cholera.

Although the diarrhea can be fatal to the host, it increases the fitness of *V. cholerae* cells in environments where human sanitation is bad. Because *V. cholerae* cells are shed in diarrhea, they can enter drinking-water supplies if systems for transporting and processing human sewage are in poor repair or nonexistent. In this way, *V. cholerae* cells that induce diarrhea are efficiently transmitted to new hosts. *Vibrio cholerae* cells that have alleles responsible for diarrhea have higher fitness than do *V. cholerae* cells that lack these alleles.

The genes that encode the pilus proteins and cholera toxin are under the control of regulatory genes called the *Tox* (for toxin) loci. As **Figure 17.16** shows, *ToxR* produces a transcription factor that responds to a signal in the intestinal environment. When the ToxR protein is activated, it stimulates transcription of the *ToxT* gene. The *ToxT* product, ToxT, is a DNA-binding protein that activates transcription of several genes required for infection, including the genes that encode the cholera toxin subunits. These genes are called *ctxA* and *ctxB*. As you might suspect, they are part of a polycistronic operon.

Biologists are currently trying to translate research on gene regulation into better tools for fighting cholera. For example, blocking the activity of the ToxR or ToxT proteins should prevent disease. This hypothesis has been supported by the observation that *ToxR*⁻ mutants are harmless. Researchers are now attempting to develop drugs that prevent ToxR from binding the signal molecule, that stop ToxR from activating *ToxT*, or that prevent ToxT from contacting DNA.

The axiom "knowledge is power" applies here: Although drugs like those proposed earlier are not yet available, research on gene regulation has provided several novel and promising drug targets.

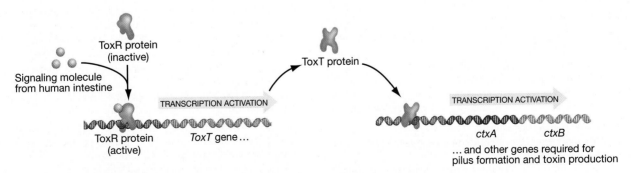

FIGURE 17.16 In the Cholera Bacterium, the Genes Required for Infection Are Regulated
QUESTION Biologists have not been able to identify the signaling molecule that triggers the cascade of events shown here. What properties must this molecule have?

CHAPTER REVIEW

Summary of Key Concepts

▪ **Changes in gene expression allow bacterial cells to respond to environmental changes.**

Transcription in bacteria is constitutive or inducible. Constitutive expression occurs in genes whose products are required at all times, such as genes that encode glycolytic enzymes. Expression of most genes is induced by environmental signals, however, so gene products are produced on demand.

Precise regulation over gene expression is important to bacterial cells. For example, some species switch between different food sources to take advantage of the nutrients currently available in their environment. Efficient use of nutrients is beneficial because cells are usually competing for access to resources. In the intestinal bacterium *E. coli*, for example, changes in gene expression allow cells to use lactose as a food source when glucose is not available.

▪ **Gene expression can be controlled at three levels: transcription, translation, or post-translation (protein activation).**

Among the three levels of gene regulation, there is a trade-off between the speed of response to changed conditions and the efficient use of resources. In post-translational control, for example, the change in gene expression is extremely fast but demands a considerable investment of resources to make and regulate the protein in question.

▪ **Transcriptional control can be positive or negative. In the case of the genes for lactose metabolism in *E. coli*, positive control occurs when a regulatory protein binds to DNA and increases**

the transcription rate. Negative control occurs because another regulatory protein must be released before transcription can occur.

Protein-coding genes involved in lactose metabolism are normally transcribed at low levels because a repressor protein binds to a DNA sequence—the operator—near their promoter. But when present, lactose binds to the repressor and induces a conformational change that causes the repressor to fall off the operator. Rapid transcription of the *lac* loci does not begin, however, unless glucose is also lacking. When glucose is scarce, cAMP levels rise in the cell. This is important because the regulatory protein CAP undergoes a conformational change when it is bound to cAMP. The CAP-cAMP complex binds to a control sequence near the *lac* promoter and increases transcription by facilitating binding by RNA polymerase.

Web Tutorial 17.1   The *lac* Operon

▪ **Many regulatory proteins bind to specific sites in DNA. Because different regulatory proteins have different amino acid sequences, they bind to different DNA sequences.**

The regulatory proteins involved in negative control and positive control are often allosterically regulated, meaning that they have the ability to switch between two different conformations in response to binding a small molecule—often a nutrient or product affected by the gene being regulated. The change in conformation affects the protein's ability to bind to DNA.

Questions

Content Review

1. Genes for enzymes in glycolysis are expressed constitutively. Why?
   a. Their expression is controlled at three levels: transcriptional, translational, and post-translational.
   b. Glycolysis occurs at all times, so they need to synthesize ATP-producing enzymes constantly.
   c. Transcription is activated only in response to signals indicating that ATP supplies are low.
   d. They do not need to be expressed when fermentation instead of cellular respiration is occurring.

2. Why do researchers frequently begin a search for mutants by exposing organisms to UV light or X-rays?
   a. The treatment causes mutants to turn yellow.
   b. The treatment triggers gene expression.
   c. The treatment exposes constitutive mutants.
   d. The treatment increases the mutation rate by damaging DNA.

3. What is the goal of a genetic screen?
   a. to identify individuals that have mutations in genes that are required for a particular process
   b. to produce a large number of mutant individuals, which can later be analyzed in detail
   c. to determine whether the mutations carried by particular individuals are in protein-coding genes or regulatory elements
   d. to determine the fitness or viability of mutant individuals

4. Why are the genes involved in lactose metabolism considered to be an operon?
   a. They occupy adjacent locations on the *E. coli* chromosome.
   b. They have a similar function.
   c. They are all required for normal cell function.
   d. They are under the control of the same promoter.

5. What is catabolite repression?
   a. a mechanism that turns off the synthesis of enzymes responsible for catabolic reactions when the product is present
   b. a mechanism that turns off the synthesis of enzymes responsible for catabolic reactions when the product is absent
   c. repression that occurs because of allosteric changes in a regulatory protein
   d. repression that occurs because of allosteric changes in a DNA sequence

6. What is a helix-turn-helix motif?
   a. a protein domain involved in folding into the active conformation
   b. a protein domain involved in induction
   c. a protein domain involved in DNA binding
   d. a protein domain involved in catalysis

Conceptual Review

1. Explain the difference between positive and negative control over transcription. Why is it advantageous for the *lac* operon in *E. coli* to be under both positive control and negative control? What would happen if only negative control occurred? What would happen if only positive control occurred?

2. In *E. coli*, rising levels of cAMP can be a considered a starvation signal. Explain.

3. Explain the role of post-translational control in the *lac* operon.

4. CAP is also known as the cAMP-receptor protein. Why?

5. Explain how the *lac* repressor binds to DNA. Then explain the relationship between the two-fold rotational symmetry of the operator and the four-part structure of the repressor protein.

6. The galactose released when β-galactosidase cleaves lactose enters the glycolytic pathway in *E. coli* once a series of enzyme-catalyzed reactions has converted the galactose to glucose-6-phosphate. Why, then, is glucose, not lactose, the preferred sugar in *E. coli*?

Group Discussion Problems

1. You are interested in using bacteria to metabolize wastes at an old chemical plant and convert them into harmless compounds. You find bacteria that are able to tolerate high levels of the toxic compounds toluene and benzene, and you suspect that it is due to the ability of the bacteria to break these compounds into less-toxic products. If that is true, these toluene- and benzene-resistant strains will be valuable for cleaning up toxic sites. How could you find out whether these bacteria have enzymes that allow them to metabolize toluene?

2. Assuming that the bacteria you examined in Group Discussion Problem 1 do have an enzymatic pathway to break down toluene, would you predict that the genes involved are constitutively expressed, under positive control, or under negative control? Why? What experiments could you conduct to test your hypothesis?

3. The *lacI* mutants are constitutive because the repressor fails to recognize and bind to the operator region. Other repressor mutants have been isolated that are called *LacI^S* mutants. These repressor proteins continue to bind to the operator, even in the presence of the inducer. How would this mutation affect the function of the *lac* operon? Specifically, how well would *LacI^S* mutants do in an environment that has lactose as its sole sugar?

4. X-gal is a colorless, lactose-like molecule that can be split into two fragments by β-galactosidase. One of these product molecules is blue. The following photograph is a close-up of *E. coli* colonies growing in a medium that contains lactose.

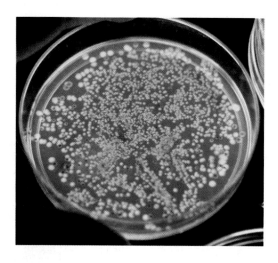

Draw a line to three colonies whose cells have functioning copies of β-galactosidase. Draw a line to three colonies whose cells have mutations in the *lacZ* locus or in one of the genes involved in regulation of the *lacZ*. Suppose you could analyze the sequence of the β-galactosidase gene from each of the mutant colonies. How would these data help you distinguish which cells are structural mutants and which are regulatory mutants?

Answers to Multiple-Choice Questions   1. b; **2.** d; **3.** a; **4.** d; **5.** a; **6.** c

18 Control of Gene Expression in Eukaryotes

KEY CONCEPTS

▦ Changes in gene expression allow eukaryotic cells to respond to changes in the environment and cause distinct cell types to develop.

▦ In eukaryotes, DNA is packaged with proteins into complex structures that must be opened before transcription can occur.

▦ In eukaryotes, transcription can be initiated only when specific proteins bind to the promoter and to regulatory sequences that may be close to the promoter or far from it.

▦ Alternative splicing allows a single gene to code for several different products.

▦ Cancer can develop when mutations disable genes that regulate cell-cycle control genes.

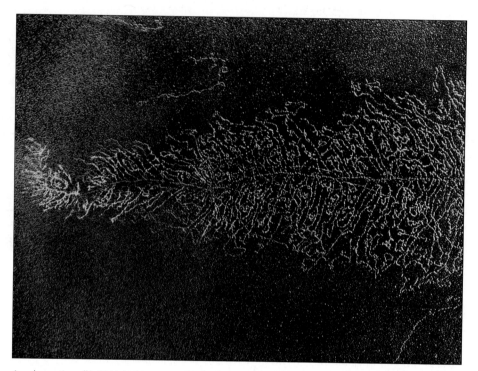

A eukaryotic cell's DNA being transcribed. The horizontal strand in the middle of the micrograph is DNA; the yellow strands above and below it are mRNAs.

In bacteria, precise regulation of genes is essential for cells to respond to changes in their environment, as Chapter 17 indicated. In *Escherichia coli*, the genes that are required to import and cleave lactose are rapidly expressed only when the cells have to rely on lactose as a source of energy—meaning glucose is absent and lactose is present. *Vibrio cholerae* can live in a wide array of environments, but the genes that allow the cells to infect a human are expressed only when the cells find themselves inside someone's small intestine.

Unicellular eukaryotes face similar challenges. Consider the yeast *Saccharomyces cerevisiae*, which is used extensively in the production of beer, wine, and bread. In nature, the cells live on the skins of grapes and other fruits. In the wild, variation in temperature and humidity can be high. In addition, the sugars that are available to the cells change dramatically in type and concentration as the fruit that they are living on ripens, falls, and rots. For yeast cells to grow and reproduce efficiently, gene expression has to be modified in response to these environmen-

tal changes. Further, the offspring of a cell that lived on a grape may be blown to a nearby orchard and take up residence on an apple. The genes that are expressed in the cells of grapes versus those expressed in the cells of apples vary markedly.

The cells that make up multicellular eukaryotes face an even greater challenge. Consider that, at one time, your body consisted of a single cell—a fertilized egg. Now you have trillions of cells, each of which has a specialized structure and function. You have heart muscle cells, lung cells, nerve cells, skin cells, and so on. Even though these cells look and behave very differently, each of them contains the same genes. Your bone cells are different from your blood cells not because they contain different genes, but because they express different genes. Your bone cells have blood cell genes—they just don't transcribe them.

How is it possible for the cells inside your body to express unique sets of genes? The answer is that your cells respond to changes in their environment, just as bacteria and unicellular eukaryotes do. But instead of responding to the presence of mole-

382

cules such as glucose or lactose in the external environment, your cells respond to the presence of signals from other cells—from an internal environment. As a human being or an oak tree develops, cells that are located in different parts of the organism are exposed to different signals. As a result, they express different genes. Differential gene expression is responsible for creating different cell types, arranging them into tissues, and coordinating their activity to form the multicellular society we call an individual.

How does all this happen? Unit 4 explores the nature of the signals that trigger the formation of muscle, bone, leaf, and flower cells in multicellular organisms. Here we need to focus on what happens once a eukaryotic cell receives a signal. Let's begin by surveying the various ways that gene expression can be controlled and then plunge into a detailed look at key points in gene regulation. The chapter closes with a look at how breakdowns in the control of gene expression can help trigger cancer.

18.1 Mechanisms of Gene Regulation— an Overview

Like bacteria, eukaryotes can control gene expression at the levels of transcription, translation, and post-translation (protein activation or inactivation). But as **Figure 18.1** shows, two additional levels of control occur in eukaryotes as genetic information flows from DNA to proteins. The first level involves the DNA-protein complex at the top of the figure. In eukaryotes, DNA is wrapped around proteins to create a protein-DNA complex called **chromatin**. Eukaryotic genes have promoters, just as bacterial genes do. Before transcription can begin in eukaryotes, DNA near the promoter must be released from tight interactions with proteins so that RNA polymerase can make contact with the promoter. To capture this point, biologists say that **chromatin remodeling** must occur prior to transcription.

The second level of regulation that is unique to eukaryotes involves **RNA processing**—the steps required to produce a mature, processed mRNA from a **primary RNA transcript**, the preliminary result of transcription. Recall from Chapter 16 that introns have to be spliced out of primary transcripts. In some cases, carefully orchestrated alterations in the splicing pattern occur. When the splicing events in a particular primary RNA transcript change, a different message emerges. The altered message, in turn, leads to the production of a different product.

How many of the five potential control points shown in Figure 18.1 are actually used? In all cells, every one of them is employed at certain times. The primary focus of this chapter is to explore three of these key transition points: (1) control of packing DNA into chromatin (control over chromatin remodeling), (2) control of transcription, and (3) control of RNA processing.

To appreciate the breadth and complexity of gene regulation in eukaryotes, let's follow the series of events that occurs as an embryonic cell responds to a developmental signal. Suppose a molecule arrives that specifies the production of a muscle-specific protein. What happens next?

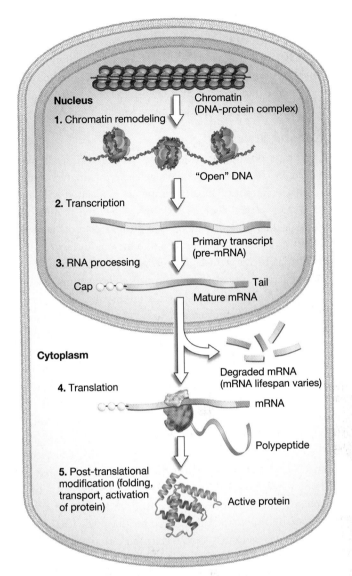

FIGURE 18.1 In Eukaryotes, Gene Expression Can Be Controlled at Many Different Steps
Numerous steps occur during protein synthesis in eukaryotes. Each step provides an opportunity for regulation of gene expression.
EXERCISE Place and label a horizontal line such that mechanisms of transcriptional control fall above the line and mechanisms of post-transcriptional control fall below it.

18.2 Eukaryotic DNA and the Regulation of Gene Expression

If the arrival of a signaling molecule from outside the cell is going to result in the transcription of a particular gene, the DNA around the target gene must be drastically remodeled. To appreciate why, consider that a typical cell in your body contains about six billion base pairs of DNA. Lined up end to end, these nucleotide pairs would form a double helix about 2 m (6.5 feet) long. But the nucleus that holds this DNA is thinner than a sheet of paper. In eukaryotes, DNA is packed inside the nucleus so tightly that RNA polymerase can't access it. Part of this packing is done by supercoiling—meaning the DNA double helix is twisted on itself many times, just as in bacterial chromosomes (see Chapter 7). The supercoiled DNA found in bacteria does not need to be altered extensively before it can be transcribed, however. Eukaryotic DNA is different—it has to undergo a series of significant physical changes before transcription can take place. The reason is that supercoiling is just part of the packing system found in eukaryotes.

Chromatin Structure

The first data on the physical nature of eukaryotic DNA were published in the early 1900s, when chemical analyses established that eukaryotic DNA is intimately associated with proteins. Later work documented that the most abundant DNA-associated proteins belong to a group called the **histones**. In the 1970s electron micrographs like the one in **Figure 18.2a** revealed that the protein-DNA complex, or chromatin, has a regular structure. In some preparations for electron microscopy, chromatin actually looked like beads on a string. The "beads" came to be called **nucleosomes**. More details emerged in 1984 when researchers determined the three-dimensional structure of eukaryotic DNA by using X-ray crystallography (a technique introduced in Chapter 4). The X-ray crystallographic data indicated that each nucleosome consists of DNA wrapped almost twice around a core of eight histone proteins. As **Figure 18.2b** indicates, between each pair of nucleosomes there is a "linker" stretch of DNA associated with a histone called **H1**.

The intimate association between DNA and histones occurs in part because DNA is negatively charged and histones are positively charged. DNA has a negative charge because of its phosphate groups; histones are positively charged because they contain many lysine and/or arginine residues (see Chapter 3).

More recent work has shown that there is another layer of complexity in eukaryotic DNA. As **Figure 18.2c** indicates, H1 histones interact with each other and with histones in other nucleosomes to produce a tightly packed structure. Based on its width, this structure is called the **30-nanometer fiber**. (Recall that a nanometer is one-billionth of a meter and is abbreviated nm.) Often, 30-nm fibers are packed into still larger structures, which organize DNA inside the nucleus.

The elaborate structure of chromatin does more than just package DNA into the nucleus. Chromatin structure has profound implications for the control of gene expression. To drive this point home, consider the 30-nanometer fiber illustrated in Figure 18.2c. If this tightly packed stretch of DNA contains a promoter, how can RNA polymerase bind to it and initiate transcription?

Evidence that Chromatin Structure Is Altered in Active Genes

Once the nucleosome-based structure of chromatin was established, biologists hypothesized that the close physical interaction between DNA and histones must be altered for RNA polymerase to make contact with DNA. More specifically, biologists hypothesized that a gene could not be transcribed until the chromatin near its promoter was remodeled. The idea was that the chromatin must be relaxed or decondensed for RNA

(a) Nucleosomes in chromatin

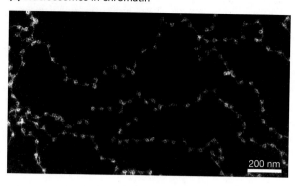

200 nm

(b) Nucleosome structure

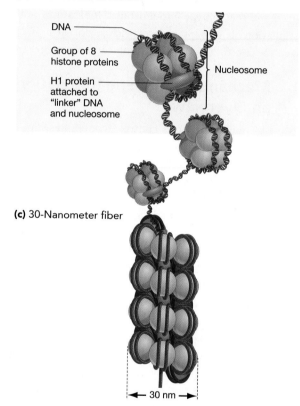

DNA

Group of 8 histone proteins

H1 protein attached to "linker" DNA and nucleosome

Nucleosome

(c) 30-Nanometer fiber

←— 30 nm —→

FIGURE 18.2 Chromatin Has Several Levels of Structure
(a) Electron micrograph of chromatin, showing that bead-like structures, called nucleosomes, are linked by short stretches of DNA. **(b)** X-ray crystallographic data revealed that nucleosomes consist of DNA wrapped around a core of eight histone proteins. The linker DNA is associated with a particular histone called H1. **(c)** A model of the structure of the 30-nanometer fiber. It proposes that H1 proteins form a core in the center of the fiber. Each ball here represents a group of eight histones, to make the overall structure of the 30-nanometer fiber clearer.

polymerase to bind to the promoter. If so, then chromatin remodeling would represent the first step in the control of eukaryotic gene expression. Two types of studies have provided strong support for this hypothesis.

The first type of evidence comes from studies with the enzyme DNase, which cuts DNA at random locations. DNase cannot cut DNA efficiently if the molecule is tightly complexed with his-

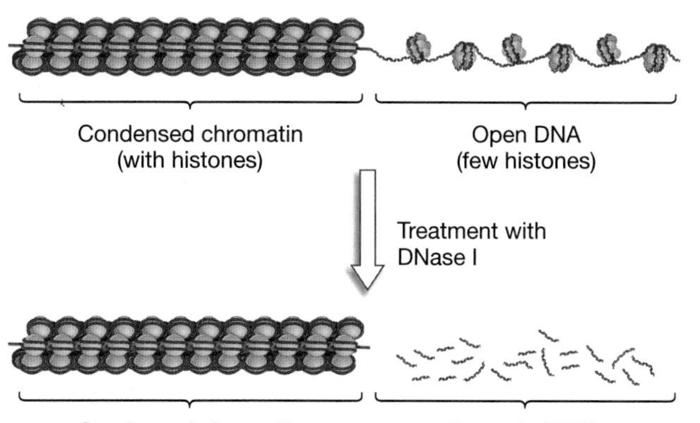

DNase assay for chromatin structure

Condensed chromatin
(with histones)

Open DNA
(few histones)

Treatment with
DNase I

Condensed chromatin

Degraded DNA

FIGURE 18.3 DNase Cannot Cut Intact Condensed Chromatin
DNase is an enzyme that cuts DNA at random locations.

tones. As **Figure 18.3** shows, the enzyme works effectively only if DNA is in the "open" configuration. Harold Weintraub and Mark Groudine used this observation to test the hypothesis that the DNA of actively transcribed genes is in an open configuration. They performed the test by comparing the structure of chromatin in two genes of the same cell type: the β-globin and ovalbumin genes of chicken blood cells. β-Globin is a protein that is part of the hemoglobin found in red blood cells; ovalbumin is a major protein of egg white. In blood cells, the β-globin gene is transcribed at high levels. The ovalbumin gene, in contrast, is not transcribed at all. After treating blood cells with DNase and then analyzing the state of the β-globin and ovalbumin genes, the researchers found that DNase cut up the β-globin gene much more readily than the ovalbumin gene. They interpreted this finding as evidence that chromatin in blood cells was in an open configuration at the β-globin gene but closed at the ovalbumin gene. Analogous studies using DNase on different genes and in different cell types yielded similar results.

The second type of evidence in support of the chromatin-remodeling hypothesis comes from studies of mutant brewer's yeast cells that do not produce the usual complement of histones. Researchers found that many yeast genes that are normally never transcribed are instead transcribed at high levels at all times in these mutant cells. To interpret this finding, biologists hypothesized that the lack of histone proteins prevented the assembly of normal chromatin. If the absence of normal histone-DNA interactions promotes transcription, then the presence of normal histone-DNA interactions must prevent it.

Taken together, the data suggest that the normal or default state of eukaryotic genes is to be turned off. If so, then gene expression depends on chromatin being opened up in the promoter region.

How Is Chromatin Altered?

Although much remains to be learned about how chromatin is opened up prior to transcription, researchers have made impor-

tant progress recently. One key finding is that two major types of protein are involved in modifying chromatin structure. One group of proteins creates structures called **chromatin-remodeling complexes**, which reshape chromatin through a series of reactions that are dependent on ATP. It is not known exactly how these ATP-dependent enzymes work, however. The second type of chromatin-modifying protein works by adding small molecules such as acetyl (CH_3COOH) or methyl (CH_3) groups to histones. These processes are known as **acetylation** and **methylation**, respectively.

Figure 18.4 shows how the enzymes called **histone acetyl transferases (HATs)** affect chromatin structure. HATs acetylate the positively charged lysine residues in histones. When a HAT adds an acetyl group to selected histones, the number of positive charges on the histones is reduced. The result is less electrostatic attraction between histones and the negatively charged DNA. The association between nucleosomes and DNA is loosened and the chromatin decondenses.

Figure 18.4 also shows how chromatin is "recondensed" by a group of enzymes called **histone deacetylases (HDACs)**, which remove the acetyl groups added by HATs. Histone deacetylase activity reverses the effects of acetylation, returning chromatin to its default condensed state. If HATs are an on switch for transcription, HDACs are the off switch.

What happens once a section of DNA is opened up and exposed to RNA polymerase? In bacteria, the protein sigma binds to RNA polymerase and allows it to contact the promoter. Then transcription begins if any required positive regulators, such as CAP, are in place and negative regulators, such as the repressor, are absent. Does anything similar happen in eukaryotes?

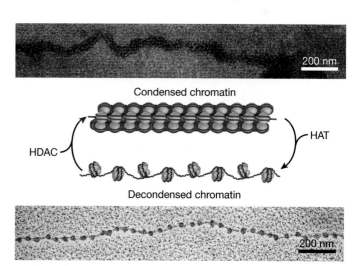

200 nm

Condensed chromatin

HDAC

HAT

Decondensed chromatin

200 nm

FIGURE 18.4 Chromatin Decondensation and Condensation Are Driven by Histone Acetyl Transferases and Histone Deacetylases
Histone acetyl transferases (HATs) cause chromatin to decondense, and histone deacetylases (HDACs) cause it to condense. EXERCISE In decondensed chromatin, each nucleosome has an acetyl group attached to it. Indicate this by adding "Ac" to several nucleosomes in the bottom part of the drawing.

✓ CHECK YOUR UNDERSTANDING

Eukaryotic DNA is wrapped tightly around histones, forming nucleosomes, which are then coiled into dense structures called 30-nm fibers. Before transcription can begin, the chromatin (DNA-protein complex) must be relaxed by chromatin-remodeling complexes and HATs so that RNA polymerase can contact the promoter. You should be able to (1) draw and label the structure of a nucleosome and a 30-nm fiber and explain their function, and (2) compare and contrast the structure of DNA in bacteria versus that in eukaryotes.

18.3   Regulatory Sequences and Regulatory Proteins

Chapter 16 introduced the **promoter**—the site in DNA where RNA polymerase binds to initiate transcription. Eukaryotic promoters are located about 30 nucleotides upstream of the point where RNA polymerase begins transcription. Many contain a specific base sequence, known as a **TATA box**, where a sigma-like protein binds and allows RNA polymerase to contact DNA. Based on their position and function, eukaryotic promoters are much like bacterial promoters. Recall from Chapter 16 that bacterial promoters may vary in sequence and bind different types of sigma proteins. Eukaryotic genes that are transcribed by RNA polymerase II also have promoters that vary in sequence, but all eukaryotic promoters are bound by the same protein: the **TATA-binding protein (TBP)**.

If eukaryotic genes have promoters that interact with the same promoter-binding protein, how can transcription be controlled? The answer lies in interactions between regulatory proteins and regulatory sequences. Eukaryotic regulatory proteins are analogous to *E. coli*'s CAP and the repressor protein analyzed in Chapter 17. **Regulatory sequences** are sections of DNA that are involved in controlling the activity of genes, similar to the CAP site and operators introduced in Chapter 17.

As it turns out, some regulatory sequences found in eukaryotes are very similar to those observed in bacteria. Others are radically different. Let's take a closer look.

Promoter-Proximal Elements

The first regulatory sequences in eukaryotic DNA were discovered in the late 1970s, when Yasuji Oshima and co-workers developed an important model system to study control of gene expression in eukaryotes. Much as François Jacob and Jacques Monod had focused on lactose metabolism in *E. coli*, Oshima's team set out to understand how yeast cells control the metabolism of the sugar galactose. When galactose is absent, yeast cells produce tiny quantities of the enzymes required to metabolize it. But when galactose is present, transcription of the genes encoding these enzymes increases by a factor of 1000.

Oshima and co-workers focused their early work on mutants that were unable to use galactose. The team's first major result was the discovery of mutant cells that failed to produce any of the five enzymes required for galactose metabolism, even

if galactose was present. To interpret this observation, they hypothesized that the cells had a knock-out (loss-of-function) mutation that completely disabled a regulatory protein. Like CAP, this hypothesized protein was thought to exert positive control over the five genes. The team called the gene *GAL4* and the protein product GAL4. (In *S. cerevisiae*, the names of genes are written in full caps and italicized. As in other species, gene names are italicized but protein names are not.)

This hypothesis was supported when other researchers isolated the regulatory protein and found that it has a DNA-binding domain, analogous to the helix-turn-helix motif introduced in Chapter 17. The domain binds to a 20-base-pair stretch of DNA located just upstream from the promoter for the five genes that GAL4 regulates. The location and structure of this regulatory sequence are comparable to those of the CAP binding site in the *lac* operon of *E. coli*. Similar regulatory sequences have now been found in a wide array of eukaryotic genes and species. Because such sequences are located close to the promoter and bind regulatory proteins, they are termed **promoter-proximal elements**.

As **Figure 18.5** shows, promoter-proximal elements are just upstream from the promoter and the gene's start site; the exons and introns that make up the actual gene are farther downstream. Unlike the promoter itself, promoter-proximal elements have sequences that are unique to specific genes. In this way, they furnish a mechanism for eukaryotic cells to exert precise control over transcription.

The discovery of positive control and promoter-proximal elements provided a satisfying parallel between gene regulation in bacteria and in eukaryotes. This picture changed, however, when researchers discovered a new class of eukaryotic DNA regulatory sequences—sequences unlike anything in bacteria.

Enhancers

Susumu Tonegawa and colleagues made a discovery that may rank as the most startling in the history of research on gene expression. Tonegawa's group was exploring how human immune system cells regulate the genes involved in the production of antibodies. **Antibodies** are proteins that bind to specific sites on other molecules. In your immune system, antibodies bind to viruses and bacteria and mark them for destruction by other

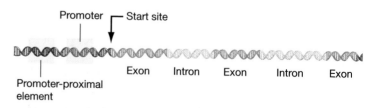

FIGURE 18.5 Promoter-Proximal Elements Regulate the Expression of Some Eukaryotic Genes

All eukaryotic genes have promoters, the sites where RNA polymerase initially is in contact with the gene. Exons and introns are not drawn to scale throughout this chapter. They are typically very large compared with promoters and promoter-proximal (regulatory) elements.

cells. The antibody gene that the researchers were working with is broken into many introns and exons. Introns are DNA sequences spliced out of the primary mRNA transcript; exons are regions of eukaryotic genes that are included in the mature RNA once splicing is complete. The biologists used techniques that will be introduced in Chapter 19 to place copies of an intron in new locations and had found that, when they placed the intron close to a gene, the gene's transcription rate increased. Based on this observation, Tonegawa and co-workers hypothesized that the intron contained some sort of regulatory sequence.

To test this hypothesis, the biologists performed what is now considered a classic experiment (**Figure 18.6**). The protocol was simple in concept: Starting with a human antibody-producing gene that included an intron flanked by two exons, as shown in step 1 of Figure 18.6, the team removed different portions of the intron and observed whether transcription of the gene still took place. The researchers used enzymes to cut out specific pieces of the intron in multiple copies of the gene (step 2) and to ligate the resulting sections back together (step 3). Each of the modified genes was missing a different section of the intron. Then the team

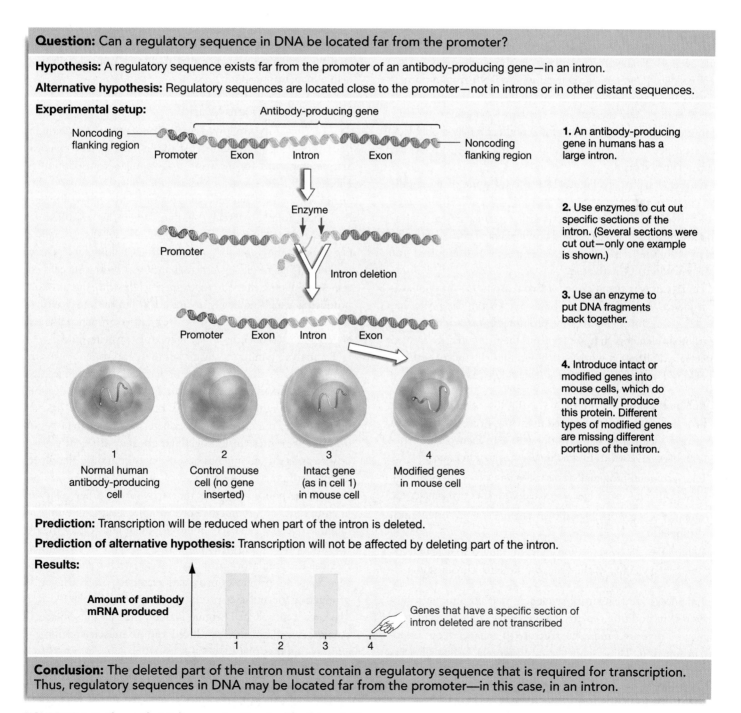

FIGURE 18.6 Evidence That Enhancers Are Required for Transcription

QUESTION Why did the researchers assess mRNA production in human cells that did *not* receive a modified or intact gene?

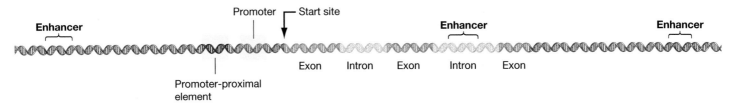

FIGURE 18.7 Enhancers Are Far from the Genes They Regulate
Eukaryotic genes often have more than one enhancer. Enhancers can be located either 5' or 3' to the gene they regulate or in introns and can be located tens of thousands of base pairs away from the promoter.

inserted different versions of the gene into mouse cells, which do not normally contain the antibody-producing gene. Some mouse cells received normal copies of the gene; some received copies of the gene missing different portions of the intron; some received no gene (step 4). Finally, the researchers analyzed the mRNAs that were produced and compared them with mRNA from a normal human cell and from mouse cells that received the normal gene. If their hypothesis was correct and a regulatory sequence was located inside the intron, then some of the modified genes should lack that sequence. Transcription should be reduced in genes lacking the regulatory sequence.

As the "Results" of Figure 18.6 shows, some of the modified copies of the gene were not transcribed at all. The message is clear: The gene is not transcribed if certain parts of the intron are missing. Tonegawa and co-workers proposed that the intron contains a regulatory sequence that is required for transcription to be activated.

This result was remarkable for two reasons: (1) the regulatory sequence was thousands of bases away from the promoter, and (2) it was downstream of the promoter instead of upstream. Regulatory elements that are far from the promoter are termed **enhancers**. Follow-up work has shown that enhancers occur in all eukaryotes and that they have several key characteristics:

- Enhancers can be more than 100,000 bases away from the promoter. They can be located in introns or in untranscribed 5' or 3' sequences flanking the gene (see **Figure 18.7**). Researchers have yet to find enhancers located in exons.

- Like promoter-proximal elements, many types of enhancers exist. Different genes are associated with different enhancers.

- Enhancers can work even if their normal 5' → 3' orientation is flipped.

- Enhancers can work even if they are moved to a new location in the vicinity of the gene, on the same chromosome.

Enhancers are regulatory sequences that are unique to eukaryotes. In addition, eukaryotic genomes contain regulatory sequences that are similar in structure to enhancers but opposite in function. These sequences are **silencers**. When silencers are active, they shut down transcription.

Once enhancers and silencers had been characterized, it was clear that they represented a type of regulatory sequence that worked very differently than promoter-proximal elements did.

For example, there was good evidence that GAL4 and other types of regulatory proteins bind to promoter-proximal elements and interact directly with TBP or RNA polymerase. These interactions stabilize binding and promote transcription. But how could a regulatory sequence that is distant from the promoter help initiate transcription?

How Do Enhancers Work?

An experiment that followed up on Tonegawa's work with antibody-producing genes supported the hypothesis that enhancers are binding sites for proteins that regulate transcription. As **Figure 18.8** shows, researchers cut the enhancer sequence from an antibody-producing gene and spliced it into copies of the gene that codes for the protein β-globin. When this modified gene was inserted into antibody-producing cells, which normally do not express the gene for β-globin, the β-globin mRNA was produced. This result suggests that antibody-producing cells contain some factor that interacts specifically with the antibody-producing gene's enhancer to induce transcription. Follow-up work showed that the factor is a regulatory protein. Enhancers are regulatory sequences where regulatory proteins bind.

By analyzing mutant yeast, fruit flies, and roundworms that have defects in the expression of particular genes, biologists have identified a large number of regulatory proteins that bind to enhancers and silencers. This research program has supported one of the most general conclusions about gene regulation in eukaryotes: In multicellular species, different types of cells express different genes because they contain different regulatory proteins. These regulatory proteins, in turn, are produced in response to signals that arrive from other cells early in embryonic development. For example, a signaling molecule might arrive at a cell early in development and trigger the production of regulatory proteins that are specific to muscle cells. Because the regulatory proteins bind to specific enhancers and silencers and promoter-proximal elements, they trigger the production of muscle-specific proteins (**Figure 18.9**). If no "become a muscle cell" signal arrives, then no muscle-specific regulatory proteins are produced and no muscle-specific gene expression takes place.

To summarize, differential gene expression is based on the production of specific regulatory proteins. Eukaryotic genes are turned on when specific regulatory proteins bind to enhancers and promoter-proximal elements; the genes are turned off

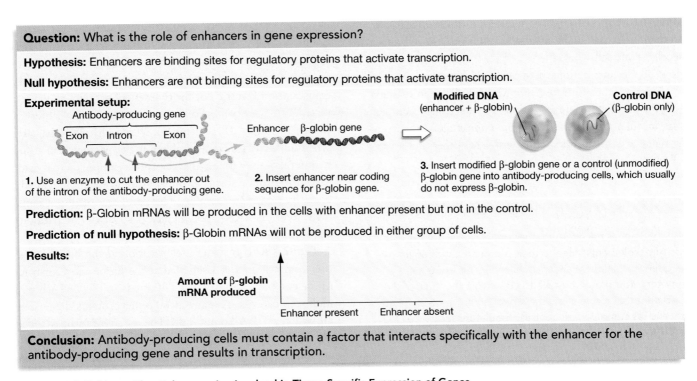

Question: What is the role of enhancers in gene expression?

Hypothesis: Enhancers are binding sites for regulatory proteins that activate transcription.

Null hypothesis: Enhancers are not binding sites for regulatory proteins that activate transcription.

Experimental setup:

1. Use an enzyme to cut the enhancer out of the intron of the antibody-producing gene.

2. Insert enhancer near coding sequence for β-globin gene.

3. Insert modified β-globin gene or a control (unmodified) β-globin gene into antibody-producing cells, which usually do not express β-globin.

Prediction: β-Globin mRNAs will be produced in the cells with enhancer present but not in the control.

Prediction of null hypothesis: β-Globin mRNAs will not be produced in either group of cells.

Results:

Conclusion: Antibody-producing cells must contain a factor that interacts specifically with the enhancer for the antibody-producing gene and results in transcription.

FIGURE 18.8 Evidence That Enhancers Are Involved in Tissue-Specific Expression of Genes
Researchers attached the enhancer sequence from an antibody-producing gene to the coding sequence for the β-globin gene. The altered gene was expressed in antibody-producing cells, which normally do not produce β-globin.

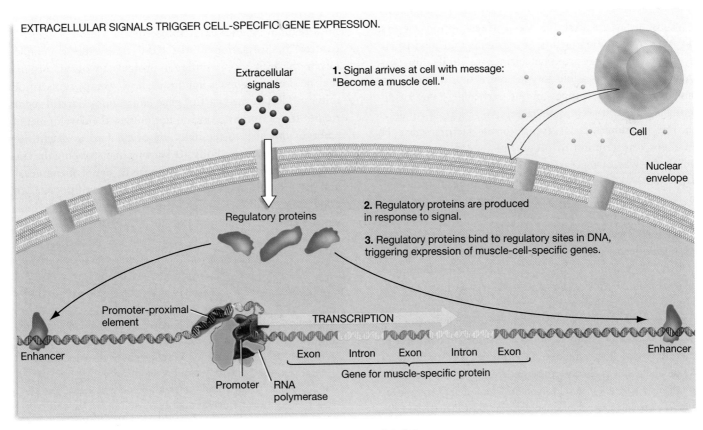

EXTRACELLULAR SIGNALS TRIGGER CELL-SPECIFIC GENE EXPRESSION.

1. Signal arrives at cell with message: "Become a muscle cell."

2. Regulatory proteins are produced in response to signal.

3. Regulatory proteins bind to regulatory sites in DNA, triggering expression of muscle-cell-specific genes.

FIGURE 18.9 Why Are Certain Proteins Produced Only in Certain Types of Cells?
Signaling molecules from outside the cell trigger the production or activation of cell-specific regulatory proteins, which then influence transcription by binding to enhancers, silencers, or promoter-proximal elements in DNA.

when regulatory proteins bind to silencers. Distinctive regulatory proteins are what make a muscle cell a muscle cell and a bone cell a bone cell.

How do the interactions between regulatory proteins and DNA sequences control transcription? As we'll see in the next section, the initiation of transcription is a marvelously complex process in eukaryotes. It requires the coordinated activity of many types of proteins interacting with each other and a variety of DNA sequences.

✓CHECK YOUR UNDERSTANDING

Eukaryotic genes have regulatory sequences called promoter-proximal elements close to their promoters and regulatory sequences called enhancers or silencers far from their promoters. You should be able to (1) explain the logic of the experiment that led to the discovery of an enhancer in an intron, and (2) compare and contrast regulatory sequences in the DNA of bacteria versus those in eukaryotes.

18.4   Transcription Initiation

After several decades of research, many questions remain about how transcription is initiated in eukaryotes. It is clear, however, that two broad classes of proteins interact with regulatory sequences at the start of transcription: (1) regulatory transcription factors and (2) basal transcription factors. **Regulatory transcription factors** are proteins that bind to enhancers and promoter-proximal elements. These transcription factors are responsible for the expression of particular genes in particular cell types and at particular stages of development. **Basal transcription factors**, in contrast, interact with the promoter. Basal transcription factors must be present for transcription to occur, but they do not provide much in the way of regulation. TBP is a basal transcription factors that is common to all genes. Other basal transcription factors are specific to promoters recognized by RNA polymerase I, II, or III.

In addition, proteins called **coactivators** are involved in starting transcription. Coactivators do not bind to DNA. Instead they link the proteins involved in initiating transcription—regulatory transcription factors and basal transcription factors.

Figure 18.10 summarizes the current model for how transcription is initiated in eukaryotes. The first step is the binding of regulatory transcription factors to DNA; these factors recruit chromatin-remodeling complexes and histone acetyl transferases (HATs). The result is chromatin remodeling, a loosening of the chromatin structure. Once the chromatin-remodeling complexes and HATs are in place, they open a broad swath of chromatin that includes the promoter region (step 2).

Once chromatin's grip is relaxed, other regulatory transcription factors can bind enhancers and promoter-proximal

elements. When these regulatory transcription factors are bound to DNA, they interact with basal transcription factors. Because enhancers are located far from the promoter, the regulatory transcription factors are normally far from the basal transcription factors. For the two sets of proteins to make contact, DNA has to loop out and away from the promoter, as shown in Figure 18.10 (step 3). When all of the basal transcription factors have assembled at the promoter in response to interactions with regulatory transcription factors and coactivators, they form a multi-protein machine called the **basal transcription complex**. The basal transcription complex then "recruits" RNA polymerase II so that transcription can begin (step 4).

Figure 18.11 provides a closer look at the basal transcription complex as it is assembled at the promoter. Construction of the complex begins when TBP binds to the TATA box. In a multistep process, as many as 60 other proteins then assemble around the DNA-bound TBP. The assembly process depends on interactions with regulatory transcription factors that are bound to enhancers, silencers, and promoter-proximal elements; the result is the formation of a basal transcription complex that can position RNA polymerase II in a way that initiates transcription. Compared with the situation in bacteria, where just 3 to 5 proteins may interact at the promoter to initiate transcription, the state of affairs in eukaryotes is remarkably complicated.

Currently, biologists are focused on understanding exactly how regulatory and basal transcription factors interact to control the formation of the basal transcription complex. Given the number of different proteins involved, progress may be slow. Progress is important, however, for transcription initiation lies at the heart of gene expression. Careful regulation of transcription is critical not only to the development of embryos but also to the daily life of eukaryotes. Right now, cells throughout your body are starting and stopping the transcription of specific genes in response to signals from nearby and distant cells. As the environment inside and outside your body continually changes, your cells continually change which genes are being transcribed.

✓CHECK YOUR UNDERSTANDING

Transcription initiation is a multistep process that begins when regulatory transcription factors bind to DNA and recruit proteins that open chromatin. Interactions between regulatory transcription factors and basal transcription factors result in the formation of the basal transcription complex and the arrival of RNA polymerase at the gene's start site. You should be able to compare and contrast the sigma proteins and regulatory proteins observed in bacteria with the basal and regulatory transcription factors found in eukaryotes.

THE ELEMENTS OF TRANSCRIPTIONAL CONTROL: A MODEL

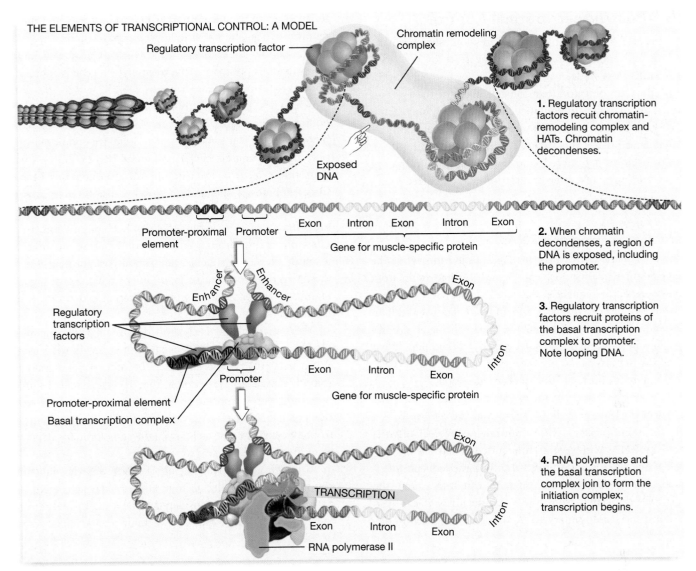

Regulatory transcription factor

Chromatin remodeling complex

1. Regulatory transcription factors recuit chromatin-remodeling complex and HATs. Chromatin decondenses.

Exposed DNA

Promoter-proximal element

Promoter

Exon Intron Exon Intron Exon

Gene for muscle-specific protein

2. When chromatin decondenses, a region of DNA is exposed, including the promoter.

Enhancer

Enhancer

Regulatory transcription factors

Exon

Promoter-proximal element

Basal transcription complex

Promoter

Exon Intron Exon

Intron

Gene for muscle-specific protein

3. Regulatory transcription factors recruit proteins of the basal transcription complex to promoter. Note looping DNA.

Exon

TRANSCRIPTION

Exon Intron Exon

Intron

RNA polymerase II

4. RNA polymerase and the basal transcription complex join to form the initiation complex; transcription begins.

FIGURE 18.10 The Elements of Transcriptional Control: A Model
According to the currently accepted model, transcription is initiated through a series of steps.

FIGURE 18.11 Many Proteins Interact at the Start Site for Transcription
Basal transcription factors associate with TBP when it binds to the promoter. The basal transcription complex also includes other basal transcription factors and RNA polymerase II.

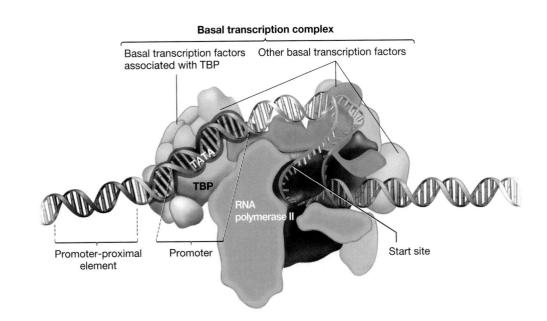

Basal transcription complex

Basal transcription factors associated with TBP

Other basal transcription factors

TATA

TBP

RNA polymerase II

Promoter-proximal element

Promoter

Start site

18.5   Post-Transcriptional Control

In the process of regulating gene expression, chromatin remodeling and transcription are just the start of the story. Once an mRNA is made, a series of events have to occur if the final product is going to affect the cell. Each of these events offers an opportunity to regulate gene expression, and each is used in some cells at least some of the time. Control points include splicing mRNAs in various ways, altering the rate at which translation is initiated, and modifying the life span of mRNAs and proteins after translation has occurred. Let's consider each in turn.

Alternative Splicing of mRNAs

Introns are spliced out of primary RNA transcripts while the message is still inside the nucleus. The mRNA that results from splicing consists of sequences that are encoded by exons and are protected by a cap on the 5' end and a long poly (A) tail on the 3' end. Recall from Chapter 16 that splicing is accomplished by the molecular machines called *spliceosomes*. What that chapter did not mention, however, is that splicing provides an opportunity for the regulation of gene expression.

During splicing, changes in gene expression are possible because selected exons, as well as introns, may be removed. As a result, the same primary RNA transcript can yield mature, processed mRNAs with several different combinations of transcribed exons. This is important, because, if the sequence of ribonucleotides in the mature mRNAs can vary, then the polypeptides translated from those mature mRNAs can also vary. When the same primary RNA transcript is spliced in different ways to produce different mature mRNAs and thus different proteins, **alternative splicing** is said to occur.

To see how alternative splicing works, consider the muscle-cell protein tropomyosin. The tropomyosin gene is expressed in both skeletal muscle and smooth muscle, which are distinct muscle types. (Skeletal muscle is responsible for moving your bones; smooth muscle lines many parts of your gut and certain blood vessels. The two types of muscle are composed of distinct types of muscle cells.) As **Figure 18.12a** shows, the primary transcript from the tropomyosin gene contains 14 exons. In skeletal muscle and smooth muscle cells, different subsets of the 14 exons are spliced together to produce two different messages for translation (**Figure 18.12b**). Each mature mRNA contains information from a different combination of exons. As a result of alternative splicing, the tropomyosin proteins found in these two cell types are distinct. One of the reasons that skeletal muscle and smooth muscle are different is that they contain different types of tropomyosin.

Alternative splicing is controlled by regulatory proteins that bind to mRNAs in the nucleus and interact with spliceosomes. When cells that are destined to become skeletal muscle or smooth muscle are developing, they receive signals leading to the production of specific proteins that are active in the regulation of splicing. Instead of transcribing different versions of the tropomyosin gene, the cells splice the same primary RNA transcript but in different ways.

Although alternative splicing of RNA was viewed as a curiosity until about a decade ago, attitudes have changed. Alternative splicing is now considered to be a major mechanism in the control of gene expression in multicellular eukaryotes. In

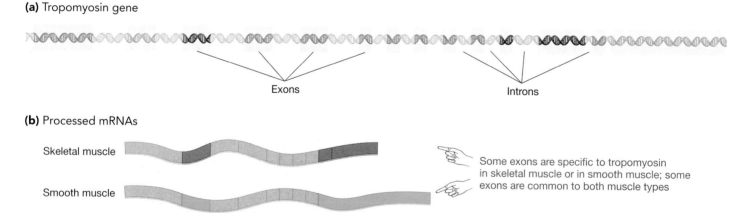

(a) Tropomyosin gene

Exons   Introns

(b) Processed mRNAs

Skeletal muscle

Smooth muscle

Some exons are specific to tropomyosin in skeletal muscle or in smooth muscle; some exons are common to both muscle types

FIGURE 18.12 Alternative Splicing Produces More than One Mature mRNA from the Same Gene
(a) The tropomyosin gene has a large series of exons and introns. The purple exons are found in skeletal muscle tropomyosin, and the blue exons are found in smooth muscle tropomyosin. The orange exons are found in the tropomyosin of both types of muscle cells. **(b)** In skeletal and smooth muscle cells, alternative splicing leads to the production of distinct mature mRNAs for the tropomyosin protein.

fruit flies, for example, the gene that determines gender is transcribed into the same primary RNA transcript in both males and females. A cascade of regulated splicing in females ultimately results in the production of a protein called *doublesex*, which represses the transcription of male-specific genes. In males, an alternative series of splicing events results in a version of the doublesex protein that represses the transcription of female-specific genes.

Alternative splicing makes it possible to produce more than one protein from a single gene. Before the importance of alternative splicing was recognized, a gene was considered to be a nucleotide sequence that encodes a single protein or RNA. Now genes have to be thought of as DNA sequences that are capable of directing the production of one or more related polypeptides or RNAs. As a result, the number of proteins that an organism can produce far exceeds the number of genes it has. For example, at least 35 percent of human genes undergo alternative splicing. Although the human genome contains an estimated 40,000 genes, the number of different proteins that your cells can produce is believed to be between 100,000 and 1 million.

The current record holder for the number of mRNAs derived from one gene is the *Dscam* gene in the fruit fly *Drosophila melanogaster*. The products of this gene are involved in guiding growing nerve cells within the embryo. Because the primary transcript is spliced into about 38,000 distinct forms of mRNA, the *Dscam* gene can produce about 38,000 different products. Researchers are testing the hypothesis that particular combinations of Dscam proteins allow particular nerve cells to connect to each other in the fly's brain.

Current research on alternative splicing is focused on two topics: (1) understanding how regulatory proteins control the process and (2) documenting the extent to which alternative splicing enriches the genetic information stored in a genome. Stay tuned.

Translational Control

Once splicing is complete and processed mRNAs are exported to the cytoplasm, several new regulatory mechanisms come into play prior to translation. The following examples illustrate the array of mechanisms that influence the timing or extent of translation:

- The life span of an mRNA in the cell can vary. For example, the mRNA for casein—the major protein in milk—is produced in the mammary gland tissue of female mammals. Normally this mRNA has a half-life of just over 1 hour, and little casein protein is produced. But when a female mouse is lactating, regulatory molecules increase the mRNA's half-life to 28.5 hours—leading to a huge increase in the production of casein protein.

- In many species of animals, eggs are loaded with mRNAs that are not translated until fertilization occurs. The proteins produced from these messages are involved in directing the early development of the embryo, so they are not needed until fertilization is complete. Translation of the mRNAs in the egg can be prevented when a regulatory protein binds to them, or when the message's cap or tail is modified.

- The overall rate of translation in a cell may slow or stop in response to a sudden increase in temperature or infection by a virus. The slowdown occurs because regulatory proteins add a phosphate group to a protein that is part of the ribosome. As we have pointed out in earlier chapters, phosphorylation frequently leads to changes in the shape and chemical reactivity of proteins. In the case of the phosphorylated ribosomal protein, the shape change slows or prevents translation. For the cell, this dramatic change in gene expression can mean the difference between life and death. If the danger is due to a sudden increase in ambient temperature, shutting down translation prevents the production of improperly folded polypeptides; if the insult is a virus, the cell avoids manufacturing viral proteins.

Post-Translational Control

Control of gene expression may continue even after translation occurs and a protein product is complete. Recall from Chapter 17 that, in bacteria, mechanisms of post-translational regulation are important because they allow the cell to respond to new conditions rapidly. The same is true for eukaryotes. Instead of waiting for transcription and translation to occur, the cell can respond to altered conditions by quickly activating or inactivating existing proteins. Regulatory mechanisms occurring late in the flow of information from DNA to RNA to protein involve a trade-off between speed and resource use, however, because transcription and translation require energy and materials.

A group of regulatory transcription factors that is found in mammals is called **signal transducers and activators of transcription (STATs)** and provides a good example of post-translational control. STATs participate in several signal transduction pathways. Recall from Chapter 8 that signal transduction pathways are interactions that convert a signal at the cell surface into a response by the cell. In white blood cells, STAT activity results in increased cell growth and cell division.

In white blood cells, STATs normally reside in the cytoplasm as single polypeptide chains. In this form, the STATs are inactive. But a chain of events is triggered when a signaling molecule binds to and activates a receptor on the cell surface (**Figure 18.13a**). The activated receptor adds a phosphate to a STAT polypeptide chain, triggering the formation of a **dimer** (a unit made of two parts). The activated STAT dimer then moves into the nucleus, binds to an enhancer, and activates transcription (**Figure 18.13b**). The STAT remains active until the phosphate is removed by another regulatory protein. At this point, the two polypeptide chains come apart and are transported back to the cytoplasm.

Phosphorylation is a very common mechanism of post-translational control over gene expression, particularly in signal transduction pathways. It is only one of several post-translational control mechanisms that have been documented, however. The activity of a protein may also be modified by folding or by enzymes that cleave off a portion of the molecule. But no matter what the mechanism, post-translational control is correlated with particularly rapid changes in gene expression.

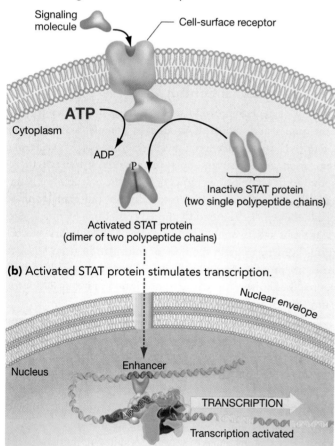

(a) Growth signal activates STAT protein.

(b) Activated STAT protein stimulates transcription.

FIGURE 18.13 STATs Activate Transcription after They Are Phosphorylated

(a) A series of events, beginning with the binding of a signaling molecule to a cell-surface receptor, leads to the phosphorylation of STAT proteins. This causes two STAT polypeptide chains to join, forming an activated dimer. **(b)** Activated STAT proteins move to the nucleus, bind to DNA, and activate the transcription of selected target genes.

A 50-Year Retrospective: How Does Gene Expression in Bacteria and in Eukaryotes Compare?

For about five decades, biologists have been studying how gene expression is controlled. Almost as soon as they knew that information in DNA is transcribed into RNA and then translated into proteins, researchers began asking questions about how that flow of information is regulated. Early studies on bacteria established the existence of promoters, the structure and function of regulatory proteins and regulatory sequences, and the principles of positive control and negative control. More recent work has highlighted the importance of chromatin structure and alternative splicing in eukaryotes.

What are the fundamental differences in how gene expression is regulated in bacteria and eukaryotes? Biologists point to four primary differences, two of which involve levels of control that exist in eukaryotes but not in bacteria:

1. *Packaging* The chromatin structure of eukaryotic DNA must be opened in order for TBP, the basal transcription complex, and RNA polymerase to gain access to genes and initiate transcription. A key insight here is that, because eukaryotic DNA is packaged so tightly, the default state of transcription in eukaryotes is off. In contrast, the default state of transcription in bacteria, with their supercoiled DNA and freely accessible promoters, is on.

2. *Alternative splicing* Prior to translation, primary transcripts in eukaryotes must be spliced—an event that does not occur in bacteria. The fundamental consequence is that the one-to-one correspondence between the number of genes and the number of gene products observed in bacteria is not seen in eukaryotes. Instead, each eukaryotic gene may code for one to thousands of distinct products.

3. *Complexity* Transcriptional control is much more complex in eukaryotes than in bacteria. The function of sigma proteins in bacteria is analogous to the role of the basal transcription complex in eukaryotes. Likewise, the function of CAP, the repressor, and other regulatory proteins is analogous to the role of regulatory transcription factors in eukaryotes. But the sheer number of eukaryotic proteins involved in regulating transcription—and the complexity of their interactions—dwarfs those in bacteria.

4. *Coordinated expression* In bacteria, genes that are involved in the same cellular response are organized into operons controlled by a single promoter. Because their polycistronic mRNAs are translated together, several proteins are produced in a coordinated fashion. In contrast, operons are rare in eukaryotes. In these organisms, genes that are physically scattered can be expressed at the same time because a single set of regulatory transcription factors can trigger the transcription of several genes. For example, muscle-specific genes found on several different chromosomes can be transcribed in response to the same muscle-specific regulatory transcrip-

tion factor. In this way, eukaryotes coordinate the expression of functionally related genes.

To explain why these differences exist, biologists point to a fundamental contrast between unicellular and multicellular individuals. In both types of organisms, cells have to respond to environmental changes in an appropriate way. Bacteria and yeast cells change gene expression to respond to changes in the availability of sugars and other nutrients; your muscle cells change gene expression in response to changes in temperature, the availability of oxygen and sugars, and general activity. But in addition, the cells of multicellular organisms have to differentiate as an individual develops. Changes in gene expression are responsible for the differentiation of muscle cells, bone cells, leaf cells, and flower cells in response to signals from other cells. The need for each cell type to have a unique pattern of gene expression is thought to be the reason why control of gene expression is so much more complex in eukaryotes than in bacteria.

One of the two great frontiers in research on gene expression is to understand how developmental signals produce cell-specific gene expression in multicellular organisms. The other major frontier is to understand how certain defects in gene regulation result in uncontrolled cell growth and the suite of diseases called cancer.

18.6 Linking Cancer with Defects in Gene Regulation

Normal regulation of gene expression results in the orderly development of an embryo and appropriate responses to environmental changes. Abnormal regulation of gene expression, in contrast, can lead to developmental abnormalities and diseases such as cancer. Hundreds of distinct cancers exist. These diseases are enormously variable regarding the tissues they affect, their rate of progression, and their outcome. Because the underlying defects, symptoms, and consequences are so diverse, cancer is not a single disease but a family of related diseases. Cancers are related because they all result from uncontrolled cell growth.

From this perspective, the fundamental question in cancer biology is a simple one: What causes uncontrolled cell growth? The short answer is that each type of cancer is caused by a different set of defects that lead to uncontrolled cell growth. The longer and more accurate answer is based on two key observations. First, recall from Chapter 11 that cancer results from defects in the proteins that control the cell cycle. Second, cancer is associated with mutations that knock out key genes. For example, Chapter 14 linked the cancer-causing disorder xeroderma pigmentosum and hereditary nonpolyposis colorectal cancer with defects in DNA repair. This association supports the hypothesis that increased mutation rates are involved in triggering cancer. The hypothesis is also supported by the observation that people who have been exposed to **mutagens**—that is, radiation or chemicals that induce mutation—have an increased risk of developing cancer. For example, cancer rates were extremely high in the tens of thousands of people who were exposed to massive doses of radiation from the atomic bombs dropped on Hiroshima and Nagasaki, Japan, at the end of World War II. The link between mutation and cancer is so strong that suspected **chemical carcinogens** (cancer-causing compounds) are first tested by their ability to cause mutations in bacteria.

The fundamental question of cancer biology comes into sharper focus once the role of cell-cycle regulators and mutation is acknowledged. Now we can ask, Which genes, when mutated, disrupt the cell cycle and trigger uncontrolled cell growth?

Intensive research over the past two decades has shown that many cancers are associated with mutations in regulatory transcription factors or regulatory sequences in DNA. These mutations lead to cancer when they affect one of two classes of genes: (1) genes that stop or slow the cell cycle, and (2) genes that trigger cell growth and division by initiating specific phases in the cell cycle.

Genes that stop or slow the cell cycle are called **tumor suppressor genes**. Their products prevent the cell cycle from progressing unless specific signals indicate that conditions are right for moving forward with mitosis and cell division. If a mutation disrupts normal function of a tumor suppressor gene, then a key "brake" on the cell cycle is eliminated.

Genes that encourage cell growth by triggering specific phases in the cell cycle are called **proto-oncogenes** ("first-cancer-genes"). In normal cells, proto-oncogenes are required to initiate each phase in the cell cycle. They are active only when conditions are appropriate for growth, however. In cancerous cells, defects in the regulation of proto-oncogenes cause these genes to stimulate growth at all times. In cases such as this, a mutation has converted the proto-oncogene into an **oncogene**—an allele that promotes cancer development.

To gain a deeper understanding of how defects in gene expression can lead to cancer, let us consider two examples from recent research. The first, the *p53* gene, illustrates how problems with regulating tumor suppressor genes can lead to uncontrolled cell growth. The second, STAT mutations, demonstrates how defects in regulatory transcription factors can lead to the constant activation of proto-oncogenes, constant cell growth, and tumor formation.

p53—Guardian of the Genome

The gene that is most often defective in human cancers codes for a regulatory transcription factor. This gene is called *p53*, because the protein it codes for has a molecular weight of approximately 53 kilodaltons (53,000 daltons, or 53,000 amu—see Chapter 2). DNA sequencing studies have revealed that mutant, nonfunctional forms of *p53* are found in over half of all human cancers.

What is the link between a loss of p53 protein activity and cancer? A key observation suggested an answer to this question: When researchers exposed normal, noncancerous human cells to UV radiation, levels of p53 protein increased markedly. Recall from Chapter 14 that UV radiation damages DNA. Follow-up studies confirmed that there is a close correlation between DNA damage and the amount of p53 in a cell. In addition, analyses of the protein's structure showed that it contains a DNA-binding domain.

These observations inspired the hypothesis that p53 is a transcription factor that serves as the master brake on the cell cycle (**Figure 18.14a**). In this model, p53 is activated after DNA damage occurs. The hypothesis contends that the activated protein binds to the enhancers of genes that arrest the cell cycle. Once the cell cycle is arrested, the cell has time to repair its DNA before continuing to grow and divide.

Recent research has shown that this model of p53 function is correct in almost every detail. For example, consider the results of studies based on X-ray crystallography. The three-dimensional models generated by this technique confirmed that p53 binds directly to DNA (**Figure 18.14b**). In addition, researchers who mapped the location of mutations that make p53 defective found that virtually all of the cancer-causing mutations were located in p53's DNA-binding site (see the gray regions of Figure 18.14b). This observation supports the hypothesis that defective forms of the protein can't bind to enhancers. Further, investigators have documented that one of the genes induced by p53 encodes a protein that binds to the cyclin-Cdk complexes introduced in Chapter 11. Binding by this protein prevents the cyclin-Cdk complexes from triggering M (mitosis) phase in the cell cycle.

Recent research has also shown that when a cell's DNA is extensively damaged and cannot be repaired, p53 activates the transcription of genes that cause the cell to take its own life via apoptosis (see Chapter 11). But if mutations in the *p53* gene make the protein product inactive, then damaged cells are not shut down or killed. They continue to move through the cell cycle, except now they are likely to contain many mutations because of the DNA damage they have left unrepaired. If these mutations create oncogenes or inactivate genes for tumor suppressors, the cells have taken a key step on the road to cancer.

To summarize, *p53* functions as a tumor suppressor gene. It prevents cancer onset by stopping the cell cycle when DNA is damaged. When *p53* is functioning normally, mutations that produce oncogenes are repaired or eliminated before uncontrolled cell growth can begin. The role of *p53* in preventing cancer is so fundamental that biologists call this gene "the guardian of the genome." Currently, research on the p53 protein is forging ahead on two fronts: Biologists are striving to identify more of the genes that are regulated by this protein and to find molecules that could act as anticancer drugs by mimicking p53's shape and activity.

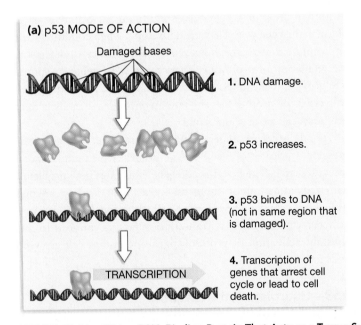

(a) p53 MODE OF ACTION

Damaged bases

1. DNA damage.

2. p53 increases.

3. p53 binds to DNA (not in same region that is damaged).

TRANSCRIPTION

4. Transcription of genes that arrest cell cycle or lead to cell death.

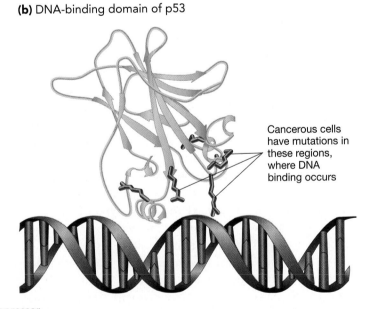

(b) DNA-binding domain of p53

Cancerous cells have mutations in these regions, where DNA binding occurs

FIGURE 18.14 p53 Is a DNA-Binding Protein That Acts as a Tumor Suppressor
(a) Summary of the current model for p53's mode of action. **(b)** Ribbon model showing the DNA-binding domain of p53. In cancer patients, the amino acids in the gray portions of the molecule are often different than the normal versions.   **QUESTION** Why would substituting one amino acid for another at one of the locations highlighted in part (b) affect a protein's ability to bind to DNA? If the model in part (a) is correct, what happens when p53 cannot bind to DNA efficiently?

STAT Mutations in Cancer

The family of transcription activators called STATs is another type of protein implicated in cancer. Recall from Section 18.5 that, in several types of cells, STATs activate genes that promote cell growth in response to signals from outside the cell. In normal cells, STATs must be phosphorylated before they can bind to regulatory sequences in DNA and activate transcription. But in some human cancers, STATs bind to DNA all the time, or constitutively—not just when they are phosphorylated. When this happens, STATs constantly stimulate cell growth and division. These observations suggest that mutations can convert STATs into oncogenes.

What type of mutations could turn STATs into oncogenes? To answer this question, recall that the activated form of a STAT is a dimer—a combination of two polypeptides. For a dimer to form, an amino acid residue at the point of contact between the two STAT proteins must be phosphorylated (**Figure 18.15a**).

These observations inspired a hypothesis that STATs turn into oncogenes if mutations cause them to form dimers without being phosphorylated—that is, without being stimulated by a normal growth signal. More specifically, researchers suggested that if mutations caused cysteine residues to be substituted in the portion of the protein where dimer formation takes place, then disulfide bonds might form between STAT3 monomers, creating constitutively activated dimers (**Figure 18.15b**).

To test this hypothesis, researchers created mutant versions of the STAT3 protein with cysteines in the appropriate location. The researchers expressed the protein in normal (control) cells and analyzed the amount of protein produced by a gene regulated by STAT3. Compared with cells with normal STAT3 proteins, the cysteine-containing mutant cells produced 10 times as much of the target gene product (**Figure 18.15c**). This observation was consistent with the biologists' prediction that cysteine mutants form dimers and bind to DNA constitutively.

To confirm their result, the biologists injected mice either with cells that expressed cysteine-mutant STAT3 or with cells that expressed normal STAT3. After just four weeks, all of the

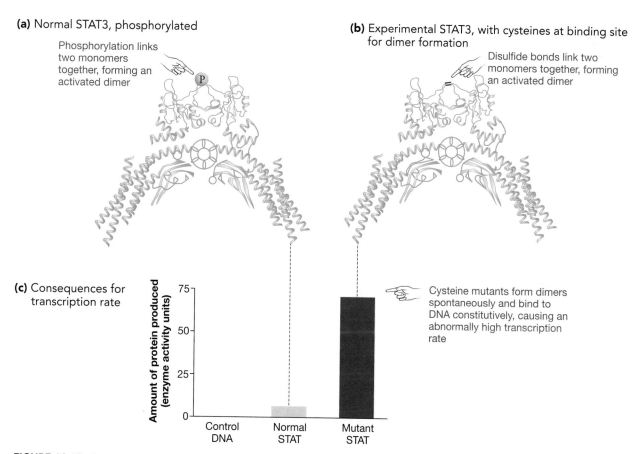

(a) Normal STAT3, phosphorylated

Phosphorylation links two monomers together, forming an activated dimer

(b) Experimental STAT3, with cysteines at binding site for dimer formation

Disulfide bonds link two monomers together, forming an activated dimer

(c) Consequences for transcription rate

Cysteine mutants form dimers spontaneously and bind to DNA constitutively, causing an abnormally high transcription rate

Amount of protein produced (enzyme activity units)

Control DNA Normal STAT Mutant STAT

FIGURE 18.15 Generating a Cancer-Causing Mutation Experimentally
(a) Phosphorylated sections of STAT3 proteins bind to one another, forming a dimer. **(b)** Substitution of cysteines at the locations shown in red result in the formation of disulfide bonds between STAT3 proteins. As a result, the STAT proteins are activated at all times. **(c)** Histogram showing the transcriptional activity of a gene that is normally activated by STAT3. The data come from cells that received either a control DNA (a gene that does not encode a transcription activator), a normal *STAT3* gene, or a copy of the cysteine-mutant version of the *STAT3* gene.

mice that had been injected with mutant cells had developed tumors at the site of injection. In contrast, none of the mice that had been injected with normal cells showed evidence of tumors, even after eight weeks.

These experiments have an important message: Any mutation that constitutively forms dimers of STAT proteins causes constitutive activation of STAT target genes. As a result, these mutations set the stage for cancer. An even more global message is that constant activation of a regulatory transcription factor can lead to cancer. Cancers originate with mutations that disrupt the normal regulation of cell-cycle control genes.

✓ CHECK YOUR UNDERSTANDING

Cancer is associated with mutations and with loss of control over the cell cycle. These two observations are connected, because mutations can cause cancer when they occur in genes that are involved in regulating cell-cycle control genes. For example, when a protein that activates the cell cycle is activated constitutively because of a mutation in a regulatory gene, then uncontrolled cell growth may occur. You should be able to explain why knock-out (loss-of-function) mutations in *p53* occur in so many cancers.

ESSAY Fly Eyes and Other Strange Tales of Gene Regulation

Imagine a 1950s-vintage science-fiction movie. As thunder cracks and lightning flashes, a mad-eyed scientist working in the bowels of a sinister-looking laboratory inserts a mouse gene into a fruit fly's genome. The next day, the scientist cackles with delight as she discovers eyes sprouting in bizarre locations on the fly. But then the fly escapes. It mates with a normal fly and sires offspring with hundreds of eyes. Thanks to their remarkable 360° vision, no predator or flyswatter can stop the little monsters. Clouds of them soar off, bent on global domination.

Preposterous? A team of not-so-mad-eyed scientists has inserted a mouse gene into fruit flies and observed the formation of eyes on the adult fly's legs, antennae, and torso (**Figure 18.16**). The gene in question is a regulatory transcription factor. When the gene product binds to enhancers, it sets off a cascade of gene expression events that are required for eye development.

Walter Gehring and colleagues were first to isolate and characterize the gene involved. It was called *eyeless*, because it was originally discovered as a mutant allele that causes flies to develop without eyes. When Gehring's team created *Drosophila* embryos that expressed the normal form of the gene in locations that naturally give rise to appendages, eyes developed on the legs and antennae. This result underscores an important conclusion: During development, the fate of a cell depends on the activity of regulatory transcription factors. When *eyeless* is activated in leg cells, the leg cells give rise to eyes.

> *When eyeless is activated in leg cells, the leg cells give rise to eyes.*

When the researchers looked for genes in other organisms that were similar to *eyeless* in DNA sequence, the team found them in a wide array of species. Consider a mouse gene called *Small Eye* or *pax6*, which is similar in sequence to *eyeless*. The name *Small Eye* was inspired by the observation that a mutant allele of this gene leads to reduced eye development. Because *eyeless* and *Small Eye* are similar in structure and function, Gehring hypothesized that they are **homologous**—that the genes are descended from a similar gene in a common ancestor of flies and mice, which, according to the fossil record, lived some 550 million years ago. To test this hypothesis, Gehring inserted the mouse gene into *Drosophila* embryos. *Small Eye* induced the development of fly eyes essentially anywhere in the embryo that it was expressed. He got the same result when he inserted a squid gene that was similar to *eyeless*.

What does this tell us about biology? First, it is reasonable to conclude that *eyeless*, *Small Eye*, and like genes in other species are similar in structure and function because they are homologous. Second, these regulatory transcription factors and their recognition sequences in DNA have been so conserved that the proteins are nearly interchangeable among organisms separated by 550 million years of evolution. Third, one key regulatory transcription factor can unleash a cascade of responses that results in a structure as complex as an eye. Fourth, the same starting point—expression of the *eyeless* gene—can lead to very different ends, depending on the context. In flies, the mouse gene *Small Eye* leads to a fly eye, not a mouse eye.

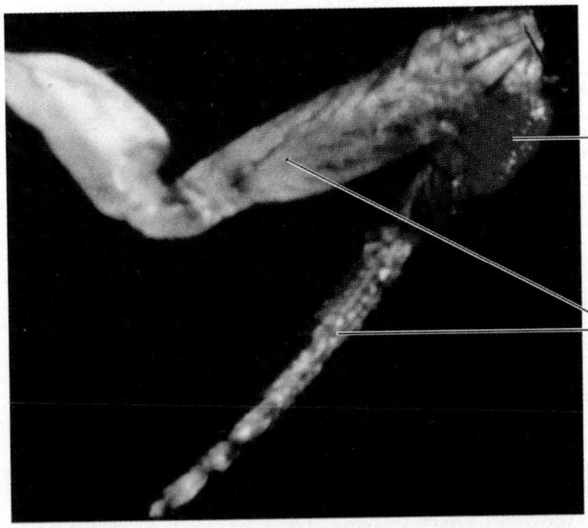

FIGURE 18.16 *Eyeless* **Encodes a Transcription Factor That Induces Eye Development**
When *Drosophila* embryos are made to express the *eyeless* gene in cells that gave rise to the legs, adults develop with eyes on their legs.

Fly eye

Fly leg

CHAPTER REVIEW

Summary of Key Concepts

■ **Changes in gene expression allow eukaryotic cells to respond to changes in the environment and cause distinct cell types to develop.**

In a multicellular eukaryote, different types of cells are radically different in size, shape, and function—even though they have the same DNA. Cells are different not because they have different genes but because they express different genes.

In embryos, cells begin expressing specific genes in response to signals from other cells. Because different cells receive different signals, one cell might begin expressing muscle-specific genes while a nearby cell expresses bone-specific genes. Once cells mature, they continue to receive signals that are released by other cells and that carry information about changes in the environment. In response, changes in gene expression may occur.

In both embryonic cells and mature cells, several types of changes in gene expression may occur. The transcription of specific genes may be initiated or repressed, mRNAs may be spliced in different ways to produce a different product, the life span of specific mRNAs may be extended or shortened, and the life span or activity of particular proteins may be altered.

■ **In eukaryotes, DNA is packaged with proteins into complex structures that must be opened before transcription can occur.**

Eukaryotic DNA is wrapped around histone proteins to form a bead-like nucleosome, which is then coiled into 30-nm fibers and higher-order chromatin structures. Transcription cannot be initiated until the interaction between DNA and histones in chromatin is relaxed. These changes depend on the acetylation of histones and the action of chromatin-remodeling complexes.

■ **In eukaryotes, transcription can be initiated only when specific proteins bind to the promoter and to regulatory sequences that may be close to the promoter or far from it.**

Regulatory transcription factors are proteins that bind to regulatory sequences called enhancers and silencers, which are often located at a distance from the gene in question, or to promoter-proximal sequences near the start of the coding sequence. The first regulatory transcription factors that bind to DNA recruit proteins that loosen the histones' grip on the gene, making the promoter accessible to basal transcription factors. Interactions between regulatory and basal transcription factors lead to the formation of the basal transcription complex. Once this large, multi-protein complex is intact, RNA polymerase is recruited to the site and transcription begins.

Web Tutorial 18.1 Transcription Initiation in Eukaryotes

■ **Alternative splicing allows a single gene to code for several different products.**

Once a message has been transcribed, several other regulatory events come into play. Alternative splicing allows a single locus to produce more than one mRNA and more than one protein. Mechanisms of post-translational control, such as the activation of transcription factors by phosphorylation, trigger rapid responses to signals from outside the cell.

■ **Cancer can develop when mutations disable genes that regulate cell-cycle control genes.**

The cell cycle is controlled by specific genes. But if mutations alter the regulatory sequences of cell-cycle control genes or the genes for their regulatory transcription factors, then uncontrolled cell growth and tumor formation may result. For example, the cell-cycle regulator *p53* is responsible for stopping the cell cycle when DNA is damaged. If *p53* is mutated in a way that prevents its protein product from binding to DNA, then the cell cycle is not arrested and the damaged DNA is not repaired—leading to mutations. If the mutations alter genes that control the cell cycle, such as STAT genes, cancer may occur. Mutant STAT proteins that promote cancer bind to DNA at all times and trigger the expression of genes that activate the cell cycle, leading to uncontrolled cell growth.

Questions

Content Review

1. What is chromatin?
   a. the protein core of the nucleosome, which consists of histones
   b. the 30-nm fiber
   c. the DNA-protein complex found in eukaryotes
   d. the histone *and* non-histone proteins in eukaryotic nuclei

2. What is a tumor suppressor?
   a. a gene associated with tumor formation when its product does not function
   b. a gene associated with tumor formation when its product functions normally
   c. a gene that accelerates the cell cycle and leads to uncontrolled cell growth
   d. a gene that codes for a transcription factor involved in tumor formation

3. Which of the following statements about enhancers is correct?
   a. They contain a unique base sequence called a TATA box.
   b. They are located only in 5′-flanking regions.
   c. They are located only in introns.
   d. They are found in a variety of locations and are functional in any orientation.

4. In eukaryotes, why are certain genes expressed only in certain types of cells?
   a. Regulatory transcription factors vary from cell to cell.
   b. The promoter sequence varies from cell to cell.
   c. The location of enhancers varies from cell to cell.
   d. The type of RNA polymerase varies from cell to cell.

5. What is alternative splicing?
   a. the phosphorylation events that lead to different types of post-translational regulation
   b. mRNA processing events that lead to different combinations of exons being spliced together
   c. folding events that lead to proteins with alternative conformations
   d. action by regulatory proteins that leads to changes in the life span of an mRNA

6. What types of proteins bind to promoter-proximal elements?
   a. the basal transcription complex
   b. the basal transcription complex plus RNA polymerase
   c. basal transcription factors
   d. regulatory transcription factors

Conceptual Review

1. Compare and contrast (a) enhancers and the CAP site; (b) promoter-proximal elements and the *lac* operon operator; (c) basal transcription factors and sigma; and (d) eukaryotic promoters and bacterial promoters.

2. Why does chromatin need to be "remodeled" for transcription to occur? Why was it logical to observe that DNA-protein interactions can be loosened by the addition of acetyl groups to histones?

3. Compare and contrast (a) enhancers and silencers; (b) promoter-proximal elements and enhancers; (c) transcription factors and coactivators; and (d) regulatory transcription factors and basal transcription factors.

4. Explain the logic behind the experiment by Tonegawa and colleagues that was responsible for the discovery of enhancers. Which treatments in the experiment served as controls, and what did they control for?

5. When the human genome was first sequenced, researchers estimated that only about 30,000 genes were present. Previously, most investigators expected that the human genome held at least 100,000 genes. Explain how alternative splicing can explain the discrepancy between the observed and predicted numbers of genes.

6. Explain why mutations in regulatory proteins and regulatory sequences can lead to loss of control over the cell cycle and the development of cancer. Include a specific example.

Group Discussion Problems

1. Histone proteins have been extremely highly conserved during evolution. The histones found in fruit flies and humans, for example, are nearly identical in amino acid sequence. Offer an explanation for this observation. (Hint: What are the consequences of a mutation in a histone?)

2. Cancers are most common in tissues where cell division is common, such as blood cells and cells in the lining of the lungs or gut. Why is this observation logical?

3. Levels of p53 protein in the cytoplasm increase after DNA damage. Design an experiment to determine whether this increase is due to increased transcription of the *p53* gene or to activation of preexisting p53 proteins via a post-translational mechanism such as phosphorylation.

4. The frequency of cancer was elevated in individuals who were exposed to the nuclear bombs that were dropped on Japan in World War II. How are these data related to the observation that individuals who experience severe sunburn or deep tanning are more likely to develop skin cancer?

Answers to Multiple-Choice Questions   1. c; 2. a; 3. d; 4. a; 5. b; 6. d

Analyzing and Engineering Genes

19

KEY CONCEPTS

▓ The discovery of enzymes that cut DNA at specific locations, along with enzymes that piece DNA segments back together, gave biologists the ability to move genes from one location to another.

▓ Biologists can determine a gene's base sequence once they have obtained many copies of the gene—by inserting it into loops of DNA called plasmids in bacterial cells and then allowing the cells to grow or by performing a polymerase chain reaction.

▓ Researchers use several strategies to find and characterize the genes responsible for specific traits, such as the alleles associated with certain genetic diseases.

▓ In some cases, it has been possible to insert genes into humans to cure genetic diseases or into plants to provide them with novel traits, such as the ability to resist insect attacks.

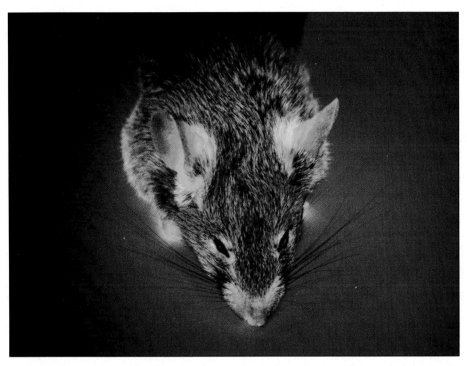

A mouse that has been genetically engineered to express a gene from jellyfish. The jellyfish gene codes for a protein that naturally luminesces, or emits light. The gene's product is called green fluorescent protein, or GFP.

Biologists engineer DNA sequences by removing them from a particular organism, manipulating them, and inserting the sequences into another organism of the same species or into one of a different species. These efforts have recently been applied to humans and crop plants in the hope of curing devastating genetic diseases and increasing the quality and quantity of food available to people in poverty-stricken areas of the globe. But manipulating the genetic makeup of organisms raises ethical concerns and has the potential to create harm. Efforts to manipulate DNA sequences in organisms are often referred to as **genetic engineering**. Genetic engineering became possible after the discovery of bacterial enzymes that cut DNA at specific sites and other enzymes that paste DNA sequences together. By mixing and matching sequences from different sources, biologists can create "designer genes." Efforts to manipulate genes usually result in novel combinations of genes along chromosomes, or *genetic recombination* (Chapter 12). Appropriately enough, techniques used to engineer genes are often referred to as *recombinant DNA technology*.

The enzymes used to cut and paste DNA were initially discovered and used during the 1960s and 1970s. Over the subsequent two decades, researchers followed up by developing systems for transferring recombinant genes into various types of organisms and by working out techniques for controlling the expression of the introduced DNA. Since then, research has concentrated on applying these techniques to solve problems in medicine, industry, and agriculture. The goals of genetic engineering are to (1) improve our understanding of how genes work and (2) advance **biotechnology**—the manipulation of organisms to create products or cure disease.

The goal of this chapter is to introduce some of the key techniques and conceptual issues involved in genetic engineering, using a case history approach. Our first example represents one of the initial efforts to use recombinant DNA technology to cure an inherited disease in humans called pituitary dwarfism. The research relied on techniques for identifying genes, copying them, moving them into a new host organism, and then controlling their expression. The remaining sections introduce how biologists determine the DNA sequence of genes, how they located the gene responsible for Huntington's disease in humans, and how novel genes are being placed into humans and crop plants.

In addition to exploring the techniques used to manipulate DNA, it is essential to consider the ethical, economic, ecological, and political issues involved. *Gene therapy* (the introduction of normal alleles to cure genetic defects) and the release of genetically modified plants are under intense scrutiny in the press, and lawmaking bodies around the world are debating legislation to regulate both efforts. What are the potential perils and benefits of introducing recombinant genes into human beings, food plants, and other organisms? This question represents one of the great ethical challenges of the twenty-first century.

19.1 Using Recombinant DNA Techniques to Manufacture Proteins: The Effort to Cure Pituitary Dwarfism

To understand the basic techniques and tools that are used by genetic engineers, let's consider the effort to treat **pituitary dwarfism** in humans. The pituitary gland is a structure at the base of the mammalian brain that produces several important molecules, including a protein that stimulates growth. The molecule, which was found to be just 191 amino acids long, was named **growth hormone**. The gene that codes for this protein is called *GH1*.

Once growth hormone was discovered, researchers immediately suspected that at least some forms of inherited dwarfism might be due to a defect in the GH1 protein. This hypothesis was confirmed when it was established that people with certain types of dwarfism produce little growth hormone or none at all. These people have defective copies of *GH1* and exhibit pituitary dwarfism, type I.

By studying the pedigrees of families in which dwarfism was common, several teams of researchers established that pituitary dwarfism, type I, is an autosomal recessive trait (see Chapter 13). Stated another way, affected individuals have two copies of the defective allele. People who have only one defective allele are carriers of the trait, meaning that they can transmit the defective allele to their offspring but are not themselves affected. Affected individuals grow more slowly than average people, reach puberty from two to ten years later than average, and are short in stature as adults—typically no more than 120 cm (4 feet) tall.

Why Did Early Efforts to Treat the Disease Fail?

Once the molecular basis of pituitary dwarfism was understood, physicians began treating the disease with injections of growth hormone. This approach was inspired by the spectacular success that physicians had achieved in treating type I diabetes mellitus. Diabetes mellitus is due to a deficiency of the peptide hormone insulin, and clinicians had been able to alleviate the disease's symptoms by injecting patients with insulin from pigs. Early trials showed that people with pituitary dwarfism could be treated successfully with growth hormone therapy, but only if the protein came from humans. Growth hormones isolated from pigs, cows, or other animals were ineffective. Until the 1980s, however, the only source of human growth hormone was pituitary glands dissected from human cadavers. Up to 20,000 pituitaries had to be collected from cadavers to supply enough growth hormone to treat the population of affected individuals. As a result, the drug was extremely scarce and expensive.

Meeting demand turned out to be the least of the problems with growth hormone therapy, however. To understand why, recall from Chapter 3 that certain degenerative brain disorders in mammals are caused by infectious proteins called **prions**. Kuru and Creutzfeldt-Jakob disease are prion diseases that affect humans. Certain prion diseases, including some forms of Creutzfeldt-Jakob disease, are hereditary. In most cases, however, people and other animals become infected with prion diseases by directly ingesting prion proteins in food. For example, humans can contract a variant of mad cow disease by ingesting beef products that are contaminated with prion proteins. When physicians found that some of the children treated with human growth hormone developed Creutzfeldt-Jakob disease in their teens and twenties, the physicians realized that the supply of growth hormone was contaminated with a prion protein from the brains of the cadavers supplying the hormone. As a result, the use of growth hormone isolated from cadavers was banned in 1984. What happened next?

Using Recombinant DNA Technology to Produce a Safe Supply of Growth Hormone

Even as the problems with traditional hormone therapy for pituitary dwarfism became apparent, researchers began to work out a genetic engineering approach for producing growth hormone. The idea was to find the gene that encodes the human growth hormone protein and insert it into bacterial cells or yeast cells. The hope was that huge quantities of recombinant *Escherichia coli* or *Saccharomyces cerevisiae* cells could be grown and that the recombinant cells would produce uncontaminated growth hormone in sufficient quantities to meet demand at an affordable price.

To meet this goal, investigators used the steps outlined in **Figure 19.1** to look for the human growth hormone. The first task was to isolate the mRNAs that are produced in the pituitary gland. One of these mRNA transcripts would encode growth hormone. Next, the biologists used the enzyme

WEB TUTORIAL 19.1 Producing Human Growth Hormone

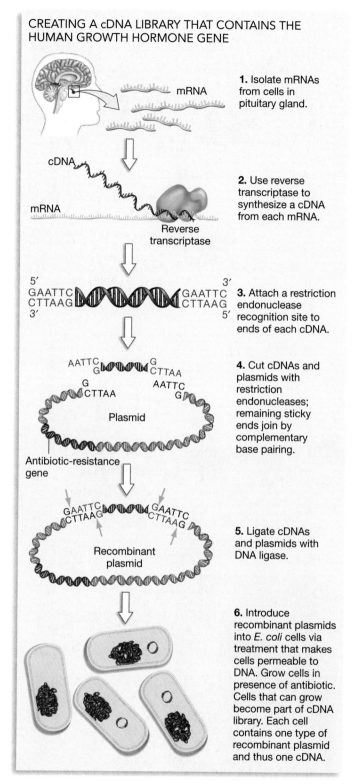

CREATING A cDNA LIBRARY THAT CONTAINS THE HUMAN GROWTH HORMONE GENE

mRNA

1. Isolate mRNAs from cells in pituitary gland.

cDNA

mRNA

Reverse transcriptase

2. Use reverse transcriptase to synthesize a cDNA from each mRNA.

5'
GAATTC   GAATTC
CTTAAG   CTTAAG
3'   5'
                      3'

3. Attach a restriction endonuclease recognition site to ends of each cDNA.

AATTC   G
G   CTTAA
      G   AATTC
   CTTAA   G

Plasmid

Antibiotic-resistance gene

4. Cut cDNAs and plasmids with restriction endonucleases; remaining sticky ends join by complementary base pairing.

GAATTC   GAATTC
CTTAAG   CTTAAG

Recombinant plasmid

5. Ligate cDNAs and plasmids with DNA ligase.

6. Introduce recombinant plasmids into *E. coli* cells via treatment that makes cells permeable to DNA. Grow cells in presence of antibiotic. Cells that can grow become part of cDNA library. Each cell contains one type of recombinant plasmid and thus one cDNA.

FIGURE 19.1 Complementary DNA (cDNA) Libraries Represent the mRNAs in a Cell
The hunt for growth hormone gene began with the creation of a cDNA library from pituitary-gland cells. Because growth hormone is produced in this tissue, the cDNA for growth hormone should exist somewhere in the library. **QUESTION** Would each type of cDNA in the library be represented just once? Why or why not?

reverse transcriptase, which catalyzes the synthesis of a DNA strand from an RNA template. By means of complementary base pairing, reverse transcriptase made a **complementary DNA (cDNA)** of each pituitary mRNA. DNA polymerase was then used to synthesize the complementary strand and yield double-stranded DNA. The product of this step—shown in step 2 of Figure 19.1—was thousands of different cDNAs. But which of these cDNAs encoded the growth hormone protein?

The key to answering this question was to isolate each of the many cDNAs and then obtain enough copies of them so that each of the isolated sequences could be analyzed. The process of producing many identical copies of a gene is called **genetic cloning**. Steps 3 through 6 of Figure 19.1 show how the gene for growth hormone was cloned. These steps also illustrate some of the most basic and widely used techniques in genetic engineering. Let's take a closer look.

How Are Plasmids Used in Cloning? The small, circular DNA molecules called **plasmids** were initially discovered in bacteria. Recall from Chapter 7 that plasmids are physically distinct from the bacterial chromosome. They are not required by the cell for normal growth and reproduction, and most replicate independently of the chromosome. Some plasmids carry genes for antibiotic resistance or other traits that increase the cell's ability to grow in a particular environment. In natural populations of bacteria, plasmids are sometimes copied and transferred to another bacterial cell via the mechanism explained in **Box 19.1** (page 404). Plasmids have also been discovered in the nuclei of certain unicellular eukaryotes, including baker's yeast.

Researchers quickly realized that if they could splice a loose piece of DNA into a plasmid and insert the modified plasmid into a bacterial cell, then the engineered plasmid would be replicated and passed on to daughter cells as the bacterium grew and divided. If this recombinant bacterium were placed in a liquid nutrient broth and allowed to grow overnight, billions of copies of the plasmid DNA would be produced. Biologists could harvest the recombinant genes by breaking the bacteria open, releasing the DNA, and separating out the plasmids.

Over the past several decades, researchers have modified naturally occurring plasmids to use them as carriers of recombinant genes. For example, the plasmids used in genetic engineering experiments carry at least one gene that readily allows researchers to identify and isolate the bacterial cells that carry these plasmids. In many cases, the gene involved codes for a molecule that confers resistance to a specific antibiotic. Thus, when bacterial cells are grown on plates containing that antibiotic, only cells carrying the plasmid are able to survive. Now, how could researchers cut a plasmid open and insert a specific DNA sequence to create a recombinant plasmid that could be copied and transferred?

BOX 19.1   How Do Plasmids Move between Cells Naturally?

In natural populations of bacteria, plasmids can be transferred from one cell to another through a process called **conjugation**. In *E. coli*, conjugation is possible if a cell has a particular type of plasmid called the **F-plasmid** (the "F" stands for fertility). Cells that have an F-plasmid are referred to as F⁺; cells that lack an F⁻ plasmid are F⁻.

Conjugation begins when tubules on the surface of an F⁺ cell (the donor) make contact with an F⁻ cell (the recipient; **Figure 19.2a**, step 1; **Figure 19.2b**). After the tubule attaches, it retracts. As a result, the cell walls of the two cells touch, and a structure called a *conjugation tube* forms between them (step 2 of Figure 19.2a). Then the double-stranded plasma DNA in the F⁺ cell is cut, the strands separate, and a single strand of that DNA passes through the conjugation tube into the F⁻ cell (steps 2 and 3). The single strand of plasmid DNA that is now in each cell serves as a template for the formation of double-stranded DNA, resulting in the formation of two F⁺ cells (step 4).

If the F-plasmid involved happens to include a stretch of DNA called an *insertion sequence* (IS), the plasmid can integrate into the bacterial chromosome. Cells in which an F-plasmid has been incorporated into their main chromosome are called **Hfr strains** (for *h*igh *f*requency of *r*ecombination). This name was inspired by the observations that (1) the plasmid DNA retains the ability to mobilize and travel through the conjugation tube; (2) when conjugation occurs between an Hfr cell and an F⁻ cell, segments of the Hfr cell's chromosomal DNA frequently accompany the plasmid DNA during the transfer; and (3) if the new alleles are inserted into the recipient cell's main chromosome, a recombinant cell—one with a new sequence of genes along the chromosome—results.

(a) CONJUGATION

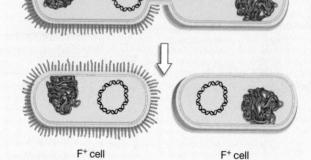

Donor F⁺ cell   Recipient F⁻ cell

F-plasmid
(double-stranded DNA)

Bacterial
chromosome

Conjugation
tube

F⁺ cell   F⁺ cell

1. Conjugation begins when tubules on the surface of an F⁺ cell make contact with an F⁻ cell.

2. After the tubule attaches, it retracts, pulling the two cells into contact. Where they touch, a conjugation tube forms. A single strand of plasmid DNA passes through the tube.

3. The single strand of plasmid DNA in each cell serves as a template for the formation of double-stranded DNA.

4. Both cells have a double-stranded plasmid. Both are F⁺.

(b) An F⁺ cell contacts an F⁻ cell.

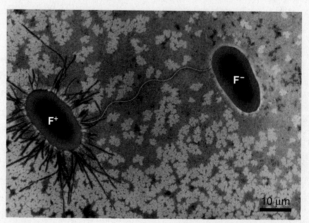

FIGURE 19.2 In *E. coli*, F-Plasmids Are Transferred via Conjugation
(a) After the conjugation tube has formed, one strand of the F-plasmid's DNA is cut and the cut strand travels through the conjugation tube. The transferred strand in the F⁻ cell and the remaining strand in the F⁺ cell are used as templates for DNA replication. **(b)** An F⁺ cell and an F⁻ cell make contact. **EXERCISE** In part (b), label the tubule connecting the two cells.

Using Restriction Endonucleases to Cut DNA   In addition to containing a gene for antibiotic resistance, the plasmids used in genetic engineering contain base sequences that can be cut by enzymes called *restriction enzymes* or (more precisely) restriction endonucleases. A **restriction endonuclease** is a bacterial enzyme that cuts DNA molecules at a specific base sequence. In nature, restriction endonucleases help bacterial cells fend off attacks by viruses. When a virus infects a bacterial cell, the virus injects its genetic material into the cell. In many cases, the bacterial cell's restriction endonucleases cut the viral DNA, inactivating it and rendering the virus harmless. The cell's own DNA is protected from being cut because the specific sequence of bases that is targeted by the enzyme—also called its **recognition site**—has methyl (—CH₃) groups attached (**Figure 19.3**).

Over 400 restriction endonucleases have been identified from a wide variety of bacterial species. The most commonly used enzymes cut DNA only at sites that form palindromes. In English, a word or sentence is a palindrome if it reads the same way backward as it does forward. (*Madam, I'm Adam* is an example.) In biology, a stretch of double-stranded DNA forms a palindrome if the 5′ → 3′ sequence of one strand is identical to the 5′ → 3′ sequence on the antiparallel, complementary strand. Most of the known restriction endonucleases recognize a unique type of palindromic DNA sequence.

To pull these observations together, consider how researchers inserted pituitary-gland cDNAs into plasmids. To begin, the researchers attached the same palindromic recognition site to the ends of each cDNA produced from mRNAs in the pituitary gland. Step 3 of Figure 19.1 shows the resulting molecules. Note that the 5′ → 3′ sequence on one strand of the endonuclease recognition site reads GAATTC and that the 5′ → 3′ sequence on the complementary strand also reads GAATTC.

The researchers' next step was to cut the recognition sites at the ends of each cDNA with a restriction endonuclease called EcoRI (for *Escherichia coli* restriction I—this was the first restriction endonuclease discovered in *E. coli*). As the top part of step 4 in Figure 19.1 shows, however, the enzyme involved did not break its recognition sequence neatly in two. Like most restriction endonucleases, it made a staggered cut in the palindrome. Because a staggered cut was made in a palindromic sequence, the DNA fragments that resulted had **sticky ends**. The ends of the fragments are said to be sticky because the single-stranded bases on one fragment are complementary to the single-stranded bases on the other fragment. As a result, the two ends will tend to pair up and hydrogen bond to one another.

When researchers realized that restriction endonucleases created sticky ends in DNA, they had a fundamentally important insight: If restriction sites in different parts of the genome were cut with the same restriction endonuclease, the presence of sticky ends would allow the resulting fragments to be spliced together. This is the essence of **recombinant DNA technology**—the ability to create novel DNA sequences by cutting specific sequences and pasting them into new locations.

Step 5 of Figure 19.1 demonstrates this aspect of genetic engineering. The sticky ends of the plasmid and of the cDNAs bind to each other via complementary base pairing. Researchers then used DNA ligase—the enzyme introduced in Chapter 14 that connects Okazaki fragments during DNA replication—to seal the recombinant pieces of DNA together. The result was a set of recombinant plasmids. Each contained a different cDNA from the human pituitary gland.

Transformation: Introducing Recombinant Plasmids into Bacterial Cells   Plasmids serve as a **vector**, meaning a vehicle for transferring recombinant genes to a new host. If a recombinant plasmid can be inserted into a bacterial (or yeast) cell, the foreign DNA will be copied and transmitted to new cells as the

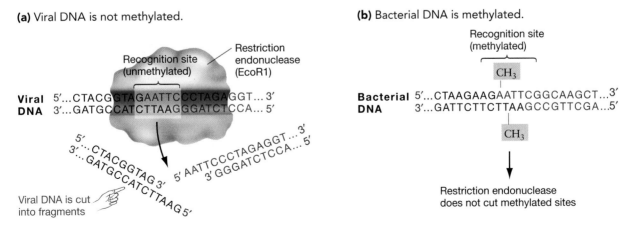

(a) Viral DNA is not methylated.

Recognition site (unmethylated)

Restriction endonuclease (EcoR1)

Viral DNA
5′...CTACG**GTAGAATTCCCTAGA**GGT...3′
3′...GATGC**CATCTTAAGGGATCT**CCA...5′

5′...CTACGGTAG 3′
3′...GATGCCATCTTAAG 5′

5′ AATTCCCTAGAGGT...3′
3′ GGGATCTCCA...5′

Viral DNA is cut into fragments

(b) Bacterial DNA is methylated.

Recognition site (methylated)

CH₃

Bacterial DNA
5′...CTAAGAAGAATTCGGCAAGCT...3′
3′...GATTCTTCTTAAGCCGTTCGA...5′

CH₃

Restriction endonuclease does not cut methylated sites

FIGURE 19.3 Restriction Endonucleases Protect Bacteria from Viruses
(a) Viral DNA is not methylated, so it is cut and destroyed by restriction endonucleases. **(b)** Restriction sites in bacteria are not cut by restriction endonucleases, because the sites are methylated.

host cell grows and divides. In this way, researchers can obtain thousands or millions of copies of specific genes.

How can recombinant plasmids be introduced into cells? The natural process of plasmid transfer introduced in Box 19.1 is too inefficient to be useful in a laboratory setting. And even though some species of bacteria have membrane proteins that allow the cells to take up DNA directly from their environment, most do not. Cells that take up DNA from the environment and incorporate it into their genomes are said to undergo **transformation**. In order for most cells to take up foreign DNA efficiently, the permeability of their cell membranes must be increased by means of a specific chemical treatment or an electrical shock. Typically, just a single plasmid enters the cell. Cells that are subjected to this treatment are subsequently spread out on a gelatinous medium (agar) that contains antibiotic and are allowed to grow. Only bacterial cells that have successfully taken up a plasmid carrying the antibiotic-resistance gene are able to grow in the presence of the antibiotic. The resulting collection of transformed bacterial cells represents a cDNA library (step 6 of Figure 19.1). A **cDNA library** is a collection of bacterial cells, each containing a vector with one cDNA from a particular cell type or tissue.

Using Nucleic Acid Hybridization to Find a Target Gene

Which of the cDNAs in the pituitary-gland cDNA library encodes growth hormone? Investigators used a multistep process to answer this question and find the gene. They began by using the genetic code to infer the approximate DNA sequence of the growth hormone gene. This was possible because the sequence of amino acids in the polypeptide was known. Thus, the researchers could infer which mRNA codon and DNA sequences coded for each amino acid. But recall from Chapter 15 that the genetic code is redundant, so it is possible for more than one codon to code for an amino acid. As a result, their inferred sequence for the growth hormone gene was approximate—not exact.

The next step was to synthesize many copies of a short, single-stranded stretch of DNA that was complementary to the inferred sequence. Because these molecules would bind to single-stranded fragments from the actual gene via complementary base pairing, they could act as a probe. A **probe** is a single-stranded fragment of a labeled, known gene that binds to a complementary sequence in the sample being analyzed. Probes are a way to find one specific sequence in a collection of many different sequences. Probes must be labeled in some way so that they can be found after they have bound to the complementary sequence. In this case, researchers attached a radioactive atom to the probe. The labeled DNA was then added to a copy of the cDNA library (**Figure 19.4**). As predicted, the radioactively labeled DNA did bind to its complementary sequence in the cDNA library. The probe had identified the recombinant cell that contained the human growth hormone.

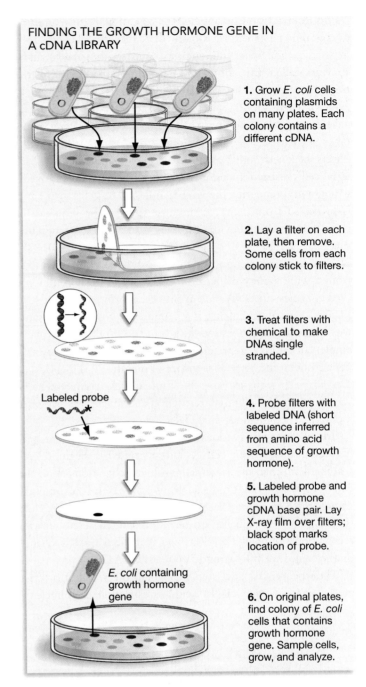

FINDING THE GROWTH HORMONE GENE IN A cDNA LIBRARY

1. Grow *E. coli* cells containing plasmids on many plates. Each colony contains a different cDNA.

2. Lay a filter on each plate, then remove. Some cells from each colony stick to filters.

3. Treat filters with chemical to make DNAs single stranded.

Labeled probe

4. Probe filters with labeled DNA (short sequence inferred from amino acid sequence of growth hormone).

5. Labeled probe and growth hormone cDNA base pair. Lay X-ray film over filters; black spot marks location of probe.

E. coli containing growth hormone gene

6. On original plates, find colony of *E. coli* cells that contains growth hormone gene. Sample cells, grow, and analyze.

FIGURE 19.4 Finding Specific Genes by Probing a cDNA Library
Because the amino acid sequence of the growth hormone protein was known, researchers could infer the DNA sequence of the cDNA and use this sequence to create a DNA probe. Once bacteria with the growth hormone cDNA were isolated from the cDNA library, researchers could produce virtually unlimited quantities of the cDNA.

Mass-Producing Growth Hormone To accomplish their goal of producing large quantities of the human growth hormone, the investigators used recombinant DNA techniques to transfer the growth hormone cDNA to a new plasmid. The plasmid in question was one containing a promoter sequence, like those introduced in Chapter 16, that is recognized by *E. coli*'s

RNA polymerase holoenzyme. The recombinant plasmids were then introduced into *E. coli* cells.

The *E. coli* cells that resulted from these steps contained a gene for human growth hormone attached to an *E. coli* promoter. As a result, the transformed *E. coli* cells began to transcribe and translate the human growth hormone gene. (Note that because the cDNAs are synthesized from mature mRNAs, which do not contain introns, cDNAs do not contain introns either. Thus, bacterial cells can translate an mRNA transcribed from a cDNA directly—no splicing is required.)

The human growth hormone accumulated in the cells and could be isolated and purified. Bacterial cells containing the human growth hormone gene are now grown in huge quantities. These cells have proved to be a safe and reliable source of the human growth hormone protein.

Ethical Concerns over Recombinant Growth Hormone

As supplies of growth hormone increased, physicians used it in treating not only people with pituitary dwarfism but also children who were short, although they had no actual growth hormone deficiency. Even though the treatment requires several injections per week until adult stature is reached, growth hormone therapy was popular because it often increased the height of these children by a few centimeters. In essence, the drug was being used as a cosmetic—a way to improve appearance in cultures where height is deemed attractive. But should a parent request growth hormone treatment for genetically normal children to change their appearance? If short people are discriminated against in a culture, is a medical treatment a better solution than education and changes in attitudes? And what if parents wanted a tall child to be even taller to enhance her potential success as, say, a basketball player? Currently, the U.S. Food and Drug Administration has approved the use of human growth hormone for only the shortest 1.2 percent of children, who are projected to reach adult heights of less than 160 cm (5′3″) in males and 150 cm (4′11″) in women.

Growth hormone has also been found to enhance the maintenance of bone density and muscle mass. As a result, it has become a popular performance-enhancing drug for athletes. Part of its popularity stems from the fact that it is currently undetectable in the drug tests administered by governing bodies. Should athletes be able to enhance their physical skills by taking hormones or other types of drugs? And is the drug safe at the dosages being used by athletes? These types of questions are being debated by physicians and researchers, agencies that govern sports, and legislative bodies. Even though these questions are still unanswered, it is clear that, while solving one important problem, recombinant DNA technology created others. Throughout this chapter, an important theme is that genetic engineering has costs that must be carefully weighed against its benefits.

✔ CHECK YOUR UNDERSTANDING

The essence of recombinant DNA technology is to cut DNA into fragments with a restriction endonuclease, paste specific sequences together via complementary base pairing of sticky ends and the action of DNA ligase, and insert the resulting recombinant genes into a bacterial (or yeast) cell so that the genes are expressed. You should be able to (1) make a diagram showing how certain restriction endonucleases create DNA fragments with sticky ends, (2) make a diagram showing how DNA ligase can be used to splice sequences with sticky ends into a plasmid or other type of DNA sequence, and (3) explain how a cDNA library is constructed and probed with a known sequence.

19.2 Analyzing DNA: Did Our Ancestors Mate with Neanderthals?

Biologists reached the goal of producing human growth hormone on a commercial scale without knowing the exact sequence of the entire growth hormone gene. This is unusual. Determining the sequence of a gene is usually one of the first things that investigators wish to do. Understanding a gene's sequence is valuable for a variety of reasons:

- Once a gene's sequence is known, it is straightforward to infer the amino acid sequence of its product from the genetic code. Understanding a protein's primary structure, in turn, is usually required to determine its three-dimensional structure and normal function.

- Sequence differences among alleles can be analyzed to understand why some versions of the gene function better or differently than others do.

- A gene sequence can be compared with sequences of genes that have the same function in other species. These comparisons are often interesting. For example, bacteria, yeast, and humans are about as different as organisms can be, yet all three contain DNA polymerase genes with sections that are nearly identical in base sequence. Biologists explain this similarity by hypothesizing that the bacteria inhabiting your gut, the yeast in the bread you eat, and you are descended from a common ancestor that had DNA polymerase with the same sequence. Genes that are similar due to descent from a common ancestor are said to be homologous.

- Gene sequences can be compared to infer how closely related various species are. For example, analyzing DNA sequences allowed researchers to evaluate whether our species, *Homo sapiens*, ever interbred with the species of human called *Homo neanderthalensis*, Neanderthals.

Given that sequence data are important, how can biologists obtain them? This section introduces two common techniques

(a) Primers are required to run PCR.

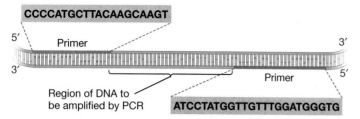

FIGURE 19.5 The Polymerase Chain Reaction (PCR) Is a Method for Producing Many Copies of a Specific Sequence
(a) The orange sequences indicate a set of single-stranded primers, which bracket the region of DNA to be amplified. (These are the actual sequences used in a PCR experiment explained later in the chapter.) **(b)** Each PCR cycle (denaturation, primer annealing, and extension) results in a doubling of the number of copies of the sequence between and including the primers. EXERCISE In part (b), draw in the events that produced the sequences in step 5.

used to analyze DNA sequences in the laboratory: the polymerase chain reaction and dideoxy sequencing. Both are variations of in vitro DNA synthesis reactions. The *polymerase chain reaction* (PCR) is a way to create many copies of a particular stretch of DNA. Once many copies of a gene are available, *dideoxy sequencing* allows researchers to determine the exact sequence of bases.

PCR and dideoxy sequencing rank among the greatest of all technological advances in the history of biological science. Their impact is comparable to the development of the light microscope, the electron microscope, and recombinant gene technology.

The Polymerase Chain Reaction (PCR)

Researchers have to obtain many copies of a gene in order to analyze its molecular structure. Inserting a gene into a bacterial plasmid is one method for producing many copies of a particular gene. The polymerase chain reaction is another technique for generating these copies. The **polymerase chain reaction (PCR)** is an in vitro DNA synthesis reaction in which a specific section of DNA is replicated over and over to amplify the number of copies of that sequence.

Figure 19.5 illustrates the reaction protocol, which was originally designed by Kary Mullis in 1983. As Figure 19.5a shows, the reaction mix includes two short sequences of single-stranded DNA. These sequences act as the primers that are required for DNA polymerase to work (see Chapter 14). The primers used in a PCR reaction have to bracket the region that will be copied. One primer is complementary to a sequence on one strand of the target DNA; the other primer is complementary to a sequence on the other strand. This is a key feature of PCR. To perform the reaction, a researcher must know the sequences on

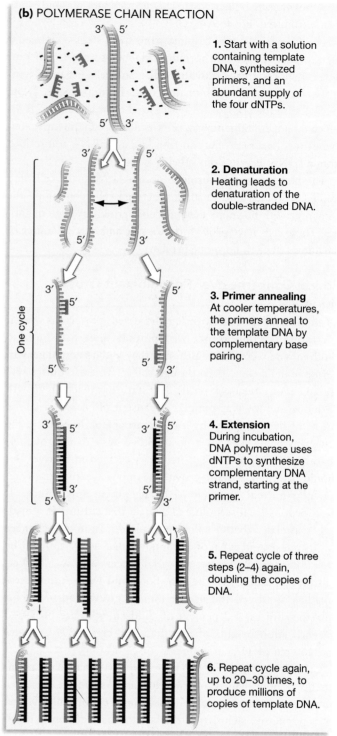

(b) POLYMERASE CHAIN REACTION

One cycle

1. Start with a solution containing template DNA, synthesized primers, and an abundant supply of the four dNTPs.

2. Denaturation
Heating leads to denaturation of the double-stranded DNA.

3. Primer annealing
At cooler temperatures, the primers anneal to the template DNA by complementary base pairing.

4. Extension
During incubation, DNA polymerase uses dNTPs to synthesize complementary DNA strand, starting at the primer.

5. Repeat cycle of three steps (2–4) again, doubling the copies of DNA.

6. Repeat cycle again, up to 20–30 times, to produce millions of copies of template DNA.

either side of the region of interest so that the correct primers can be synthesized.

To begin the reaction, Mullis added the primers to a solution containing an abundant supply of the four deoxynucleoside triphosphates (dNTPs; see Chapter 14) and copies of the template DNA—that is, a DNA sample that includes the gene of interest (Figure 19.5b, step 1). Then he heated the mixture to 95°C so that the double-stranded template DNA would *denature*—meaning that the two strands would separate, forming single-stranded templates (step 2). As the mixture cooled to about 60°C, some of the denatured DNA strands re-formed double helices. But some of the primers bonded to complementary portions of the single-stranded template DNA (step 3). This step is called *primer annealing*. The next step (4) was to add DNA polymerase I and incubate the mixture at 37°C, so that the enzyme could synthesize the complementary DNA strand, starting at the primer. This step is called *extension*.

The three steps denaturation, primer annealing, and extension constitute a single PCR cycle. If one copy of the template sequence existed in the sample originally, then there would be two copies at the end of the first cycle. Mullis added more DNA polymerase and repeated the cycle. The two copies present at the start of the second cycle acted as templates, so four copies existed at the end of cycle number 2 (step 5 of Figure 19.5b). As he repeated the cycle again and again, the amount of template sequence doubled each time (step 6). Doubling occurs because each newly synthesized segment of DNA serves as a template in the subsequent cycle. Starting with a single copy, successive cycles resulted in the production of 2, 4, 8, 16, 32, 64, 128, 256 copies, and so on. (A total of n cycles generates 2^n copies.) In just 20 cycles, one sequence can be amplified to over a million copies. By performing up to 30 cycles, Mullis obtained enormous numbers of copies of the template sequence.

Improving the Protocol   Mullis's PCR protocol has been modified in several important ways. Current methods were inspired by the discovery of the bacterium *Thermus aquaticus* in a hot spring inside Yellowstone National Park, Wyoming. This organism contains a heat-stable form of DNA polymerase known as *Taq* polymerase. Although most DNA polymerases are denatured and destroyed when heated to temperatures high enough to denature DNA, *Taq* polymmerase continues to function normally when heated to 95°C. (This observation is not surprising, because 95°C is the normal temperature at which the species lives.) By switching to *Taq* polymerase, researchers no longer had to add fresh DNA polymerase at each cycle. In current protocols, then, each PCR cycle consists of three steps:

1. Denaturation of DNA (94°C);
2. Annealing of primers to single-stranded DNA (50–60°C);
3. Extension of the complementary strands from the primers (72°C).

A single application of heat-stable *Taq* polymerase at the start of the reaction supports as many cycles as needed. The other major change in Mullis's protocol is that the temperature changes required in each cycle are now automated by PCR machines.

Using PCR to Study Fossil DNA   To understand why PCR is so valuable, consider a recent study by Svante Pääbo and colleagues. These biologists wanted to analyze DNA recovered from the 30,000-year-old bones of a fossilized human of the species *Homo neanderthalensis*. The individual involved was actually the first Neanderthal fossil ever discovered, from the Neander valley of Germany. Pääbo's goal was to determine the sequence of bases in the ancient DNA, compare it with DNA from modern humans (*Homo sapiens*), and analyze how similar the two species are. If the base sequences found in Neanderthals were similar enough to those found in modern humans, it would support the hypothesis that some of the DNA in our cells is descended from sequences that existed in Neanderthals. That could happen only if *Homo sapiens* and *Homo neanderthalensis* interbred while they coexisted in Europe about 30,000 years ago. This would mean that some of our ancestors were Neanderthals.

The Neanderthal bone was so old, however, that most of the DNA in it had degraded into tiny fragments too small to sequence. The biologists could recover only a minute amount of DNA that was still in moderate-sized pieces. To have any hope of analyzing the sample, they needed to generate many copies. To accomplish this, they turned to PCR.

Which Neanderthal gene should they try to amplify? Pääbo's group reasoned that, because human cells contain up to a thousand mitochondria, and because each mitochondrion in a cell contains many copies of mitochondrial DNA, there might still be some intact mitochondrial DNA somewhere in the Neanderthal DNA sample. More specifically, the team set out to amplify a locus called the control region of mitochondrial DNA.

To amplify the control region, the team added two single-stranded DNA primers, each about 20 base pairs long, to the sample. These primers had a sequence identical to sequences on either side of the control region in *Homo sapiens*. Using these primers allowed the researchers to amplify just the sequences of the Neanderthal control region, along with the primer sequences. After adding the primers to the reaction mix, the researchers added the four types of dNTPs and *Taq* polymerase, and used a PCR machine to change the temperature of the mix through a large number of cycles automatically. Using PCR, Pääbo's group was able to synthesize many copies of the control region. Obtaining many copies of the gene was essential to meeting their goal of determining the exact DNA sequence of the region.

Dideoxy Sequencing

Once researchers have many copies of a specific stretch of DNA, how can they determine the exact sequence of bases? In 1975 Frederick Sanger developed **dideoxy sequencing** as a clever variation on the basic in vitro DNA synthesis reaction to determine exact base sequences. But saying "clever" is an understatement. Sanger had to link three important insights to make his sequencing strategy work:

1. Sanger chose the name "dideoxy" because he used monomers called dideoxyribonucleoside triphosphates, or ddNTPs, along with deoxyribonucleoside triphosphates (dNTPs). These ddNTPs are identical to the dNTPs found in DNA, except they lack a hydroxyl group at their 3′ carbon (**Figure 19.6a**). Sanger realized that if a ddNTP were added to a growing DNA strand, it would terminate synthesis. Why? A ddNTP has no hydroxyl group available on its 3′ carbon to link to the 5′ carbon on an incoming dNTP monomer. As a result, DNA polymerization stops once a ddNTP is added.

2. Sanger linked this property of ddNTPs to a second fundamental insight. Suppose, he reasoned, that a radioactive primer could be attached to a template DNA, like the control region in Neanderthals that was studied by Pääbo's group. If this labeled template were incubated with DNA polymerase, the four dNTPs, and a small amount of ddGTP, along with a large amount of dGTP, the resulting daughter strands would constitute a limited set of lengths, each ending with a ddGTP. To understand why, consider that the synthesis of each daughter molecule would start at the same point—the primer—but end whenever a ddGTP happened to be incorporated into the growing strand opposite a C in the template. The addition of a ddGTP would stop further elongation. The collection of newly synthesized strands would vary in length as a result, with each specifying the distance from the labeled primer to successive C's in the template strand (**Figure 19.6b**). Analogous reactions done using ddTTP, ddATP, and ddCTP would give the distances between successive A's, T's, and G's, respectively.

3. Finally, Sanger realized that when the fragments produced by the four reactions are lined up by size, they reveal the sequence of bases in the template DNA. For example, when Pääbo and co-workers implemented Sanger's protocol and ran all four reactions, they had a collection of fragments identifying where every A, T, C, and G occurred in the control-region sequence of Neanderthals. To line these fragments up in order of size, they could be separated by gel electrophoresis. As **Figure 19.6c** shows, Pääbo's team could then read the sequence directly from the resulting autoradiograph.

Their result? The control region sequences found in Neanderthal DNA are highly distinct from the control region sequences observed in living humans. Thus, there is no evidence that we inherited some of our DNA from a Neanderthal ancestor. Based on these data, Pääbo and co-workers doubt the hypothesis that Neanderthals and members of our own species interbred. More recent studies, of Neanderthal DNA from Croatia, the Caucasus region of Russia, and a second German individual, support the same conclusion.

(a) ddNTPs terminate DNA synthesis.

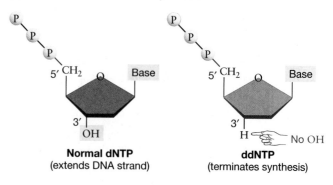

Normal dNTP
(extends DNA strand)

ddNTP
(terminates synthesis)

(b) Daughter strands of different lengths can be produced by using a mix of dNTPs and ddNTPs.

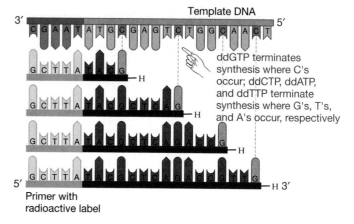

ddGTP terminates synthesis where C's occur; ddCTP, ddATP, and ddTTP terminate synthesis where G's, T's, and A's occur, respectively

Primer with radioactive label

(c) Different-length strands can be lined up by size to determine DNA sequence.

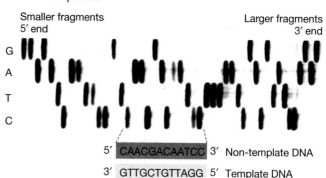

5′ CAACGACAATCC 3′ Non-template DNA

3′ GTTGCTGTTAGG 5′ Template DNA

FIGURE 19.6 Dideoxy Sequencing Is a Method for Determining the Base Sequence of DNA
(a) Unlike dNTPs, ddNTPs have no hydroxyl (OH) group on their 3′ carbon, so DNA synthesis stops if they are inserted into a strand. **(b)** If a small amount of ddGTP is added to an in vitro DNA synthesis reaction along with dNTPs and a short, single-stranded primer that is complementary to the 3′ end of a target sequence, then DNA polymerase will synthesize fragments of varying lengths. **(c)** The fragments that result from separate reactions using ddGTP, ddTTP, ddATP, and ddCTP can be run out on a gel. (The labels along the left-hand side indicate which ddNTP was used for the reaction loaded in each lane.) The DNA sequence of the strand complementary to the template can be read directly off the gel as shown. **EXERCISE** Starting at the far left end of the gel pictured in part (c), write down the entire sequence shown. Hint: The sequence starts with GG.

Recent innovations in sequencing techniques include the use of fluorescent markers instead of radioactive atoms on ddNTPs. The resulting fluorescently labeled fragments can be read by machine, rather than by eye, to determine the DNA sequence of the template region. Chapter 20 explains how automated sequencing machines are currently being used for large-scale sequencing projects. Even though automation allows today's gene sequencing centers to analyze hundreds of thousands of base pairs of DNA per day, the basic sequencing reaction that is used is still based on Sanger's dideoxy method.

✓ CHECK YOUR UNDERSTANDING

PCR is a technique for amplifying a specific region of DNA into millions of copies, which can then be sequenced or used for other types of analyses. You should be able to (1) explain the steps in a PCR cycle; (2) write down the sequence of a DNA strand 50 base pairs long, then design primers that would allow you to amplify the segment via PCR; and (3) explain the logic behind the dideoxy sequencing method and read a base sequence off a sequencing gel.

19.3 Gene Hunting Based on Pedigree Analysis: The Huntington's Disease Story

One of the great promises of DNA sequencing technology is the ability to analyze genes that are associated with inherited diseases of humans and thereby understand the nature of the molecular defect that causes illness. Once the difference between normal and dysfunctional alleles is understood, investigators hope to be able to introduce working copies of alleles into patients with defective alleles and cure these individuals. As the next two sections of this chapter will show, the simple-sounding goal of replacing defective alleles with functioning alleles has been tremendously difficult to achieve.

The effort to cure genetic diseases via recombinant alleles starts with the ability to find the disease genes. As an example of how this type of gene hunt is done, let's consider the first successful effort ever conducted in humans: the search for the gene associated with Huntington's disease.

Huntington's disease is a rare but devastating illness. Typically, affected individuals first show symptoms between the ages of 35 and 45. At onset, an individual appears to be clumsier than normal and tends to develop small tics and abnormal movements. As the disease progresses, uncontrollable movements become more pronounced. Eventually the affected individual twists and writhes involuntarily. Personality and intelligence are also affected—to the extent that the early stage of this disease is sometimes misdiagnosed as the personality disorder schizophrenia. The illness may continue to progress for 10 to 20 years and is eventually fatal.

Because Huntington's disease appeared to run in families, physicians suspected that it was a genetic disease. Recall from Chapter 13 that an analysis of pedigrees from families affected by Huntington's disease suggested that the trait was due to a single, autosomal dominant allele. To test this hypothesis, researchers set out to identify the gene or genes involved and to document that one or more genes are altered in affected individuals. Reaching this goal took over 10 years of intensive effort.

How Was the Huntington's Disease Gene Found?

The search for the Huntington's disease gene was led by Nancy Wexler, whose mother had died of the disease. If the trait was indeed due to an autosomal dominant allele, it meant that there was a 50 percent chance that Wexler had received the allele from her mother and would begin to show symptoms when she reached middle age. Identifying the gene would make it possible for Wexler's research team to reach several important objectives:

- *Genetic Testing* Because the symptoms of Huntington's disease do not appear until middle age, many affected people have married and had children by the time they develop symptoms. If members of families with Huntington's patients could be tested for the defective allele early in life, they might make different decisions about starting their own families and passing on the illness.

- *Therapy* Once the sequence of a gene is known, researchers can better understand the structure and function of its protein product. Information on a protein's mode of action is extremely helpful in developing drugs or other treatments. As we'll see in Section 19.4, finding a gene also makes it possible to consider replacing defective alleles with normal copies.

- *Improved Understanding of the Phenotype* Several fundamental questions about Huntington's disease might be resolved if the causative gene were found. How can one defective protein lead to abnormal movements *and* personality changes? Why does it take so long for symptoms to appear, and why is the illness progressive?

Using Genetic Markers   To locate disease-causing genes, researchers start with a **genetic map**, also known as a **linkage map** or **meiotic map** (see Chapter 13). Recall that in fruit flies and other organisms, the relative positions of genes on the same chromosome can be determined by analyzing the frequency of recombination between pairs of genes. A gene that has been mapped can serve as a **genetic marker**, particularly if that gene has two or more alleles present in the population being studied. A genetic marker provides a physical landmark at a known position along a chromosome.

To understand how genetic markers can be used to locate the positions of unknown genes, suppose that you knew the position of a hair-color gene in humans and that various alleles of this gene contributed to the development of black hair, red hair, blond hair, and brown hair in the population that you were studying. Now suppose that you analyzed the pedigree of a large family in which cystic fibrosis was common and you noticed that individuals who had cystic fibrosis almost always had black hair—even though they were just as likely as unaffected individuals to have any other trait. If you observe that two particular alleles are almost always inherited together, it is logical to conclude that the genes involved are close to each other on the same chromosome—meaning that they are closely linked. If they were not closely linked, then crossing over between them would be common and they would not be inherited together. Based on these data, you could infer that the gene for cystic fibrosis is very close to the hair-color gene.

In essence, then, disease-gene hunting boils down to this: Researchers have to find a large number of closely related people who are affected and unaffected, and then attempt to locate a genetic marker with an allele that almost always occurs in the affect-

ed individuals but not in the unaffected people. If such a marker is found, the disease gene is almost guaranteed to be nearby.

The types of genetic markers used in gene hunts have changed over time. In the late 1970s and early 1980s, when biologists were searching for the Huntington's disease gene, the best genetic markers available were short stretches of DNA where specific restriction endonucleases cut the double helix. These sequences are known as *restriction endonuclease recognition sites*. (Chapter 20 introduces the types of markers in common use today. Although the nature of the markers employed has changed over time, the logic of how they are used is the same.)

How do restriction endonuclease recognition sites work as genetic markers? The key idea is that the number of recognition sites varies among individuals because the base sequence of their DNA varies. For example, if a mutation alters the DNA sequence of a recognition site, the restriction endonuclease will

---

BOX 19.2 Southern Blotting

Southern blotting was named after its inventor, Edwin Southern, and it is among the most basic techniques in molecular biology. It is a multistep procedure that allows researchers to identify and characterize specific genes within an organism's genome. If you have a DNA sample from an individual and wish to find a specific gene in that sample, Southern blotting is a way to do it.

The first step in Southern blotting is to obtain DNA from the cells of the or-

ganisms being studied and digest it with restriction endonucleases (**Figure 19.7**, step 1). The DNA fragments generated in this way can then be separated by gel electrophoresis (steps 2 and 3). Once the fragments are sorted by size, they are treated with chemicals that break the hydrogen bonds between base pairs, resulting in the formation of single-stranded DNA (step 4). Next the fragments of single-stranded DNA are transferred from the gel to a piece of nitrocellulose

or nylon filter, using the blotting technique illustrated in step 5.

The product of these steps is a series of single-stranded DNA fragments, sorted by size and permanently bound to the membrane. In some cases, these fragments represent the entire genome of the organism being studied. To find a certain gene in this collection of fragments, a researcher must have a DNA sequence that is complementary to some region of that gene. Recall that such a sequence is called a

SOUTHERN BLOTTING

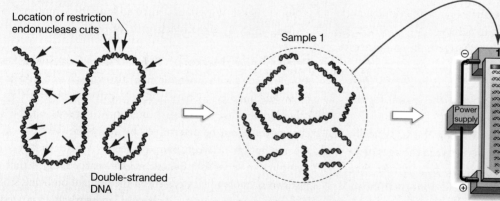

1. Restriction endonucleases cut DNA sample into fragments of various lengths. Each type of restriction endonuclease cuts a specific sequence of DNA.

2. A sample consists of all the DNA fragments of various lengths. The sample is loaded into a gel for electrophoresis.

3. Electrophoresis. Use voltage difference to separate DNA fragments by size. Small fragments run faster.

FIGURE 19.7 Southern Blotting Protocol
Southern blotting is a technique for locating a specific gene in a DNA sample containing many genes.

no longer cut at that location. Thus, each recognition site has two alleles: The site is either present or absent.

When an investigator cuts DNA from many individuals with the same restriction endonuclease, the size of the resulting fragments will vary. These size variants are called **restriction fragment length polymorphisms (RFLPs)**. To determine which RFLPs occur in a particular individual, a researcher performs the following four steps, which are explained in more detail in **Box 19.2**:

1. Obtain a DNA sample from individuals in the population being studied (via blood samples in humans). Divide the purified DNA into several subsamples. Allow one of these subsamples to be cut with a particular restriction endonuclease.

2. The fragments from a series of individuals are sorted by size via gel electrophoresis. The separated fragments are then treated chemically to make them single stranded.

3. The fragments are probed with a labeled, single-stranded segment of DNA from a particular genetic marker—meaning a specific location on a chromosome. The probe binds to complementary sequences in the samples being analyzed. The act of binding identifies fragments that include a particular genetic marker or set of markers.

4. The restriction fragments that bind to the probe are visualized via autoradiography or other techniques.

When these steps are complete, a researcher has documented the "phenotype" of a person at a particular genetic marker. Stated another way, the fragment patterns show which allele the person has (recognition site present or absent) at one or more markers.

The bands on the gel are a genetic trait, just like height or hair color or the presence or absence of Huntington's disease. Because

probe. The probe DNA is labeled—typically with a radioactive atom or fluorescent marker—and made single stranded by heating. The probe is then added to a solution bathing the filter (step 6). During incubation, the labeled probe binds to the fragment or fragments on the nylon that have complementary base pairs. This step is called hybridization. In this way, the probe identifies the gene of interest.

To visualize which fragments hybridized with a radioactively labeled probe, a researcher lays X-ray film over the nylon filter. As step 7 of Figure 19.7 shows, radioactive emissions from the probe DNA expose the film. The black band that results identifies the target gene. Fluorescent markers can be visualized and photographed under the appropriate wavelengths of light.

A variation on Southern blotting is based on separating RNAs via gel electrophoresis, transferring them to a filter paper, and probing them with a single-stranded and radioactively labeled DNA probe. This technique is used to identify the RNA fragments produced by a particular gene. It is called **Northern blotting**, in a lighthearted tribute to the protocol from which it was derived. The variation called **Western blotting** involves separating proteins via electrophoresis and then probing the resulting filter with an antibody probe that binds to the protein of interest. The use of antibodies in research is explored in detail in later chapters.

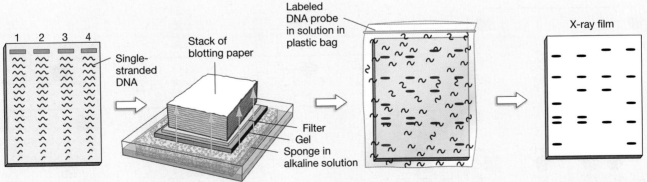

4. The DNA fragments are treated to make them single stranded.

5. Blotting. An alkaline solution wicks up into blotting paper, carrying DNA from gel onto nylon filter, where it is then permanently bound.

6. Hybridization with radioactive probe. Incubate the nylon filter with a solution containing labeled probe DNA. The radioactive probe binds to the fragments containing complementary sequences.

7. Autoradiography. Place filter against X-ray film. Radioactive DNA fragments expose film, forming black bands that indicate location of target DNA.

banding patterns are due to particular sequences in DNA, they are passed from parent to offspring just like any other DNA sequence.

By cutting DNA with different restriction endonucleases and probing the resulting gels with sequences from different locations in the genome, researchers gradually build a large catalog of recognition sites that vary among individuals.

For these RFLPs or other types of molecular markers to be used in a gene hunt, they have to be placed on a map of their species' chromosomes. Genetic markers can be placed on a genetic map by analyzing the frequency of recombination between them. In contrast, a **physical map** of the genome records the absolute position—in numbers of base pairs—along a chromosome. Techniques for creating physical maps are introduced briefly in Chapter 20.

Using a Pedigree   Once a genetic map containing many restriction endonuclease recognition sites or other genetic markers has been assembled, biologists need help from families affected by an inherited disease to find the gene in question. Recall that the fundamental goal is to find a genetic marker that is almost always inherited along with the disease-causing allele. Gene hunts are more likely to be successful if large families are involved. Large sample sizes minimize the probability that researchers will observe an association between one or more markers and the disease just by chance—not because they are closely linked. Wexler's Huntington's disease team was fortunate to find a large, extended family affected with the disease living along the shores of Lake Maracaibo, Venezuela.

From historical records, the researchers deduced that the Huntington's disease allele was introduced to this family by a European sailor or trader who visited the area in the early 1800s. When family members agreed to participate in the study, there were over 3000 of his descendants living in the area. One hundred of these people had been diagnosed with Huntington's disease. To help in the search for the gene, family members agreed to donate skin or blood samples for DNA analysis and to furnish information on who was related to whom.

This research resulted in the pedigree shown in **Figure 19.8**. Note that the diagram includes information about the disease phenotype and the particular RFLP pattern observed in each family member. To find the Huntington's disease gene, researchers looked for fragment patterns that were present in affected individuals but absent from unaffected individuals. Several restriction endonuclease recognition sites appeared to be inherited along with Huntington's disease. Were these associations due to chance or to a close physical association between the recognition site and the defective allele? Stated another way, were the recognition site and Huntington's disease gene close to one another on the DNA double helix (**Figure 19.9**)?

To answer this question, the researchers used a statistical analysis to determine the likelihood that an apparent association between a genetic marker and the disease phenotype was due to chance. The analysis supported the hypothesis that the region shown in **Figure 19.10a**—which happened to be on chromosome 4—was close to the Huntington's disease gene. Note that two of the seven restriction endonuclease recognition sites in this region (marked by arrows) are polymorphic, meaning some individuals have one or both sites and some do not. As **Figure 19.10b** shows, four combinations of restriction endonuclease cuts are possible at these two sites. Thus, all four fragment patterns A through D shown in **Figure 19.10c** are possible.

The key finding was that pattern C shown in Figure 19.10c was inherited along with the defective allele often enough to

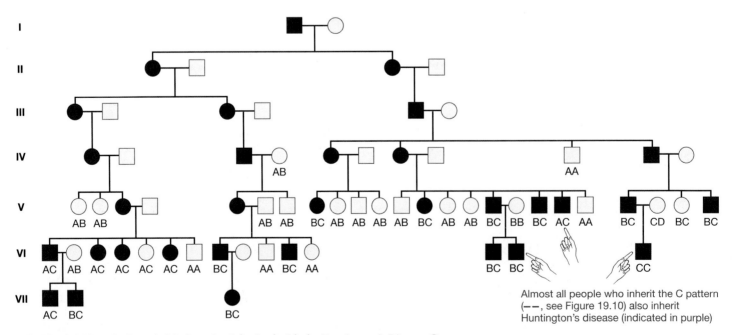

Almost all people who inherit the C pattern (——, see Figure 19.10) also inherit Huntington's disease (indicated in purple)

FIGURE 19.8 Certain Genetic Markers Are Inherited with the Huntington's Disease Gene
Part of a pedigree from seven generations (I–VII) of a large extended family. Individuals affected by Huntington's disease inherited the defective allele from a common ancestor. The letters A–D represent the restriction fragment patterns found in each individual.

QUESTION In generation VI, an individual with the genotype AC does not have Huntington's disease. How is this possible?

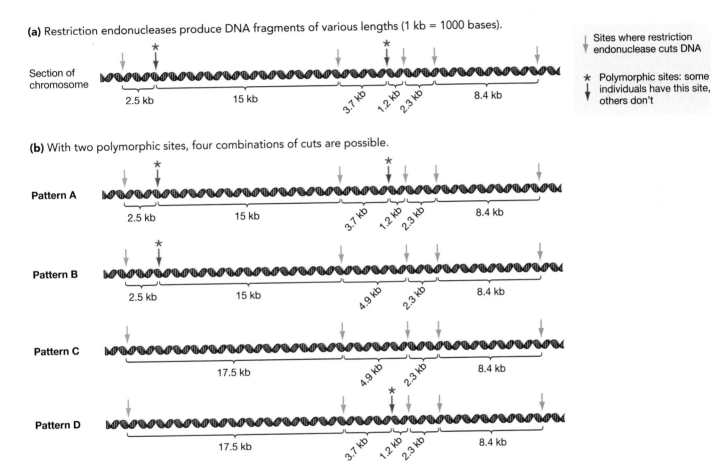

FIGURE 19.9 Genetic Markers and Disease Alleles Are Inherited Together if They Are Closely Linked
To find a gene, researchers look for alleles of a genetic marker that almost always occur in individuals with a specific allele of the gene.

(a) Restriction endonucleases produce DNA fragments of various lengths (1 kb = 1000 bases).

(b) With two polymorphic sites, four combinations of cuts are possible.

(c) Four banding patterns are possible when fragments from homozygous individuals are run out on a gel.

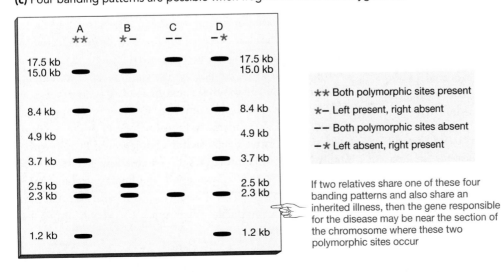

✱✱ Both polymorphic sites present

✱– Left present, right absent

– – Both polymorphic sites absent

– ✱ Left absent, right present

If two relatives share one of these four banding patterns and also share an inherited illness, then the gene responsible for the disease may be near the section of the chromosome where these two polymorphic sites occur

FIGURE 19.10 Restriction Endonuclease Recognition Sites Can Be Used as Genetic Markers
(a) A DNA segment with five restriction sites (green arrows) that occur in all individuals in a given population and two sites (blue starred arrows) that occur in some individuals but not in others. Here "kb" stands for kilobase; 1 kb = 1000 bases. **(b)** Because there are two polymorphic sites, four possible combinations of cuts can occur. **(c)** Four banding patterns predicted for individuals who are homozygous for each pattern.

indicate that the association was not due to chance. Apparently, the English sailor who introduced the Huntington's disease allele also had the "C" allele in the RFLP pattern. The two loci are so close together that recombination between them—which would put an "A" or "B" or "D" allele next to the Huntington's disease allele in his descendants—has been extremely rare. By examining associations between other markers on chromosome 4 with the Huntington's disease phenotype, the team succeeded in narrowing down the location of the Huntington's disease gene to a region about 500,000 base pairs long. Because the haploid human genome contains over 3 billion base pairs, this was a significant accomplishment.

Pinpointing the Defect Once the general location of the Huntington's disease gene was known, biologists looked in that region for exons that encode a functional mRNA. They did this by sequencing exons from diseased and normal individuals, comparing the data, and pinpointing specific bases that were different between the two groups of individuals. In this way, dideoxy sequencing played a key role in the gene hunt.

When this analysis was complete, the research team found that individuals with Huntington's disease have an unusual number of CAG codons near the 5′ end of a particular gene. CAG codes for glutamine. Healthy individuals have 11–25 copies of the CAG codon, while affected individuals have 42 or more copies. Although many genetic diseases are caused by single base changes that alter the amino acid sequence of a protein, diseases caused by the expansion of a particular codon repeat have also been observed. For example, the most common form of inherited mental retardation, **fragile-X syndrome**, is caused by an increase in the number of copies of a CGG codon at a specific site on the X chromosome.

When the Huntington's disease research team confirmed that the increase in the CAG codon was always observed in affected individuals, the team concluded that the long search for the Huntington's disease gene was over. They named the newly discovered locus *IT15* and its protein product huntingtin. In both affected and normal individuals, the huntingtin protein is involved in the early development of nerve cells. It is only later in life that mutant forms of the protein cause disease.

New Approaches to Therapy How has the effort to locate the Huntington's disease gene helped researchers and physicians understand and treat the illness? Biomedical researchers point to three major benefits of a successful gene hunt:

1. The discovery of *IT15* allowed investigators to understand the molecular nature of Huntington's disease for the first time. Autopsies of Huntington's patients showed that their brains actually decrease in size due to the death of neurons and that the brain tissue contains insoluble aggregates of the huntingtin protein. These aggregates are thought to be a direct consequence of the changes in the number of CAG repeats in the *IT15* gene and thus the number of glutamine residues in the protein. Long stretches of polyglutamine (a polymer of glutamine residues) are known to result in the formation of protein aggregates. The leading hypothesis to explain Huntington's disease proposes that a gradual buildup of the huntingtin protein aggregates triggers neurons to undergo apoptosis, or programmed cell death.

2. The discovery of the Huntington's disease gene has allowed scientists to introduce the defective allele into mice via the types of genetic engineering techniques that we will discuss in Section 19.4. These mice are called **transgenic** ("across-genes"), because they have alleles that have been modified by genetic engineering.

   The goal of creating transgenic mice with the defective *IT15* allele was to stimulate research on Huntington's disease. Laboratory animals with disease symptoms that parallel those of a human disease are said to provide an **animal model** of the disease. Transgenic mice that produce defective versions of the huntingtin protein develop tremors and abnormal movements, exhibit higher-than-normal levels of aggression toward litter and cage mates, and experience a loss of neurons in the brain. The transgenic mice get a version of Huntington's disease.

   Animal models are valuable in disease research, because they can be used to test potential treatments before investigators try them on human patients. Now that an animal model for Huntington's disease is available, research groups have begun testing drugs that appear to prevent or reduce the aggregation of the huntingtin protein. Other biologists are attempting to design drugs that prevent the death of cells containing aggregated huntingtin proteins.

3. A third benefit of finding a disease gene is that it makes genetic testing possible. What is genetic testing, and how is it beneficial?

Genetic Testing

When the Huntington's gene was found, biologists used the information to develop a test for the presence of the defective allele. The test involves obtaining a DNA sample from an individual and using the polymerase chain reaction to amplify the chromosome region that contains the CAG repeats responsible for the disease. If the number of CAG repeats is 35 or less, the individual is considered to be normal. Forty or more repeats results in a positive diagnosis for Huntington's. People with a family history of Huntington's disease can now be tested.

As more becomes known about the human genome, more and more tests like this will be developed for diseases that have some genetic component. What types of genetic testing are done currently?

- *Carrier Testing* Before starting their own family, people from families affected by a genetic disease frequently want to know whether they carry the allele responsible. That is especially true for diseases, such as cystic fibrosis (CF), that are due to recessive alleles. If only one of the prospective parents has the allele, then none of the children they have together should develop CF. But if both prospective parents carry the allele, then each child they produce has a 25 percent chance of having CF. Carrier testing can determine whether an individual is a carrier for a genetic disease.

- *Prenatal Testing* Suppose that two parents, both carrying the CF allele, decide to have children but do not want to pass along that allele. Once the mother is pregnant, a physician can obtain fetal cells early in gestation. The cells can be cultured and DNA isolated. The CF allele can then be amplified by PCR and sequenced. Based on the test results, the couple may choose to continue or terminate a pregnancy.

- *Adult Testing* Huntington's disease isn't the only trait that appears in adulthood. For example, about 5 percent of the women who develop breast cancer have inherited a faulty gene. To date, three genes are known to create a predisposition to breast cancer. Adult women with a family history of breast cancer can be tested for the presence of these mutant genes.

It is important to note, though, that the results of genetic tests can be difficult to interpret. If a person has an expanded CAG repeat in the *IT15* gene, it is virtually certain that he or she will develop Huntington's disease. But if a woman has one of the alleles associated with breast cancer, there is just an 80 percent chance that she will develop the illness. In such cases, the benefit of testing is to make the person involved more vigilant about getting regular checkups. Then if breast cancer does develop, it can be caught at an early and potentially curable stage.

Ethical Concerns over Genetic Testing

Genetic testing raises controversial ethical issues. For example, some individuals maintain that it is morally wrong to terminate any pregnancy, even if the fetus involved is guaranteed to be born with a debilitating genetic disease. Also, consider Nancy Wexler's position soon after the discovery of *IT15*. Would you choose to be tested for the defective allele and risk finding out that you were almost certain to develop Huntington's disease? Should physicians agree to test people for genetic diseases that have no cure? Should it be legal for health insurance companies to test clients for genetic diseases that will require expensive treatments so that the companies can refuse coverage?

These questions are still being debated by political and religious leaders, health-care workers, philosophers, and members of the general public. Difficult as they are, the issues raised by genetic testing pale in comparison with some of the questions raised by the effort to create transgenic humans.

✓CHECK YOUR UNDERSTANDING

Genes for particular traits can be located if they are closely linked to a known genetic marker and are thus inherited along with the marker. You should be able to draw a pedigree showing a close association between a particular genetic illness and a specific genetic marker.

19.4 Can Gene Therapy Cure Inherited Diseases in Humans? Research on Severe Immune Disorders

For biomedical researchers interested in curing inherited diseases such as Huntington's, sickle-cell anemia, and cystic fibrosis, the ultimate goal is to replace or augment defective copies of the gene with normal alleles. This approach to treatment is called **gene therapy**.

For gene therapy to succeed, two crucial steps must be completed. First, the wild-type allele for the gene in question and its regulatory sequences must be sequenced and understood. Second, there must be a method for introducing a copy of the normal allele into affected individuals. The DNA has to be introduced in a way that ensures expression of the gene in the correct tissues, in the correct amount, and at the correct time. If the defective allele is dominant, then the introduction step may be even more complicated: In at least some cases, the introduced allele must physically replace or block the expression of the existing dominant allele.

Chapter 18 discussed aspects of gene regulation in humans and other eukaryotes; here, we focus on techniques for introducing DNA into humans. As a case study, let's consider the first successful application of gene therapy in humans.

How Can Novel Alleles Be Introduced into Human Cells?

Section 19.1 reviewed how recombinant DNA sequences are packaged into plasmids and taken up by *E. coli* cells. Humans and other mammals lack plasmids, however, and their cells do not take up foreign DNA in response to chemical or electric treatments as efficiently as bacterial cells do.

How can foreign genes be introduced efficiently into human cells? To date, researchers have focused on packaging foreign DNA into viruses for transport into human cells. A viral infection begins when a virus particle enters or attaches to a host cell and inserts its genome into that host cell. In some cases the viral DNA becomes integrated into a host-cell chromosome. As a result, viruses that infect human cells can be used as vectors to carry engineered alleles into the chromosomes of target cells. Potentially, the alleles delivered by the virus could be expressed and produce a product capable of curing a genetic disease.

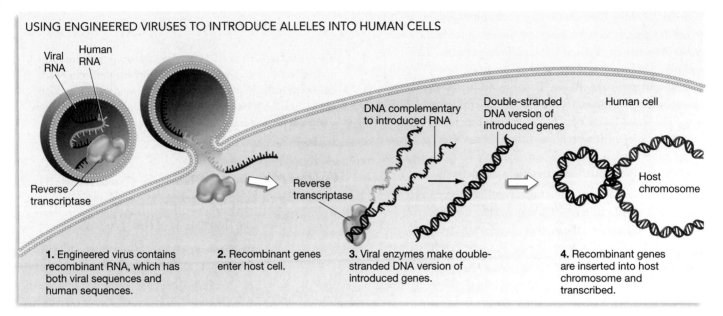

USING ENGINEERED VIRUSES TO INTRODUCE ALLELES INTO HUMAN CELLS

FIGURE 19.11 Retroviruses Insert Their Genes into Host-Cell Chromosomes
Retroviruses can be used as vectors to carry specific genes into cells and to insert these genes into the chromosomes of target cells. **QUESTION** What happens if the recombinant DNA is inserted in the middle of a gene that is critical to the cell's normal function?

Two major classes of viruses have proven to be effective vectors for introducing recombinant DNA: (1) retroviruses and (2) adenoviruses. Retroviruses have an RNA genome. The genomes include the gene for the enzyme reverse transcriptase, which catalyzes the production of a double-stranded DNA molecule from a single-stranded RNA template. (Recall from Section 19.1 that researchers use reverse transcriptase to produce cDNAs.) When a retrovirus infects a human cell, reverse transcriptase catalyzes the production of a DNA copy of the viral genome. Other viral enzymes catalyze the insertion of the viral DNA into a host-cell chromosome. If human genes can be packaged into a retrovirus, then the virus is capable of inserting the human alleles into a chromosome in a target cell (**Figure 19.11**).

The adenoviruses have a DNA genome and have also been used to carry recombinant alleles into human cells. Certain adenoviruses infect human lung cells and other tissues in the respiratory tract and cause colds or flu-like symptoms. Because they are specialized for infecting cells in the human respiratory tract, adenoviruses have been tested as vectors for treating individuals who suffer from inherited disorders of the lung, such as cystic fibrosis. Unfortunately, the genes of adenoviruses do not integrate into host-cell chromosomes. As a result, they are expressed in target cells for only a relatively short time. For such reasons, most gene therapy research is focused on retroviruses.

Although viruses would appear to be ideal vectors for the delivery of normal human alleles, there are serious problems associated with their use. Viruses usually cause disease. The retroviruses include the human immunodeficiency virus (HIV), which causes AIDS, as well as an array of viruses that cause cancer in animals. Using these agents for gene therapy requires that sequences responsible for causing disease be inactivated or removed from their genomes. Even if they have been inactivated, the altered particles may be able to recombine with viral DNA that already exists in the individual being treated and lead to the formation of a new infectious strain. In addition, viral proteins trigger a response by the immune system that can cause dangerous side effects during treatment. Finally, if viral genes happen to insert themselves in a position that disrupts the function of an important gene in the target cell, the consequences can be serious. Despite these risks, viruses are still the best vectors currently available for human gene therapy.

Using Gene Therapy to Treat X-Linked Immune Deficiency

Recently, a research team reported the successful treatment of a disease by means of gene therapy. The illness is called **severe combined immunodeficiency (SCID)**. Children who are born with SCID lack a normal immune system and are unable to fight off infections.

The type of SCID treated by the group is designated SCID-X1, because it is caused by mutations in a gene on the X chromosome. The gene is required for the development of immune-system cells called *T cells*. T cells develop in bone marrow, from undifferentiated cells called **stem cells**. Unlike mature T cells, stem cells divide continuously.

The gene responsible for SCID-X1 codes for a polypeptide called γc ("gamma-c"), which is a key component of several

FIGURE 19.12 A "Bubble Child"
Children who are born with SCID cannot fight off bacterial or viral infections. As a result, such children must live in a sterile environment.

plasma membrane proteins. The function of these proteins is to receive signals from growth factors that stimulate the development of T cells from stem cells. Without functioning copies of the receptor proteins, the growth signals are not received and T cells do not mature properly. Without a large supply of mature T cells, infants are helpless to ward off bacterial and viral infections. SCID-X1 is fatal.

Traditionally, physicians have treated SCID-X1 by keeping the patient in a sterile environment, isolated from any direct human contact until the patient could receive bone-marrow tissue transplanted from a close relative (**Figure 19.12**). In most bone-marrow transplants, the T cells that the patient needs then differentiate from the transplanted bone-marrow cells and allow the individual to live normally. In some cases, though, no suitable donor is available. One boy born with SCID had to live in a germ-

free "bubble" for over 12 years; when a bone-marrow transplant was finally attempted, it failed. Could gene therapy cure this disease by furnishing functioning copies of the defective gene?

Before any protocol for gene therapy can be implemented in humans, researchers must investigate its safety by showing that the proposed strategy works in animals. To begin their work, then, biologists introduced the γc gene into a retrovirus vector and used the recombinant virus to infect bone-marrow cells from dogs. Follow-up analyses showed that the dog cells that took up the vector were long lived and produced the receptor protein. Further, the receptor recognized the proper growth factor and stimulated the development of T cells. Another set of experiments was carried out on an animal model for SCID—mice that lacked functioning copies of the γc gene. When the team transferred a normal copy of the gene into the marrow cells of these mice, immunodeficiency was prevented.

Based on these results, the team gained approval to treat ten boys, each less than one year old, who had SCID-X1. No suitable bone-marrow donor was available to treat these patients. Even though they'd been isolated in germ-free rooms, several had developed illnesses such as pneumocystic pneumonia and thrush (oral yeast infections), which appear in people with malfunctioning immune systems.

To implement gene therapy, the researchers removed bone marrow from each patient and collected the stem cells that produce mature T cells (**Figure 19.13**). Each day for three days, researchers infected the stem cells with an engineered retrovirus that carried the normal γc gene. About 30 percent of the cells took up the γc gene and manufactured the receptor protein. This recombination step was done in vitro to minimize patient discomfort and to allow researchers to monitor the procedure's success. The stem cells that were producing normal receptor protein were then isolated and transferred back into the patients.

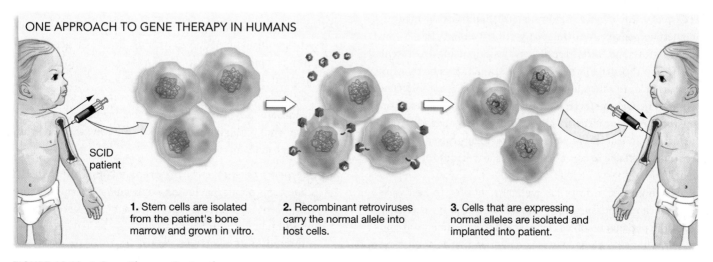

ONE APPROACH TO GENE THERAPY IN HUMANS

SCID patient

1. Stem cells are isolated from the patient's bone marrow and grown in vitro.

2. Recombinant retroviruses carry the normal allele into host cells.

3. Cells that are expressing normal alleles are isolated and implanted into patient.

FIGURE 19.13 A Gene Therapy Protocol
For gene therapy to work, copies of a normal allele have to be introduced into a patient's cells and be expressed.

Within four months after reinsertion of the transformed marrow cells, nine of the boys had normal levels of functioning T cells. T-cell concentrations in the nine children continued to increase for eight months. At that time, the research team had enough confidence in the boys' immune systems to inoculate each boy with polio and tetanus vaccines. To their great relief, they found that the boys' immune systems succeeded in producing antibodies to the polio and tetanus vaccines. Antibodies are the body's normal response proteins. Their production is crucial to fighting off infections. The children now had functioning immune systems. The infections that existed prior to gene therapy also began to clear up.

About three months after the experimental treatment began, the nine patients were removed from germ-free isolation rooms and began residing at home, where they grew and developed normally. During a routine check-up 30 months after gene therapy, however, the research team discovered that two of the boys had developed a type of cancer characterized by unchecked growth of T cells. Follow-up analyses of their bone-marrow cells showed that, in both boys, a viral-borne γc gene had been inserted near the gene for a regulatory transcription factor that triggers the growth of T cells. The viral sequences apparently acted as an enhancer and led to constitutive expression of the transcription factor. As this book goes to press, the two boys have responded to cancer chemotherapy and are healthy.

The tenth boy to receive gene therapy never succeeded in producing T cells at all. For unknown reasons, his recombinant stem cells failed to function normally when they were transplanted back into his bone marrow. Fortunately, physicians were later able to find a bone marrow donor whose cells were a close enough match to the boy's cells to make a successful transplant possible.

Ethical Concerns over Gene Therapy

Although the treatment of SCID-X1 qualifies as an important success story, gene therapy is still highly experimental, extremely expensive, and intensely controversial. Just a few months before the initial report on the promising treatment of SCID-X1 appeared in the year 2000, a major gene therapy center was closed when a patient being treated for a different enzyme deficiency disease died after gene therapy based on the use of an adenovirus. In this case, the death was directly attributable to the treatment. The incident raised serious questions about the use of adenoviruses as vectors for recombinant DNA.

Throughout the history of medicine, efforts to test new drugs, vaccines, and surgical protocols have always carried a risk for the patients involved. Gene therapy experiments are no different. The researchers who run gene therapy trials must explain the risks clearly and make every effort to minimize them.

The initial report on the development of cancer in the boys who received gene therapy for SCID-X1 concluded with the following statement: "We have proposed ... a halt to our trial until further evaluation of the causes of this adverse event and a careful reassessment of the risks and benefits of continuing our study of gene therapy." Although recombinant gene therapy holds great promise for the treatment of a wide variety of devastating inherited diseases, fulfilling that promise is almost certain to require many years of additional research and testing, as well as the refinement of legal and ethical guidelines.

19.5   Biotechnology in Agriculture: The Development of Golden Rice

Progress in human gene therapy has been slow, but progress in transforming crop plants with recombinant genes has been rapid. In the United States alone, genetically modified or transgenic strains of 15 major crop species are now approved for commercial use. Farmers growing maize (corn) in the United States currently have over 15 distinct transgenic varieties to choose from. In Mississippi, 97 percent of the land area devoted to cotton is now planted with recombinant strains. Globally, about 67.7 million hectares (167 million acres) of transgenic crops were grown in 2003. The use of genetically modified plants has become a major element of modern agriculture.

Recent efforts to develop transgenic plants have focused on three general objectives:

1. *Reducing losses to herbivore damage* The most popular strategy for reducing losses caused by plant-eating organisms is based on introducing the gene for a naturally occurring insecticide. For example, researchers have transferred a gene from the bacterium *Bacillis thuringiensis* into corn; the presence of the "Bt toxin" encoded by this gene protects the plant from corn borers and other caterpillar pests.

2. *Reducing competition with weeds* Competition with weeds is usually reduced by introducing a gene whose product makes the crop plant resistant to a herbicide (a chemical that destroys plants). For example, soybeans have been genetically engineered for resistance to the herbicide glyphosate. Soybean fields with the engineered strain can be sprayed with glyphosate to kill weeds without harming the soybeans.

3. *Improving the quality of the product consumed by people* Some crop plants, including soybeans and canola plants, have been engineered to produce a higher percentage of unsaturated fatty acids relative to saturated fatty acids. Saturated fatty acids can contribute to heart disease, so crops with less of them are healthier to eat.

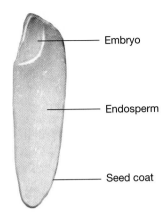

FIGURE 19.14 Endosperm Is a Nutritive Tissue in Seeds
Rice seeds consist of the three regions shown here.

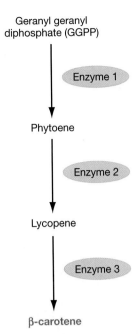

FIGURE 19.15 Synthetic Pathway for β-Carotene
GGPP is a molecule found in rice seeds. Three enzymes are required to produce β-carotene from GGPP.

To gain a better understanding of genetic engineering in plants, let's take a detailed look at efforts to produce transgenic rice with improved nutritional qualities.

Rice as a Target Crop

Although almost half the world's population depends on rice as its staple food, this grain is an extremely poor source of certain vitamins and essential nutrients. For example, rice contains no vitamin A. This is a serious issue, because vitamin A deficiency causes blindness in 250,000 Southeast Asian children each year. Vitamin A deficiency also renders children more susceptible to diarrhea, respiratory infections, and childhood diseases such as measles.

Humans and other mammals synthesize vitamin A from a precursor molecule known as β-carotene (beta-carotene). β-Carotene belongs to a family of plant pigments called the carotenoids, which were introduced in Chapter 10. Carotenoids are orange, yellow, and red and are especially abundant in carrots. Rice plants synthesize β-carotene in their chloroplasts but not in the part of the seed that is eaten by humans. Could genetic engineering produce a strain of rice that synthesized β-carotene in the carbohydrate-rich seed tissue called endosperm (**Figure 19.14**)? A research team set out to answer this question. If successful, their research could help to solve an important global health problem.

Synthesizing β-Carotene in Rice

The research team began their work by searching for compounds in rice endosperm that could serve as precursors for the synthesis of β-carotene. They found that maturing rice endosperm contains a molecule called geranyl geranyl diphosphate (GGPP), which is an intermediate in the synthetic pathway that leads to the production of carotenoids. As **Figure 19.15** shows, three enzymes are required to produce β-carotene from geranyl geranyl diphosphate. If genes that

encode these enzymes could be introduced into rice plants along with regulatory sequences that would trigger their synthesis in endosperm, the researchers could produce a transgenic strain of rice that would contain β-carotene.

Fortunately, genes that encode two of the required enzymes had already been isolated from daffodils, and the gene for the third enzyme had been purified from a bacterium. Because the sequences had been inserted into plasmids and grown in bacteria, many copies were available for manipulation. To each of the coding sequences in the plasmids, biologists added the promoter region from an endosperm-specific protein. This segment included a regulatory site that would promote transcription of the recombinant sequences in endosperm cells.

To develop transgenic rice strains that are capable of producing β-carotene, the three sets of sequences had to be inserted into rice plants. As you know from reading this chapter, introducing recombinant DNA is fairly straightforward into *E. coli* but difficult in humans. How are foreign genes introduced into plants?

The *Agrobacterium* Transformation System

Agrobacterium tumefaciens is a bacterium that infects plants. Plant tissues that become infected with this parasite form a tumorlike growth called a **gall**. Such plants are said to have crown gall disease (**Figure 19.16a**, page 422). When researchers looked into how these infections occur, they found that a plasmid carried by *Agrobacterium* cells, called a

Ti (tumor-inducing) **plasmid**, plays a key role (**Figure 19.16b**, step 1). Ti plasmids contain several functionally distinct sets of genes. One set encodes products that allow the bacterium to bind to the cell walls of a host. Another set, referred to as the virulence genes, encodes the proteins required to transfer part of the Ti DNA, called T-DNA (transferred DNA), into the interior of the plant cell. The T-DNA then travels to the plant cell's nucleus and integrates into its chromosomal DNA (step 2). When transcribed, T-DNA induces the infected cell to grow and divide. The result is the formation of a gall that encloses a growing population of *Agrobacterium* cells (step 3).

Researchers soon realized that the Ti plasmid offered an efficient way to introduce recombinant genes into plant cells. Follow-up experiments confirmed that recombinant genes could be added to the T-DNA that integrates into the host chromosome, that the gall-inducing genes could be removed and that the resulting sequence is efficiently transferred and expressed in its new host plant.

Researchers have also been able to transfer foreign genes into crop plants directly. In this approach, tiny tungsten particles are coated with foreign genes and then shot into plant cells by means of an instrument called a "gene gun."

Golden Rice

To generate a strain of rice that produces all three enzymes needed to synthesize β-carotene in endosperm, the researchers exposed embryos to *Agrobacterium* cells containing genetically modified Ti plasmids (**Figure 19.17a**). The transformed plants were then grown in the greenhouse. When the transgenic individuals had matured and set seed, the researchers found that some rice grains contained so much β-carotene that they appeared yellow. This "golden rice" is compared with unmodified rice in **Figure 19.17b**. Testing is under way to determine whether the transgenic rice contains enough β-carotene to relieve vitamin A deficiency in children.

While the production of golden rice in the laboratory is promising, there is still much to do before the advance has an impact on public health. The transgenic rice must be grown successfully under normal field conditions. In addition, most or all of the rice strains grown in different parts of the world will need to be crossed with transgenic stock to acquire the appropriate genes. Finally, the transgenic strains must be made available to poor farmers at an affordable price.

In addition to the possibility of alleviating vitamin A deficiency, transgenic strains of rice may also be able to ease other nutritional problems in rice-eating people—specifically, protein deficiency and iron deficiency. Early attempts to develop transgenic rice strains with augmented amounts of stored protein and available iron have been promising.

As the examples reviewed in this chapter show, recent advances in agricultural and medical biotechnology promise to increase the quality and quantity of food and to alleviate at least some inherited diseases that plague humans. Each solution offered by genetic engineering tends to introduce new issues, however. As this chapter's essay points out, researchers and consumer advocates have expressed concerns about the increased numbers and types of genetically modified foods. During this debate, biology students and others who are well informed about the techniques and issues involved will be an important source of opinions and information for the community.

(a) Plant with crown gall disease

(b) *A. TUMEFACIENS* INDUCES GALL FORMATION.

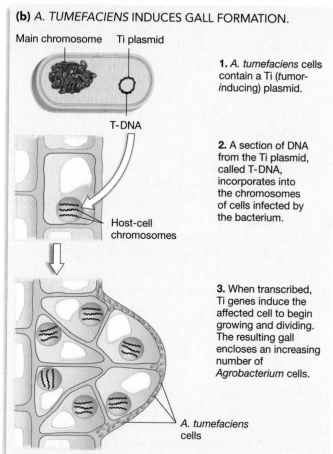

Main chromosome   Ti plasmid

T-DNA

Host-cell chromosomes

1. *A. tumefaciens* cells contain a Ti (tumor-inducing) plasmid.

2. A section of DNA from the Ti plasmid, called T-DNA, incorporates into the chromosomes of cells infected by the bacterium.

3. When transcribed, Ti genes induce the affected cell to begin growing and dividing. The resulting gall encloses an increasing number of *Agrobacterium* cells.

A. tumefaciens cells

FIGURE 19.16 *Agrobacterium* **Infections Introduce Foreign Genes into a Host-Cell Chromosome**
(a) The gall, or tumorlike growth at the bottom of this plant's stem, is due to an infection by the bacterium *A. tumefaciens*. **(b)** Ti plasmids of *A. tumefaciens* cells induce gall formation.

(a) GENETIC ENGINEERING OF Ti PLASMIDS

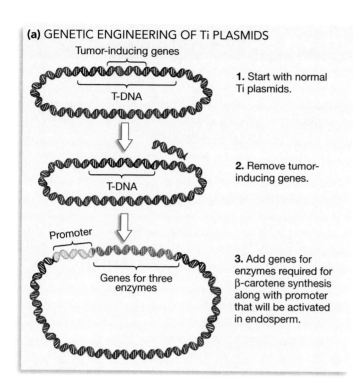

Tumor-inducing genes

T-DNA

1. Start with normal Ti plasmids.

T-DNA

2. Remove tumor-inducing genes.

Promoter

Genes for three enzymes

3. Add genes for enzymes required for β-carotene synthesis along with promoter that will be activated in endosperm.

(b) Rice plants infected with transformed *Agrobacterium* produce β-carotene in their seeds.

FIGURE 19.17 "Golden Rice" Is a Transgenic Strain
(a) The construction of a Ti plasmid to produce a strain of rice capable of synthesizing β-carotene in the endosperm of their seeds. **(b)** Seeds from individuals that were transformed with the Ti plasmids illustrated in part (a). Chemical analyses confirmed that the golden color in the transformed seeds is due to large quantities of β-carotene.

ESSAY Controversies over Genetically Modified Foods

Enthusiasm about the potential benefits of golden rice and other genetically modified foods is tempered by concern about the risks of releasing large numbers and types of recombinant crop plants. In Europe, public opinion has been extremely skeptical about the wisdom of developing genetically engineered crops. Protesters in Britain, the United States, and elsewhere have even vandalized test plantings of genetically modified crops.

Proponents of genetic engineering in agriculture can point to several important success stories. In 1998 U.S. farmers applied 8.2 million fewer pounds of active pesticide ingredient than they would have in the absence of genetically modified crops, in part because glyphosate-resistant and insecticide-producing crop strains were coming into widespread use. Recall that glyphosate is a herbicide. If crop plants are glyphosate resistant, then the compound can be applied in small amounts to kill weeds after growth starts. If they are not, then other weed killers must be applied in larger amounts earlier in the growing season. In addition, genetic engineering made it possible to develop soybeans and corn strains that produce higher concentrations of the amino acid lysine, improving their nutritional value.

Opponents of genetically modified foods can point to several important concerns, however:

- Research has shown that corn plants genetically engineered to produce the Bt toxin leak the molecule into the soil, where it might kill untargeted organisms.

- Widespread adoption of insecticide-producing crop strains is almost certain to lead to the evolution of pests that are resistant to the compounds. For example, consider an insect species that eats transgenic corn plants producing Bt toxin. If one insect in the population happened to have an allele that reduced the toxicity of the toxin, it would be able to produce many more offspring than could individuals without the allele. As a result, the allele for resistance would spread rapidly.

Are genetically modified crops good or bad?

- Research also shows that crop plants genetically engineered for glyphosate resistance produce pollen carrying the recombinant allele, and that pollen can subsequently be transferred to closely related weeds. Based on these data, there is legitimate concern that the use of these crop strains could lead to the more rapid evolution of herbicide-resistant weeds.

Are genetically modified crops good or bad? There is no easy or pat answer to this question. Like any technology, genetic engineering has costs and benefits. For citizens, the challenge is to become informed about the issues and insist that regulatory agencies strictly enforce measures to maximize the benefits of the technology but minimize the risks involved. For biologists, the challenge is to quantify those benefits and risks as carefully as possible and communicate their results clearly so that good decisions can be made.

CHAPTER REVIEW

Summary of Key Concepts

■ **The discovery of enzymes that cut DNA at specific locations, along with enzymes that piece DNA segments back together, gave biologists the ability to move genes from one location to another.**

Genetic engineering is based on inserting a modified version of an allele into an organism. Biologists were able to manipulate genes after the discovery of restriction endonucleases, DNA ligase, and plasmids. Restriction endonucleases allow researchers to cut DNA at specific locations and insert it into plasmids or other vectors with the help of DNA ligase.

Once the gene for the human growth hormone was isolated, it was cloned by introducing it into a plasmid that was taken up by *E. coli* cells. There the gene was expressed. As a result of genetic engineering, pharmaceutical companies have been able to produce large supplies of human growth hormone at an affordable price.

Web Tutorial 19.1   Producing Human Growth Hormone

■ **Biologists can determine a gene's base sequence once they have obtained many copies of the gene—by inserting it into loops of DNA called plasmids in bacterial cells and then allowing the cells to grow or by performing a polymerase chain reaction.**

The most common alternative to cloning genes in plasmids is an in vitro DNA synthesis reaction called the polymerase chain reaction (PCR). The PCR technique is based on primers that bracket a target stretch of DNA. With the use of a heat-stable form of DNA polymerase called *Taq* polymerase, a single target DNA sequence can be amplified to millions of copies.

The most common way to analyze the gene copies produced by plasmid-based cloning or PCR is to sequence them by the dideoxy sequencing method. This is an in vitro synthesis reaction that employs dideoxyribonucleotides to stop DNA replication at each base in the sequence. By running the resulting DNA fragments out on a gel, the sequence of nucleotides in a gene can be determined.

Web Tutorial 19.2   The Polymerase Chain Reaction

■ **Researchers use several strategies to find and characterize the genes responsible for specific traits, such as the alleles associated with certain genetic diseases.**

To isolate the gene for human growth hormone, investigators created a cDNA library from cells that were transcribing growth hormone mRNA and probed the library with a DNA sequence that encoded a short stretch of the known amino acid sequence.

Locating the gene responsible for Huntington's disease was difficult, because only the disease phenotype and the pattern of inheritance were known at the time the research was started. To find the gene, investigators analyzed a large number of genetic markers and a large pedigree of an affected family. Their goal was to find a marker that was inherited along with the allele responsible for the disease. Once this strategy pinpointed the general area where the gene was located, biologists sequenced exons from normal and diseased individuals to determine exactly where the defect occurred.

■ **In some cases, it has been possible to insert genes into humans to cure genetic diseases or into plants to provide them with novel traits, such as the ability to resist insect attacks.**

Once genes are located and characterized, they can be introduced into other individuals or species in an effort to change their traits. Genetic transformation can occur in several ways, depending on the species involved. In humans, recombinant DNA must be introduced by viruses. Because introducing foreign genes into humans is difficult, and because complex ethical and safety issues are involved, progress in human gene therapy has been slow. In bacteria and certain eukaryotes, foreign genes can be inserted into plasmids and the recombinant plasmids can be inserted back into a host cell. An important variation on this approach has made genetic engineering feasible for crop plants. Certain bacteria that infect plants have plasmids that integrate their genes into the host-plant genome. By adding recombinant alleles to these plasmids, researchers have been able to introduce alleles that improve the nutritional quality of crops, make them resistant to herbicides, or allow them to produce insecticides.

Questions

Content Review

1. What do restriction endonucleases do?
   a. They cleave bacterial cell walls and allow viruses to enter the cells.
   b. They join pieces of DNA by catalyzing the formation of phosphodiester bonds between them.
   c. They cut stretches of DNA at specific sites known as recognition sequences.
   d. They act as genetic markers in the chromosome maps used in gene hunts.

2. What is a plasmid?
   a. an organelle found in many bacteria and certain eukaryotes
   b. a circular DNA molecule that in some cases replicates independently of the main chromosome(s)
   c. a type of virus that has a DNA genome and that infects certain types of human cells, including lung and respiratory tract tissue
   d. a type of virus that has an RNA genome, codes for reverse transcriptase, and inserts a cDNA copy of its genome into host cells

3. When present in a DNA synthesis reaction, a ddNTP molecule is added to the growing chain. No further nucleotides can be added afterward. Why?
   a. There are not enough dNTPs available.
   b. A ddNTP can be inserted at various locations in the sequence, so fragments of different length form—each ending with a ddNTP.
   c. The 5′ carbon on the ddNTP lacks a hydroxyl group, so no phosphodiester bond can form.
   d. The 3′ carbon on the ddNTP lacks a hydroxyl group, so no phosphodiester bond can form.

4. Once the gene that causes Huntington's disease was found, researchers introduced the defective allele into mice to create an animal model of Huntington's disease. Why was this model valuable?
   a. It allowed them to test potential drug therapies without endangering human patients.
   b. It allowed them to study how the gene is regulated.
   c. It allowed them to make large quantities of the huntingtin protein.
   d. It allowed them to study how the gene was transmitted from parents to offspring.

5. To begin the hunt for the human growth hormone gene, researchers created a cDNA library from cells in the pituitary gland. What did this library contain?
   a. only the sequence encoding growth hormone
   b. DNA versions of all the mRNAs in the pituitary-gland cells
   c. all of the coding sequences in the human genome, but no introns
   d. all of the coding sequences in the human genome, including introns

6. What does it mean to say that a genetic marker and a disease gene are closely linked?
   a. The marker lies within the coding region for the disease gene.
   b. The sequence of the marker and the sequence of the disease gene are extremely similar.
   c. The marker and the disease gene are on different chromosomes.
   d. The marker and the disease gene are in close physical proximity and tend to be inherited together.

Conceptual Review

1. Explain how restriction endonucleases and DNA ligase are used to insert foreign genes into plasmids and create recombinant DNA. Illustrate some of the key events involved—including an explanation of why sticky ends are important.

2. How do researchers get foreign DNA into cells? Explain how recombinant plasmids are introduced into *E. coli* cells, how viruses are used to transport genes into human cells, and how the Ti plasmid in *Agrobacterium* cells is used to transport genes into plants.

3. What is a cDNA library? How is one created? Give an example of how a cDNA library can be used in research.

4. What are genetic markers, and how are they used to create a genetic map? Explain how researchers combine analyses of pedigrees and genetic markers to narrow down the location of disease genes.

5. Researchers added the promoter sequence from an endosperm-specific gene to the Ti plasmids used in creating golden rice. Why was this step important? Comment on the roles of promoter and enhancer sequences in genetic engineering in eukaryotes.

6. List the molecules that are required for a typical PCR reaction. Then list the function of each molecule required for the replication of DNA in vivo. For each molecule on the in vivo list, explain why it is or is not required for the corresponding in vitro reaction.

Group Discussion Problems

1. The text posed the following questions about Huntington's disease: How can one defective protein lead to abnormal movements *and* personality changes? Why does it take so long for symptoms to appear, and why is the illness progressive? Based on the molecular nature of the disease, offer hypotheses to answer these questions.

2. Discuss some of the ethical issues involved in human gene therapy. Specifically, should therapy be restricted to somatic cells, or should individuals be able to alter their germ-line cells (meaning that they would alter the alleles they pass on to their offspring)? Should gene therapy be approved for disease-causing alleles only, or should parents also be able to pay to transform their children with alleles associated with height, intelligence, hair color, eye color, athletic performance, musical ability, or similar traits?

3. Several organizations are actively trying to stop the development of transgenic crops and implement bans on the marketing of genetically modified foods. Suppose a representative from one of these organizations comes to your door seeking membership support. Explain why you would or would not support the group's efforts.

4. A friend of yours is doing a series of PCR reactions and comes to you for advice. She purchased three sets of primers, hoping that one set would amplify the template sequence shown below. (The dashed lines in the template sequence stand for a long sequence of bases.) None of the three primer pairs produced any product DNA, however.

	Primer a	Primer b
Primer Pair 1:	5′ GTCCAGC 3′ &	5′ CCTGAAC 3′
Primer Pair 2:	5′ GGACTTG 3′ &	5′ GCTGCAC 3′
Primer Pair 3:	5′ GTCCAGG 3′ &	5′ CAAGTCC 3′

Template

5′ ATTCGGACTTG——————GTCCAGCTAGAGG 3′
3′ TAAGCCTGAAC——————CAGGTCGATCTCC 5′

   a. Explain why each primer pair didn't work. Indicate whether both primers are at fault or just one primer is the problem.
   b. Your friend doesn't want to buy new primers. She asks you whether she can salvage this experiment. What do you tell her to do?

20 Genomics

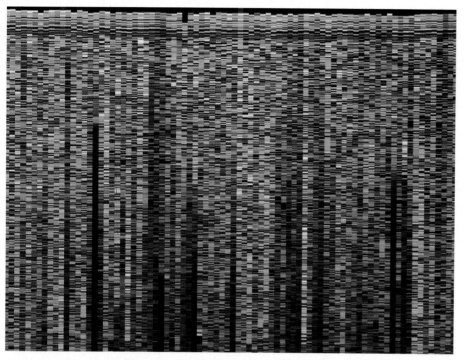

Output from an automated genome sequencing machine, representing about 48,000 bases from the human genome. Each vertical stripe represents the sequence of a stretch of DNA.

KEY CONCEPTS

- Once a genome has been completely sequenced, researchers use a variety of techniques to identify which sequences code for products and which act as regulatory sites.

- In bacteria and archaea, there is a positive correlation between the number of genes in a species and the species' metabolic capabilities.

- In eukaryotes, genomes are dominated by parasitic sequences that do not contribute to the fitness of the organism.

- Data from genome sequencing projects are now being used in the development of new drugs and vaccines.

The first data sets describing the complete DNA sequence, or **genome**, of humans were published in February 2001 as part of the **Human Genome Project**. The achievement was immediately hailed as a landmark in the history of science. It is important to recognize that the Human Genome Project is part of a much larger and ongoing effort to sequence genomes from an array of other eukaryotes, hundreds of bacteria, and dozens of archaea. The effort to sequence, interpret, and compare whole genomes is referred to as **genomics**. The pace of research in this field is nothing short of explosive.

The sudden arrival of genomics at the forefront of biological science prompts a host of questions: Why is whole-genome sequencing attracting so much attention? How do researchers sequence the entire gene complement of an organism? What have biologists learned from the first genomes that have been completed? And what might the future hold in the direction of research? These questions are the foundation of this chapter.

As an introductory biology student, you are part of the first generation trained in the genome era. Genomics is revolutionizing several fields within biological science and will almost certainly be an important part of your personal and professional life. Let's look at what genomics is and why it's being done.

20.1 Whole-Genome Sequencing

Genomics has moved to the cutting edge of research in biology largely because technological advances have increased the speed and driven down the cost of sequencing DNA. Since the Human Genome Project began in 1988, for example, advances in automation and revised experimental protocols have reduced the cost of sequencing by a factor of two every year and a half. As data became less expensive and were obtained faster, the pace of whole-genome sequencing accelerated. The result is that an almost mind-boggling number of sequences are now being generated. As this book goes to press, the primary international repository for DNA sequence data contains over 38 *billion* nucleotides. (With about 3 billion bases, humans have the largest haploid genome sequenced to date.) The size of this database—stored in a publicly funded online service called GenBank—doubles every year.

Which genomes are being sequenced, and why? A little history can shed some light on these questions. The first organismal genome to be sequenced came from a bacterium that lives in the human upper respiratory tract. This bacterium, *Haemophilus influenzae*, has one circular chromosome and a total of 1,830,138 base pairs of DNA. It was an important genome to explore, because this organism causes earaches and respiratory tract infections in children and because one particular strain is capable of infecting the membranes surrounding the brain and spinal cord, causing meningitis. The genome of *H. influenzae* was also small enough to sequence completely with a reasonable amount of time and money, given the technology available in the early 1990s.

The publication of the *H. influenzae* genome in 1995 was quickly followed by the completion of genome sequences from an assortment of bacteria and archaea. The first eukaryotic genome, from the yeast *Saccharomyces cerevisiae*, was finished in 1996. After that breakthrough, complete genome sequences were published from a variety of protists, plants, and animals. To date, genomes have been or are being sequenced from over 110 bacterial species, 16 archaeal species, and 20 eukaryotic species.

Several of the early sequencing projects focused on the need to work out techniques for the Human Genome Project. In other cases, organisms were selected for whole-genome sequencing because they cause disease or have interesting biological properties. For example, genomes of bacteria and archaea that inhabit extremely hot environments have been sequenced in the hopes of discovering enzymes useful for high-temperature industrial applications and understanding how proteins can work under those conditions. Other bacteria and archaea were chosen for sequencing because they produce compounds such as methane (CH_4, natural gas) as a by-product of cellular respiration. The rice genome was sequenced because rice is the main food source for most humans. Finally, species such as the fruit fly *Drosophila melanogaster*, the roundworm *Caenorhabditis elegans*, the house mouse *Mus musculus*, and the mustard plant *Arabidopsis thaliana* were analyzed because they serve as model organisms in biology and because data from well-studied organisms promised to help researchers interpret the human genome.

Before exploring the results of these studies in more detail, it will be helpful to review some of the technological innovations that have made whole-genome sequencing possible.

Recent Technological Advances

Recall from Chapter 19 that researchers use an in vitro DNA synthesis reaction called *dideoxy sequencing* to determine the sequence of bases in a segment of DNA. The gene being sequenced serves as the template, and the sequencing reactions

produce a series of DNA fragments that end with an A, a C, a T, or a G. These fragments are then separated via electrophoresis and visualized by autoradiography or by other techniques. This method is still the basis of genome sequencing projects, though important modifications have been adopted. For example, recall that researchers traditionally attached a radioactive isotope to the primer in the reaction mix to label the fragments of DNA produced by a sequencing reaction. Now fluorescent markers are bonded to the dideoxyribonucleoside triphosphates (ddNTPs) used in a sequencing reaction—a different color for each ddNTP. As **Figure 20.1** shows, the switch to fluorescent markers was important for two reasons: (1) Each stretch of DNA could be sequenced with one dideoxy reaction instead of four separate ones, and (2) machines were developed

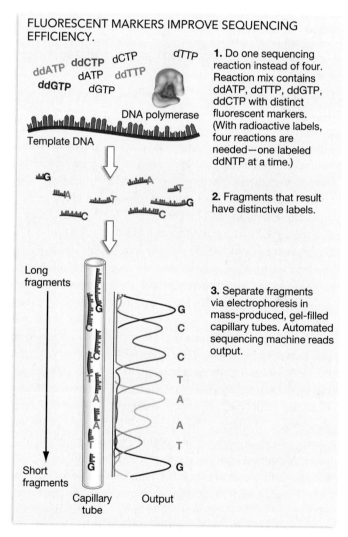

FIGURE 20.1 Use of Fluorescent Markers Simplifies Sequencing Researchers made DNA sequencing more efficient by attaching distinctive fluorescent markers to the ddATP, ddGTP, ddCTP, and ddTTP used in sequencing reactions. Fluorescent markers are safer than radioactive markers and can be read by a machine.

EXERCISE List two ways that this protocol differs from sequencing methods that rely on radioactive markers.

to detect the fluorescence produced by each fragment and read the output of the sequencing reaction. In addition, instead of separating the fragments generated by sequencing reactions through electrophoresis in hand-prepared gels that are poured between glass plates, researchers now perform the electrophoresis step with mass-produced, gel-filled capillary tubes.

Based on these and other innovations, researchers used automated sequencing machines to establish factory-style DNA sequencing centers such as the one shown at the start of this chapter. Over 20 such centers have now been established in the United States, the United Kingdom, Germany, France, Japan, and China. Some of these laboratories employ dozens of biologists and can conduct 100,000 sequencing reactions daily. **Box 20.1** introduces the way researchers design these reactions so that the resulting

data can be assembled into a completed genome sequence for a particular species. That discussion also emphasizes the importance of computer programs in analyzing DNA sequences. The vast quantity of data generated by genome sequencing centers has made **bioinformatics**—the effort to manage, analyze, and interpret information in biology—a key to continued progress in the field.

Which Sequences Are Genes?

Obtaining raw sequence data is just the beginning of the effort to understand a genome. As researchers point out, raw sequence data are analogous to the parts list for a house, except that the list has no punctuation: "windowwabeborogovestaircasedoorjubjub . . . ." Where do the genes for "window," "staircase," and "door" start and end? Are the segments that read

---

BOX 20.1   Shotgun Sequencing and Bioinformatics

Genomes range in size from a few million base pairs to several billion. How do investigators generate the stretches of DNA, approximately 1000 base pairs long, that can be analyzed in a single sequencing reaction? When researchers decide to sequence a large genome, this is a crucial question. In the Human Genome Project, a total of 3.2 billion bases had to be broken up into 1000-base-pair-long pieces, the sequence of each piece determined, and the pieces then put back together in the correct order. How was this done?

Current genome sequencing projects are based on breaking large genomes into tens of thousands of relatively tiny pieces, which are easier to sequence. Biologists call this approach **shotgun sequencing**. As step 1 of **Figure 20.2** shows, shotgun sequencing begins by using restriction endonucleases to cut a genome into pieces about 160 kilobases (kb) long (1 kb = 1000 bases). Next, each piece is inserted into a circular DNA molecule called a **bacterial artificial chromosome (BAC)**. Each BAC is then inserted into a different *Escherichia coli* cell, creating what researchers call a **BAC library** (step 2). A BAC library is an example of a **genomic library**: a set of all

the DNA sequences in a particular genome, split into small segments and inserted into a vector. When the *E. coli* cells that contain BACs are allowed to grow, researchers can isolate large numbers of each 160-kb fragment.

Once a BAC library has been constructed, each 160-kb segment is isolated. Different restriction endonucleases are used to cut the segments into an array of pieces, each about 1000 base pairs long (step 3). These small fragments are then inserted into the circular DNA molecules called plasmids and placed inside bacterial cells (step 4). In this way, a genome is broken down into two manageable levels: 160-kb fragments in BACs and 1-kb segments in plasmids. The plasmids are copied many times as the bacterial cells grow into a large population. Large numbers of each 1000-base-pair fragment are then available for sequencing reactions (step 5).

To put together the data from thousands of sequencing reactions, computer programs analyze regions where the ends of each 1000-base-pair segment overlap (step 6). Overlaps occur because different restriction endonucleases were used to fragment the 160-kb segments. The computer mixes and matches seg-

ments until an alignment consistent with all available data is obtained. Then the ends of each BAC are analyzed in a similar way (step 7). The computer's goal is to arrange each 160-kb segment in its correct position along the chromosome, based on regions of overlap.

Once complete genome sequences became available, databases that could hold completed sequence information had to be created and managed in a way that made the raw data and a variety of annotations available to the international community of researchers. These sequence databases also had to be searchable, so that investigators could evaluate how similar newly discovered genes were to genes that had been studied previously.

Because the amount of data involved is so large, the computational challenges involved in genomics are formidable. Thus far, sophisticated algorithms and continually improving computer hardware have allowed specialists in bioinformatics to keep pace with the rate of data acquisition. Individuals with a background in biology and experience with software design or development are in high demand in laboratories around the world.

"wabeborogove" and "jubjub" important in gene regulation, or are they simply spacers or other types of sequences that have no function at all?

The most basic task in annotating a genome is to identify which bases constitute genes—segments of DNA that code for an RNA or a protein product. In bacteria and archaea, identify-ing genes is relatively straightforward. Biologists begin with computer programs that scan the sequence of a genome in both directions. Such programs test each reading frame that is possi-ble on the two strands of the DNA. (Recall from Chapter 15 that a reading frame is the sequence in which codons are read.) With codons consisting of three bases, three reading frames are

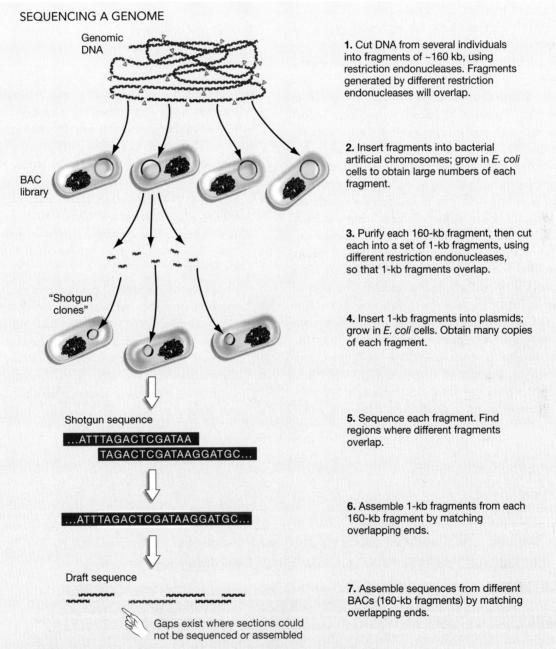

SEQUENCING A GENOME

Genomic DNA

BAC library

"Shotgun clones"

Shotgun sequence

...ATTTAGACTCGATAA
TAGACTCGATAAGGATGC...

...ATTTAGACTCGATAAGGATGC...

Draft sequence

Gaps exist where sections could not be sequenced or assembled

1. Cut DNA from several individuals into fragments of ~160 kb, using restriction endonucleases. Fragments generated by different restriction endonucleases will overlap.

2. Insert fragments into bacterial artificial chromosomes; grow in *E. coli* cells to obtain large numbers of each fragment.

3. Purify each 160-kb fragment, then cut each into a set of 1-kb fragments, using different restriction endonucleases, so that 1-kb fragments overlap.

4. Insert 1-kb fragments into plasmids; grow in *E. coli* cells. Obtain many copies of each fragment.

5. Sequence each fragment. Find regions where different fragments overlap.

6. Assemble 1-kb fragments from each 160-kb fragment by matching overlapping ends.

7. Assemble sequences from different BACs (160-kb fragments) by matching overlapping ends.

FIGURE 20.2 Shotgun Sequencing Breaks Large Genomes into Many Short Segments
Shotgun sequencing is a multistep process. A genome is initially fragmented into 160-kb sections with restriction endonucleases and cloned into bacterial artificial chromosomes (BACs). Each 160-kb section is then cut into 1000-base-pair fragments that can be cloned into plasmids, grown to a large number of copies, and sequenced.

possible on each strand, for a total of six possible reading frames (**Figure 20.3**). Because randomly generated sequences contain a stop codon about one in every 20 codons on average, a long stretch of codons that lacks a stop codon is a good indication of a coding sequence. The computer program highlights any "gene-sized" stretches of sequence that lack a stop codon but are flanked by a stop codon and a start codon. Because polypeptides range in size from 50 amino acids to many hundreds of amino acids, gene-sized stretches of sequence range from several hundred bases to over a thousand bases. In addition, the computer programs look for sequences typical of promoters, operators, or other regulatory sites. DNA segments that are identified in this way are called **open reading frames**, or **ORFs**.

Once an ORF is found, a computer program compares its sequence with the sequences of known genes from model organisms or other well-studied species. If the ORF appears to be a gene that has not yet been described in any other species, it is a promising candidate for follow-up studies to test the hypothesis that it actually codes for a product. A "hit," in contrast, means that the ORF shares a significant amount of sequence with a known gene from another species. Similarities between genes in different species are usually due to **homology**. If genes are homologous, it means that they are similar because they are related by descent from a common ancestor. Homologous genes usually have similar base sequences and the same or a similar function. For example, consider the genes introduced in Chapter 14 that code for enzymes involved in repairing mismatches in DNA. Recall that the mismatch repair genes in *E. coli*, yeast, and humans are extremely similar in structure, DNA sequence, and function. To explain this similarity, biologists hypothesize that the common ancestor of all cells living today had

mismatch repair genes—thus, the descendants of this ancestral species also have versions of these genes.

Finding and analyzing genes by identifying ORFs does not work well in eukaryotes, however. Most coding regions in eukaryotes are broken up by introns, and most regions in eukaryotic genomes do not actually code for a product. In the human genome, for example, it is estimated that less than 2 percent of the DNA present actually codes for proteins, tRNAs, ribosomal RNAs, or other types of products. As a result, scanning eukaryotic genomes for open reading frames is not as useful as it is in bacteria and archaea.

To find coding regions in eukaryotic genomes, researchers pursue several strategies:

- Computer programs can be written to search for homologs by scanning the genome sequence for regions that are similar to well-studied genes from fruit flies and other model organisms. The search programs translate the sequenced DNA into an amino acid sequence, and then they attempt to match the inferred amino acid sequences to the amino acid sequences of known genes. If two amino acid sequences are similar, then researchers suspect that the corresponding stretch of DNA in the newly sequenced genome codes for a gene whose function is similar to the function of the known gene.

- As Chapter 19 showed, investigators can isolate mRNAs from the organism being studied and then use enzymes to make the complementary DNAs (cDNAs). Once an entire genome is sequenced, a computer program can scan it and pinpoint where each cDNA is located. This approach allows researchers to identify genes that are expressed in certain cell types—the tissues where the original mRNA was found.

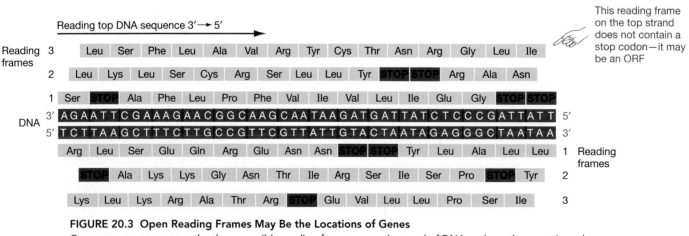

FIGURE 20.3 Open Reading Frames May Be the Locations of Genes
Computer programs scan the three possible reading frames on each strand of DNA and use the genetic code to translate each codon. A long stretch of codons that lacks a stop codon may be an open reading frame (ORF)—a possible gene.

- If genome sequences from closely related species are available, computers are used to compare them and identify sequences that are similar in all of the species involved. Sequences that are shared by closely related species are hypothesized to be located in genes—not in noncoding regions. The logic behind this claim runs as follows: Coding and regulatory sequences are expected to change much more slowly over time than noncoding regions are. This logic is valid because most genes are required for organisms to function and because most proteins work less efficiently when their amino acid sequences change by mutation. Thus, it is logical to expect that natural selection eliminates most mutations in coding sequences and that coding sequences should change slowly over time. Because changes in noncoding sequences do not change gene products, they do not affect the organism's phenotype. Mutations in noncoding regions are much less likely to be eliminated by natural selection, so noncoding regions change relatively rapidly over time.

Although each of these gene-finding strategies has been productive, it will probably be many years before biologists are convinced that they have identified all of the coding regions in even a single eukaryotic genome. As that effort continues, though, researchers are mining the data and making some remarkable observations. Let's first consider what genome sequencing has revealed about the nature of bacterial and archaeal genomes and then move on to eukaryotes. Is the effort to sequence whole genomes paying off?

20.2 Bacterial and Archaeal Genomes

By the time you read this paragraph, the genomes of over 110 bacterial species and 16 archaeal species will have been completely sequenced, and projects focused on sequencing dozens of additional prokaryotic genomes will be under way. In addition to this impressive array of different species, complete genome sequences are now available for several strains of the same bacterial species. For example, researchers have sequenced the genome of a harmless laboratory strain of *Escherichia coli*—derived from the harmless strain that lives in your gut—as well as the genome of a form that causes severe disease in humans. As a result, researchers can now compare the genomes of closely related cells that have different ways of life.

This section focuses on a simple question: Based on data published between 1995 and 2004, what general observations have biologists been able to make about the nature of bacterial and archaeal genomes?

The Natural History of Prokaryotic Genomes

In a sense, biologists who are working in genomics can be compared with the naturalists of the eighteenth and nineteenth centuries. The early naturalists explored the globe, collecting the plants and animals they encountered. Their goal was to describe what existed. Similarly, the first task of a genome sequencer is to catalog what is in a genome—specifically, the number, type, and organization of genes. Several interesting conclusions can be drawn from relatively straightforward observations about the data obtained thus far:

- In bacteria, there is a general correlation between the size of a genome and the metabolic capabilities of the organism. In general, parasites have much smaller genomes than nonparasitic organisms do. **Parasites** live off a host, to the detriment of the host. The smallest genomes are found in parasitic bacteria from the genus *Mycoplasma*. (*Mycoplasma pneumoniae* causes pneumonia in humans.) These bacteria live and multiply inside host cells. Because *Mycoplasma* acquire almost all of their nutrients from their hosts, it is not surprising to observe that they lack the enzymes required to manufacture many essential compounds—these bacteria don't need genes that are being expressed by their hosts. In contrast, the genomes of nonparasitic strains of the bacteria *E. coli* and *Pseudomonas aeruginosa* are 8 to 10 times larger. Their genes code for enzymes that synthesize virtually every molecule needed by the cell. Based on this observation, it is not surprising that *E. coli* is able to grow under a wide variety of environmental conditions. Using similar logic, researchers hypothesize that the large genome of *Pseudomonas* explains why it is able to occupy a wide array of soil types, including marine and marshy habitats, as well as human tissues, where it can cause illness.

- Biologists still do not know the function of many of the genes that have been identified. Although *E. coli* probably qualifies as the most intensively studied of all organisms, the function of over 30 percent of its genes is unknown.

- There is tremendous genetic diversity among prokaryotes. About 15 percent of the genes in each prokaryotic genome appear to be unique to its own species. That is, about one in six genes in a prokaryotic species is found nowhere else.

- Redundancy among genes is common. For instance, the genome of *E. coli* has 86 pairs of genes whose DNA sequences are nearly identical—meaning that the proteins they produce are nearly alike in structure and presumably in function. Although the significance of this redundancy is unknown, biologists hypothesize that slightly different forms of the same protein are produced in response to slight changes in environmental conditions.

- Multiple chromosomes are more common than anticipated. Several species of prokaryotes have two different circular chromosomes instead of one.

- Many species contain the small, extrachromosomal DNA molecules called plasmids. Recall from Chapter 19 that plasmids contain a small number of genes, though not genes that are absolutely essential for growth. In many cases, plasmids can be exchanged between cells of the same species or even of different species.

Perhaps the most surprising observation of all is that in many bacterial and archaeal species, a significant proportion of the genome appears to have been acquired from other, often distantly related, species. In some bacteria and archaea, 15–25 percent of the genetic material appears to be "foreign." This is a remarkable claim. What evidence backs up the assertion that prokaryotes acquire DNA from other species? How could this happen, and what are the consequences?

Evidence for Lateral Gene Transfer

Biologists use two general criteria to support the hypothesis that sequences in bacterial or archaeal genomes originated in another species: (1) when stretches of DNA are much more similar to genes in distantly related species than to those in closely related species and (2) when the proportion of G-C base pairs to A-T base pairs in a particular gene or series of genes is markedly different from the base composition of the rest of the genome. In many cases, the proportion of G-C bases in a genome is characteristic of a genus or species.

How can genes move from one species to another? In at least some cases, plasmids appear to be involved. For example, most of the genes that are responsible for conferring resistance to antibiotics are found on plasmids. Although the physical mechanism involved is not known, researchers have recently documented the transfer of antibiotic-resistance genes between very distantly related species of disease-causing bacteria, by means of plasmids. In some cases, genes from plasmids become integrated into the main chromosome of a bacterium, resulting in genetic

recombination (see Chapter 12). The movement of DNA from one species to another species is called **lateral gene transfer** (**Figure 20.4**).

Some biologists hypothesize that lateral gene transfer also occurs when bacteria and archaea take up raw pieces of DNA from the environment—perhaps in the course of acquiring other molecules. This may have occurred in the bacterium *Thermotoga maritima*, which occupies the high-temperature environments near deep-sea hot springs. Almost 25 percent of the genes in this species are extremely closely related to genes found in archaea that live in the same habitats. The archaea-like genes occur in distinctive clusters within the *T. maritima* genome, which supports the hypothesis that the sequences were transferred in large pieces from an archaean to the bacterium.

Similar types of direct gene transfer are hypothesized to have occurred in the bacterium *Chlamydia trachomatis*. This organism is a major cause of blindness in humans from Africa and Asia; it also causes chlamydia, the most common sexually transmitted bacterial disease in the United States. The *C. trachomatis* genome contains 35 genes that resemble eukaryotic genes in structure. Because *C. trachomatis* lives inside the cells that it parasitizes, the most logical explanation for this observation is that the bacterium occasionally takes up DNA directly from its host cell, resulting in a eukaryote-to-bacterium transfer.

In addition to being transferred between species by means of plasmids or DNA fragments, genes can be transported by viruses.

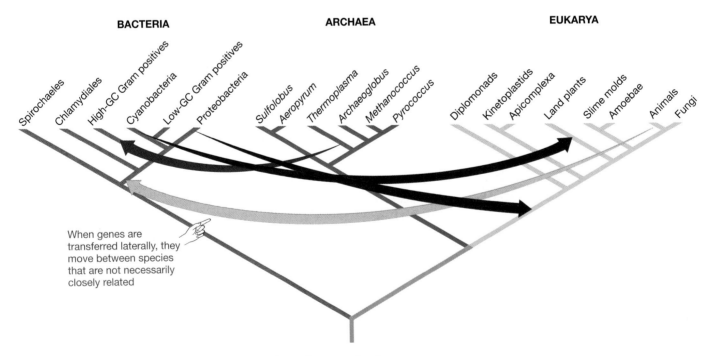

FIGURE 20.4 Lateral Gene Transfer Is Movement of DNA between Species
Lateral gene transfer can occur between very distantly related organisms. EXERCISE List the mechanisms responsible for lateral gene transfer.

For example, investigators who compared the sequences of laboratory and disease-causing strains of *E. coli* found that the pathogenic cells have almost 1400 "extra" genes. Compared with the rest of the genome, most of these genes have a distinctive G-C to A-T ratio. In addition, many are extremely similar to sequences isolated from viruses that infect *E. coli*. Based on these observations, most researchers support the hypothesis that at least some of the disease-causing genes in *E. coli* were brought in by viruses.

To summarize, mutation and gene transfer within species are not the only source of genetic variation in bacteria and archaea. Over the course of evolution, lateral gene transfer has been an important source of new genes and allelic diversity in these domains. This insight would not have been possible without data from whole-genome sequencing. Have efforts to sequence eukaryotic genomes led to similar types of insights?

✓ CHECK YOUR UNDERSTANDING

The major surprise that came out of genome sequencing projects involving bacteria and archaea was the extent and importance of lateral gene transfer—that is, the movement of DNA from one species to another. You should be able to (1) summarize evidence that lateral gene transfer is responsible for the presence of a particular DNA sequence and (2) summarize evidence that the size of a prokaryotic genome is correlated with the organism's metabolic capabilities.

20.3 Eukaryotic Genomes

Sequencing eukaryotic genomes presents two daunting challenges. The first is sheer size. Compared with the genomes of bacteria and archaea, which range from 580,070 base pairs in *Mycoplasma genitalium* to over 6.3 million base pairs in *Pseudomonas aeruginosa*, eukaryotic genomes are large. The haploid genome of *Saccharomyces cerevisiae* (baker's yeast), a unicellular eukaryote, contains over 12 million base pairs. The roundworm *Caenorhabditis elegans* has a genome of 97 million base pairs; the fruit-fly genome contains 180 million base pairs; the mustard plant *Arabidopsis thaliana*'s genome has 130 million base pairs; and humans, rats, mice, and cattle contain roughly 3 billion base pairs each.

The second great challenge in sequencing eukaryotic genes is coping with noncoding sequences that are repeated many times. Many eukaryotic genomes are dominated by repeated DNA sequences that occur between genes and that do not code for products used by the organism. These repeated sequences pose serious problems in aligning and interpreting sequence data. What are they? If such sequences don't code for a product, why do they exist?

Natural History: Types of Sequences

In many eukaryotic genomes, the exons, introns, and regulatory sequences associated with genes make up a relatively small percentage of the genome. Recall from Section 20.1 that, in humans, exons constitute less than 2 percent of the total genome while repeated sequences account for well over 50 percent. In contrast, over 90 percent of the DNA sequences in a bacterial or archaeal genome code for a product used by the cell.

When noncoding and repeated sequences were discovered, they were initially considered "junk DNA" that was nonfunctional and probably unimportant and uninteresting. But subsequent work has shown that most of the repeated sequences observed in eukaryotes are actually derived from sequences known as transposable elements. **Transposable elements** are parasitic segments of DNA that are capable of moving from one location to another in a genome. They are similar to viruses, except that they are not transmitted from one host cell or host individual to another by infection. Instead, transposable elements transmit copies of themselves to additional locations within the host genome and are passed on to offspring along with the rest of the genome. Viruses leave a host cell and find a new cell to infect. But transposable elements never leave their host cell—they simply make copies of themselves and move to new locations in the genome. They are examples of what biologists call **selfish genes**—DNA sequences that survive and reproduce but that do not increase the fitness of the host genome.

How Do Transposable Elements Work?   Transposable elements come in a wide variety of types and spread through genomes in a variety of ways. Different species typically contain distinct types of transposable elements. The genomes of fruit flies, yeast, and humans each contain distinctive kinds of parasitic sequences.

As an example of how these selfish genes work, let's consider a type called a **long interspersed nuclear element** (LINE), found in humans and other eukaryotes. A LINE is similar to the retroviruses, introduced in Chapter 19. Each LINE consists of a stretch of DNA that contains a gene that codes for the enzyme reverse transcriptase, a gene that codes for the enzyme integrase, and a single promoter that is recognized by RNA polymerase II. Your genome contains tens of thousands of LINEs, each between 1000 and 5000 bases long.

A LINE contains all the sequences required for it to make copies of itself and insert them into a new location in the genome. When a LINE is expressed, reverse transcriptase is translated from the LINE mRNA. This enzyme makes a cDNA version of the LINE mRNA. The second LINE product, integrase, then inserts the newly synthesized LINE DNA into a new location in the genome. In this way, the parasitic sequence reproduces. The current model for how this happens is shown in

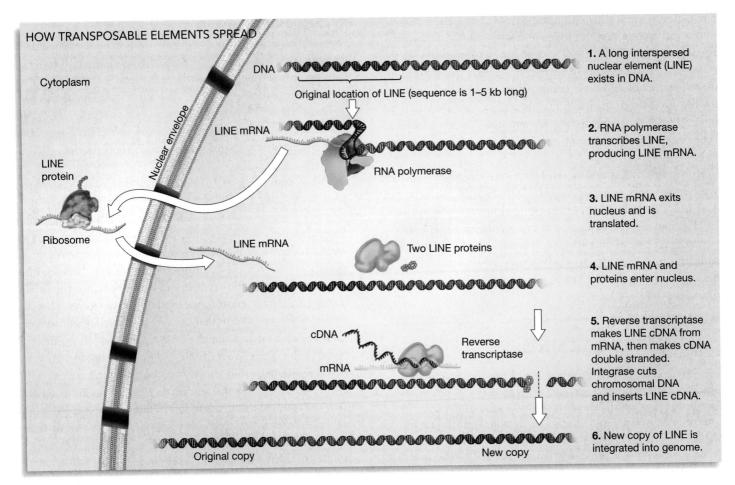

HOW TRANSPOSABLE ELEMENTS SPREAD

Cytoplasm

Nuclear envelope

DNA

Original location of LINE (sequence is 1–5 kb long)

1. A long interspersed nuclear element (LINE) exists in DNA.

LINE mRNA

RNA polymerase

2. RNA polymerase transcribes LINE, producing LINE mRNA.

LINE protein

Ribosome

3. LINE mRNA exits nucleus and is translated.

LINE mRNA

Two LINE proteins

4. LINE mRNA and proteins enter nucleus.

cDNA

mRNA

Reverse transcriptase

5. Reverse transcriptase makes LINE cDNA from mRNA, then makes cDNA double stranded. Integrase cuts chromosomal DNA and inserts LINE cDNA.

Original copy

New copy

6. New copy of LINE is integrated into genome.

FIGURE 20.5 Transposable Elements Spread within a Genome
This sequence of events is the leading hypothesis for how LINEs spread.

Figure 20.5. If the transposition event occurs in reproductive cells that go on to form eggs or sperm, the copied LINE will be passed on to offspring. If the LINE happens to insert itself inside a gene or a regulatory sequence, it causes a mutation that is almost certain to reduce the host's fitness.

Most of the LINEs observed in the human genome are defective, however. Specifically, most of the LINEs in your cells don't contain a promoter or the genes for either reverse transcriptase or integrase. To make sense of this observation, researchers hypothesize that the insertion process illustrated in Figure 20.5 is usually disrupted in some way. Analyses of the human genome have revealed that only a handful of our LINEs appear to be complete and potentially active.

Virtually every prokaryotic and eukaryotic genome examined to date contains at least some transposable elements. They vary widely in type and number, however; bacterial and archaeal genomes have relatively few transposable elements. This observation has inspired the hypothesis that bacteria and archaea either have efficient means of removing parasitic sequences or can somehow thwart insertion events. To date, however, this hypothesis has yet to be tested rigorously.

Research on transposable elements and lateral gene transfer has had a huge impact on how biologists view the genome.

Many genomes are riddled with parasitic sequences, and others have undergone radical change in response to lateral gene transfer events. The general picture is that genomes are dynamic. Their size and composition can change dramatically over time.

Repeated Sequences and DNA Fingerprinting In addition to containing repeated sequences from transposable elements, eukaryotic genomes have several hundred loci called **simple tandem repeats**. Repeating units that are just 1 to 5 bases long are known as **microsatellites** or **simple sequence repeats**; repeating units that are 6 to 500 bases long are called **minisatellites** or **variable number terminal repeats** (VNTRs). Both types of repeated sequences make up 3 percent of the human genome. The most common type is a repeated stretch of the dinucleotide AC, giving the sequence ACACACAC .... Both types of satellite sequences are thought to originate when DNA polymerase skips or mistakenly adds extra bases during replication.

Soon after these sequences were first characterized, Alec Jeffreys and co-workers established that microsatellite and minisatellite loci are "hypervariable"—they vary among individuals much more than any other type of sequence does. **Figure 20.6** illustrates one hypothesis for why minisatellites have so many different alleles: These highly repetitive stretches often align incorrectly when

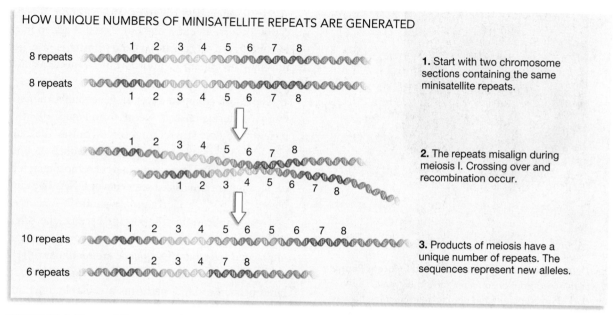

FIGURE 20.6 Unequal Crossover Changes the Numbers of Simple Sequence Repeats

The alignment of homologous chromosomes during prophase of meiosis I is driven by sequence similarity between homologs. Because simple sequence repeats are so similar, they are likely to misalign during synapsis.

homologous chromosomes synapse and cross over during prophase of meiosis I. Instead of lining up in exactly the same location, the two chromosomes pair in a way that matches up bases in different repeated segments. Due to this misalignment, **unequal crossover** occurs. Chromosomes produced by unequal crossover contain different numbers of repeats. The key observation is that, if a particular microsatellite or minisatellite locus has a unique number of repeats, it represents a unique allele. Each of these alleles has a unique length. As with any allele, microsatellite and minisatellite alleles are transmitted from parents to offspring.

Misalignment or errors by DNA polymerase are so common at these loci that, in most populations of eukaryotes, the genome of virtually every individual has at least one new allele. This variation in repeat number among individuals is the basis of DNA fingerprinting. **DNA fingerprinting** refers to any technique for identifying individuals on the basis of unique features of their genomes. Because microsatellite and minisatellite loci vary so much among individuals, they are now the loci of choice for DNA fingerprinting. To fingerprint an individual, researchers obtain a DNA sample and perform the polymerase chain reaction (PCR), using primers that flank a region containing microsatellites or minisatellites. Once many copies of the region are available, they can be analyzed to determine the number of repeats present.

Research on repeated sequences has revealed that the probability of getting a new allele is higher for shorter repeats than for longer repeats. For some two-base-pair repeats, the number of repeats present changes so quickly over time that only very close relatives are likely to share any of the same alleles. This observation has important practical implications. For example, because parents and offspring tend to share alleles at these loci, analyzing microsatellite and minisatellite sequences offers an accurate

way to assign paternity in birds, humans, and other species that have these types of sequences (**Figure 20.7**).

DNA fingerprinting has also been enormously useful in **forensic science**—the use of scientific analyses to solve crimes. As an example, consider the case of Lonnie Erby, who was convicted of raping three women in 1985. After Erby had served 17 years of a 115-year prison sentence, a DNA fingerprinting analysis of preserved semen taken from the victims

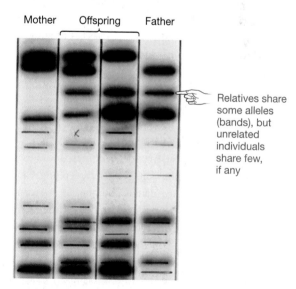

Relatives share some alleles (bands), but unrelated individuals share few, if any

FIGURE 20.7 Close Relatives Have Similar DNA Fingerprints

A gel that contains DNA fragments from parents and offspring and has been probed with sequences from several minisatellite loci. Different individuals have fragments of various lengths if minisatellite loci contain different numbers of repeats. Related individuals tend to share fragment patterns.

FIGURE 20.8 DNA Fingerprinting Data Has Freed Innocent People
Lonnie Erby was released from prison after serving 17 years. DNA fingerprinting data showed that he did not commit the crime he was accused of.

was performed. Several microsatellite and minisatellite loci were amplified from the preserved DNA samples, via PCR, and the alleles present were compared with those in Erby's DNA. They did not match. The data showed that there was virtually no chance that Erby had committed the crimes. Once the result was confirmed, he was released (**Figure 20.8**).

A nonprofit group that is coordinating DNA testing of convicted criminals in the United States has documented a total of 136 prisoners exonerated and released to date. Some of the individuals involved had been incarcerated even longer than Erby, for crimes ranging from rape to murder. It is no exaggeration to claim that DNA fingerprinting has given these innocent men and women their lives back.

Now that we've reviewed the characteristics of some particularly prominent types of noncoding sequences in eukaryotes, it's time to consider the nature of the coding sequences in these genomes. Let's start with the most basic question of all: Where do eukaryotic genes come from?

What Is a Gene Family?

In both prokaryotic and eukaryotic genomes, biologists routinely find groups of similar genes clustered along the same chromosome. The genes are usually similar in structural aspects, such as the arrangement of exons and introns, and in their base sequence. The degree of sequence similarity among these clustered genes varies. In the genes that code for ribosomal RNAs (rRNAs) in vertebrates, the sequences are virtually identical—meaning that each individual has many exact copies of the same gene. In other cases, though, the proportion of bases that are identical is 50 percent or less.

Genes that are extremely similar to each other are considered to be part of the same **gene family**. Genes that make up gene families are hypothesized to have arisen from a common

ancestral sequence through **gene duplication**. When gene duplication occurs, an extra copy of a gene is added to the genome. In eukaryotes, where lateral gene transfer is rare, gene duplication may be the most important of all mechanisms that generate new genes.

The most common type of gene duplication results from crossover during meiosis. Recall from Figure 20.6 that misalignment of homologous chromosomes causes the duplication or deletion of a small section of DNA sequence. When unequal crossover involves large sections of chromosomes, it can produce duplications of gene-sized segments of DNA. The duplicated segments that result are arranged in tandem—one after the other.

Gene duplication is important because the original gene is still functional and produces a normal product. As a result, the duplicated stretches of sequence are redundant. If mutations in the duplicated sequence alter the protein product, and if the altered protein product performs a valuable function in the cell, then an important new gene has been created. The duplicated gene may also be regulated in a different way, so that it is expressed in novel locations or at novel times. In either case, the duplicated sequences represent new genes and can lead to the evolution of novel traits. Gene duplication produces new genes and creates gene families.

Alternatively, mutations in the duplicated region may make expression of the new locus impossible. For example, a mutation could produce a stop codon in the middle of an exon. When a gene is disabled in some way, the resulting sequence is referred to as a pseudogene. A **pseudogene** is a sequence that closely resembles a working gene but does not produce a working product. Pseudogenes have no function.

As an example of a gene family, consider the human **globin** genes diagrammed in **Figure 20.9**. These genes code for proteins that form part of hemoglobin—the oxygen-carrying molecule in your red blood cells. Analyzing the globin loci illustrates several important points about gene families. In humans, the globin gene family contains several pseudogenes, along with several genes that code for oxygen-transporting proteins. The various coding genes in the family serve slightly different functions. For example, some genes are active only in the fetus or the adult. Follow-up work showed that the proteins encoded by the fetal genes have an exceptionally high affinity for oxygen, compared with the proteins expressed in adults. As a result, oxygen is able to move from the mother's blood to the fetus's blood.

In addition to the gene duplication events that result from unequal crossover, an entire complement of chromosomes may be duplicated as a result of a mistake in either mitosis or meiosis. In this case, the resulting organism contains double the normal complement of chromosomes. Recall from Chapter 12 that species with duplicated chromosome complements are said to be **polyploid**. When polyploidy occurs, every gene in the duplicated genome is free to mutate and possibly to acquire a new function. Polyploidy has been a particularly important source of new genes in plants.

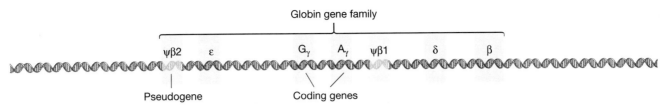

FIGURE 20.9 Gene Families Are Clusters of Closely Related Genes
Genes within the globin family. Red segments represent functioning genes, and yellow segments are pseudogenes. The members of a gene family are arranged one after the other, or in tandem. EXERCISE Add labels to each functioning gene, indicating when it is expressed: ε, 0–6 weeks postconception; G_γ and A_γ, highest 8–36 weeks postconception; δ, very low levels after birth through adulthood; β, low levels 12–36 weeks postconception, increasing to high levels from 6 weeks after birth through adulthood.

✔CHECK YOUR UNDERSTANDING

Eukaryotic genomes are riddled with parasitic sequences that do not contribute to the fitness of the organism. Simple repeated sequences are also common, and many of the coding sequences are organized into families of genes with related functions. You should be able to explain (1) why transposable elements are considered selfish genes, (2) why simple sequence repeats make DNA fingerprinting possible, and (3) how unequal crossover leads to duplicated sequences.

20.4 Comparative Genomics

Analyzing the genomes of individual bacterial, archaeal, and eukaryotic species has provided a wealth of insights, ranging from the size of specific gene families to the nature of transposable elements. Recently, however, biologists have begun to compare the genomes of different species. As the number of completely sequenced genomes expands, and as a wider diversity of species is studied, comparative genomics becomes more and more productive. We've already seen how comparing genome sequences across species led to the realization that there is a positive correlation between genome size and metabolic capabilities in bacteria and archaea, and that lateral gene transfer has played a large role in the evolution of these lineages. Let's delve into comparative analyses, starting with what comparative genomics has to say about the evolution of disease and the importance of changes in gene regulation during human evolution.

Understanding Virulence

Currently, dozens of whole-genome sequencing projects are focused on disease-causing bacteria. As each of these projects is completed, biologists begin comparing the genomes of harmless strains of bacteria with those of pathogenic strains. The goal is to achieve a much more detailed understanding of the genetic basis of **virulence**—the tendency for a parasite to harm its host.

As an example of this work, let's look more closely at what biologists found when they compared the genomes of benign (harmless) and pathogenic strains of *E. coli*. Biomedical researchers are interested in the genome of the disease-causing strain called O157:H7, because it is a common cause of food poi-

soning. People who eat undercooked meat may develop serious, and occasionally fatal, infections by O157:H7. Cells of this strain have a large number of genes not found in benign strains. The genes found in this strain that are associated with disease originally may have been introduced into the bacterial cells by viruses.

When biologists analyzed the novel genes in O157:H7, they found that several of those genes were extremely similar to sequences from a closely related pathogenic bacterium called *Shigella*. In countries where public water supplies are untreated, *Shigella* is an important cause of dysentery. Some of the *Shigella*-like genes found in O157:H7 code for proteins that poison cells in the human intestine and cause severe diarrhea. This observation suggests that the same genes may be responsible for causing food poisoning by *E. coli* and dysentery by *Shigella*. More specifically, biologists hypothesize that *E. coli* obtained the virulence genes by lateral gene transfer from *Shigella*, via a virus that infects both species. In addition to being intrinsically interesting, this research has an important practical impact: If drugs that neutralize *Shigella* toxins can be developed, they should also be effective against infections by *E. coli* O157:H7.

The data available indicate that the *Shigella*–*E. coli* story may be typical. In the disease-causing organisms whose genomes have been sequenced thus far, biologists have found that the genes responsible for causing illness or antibiotic resistance tend to occur in discrete clusters. This observation supports the hypothesis that the genes originated in lateral gene transfer events.

The ability to identify genes that are specific to disease-causing strains of bacteria gives investigators a host of new drug targets. If drugs that knock out the protein products of these genes can be developed, they would inhibit disease-causing bacterial strains while leaving helpful forms intact.

Studying Adaptation

Comparing the genomes of different strains or species has been a productive way to identify genes that are responsible for adaptations to particular environments or ways of life. For example, disease-causing organisms such as the cholera bacterium introduced in Chapter 17, *E. coli* O157:H7, and *Shigella* thrive in environments where human populations are dense and sanitation is poor. The presence of virulence genes is thought to increase the ability of these cells to survive and reproduce in those habitats, compared with strains that are not virulent. To use the terms introduced in Chapter 1, virulence genes are thought to increase the fitness of cells in certain types of habitats.

Virulence genes are just one case in which genome sequencing has linked specific genes and specific adaptations. To drive this point home, consider the following examples:

- *Anopheles gambiae* is the mosquito species that transmits the malaria parasite to humans. Female *Anopheles* mosquitoes must obtain a blood meal prior to laying eggs. Thus, it was logical to find that their genome contains 58 genes for fibrinogen-like proteins. In humans, *fibrinogen* forms clots that reduce blood loss from a wound. In contrast, the fibrinogen-like molecules in mosquitoes are thought to prevent blood from clotting when females are sucking blood. This hypothesis is supported by the observation that fruit flies have just 13 genes for fibrinogen-like proteins. Fruit flies are also insects, but they eat rotting fruit instead of blood.

- The single-celled eukaryotes that parasitize vertebrates and cause malaria are in the genus *Plasmodium*. During a *Plasmodium* infection, the host's immune system cells interact with proteins on the surface of the parasitic cells. If host cells recognize *Plasmodium* as foreign, other immune system cells are called in to destroy the parasites. But when the *Plasmodium* genome was sequenced, researchers found that it has 600 to 1000 distinct copies of genes that code for cell-surface proteins—many more than observed in other species. The *Plasmodium* genes are similar in sequence, clustered in the same chromosomal region, and repeated in tandem, suggesting that they are part of the same gene family. To explain this observation, biologists suggest that natural selection has favored the evolution of many cell-surface proteins in *Plasmodium*. The hypothesis is that parasitic cells with many different cell-surface proteins are difficult for the immune system to detect and destroy.

- *Prochlorococcus* is a very abundant photosynthetic bacterium in the oceans. The genomes of three strains have been sequenced. One of the strains is adapted to high light levels and lives near the surface. The other two strains are adapted to moderate or very low light levels and live in deeper water. Researchers who compared the number of genes that code for chlorophyll-binding proteins in the antenna complexes of photosystems I and II (see Chapter 10) found a strong correlation with light availability. The high-light strain contains just one of these genes. The moderate-light strain has two genes for these proteins; the low-light strain has eight. Apparently, natural selection has favored the evolution of particularly large and complex antenna systems in organisms that live in low-light habitats.

Gene Number and Alternative Splicing

Of all observations about the nature of eukaryotic genomes, perhaps the most striking is that particularly complex organisms do not appear to have particularly large numbers of genes. **Table 20.1** indicates the estimated number of genes

TABLE 20.1 Number of Genes in Selected Genomes

Species	Description	Genome Size (Millions of Base Pairs)	Estimated Number of Genes
Mycoplasma genitalium	Parasitic bacterium; causes urogenital-tract infections in humans	0.58	517
Haemophilus influenzae	Bacterium; common resident of human upper respiratory tract; can cause earache and meningitis	1.83	1743
Thermotoga maritima	Bacterium that lives in extremely hot environments	1.89	1877
Vibrio cholerae	Bacterium that lives in saltwater marshes and coastal environments; causes cholera	4.0	3885
Escherichia coli (laboratory strain)	Bacterium that is an important model organism in biochemistry and genetics	4.6	4288
Saccharomyces cerevisiae	Baker's and brewer's yeast; a unicellular fungus; an important model organism in biochemistry and genetics	12	6000
Plasmodium falciparum	Single-celled, parasitic eukaryote; causes malaria in humans	30	6500
Drosophila melanogaster	Fruit fly; an important model organism in genetics and developmental biology	180	13,600
Caenorhabditis elegans	A roundworm; an important model organism in developmental biology	97	19,000
Mus musculus	House mouse; an important model organism in genetics and developmental biology	2500	~30,000
Arabidopsis thaliana	A mustard plant; an important model organism in genetics and developmental biology	119	26,000
Homo sapiens	Humans	3000	~25,000

found in selected genomes. Note that the total number of genes in *Homo sapiens*, which is considered a particularly complex organism, is not that much higher than the total number of genes in fruit flies and roundworms. In fact, gene number in humans is about the same as in mice and in the weedy mustard plant *Arabidopsis*. Before the human genome was sequenced, many biologists expected that humans would have at least 100,000 genes. We now know that we have only 20,000–25,000.

How can this be? Recall that, in prokaryotes, there appears to be a rough correlation between genome size, gene number, a cell's metabolic capabilities, and the cell's ability to live in a variety of habitats. Similarly, it is logical to observe that plants have exceptionally large numbers of genes because they synthesize so many different and complex molecules from just carbon dioxide, nitrate ions, phosphate ions, and other simple nutrients. The idea is that large numbers of genes enable plants to produce large numbers of enzymes. Why isn't there a stronger correlation between gene number and morphological and behavioral complexity in animals?

The leading hypothesis focuses on alternative splicing. Recall from Chapter 18 that the exons of a particular gene can be spliced in ways that produce distinct mature mRNAs. As a result, a single eukaryotic gene can code for multiple transcripts and thus multiple proteins. The alternative-splicing hypothesis claims that at least certain multicellular eukaryotes do not need enormous numbers of distinct genes. Instead, alternative splicing creates different proteins from the same gene. The alternative forms might be produced at different developmental stages or in response to different environmental conditions.

In support of the alternative-splicing hypothesis, researchers have analyzed the mRNAs produced by human genes and have estimated that each gene produces an average of slightly more than three distinct transcripts. If this result is valid for the rest of the genome, the actual number of different proteins that can be produced is more than triple the gene number. Humans have only 20,000–25,000 genes, but these genes may have the ability to produce at least 100,000 different transcripts.

Based on such results, gaining a better understanding of how alternative splicing is regulated has become an urgent research priority. Gene regulation has, in fact, been an increasingly important theme in comparisons of closely related species.

Sequence Similarity and the Importance of Regulatory Sequences

Comparing the numbers of genes found in humans and in mice created a paradox that may be resolved by the alternative-splicing hypothesis. Comparing the base sequences of genes in humans and in other species has created an analogous paradox.

Here is the issue: At the level of base sequences, human beings and chimpanzees are 98.8 percent identical on average. If humans and chimps are so similar genetically, why do they appear to be so different in their morphology and behavior?

The leading hypothesis to resolve this paradox focuses on the importance of regulatory genes and regulatory sequences. Recall from Chapter 18 that a **regulatory sequence** is a section of DNA involved in controlling the activity of other genes; it may be a promoter, a promoter-proximal element, an enhancer, or a silencer. The term **structural gene**, in contrast, refers to a sequence that codes for a tRNA, rRNA, protein, or other type of product.

To resolve the sequence-similarity paradox, biologists propose that even though many structural genes in closely related species, such as humans and chimps, are identical or nearly identical, regulatory sequences and genes that code for regulatory transcription factors might have important differences between the two species. Suppose that the structural gene for human growth hormone and chimp growth hormone are identical in base sequence. But if subtle differences in transcription factors or regulatory sequences change the pattern of expression of that gene—perhaps turning it on later and longer in humans than in chimps—then height and other characteristics will change even though the structural gene is the same. Based on current analyses, biologists suggest that the human genome contains about 3000 different regulatory transcription factors. Subtle mutations in these proteins could have a significant effect on gene expression and thus on the phenotype.

The regulatory hypothesis is logical, but it remains to be tested rigorously. To date, there are no specific examples of changes in regulatory sequences that are responsible for phenotypic differences in humans and chimps or other closely related species.

20.5 Future Prospects

To explain the impact of genomics on the future of biological science, Eric Lander has compared the sequencing of the human genome to the establishment of the periodic table of the elements in chemistry. Once the periodic table was established and validated, chemists focused on understanding how the elements combine to form molecules. Similarly, biologists now want to understand how the elements of the human genome combine to produce an individual.

In essence, a genome sequence is a parts list. Once that list is assembled, researchers delve deeper to understand how genes interact to produce an organism. Let's explore some of the ways in which researchers use whole-genome data to answer fundamental questions about how organisms work.

Functional Genomics and Proteomics

Biologists who are interested in the research field called **functional genomics** look at the lists of protein-coding regions in a genome and ask, How do all these gene products interact? After all, the products of genes do not exist in a vacuum. Instead, groups of proteins act together to respond to environmental challenges such as extreme heat or drought. Similarly, distinct groups of genes are transcribed at different stages as a multicellular eukaryote grows and develops.

Whole-genome sequencing has inspired two important new approaches to studying how genes interact. One method, using DNA microarrays, focuses on analyzing changes in gene expression; the other, proteomics, seeks to understand protein-protein interactions. Let's take a closer look.

DNA Microarrays DNA-based microarrays are beginning to revolutionize how researchers analyze changes in gene activity. A **DNA microarray** consists of a large number of single-stranded DNAs permanently affixed to a glass slide. By probing DNA microarrays with the mRNAs from a particular cell, researchers can determine which genes are being expressed. Using microarrays is exciting, because researchers no longer have to study changes in the expression of genes one at a time. Instead, they can study the expression of thousands of genes at a time and identify which sets of genes are expressed in concert.

As **Figure 20.10** shows, the thousands of different short, single-stranded DNA sequences that make up a microarray are affixed to a glass plate in tiny but separate spots. In many instances, the DNAs on the plate represent short stretches from each exon in a genome. If so, then each spot in the microarray represents a different gene or exon. These types of microarrays give researchers a plate that contains segments from virtually every gene in an organism.

Once such a plate has been manufactured, a variety of experiments are possible. For example, mRNAs can be isolated at different stages in development and converted to cDNAs with reverse transcriptase. The resulting cDNAs are then used as probes. The sequences are labeled with a fluorescent marker, made single stranded, and hybridized with the microarray. Because the labeled cDNAs bind to the complementary sequences on the plate, the results give an accurate picture of which genes are being expressed at a particular stage of development. In this way, researchers can establish which genes are transcribed in response to heat, to the presence of an antibiotic, or to a viral infection.

The sequences in a microarray are attached to the slide via covalent bonds, so they remain affixed after the probe cDNAs are washed off. As a result, the microarray slide can be reused and re-probed—perhaps with cDNAs from a later stage in development or during an attack by a different virus. In this way, researchers collect data on which genes are expressed at different intervals as the organism matures or responds to different types of environmental conditions.

Proteomics **Proteomics** is the large-scale study of protein function. Now that biologists have a catalog of all protein-coding genes in an organism, researchers are working out new techniques for examining protein-protein interactions on a massive scale. One approach is similar to the use of DNA microarrays, except that large numbers of proteins, rather than DNA sequences, are affixed on a glass plate. This microarray of proteins is then treated with an assortment of proteins produced by the same organism. These proteins are labeled with a fluorescent or

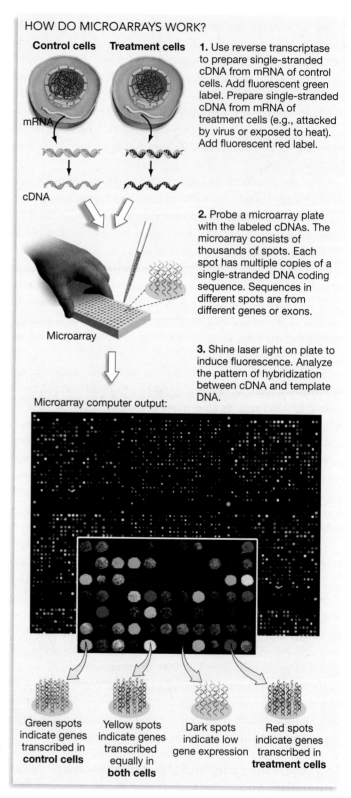

HOW DO MICROARRAYS WORK?

Control cells **Treatment cells**

mRNA

cDNA

1. Use reverse transcriptase to prepare single-stranded cDNA from mRNA of control cells. Add fluorescent green label. Prepare single-stranded cDNA from mRNA of treatment cells (e.g., attacked by virus or exposed to heat). Add fluorescent red label.

Microarray

2. Probe a microarray plate with the labeled cDNAs. The microarray consists of thousands of spots. Each spot has multiple copies of a single-stranded DNA coding sequence. Sequences in different spots are from different genes or exons.

3. Shine laser light on plate to induce fluorescence. Analyze the pattern of hybridization between cDNA and template DNA.

Microarray computer output:

Green spots indicate genes transcribed in **control cells**

Yellow spots indicate genes transcribed equally in **both cells**

Dark spots indicate low gene expression

Red spots indicate genes transcribed in **treatment cells**

FIGURE 20.10 DNA Microarrays Represent Every Gene in a Genome

To create a DNA microarray, investigators spot thousands of short, single-stranded DNA sequences from coding sequences onto a glass plate. By probing this microarray with labeled cDNAs synthesized from mRNAs, researchers can identify which coding sequences are being transcribed. Here mRNAs from cells treated with a hormone are red, while mRNAs from untreated (control) cells are green.

radioactive tag. If any labeled proteins bind to the proteins in the microarray, the two molecules may also interact in the cell.

Medical Implications

With the advent of microarray technology, the "periodic tables" provided by genome projects are having an important impact on research into gene expression and protein-protein interactions. But the governments and corporations that fund genome projects have undertaken the expense primarily because of the potential benefits to biomedical research. In this respect, is genomics living up to its promise?

Even though genome data became available only recently, the prospect of improving human health and welfare is clear. To support this statement, consider how whole-genome analyses are being used in drug and vaccine development.

Identifying Potential Drug Targets   Many drugs act by disabling a particular enzyme. Because genome sequencing provides a comprehensive catalog of the proteins produced by an organism, it offers a comprehensive list of targets for drug therapy. In addition, comparative genomics can identify which drug targets are found in the pathogen but not in the host. If an enzyme is unique to a disease-causing organism, taking a drug that disables it is less likely to damage nontarget species or the host.

For example, malaria researchers have now isolated and characterized all of the proteins produced by the malaria parasite *Plasmodium* at four stages of its life cycle. The genes that code for these proteins have been identified, and many of the proteins are now slated for drug testing. New drug therapies against malaria are needed urgently, because most *Plasmodium* populations have evolved resistance to the array of drugs currently available.

Designing Vaccines   Although efforts to exploit genome data in drug design are still in their infancy, genomics has already inspired important advances in vaccine development. To illustrate how this work is proceeding, consider recent research on the bacterium *Neisseria meningitidis*. This species is a major cause of meningitis and blood infections in children and was one of the first prokaryotes to have its genome sequenced. Although antibiotics can treat *N. meningitidis* infections effectively, the organism grows so quickly that it often injures or even kills the victim before a diagnosis can be made and drugs administered. As a result, biomedical researchers have been interested in developing a vaccine that would prime the immune system and allow children to ward off infections.

Vaccine development has been difficult in this case, however. The immune system usually responds to molecules on the outer surface of bacteria or viruses, and *N. meningitidis* is covered with a polysaccharide that is identical to a compound found on the surface of brain cells. Immune system cells normally do not attack compounds found on the body's own cells, so a vaccine composed of the *N. meningitidis* polysaccharide would elicit no response.

To circumvent this problem, biologists analyzed the genome sequence of *N. meningitidis* and tested 600 open reading frames for the ability to encode vaccine components. The researchers inserted the 600 DNA sequences into *E. coli* cells, following the steps shown in **Figure 20.11**. Later they succeeded in isolating 350 different *N. meningitidis* proteins from the transformed cells. Then the biologists injected these proteins into

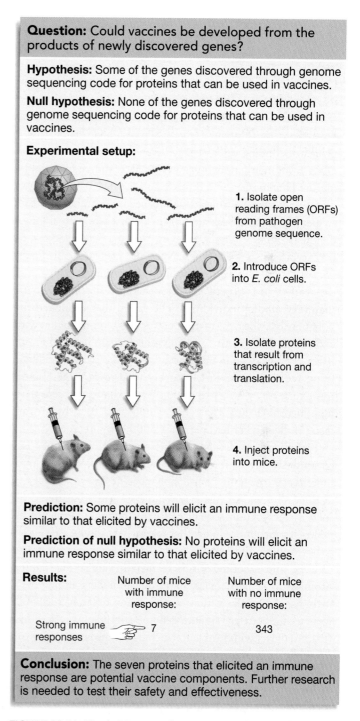

Question: Could vaccines be developed from the products of newly discovered genes?

Hypothesis: Some of the genes discovered through genome sequencing code for proteins that can be used in vaccines.

Null hypothesis: None of the genes discovered through genome sequencing code for proteins that can be used in vaccines.

Experimental setup:

1. Isolate open reading frames (ORFs) from pathogen genome sequence.

2. Introduce ORFs into *E. coli* cells.

3. Isolate proteins that result from transcription and translation.

4. Inject proteins into mice.

Prediction: Some proteins will elicit an immune response similar to that elicited by vaccines.

Prediction of null hypothesis: No proteins will elicit an immune response similar to that elicited by vaccines.

Results:

	Number of mice with immune response:	Number of mice with no immune response:
Strong immune responses	7	343

Conclusion: The seven proteins that elicited an immune response are potential vaccine components. Further research is needed to test their safety and effectiveness.

FIGURE 20.11 Newly Discovered Proteins Can Be Tested for Vaccine Development
Because all potential genes are identified after whole-genome sequencing, virtually all proteins produced by a pathogen can be tested for their ability to provoke an immune response and act as a vaccine.

mice and analyzed whether an immune response occurred. Their results show that seven of the proteins tested evoked a strong immune response and represent promising vaccine components. Follow-up work is now under way to determine whether one or more of these proteins could act as a safe and effective vaccine in humans.

The general message of these case histories is that genome sequencing data are providing biomedical researchers with new drug targets and candidate vaccine components. Although early results are promising, only time will tell if genomics ushers in what some have claimed will be an entirely new era of medical research and practice.

ESSAY Genomics and Genetic Discrimination

Chapter 19 explained how gene hunters analyzed the genetic markers called *restriction fragment length polymorphisms* (RFLPs) to find the allele responsible for Huntington's disease. Data from the Human Genome Project have made this approach to finding disease genes even more powerful. Because DNA from many individuals was used as source material during the Human Genome Project, and because overlapping segments of genes were routinely sequenced, researchers were able to identify 1.42 million sites where single bases vary among individuals. Where you might have a "C" at a particular site, others may have a "T." These variable sites are called *single nucleotide polymorphisms*, or SNPs (pronounced "snips"), and they can serve as genetic markers—mapped sites in the genome that vary among individuals. The group that cataloged these polymorphic sites estimates that 60,000 of them lie within exons and that at least one polymorphic site lies within 5000 bases of every exon in the genome. To realize why this is significant, recall that the RFLP marker used to find the Huntington's disease gene was 500,000 bases away from the Huntington's locus. That gene would have been found much faster if the marker had been just 5000 bases away.

For disease gene hunters, this new catalog of SNPs is an enormously powerful resource. The possibility of analyzing the inheritance of millions of polymorphic sites all over the genome—instead of just a few hundred widely scattered ones such as recognition sequences for restriction endonucleases—makes it much more likely that researchers can locate genes associated with illnesses such as Alzheimer's disease, certain cancers, and inherited forms of mental illness. The database promises to be particularly important in understanding the genetic basis of diseases that involve many different loci. As efforts to annotate the human genome continue, it is very likely that biologists will be able to track down the genes responsible for many or even most inherited diseases.

Suppose that this prediction is correct, and researchers succeed in identifying alleles that make people susceptible to disorders such as heart attack, high blood pressure, diabetes, certain cancers, and mental illness. Now suppose that, in the near future,

technology advances to the point at which you will be able to walk into a clinic, request that your DNA be sequenced, and then receive a report detailing the disease-related alleles that you carry. Would you want to know which of your alleles make you suscep-

Would you want to know which of your alleles make you susceptible to certain diseases?

tible to certain diseases? If you had a family history of the cancer called HNPCC (see Chapter 14), the answer might be yes. The allele associated with this cancer has been identified, so people can learn whether they have a genetic predisposition to the disease. If so, they can change their diets to make tumor development less likely, and they can be screened regularly for the onset of the disorder. But suppose that health insurance companies, life insurance companies, your friends, and even prospective spouses or employers had access to these data. Would you want them to know?

Early in the debate about whether government agencies should support the Human Genome Project, questions like these were discussed extensively. As part of the funding proposals that allowed the project to move forward, money was set aside to study the ethical and legal implications of the project. During the rush to sequence the human genome, however, these questions received somewhat less attention than they had earlier. Now that the genome is complete, the discussion over the ethical and legal ramifications of the data is starting anew.

Although actual cases of genetic discrimination have been extremely rare, most U.S. states have passed laws prohibiting the use of genetic information in evaluating applications for health insurance. As this book goes to press, the U.S. Congress is considering legislation that would ban discrimination on the basis of genotype when individuals apply for insurance or employment. But the efforts to understand the social and ethical ramifications of genome sequencing and to regulate the use of the data are still in their infancy. It will be important for you, as a biology student, to help educate the general public and contribute an informed opinion as issues arise.

CHAPTER REVIEW

Summary of Key Concepts

■ **Once a genome has been completely sequenced, researchers use a variety of techniques to identify which sequences code for products and which act as regulatory sites.**

Recent technical advances have allowed investigators to sequence DNA much more rapidly and cheaply than before, resulting in a flood of genome data. Researchers annotate genome sequences by finding genes and determining their function. To identify genes in bacteria and archaea, researchers use computers to scan the genome for start and stop codons that are in the same reading frame and that are separated by gene-sized stretches of sequence. Finding such open reading frames (ORFs) is difficult in eukaryotes, because exons are interrupted by introns and because most eukaryotic DNA does not code for a product. One approach to finding eukaryotic genes is to analyze the sequences of complementary DNAs (cDNAs) synthesized from mRNAs and then match these sequences to DNA found in the genome itself. Sequences that are highly conserved among species are also hypothesized to indicate the locations of genes.

Web Tutorial 20.1   Human Genome Sequencing Strategies

■ **In bacteria and archaea, there is a positive correlation between the number of genes in a species and the species' metabolic capabilities.**

Species of bacteria and archaea are usually targeted for whole-genome sequencing because they cause disease or have interesting metabolic abilities. In these groups, a general correlation seems to exist between the size of an organism's genome and its morphological complexity or biochemical capabilities. Parasites tend to have small genomes; organisms that live in a broad array of habitats or that use a wide variety of nutrients tend to have larger genomes. Many of the genes identified in bacteria and archaea still have no known function, however, and a significant percentage of them are extremely similar to other genes in the same genome. Another generalization about prokaryotic genomes is that genes are frequently transferred laterally, or between species. Lateral gene transfer appears to be common in genes responsible for causing disease.

■ **In eukaryotes, genomes are dominated by parasitic sequences that do not contribute to the fitness of the organism.**

Compared with prokaryotic genomes, eukaryotic genomes are large and contain a high percentage of transposable elements, repeated sequences, and other noncoding sequences. There is no obvious correlation between morphological complexity and gene number in eukaryotes, although the number of distinct transcripts produced may be much larger than the actual gene number in certain species as a result of alternative splicing. By analyzing genes that belong to closely related gene families, researchers have confirmed that gene duplication and polyploidy have been important sources of new genes in eukaryotes.

■ **Data from genome sequencing projects are now being used in the development of new drugs and vaccines.**

The availability of whole-genome sequences is inspiring new research programs. Biologists are affixing small amounts of proteins or cDNAs to microarrays in order to study protein-protein interactions or changes in gene expression and in mRNA populations over time. In addition, the availability of whole-genome data has allowed investigators to find novel genes that are closely related to existing drug targets and to find new proteins that may serve as vaccine candidates.

Questions

Content Review

1. What is an open reading frame?
   a. a gene whose function is already known
   b. a DNA section that is thought to code for a protein because it is similar to a complementary DNA (cDNA)
   c. a DNA section that is thought to code for a protein because it has a start codon and a stop codon flanking hundreds of base pairs
   d. any member of a gene family

2. What best describes the logic behind shotgun sequencing?
   a. Break the genome into tiny pieces. Sequence each piece. Use overlapping ends to assemble the pieces in the correct order.
   b. Start with one end of each chromosome. Sequence straight through to the other end of the chromosome.
   c. Use a variety of techniques to identify genes and ORFs. Sequence these segments—not the noncoding and repeated sequences.
   d. Break the genome into pieces. Map the location of each piece. Then sequence each piece.

3. What are minisatellites and microsatellites?
   a. small, extrachromosomal loops of DNA that are similar to plasmids
   b. parts of viruses that have become integrated into the genome of an organism
   c. incomplete or "dead" remains of transposable elements in a host cell
   d. short and simple repeated sequences in DNA

4. What is the leading hypothesis to explain the paradox that large, complex eukaryotes such as humans have relatively small numbers of genes?
   a. lateral transfer of genes from other species
   b. alternative splicing of mRNAs
   c. polyploidy, or the doubling of the genome's entire chromosome complement
   d. expansion of gene families through gene duplication

5. What evidence do biologists use to infer that a gene is part of a gene family?
   a. Its sequence is exactly identical to that of another gene.
   b. Its structure—meaning its pattern of exons and introns—is identical to that of a gene found in another species.
   c. Its composition, in terms of percentage of A-T and G-C pairs, is unique.
   d. Its sequence, structure, and composition are similar to those of another gene in the same genome.

6. What is a pseudogene?
   a. a coding sequence that originated in a lateral gene transfer
   b. a gene whose function has not yet been established
   c. a polymorphic gene—meaning that more than one allele is present in a population
   d. a gene whose sequence is similar to that of functioning genes but does not produce a functioning product

Conceptual Review

1. Explain how open reading frames are identified in the genomes of bacteria and archaea. Why is it more difficult to find open reading frames in eukaryotes?

2. Why is the observation that parasitic organisms tend to have relatively small genomes logical?

3. Review how a LINE sequence transmits a copy of itself to a new location in the genome. Why are LINEs and other repeated sequences referred to as "genomic parasites"?

4. How does DNA fingerprinting work? Stated another way, how does variation in the size of microsatellite and minisatellite loci allow investigators to identify individuals?

5. Researchers can create microarrays of short, single-stranded DNAs that represent many or all of the exons in a genome. Explain how these microarrays are used to document changes in the transcription of genes over time or in response to environmental challenges.

6. Explain the concept of homology and how identifying homologous genes helps researchers identify the function of unknown genes. Are duplicated sequences that form gene families homologous? Explain.

Group Discussion Problems

1. Parasites lack genes for many of the enzymes found in their hosts. Most parasites, however, have evolved from free-living ancestors that had large genomes. Based on these observations, W. Ford Doolittle claims that the loss of genes in parasites represents an evolutionary trend. He summarizes his hypothesis with the quip "use it or lose it." What does he mean?

2. According to eyewitness accounts, communist revolutionaries executed Nicholas II, the last czar of Russia, along with his wife and three children, the family physician, and several servants. Many decades after this event, a grave purporting to hold the remains of the royal family was identified. Biologists were asked to analyze DNA from each adult and juvenile skeleton and determine whether the bodies were indeed those of several young siblings, two parents, and several unrelated adults. If the grave was authentic, describe how similar the DNA fingerprints of each skeleton would be relative to the fingerprints of other individuals in the grave.

3. The human genome contains a gene that encodes a protein called syncytin. This gene is expressed in placental cells during pregnancy. The syncytin gene is nearly identical in DNA sequence to a gene in a virus that infects humans. In this virus, the syncytin-like gene codes for a protein found in the virus's outer envelope. State a hypothesis to explain the similarity between the two genes.

4. A recent study used microarrays to compare the patterns of expression of genes that are active in the brain, liver, and blood of chimpanzees and humans. Although the overall patterns of gene expression were similar in the liver and blood of the two species, expression patterns were strikingly different in the brain. How does this study relate to the hypothesis that most differences between humans and chimps involve changes in gene regulation?

Answers to Multiple-Choice Questions   1. c; 2. a; 3. d; 4. b; 5. d; 6. d

www.prenhall.com/freeman is your resource for the following: Web Tutorials; Online Quizzes and other Online Study Guide materials; Answers to Conceptual Review Questions; Solutions to Group Discussion Problems; Answers to Figure Caption Questions and Exercises; and Additional Readings and Research.

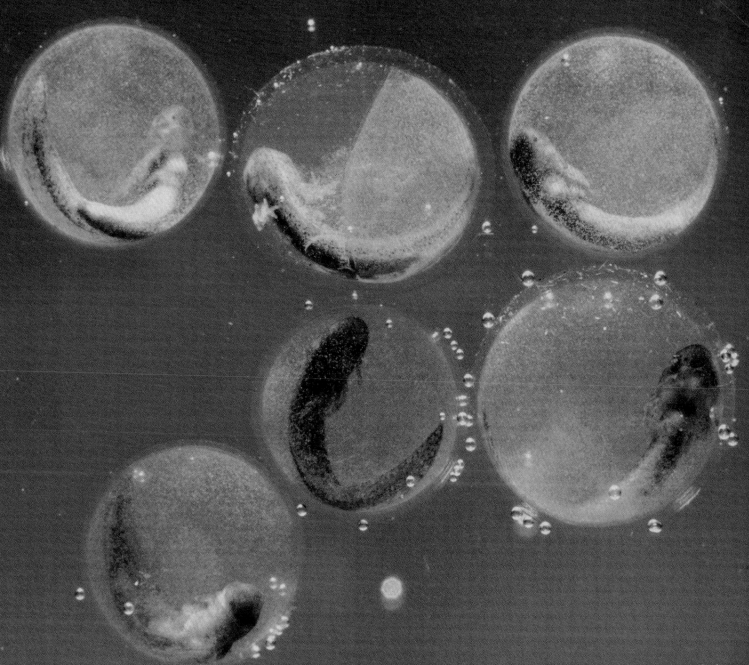

These spotted salamander embryos are growing in association with green algae. The oxygen that the algae produce as a by-product of photosynthesis is used by the embryos in cellular respiration.

21 Early Development

KEY CONCEPTS

- In most animals, fertilization depends on specific interactions among proteins on the plasma membranes of sperm and egg.

- In animals, the earliest cell divisions divide the fertilized egg into a mass of cells. These cells are not all alike; eventually they give rise to different types of cells and tissues as development proceeds.

- In animals as well as plants, the early events in development result in the production of three embryonic tissues and the establishment of the major body axes.

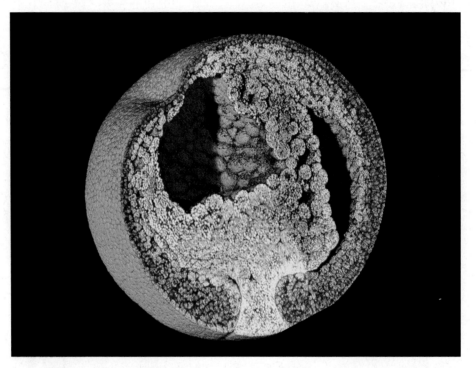

Scanning electron micrograph of a sea urchin embryo, cut in half. As development continues, the cells shown will form the individual's gut, muscle, and skin.

In 1859 Charles Darwin referred to the origin of species as "the mystery of mysteries." At that time the most urgent task confronting biologists was to explain how species come to be. Today, however, Darwin's theory of evolution by natural selection explains many of the most fundamental questions about how species form and how they change through time. What question qualifies as the current mystery of mysteries in biological science? Although there are many candidates, one of the most compelling is the question of how a multicellular individual develops from a single cell—the fertilized egg.

Biologists in the 1800s first began to study how plants and animals develop. Early investigators sought to catalog the events that occur as an individual develops from a fertilized egg to its juvenile and then adult forms. This early work was observational and descriptive in nature and was carried out by means of microscopes.

Based on these studies, development appeared to be highly variable among organisms. In humans, a fertilized egg develops into a newborn child composed of hundreds of cell types and trillions of cells. This process takes nine months. Rapid growth and development continue for another 15 to 18 years, until individuals become adults. In contrast, a fertilized egg from a burr oak tree takes about four months to grow into an **embryo**, or newly developing individual, that is encased in an acorn. After spending the winter in a dormant state, the embryo grows into a seedling. Over the next 250 to 300 years, the tree will continue to grow and develop new leaves, roots, and flowers. At the other extreme, a fertilized egg of the fruit fly *Drosophila melanogaster* transforms into a feeding larva in just a day. About five days later, the larva stops feeding and forms a pupa. In another four days, an adult emerges. The adult fly flies, feeds, courts, mates, and starts the cycle anew.

How did biologists start to make sense of all this variability in the rate and pattern of development? By the late nineteenth century, researchers had started to use experimental approaches to study embryos. Biologists focused on organisms that participate in external fertilization (outside the body) rather than internal fertilization (inside the body), so that embryos could be easily collected and observed. They also chose organisms that developed rapidly and produced large numbers of eggs that were easy to manipulate. Then they removed cells at various stages of development to see which were required for the formation of certain structures, injected dyes into cells to follow their fate through time, and analyzed mutant individuals that developed abnormally. After a century of poking and prodding embryos with these and other techniques, biologists have concluded that several unifying mechanisms and themes underlie the wonderful variety of plant and animal developmental programs.

The goal of this chapter is to explore how common genetic and cellular mechanisms unify the diversity of developmental patterns seen in different groups of plants and animals. We begin by examining *gametogenesis*, the process responsible for producing gametes. Recall from Chapter 12 that **gametes** are haploid reproductive cells; when they differ in size, they are called **sperm** (for males) and **eggs** (for females). As **Figure 21.1** shows, gametogenesis is followed by *fertilization*, the union of a sperm and an egg. In animals, a fertilized egg undergoes a series of rapid cell divisions called *cleavage*, forming a mass of cells known as a *blastula*. In the blastula, massive rearrangements of cells during *gastrulation* result in the formation of a *gastrula*. These processes are described in Sections 21.1 through 21.3.

They set the stage for the development of specialized tissues and organs during a process called *organogenesis*. In Section 21.4 we'll survey the diversity of developmental schemes by comparing development in four representative organisms. Let's delve in.

21.1 Gametogenesis

The development of a new individual begins with **gametogenesis**, which is the formation of gametes—sperm and egg—in the reproductive organs of adult organisms. The DNA and cytoplasm in these reproductive cells are the initial components of the new individual. Both sperm and egg contribute an equal number of chromosomes to the offspring—usually a haploid genome containing one copy of each gene. But because egg cells are routinely hundreds or thousands of times larger than sperm cells, an egg contributes much more cytoplasm than does a sperm.

The sequence of mitotic and meiotic cell divisions that lead up to the production of sperm and eggs are discussed in Chapters 40 and 48. Here, let's focus on the structure of the mature reproductive cells. Understanding how these specialized cells are put together will lay the groundwork for addressing questions about fertilization and early development.

Sperm Structure and Function

A mammalian sperm cell has four main compartments: (1) a head, (2) a neck, (3) a midpiece, and (4) a tail. The head region contains the nucleus and an enzyme-filled structure called the **acrosome**; the neck encloses a centriole; the midpiece is packed

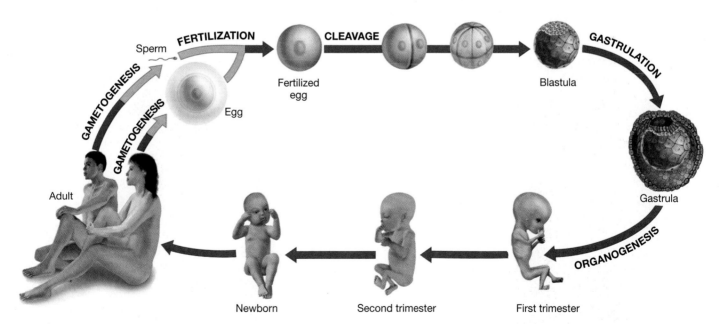

FIGURE 21.1 Development Proceeds in Ordered Phases
In animals, the development of a new individual starts with the formation of gametes and continues with fertilization and early cell divisions (cleavage), resulting in a blastula. Gastrulation then rearranges the blastula into a gastrula. The events through the formation of the gastrula are discussed in this chapter; Chapter 22 explores what happens next—organogenesis.

with mitochondria; and the tail region consists of a flagellum (see Chapter 7). **Figure 21.2a** shows the basic structure of animal sperm. A sperm cell is so highly specialized that it has been called "DNA with a propeller." The propeller is the flagellum, which is powered by ATP manufactured in the mitochondria.

Sperm cells develop after meiosis has resulted in the production of a haploid nucleus. As a sperm cell matures, its chromosomes are condensed into an even more densely packed arrangement than in other cells. The acrosome develops from the Golgi apparatus, and the flagellum arises from the centrioles. Once the sperm reaches an egg, the enzymes in the acrosome are responsible for digesting the outer coverings of the egg cell so that the two cells' plasma membranes can make contact and fuse. In most animal species, the sperm's nucleus, centriole, and mitochondria enter the egg after fusion occurs.

As **Figure 21.2b** shows, the sperm of flowering plants are even simpler in form than the sperm of animals. In most plants, sperm develop from **pollen grains**, which consist of several haploid cells derived from a product of meiosis. An insect or the wind carries a pollen grain to the top of a structure that contains an *ovule*, the female reproductive structure in which a plant's egg cell forms. One of the cells inside the pollen grain then divides by mitosis, producing two **sperm nuclei**. These nuclei move down the **pollen tube**, a structure that grows out of the pollen grain toward the egg cell. Once the pollen tube reaches the egg inside the ovule, the sperm nuclei bud off and move toward the egg. As we'll see in Section 21.2, both sperm cells will participate in fertilization.

Egg Structure and Function

Sperm are small and motile reproductive cells; eggs are large and nonmotile. Eggs are large mainly because they contain the nutrients required for the embryo's early development. Even among mammals, whose embryos start to obtain nutrition through a maternal organ called the *placenta* within a week or two after fertilization, the egg itself must supply the nutrients for early development. In species in which eggs are laid into the environment, stores in the egg are the *only* source of nutrients until organs have formed and a larva or juvenile hatches and begins to feed. In these species, the nutrients required for early development are provided by a **yolk**, the fat- and protein-rich cytoplasm that is loaded into egg cells as they mature. Yolk may be present as one large mass or as many small granules. Plants also provide their embryos with nutritive tissue, which the offspring use until they are large enough to initiate photosynthesis and feed themselves. In contrast to the nutrient-rich yolk of animals, though, these nutrients are not synthesized and stored until after fertilization has occurred. As we'll see in Section 21.2, a flowering plant embryo's nutrient supply is created during seed maturation—not during gametogenesis.

Yolk is not the only specialized material or structure found in animal eggs. Many eggs also contain organelles called **cortical granules**, which are small vesicles filled with enzymes that are involved in fertilization in egg-laying animals. As the egg matures, cortical granules are synthesized, transported to the cell surface, and bound to the inner surface of the plasma membrane. Just outside the plasma membrane, a fibrous, mat-like sheet of

(a) Animal sperm swim to the egg.

(b) Plant sperm move to the egg through a pollen tube.

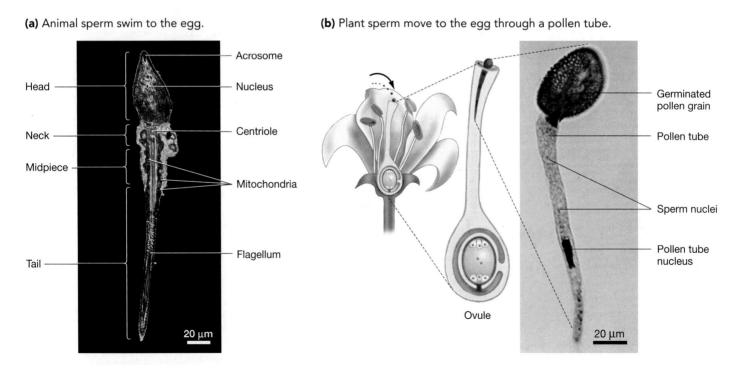

FIGURE 21.2 The Structures of Animal and Plant Sperm
(a) The morphology of human sperm is typical of many animal species. **(b)** Two sperm cells are produced when pollen germinates. They travel down the pollen tube toward the ovule, which contains the egg.

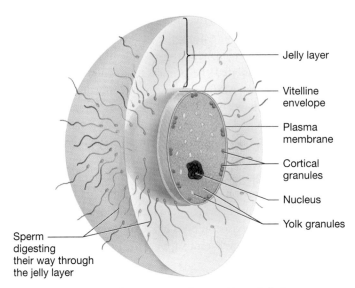

Jelly layer

Vitelline envelope

Plasma membrane

Cortical granules

Nucleus

Yolk granules

Sperm digesting their way through the jelly layer

FIGURE 21.3 Sea Urchin Eggs Are Covered by a Jelly Layer
Animals that lay their eggs in water frequently surround their eggs with a jelly layer, a gelatinous coat.

glycoproteins called the **vitelline envelope** forms and surrounds the egg (**Figure 21.3**). (In humans, the equivalent structure is the unusually thick *zona pellucida*.) In some species, a large gelatinous matrix known as a **jelly layer** also encloses the egg. How does a sperm cell penetrate these coatings to reach the egg's plasma membrane and fertilize the egg?

21.2 Fertilization

Fertilization seems like a simple process: A sperm cell fuses with an egg cell to form a diploid cell known as a **zygote** (a fertilized egg). Upon reflection, though, you should begin to appreciate that the process is extraordinarily complex. For fertilization to take place, sperm and egg cells must be in the same place at the same time and then must recognize and bind to each other. Next they must fuse—even though most of the other cells in the body do not fuse with cells they contact. In most species, fusion must also be limited to a single sperm so that the zygote does not receive extra chromosomes. Finally, the fusion of the two gametes has to trigger the onset of development. This complexity has made fertilization a fascinating research topic. The contact that takes place between sperm and egg at fertilization qualifies as the best studied of all cell-to-cell interactions.

Fertilization in Sea Urchins

Work on fertilization began in earnest early in the twentieth century, when biologists started to study the sperm-egg interaction in sea urchins. Sea urchins are marine invertebrates that perform fertilization externally (**Figure 21.4a**). To maximize the probability that sperm and egg will meet, male and female sea urchins secrete huge quantities of gametes. As a result, this organism provided researchers with large numbers of gametes that could be studied in a test tube or culture dish filled with

seawater. Sea urchins continue to be an intensively studied model system in research on fertilization.

In sea urchins, the jelly layer that surrounds the egg contains a molecule that attracts sperm. The attractant is a small peptide that diffuses away from the egg and into the surrounding seawater. Sperm respond to the attractant by swimming toward areas where its concentration is high. The same type of interaction between sperm and eggs has been observed in a wide variety of animals that lay their eggs in seawater.

When a sea urchin sperm and egg meet, the head of the sperm initially encounters the jelly layer of the egg cell (**Figure 21.4b**). To reach the plasma membrane of the egg cell, the sperm head must

(a) Sea urchin releasing gametes

(b) FERTILIZATION SEQUENCE

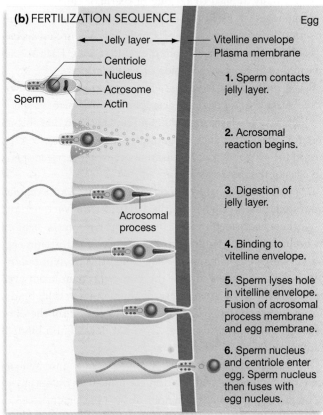

Jelly layer

Egg

Vitelline envelope
Plasma membrane

Centriole
Nucleus
Acrosome
Actin

Sperm

1. Sperm contacts jelly layer.

2. Acrosomal reaction begins.

3. Digestion of jelly layer.

Acrosomal process

4. Binding to vitelline envelope.

5. Sperm lyses hole in vitelline envelope. Fusion of acrosomal process membrane and egg membrane.

6. Sperm nucleus and centriole enter egg. Sperm nucleus then fuses with egg nucleus.

FIGURE 21.4 Fertilization in Sea Urchins
(a) Sea urchins release their gametes into the ocean. **(b)** Fertilization involves a complex sequence of events.

digest its way through the egg's jelly layer and vitelline envelope. This process begins with an **acrosomal reaction,** a response that is triggered by contact between the sperm's head and the jelly layer. In the first part of this reaction, the contents of the acrosome—the structure at the tip of the sperm's head—are expelled. The enzymes that are released digest a pathway through the egg's jelly layer and allow the sperm to reach the vitelline envelope of the egg. The second part of the acrosomal reaction involves the polymerization of actin into microfilaments that form the **acrosomal process,** a protrusion that extends until it makes contact with the vitelline envelope. Finally, the plasma membranes of the egg and sperm fuse. The sperm nucleus, mitochondria, and centriole enter the egg, and the sperm and egg nuclei fuse to form the zygote nucleus. Fertilization is complete.

Frank Lillie was the first to recognize that an important question was hidden in this sequence of events: How do gametes from the same species recognize each other? After all, in many habitats sperm and eggs from a particular sea urchin species float in seawater along with eggs and sperm from other sea urchin species and many other organisms. What prevents cross-species fertilization and the production of dysfunctional hybrid offspring?

Species Recognition in Sea Urchins Lillie was the first researcher to identify a substance on the surface of egg cells that appeared to be involved in binding sperm. He called the compound *fertilizin* and showed that if the purified molecule were experimentally applied to a group of sperm, it caused the cells to clump together. He also showed that clumping occurred only when fertilizin from the eggs of a particular sea urchin species was combined with sperm from the same species. Each species appeared to have its own version of fertilizin.

Based on these observations, Lillie proposed that fertilizin on the surface of an egg interacted with a substance on sperm in a lock-and-key fashion. Further, Lillie suggested that this interaction was a product of natural selection, because it increased the probability of fertilization occurring between sperm and eggs of the same species.

Decades passed before Lillie's hypothesis was extended and tested, however. Then, in the 1970s, Victor Vacquier and co-workers succeeded in identifying a protein on the head of sea urchin sperm that binds to the surface of sea urchin eggs in a species-specific manner. They called this protein *bindin*. Follow-up work showed that the bindin proteins from even very closely related species are distinct. As a result, bindin should ensure that a sperm binds only to eggs from the same species. The next question was, To what does bindin bind? If bindin acts as a key, what acts as the lock?

To address these questions, Kathleen Foltz and William Lennarz hypothesized that fertilizin is a protein on the surface of sea urchin eggs that binds to bindin. To find this bindin-receptor protein, the researchers attempted to isolate the part of the receptor that is exposed on the outside of the egg. They predicted that this region of the protein interacts with the bindin on sperm.

Figure 21.5 illustrates Foltz and Lennarz's experimental approach. They began by treating the surface of sea urchin eggs with a **protease**—an enzyme that cleaves peptide bonds. When the investigators isolated the protein fragments that were released from the egg surface, they found one that bound to sperm and to isolated bindin molecules. Further, this binding occurred in a species-specific manner. A protein fragment from the eggs of one species bound to sperm of its own species, but did not bind to sperm of different species. Based on these observations, the biologists claimed that they had found the outward-facing portion of the egg-cell receptor for sperm.

These experiments provided convincing evidence that an egg receptor for sperm exists on the surface of sea urchin eggs. The work also provided important support for Lillie's lock-and-key

---

Question: If bindin is the protein on the surface of sperm that acts like a key, what is the "lock"?

Hypothesis: The lock is a protein on the surface of sea urchin eggs that binds to bindin.

Null hypothesis: The lock is not a protein on the surface of sea urchin eggs that binds to bindin.

Experimental setup:

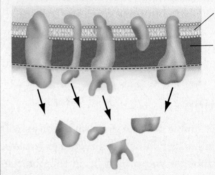

Egg cell's plasma membrane

Vitelline envelope

1. Use protease to release protein fragments from egg surface.

2. Isolate each type of protein fragment and see which ones bind to sperm and to isolated bindin molecules.

Prediction: One of the protein fragments isolated from egg surface will bind to sperm and to isolated bindin molecules.

Prediction of null hypothesis: The protein fragments isolated from egg surface will not bind to sperm or to isolated bindin molecules.

Results:

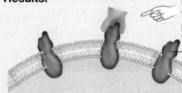

Fertilizin (protein fragment from egg) binds to bindin protein on sperm if sperm is from same species as egg

Bindin

Sperm cell's plasma membrane

Conclusion: Fertilizin is a "lock" protein on the surface of eggs that binds to bindin (the "key") in a species-specific manner.

FIGURE 21.5 The Egg-Cell Receptor for Sperm Was Isolated and Characterized
By treating sea urchin egg cells with a protease, researchers isolated the fragment of the egg-cell receptor that binds to sperm.

hypothesis. During sea urchin fertilization, species-specific bindin molecules on sperm interact with species-specific fertilizin receptors on the surface of the egg. This interaction is required for the plasma membranes of sperm and egg to fuse. As a result, cross-species fertilization is rare.

Blocking Polyspermy Early in the history of studies on sea urchin fertilization, researchers noticed that only one sperm succeeded in fertilizing the egg, even when dozens or even hundreds of sperm were clustered around the vitelline envelope. From the standpoint of producing a viable offspring, this observation was logical. If multiple fertilization, or **polyspermy**, occurred, the resulting zygote would have more than two copies of each chromosome. Sea urchin embryos with more than two copies of each chromosome die. If many sperm are present, how is polyspermy avoided?

Research over the past 70 years has revealed a wide array of mechanisms that block polyspermy in various animal species. In sea urchins, for example, fertilization results in the erection of a physical barrier to sperm entry. As **Figure 21.6a** shows, the entry of a sperm causes calcium ions (Ca^{2+}) to be released from storage areas inside the egg. A wave of ions starts at the point of sperm entry and propagates throughout the egg. In response to this dramatic increase in Ca^{2+} concentration, a series of events occurs in the egg. For example, the cortical granules located just inside the membrane fuse with the egg cell's plasma membrane and release their contents to the exterior. The contents of the cortical granules include proteases that digest the exterior-facing fragment of the egg-cell receptor for sperm. In addition, other compounds from the cortical granules are trapped between the egg cell's plasma membrane and the vitelline envelope and cause water to flow into the space by osmosis. The influx of water then causes the envelope matrix to lift away from the cell and form a **fertilization envelope** (**Figure 21.6b**). The fertilization envelope, in turn, keeps additional sperm from contacting the egg membrane.

To summarize, a century of research on sea urchin fertilization has illuminated one of the most important cell-to-cell interactions in nature. The interaction between sperm and egg triggers a series of remarkable events, including the acrosomal reaction and mechanisms for blocking polyspermy.

(a) A WAVE OF Ca^{2+} SPREADS FROM THE SITE OF SPERM ENTRY.

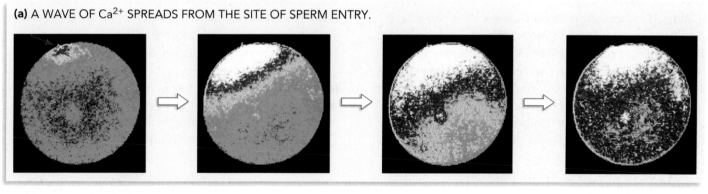

(b) THE FERTILIZATION ENVELOPE LIFTS AND BLOCKS EXCESS SPERM.

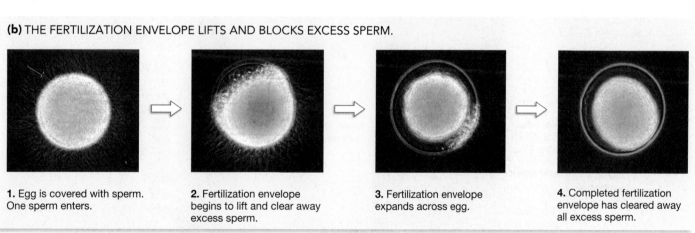

1. Egg is covered with sperm. One sperm enters.

2. Fertilization envelope begins to lift and clear away excess sperm.

3. Fertilization envelope expands across egg.

4. Completed fertilization envelope has cleared away all excess sperm.

FIGURE 21.6 A Physical Barrier Erected after Fertilization Prevents Polyspermy
(a) During fertilization, a wave of Ca^{2+} begins at the point of sperm entry (arrow) and spreads throughout the egg. The white dots are from a reagent that reacts with calcium ions. **(b)** In response to increased Ca^{2+} concentrations, cortical granules fuse with the egg cell's plasma membrane and release their contents to the exterior, causing a fertilization envelope to rise and clear away excess sperm.

Fertilization in Mammals

Although sea urchins have acted as a productive model system, an important question remains: How relevant are these findings to other species? In particular, how similar is sea urchin fertilization to events in mammals and plants? Unlike what occurs in sea urchins, fertilization in humans and other mammals occurs internally. This fact makes mammalian fertilization much more difficult to study than fertilization in sea urchins and similar species. With the advent of the *in vitro fertilization* (IVF) methods highlighted in this chapter's essay, however, biologists have finally acquired the ability to study mammalian fertilization under laboratory conditions. What have they found?

In most cases, females actively choose a mate prior to the sperm-egg interaction. As a result, species recognition is not as much of an issue in mammals and other species with internal fertilization as it is in sea urchins and other species with external fertilization. The acrosomal reaction still occurs, however, after the sperm's head reaches the equivalent of the sea urchin vitelline envelope—a gelatinous extracellular matrix known as the **zona pellucida**. The enzymes released from the acrosome digest the zona pellucida. As a result, the sperm's head is able to reach the egg cell's plasma membrane and fuse with it.

Although as yet there is no convincing evidence for a bindin-like protein on the sperm head, biologists have recently presented data suggesting that egg cells have a binding site for sperm. To search for this site in mouse eggs, the researchers analyzed the three glycoproteins found in the zona pellucida. They found that one of the three glycoproteins, called ZP3, binds to the heads of sperm. But as predicted, the binding of mammalian sperm to ZP3 is not species specific. Finally, researchers have found that enzymes released from cortical granules modify ZP3 in a way that prevents binding by additional sperm.

Fertilization in Flowering Plants

It should not be a surprise that marked differences occur in plant and animal fertilization. Considerable variation occurs in developmental processes among groups that have evolved independently for long periods.

Fertilization in flowering plants has been exceptionally difficult to study, because the process takes place inside a female reproductive structure called the **ovule**. Two important observations are worth noting, however. First, an event known as **double fertilization** takes place in flowering plants. As **Figure 21.7a** shows, two sperm enter the ovule via the pollen tube. In the ovule, they encounter the egg cell and a cell that in many species contains two haploid nuclei. One sperm fuses with the egg to form the zygote, while the other sperm fuses with the cell with two haploid nuclei to form a triploid ($3n$) cell. The triploid cell divides repeatedly to form a nutritive tissue called endosperm. **Endosperm** ("inside-seed") provides the proteins, carbohydrates, and fats or oils required for embryonic development, **germination** (resumptions of seed growth), and early seedling growth. In species with large seeds, the endosperm grows into a sizeable nutrient reservoir as the ovule matures (**Figure 21.7b**). When you eat wheat, rice, corn, almonds, or other grains or nuts, you are eating primarily endosperm. Functionally, endosperm is equivalent to the yolk found in animal eggs.

A second major observation about plant fertilization involves the start of the process. The development of a new plant begins when a pollen grain germinates at the top of the female reproductive structure. Once a pollen grain germinates, sperm are produced and begin to move toward the ovule, at

(a) How does double fertilization occur?

(b) Products of double fertilization in wheat seed

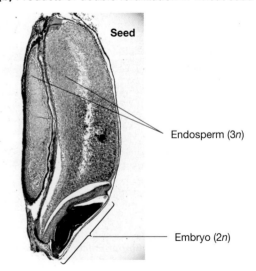

Ovule

Pollen tube carries two sperm (*n*) to ovule

Double fertilization

Two maternal nuclei and one sperm nucleus fuse to form **endosperm** (3*n*)

Egg and one sperm fuse to form **zygote** (2*n*)

Seed

Endosperm (3*n*)

Embryo (2*n*)

FIGURE 21.7 Double Fertilization Leads to the Formation of Endosperm
(a) In flowering plants, a sperm (*n*) fuses with two haploid maternal nuclei (*n*) near the egg to form triploid (3*n*) endosperm tissue. **(b)** Endosperm is a nutritive tissue packed with proteins and carbohydrates and fats or oils.

the base of that structure. Recent research has shown that the germination of a pollen grain involves complex interactions between proteins on the surface of the pollen grain and proteins on the walls and membranes of cells at the top of the female reproductive structure. Some of these interactions are involved in species recognition and ensure that pollen from only the same species is allowed to germinate. But in some species, the interactions also prevent pollen from the same individual from germinating on its own female reproductive parts. In this way, self-fertilization and inbreeding are avoided.

Later chapters will explore the molecular mechanisms involved in self-self recognition in plants, as well as the damaging consequences of inbreeding. For now, let's turn to the cell divisions that take place immediately after fertilization.

✓CHECK YOUR UNDERSTANDING

Fertilization is based on specific interactions between proteins in the plasma membranes of animal sperm and egg cells or on the surfaces of plant pollen grains and female reproductive structures. You should be able to (1) describe the protein-protein interactions in sea urchins, mammals, and flowering plants; and (2) explain why sea urchin eggs have such a sophisticated mechanism to block polyspermy.

21.3 Cleavage and Gastrulation

Cleavage refers to the rapid cell divisions that take place in animals after fertilization. Cleavage partitions the egg cytoplasm without additional cell growth taking place. As **Figure 21.8a** shows, the zygotes of many species simply divide into two cells, then four cells, then eight, and so on, without concurrent growth overall. The cells that are created by cleavage divisions are called **blastomeres**. When cleavage is complete, the embryo consists of a sphere of cells called a **blastula**. The blastula then undergoes *gastrulation*—the coordinated cell movements that result in the formation of a layered body structure. In animals, both cleavage and gastrulation are part of **embryogenesis**, the process by which a single-celled zygote becomes a multicellular embryo.

The exact pattern of cleavage varies widely among species, however. Cells can divide at right angles to each other so that they form tiers, as in Figure 21.8a; this pattern is **radial cleavage**. Instead they may divide at oblique angles so that they pile up as shown in **Figure 21.8b**; this pattern is **spiral cleavage**. In birds, fish, and other species whose eggs have large, membrane-bound structures filled with yolk, cleavage does not split the egg completely but produces a mound of cells around the yolk or on top of it, as shown in **Figure 21.8c**. How are these patterns controlled?

(a) Cleavage divides up the egg cytoplasm.

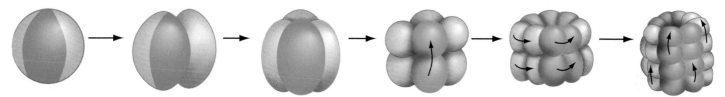

(b) In some species, cleavage creates a spiral group of cells.

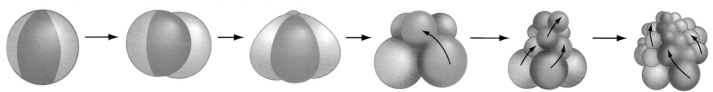

(c) In some species, a disk of small cells forms on top of a large cell containing the yolk.

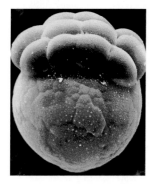

FIGURE 21.8 Cleavage Patterns Vary Widely among Species
(a) Radial cleavage. During cleavage, cells divide but growth does not occur. Some of these cells are colored green simply to help you follow their fate. **(b)** Spiral cleavage. **(c)** Distinctive cleavage patterns occur in fishes, reptiles, birds, and other species with extremely large amounts of yolk.

Controlling Cleavage Patterns

In 1894 H. E. Crampton published a remarkable observation that provided an important clue about how cleavage is controlled. Crampton was studying coiling patterns in the shells of snails. Snail shells coil either to the left or to the right of the opening where the individual's body emerges. Usually, all members of the same species have shells that coil in the same direction. Occasionally, though, Crampton found mutant individuals whose shells coiled in the direction opposite the norm for their species. The most remarkable observation he made is that the direction of cleavage and shell coiling is the same. As **Figure 21.9a** shows, normal members of a given snail species with left-handed coiling undergo cleavage in a way that leads to a left-handed coil of cells in blastulas of that species. But in mutant individuals of the same species, cleavage leads to a right-handed coil of cells as illustrated in **Figure 21.9b**. The mutants also had shells with right-handed coiling.

Another researcher followed up on this observation by arranging matings between right-handed snails and left-handed snails and analyzing the phenotypes of their offspring. He found that the pattern of inheritance could be explained by hypothesizing that a single gene with two alleles controlled the coiling pattern. He called the coiling alleles *D* and *d* and determined that *D* is dominant. Some crosses gave particularly interesting results, however. For example, when *DD* fe-

males were crossed with *dd* males, all of the offspring had the *Dd* genotype and had right-handed shells. But when *dd* females were crossed with *DD* males, all of the offspring had the *Dd* genotype but had left-handed shells. To explain these results, the biologist proposed that the direction of cleavage and coiling depends on the mother's genotype and not on the offspring's genotype. That is, mothers with *dd* genotypes produce left-coiling shells in their offspring. Mothers with a *Dd* or *DD* genotype, however, produce offspring with a right-coiled shell.

How is this "maternal effect" possible? The most likely answer is that the cleavage pattern in snails, and thus the coiling pattern of their shells, is established by a molecule that is present in the egg. A molecule that exists in eggs and that helps direct early development is called a **cytoplasmic determinant**. In this case, researchers hypothesized that a molecule produced by the mother and localized in the egg cytoplasm determines the orientation of the mitotic spindle, as shown in Figure 21.9. The mother's genotype determines the coiling pattern because the gene responsible for the pattern is expressed during egg maturation—not during embryo development. This result suggests that the orientation of cleavage is controlled primarily by cytoplasmic determinants from the mother and not by proteins produced in the zygote. What other effects do cytoplasmic determinants have?

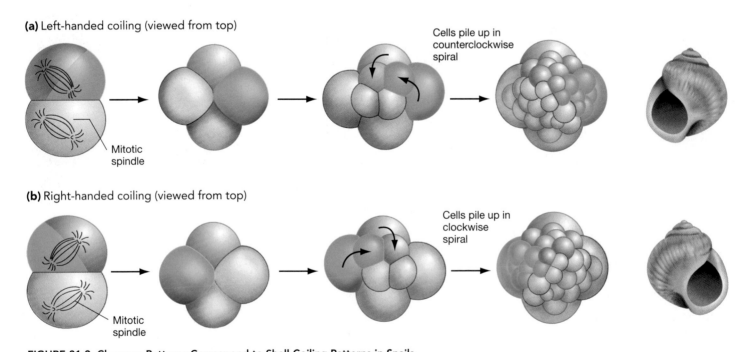

(a) Left-handed coiling (viewed from top)

Mitotic spindle

Cells pile up in counterclockwise spiral

(b) Right-handed coiling (viewed from top)

Mitotic spindle

Cells pile up in clockwise spiral

FIGURE 21.9 Cleavage Patterns Correspond to Shell-Coiling Patterns in Snails
(a) In snails, a cytoplasmic determinant sets up the orientation of the mitotic spindles during cleavage.
(b) Mutant forms of this cytoplasmic determinant switch the orientation of the mitotic spindles, leading to blastomeres and shells that coil in a direction opposite of normal.

The Role of Cytoplasmic Determinants in Development

Some of the earliest insights into the nature of cytoplasmic determinants came from studies on a sea squirt (**Figure 21.10a**). Sea squirts are marine organisms that are closely related to the vertebrates. In his experiments, Edwin Conklin focused on sea squirts because their embryos are transparent enough to make individual cells in live specimens visible with just the light microscope. In addition, the cytoplasm inside each sea squirt embryo contains differently pigmented regions. Conklin realized that the presence of different pigments might mark individual cells and allow him to follow their fate as development progressed. When biologists refer to an embryonic cell's **fate**, they mean what that cell is likely to become in the adult individual. Conklin wanted to know what pigmented cells in the embryos of sea squirts were likely to become in the corresponding adult sea squirts.

Conklin began this study by focusing on a band of yellow pigment that is present in fertilized eggs. As **Figure 21.10b** shows, certain embryonic cells inherit the pigment during cleavage. This is a key point: Cleavage divides up the egg cytoplasm in a specific way. As a result, not all blastomeres are identical. Because of their position in the embryo, different blastomeres contain different cytoplasmic determinants or other maternal factors.

By following the yellow-pigmented cells over time, Conklin was able to determine that they were destined to become muscle cells in the **larva**—the sexually immature phase in the life cycle. The yellow pigment, or a molecule associated with it, appeared to be acting as a cytoplasmic determinant. This was a crucial observation, because it showed that cytoplasmic determinants are involved in the process known as differentiation. **Differentiation** is the generation of different cell types from a single cell—the fertilized egg. The muscle cells of sea squirts, for example, have a specialized structure and function and express muscle-specific genes. Conklin's result suggested that a cytoplasmic determinant associated with the yellow pigment was responsible for the differentiation of muscle cells.

By doing the same type of analysis on other pigments in the fertilized egg, Conklin was able to construct a **fate map**—that is, a description of what each cell in the embryo is destined to become. The fate map showed how the positions of cells early in development correlated with their specialized state in larvae.

Follow-up work showed that the role of cytoplasmic determinants is highly variable among species. Cytoplasmic determinants and other types of maternal factors do not appear to play a major role in the early development of mammalian embryos, and it is unclear at present whether they are important in plants. But in millions of animal species, molecules in the egg's cytoplasm direct cleavage and influence the fate of embryonic cells.

The Role of Cell-to-Cell Signals in Development

In early studies on cytoplasmic determinants and other maternal factors, researchers turned to the eggs of the African clawed frog *Xenopus laevis*. In eggs of this species, an array of key molecules is asymmetrically distributed. The most obvious of these molecules are components of yolk. About 75 percent of the yolk is localized to just half of the egg. The yolk-rich region is

(a) The sea squirt *Styela partita*

1 cm

(b) Mapping the fate of pigment in cytoplasm

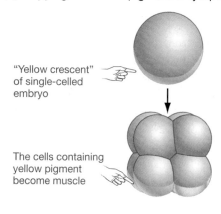

"Yellow crescent" of single-celled embryo

The cells containing yellow pigment become muscle

FIGURE 21.10 Molecules in the Egg's Cytoplasm Influence Development
(a) The sea squirt *Styela partita* has eggs that contain distinct pigments. **(b)** The yellow pigment found in sea squirt eggs is associated with the development of muscle cells.

(a) In frogs, yolk is concentrated in the vegetal hemisphere.

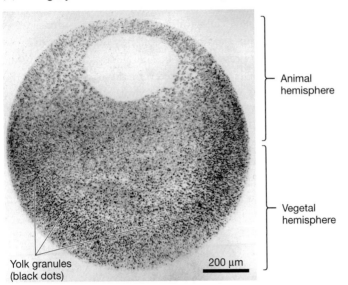

Animal
hemisphere

Vegetal
hemisphere

Yolk granules
(black dots)

200 µm

(b) *Vg1* mRNA is localized to the vegetal pole.

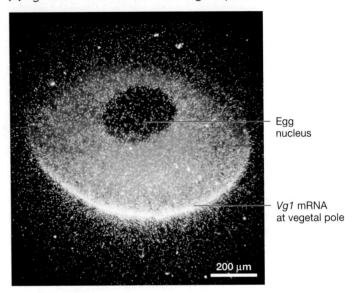

Egg
nucleus

Vg1 mRNA
at vegetal pole

200 µm

FIGURE 21.11 Eggs Are Highly Asymmetrical
(a) In frogs and in many other species, the yolk is concentrated in half of the egg, called the vegetal hemisphere.
(b) Bright light spots mark the location of *Vg1* mRNA (see Box 21.1).

known as the **vegetal hemisphere**; the yolk-poor half is called the **animal hemisphere** (Figure 21.11a).

In the 1980s Douglas Melton discovered a less obvious asymmetrically distributed molecule in frog eggs—an mRNA called *Vg1* (for "vegetal 1"). Using techniques introduced in **Box 21.1**, Melton was able to visualize the mRNA and confirm that it is localized to the vegetal pole of *Xenopus* eggs (**Figure 21.11b**). Researchers in other labs showed that another mRNA, called *VegT*, is also localized in the vegetal pole.

Subsequent work showed that both the *Vg1* and *VegT* mRNAs stay at the vegetal pole during cleavage and that cells inheriting these molecules become a specific type of embryonic tissue. Most animal embryos develop three types of tissues, called **germ layers**, early in development: (1) ectoderm ("outside skin"), (2) mesoderm ("middle skin"), and (3) endoderm ("inner skin"). Embryonic **ectoderm** forms the outer covering and nervous system; **mesoderm** gives rise to muscle, internal organs, and connective tissues such as blood and cartilage; and **endoderm** produces the lining of the digestive tract or gut, along with some of the associated organs. Cells that inherit *Vg1* and *VegT* mRNA become endoderm. Differentiation is triggered by the VegT protein, which is a regulatory transcription factor (Chapter 18). When the VegT protein binds to DNA, it triggers the expression of endoderm-specific genes. VegT is a cytoplasmic determinant.

What does the *Vg1* mRNA do? When the message is translated inside cells that are becoming endoderm, the resulting Vg1 protein is secreted from the cells, via the exocytosis pathway introduced in Chapter 7. In response to Vg1, nearby cells become mesoderm.

Based on these observations, the researchers concluded that the Vg1 protein functions as a **cell-to-cell signal**—a molecule that is secreted from a cell and received by a nearby cell, which changes its activity in response. To capture this point, biologists say that induction has occurred. **Induction** is an interaction between a signaling cell and a nearby recipient cell that changes the activity or fate of the recipient cell. The Vg1 protein produced by endodermal cells induces mesoderm. More specifically, the cells that respond to Vg1 become mesoderm on the **dorsal** (back) side of the embryo.

To summarize, cleavage is organized by molecules that are stored in the egg's cytoplasm. As a result of cleavage, different cytoplasmic determinants and other maternal factors are localized in specific blastomeres. The cytoplasmic determinants found in a cell cause them to differentiate into a particular embryonic tissue. In addition, the cell-to-cell signals produced by certain blastomeres can induce changes in neighboring cells.

Activating the Zygotic Genome

In most animals besides mammals, genes that are present in the zygote are not expressed during cleavage. Instead, early cleavage is directed by cytoplasmic determinants in the egg. The zygotic genome is transcribed for the first time after cleavage is well under way.

These points emerged from a series of studies on the African clawed frog that were published in the 1980s. Researchers injected *Xenopus* embryos with α-amanitin, which is derived from poisonous mushrooms. Recall from Chapter 16 that α-amanitin inhibits transcription. Although the presence of α-amanitin prevented the genomes of the experimental

BOX 21.1 Visualizing mRNAs by in Situ Hybridization

Where are the mRNAs for a particular cytoplasmic determinant localized in an egg? Once the zygote's genome is activated, when and where are certain genes transcribed in the embryo? Answering questions like these became possible when biologists developed a technique called in situ ("in place") hybridization. **In situ hybridization** allows researchers to visualize the location of specific mRNAs in organisms. As a result, biologists are able to infer where particular genes are being transcribed at various times during development.

To perform in situ hybridization for a particular mRNA molecule, a researcher obtains single-stranded DNA or RNA molecules that are complementary in nucleotide sequence to the RNA in question (**Figure 21.12**, step 1). For the DNA or RNA to be used as a probe, some sort of label must be attached to it. The label gives researchers a way of detecting the molecule's location. Radioactive atoms were used as labels when this technique was first developed, but now most researchers prefer to attach enzymes that catalyze color-producing reactions (step 2) for their visibility.

Because in situ hybridization cannot be done in living organisms, the labeled probe is added to a preparation containing preserved tissue. Prior to this hybridization step, the preserved tissue is treated with detergents or other agents that make plasma membranes permeable to small DNA molecules (step 3). As the probe enters the cells, it binds to mRNA molecules whose sequence is complementary (step 4). In this way, the single-stranded, labeled probe identifies a specific mRNA. Excess labeled probe that does not bind to complementary mRNAs is washed away. When the in situ hybridization is complete, the location and intensity of the signal from the label indicate the location and amount of mRNA present in particular regions of the organism (step 5).

VISUALIZING mRNAs BY IN SITU HYBRIDIZATION

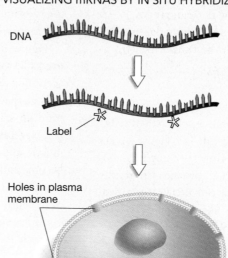

DNA

Label

1. Start with a single-stranded DNA or RNA probe, complementary in sequence to target mRNA.

2. Add label to probe (an enzyme that catalyzes a color-producing reaction).

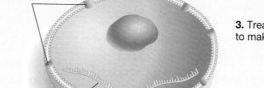

Holes in plasma membrane

Target mRNA

3. Treat preserved cells or tissues to make them permeable to probe.

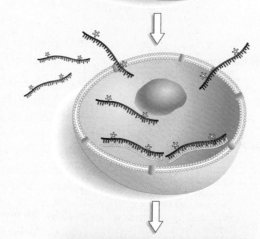

4. Add many copies of probe. Probe binds to target mRNA.

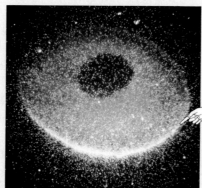

5. In this case, target mRNAs are concentrated in one end of the cell.

The label shows up as white in this image

FIGURE 21.12 In Situ Hybridization Allows Researchers to Pinpoint the Location of Specific mRNAs

QUESTION In situ hybridization is usually used to identify cells that are expressing a particular gene. Why is this method valid?

embryos from being expressed, the embryos progressed through early cleavage normally. In follow-up experiments, the researchers showed that new RNAs start to be produced in *Xenopus* after the twelfth cleavage division. Until then, development is directed by cytoplasmic determinants.

Mammals are an important exception to this rule. In mice and other mammals, mRNAs begin to be transcribed from the zygotic genome at the two-cell stage.

Gastrulation

As cleavage continues, the embryo becomes a solid ball of cells, which then hollows out and forms a blastula (see Figure 21.1). As this point, cell division slows. Cleavage is complete, and cells start actively making mRNAs and proteins from the embryo's genes. Next a series of dramatic cell movements called gastrulation, a key event in animal development, begins. **Gastrulation** radically rearranges cells and results in the formation of a **gastrula**, which contains the three embryonic tissue types. As **Figure 21.13** shows, these embryonic tissues give rise to the many types of tissues found in adults.

Research on gastrulation started in the 1920s with efforts to map the fate of individual cells in newt and frog embryos. These early experiments were based on soaking tiny blocks of

agar (a gelatinous compound) with a nontoxic dye. The dyed blocks were then pressed against the surface of blastula-stage embryos, so that a small number of blastomeres became marked with dye. By allowing marked embryos to develop and then examining them at intervals during gastrulation, researchers were able to follow the movement of cells.

Figure 21.14 summarizes the results of experiments with dyed blastomeres in frogs. The drawings identify blastomeres according to the type of embryonic tissue they will become. The cells at the animal pole are destined to become ectoderm and produce skin and nerve tissues; the cells between the animal and vegetal poles will develop into mesoderm and generate muscle and connective tissue; the cells at the vegetal pole will become endoderm and form the gut and associated organs. As Figure 21.14 shows, each of these three populations of cells moves into a new position during gastrulation, forming layers.

Although the gastrulation pattern varies among species almost as much as do cleavage patterns, certain features are shared:

- In most species, gastrulation begins when an invagination appears in the blastula, similar to the indentation that forms when you push a finger into a balloon. In frogs this invagination becomes round and is called the **blastopore**; it appears about two-thirds of the way between the embryo's animal and vegetal poles and surrounds a plug of yolk cells.

- Cells from the periphery move to the interior of the embryo through the blastopore.

- By the end of gastrulation, (1) the three embryonic tissues are arranged in layers, (2) the gut has formed, and (3) the major body axes have become visible. In frogs the blastopore becomes the anus, and the region just above the blastopore (as drawn in Figure 21.14) becomes the dorsal side of the embryo. In this way, the head-to-tail and back-to-belly axes of the body (discussed next) become apparent.

The discovery of cytoplasmic determinants and cell-to-cell signals clarified how blastomeres become differentiated into the three embryonic tissues. Current research on gastrulation is focused on understanding how cells move in such an organized way. But what about the other major result of gastrulation—the formation of the major body axes?

Organizing the Major Body Axes

Formation of the major body axes is among the most important events in all of development. There are three of these axes, illustrated in **Figure 21.15**:

1. One axis runs **anterior** (toward the head) to **posterior** (toward the tail);

2. One axis runs **ventral** (toward the belly) to **dorsal** (toward the back);

3. One axis runs **proximal** (toward the center of the body) to **distal** (away from the center of the body).

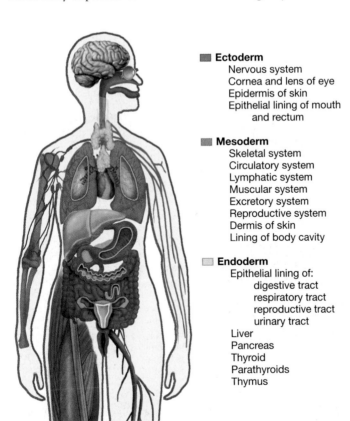

■ **Ectoderm**
  Nervous system
  Cornea and lens of eye
  Epidermis of skin
  Epithelial lining of mouth
    and rectum

■ **Mesoderm**
  Skeletal system
  Circulatory system
  Lymphatic system
  Muscular system
  Excretory system
  Reproductive system
  Dermis of skin
  Lining of body cavity

■ **Endoderm**
  Epithelial lining of:
    digestive tract
    respiratory tract
    reproductive tract
    urinary tract
  Liver
  Pancreas
  Thyroid
  Parathyroids
  Thymus

FIGURE 21.13 The Three Embryonic Tissues Give Rise to Different Adult Tissues and Organs
QUESTION In a gastrula, mesoderm is sandwiched between ectoderm on the outside of the embryo and endoderm on the inside. Relate this observation to the positions of adult tissues and organs derived from mesoderm, ectoderm, and endoderm.

DURING GASTRULATION, EMBRYONIC TISSUES FORM DISTINCT LAYERS.

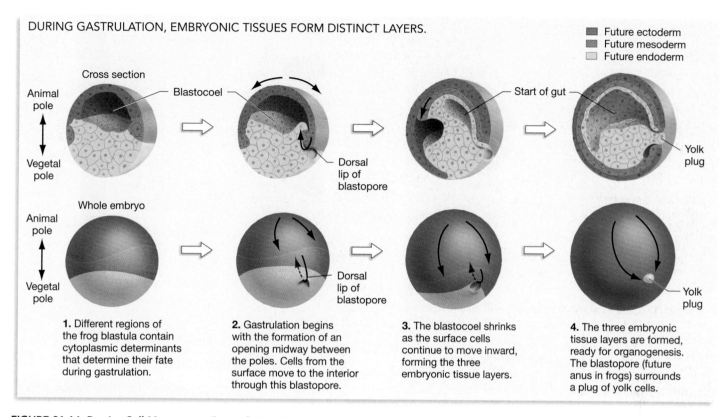

Future ectoderm
Future mesoderm
Future endoderm

Cross section

Animal pole — Blastocoel
Vegetal pole

Dorsal lip of blastopore

Start of gut

Yolk plug

Whole embryo

Animal pole
Vegetal pole

Dorsal lip of blastopore

Yolk plug

1. Different regions of the frog blastula contain cytoplasmic determinants that determine their fate during gastrulation.

2. Gastrulation begins with the formation of an opening midway between the poles. Cells from the surface move to the interior through this blastopore.

3. The blastocoel shrinks as the surface cells continue to move inward, forming the three embryonic tissue layers.

4. The three embryonic tissue layers are formed, ready for organogenesis. The blastopore (future anus in frogs) surrounds a plug of yolk cells.

FIGURE 21.14 Precise Cell Movements Occur during Gastrulation
EXERCISE In frogs, the initial opening of the blastopore is on what will become the dorsal side of the larva and adult. The blastopore becomes the anus of the larva and adult. On the last drawing, mark where the head, tail, back, and belly will develop.

How are these axes established so that the head, tail, back, and belly of the embryo develop in the correct relationship to each other?

The Spemann-Mangold Experiments

In 1924 Hans Spemann and Hilde Mangold published the results of a dramatic experiment on newt and salamander embryos. These researchers wanted to understand how the body axes of the embryo are established. From fate-mapping studies, they knew that the cells on the animal-pole side of the blasto-

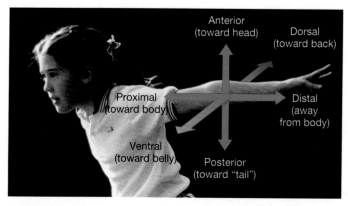

Anterior (toward head)
Dorsal (toward back)
Proximal (toward body)
Distal (away from body)
Ventral (toward belly)
Posterior (toward "tail")

FIGURE 21.15 Most Animals Have Three Major Body Axes

pore became part of the animal's back. Spemann and Mangold predicted that if these cells were designated prior to gastrulation to become back tissue, then the cells should still form back tissue if they were transplanted to a different location in the embryo.

To test this prediction, the investigators transplanted embryonic tissues between embryos of two differently pigmented species of newt. The color differences allowed them to follow the fate of the transplanted cells. For example, **Figure 21.16** (page 460) illustrates an experiment in which they took cells from the dorsal side of a blastopore in a nonpigmented embryo and added those cells to the ventral side of a pigmented embryo. Much to the researchers' amazement, the transplanted cells did not just become back tissue. Instead, a second, "twinned" embryo developed that was joined to the normal host embryo. The second embryo contained both pigmented and nonpigmented cells.

To interpret this result, Spemann and Mangold reasoned that the transplanted cells must be able to alter the fates of host tissues. For example, cells that would otherwise have become part of the belly become part of the back, in response to the transplanted cells. The logic was that, because the second embryo contained both donor and host cells, the transplanted cells must have recruited host cells to form part of the second embryo. Later, Spemann began using the term **organizer** to refer to

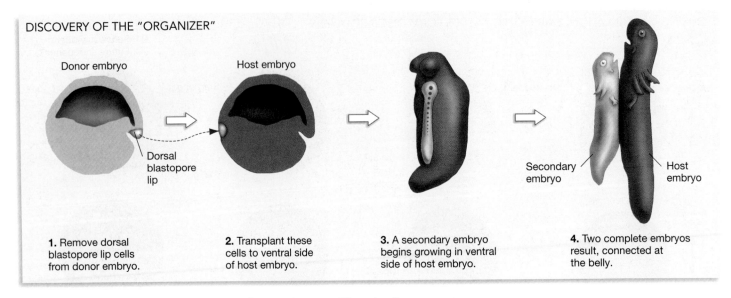

DISCOVERY OF THE "ORGANIZER"

Donor embryo

Host embryo

Dorsal blastopore lip

Secondary embryo

Host embryo

1. Remove dorsal blastopore lip cells from donor embryo.

2. Transplant these cells to ventral side of host embryo.

3. A secondary embryo begins growing in ventral side of host embryo.

4. Two complete embryos result, connected at the belly.

FIGURE 21.16 Cells from the Blastopore Lip of Newts Act as an "Organizer"
QUESTION Why do the twinned embryos end up belly to belly? Why do they have the same head-to-tail orientation?

the cells that do the recruiting, which come from a specific region on the embryo's dorsal side. The name was logical because these cells have the ability to organize host and donor tissues into a second embryo. Spemann also coined the term *induction* to describe the ability of organizer cells to direct the development of other cells. His idea was that the organizer somehow induces the formation of a second embryo.

The organizer appeared to be the key to the movement of cells during gastrulation and the formation of the body axes. Further research has demonstrated that embryos from other vertebrate species have an organizer-like region where gastrulation is initiated. How does this group of cells form, and how does it function?

Forming the Organizer   An unusual observation about frog eggs provided an initial clue about the formation of the organizer. Just after fertilization, a region of gray cytoplasm appears in the zygotes of some frog species opposite the point of sperm entry. Because of its shape and color, the region is called the **gray crescent**. Several lines of evidence indicate that the gray crescent corresponds to the region where the organizer forms. For example, fate-mapping studies showed that cells derived from this region become the area where the organizer forms. In addition, Spemann found that when he artificially separated blastomeres at the two-cell stage, blastomeres that lacked the gray crescent developed as a mass of unorganized tissues. In contrast, blastomeres containing the gray crescent developed normally.

How does fertilization lead to the formation of the gray crescent and the organizer? Investigators have shown that, in frogs, fertilization triggers a dramatic rearrangement of the egg cytoplasm. When the cytoplasm moves, a protein called β-catenin in the gray crescent region is activated. Recent work suggests

that β-catenin works in conjunction with the Vg1 protein to stimulate the transcription of organizer-specific genes. What are these genes?

Molecules Involved in Organizer Function   Cells in the organizer can induce changes in other cells. Based on this observation, biologists hypothesized that cells in the organizer must produce and secrete one or more cell-to-cell signals. Several teams of biologists have been actively searching for the molecules involved in the organizer function and the genes that encode them. This research is driven by the hope that the molecules that are active in frogs will lead to the discovery of similar molecules active in the development of other species with organizer-like regions—including humans.

In frogs, the most common experimental strategy for identifying important organizer molecules is to inject blastulas with particular compounds and observe the response. For example, one research team began a study by isolating the mRNAs found in the organizer regions of frog gastrulas. Then the researchers injected these candidate mRNAs one by one into embryos that had been irradiated with ultraviolet light. This was a clever experimental strategy, because researchers had found that UV light inhibits the egg's cytoplasm from rotating at fertilization. As a result, irradiated embryos do not normally form a gray crescent or organizer. The researchers' goal was to find an mRNA that could restore the ability of these eggs to form an organizer and undergo gastrulation.

Among the many mRNAs that they tested, the biologists found one that resulted in the formation of a complete embryo. This mRNA appeared to code for a protein that is an organizer product (**Figure 21.17**). To support this hypothesis, the researchers used in situ hybridization techniques (see Box 21.1) to locate where the mRNA is produced in intact embryos. As predicted, the mRNA was found only in the organizer region. The investigators also found that if they injected an excess of

this mRNA, the treated embryos developed as an enlarged head region with no trunk or tail. As a result, the biologists named the corresponding gene *noggin*. They hypothesized that the noggin protein functions in setting up the major body axes of the embryo. Follow-up experiments have shown that noggin works by binding to signaling molecules called *BMPs* and activating them.

Similar studies have identified other signaling molecules produced by the organizer region. Currently, research is focused on understanding exactly how noggin and these other organizer products act on target cells during and after gastrulation. Noggin and other organizer products are among the most important molecules in all of development. In addition to establishing the major body axes, they set in motion a chain of events that leads to the formation of specialized cells and their organization into tissues and organs.

✓CHECK YOUR UNDERSTANDING

Cleavage results in a sphere of cells that then undergo a massive rearrangement during gastrulation. At the end of gastrulation, the three embryonic tissues are arranged in layers and the major body axes are established. You should be able to give an example of how cytoplasmic determinants or cell-to-cell signals direct the pattern of cleavage, the fate of the blastomeres that are created by cleavage, and the location of the initial invagination in gastrulation.

21.4 Patterns of Development

Fertilization, cleavage, and gastrulation produce a multicellular embryo that is organized into several distinct embryonic tissues. Before we go on to explore additional details on how the major body axes are established and how cells become differentiated into specialized types, it will be helpful to review how the overall pattern of development varies among species.

To step back and sample the diversity of developmental patterns observed in plants and animals, let's consider embryonic development in a plant, an invertebrate, and a couple of vertebrates. More specifically, we'll look briefly at the earliest events in the life of the wild mustard plant *Arabidopsis thaliana*, the fruit fly *Drosophila melanogaster*, frogs, and humans. Two vertebrates are featured to illustrate the differences in developmental patterns between species that lay eggs (represented by frogs) and those that give live birth (represented by humans).

Arabidopsis thaliana, D. melanogaster, and frogs such as *Xenopus laevis* are among the most important model organisms in developmental biology. These species are relatively easy to rear in the laboratory, and the genetic makeup of individuals can be manipulated through selective breeding or recombinant DNA technology. Experimental results from these species have provided insight into how development works in organisms that are difficult to study experimentally, including humans.

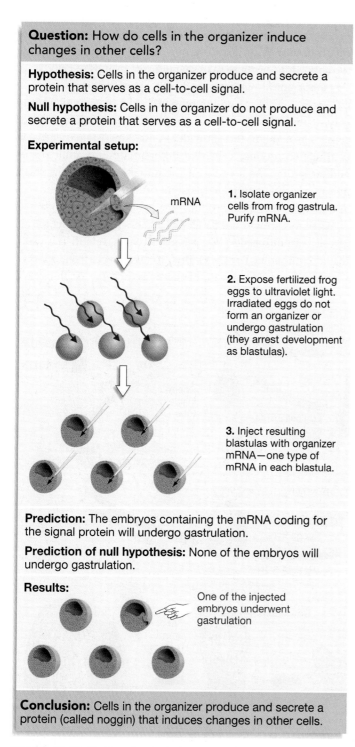

Question: How do cells in the organizer induce changes in other cells?

Hypothesis: Cells in the organizer produce and secrete a protein that serves as a cell-to-cell signal.

Null hypothesis: Cells in the organizer do not produce and secrete a protein that serves as a cell-to-cell signal.

Experimental setup:

mRNA

1. Isolate organizer cells from frog gastrula. Purify mRNA.

2. Expose fertilized frog eggs to ultraviolet light. Irradiated eggs do not form an organizer or undergo gastrulation (they arrest development as blastulas).

3. Inject resulting blastulas with organizer mRNA—one type of mRNA in each blastula.

Prediction: The embryos containing the mRNA coding for the signal protein will undergo gastrulation.

Prediction of null hypothesis: None of the embryos will undergo gastrulation.

Results:

One of the injected embryos underwent gastrulation

Conclusion: Cells in the organizer produce and secrete a protein (called noggin) that induces changes in other cells.

FIGURE 21.17 Finding the Organizer Product

QUESTION Cells that receive cell-to-cell signals secreted by the organizer give rise to a wide array of tissues and structures along the dorsal side of an individual. Do you agree with the hypothesis that the organizer produces a single product? Explain why or why not.

Embryonic Development in *Arabidopsis thaliana*

Arabidopsis thaliana is a flowering plant whose short life span and ease of culture has made it a favorite experimental subject for biologists interested in development and genetics (**Figure 21.18a**). This plant's entire life cycle can take just 6 weeks under laboratory conditions. The cycle begins with gametogenesis (**Figure 21.18b**). Sperm formation occurs in pollen grains, while egg cell formation occurs inside the ovule, within the flower.

After fertilization, the zygote undergoes mitosis and its daughter cells continue dividing mitotically. As a result of these early cell divisions, an embryo begins to develop inside the ovule. As embryogenesis continues, the ovule's covering develops into a **seed coat**, a protective layer that encases both the embryo and the endosperm. The mature seeds are usually dispersed away from the parent plant by wind, an animal, or water. When the seed germinates, the seed coat breaks and the embryo emerges to become a **seedling**. Mature embryos and seedlings have three prominent parts: a **root**, the belowground portion that anchors the plant and absorbs nutrients; initial leaves called **cotyledons**; and a stem-like structure called the **hypocotyl**, which joins the roots to the cotyledons (see Figure 21.18b). Together, the leaves and stem—the aboveground portion of the plant—making up the **shoot**.

How do these embryonic structures form? As **Figure 21.18c** shows, the cells that result from the fertilized egg's first division are asymmetrical in size, orientation, and their eventual function or fate. The bottom, or *basal cell*, is large and gives rise to a column of cells called the *suspensor*, which anchors the embryo as it develops. The small cell above the basal cell, called the *apical cell*, in contrast, is the progenitor of the mature embryo. Initially it gives rise to a simple ball of cells at the tip of the suspensor. As this group of cells grows and develops, the cotyledons begin to take shape. Later, groups of cells called the **shoot apical meristem** and **root apical meristem** form. A **meristem** consists of undifferentiated cells that divide repeatedly, with some of their daughter cells becoming specialized cells. Throughout the individual's life, meristematic tissues continue to produce cells that differentiate into adult tissues and structures. As a result of the initial growth of the shoot apical meristem and root apical meristem, the embryonic root and hypocotyl elongate to form the main axis of the body.

All of this growth and development takes place without the aid of cell rearrangements. Because plant cells have stiff cell walls, they do not move. For the embryo to take shape, then, cell divisions must occur in precise orientations, and the resulting cells

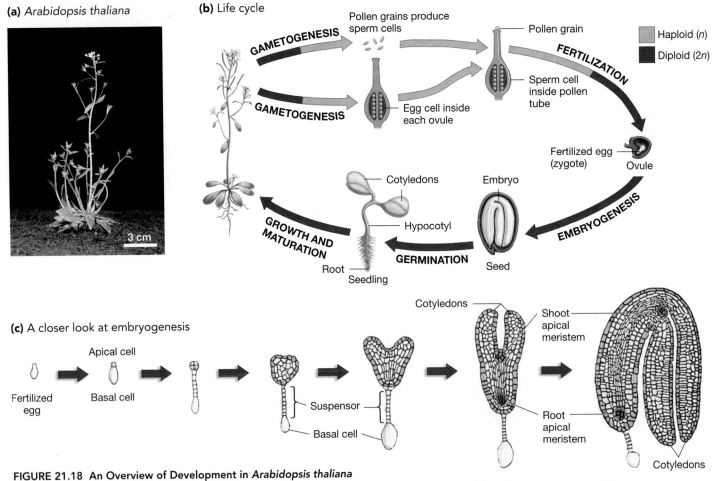

FIGURE 21.18 An Overview of Development in *Arabidopsis thaliana*
(a) *Arabidopsis thaliana* is an important model organism in developmental biology and genetics. (b) Its life cycle can take as little as 6 weeks. Fertilization and early embryonic development take place inside the ovule, which matures into a seed. (c) During embryogenesis, the root and shoot apical meristems form in addition to the first leaves (cotyledons), the root, and a hypocotyl.

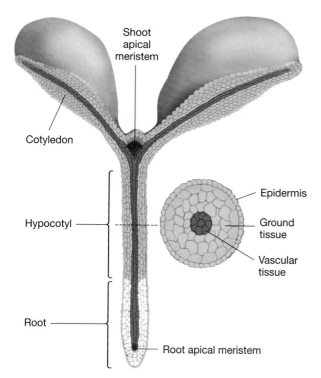

FIGURE 21.19 *Arabidopsis* **Has Three Embryonic Tissues**
Embryogenesis in plants produces three embryonic tissues: epidermis, ground tissue, and vascular tissue, organized in a radial fashion.

must exhibit differential growth—meaning that some cells grow larger than others. In contrast, animal embryos take shape largely as the result of directed cell movements during gastrulation, with each cell approximately the same size as its neighbors.

The root, hypocotyl, and cotyledons are organs that form along the main axis of the embryo (**Figure 21.19**). The cross-sectional view in the figure shows that early development in *A. thaliana* also produces three embryonic tissues. Note that an

embryo contains an **epidermis**—an outer covering of cells, called *epidermal cells*, that protect the individual. Inside this layer of cells is **ground tissue**, a mass of tissue that may later differentiate into cells that are specialized for photosynthesis, food storage, or other functions. The **vascular tissue** in the center of the plant will eventually differentiate into specialized cells that transport food and water between root and shoot. In the embryo, the three tissue systems are arranged in a radial pattern.

Although this pattern of development is typical of many plants, the sequence of events and the structures and tissues that form are very different from those observed during the development of animals. These differences are not surprising, given that both the fossil record and evolutionary trees show that plants and animals evolved from different single-celled ancestors. Stated another way, multicellular bodies evolved independently in plants and animals. As a result, their developmental pathways are distinct. Let's begin an analysis of animal development by looking at what may be the best studied of all multicellular organisms—the fruit fly.

An Invertebrate Model: *Drosophila melanogaster*

In the fruit fly, *Drosophila melanogaster*, gametogenesis leads to the production of sperm and eggs in different individuals—not the same individual, as in *Arabidopsis*. But fertilization occurs inside female reproductive structures, as it does in *Arabidopsis*. A female fly then lays the fertilized eggs in rotting fruit or a similar medium, so they develop outside her body. (In contrast, recall that in *Arabidopsis* the early stages of embryo formation occur while the seed is maturing and still attached to the parent plant.)

As **Figure 21.20** shows, the nucleus of a fertilized fly egg repeatedly undergoes mitosis without cytokinesis occurring. The result is an embryo with many nuclei scattered throughout a

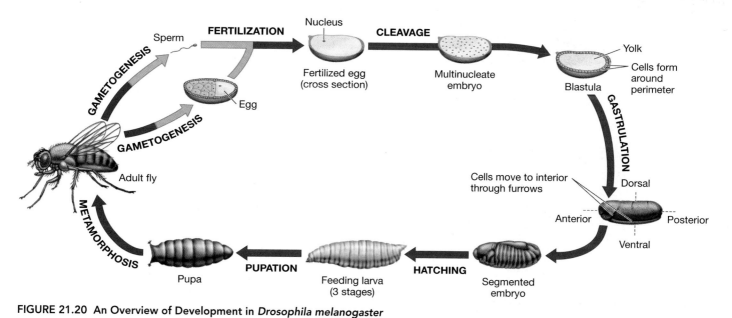

FIGURE 21.20 An Overview of Development in *Drosophila melanogaster*
When fruit-fly embryogenesis is complete, the segmented embryo hatches into a feeding larva. After feeding for several days, the larva stops moving, secretes a case, and becomes a pupa. The entire body is remodeled into the adult form via metamorphosis.

cytoplasm filled with nutrient-rich yolk. At the blastula stage of development, most nuclei migrate to the periphery of the embryo and become surrounded by a plasma membrane.

Gastrulation begins when a long furrow forms on one side of the embryo, and cells on the periphery begin to move through the furrow to the interior. As gastrulation continues, another furrow forms and defines the head region; soon after, furrows form along the length of the body and demarcate the series of body regions called **segments**. A segment is a region of the body that is repeated. Segments contain similar structures and have similar functions.

When gastrulation is complete and the larva matures, the larva has a functioning mouth and digestive tract and begins feeding on rotting fruit or other food sources. Then the fly undergoes **pupation**, the secretion of a hard case around the larva. The result is the formation of a **pupa**, a nonfeeding stage in which the fly's body is remodeled to form an adult. This remodeling process is called **metamorphosis** ("change-form"). It is important to appreciate how remarkably different the larval and adult forms are in many species that undergo metamorphosis. Contrast a maggot and a fly or a caterpillar and a butterfly. In these species, juvenile and adult forms have completely different bodies, live in different habitats, and exploit different food sources.

As **Figure 21.21** indicates, there is a one-to-one correspondence between the segments that appear in the larva and the segments found in the adult. The segments that form in fruit flies are grouped into regions called the *head, thorax,* and *abdomen*. Despite the massive changes in shape and tissue organization that occur during metamorphosis, segmentation is

fundamental to the organization of both the larval body and the adult body.

Finally, it is important to note that following metamorphosis, gastrulation is responsible for segregating cells into the three embryonic tissues of ectoderm, mesoderm, and endoderm.

A Vertebrate Model: The Frog

How does fly development compare with the development of vertebrates such as the frog *Xenopus laevis*? Although the fossil record indicates that fruit flies and frogs last shared a common ancestor more than 600 million years ago, some features of fly and frog development are similar. Features such as asymmetry, segmentation, and three embryonic tissues are also found in frog embryos. These are remarkable observations. Even though many features of development vary among animal species, certain aspects have been highly conserved during evolution. Recent research has shown that many of the underlying genetic mechanisms responsible for development are also shared among some species.

Figure 21.22 shows some prominent stages in the development of a frog. The result of gastrulation in the frog is fundamentally similar to the result of gastrulation in the fly, even though there are many differences in the outward appearance of the embryos produced by these two species. Embryonic tissues separate into distinct layers or compartments, and the major body axes take shape. Notice that as gastrulation ends, a structure called the neural tube begins to form along the embryo's back. The **neural tube** will eventually become the spinal cord and brain of the larva and adult frog. In this way, the dorsal-to-ventral (back-to-belly) axis of the body takes shape. The formation of the neural tube is the key event in preparation for the developmental stage called **organogenesis**, which occurs after gastrulation and results in the formation of specialized cells, tissues, and organs. (An **organ** is a group of tissues that are integrated into a structure with a specific function, such as the heart or lungs.) At this stage, the frog embryo is called a *neurula*. The embryo will hatch into a larva, known as a *tadpole*, which will undergo metamorphosis to become an adult frog.

Frogs, like flies, lay their eggs so that development takes place in the surrounding environment. But in some lizards, sharks, and fish—and in the vast majority of mammals, including humans—early development takes place inside the mother's body. Despite this difference, the early stages of human development are also marked by cleavage, gastrulation, organogenesis, and the formation of body segments.

Early Development in Humans

Like fruit flies and frogs, humans have separate sexes. Gametogenesis in males results in the production of the tiny, motile gametes called sperm; the analogous process in females results in larger, nonmotile eggs. As **Figure 21.23** shows, fertilization and cleavage occur inside the reproductive tract of human mothers, in the fallopian tube. The **fallopian tube** is a structure that connects the ovary and the uterus. The **ovary** is the organ in which the egg matures; release of the egg from the ovary is known as

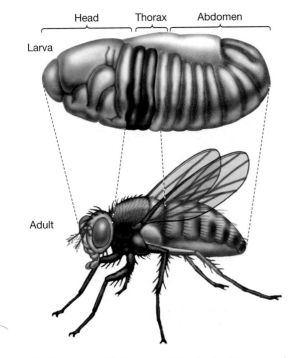

FIGURE 21.21 *Drosophila* **Larvae and Adults Are Segmented**
The segmented body of insects is organized into head, thorax, and abdomen regions.

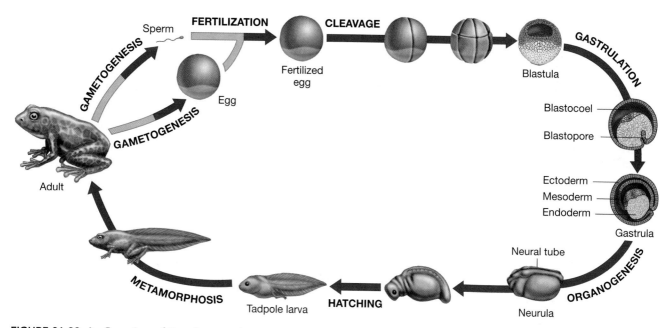

FIGURE 21.22 An Overview of Development in a Frog
In vertebrates, early embryonic development consists of cleavage, gastrulation, and organogenesis. Frogs hatch into a larval stage called a tadpole. Although frogs undergo metamorphosis, they do not form a pupa.

ovulation. The **uterus** is the organ in which the embryo develops. Cleavage occurs as the embryo travels down the length of the fallopian tube.

When cleavage is complete, the embryo, now known as a *blastocyst*, breaks out or hatches from the zona pellucida and implants into the uterine wall. Once this process, called **implantation**, occurs, the placenta begins to form. The **placenta** is an organ derived from a mixture of maternal and embryonic cells. It allows nutrients and wastes to be exchanged between the mother's blood and the embryo's blood. For this reason, human eggs are small compared with the eggs of frogs and largely lack

a source of nutrients such as the endosperm of seed plants and the yolk of fly or frog eggs. Gastrulation and the formation of the neural tube occur after the embryo has become implanted into the uterine wall. Later events in pregnancy and birth are detailed in Chapter 48.

Making Sense of Developmental Variation and Similarity

Early descriptive studies of plant and animal development established that the sequence of events varies a great deal among species. At first glance, it seems puzzling that so much variation

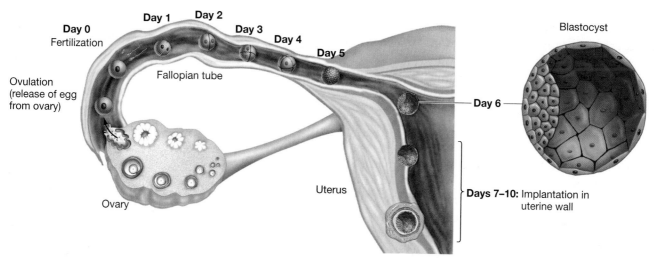

FIGURE 21.23 In Humans, Cleavage Occurs before Implantation into the Uterus
After the blastocyst produced by cleavage is implanted in the uterine wall, gastrulation and organogenesis occur, and the placenta begins to form from a combination of embryonic and maternal cells.

exists in processes as fundamental as gametogenesis, fertilization, and early development. Why do so many differences exist?

Part of the variation observed in developmental patterns is undoubtedly due to natural selection. Recall from Chapter 1 that natural selection occurs when individuals with certain genetically based traits produce more offspring than do individuals without those traits. In the case of development, natural selection can favor certain patterns or events over others. For example, the evolution of larval forms in groups such as insects and frogs was probably favored by natural selection because it allowed juveniles and adults of the same species to exploit different food sources and thus avoid competing with each other. The evolution of the placenta and pregnancy is thought to have been advantageous because of the protection they offer the developing embryo. Similarly, the seed is seen as a trait that protects the embryo during early development and allows it to be dispersed to a location away from the parent plant prior to germination.

Amid this variation, though, several common themes emerge about developmental patterns:

- In both plants and animals, the earliest events in development result in asymmetries in the embryo that establish fundamental aspects of the adult body plan. In plants, an asymmetrical initial division sets up the root-shoot body axis. In most animals, either cleavage or gastrulation makes the head-to-tail and back-to-belly axes of the embryonic and adult bodies visible.

- Although the details of how cleavage occurs vary among animal species, the outcome is the same. Cleavage divides the maternal cytoplasm, produces blastomeres that contain specific cytoplasmic determinants and other maternal factors, and creates a sphere of cells that undergoes gastrulation.

- In both plants and animals, early development results in the formation of three embryonic tissues, which give rise to all the different adult tissues and structures. In animals, gastrulation is responsible for organizing these three populations of cells into distinct layers.

ESSAY   Treating Human Infertility

In 1978 the world welcomed the first "test-tube baby" when Louise Brown was born in Britain. Unlike babies born previously, Louise did not develop from a zygote produced by natural conception. Instead, she developed from a zygote created by fertilization in a laboratory dish. Since then, over 100,000 babies have been produced by the technique called **in vitro fertilization (IVF)**.

Many human couples are **infertile**—they are unable to conceive children naturally. For them, IVF can be an important alternative strategy for becoming pregnant. In IVF, egg cells isolated from the woman's ovaries are combined with sperm cells collected from the man's semen.

The gametes are placed in small dishes, under conditions that attempt to duplicate the environment inside the female reproductive tract. Typically, 50,000 to 5 million sperm are used per egg. As in natural conception, fertilization occurs when a healthy sperm penetrates the zona pellucida and fuses with the egg. The resulting embryos are then surgically implanted in the mother's uterus.

IVF is often successful for couples whose infertility results from problems in the woman's reproductive system, such as damaged or absent fallopian tubes. But for up to 40 percent of infertile couples, the problem stems from low sperm counts, poor sperm motility, deformed sperm, or other male problems. Traditional IVF seldom works with such couples, because it requires large numbers of high-quality sperm. In the early 1990s, though, improvements to IVF were introduced that provided new hope for infertile men. In one technique, called **subzonal injection,** or SUZI for short, a concentrated sample of sperm is examined under a microscope. Five to 10 of the healthiest sperm are drawn into a microneedle and then injected directly into the zona pellucida. Depositing the sperm closer to the egg increases the chances of sperm-egg fusion. SUZI may not help men who produce very few sperm, however, and it leads to polyspermy and inviable embryos in about one-third of attempts. A more recent technique called **intracytoplasmic sperm injection** (ICSI, pronounced "*ICK-see*") overcomes these problems. In ICSI, a single sperm is drawn into a microneedle and injected into the cytoplasm of the egg (**Figure 21.24**). In a significant fraction of attempts, ICSI results in the fusion of sperm and egg nuclei and a viable embryo.

Even if IVF, SUZI, or ICSI is successful, only a minority of couples achieve pregnancy when an embryo is transferred to the uterus.

Even if IVF, SUZI, or ICSI is successful, only a minority of couples achieve pregnancy when an embryo is transferred to the uterus. In some cases, failure results from problems with implantation in the uterine wall. To develop in the mother's womb, the embryo must break out or hatch from the zona pellucida and implant into the uterine wall. Eggs from women over age 40 often have problems hatching. To help overcome these problems, a physician may assist hatching by using a micropipette to deliver a small amount of acid or enzyme solution to the zona pellucida. This treatment facilitates hatching by dissolving the treated region of the zona.

Because of IVF and other assisted reproductive technologies, the chances of naturally infertile couples conceiving their own children have greatly increased. An intensive research program undertaken over the past three decades has led to huge advances in our understanding of human fertilization and early development, and has allowed many couples to experience the joy of raising their own children.

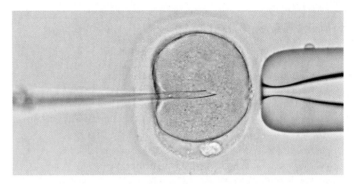

FIGURE 21.24 Human Sperm Can Be Injected Directly into an Egg
A human egg undergoing intracytoplasmic sperm injection (ICSI).

CHAPTER REVIEW

Summary of Key Concepts

◼ **In most animals, fertilization depends on specific interactions among proteins on the plasma membranes of sperm and egg.**

The development of an organism begins with the formation of gametes in its parents. In plants and animals, sperm cells contribute a haploid genome to the embryo; depending on the species involved, the sperm cell may also contribute a few other cell components. Eggs, in contrast, contribute a haploid genome and a large amount of cytoplasm to the embryo.

Fertilization is the best known of all cell-to-cell interactions, primarily because of observational and experimental studies on sea urchin eggs and sperm. When a sea urchin sperm contacts the jelly layer surrounding the egg, the acrosome in the sperm head releases digestive enzymes, and microfilaments polymerize to form a protrusion that extends to the vitelline envelope. When the sperm reaches the egg, a protein called bindin on the sperm-head membrane binds to a receptor on the vitelline envelope surrounding the egg. Because bindin and egg-cell receptors are species specific, cross-species fertilizations are prevented. After the sperm and egg cells' plasma membranes make contact and fuse, a wave of calcium ions is released from stores inside the egg. The increase in Ca^{2+} concentration causes cortical granules to fuse with the egg cell's plasma membrane. The contents of the cortical granules cause a fertilization envelope to lift off the egg cell's plasma membrane and protect against multiple fertilizations.

The fertilization sequence is similar in mammals, except that no bindin-like protein has yet been identified in sperm. Plants, in contrast, undergo a very different series of events. In flowering plants, two sperm nuclei leave the germinating pollen grain and migrate to the ovule. One sperm nucleus fertilizes the egg to form a zygote, while the other sperm nucleus fuses with two maternal nuclei to form the nutritive tissue called endosperm.

◼ **In animals, the earliest cell divisions divide the fertilized egg into a mass of cells. These cells are not all alike; eventually they give rise to different types of cells and tissues as development proceeds.**

Animal development begins with cleavage, a series of cell divisions that divide the egg cytoplasm into a large number of cells. During cleavage, an array of cytoplasmic determinants and other maternal factors are apportioned to different blastomeres. As a result, different blastomeres have different fates.

Once cleavage is complete, the embryo consists of a sphere of cells. Fate-mapping studies with dyes and other tools have shown that cells undergo massive movements during gastrulation. These movements arrange the three embryonic tissues in layers and make the back-to-belly and head-to-tail axes of the body visible. In vertebrates, a region of cells near the dorsal lip of the blastopore acts as an organizer during gastrulation. Cells in the organizer secrete the protein noggin and other cell-to-cell signals that induce changes in target cells.

Patterns and rates of embryonic development vary widely among species of plants and animals. Cleavage and gastrulation are nearly universal in animals, but gastrulation does not occur at all in plants. Among animals, some species develop into distinct larval and adult stages, while others do not. The time to adulthood may be extremely fast, as in *Arabidopsis thaliana* or *Drosophila melanogaster*, or slow, as in humans.

Web Tutorial 21.1 Early Stages of Animal Development

◼ **In animals as well as plants, the early events in development result in the production of three embryonic tissues and the establishment of the major body axes.**

Common themes unify the diversity of ways that plants and animals develop. For example, in both plants and animals, the earliest events in development make the major body axes visible and result in the formation of three types of embryonic tissues, which later give rise to distinct larval or adult structures.

Web Tutorial 21.2 The Gray Crescent in Frog Eggs

Questions

Content Review

1. How are the vitelline envelope of sea urchins and the zona pellucida of mammals similar?
   a. Both are gelatinous coats that protect the egg.
   b. Both hold cortical granules, which block polyspermy.
   c. Both hold stores of Ca^{2+}, which block polyspermy.
   d. Both are an extracellular matrix that sperm bind to and digest.

2. What happens during the acrosomal reaction?
   a. Bindin binds to the egg-cell receptor for sperm.
   b. The sperm and egg cells' plasma membranes fuse.
   c. Enzymes that digest the egg's jelly layer are released, and microfilaments in the tip of the sperm head polymerize to form a protrusion.
   d. The centriole released from the sperm orients microtubules in the fertilized egg and causes the cytoplasm to rotate—creating the gray crescent.

3. Many flowering plant species have elaborate mechanisms to prevent an individual's pollen from germinating on its own female reproductive parts. Why?
   a. to prevent self-fertilization and inbreeding
   b. to prevent polyspermy
   c. to prevent cross-species fertilization and the production of dysfunctional hybrid offspring
   d. to prevent double fertilization and the formation of endosperm

4. What happens during cleavage?
   a. The neural tube—precursor of the spinal cord and brain—forms.
   b. Basal and apical cells—precursors of the suspensor and embryo, respectively—form.
   c. The fertilized egg divides without growth occurring, forming a ball of cells.
   d. Massive movements of cells make the primary body axes visible and organize the three embryonic tissues.

5. What happens during gastrulation?
   a. The neural tube—precursor of the spinal cord and brain—forms.
   b. Basal and apical cells—precursors of the suspensor and embryo, respectively—form.
   c. The fertilized egg divides without growth occurring, forming a ball of cells.
   d. Massive movements of cells make the primary body axes visible and organize the three embryonic tissues.

6. In animals, which adult tissues and organs are ectoderm-derived?
   a. lining of the digestive tract and associated organs
   b. blood, heart, kidney, bone, and muscle
   c. nerve cells and skin
   d. blastopore and blastocoel

Conceptual Review

1. List the contributions to offspring that are made by a sperm versus an egg. Given a sample of gametes from a new species, how would you know which gametes were eggs and which were sperm?

2. Why is it logical that sperm-egg interactions are species specific in sea urchins but not in mammals?

3. Compare and contrast yolk and endosperm. How are they similar in structure and function, and how are they different? Be sure to compare when and how they are produced.

4. How did analyses of cleavage and shell-coiling patterns in snails support the hypothesis that cleavage is controlled by cytoplasmic determinants present in the egg? Does the observation that transcription does not start in some animals until cleavage is complete support this hypothesis or challenge it? Explain.

5. Compare and contrast cytoplasmic determinants and cell-to-cell signals. Explain why biologists would say that *Vg1* is a maternal factor that acts as a cell-to-cell signal.

6. What does it mean to say that a cell has a certain fate? Why do some cells in early embryos have different fates than other cells? Support your answer with specific examples.

Group Discussion Problems

1. Many questions about fertilization and early development remain unanswered:
   - Is there a protein on the sperm head of mammals that binds specifically to the glycoprotein ZP3 in the zona pellucida?
   - How are cytoplasmic determinants and factors localized to specific regions of the egg? How is yolk localized?
   - If the zygotic genome is transcribed at the two-cell stage in mammals, do cytoplasmic determinants play a significant role in early development?
   - What forces cause cells to move during gastrulation?

   Choose one of these questions and design an experiment that would contribute to answering it.

2. Suppose you are a physician at a fertility clinic. A couple seeks your assistance with becoming pregnant. The woman is 41 years old, and the man has a low sperm count. Which treatment would you recommend, and why?

3. In a lizard species native to Mexico, some populations lay eggs; others bear live young. How could you use these populations to test the hypothesis that bearing live young is favored by natural selection in certain environments because offspring are better protected?

4. Suppose there is a drug that leads to direct development when administered to frog eggs—meaning that the larval (tadpole) stage is skipped. How could you use this drug as an experimental tool to test the hypothesis that natural selection favors the development of larval forms because of reduced competition with adults for food?

Answers to Multiple-Choice Questions 1. d; 2. c; 3. a; 4. c; 5. d; 6. c

Pattern Formation and Cell Differentiation

22

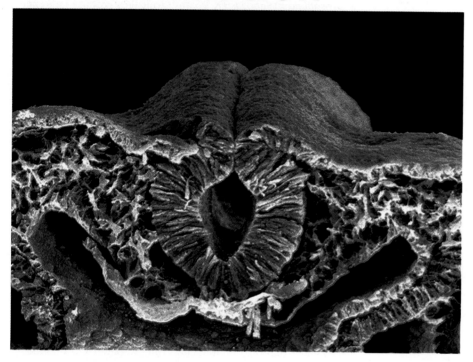

KEY CONCEPTS

▥ The fate of an embryonic cell—what it will become in the juvenile and adult organism—is established in a series of steps.

▥ At each step in development, cell-to-cell signals activate regulatory transcription factors that direct the expression of genes in cells throughout the embryo.

▥ Because virtually all cells in a multicellular organism contain the same genes, development is based on differential gene expression.

Artificially colored scanning electron micrograph of a chick embryo. Ectodermal cells along the back of the embryo (top of photograph) have folded into a neural tube. The neural tube will develop into the brain and spinal cord of the adult. This chapter explores how cells form specific structures based on their position in the embryo and then differentiate to form specialized cells such as nerve or muscle.

When the physician, researcher, and writer Lewis Thomas considered how a human being develops from a fertilized egg, he could only marvel: "You start out as a single cell derived from the coupling of a sperm and an egg, this divides into two, then four, then eight, and so on, and at a certain stage there emerges a single cell which will have as all its progeny the human brain. The mere existence of that cell should be one of the great astonishments of the earth. People ought to be walking around all day, all through their waking hours, calling to each other in endless wonderment, talking of nothing except that cell. It is an unbelievable thing, and yet there it is, popping neatly into its place amid the jumbled cells of every one of the several billion human embryos

around the planet, just as if it were the easiest thing in the world to do."[1]

Many biologists share Thomas's feelings. The experience of watching a fertilized egg go through early development inspires awe and wonder. In animals, the zygote divides during cleavage. The resulting cells then begin the massive, coordinated movements of gastrulation, and the embryo is dramatically reorganized. Just a few hours or days later, a recognizable creature with a head, eyes, back, belly, and other structures appears. In plants, the fertilized egg divides to form

[1]Thomas, L. 1979. *The Medusa and the Snail* (New York: Viking Press), p. 156. Note that Thomas was exercising some poetic license here. The brain actually arises from a group of cells in the embryo rather than from a single cell.

two simple-looking cells—one small and one large. The small cell divides repeatedly to form a seemingly disorganized clump of cells. But as this clump grows and lengthens, tiny leaves and a root appear. A mature seedling takes shape a short time later, ready to germinate and begin a life that may last a few weeks or several thousand years.

Understanding how these events happen is among the greatest of all challenges facing biologists today. Chapter 21 introduced this challenge by providing an overview of gametogenesis, cleavage, and gastrulation. It also considered key events in the development of *Arabidopsis*, *Drosophila*, frogs, and humans. As you learned in that discussion, the establishment of the major body axes and the formation of embryonic tissues are among the most important of all early developmental events. Chapter 21 also began exploring how these key events happen, by introducing the role of cytoplasmic determinants and cell-to-cell signals in establishing the fate of embryonic cells. Recall that cytoplasmic determinants are molecules that exist in eggs and that help direct early development.

This chapter begins with a look at how the overall body plan of an animal or plant takes shape. Section 22.1 introduces work on embryonic development in the fruit fly *Drosophila melanogaster*—the best studied of all multicellular organisms. Research on fruit flies has emphasized the importance of a cell's position in determining its fate. Position is important because a cell's fate is specified by localized cytoplasmic determinants and signals from other cells. The same themes carry over to Section 22.2, which examines recent research on the best studied of all plants, the mustard *Arabidopsis thaliana*.

Once the overall form of the body is established and cells are arranged in space, cells start becoming organized into tissues and organs. Eventually cells differentiate and begin to express tissue-specific genes. Section 22.3 explains how tissue-specific gene expression occurs by exploring research on vertebrates—specifically, the transformation of mesoderm into cells that express muscle-specific proteins.

The fundamental messages of this chapter are that patterns of development are established early in embryogenesis, that a single cell's fate during development depends on its position within the embryo, and that its fate is realized in a progression of steps—each mediated by specific regulatory proteins or cell-to-cell signals. For example, a cytoplasmic determinant may mark a blastomere as mesoderm. After gastrulation, other molecules may signal that the cell is now part of the mesoderm on the embryo's dorsal (back) side. Subsequent signals might indicate that this particular cell will become part of the dorsal mesoderm that contributes to muscle or bone. Finally, the cell may receive signals that trigger the production of muscle-specific proteins. A generalized embryonic cell has become a specialized mature cell. How does this process begin? If a cell's fate depends on its position, how does it "know" where it is in the body?

22.1   Pattern Formation in *Drosophila*

Biologists refer to the events that determine the spatial organization of an embryo as **pattern formation**. If a molecule signals that a target cell is in the embryo's head, or tail, or dorsal side, or ventral side, it is involved in pattern formation. Pattern formation is the first step in determining a cell's fate. Before a cell becomes part of a muscle or nerve or gut lining, it receives precise information about where it is positioned in the body.

To understand how cells get information that specifies their position inside the embryo, biologists turned to the fruit fly *Drosophila melanogaster*. Because fruit flies produce large numbers of offspring rapidly, researchers could survey laboratory populations for rare mutant embryos in which the normal spatial relationships among cells are disrupted. In other words, biologists took a genetic approach to studying development. The idea was that identifying individuals with abnormal pattern formation would make it possible to find the genes and gene products responsible for normal development. A *Drosophila* larva that lacks a head, for example, is likely to have a mutation in a gene that helps pattern the head. This approach has turned out to be extraordinarily productive. The genes that direct pattern formation in fruit flies also direct pattern formation in mice, humans, and other animals.

The effort to dissect pattern formation in *Drosophila* began in the 1970s, when Christiane Nüsslein-Volhard and Eric Wieschaus undertook a massive effort to identify pattern-formation mutants. They began by exposing adult flies to treatments that cause mutations by damaging DNA. Later, the researchers examined embryos or larvae descended from these individuals for body plan defects. After intensive effort, Nüsslein-Volhard and Wieschaus were able to identify over 100 genes that play fundamental roles in pattern formation.

The Discovery of *bicoid*

A normal fruit-fly embryo has three types of body segments: head (anterior), thoracic (middle), and abdominal (posterior) segments (**Figure 22.1a**). One of the most dramatic mutations that Nüsslein-Volhard and Wieschaus analyzed is illustrated in **Figure 22.1b**. Embryos with that mutation are missing all of the structures normally found in the anterior end. These mutants have duplicated posterior structures. The gene responsible for this phenotype is called *bicoid*, meaning "two tailed." The mutation is lethal, because larvae cannot develop beyond the stage shown in the figure. Based on its phenotype, Nüsslein-Volhard and Wieschaus suspected that the *bicoid* gene's product plays a role in pattern formation along the anterior-posterior body axis.

In a series of mating experiments with flies that carried the mutant *bicoid* allele, Nüsslein-Volhard and Wieschaus found that the gene displays **maternal effect inheritance**—a pattern of inheritance in which a gene is transcribed in the mother and affects the phenotype observed in offspring. As **Box 22.1** explains, it is the mother's genotype and not the offspring's genotype that determines the expression of the trait. Like the

(a) A normal fruit-fly embryo

(b) A *bicoid* mutant

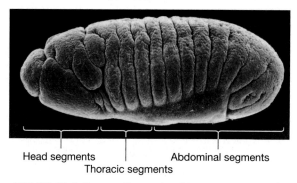

Head segments | Abdominal segments
Thoracic segments

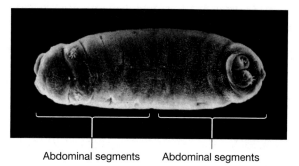

Abdominal segments   Abdominal segments

FIGURE 22.1 Pattern-Formation Mutants Have Misshapen Bodies
(a) Normal embryos have distinct head, thorax, and abdominal regions. **(b)** In *bicoid* mutants, embryos have no head or thoracic segments. Instead they have duplicated sets of the posteriormost regions, making them "two tailed."

BOX 22.1   Maternal Effect Inheritance

When researchers find a gene that appears to be important early in development, one of their first goals is to establish how the gene is inherited. Breeding experiments can answer a fundamental question about the gene: Is it expressed in the mother, meaning it codes for a cytoplasmic determinant present in the egg, or is it expressed in offspring? That is, does the gene have a maternal effect or a zygotic effect?

Crosses are set up to provide an answer. In one example, the mutant gene being considered is recessive. Investigators arrange matings between parents that are heterozygous at this locus. The two sets of crosses show the contrasting patterns of inheritance in a gene that is expressed in offspring (**Figure 22.2a**) versus a maternal effect gene (**Figure 22.2b**). The key point is that the zygotic effect hypothesis and the maternal effect hypothesis make contrasting predictions:

- *Zygotic effect hypothesis*: If the gene is expressed in offspring, then offspring will have the mutant phenotype if they are homozygous for the mutant gene—no matter what their mother's genotype.

- *Maternal effect hypothesis*: If the gene codes for a cytoplasmic determinant, then offspring will have the mutant phenotype only if their mother is homozygous for the mutant gene.

Researchers who performed this experiment with mutant and normal alleles of *bicoid* discovered that the second pattern was correct: *bicoid* displays maternal effect inheritance and has no zygotic effect.

(a) Case 1: Cross heterozygous parents; no maternal effect occurs.

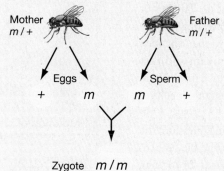

Mother *m / +*   Father *m / +*

Eggs   Sperm

+   *m*   *m*   +

Zygote *m / m*

Offspring are affected only if offspring are homozygous

Abnormal development, death

Offspring are affected only if mother is homozygous

(b) Case 2: Cross heterozygous parents; gene has maternal effect.

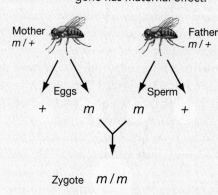

Mother *m / +*   Father *m / +*

Eggs   Sperm

+   *m*   *m*   +

Zygote *m / m*

Normal development

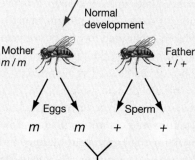

Mother *m / m*   Father *+ / +*

Eggs   Sperm

m   *m*   +   +

Zygote *m / +*

Abnormal development, death

FIGURE 22.2 Experimental Crosses Test the Hypothesis That a Gene Has a Maternal Effect
(a) The zygotic effect hypothesis predicts that offspring will have a mutant phenotype whenever the offspring are homozygous for the mutant allele.
(b) The maternal effect hypothesis predicts that offspring will have a mutant phenotype whenever the mother is homozygous for the mutant allele—no matter what the offspring's genotype is.

snail-shell-coiling gene introduced in Chapter 21, *bicoid* affects the development of an embryo because it is expressed in the mother's tissues during egg formation. The Bicoid protein is a cytoplasmic determinant. The *bicoid* mRNA is deposited in the egg along with other maternal determinants and yolk.

What Does *bicoid* Do?   To determine how the *bicoid* product works, Nüsslein-Volhard and colleagues crossed individuals carrying the mutant form of *bicoid* with individuals carrying other known genetic markers. This linkage mapping technique, introduced in Chapter 13 and explained in more detail in Chapter 19, allowed the team to map *bicoid*'s location on the *Drosophila* chromosomes. Once the location of the gene was established, the researchers were able to isolate and analyze the DNA sequence that encodes the *bicoid* product. Once they had many copies of the DNA sequence, they used the in situ hybridization technique described in Box 21.1 to locate *bicoid* mRNA in embryos and adults. The investigators added fluorescent or radioactive markers to single-stranded copies of *bicoid* DNA and then treated preserved *Drosophila* adults and embryos with the labeled copies. The probe bound to *bicoid* mRNA in cells where the gene was being transcribed.

When Nüsslein-Volhard's group treated adult flies with labeled copies of *bicoid* DNA, they found that the corresponding mRNA was located only in the mother's ovary—the part of the female reproductive tract where eggs are produced. This result was consistent with the conclusion from the breeding experiments that identified *bicoid* as a maternal effect gene. The idea is that the gene codes for a cytoplasmic determinant and that *bicoid* mRNA is loaded into eggs. According to data from in situ hybridization (**Figure 22.3a**), the mRNA is highly localized in one end of the developing egg.

What is the fate of this mRNA? To answer this question, the biologists set out to determine where and when the Bicoid protein is produced. As **Figure 22.3b** shows, they began by making the Bicoid protein from *bicoid* DNA (step 1). Using techniques introduced later in the text, they produced antibodies that bind specifically to the Bicoid protein (step 2). Recall that antibodies are proteins that bind to specific segments of a molecule. When researchers attach a fluorescent or radioactive compound to an antibody, it can be used as a labeled probe to mark the location of a specific protein (step 3). As Figure 22.3b shows, this antibody-staining experiment allowed researchers to document the location and quantity of Bicoid protein in embryos.

By doing the experiment outlined in Figure 22.3b at different stages of development, Nüsslein-Volhard and co-workers found that Bicoid protein first appears at fertilization. Further, their data indicated that the protein diffused away from the site of translation at the anterior end of the embryo. As the "Results" photograph shows, a steep concentration gradient is produced: The protein is abundant in the anterior end but declines to low concentrations in the posterior end. Recall from Chapter 21 that at this stage in fly development, nuclei

(a) Where is *bicoid* mRNA located?

In the fly ovary, *bicoid* mRNA is localized in the anterior end of the developing egg ☞

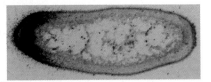

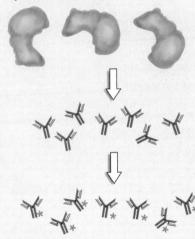

(b) Question: Is Bicoid protein as highly localized as *bicoid* mRNA?

Hypothesis: Bicoid protein is found only in the same location as *bicoid* mRNA.

Null hypothesis: Bicoid protein is not found in the same location as *bicoid* mRNA.

Experimental setup:

1. Make many copies of Bicoid protein from *bicoid* DNA.

2. Using Bicoid protein, make antibodies that bind specifically to Bicoid protein.

3. Attach a label to antibodies. Add labeled antibodies to embryo, where they will bind to Bicoid protein.

Prediction: Bicoid protein is found only at the anterior end.

Prediction of null hypothesis: Bicoid protein is not found at the anterior end.

Results:

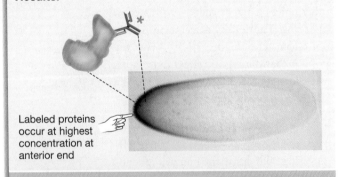

Labeled proteins occur at highest concentration at anterior end ☞

Conclusion: Both hypotheses are partially correct: Bicoid protein forms an anterior-posterior concentration gradient.

FIGURE 22.3 The *bicoid* mRNA Is Localized, and Bicoid Protein Forms a Concentration Gradient
(a) In situ hybridization shows that *bicoid* mRNA is sequestered in the anterior portion of the egg. **(b)** Bicoid protein forms an anterior-posterior concentration gradient.

are scattered throughout the embryo after repeated mitoses occur without cytokinesis. When these nuclei become surrounded by plasma membranes, the Bicoid protein is trapped in the newly formed cells. In other animal and plant species, cytoplasmic determinants like *bicoid* are localized in particular cells during cleavage.

To pull these observations together, Nüsslein-Volhard and co-workers hypothesized that high concentrations of Bicoid protein inside cells lead to the formation of anterior structures such as the head, with progressively lower concentrations giving rise to thoracic segments and the first abdominal segments. Absence of Bicoid, in contrast, results in formation of posterior structures. This hypothesis provided an explanation for the original *bicoid* mutants that the researchers had observed. Because the eggs of mothers homozygous for the mutant form of *bicoid* lack functional mRNAs, no Bicoid protein is produced anywhere in the embryo. The result is a larva in which anterior structures have been replaced by duplicated posterior structures.

Testing the Gradient Hypothesis The hypothesis that a Bicoid gradient sets up the head-to-tail axis of the fly embryo is consistent with the data from antibody-staining experiments. To test the hypothesis more rigorously, Nüsslein-Volhard and colleagues isolated *bicoid* mRNA from wild-type individuals and injected it into eggs from females that lacked a functioning copy of the *bicoid* gene. Because the treated eggs developed normally, the researchers could infer that *bicoid* is responsible for establishing a normal anterior-posterior gradient.

In addition to "rescuing" mutant larvae by injecting *bicoid* mRNA, the researchers tested the gradient hypothesis by artificially altering the distribution of Bicoid protein in developing embryos and determining the effects on the body plan. For example, the biologists produced purified *bicoid* mRNA and injected it into wild-type embryos. The treated individuals developed head structures at the site of injection—even at the posterior pole. Observations such as these provided strong support for the gradient hypothesis.

Bicoid Is a Regulatory Transcription Factor The Bicoid protein appears to provide cells with information about where they are along the anterior-posterior, or head-to-tail, axis of the embryo. In addition, the in situ hybridization and antibody-staining experiments indicated that Bicoid was active when the fly embryo consisted of many nuclei scattered throughout the egg cell's cytoplasm. How does Bicoid act on these target nuclei?

DNA sequencing studies helped answer this question. The *bicoid* gene contains sequences that are typical of regulatory transcription factors—the proteins that bind to enhancers or other regulatory sequences in DNA and control the transcription of specific genes (see Chapter 18). Based on this observation, biologists concluded that the Bicoid protein must enter nuclei, bind to DNA, and either increase or decrease the expression of specific genes. Bicoid turns on genes responsible for forming anterior structures, shuts down the transcription of genes responsible for producing posterior structures, or both.

To summarize, a gradient in Bicoid protein concentration provides cells with information about their position along the anterior-posterior body axis. Because it is sequestered in a precise location in the egg and because it affects the fate of cells later in development, the *bicoid* gene product can be considered a cytoplasmic determinant. The absence of the *bicoid* product leads to a deformed body and death. As other research showed, however, *bicoid* is just one of dozens of pattern-formation genes found in fruit flies.

The Discovery of Segmentation Genes

Among the dozens of mutants isolated by Nüsslein-Volhard and Wieschaus, one called *hunchback* attracted a great deal of attention. Embryos with mutant forms of *hunchback* resemble *bicoid* mutants. Both types of embryos lack a series of consecutive segments from the anterior end. Breeding experiments showed that the *hunchback* mRNA responsible for these mutations is produced in the embryo itself as well as in the mother, however. Unlike *bicoid*, *hunchback* is not a cytoplasmic determinant.

To understand what *hunchback* does, recall from Chapter 21 that the bodies of fly larva and adults are partitioned into a series of segments. The *hunchback* gene turned out to be one of many zygotic genes that affect the identity or arrangement of these segments. Nüsslein-Volhard and Wieschaus were eventually able to identify three general classes of **segmentation genes**—genes that affect segmentation. Except for *hunchback*, all of these genes are transcribed for the first time in the embryo and not in the mother. The three types of segmentation genes are called **gap genes**, **pair-rule genes**, and **segment polarity genes**. Gap gene mutants (such as those with *hunchback*) lack several consecutive segments. Pair-rule mutants lack alternate segments and have just half the normal total number of segments. Segment polarity mutants lack portions of each segment. In many segment polarity mutants, the missing portion is replaced by a mirror-image duplication of the intact part of the same segment.

Figure 22.4 (page 474) shows the patterns of mRNA expression for segmentation genes in normal larvae. According to these in situ hybridizations, gap genes are expressed in broad regions along the head-to-tail axis (Figure 22.4a). Pair-rule genes, in contrast, are expressed in alternate segments (Figure 22.4b), while segment polarity genes are expressed in restricted regions of each segment (Figure 22.4c). If a segmentation gene does not function, the mutant individual lacks segments or parts of segments where the mRNA is found in normal individuals. Based on these data, Nüsslein-Volhard and Wieschaus and co-workers concluded that the segmentation genes are responsible for defining the segmented body plan of the fruit fly.

(a) Gap genes: expressed in groups of segments

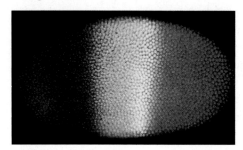

(b) Pair-rule genes: expressed in alternate segments

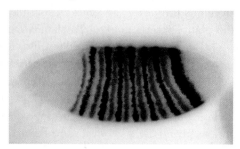

(c) Segment polarity genes: expressed in portions of segments

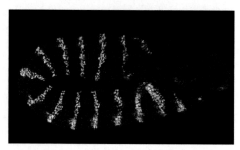

FIGURE 22.4 A Series of Genes Demarcates Body Segments in Fruit Flies
In situ hybridizations show the location of segmentation gene products in fly embryos. The embryos in **(a)** and **(b)** two were stained with antibodies for two different gene products. The embryo in **(c)** was stained for one gene product. **QUESTION** From (a) to (c), these photographs show embryos at progressively later stages of development. Why is the sequence logical?

Do Segmentation Genes Interact? Researchers who first identified the segmentation genes and described when and where they are expressed noticed that they are expressed in sequence. The gap genes are expressed first, followed by pair-rule genes, followed by segment polarity genes. This observation was logical, because gap, pair-rule, and segment polarity genes are expressed in increasingly restricted regions. The observation also suggested that the segmentation genes might interact in some way. Specifically, investigators hypothesized that the gap genes may activate the transcription of pair-rule genes, which in turn may trigger the expression of other pair-rule genes and segment polarity genes. In this way, the pattern of an embryo's segments would form in a step-by-step fashion.

Although research on interactions among segmentation gene products is continuing, an enormous amount of progress has already been made. One of the key early discoveries was that the Bicoid protein directly activates transcription of *hunchback*. The *hunchback* gene is not expressed until the Bicoid protein binds to its regulatory DNA sequences. As a result, the *hunchback* mRNA protein is found in high concentration in the anterior part of the embryo, where the Bicoid protein is abundant. Subsequent work showed that *hunchback* and other gap genes encode transcription factors that regulate the expression of pair-rule genes. Each gap gene regulates the production of a defined set of pair-rule proteins. In addition, several of the gap genes' products regulate the expression of other gap genes or even their own expression. The pair-rule genes activated by the gap genes also encode regulatory transcription factors, and some of the genes they regulate are segment polarity genes. Based on these data, the step-by-step hypothesis for establishing segment identity appears to be correct.

As **Figure 22.5** shows, this sequence of events can be described as a **regulatory cascade** or hierarchy. The *bicoid* gene, gap genes, pair-rule genes, and segment polarity genes each define a level in the cascade. As development proceeds, genes at levels farther down in the cascade are progressively activated or repressed. The products of these regulatory genes may also interact with genes at the same level.

To appreciate just how complex this regulatory cascade is, consider that the pair-rule gene *even-skipped* has at least five distinct promoters that control its expression. One of these promoters has six binding sites for Krüppel protein, three binding sites for Hunchback protein, three binding sites for Giant protein, and five binding sites for Bicoid protein. For the *even-skipped* gene to be expressed at the maximum level, each of these regulatory transcription factors must be present at a high concentration and bound to the promoter.

The regulatory cascade described in Figure 22.5 results in the progressive subdivision of the body's anterior-posterior axis into smaller and smaller units. A cell first receives information about its position on the body's anterior-posterior axis. Then gap gene products identify the group of segments that will form around the cell. Pair-rule gene products identify the particular segment that will form at that location, while segment polarity gene products provide information on position within that segment.

At each step in the developmental sequence, different cells receive a different combination of signals, and a different set of genes in the cascade is activated or shut down. In early fly embryos, hundreds of different regulatory transcription factors and dozens of different cell-to-cell signals affect the fate of cells. The combination of signals that a particular cell receives depends on its position in the embryo, and each signal causes a change in gene expression. Thus, long before a cell begins expressing tissue-specific proteins that identify it as a muscle cell or a nerve cell, it expresses genes that identify it as a cell in a particular segment or region of the body. This is the essence of pattern formation.

The Nature of Developmental Signals As the in situ hybridization and antibody-staining results in Figures 22.3 and 22.4 show, the gene products that regulate development often

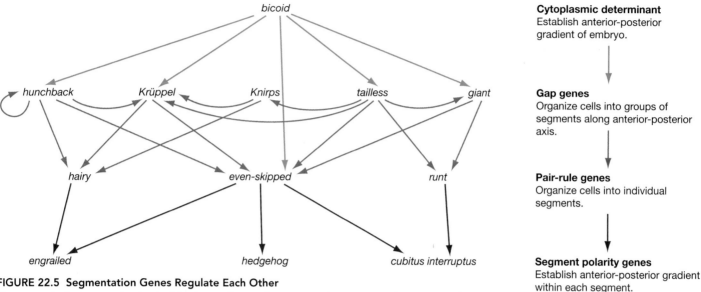

FIGURE 22.5 Segmentation Genes Regulate Each Other
The arrows indicate a small subset of the known interactions among the segmentation genes and their products. In each case, the outcome depends on the concentration of each protein involved.

Cytoplasmic determinant
Establish anterior-posterior gradient of embryo.

Gap genes
Organize cells into groups of segments along anterior-posterior axis.

Pair-rule genes
Organize cells into individual segments.

Segment polarity genes
Establish anterior-posterior gradient within each segment.

form gradients in specific regions of the embryo. This observation suggests that two types of information are being delivered to target nuclei: Both the type and concentration of signals received by cells or nuclei are important. Further, both the type and concentration of signaling molecules vary with respect to position in the embryo.

To grasp why these points are important, think back to the mechanisms of gene regulation introduced in Chapters 17 and 18. Recall that, for a eukaryotic gene to be transcribed efficiently, the regulatory transcription factors that bind to enhancers and activate transcription must be present while transcription factors that bind to silencers must be absent. Differential gene expression occurs because many of the developmental signals received by a cell alter the array of regulatory transcription factors that are present in the nucleus and are active. These transcription factors bind to promoters, enhancers, silencers, or other types of regulatory sequences in DNA and affect the amount and rate of gene expression (**Figure 22.6**). Other developmental signals may lead to alterations in the mRNAs found in target cells or induce changes in proteins that are already present. Which genes are expressed by developing cells at particular places and times? The answer depends on complex interactions among an array of developmental signals and thus transcription factors. The differential gene expression triggered by regulatory proteins shapes embryos.

Given that segmentation genes are responsible for organizing cells and tissues into distinct segments, let's consider the next level of pattern formation. What genes are responsible for producing segment-specific structures?

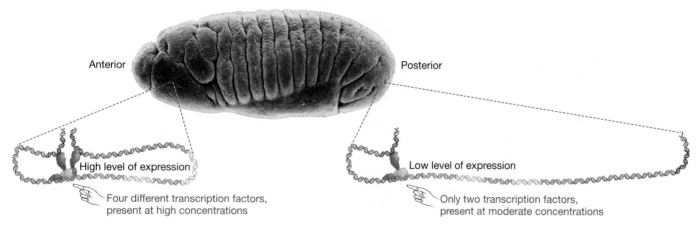

Anterior

Posterior

High level of expression

Four different transcription factors, present at high concentrations

Low level of expression

Only two transcription factors, present at moderate concentrations

FIGURE 22.6 Differential Gene Expression Is Based on the Presence of Different Regulatory Transcription Factors
The genes involved in pattern formation code for cell-to-cell signals or regulatory transcription factors that direct gene expression.

The Discovery of Homeotic Genes

During the 1940s Edward Lewis began a long series of studies that helped researchers pinpoint the genes that identify particular segments. Lewis studied mutations that alter the body pattern of adult flies, much as Nüsslein-Volhard and Wieschaus searched for mutations that affect the body pattern of embryos and larvae. Some of the mutants Lewis and others studied had bizarre phenotypes. As **Figure 22.7** shows, some flies had legs where their antennae should be; others had two sets of wings instead of one.

Lewis realized that these mutants possessed an entire segment or structure that had been transformed into a related segment or structure. This phenomenon had been observed in plants by a series of workers starting in the late 1800s. The replacement of one structure by another is termed **homeosis** ("like-condition"). For example, a mutation or series of mutations could transform thoracic segment number 3 into thoracic segment 2. Instead of bearing a pair of small stabilizer structures called *halteres*, the transformed segment would bear a pair of wings. As a result, the mutant would have four wings instead of two.

The existence of homeotic mutants meant that there must be **homeotic genes**—genes that specify a location and lead to the de-

velopment of structures appropriate to that location. (Later work showed that homeotic genes interact directly with segmentation genes.) To explain the existence of a four-winged fly, for example, Lewis suggested that the homeotic gene responsible for identifying thoracic segment number 3 was defective. Consequently, the cells in what would normally be thoracic segment 3 developed as if they were in thoracic segment 2. Stated another way, Lewis hypothesized that the products of homeotic genes specify the identity of segments along the body's anterior-posterior axis.

The Homeotic Complex   When techniques became available for identifying and sequencing genes, researchers confirmed the homeotic gene hypothesis in spectacular fashion. Studies conducted during the 1970s and 1980s initially identified eight genes in the *Drosophila* genome that lead to homeosis when they are defective. As **Figure 22.8a** shows, the eight genes are found in two clusters on the same chromosome. Investigators who explored where these genes are expressed in *Drosophila* embryos found that five are activated in the anterior part of the embryo while three are expressed in the posterior sections. The five genes expressed in the anterior part of the embryo are known as the *Antennapedia complex*; the three ex-

(a) Normal fruit flies   **(b)** Homeotic mutant fruit flies

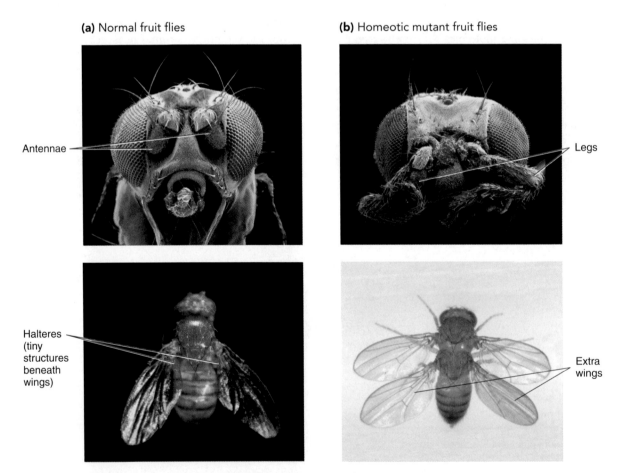

FIGURE 22.7 **Homeotic Mutants in *Drosophila* Have Structures in the Wrong Locations**
(a) The head and top view of a normal fruit fly. (b) Among the most spectacular homeotic mutants of fruit flies are individuals with legs growing where antennae should be and individuals with wings growing where small, stabilizing structures called halteres should be.

(a) Two clusters of genes form the homeotic complex.

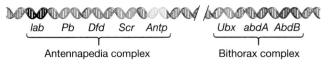

lab	Pb	Dfd	Scr	Antp		Ubx	abdA	AbdB

Antennapedia complex   Bithorax complex

(b) The sequence of homeotic genes on the chromosome correlates with where they are expressed in the embryo.

Anterior segments   Posterior segments

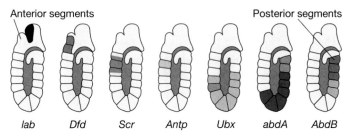

lab   Dfd   Scr   Antp   Ubx   abdA   AbdB

FIGURE 22.8 Organization and Expression of the Homeotic Complex
(a) The arrangement of *HOM-C* genes on the chromosome
(b) Where genes in the homeotic complex are expressed.

pressed in the posterior regions are called the *Bithorax complex*. As a group, the genes became known as the **homeotic complex** (abbreviated ***HOM-C***).

Studies on the timing of gene expression revealed the remarkable pattern illustrated in **Figure 22.8b**. In the embryo, genes in the homeotic complex are expressed in the same sequence as they are found along the chromosome. The gene called *lab*, for example, occurs at one end of the complex and is turned on in the most anterior segment of the embryo. Each subsequent gene in the complex is expressed in slightly more posterior segments, so these genes are expressed in the same "spatial" sequence in the embryo as their order on the chromosome. Why this pattern occurs is currently being debated. The mechanism of gene *HOM-C* action, in contrast, is well understood.

Genes in the Homeotic Complex Encode Proteins That Regulate Gene Expression Sequencing studies revealed that each gene in the homeotic complex has a 180-base-pair DNA sequence called the homeobox. The **homeobox** codes for a DNA-binding domain in the proteins produced by these genes. More specifically, the homeobox codes for a helix-loop-helix motif, which binds to specific sites in the major groove of DNA (see Chapter 17). Based on this observation, biologists hypothesized that *HOM-C* genes code for regulatory transcription factors that bind to DNA and regulate the activity of other genes. More specifically, products from the homeotic complex are thought to activate genes involved in the production of segment-specific structures such as legs, wings, antennae, and halteres. Subsequent work has identified the gene *Distal-less*, which is involved in the formation of legs and other appendages, as an example of a gene regulated by

the *HOM-C* loci. Biologists are now working to document other genes targeted by *HOM-C* products.

Where Do Homeotic Genes Fit in the Regulatory Cascade?
The segmentation genes introduced earlier in this chapter demarcate segments along the anterior-posterior axis of the fly embryo. The products of *HOM-C* genes trigger the production of structures in particular segments of adults. Is there a direct relationship between the segmentation genes and *HOM-C* genes? The answer is yes. In flies that carry a mutation in a gap gene or a pair-rule gene, the expression of one or more genes in the homeotic complex is abnormal. Based on this observation, researchers have concluded that the gap and pair-rule genes regulate the transcription of the homeotic loci. The eight *HOM-C* genes represent the next step of the regulatory cascade drawn in Figure 22.5.

Before we leave the topic of homeotic gene action and regulation, it's essential to note that not all homeotic loci are part of the Antennapedia or Bithorax complexes. Many other genes are involved in specifying the structures that develop in particular parts of the fly body; some of these genes cause homeosis when they are defective. The genes discussed in detail here are only part of a large and complicated network of genes that regulate the later stages of pattern formation.

Drosophila as a Model Organism

Based on the experiments reviewed thus far, it is clear that biologists have invested an enormous amount of time and effort in understanding pattern formation in *Drosophila*. Why? In addition to being fascinating and important in its own right, research on *Drosophila* development was intended to be applicable to other species that are difficult to manipulate in the lab. *Drosophila* was used as a model organism—an organism that is studied intensively in the hope that knowledge learned about it can be applied to other organisms.

A variety of model organisms are featured in other chapters. Research on *Escherichia coli* allowed biologists to understand the mechanisms of DNA synthesis and repair in plants and animals. Similarly, work on the yeast *Saccharomyces cerevisiae* led to an understanding of how glycolysis works in virtually every organism on Earth and how the cell cycle is controlled in multicellular species. In the same way, research on pattern formation in *Drosophila* has led to a better understanding of how pattern formation works in many other animal groups. For example, researchers have found that a homeotic complex called the **Hox gene** complex occurs in frogs, crustaceans (crabs and their relatives), birds, various types of worms, mice, and humans. Although the number of *Hox* loci varies widely among species, their chromosomal organization is similar to that of *HOM-C* genes of flies, shown in Figure 22.8a.

Recent studies of *Hox* genes have shown that they are expressed along the head-to-tail axis of the mouse embryo in the same sequence as in fruit flies. In addition, experiments have shown that when mouse *Hox* genes are altered by mutation, defects in pattern formation result. Based on these data, biologists conclude that, in flies, mice, and probably humans and most other animals, the *HOM-C* and *Hox* genes appear to play a key role in defining the position of cells along the head-to-tail axis of the body. This conclusion was supported in spectacular fashion when researchers in William McGinnis's lab introduced the *Hoxb6* gene from mice into fruit-fly eggs. The *Hoxb6* gene in mice is similar in structure and sequence to the *Antp* gene of flies. Because it was introduced without its normal regulatory sequences, the *Hoxb6* gene was expressed throughout the treated fly embryos. The resulting larvae had defects identical to those observed in naturally occurring fly mutants in which the *Antp* gene is mistakenly expressed throughout the embryo. This is a stunning result: A mouse allele not only affected the development of a fly but also mimicked the effect of a specific fly allele.

To interpret these observations, biologists hypothesize that the genes in *HOM-C/Hox* complexes are *homologous*—meaning that they are similar because they are descended from genes in a common ancestor. The hypothesis of homology contends that the first *HOM-C/Hox* loci arose before the origin of animals. Since then, the number of *HOM-C/Hox* genes has changed dramatically due to the types of gene duplication events described in Chapter 20, creating a family of related genes. But over this span of about a billion years, the same types of gene products have been involved in directing cells to develop according to their position inside the embryo. In other words, at least some of the molecular mechanisms of pattern formation have been highly conserved during animal evolution. The discovery of these shared mechanisms is one of the most significant results to have emerged from animal development studies.

Even more remarkably, researchers have discovered genes that contain homeoboxes in fungi and plants. No genes similar to those found in the *HOM-C* or *Hox* complexes exist in these groups, however. In plants, the genes involved in pattern formation are different from the segmentation and homeotic genes of animals. Why? Recall from Chapter 21 that multicellular bodies evolved independently in plants and animals. Based on this observation, it is logical to predict that the mechanisms for specifying the positional identity of cells differ between the groups. How does pattern formation occur in plants?

✔ CHECK YOUR UNDERSTANDING

Pattern formation occurs as embryonic cells respond to a regulatory cascade of cell-to-cell signals and regulatory transcription factors. In *Drosophila*, this cascade starts with the Bicoid protein and continues with the products of gap genes, pair-rule genes, segment polarity genes, and homeotic genes. At the end of this sequence, each cell in the embryo is expressing genes specific to its exact position in the body. You should be able to describe how an individual cell in a *Drosophila* embryo comes to express genes typical of a specific location—for example, the anterior portion of thoracic segment 2.

22.2 Pattern Formation in *Arabidopsis*

Historically, the field of developmental biology has been dominated by the study of animals. The reason is simple: Plant embryos develop inside the mother plant, but sea urchins, frogs, flies, and many other animals lay eggs outside their bodies. As a result, these animal embryos are accessible and relatively easy to observe and study.

Recently, however, biologists have begun studying how pattern formation occurs in plants by analyzing mutant individuals. The genetic approach to development that was pioneered with research on *Drosophila melanogaster* has proven to be a powerful way of studying plant development. Although the specific genes involved in plant development have turned out to be different from those involved in animal development, the basic logic of pattern-formation mechanisms is often quite similar in plants and animals.

Most work on pattern formation in plants has been done on the weedy mustard plant *Arabidopsis thaliana*. Recall from Chapter 21 that *Arabidopsis* seedlings grow from a seed into a mature plant capable of producing gametes in as little as six weeks. Even though their life span is short, individuals are complex enough to be interesting to study.

An important advantage of studying plants is that pattern formation is not limited to the period of early development. Instead, complex structures such as leaves, roots, branches, and flowers are produced throughout an individual's life. Recall from Chapter 21 that these structures are derived from the apical meristems located in roots and shoots. Stems and roots elongate as cell division takes place in the meristems located at their tips. As these structures grow, certain populations of cells become patterned into leaves, flowers, tubers, or other specialized structures.

To investigate how pattern formation occurs in embryonic and adult plants, let's consider two questions: (1) How is the root-to-shoot body axis established during embryogenesis, and (2) what is the mechanism of pattern formation during flower development? Biologists are addressing both questions by analyzing mutant individuals with defective roots, shoots, or flowers.

The Root-to-Shoot Axis of Embryos

Gerd Jurgens and colleagues set out to identify genes that are transcribed in the zygote or embryo of *Arabidopsis* and that are involved in establishing the root-to-shoot axis of the body. It's no surprise that this effort was similar to the project that Nüsslein-Volhard and Wieschaus had undertaken with *Drosophila*—Jurgens had participated in the work with flies.

The biologists' initial goal was to identify individuals with defects in pattern formation at the seedling stage. More specifically, they were looking for mutants that lacked particular regions along the root-to-shoot axis. The team succeeded in finding several bizarre-looking mutants (**Figure 22.9**). Certain individuals (*apical mutants*) lacked the first leaves, or cotyledons; some individuals (*central mutants*) lacked the embryonic stem, or hypocotyl; and others (*basal mutants*) lacked roots.

To interpret these results, the researchers suggested that each type of *Arabidopsis* mutant had a defect in a different gene and that each gene was involved in specifying the position of cells along the root-to-shoot axis of the body. The idea was that these genes are analogous to gap genes, which specify the identity of cells within well-defined regions along the head-to-tail axis of fruit flies.

What are these *Arabidopsis* genes, and what do they do? To answer these questions, consider the gene responsible for the mutants lacking hypocotyls and roots. This gene has been mapped and sequenced and named *monopterous*. Because its DNA sequence indicates that this gene has a DNA-binding sequence, *monopterous* is hypothesized to encode a regulatory transcription factor that regulates the activity of target genes. The monopterous protein, in turn, is manufactured in response to signals from **auxin**, a cell-to-cell signal molecule that is produced in the apical meristem.

The genes and proteins involved in setting up the root-to-shoot axis of *Arabidopsis* are not yet understood in as much detail as the pattern-formation genes of *Drosophila*. Current evidence suggests that there are strong similarities, however. In both plants and animals, pattern formation is based on cell-to-cell signals. Concentration gradients of regulatory transcription factors are important, and regulatory cascades result in the step-by-step specification of a cell's position and fate.

Many questions remain about pattern formation in plant embryos. How is auxin production turned on as the apical meristem first begins to form in embryos? Once production of the *monopterous* gene product begins, what target genes are affected? What genes other than *monopterous* are found in the regulatory cascade responsible for development along the root-to-shoot axis? Are any of them active in adults as well? Research on pattern formation in plant embryos presents a host of interesting challenges.

Flower Development in Adults

When receptor proteins in *Arabidopsis* sense that days are getting longer and the temperature is favorable, the shoot apical meristem is stimulated to act as a floral meristem. The **floral meristem** is a modified meristem that produces flowers containing the individual's reproductive organs. During flower development, the floral meristem produces four organs, arranged as shown in **Figure 22.10**: (1) sepals, (2) petals, (3) stamens, and (4) the carpel. **Sepals** are leaflike structures, located around the outside of the flower, that protect the flower. Inside the sepals is a *whorl*, or circular arrangement, of **petals**, modified leaves that enclose the male and female reproductive organs. If insects or other animals pollinate the species in question, the petals may be colored to help advertise the reproductive structures. The pollen-producing organs, or **stamens**, are located in a whorl inside the petals. In the center of the entire structure is the egg-producing reproductive organ, or **carpel**. The question is, How does the floral meristem produce these four organs in the characteristic pattern of whorls within whorls?

The first hint of an answer came in the late 1800s, when researchers realized that several types of mutant flowering plants—including some that were popular garden plants—were homeotic. In the mutant individuals, one kind of floral organ was replaced by another. Instead of having sepals, petals, stamens, and carpels, the flowers had sepals, petals, another ring of petals, and carpels. The mutant phenotype was similar to the transformation of segments later observed in *Drosophila* homeotic mutants.

Over 100 years later, Elliot Meyerowitz and colleagues assembled a large collection of *Arabidopsis* individuals with homeotic mutations in flower structure. The researchers' goal was to identify and characterize the genes responsible for specifying the four floral organs.

Meyerowitz's group found that the mutants could be sorted into three general classes according to the type of homeotic transformation that occurred. Some mutants had only carpels

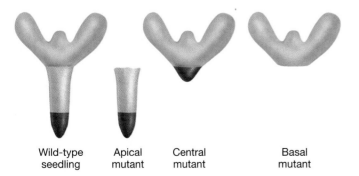

FIGURE 22.9 Pattern-Formation Mutants in *Arabidopsis* Embryos Have Misshapen Bodies
Researchers have identified *Arabidopsis* mutant individuals missing certain defined sections of the body along the root-to-shoot axis.

Wild-type seedling   Apical mutant   Central mutant   Basal mutant

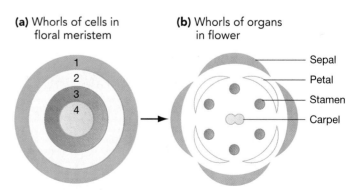

(a) Whorls of cells in floral meristem   **(b)** Whorls of organs in flower

1
2
3
4

Sepal
Petal
Stamen
Carpel

FIGURE 22.10 Flowers Are Composed of Four Organs
(a) In a floral meristem, whorls of cells produce **(b)** the four whorls of floral organs.

and stamens, as **Figure 22.11a** shows. Others had only sepals and carpels. Still others had only petals and sepals. The key observation was that each type of mutant lacked the elements found in two of the four whorls.

What was going on? Presumably, each of the three classes of homeotic mutation was due to a defect in a single gene. Meyerowitz realized that if three genes are responsible for setting up the pattern of a flower, then the mutants suggested a hy-

pothesis for how the three gene products interact. Because he referred to the three hypothetical genes as *A*, *B*, and *C*, his hypothesis is called the ABC model.

The ABC Model   As **Figure 22.11b** shows, there are three basic ideas behind the ABC model of pattern formation in flowers. The first idea is that each of the three genes involved is expressed in two adjacent whorls. The second idea is that

(a) Distinct types of homeotic mutants occur in *Arabidopsis* flowers.

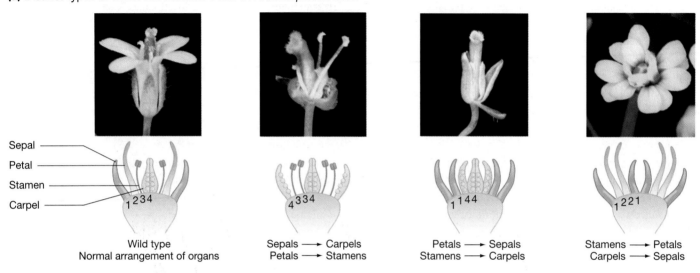

(b) A model to explain why the homeotic mutations occur

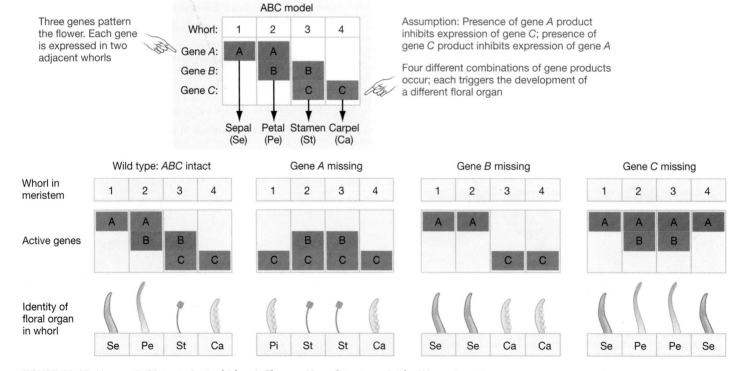

FIGURE 22.11 Homeotic Mutants in *Arabidopsis* Flowers Have Structures in the Wrong Locations
(a) Normal (wild-type) individual and three types of mutants. In each type of homeotic floral mutant, two adjacent whorls of organs are transformed into different organs. Either sepals and petals, petals and stamens, or stamens and the carpel are transformed. **(b)** The ABC model is a hypothesis to explain why three types of homeotic mutants exist.

four different combinations of gene products result from this pattern of expression. The final idea is that each of these four combinations of gene products triggers the development of a different floral organ. Specifically, Meyerowitz proposed that (1) the A protein alone causes cells to form sepals, (2) a combination of A and B proteins sets up the formation of petals, (3) B and C combined specify stamens, and (4) the C protein alone designates cells as the precursors of carpels.

Does this model explain how the three classes of homeotic mutants occur? The answer is yes, if we assume that the presence of the A protein inhibits the production of the C protein and that the presence of the C protein inhibits the production of the A protein. Then the patterns of gene expression and mutant phenotypes in Figure 22.11b correspond. For example, if the *A* gene is disabled by mutation, then it no longer inhibits the expression of the *C* gene and all cells produce the C protein. As a result, cells in the outermost whorl express only C protein and develop into carpels, while cells in the whorl just to the inside produce B and C proteins and develop into stamens.

Although the model is plausible and appeared to explain the data, it needed to be tested directly. To accomplish this, Meyerowitz and co-workers mapped the genes responsible for the mutant phenotypes and identified the DNA sequences involved. Once they had isolated the genes, they were able to obtain and use single-stranded DNAs to perform in situ hybridizations. The goal was to document the pattern of expression of the *A*, *B*, and *C* genes and see if that pattern corresponded to the model's predictions. As anticipated, the mRNAs for each of the three genes showed up in the sets of whorls predicted by the model. The *A* gene is expressed in the outer two whorls, the *B* gene is expressed in the middle two whorls, and the *C* gene is expressed in the inner two whorls.

This result strongly supported the validity of the ABC model. Just as different combinations of *HOM-C* gene products specify the identity of fly segments, different combinations of floral identity genes specify the parts of a flower.

How Do the Floral Identity Genes Work? When Meyerowitz and others analyzed the sequence of the floral organ identity genes, they discovered that all three genes contained a segment that coded for the DNA-binding domain of a protein. Based on this observation, the researchers hypothesized that, like the homeotic genes found in *Drosophila* and other animals, the floral genes appeared to be regulatory transcription factors. But instead of containing a homeobox, the genes that identify floral organs encode a long sequence of 58 amino acids, called the **MADS box**, that binds to DNA. Transcription factors with MADS boxes also occur in fungi and animals. As with research on *Drosophila*, biologists are currently working to identify the genes targeted by the ABC proteins.

To summarize, strong parallels exist between the process of pattern formation in animals and in plants. Although different genes are involved, the logic of how the genes act is the same. In both groups, pattern formation depends on signals that act in a

regulatory cascade. In each case, differences in cell fate are mediated by different combinations of regulatory transcription factors. These regulatory proteins are expressed in specific regions of the animal or plant and direct the development of cells located in that region. As **Box 22.2** (page 482) indicates, changes in those regulatory transcription factors can produce morphological change that is important in evolution.

✓CHECK YOUR UNDERSTANDING

In *Arabidopsis,* pattern formation in both embryonic and adult tissues is a step-by-step process that involves a regulatory cascade of cell-to-cell signals and regulatory transcription factors. You should be able to (1) describe two elements of this regulatory cascade in embryos and (2) explain why the *ABC* genes involved in flower development are considered homeotic genes.

22.3 Does the Genetic Makeup of Cells Change as Development Proceeds?

As development in *Arabidopsis*, fruit flies, frogs, and humans proceeds, cells become different from each other. Recall from Chapter 21 that **differentiation** is the process that results in the generation of diverse cell types from a fertilized egg. Cells are differentiated if they produce proteins that are specific to a cell type, such as muscle, nerve, or skin. One of the most basic questions in developmental biology is how differentiation occurs. Does differentiation involve changes in the genetic material of cells or simply in the kinds of genes that are expressed?

In plants, it has long been established that all cells carry similar genetic information. Gardeners and farmers have known for centuries that, in many plant species, new individuals can be produced from a section of root or shoot. Then, during the 1950s and 1960s, a series of researchers succeeded in showing that, in culture, single differentiated cells isolated from mature plants can develop into independent mature plants that function normally. Was this also the case in animals?

Are Differentiated Animal Cells Genetically Equivalent?

In the 1950s Robert Briggs and Thomas King set out to test the hypothesis that different types of differentiated animal cells have the same genetic makeup. To do this, they tested a prediction: If differentiated cells have all the genes that were present in the zygote, then a nucleus from a differentiated cell should be able to direct the development of a complete individual.

Briggs and King developed techniques for isolating nuclei from leopard frog cells and transferring them into unfertilized eggs whose nuclei had been removed. In a series of experiments, the biologists found that nuclei from an embryo undergoing cleavage or organogenesis could direct the development of complete tadpoles. In contrast, donor nuclei taken from

BOX 22.2 Regulatory Evolution

When biologists realized that the homeotic genes found in animals and plants regulate the formation of segments, limbs, flowers, and other key structures, another important realization followed: Changes in the genes responsible for pattern formation could change the size, shape, or number of segments, limbs, flowers, and other structures found in a particular species. As a result, mutations in the genes that regulate development could result in novel types of structures. In turn, the existence of novel structures in a population of organisms could trigger important evolutionary changes.

The link between genes that regulate embryonic development and evolutionary change has been termed **regulatory evolution**. The key idea is that most of the major innovations that have occurred over the course of evolution have been caused by mutations in homeotic genes and other regulatory transcription factors that are active early in development.

As an example of how researchers are exploring regulatory evolution, consider recent work on the leaf shape of tomato plants. As **Figure 22.12a** shows, tomatoes have *compound leaves*—meaning a single leaf blade divided into smaller units called *leaflets*. But leaves of other plant species vary widely in shape. *Simple leaves* consist of a single blade; *palmate leaves* have a series of leaflets that radiate from a single point.

When a new leaf begins to develop, pattern formation is controlled in part by the product of a gene called *PHANTISTICA* (abbreviated *PHAN*). The protein product of *PHAN* has a DNA-binding domain and acts as a regulatory transcription factor. It triggers the expression of genes that cause cells to form the upper surface of leaves. To explore whether changes in the regulation of *PHAN* might have a role in the evolution of various leaf shapes, a team of biologists created transgenic tomato plants

in which the *PHAN* gene product was blocked to a moderate or large extent. As **Figure 22.12b** shows, leaf shape in the transgenic individuals changed dramatically. Some of the individuals had cup-shaped (simple) leaves, while others had leaflets radiating from a single point (palmately compound).

These results have inspired the hypothesis that evolutionary changes in leaf size and shape are due to mutations that created new alleles of genes that regulate *PHAN* expression. Alleles that resulted in lowered *PHAN* expression might lead to simple leaves with a single blade; alleles that increased the extent of *PHAN* expression might result in compound leaves like those of normal tomatoes. By altering the genes that regulate development, researchers are beginning to understand the genetic changes that led to novel types of leaves, segments, limbs, and flowers.

(a) Normal leaves—compound

(b) Transgenic leaves (reduced *PHAN* expression)—simple or palmately compound

FIGURE 22.12 Changes in Pattern-Formation Genes Change the Size and Shape of Key Structures
(a) Normal tomato leaves consist of a series of leaflets. (b) When the expression of *PHAN* is reduced experimentally, leaves become cup shaped (left) or have leaflets arising from a single point. [Reprinted by permission from *Nature* (Vol. 424, July 24, 2003, pg. 439, by Kim et al.) ©2004 Macmillan Publishers Ltd. Photographs supplied by Dr. Neelima Sinah.] QUESTION Suppose that *PHAN* were overexpressed. What type of leaf do you predict would form?

tadpoles were unable to do so. These results suggested that cells may lose at least some genetic information late in development, as they become more and more specialized.

The Briggs and King experiment was criticized, however, because the failure of tadpole nuclei to direct development represented a negative result. A **negative result** is the failure to observe a particular pattern or process. Negative results can always be criticized on the grounds that something was wrong

with the experimental technique. For example, some researchers contended that the older nuclei used by Briggs and King might have complete genetic information but fail to support early development because they were more easily damaged during transfer.

These criticisms were addressed in a series of experiments that John Gurdon and co-workers carried out during the 1960s using the African clawed frog *Xenopus laevis*. In one dramatic experi-

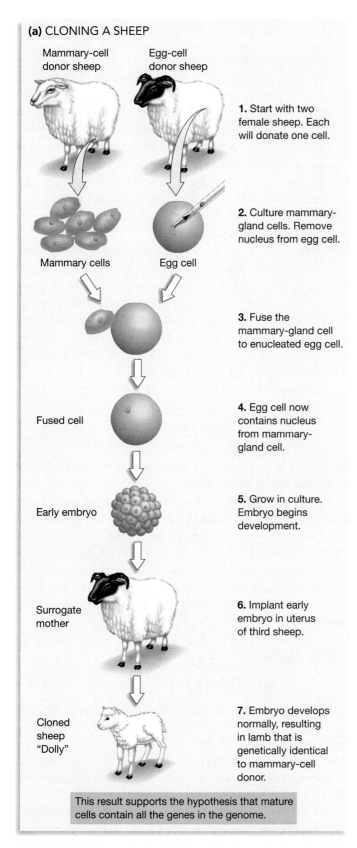

(a) CLONING A SHEEP

Mammary-cell donor sheep

Egg-cell donor sheep

1. Start with two female sheep. Each will donate one cell.

Mammary cells

Egg cell

2. Culture mammary-gland cells. Remove nucleus from egg cell.

3. Fuse the mammary-gland cell to enucleated egg cell.

Fused cell

4. Egg cell now contains nucleus from mammary-gland cell.

Early embryo

5. Grow in culture. Embryo begins development.

Surrogate mother

6. Implant early embryo in uterus of third sheep.

Cloned sheep "Dolly"

7. Embryo develops normally, resulting in lamb that is genetically identical to mammary-cell donor.

This result supports the hypothesis that mature cells contain all the genes in the genome.

ment, Gurdon's group showed that nuclei from the intestinal lining of young tadpoles could direct the development of complete tadpoles. In some cases, the tadpoles succeeded in metamorphosing into fertile, adult frogs. Although this experiment seemed to show conclusively that differentiation occurs without changes to the genetic makeup of cells, critics were not satisfied. Some researchers claimed that Gurdon and co-workers had not been careful enough in defining the source of the donor nuclei. This was a legitimate concern, because cells that produce eggs or sperm, which do not undergo differentiation, can be closely associated with the intestines of tadpoles. Other investigators contended that intestinal cells from adults, rather than tadpoles, should have been used. To test more rigorously the hypothesis that differentiation occurs without any loss of genetic information, Gurdon's group cultured skin cells that were actively producing a skin-specific protein called *keratin*—the same protein that makes up the bulk of human hair. When the researchers succeeded in producing tadpoles from eggs that received skin-cell nuclei, the result provided stronger evidence that all mature cells are genetically equivalent.

In 1997 Ian Wilmut and colleagues reinforced this conclusion through nuclear transfer experiments in sheep. As shown in **Figure 22.13a**, these researchers removed mammary-gland cells from a 6-year-old pregnant female and grew them in culture (steps 1 and 2). Later they fused these cells with *enucleated* eggs—whose nuclei had been removed—so that the egg cells contained nuclei of mammary-gland cells (steps 3 and 4). As the drawings show, the eggs came from a black-faced breed of sheep, while the donor nuclei came from a white-faced breed of sheep. After developing in culture, the resulting eggs were

(b) Dolly with an offspring of her own

FIGURE 22.13 Mammals Can Be Cloned by Transplanting Nuclei from Mature Cells
(a) When the nucleus from a differentiated mammary-gland cell was transferred to an enucleated egg, normal development occurred. The resulting lamb, Dolly, resembled the breed from which the donor nucleus came, not the breed of the egg donor or the surrogate mother. DNA testing also showed that the lamb was genetically identical to the individual that donated the nucleus. **(b)** Dolly, the sheep produced by cloning, was able to produce her own lamb through normal sexual reproduction.

BOX 22.3   Human Cloning?

Biologists can harvest nuclei from the somatic cells of mice, cows, and sheep and use the nuclei to produce embryos. These nuclear transfer experiments produce offspring that are genetically identical to the individual providing the donor nucleus. A group of genetically identical organisms are clones, so the process of creating individuals by transferring nuclei into enucleated eggs is called cloning. Can humans be cloned?

Before addressing the technical feasibility of cloning humans, let's consider a more practical issue. Why would anyone want to clone a human? Proponents of this technology contend that cloning by nuclear transfer would allow otherwise infertile couples to have offspring that are genetically related—in fact, identical—to one of the parents. Opponents of this technology are concerned that, in an attempt at immortality, dictators or wealthy eccentrics could finance clones of themselves in perpetuity.

Even benign uses of human cloning create ethical dilemmas that are yet to be thoroughly debated or resolved. What is the effect on a family of having children that are genetically identical to a parent? Would cloning be limited to individuals with "desirable" characteristics? Given the number and seriousness of the questions involved, several nations have proposed a moratorium on all research related to cloning humans.

In addition to the legal and ethical issues involved, there are significant technical hurdles to overcome before human cloning is feasible. Of the species that have been successfully cloned, monkeys and mice are most closely related to humans. Less than 3 percent of nuclear transfers result in viable mice, however. Most result in aborted pregnancies; some produce individuals that die soon after birth. Currently, the success rate is similar in cloning sheep and cows, and production of deformed newborns has occurred. Most people regard a failure rate this high as unacceptable for humans, especially given the high likelihood of producing grossly deformed offspring.

As cloning techniques improve, however, human cloning may become technologically feasible. As this book goes to press, researchers are beginning to act on a proposal to attempt cloning in as many as 600 human couples. Meanwhile, the nations of the world are debating the merits of the technology and deciding whether it will be regulated or banned outright.

---

implanted in the uteri of surrogate mothers (steps 5 and 6). In one of several hundred such transfer attempts, a white-faced lamb named Dolly was born (step 7). Dolly grew into a fertile adult and, by normal mating, produced her own lamb named Bonnie (**Figure 22.13b**). In 1998 other research groups reported similar results in mice and cows; subsequently cats, horses, a monkey, and individuals of several other species have been cloned. Recall from Chapter 12 that a **clone** is a genetically exact copy of a parent.

Although cloning protocols have been successful in animals, they are not free of complications. For example, sheep normally live about 12 years, but Dolly had to be euthanized in 2003 at age 6 because she began to suffer from debilitating diseases typical of old sheep. Researchers now hypothesize that cloned mammals may develop disorders associated with advanced age much sooner than normal. To understand why, recall from Chapter 14 that the enzyme telomerase, which prevents chromosome ends from shortening during DNA synthesis, is not active in somatic cells (cells that are not involved in gamete formation). Recall also that chromosome shortening is thought to be a signal that helps cells stop dividing as they mature. Because Dolly started life with chromosomes from a mature cell, her telomeres (the regions at the ends of linear chromosomes) were much shorter than normal. Telomere shortening is thought to contribute to the aging of cells. These types of complications are one of the many reasons that efforts to clone hu-

mans have been intensely controversial. (See **Box 22.3** for a discussion of human cloning.)

Taken together, work on cloning plants and animals has shown that, in most species, the process of cellular differentiation does not involve changes in the genetic makeup of cells. But there is an important exception to this rule. In humans and other mammals, small stretches of DNA are rearranged in certain immune system cells late in development. As a result, many immune cells are genetically unique. Chapter 49 explains how this happens.

22.4   Differentiation: Becoming a Specialized Cell

If all of the cells in an individual are genetically equivalent, how do different cell types arise? The answer is that different types of cells express different genes and manufacture different proteins. Stated another way, cellular diversity is created through **differential gene expression**. Differentiation is the process of creating specialized cell types through differential gene expression. It is based on the highly regulated expression of different subsets of genes from the common and complete set of genes found in all cells of the embryo.

Now, what controls differential gene expression? In a flower petal, the genes that are responsible for synthesizing pigments are activated. In the thorax of a fruit fly, the cells that are destined to become muscles used for flight start producing muscle-specific proteins. How does this final step in development happen?

The key to answering this question is to recognize that a cell's ultimate fate is realized through a four-step series of events:

1. Pattern formation is an early step in differentiation. Recall from Section 22.1 that it serves to organize cells in space. For example, an animal cell might begin to transcribe genes that are typical of a limb in response to signals indicating that the cell is located in a region that will develop into a leg.

2. As development continues, cells become assembled into recognizable tissues and organs in a process called **morphogenesis**. Translated literally, *morphogenesis* means "the generation of form." During pattern formation and morphogenesis, changes in gene expression cause cells to become committed to a particular fate. For example, a limb cell might become grouped with other cells that will become a muscle.

3. **Determination** is the term used to indicate a cell's irreversible commitment to becoming a particular cell type. For instance, when a cell in a muscle becomes determined, that cell is no longer capable of becoming part of a bone or part of blood or cartilage.

4. Finally, cells that are determined begin expressing tissue-specific proteins and begin functioning as muscle or nerve or bone cells. At this point, a cell is differentiated. Our example cell would begin expressing *tropomyosin* and other genes typical of skeletal muscle.

Although this series of events is not strictly linear and sequential, biologists summarize the steps as a progressive process:

Pattern formation → Morphogenesis →
Determination → Differentiation

To explore how the steps after pattern formation take place, let's consider the determination and differentiation of muscle cells in vertebrates in more detail. What causes undifferentiated mesodermal cells to begin producing the muscle-specific proteins that make breathing, walking, and opening textbooks possible?

Organizing Mesoderm into Somites

To understand how muscle cells become committed to their fate and then differentiated, recall from Chapter 21 that the precursors of muscle cells are located in mesoderm and that, once gastrulation is complete, mesoderm is sandwiched in a layer between endoderm and ectoderm. As **Figure 22.14** shows, the endoderm is internal—it is located toward the interior of the mesoderm and ectoderm. As morphogenesis proceeds, endodermal cells begin to form the digestive tract. At about the same time, the ectoderm on the dorsal (back) side of the embryo folds and forms the **neural tube**, which is the precursor of the brain and spinal cord. What is happening to mesoderm as these events take place? One of the first distinctly mesodermal structures to appear is a rod-like element called the **notochord**, shown in cross section in Figure 22.14. The notochord is unique to the group of animals called the **chordates**, which includes humans and other vertebrates. Cells in or near the notochord produce signaling molecules that induce the neural tube to form; the noto-

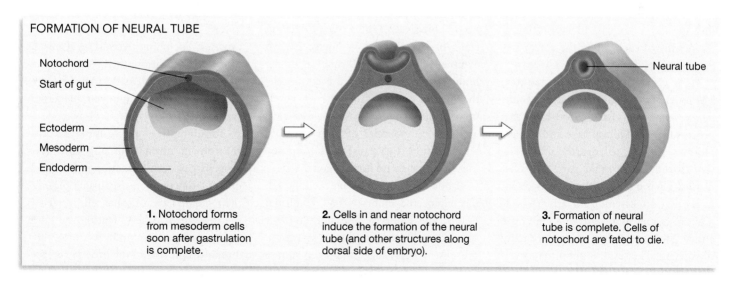

FORMATION OF NEURAL TUBE

Notochord
Start of gut
Ectoderm
Mesoderm
Endoderm

Neural tube

1. Notochord forms from mesoderm cells soon after gastrulation is complete.

2. Cells in and near notochord induce the formation of the neural tube (and other structures along dorsal side of embryo).

3. Formation of neural tube is complete. Cells of notochord are fated to die.

FIGURE 22.14 The Notochord and Neural Tube Form from Dorsal Mesoderm
In vertebrates, the notochord forms from mesoderm cells soon after gastrulation is complete. Molecules produced in or near the notochord induce the formation of the neural tube and other structures along the dorsal (back) side of the embryo.

chord itself also acts as a support as the neural tube folds into its final configuration. The notochord is a transient structure in most species, however, appearing only in embryos. As **Box 22.4** notes, many of the cells in the notochord are fated to die.

Once the neural tube forms, nearby mesodermal cells become organized into blocks of tissue called **somites**. Somites form on both sides of the neural tube, as **Figure 22.16** shows.

Fate-mapping studies have shown that the mesodermal cells in a somite are destined for a variety of structures. Cells from

---

BOX 22.4 Apoptosis: Programmed Cell Death

As animal tissues and organs take shape, certain cells are fated to die. For example, selected cells in the notochord of vertebrates die and are replaced by cells that form the vertebrae. As the human hand and foot form, cells that are initially present between the fingers and toes die. As a result, separate digits form. In these and many other cases, cell death is a normal part of development—it is carefully timed and regulated, or programmed. Programmed cell death is called **apoptosis**. Translated literally, the term means "falling away."

How does programmed cell death occur? An answer to this question emerged from studies of the roundworm *Caenorhabditis elegans*. This species is a popular research subject, because it has a complex array of organs and tissues but only about a thousand cells. In addition, the cells are transparent. As a result, biologists are able to identify individual cells in both embryos and adults. Thanks to these features, the complete fate map of this species is known. Biologists know the destiny of each cell in the embryo at every stage of development.

As a *C. elegans* individual matures, 131 of its 1090 original somatic cells undergo apoptosis. To explore how this process occurs, Hillary Ellis and Robert Horvitz set out to identify mutations that disrupt the normal pattern of cell deaths. Their initial work uncovered two genes that are essential for apoptosis. If a worm carries mutations that inactivate either gene, cells that would normally die survive. The result is abnormal development and, ironically, the death of the entire organism. Ellis and Horvitz named the genes *ced-3* and *ced-4* (for *cell death* abnormal) and proposed that they are part of a genetic program that

executes apoptosis. The researchers' hypothesis was that a cell dies if these genes are activated.

Subsequently, Horvitz and colleagues identified a gene that regulates the activity of *ced-3* and *ced-4*. Mutations that inactivate this sequence, called *ced-9*, result in the deaths of many cells that normally survive. Based on this observation, Horvitz's group suggested that the normal function of *ced-9* is to inhibit the suicidal activities of *ced-3* and *ced-4*.

By searching databases of known DNA sequences, biologists found genes in mice called *caspase-9* and *Apaf-1* that are similar to *ced-3* and *ced-4*, respectively. In addition, a mouse protein that inhibits apoptosis, called Bcl 2, is similar to the *ced-9* product. To test the hypothesis that these genes are important in mouse development, Keisuke Kuida and co-workers used genetic engineering techniques to produce mice in which both copies of the *caspase-9* gene are disrupted. In contrast to the normal embryo in **Figure 22.15a**, *caspase-9*-lacking embryos like the one in **Figure 22.15b** exhibited a severe malformation of the brain. The defect occurred because cells that would normally die early in development survived. Because the *Bcl* gene and *ced-9* are similar in both structure and function, the researchers proposed that these genes are homologous.

Kuida's group noted that even though the engineered embryos had abnormal brains, the embryos had normal digit separation. Based on this observation, the researchers suggested that mouse genes other than *caspase-9* are involved in programmed cell death during the formation of the toes.

Are these genes found in humans? The answer is yes—homologs of *Bcl*, the

(a) Normal   **(b)** Defective

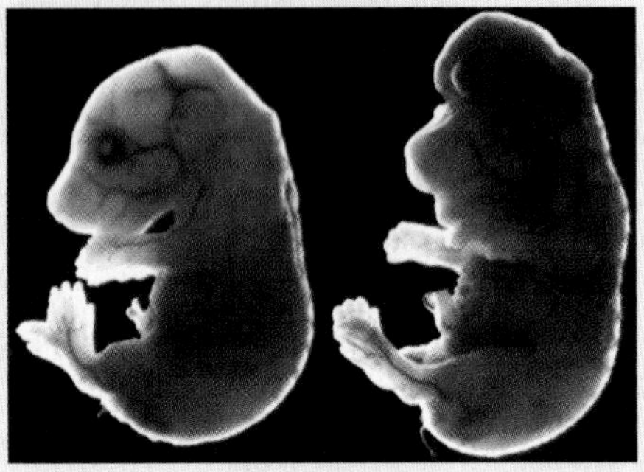

FIGURE 22.15 Defects Occur when Programmed Cell Death Fails
(a) A normal mouse embryo; **(b)** a mouse embryo that has two defective alleles of the *caspase-9* gene.

*caspase*s, and all of the other key genes that regulate apoptosis have been identified in the human genome. Normal apoptosis is important in the development of human embryos, and abnormal apoptosis has been implicated in certain diseases of adults. Inappropriate activation of programmed cell death is involved in some neurodegenerative diseases, including ALS (Lou Gehrig's disease). In addition, the failure of negative regulators such as Bcl2 has been implicated in the development of certain cancers.

To summarize, apoptosis plays a critical role in normal development. It is clear that pattern formation in the development of complex organisms requires the regulated loss of some cells as well as the growth of others. During the adult life of an organism, apoptosis also plays an important role in the orderly replacement of cells once their life span has ended. Tissue and organ stability, then, relies as much on programmed cell death as on regulated cell division. Imbalances in either process can lead to disease.

(a) Surface view of somites

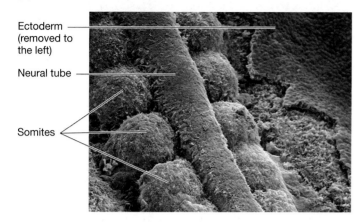

Ectoderm (removed to the left)

Neural tube

Somites

(b) Cross section of somites

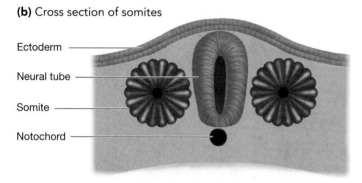

Ectoderm

Neural tube

Somite

Notochord

FIGURE 22.16 Somites Form on Both Sides of the Neural Tube
Somites are made of mesodermal cells.

somites build the vertebrae and ribs, the deeper layers of the skin that covers the back, and the muscles of the back, body wall, and limbs. By transplanting cells from one somite to another, though, researchers have found that initially any cell in a somite can become any of the somite-derived elements of the body. In other words, the cells that form the somite are not determined initially. As a somite matures, though, cells in certain sections of the structure do become committed to a specific fate. For example, biologists transplanted cells from various locations in the somite to the outer part of the somite (farthest from the neural tube) early in that structure's formation. The cells eventually became committed to form the muscles of the limbs (**Figure 22.17a**). This result supports the hypothesis that cells in a somite are initially not determined but later become committed based on their position.

Like the notochord of many species, somites are transient structures that appear relatively briefly during embryonic development. Unlike the notochord cells, however, the cells that make up somites survive and undergo differentiation. Once determination occurs, the cells that make up somites break up into distinct populations and migrate to their final location in the developing embryo. As **Figure 22.17b** shows, each population of cells is committed to becoming a different type of cell.

Recent studies have shown that distinct populations of cells in a somite become determined in response to an array of

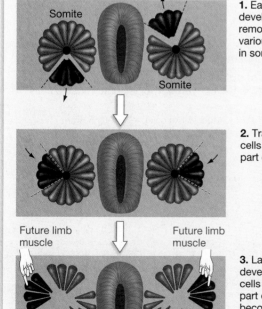

(a) CELLS IN A SOMITE ARE INITIALLY NOT DETERMINED.

Neural tube

Somite

Somite

1. Early in development, remove cells from various locations in somites.

2. Transplant these cells to the outer part of somites.

Future limb muscle

Future limb muscle

3. Later in development, cells from the outer part of somites become limb muscle, regardless of their original location.

FIGURE 22.17 Cells in the Somite Become Determined
(a) By transplanting cells early in somite formation to new locations in this structure, researchers showed that they are not determined initially. Each somite eventually breaks up into distinct populations of cells, each with a different fate. **(b)** Within a somite, the position of a cell determines its fate.

(b) Later, cells in somites become determined based on their positions.

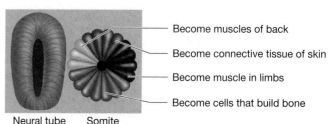

Become muscles of back

Become connective tissue of skin

Become muscle in limbs

Become cells that build bone

Neural tube   Somite

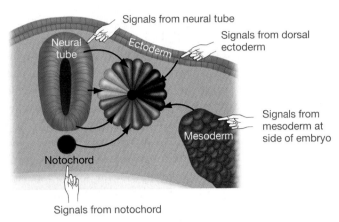

FIGURE 22.18 What Triggers Somite Determination?
Somite cells become determined in response to signals from the neural tube, notochord, and nearby ectoderm and mesoderm.

signals. **Figure 22.18** illustrates how signaling molecules diffuse away from cells in the notochord, the neural tube, and nearby ectoderm and mesoderm and act on target cells in the somite. Each type of signal acts on a distinct population of cells within the structure. In the case of the muscle cells located at the outer part of a somite, how do these signals result in determination?

Determination of Muscle Cells

In a somite, why do cells that are farthest from the neural tube become committed to producing muscle in response to the signals they receive? Harold Weintraub and colleagues answered this question by experimenting with *myoblasts*—cells that are committed to becoming muscle but that have not yet begun producing muscle-specific proteins. Specifically, Weintraub and co-workers hypothesized that myoblasts contain at least one regulatory protein that commits them to their fate. Their idea was that myoblasts begin producing this muscle-determining protein after they receive an appropriate set of signals from nearby tissues.

Figure 22.19 outlines how the biologists went about searching for this hypothetical protein. They began by isolating mRNAs from myoblasts and using reverse transcriptase to convert the mRNAs to cDNAs (steps 1 and 2). Because myoblasts transcribe only the genes required for a muscle cell to function, these cDNAs represented muscle-specific genes. Then the biologists attached a type of promoter to the cDNAs that would ensure the genes' expression in any cell (step 3). Finally, they introduced the recombinant genes into connective tissue cells called *fibroblasts* and monitored the development of the transformed cells (step 4). Just as predicted, one of the myoblast-derived cDNAs converted fibroblasts into muscle cells. Follow-up experiments showed that the same gene could convert pigment cells, nerve cells, fat cells, and liver cells into cells that produced muscle-specific proteins.

Weintraub's group called the protein product of this gene **MyoD**, for *myoblast determination*. Follow-up work showed

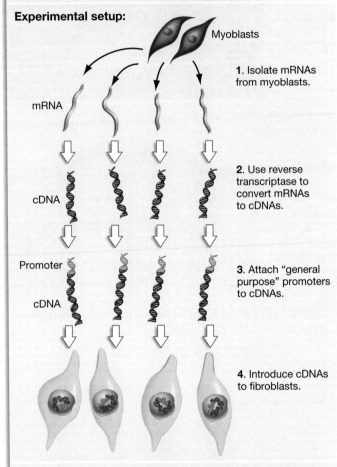

Question: Why do cells on the outside of somites become committed to producing muscle in response to the signals they receive?

Hypothesis: Myoblasts contain at least one regulatory protein that commits them to their fate.

Null hypothesis: Myoblasts do not contain a regulatory protein that commits them to their fate.

Experimental setup:

Myoblasts

1. Isolate mRNAs from myoblasts.

2. Use reverse transcriptase to convert mRNAs to cDNAs.

3. Attach "general purpose" promoters to cDNAs.

4. Introduce cDNAs to fibroblasts.

Prediction: One of the myoblast-derived cDNAs will convert fibroblasts into muscle cells.

Prediction of null hypothesis: None of the myoblast-derived cDNAs will convert fibroblasts into muscle cells.

Results:

Muscle-like cell — One of the myoblast-derived cDNAs converted fibroblasts into muscle cells

Conclusion: Cells on the outside of somites contain a regulatory protein (later called MyoD) that commits them to differentiate into muscle.

FIGURE 22.19 A Gene That Causes Muscle-Cell Differentiation
Protocol that allowed researchers to find a gene involved in the differentiation of muscle cells. QUESTION Why did the researchers have to attach a "general purpose" promoter to the cDNAs?

that *MyoD* encodes a regulatory transcription factor and that the MyoD protein binds to enhancer elements located upstream of muscle-specific genes. In addition, the MyoD protein activates expression of the *MyoD* gene. This was a key observation because it meant that once *MyoD* is turned on, it triggers its own expression—meaning that it remains on. Other researchers have found that genes closely related to *MyoD* are also required for the differentiation of muscle cells.

To summarize, cells are initially committed to become part of a vertebrate's back when determinants located in the egg change position after fertilization occurs. In response to these signals, cells become committed to become the organizer. (Recall from Section 21.3 that cells in the organizer secrete cell-to-cell signals that induce changes in target cells.) The organizer then induces mesoderm that becomes part of the notochord. Subsequently, signals from the notochord and

nearby cells induce the production of MyoD and other muscle-determining proteins in certain populations of cells from somites. In response, these target cells are committed to becoming muscle. The commitment is reinforced by MyoD's ability to enhance its own production.

As a case history, the determination and differentiation of vertebrate muscle cells illustrates several principles common to all cell types. First, it confirms that differential gene expression is at the heart of development. Second, it supports the view that the development of an individual is a stepwise process. Determination occurs when the production of a specific regulatory protein such as MyoD commits a cell to a particular fate. Differentiation occurs when the cell actually begins to produce cell-specific proteins. At that point, the process of development is largely complete. Growth takes over as the most important process in shaping the juvenile organism.

ESSAY Human Stem Cells

In most cases, the fate of a cell is sealed once it has "popped into place." At some point in the step-by-step process, development can become irreversible. When a muscle cell in a human embryo differentiates, for example, it stops growing and begins producing muscle-specific proteins. This cell remains a muscle cell for as long as it lives. It cannot change its fate and become a nerve cell.

There are important exceptions to this rule, however. The cells that make up plant meristems divide continuously and retain the ability to produce an array of cells, tissues, and organs. In many cases, even specialized plant cells retain the ability to "un-differentiate" and give rise to other types of tissues and organs. Gardeners and farmers take advantage of this trait when they take cuttings from stems or roots and use them to propagate new individuals.

Even in adult mammals, certain populations of cells retain the ability to divide and produce an array of cell types. Cells with this ability are called **stem cells**. Stem cells can be found in your muscle, skin, liver, gut lining, and elsewhere. Given that you lose an average of 100 billion cells from the lining of your intestine and from your blood each day, it is not surprising that stem cells exist.

The stem cells in your body continue to divide throughout your life. Some of their daughter cells remain as stem cells; others differentiate to replace cells in nearby tissues that have been lost or damaged. One of the key attributes of your stem cells is that they behave as though they are determined but not differentiated. The stem cells in your skin, for example, can give rise only to skin cells. Although the stem cells in your bone marrow give rise to an impressive array of different cell types, most of their descendant cells become components of blood.

In contrast, stem cells that are isolated from embryos are not as narrowly committed. Cells from the inside of a blastocyst, for

example, can give rise to virtually any cell type in the body. Recently, researchers have started to explore whether large numbers of these embryonic stem cells could be grown in culture and then implanted into adults as a treatment for diseases and injuries caused by the loss of differentiated cells. Parkinson's disease and Alzheimer's disease are caused by a loss of brain cells; the most serious form of diabetes is caused by loss of certain cells in the pancreas; muscular dystrophy results from the death of muscle cells. Biomedical researchers hope that implanted stem cells will be able to differentiate and at least partially make up for the loss of the original cells.

Research on human stem cells is intensely controversial. . . .

Research on human stem cells is intensely controversial, however. Because in vitro fertilization techniques (see the Essay in Chapter 21) frequently produce an excess of early embryos, the primary source for embryonic stem cells has been fertility clinics. If these surplus embryos are not discarded, they can be dissected and used as a source of stem cells. Opponents of this practice argue that because the embryos are sacrificed in this process, the procedure represents the taking of a human life.

Currently, nations are debating rules to govern the use of embryonic stem cells in research. Although the practice is banned in Italy and Norway, it is legal or even actively supported in most countries of the European Union. In the United States, public funding for the establishment of new cultures of embryonic stem cells has been banned, but research on cultures of embryonic stem cells that had been established prior to the year 2001 is still being supported. It remains to be seen whether stem cell research will result in important therapies and, if so, whether their use will be approved by regulatory agencies.

CHAPTER REVIEW

Summary of Key Concepts

▪ **The fate of an embryonic cell—what it will become in the juvenile and adult organism—is established in a series of steps.**

During pattern formation, cells receive information about their location in the embryo. As morphogenesis gets under way, cells change their structure and function in a way that forms an ordered arrangement of cells in space. Determination occurs when a cell becomes committed to a particular fate. A cell is differentiated when it begins producing tissue-specific proteins and functions as a mature, specialized component of the organism.

Researchers have used a genetic approach to study the mechanisms of pattern formation in fruit flies and *Arabidopsis*. By analyzing mutant embryos or adults with defects in pattern formation, biologists have been able to identify a series of genes and gene products that organize cells in space. The pattern-formation genes that have been discovered to date fall into two broad categories, based on the phenotypes of individuals with mutant forms of the genes: (1) those that cause the loss of specific regions in the embryo—for example, the segmentation genes in *Drosophila* or *monopterous* and related genes in *Arabidopsis*—and (2) those that cause homeotic transformations in adults.

The segmentation genes found in fruit flies and the *monopterous*-like genes in mustard plants are thought to define particular segments or regions in the embryo. The *Drosophila* segmentation genes generate a series of segments along the head-to-tail axis of the body; the *monopterous*-like genes of *Arabidopsis* define the cotyledons, hypocotyl, and root along the root-to-shoot axis of the body. The protein products of these genes are thought to carry signals such as "this area is a segmental boundary in the thorax" or "this area is part of the root."

In contrast, when a homeotic transformation takes place, a structure that is usually found at a different place in the body is substituted for the normal structure at a particular location. Flies with legs instead of antennae and flowers with sepals instead of petals are examples of homeotic transformations. To explain why these phenotypes occur, biologists suggest that the products of homeotic genes identify particular locations in the developing organism. For example, a *HOM-C* gene product in *Drosophila* might signal, "This is thoracic segment 2,"

whereas a MADS-box protein in *Arabidopsis* might signal, "This is the second whorl of floral organs."

Web Tutorial 22.1   Early Pattern Formation in *Drosophila*

▪ **At each step in development, cell-to-cell signals activate regulatory transcription factors that direct the expression of genes in cells throughout the embryo.**

Research on pattern-formation genes has brought to light several general principles about development. Development is a step-by-step process based on differential gene expression. As pattern formation and other aspects of development proceed, changes in gene expression occur in response to cytoplasmic determinants or signals from other cells. A cell's response depends on the types of signals it receives and their concentration. Often this response occurs through the production of regulatory transcription factors. The type and quantity of signals received by a cell depend on its position in the embryo.

▪ **Because virtually all cells in a multicellular organism contain the same genes, development is based on differential gene expression.**

The ability to clone entire plants from a cutting or a small group of cells established that mature plant cells contain all of the genes that were present in the fertilized egg. In animals, however, it was much more difficult to establish that adult cells are genetically equivalent. Eventually a series of nuclear transplant experiments culminated in the cloning of sheep, mice, and cows from somatic cell nuclei taken from adults. This result established that virtually all cells in adult plants and animals are genetically identical. In multicellular organisms, cells become specialized through the expression of specific genes—not the loss of specific genes.

One central focus of current research in developmental biology is to understand how differential gene expression leads to determination and differentiation. For example, the mesodermal cells that make up the somites found in vertebrate embryos take on one of several fates, depending on their position in the somite and thus the signals that they receive from nearby structures. Cells on the outside of somites become determined to be muscle because they receive signals that trigger the production of MyoD and other regulatory proteins. In turn, these regulatory proteins cause the cell to differentiate by triggering the transcription of muscle-specific genes.

Questions

Content Review

1. The auxin produced in *Arabidopsis* embryos and Bicoid protein produced in *Drosophila* embryos have what similar function?
   a. Both are regulatory transcription factors that serve as cell-to-cell signals.
   b. Both are cytoplasmic determinants—they are loaded into eggs by cells in the mother or by female reproductive tissues.
   c. Both trigger the transcription of homeotic loci—genes that specify the types of structures produced at a certain position.
   d. Both establish a concentration gradient that demarcates the anterior-posterior axis.

2. In combination, what do the products of gap genes, pair-rule genes, and segmentation polarity genes of fruit flies do?
   a. They trigger the reorganization of the larval body into an adult body.
   b. They define the segmented body plan of the embryo.
   c. They set up the back-to-belly axis of the larval body.
   d. They direct cell movements during gastrulation.

3. What does the ABC model of flower development attempt to explain?
   a. how different combinations of gene products trigger the formation of different floral organs
   b. why petals are found on the inside of the whorl of sepals instead of on the outside
   c. why the four types of floral organs occur in whorls
   d. why apical meristems are converted to floral meristems in response to specific cell-to-cell signals

4. What is a homeotic mutant?
   a. an individual with a structure located in the wrong place
   b. an individual with an abnormal head-to-tail axis
   c. in flies, an individual that is missing segments; in *Arabidopsis*, an individual that is missing a hypocotyl or other embryonic structure
   d. an individual with double the normal number of structures or segments

5. What evidence suggested that the MyoD protein is involved in the differentiation of muscle cells?
   a. The *MyoD* gene is expressed late—not early—in the development of somites.
   b. Expression of *MyoD* DNA can convert nonmuscle cells to muscle-like cells.
   c. *MyoD* mRNA was isolated from muscle cells.
   d. MyoD is part of a closely related family of transcription factors.

6. What is apoptosis?
   a. an experimental technique that biologists use to kill specific cells
   b. programmed cell death that is required for normal development
   c. a pathological condition observed only in damaged or diseased organisms
   d. a developmental mechanism unique to the roundworm *C. elegans*

Conceptual Review

1. What evidence suggests that at least some of the molecular mechanisms responsible for pattern formation have been highly conserved over the course of animal evolution?

2. How did researchers succeed in identifying molecules that have important roles in *Drosophila* and *Arabidopsis* pattern formation?

3. Explain why mRNA injection experiments support the hypothesis that *bicoid* sets up the normal head-to-tail axis of a *Drosophila* embryo. Why did researchers bother to do such experiments? To accept the gradient hypothesis as valid, why wasn't it enough to observe the *bicoid* mutant phenotype, the distribution of *bicoid* mRNA, and the distribution of Bicoid protein?

4. Why is it significant that many of the genes involved in pattern formation and determination encode regulatory transcription factors?

5. What is the difference between determination and differentiation? What is the relationship of these processes to pattern formation?

6. The development of in situ hybridization and antibody-staining technology had a huge impact on biologists' ability to study pattern formation and determination. Why?

Group Discussion Problems

1. Recent research has shown that the products of two different *Drosophila* genes are required to keep *bicoid* mRNA concentrated at the anterior end of the egg. In individuals with mutant forms of these proteins, *bicoid* mRNA diffuses farther toward the posterior pole than it normally does. First, predict what effect these mutations will have on segmentation of the larva. Second, suggest a hypothesis for how these proteins function at a molecular level. How would you test your hypothesis? Finally, predict whether the mutations exhibit maternal effect inheritance. Explain your rationale.

2. In 1992 David Vaux and colleagues used genetic engineering technology to introduce an active human gene for the Bcl 2 protein into embryos of the roundworm *C. elegans*. When the team examined the embryos, they found that cells that normally undergo programmed cell death survive. What is the significance of this observation?

3. Animal cells make massive movements during gastrulation. Similarly, the cells that make up somites break up into distinct populations and migrate to their final position in the embryo after determination has occurred. Plant cells, in contrast, do not move, because they are surrounded by a stiff cell wall. Discuss how this lack of movement might affect pattern formation, differentiation, or determination in plants.

4. According to the data presented in this chapter, segmentation genes set up the segmented body plan of a fruit fly while *HOM-C* genes trigger the production of segment-specific structures such as antennae, legs, and wings. Now consider the differences in the body plans of a spider, a fruit fly, a centipede, and an earthworm. Many biologists hypothesize that differences in the expression of segmentation genes or *HOM-C* genes are responsible for some of the morphological differences observed in these animal groups. Do you find this hypothesis plausible? Why or why not?

Answers to Multiple-Choice Questions   **1.** d; **2.** b; **3.** a; **4.** a; **5.** b; **6.** b

Appendix

TABLE A.1 The Metric System

Measurement	Unit of Measurement and Abbreviation	Metric System Equivalent	Converting Metric Units to English Units
Length	kilometer (km)	$1\ km = 1000\ m = 10^3\ m$	1 km = 0.62 miles
	meter (m)	1 m = 100 cm	1 m = 1.09 yards = 3.28 feet = 39.37 inches
	centimeter (cm)	$1\ cm = 0.01\ m = 10^{-2}\ m$	1 cm = 0.3937 inch
	millimeter (mm)	$1\ mm = 1000\ \mu m = 10^{-3}\ m$	1 mm = 0.039 inches
	micrometer (μm)	$1\ \mu m = 1000\ nm = 10^{-6}\ m$	
	nanometer (nm)	$1\ nm = 10^{-9}\ m$	
	angstrom (Å)	$1\ \text{Å} = 0.1\ nm = 10^{-10}\ m$	
Area	hectare (ha)	$1\ ha = 10,000\ m^2$	1 ha = 2.47 acres
	square meter (m²)	$1\ m^2 = 10,000\ cm^2$	$1\ m^2 = 1.196$ square yards
	square centimeter (cm²)	$1\ cm^2 = 100\ mm^2 = 10^{-4}\ m^2$	$1\ cm^2 = 0.155$ square inches
Mass	kilogram (kg)	1 kg = 1000 g	1 kg = 2.20 pounds
	gram (g)	1 g = 1000 mg	1 g = 0.035 ounces
	milligram (mg)	$1\ mg = 1000\ \mu g = 10^{-3}\ g$	
	microgram (μg)	$1\ \mu g = 10^{-6}\ g$	
Volume	liter (L)	1 L = 1000 mL	1 L = 1.06 quarts
	milliliter (mL)	$1\ mL = 1000\ \mu L = 10^{-3}\ L$	1 mL = 0.034 fluid ounces
	microliter (μL)	$1\ \mu L = 10^{-6}\ L$	
Temperature	†Kelvin (K)		K = °C + 273.15
	Degrees Celsius (°C)		$°F = \frac{9}{5}°C + 32$
	Degrees Fahrenheit (°F)		$°C = \frac{5}{9}(°F - 32)$

†Absolute zero is $-273.15°C = 0\ K$

TABLE A.2 Prefixes Used in the Metric System

Prefix	Abbreviation	Definition	
micro-	μ	0.000001	$= 10^{-6}$
milli-	m	0.001	$= 10^{-3}$
centi-	c	0.01	$= 10^{-2}$
deci-	d	0.1	$= 10^{-1}$
		1	$= 10^0$
kilo-	k	1000	$= 10^3$

Questions

(1) Some friends of yours just competed in a 5-kilometer run. How many miles did they run?

(2) An American football field is 100 yards long, while rugby fields are 140 meters long. In yards, how much longer is a rugby field than a football field?

(3) What is your normal body temperature in degrees Celsius? (Normal body temperature is 98.6°F.)

(4) What is your current weight in kilograms?

(5) A friend asks you to buy a gallon of milk. How many liters would you buy to get approximately the same volume?

Glossary

–10 box The six-pair base sequence (usually TATAAT) found in most prokaryotic promoters 10 bases upstream from the start of mRNA transcription. Also called Pribnow box.

30-nanometer fiber A fiber of DNA and protein found in eukaryotic nuclei and consisting of many nucleosomes wound together and tightly packed.

–35 box The six-pair base sequence (usually TTGACA) in most prokaryotic promoters 35 bases upstream from the start of mRNA transcription.

ABC model A model of pattern formation of flowers. Proposes that three genes (A, B, and C) are necessary to build a normal flower containing sepals, petals, stamens, and carpels.

abdomen The region of the body that houses most of the digestive tract. In insects, the third and hindmost of the three major body regions.

abiotic Not alive (e.g., air, water, and soil).

aboveground biomass The total mass of living plants in an area, excluding roots.

abscisic acid (ABA) A plant hormone that inhibits cell elongation and stimulates leaf shedding and dormancy.

abscission The shedding of leaves from a plant.

abscission zone The region of a petiole that thins and breaks during dropping of leaves.

absorption The uptake of the ions and small molecules from food across the lining of the intestine and into the bloodstream.

absorption spectrum A graph depicting how well a pigment absorbs various wavelengths of light.

absorptive feeding The absorption of nutrients directly from the environment, across a cell's plasma membrane.

accessory fluids The fluid portion of semen, containing nutrients, enzymes, and other substances. Added to spermatozoa by male reproductive glands.

accessory pigments Pigments, such as carotenoids, that absorb light in plant cells and pass the energy on to chlorophyll to drive photosynthesis.

acclimatization Gradual physiological adjustment of an organism to new environmental conditions.

acetylation Addition of an acetyl group (CH_3COO^-) to a molecule.

acetylcholine (ACh) A neurotransmitter that is released by nerve cells onto muscle cells to trigger muscle contraction. Also used as a neurotransmitter in the nervous system.

acetyl CoA The final product of glycolysis, produced by oxidation of pyruvate and binding to CoA. Can enter the Krebs cycle.

acid Any compound that gives up protons or accepts electrons during a chemical reaction or that releases hydrogen ions when dissolved in water.

acid-base reaction A chemical reaction that involves a transfer of protons.

acid-growth hypothesis The hypothesis that auxin's effect on plant cell elongation occurs via installation of proton pumps that make the cell wall more acidic, causing the cell wall to expand.

acid rain Rain that is substantially more acidic than normal (usually with a pH below 5), enough to adversely affect plants and other species.

acoelomate Not having a coelom (internal body cavity). Diploblasts (sponges, cnidarians, and ctenophores), flatworms, and the Acoelomorpha are all acoelomate.

acquired immune deficiency syndrome (AIDS) A human disease characterized by death of immune system cells and subsequent vulnerability to other infections. Caused by the human immunodeficiency virus (HIV).

acquired immunity Immunity to a particular pathogen conferred by antibodies produced by previous exposure to the pathogen or to a vaccine or acquired passively through the placenta or breast milk.

acrosomal process A protrusion that forms on the head of a sperm cell during fertilization in some species, such as some echinoderms.

acrosomal reaction A set of events occurring in a sperm cell upon encountering an egg cell, including release of acrosomal enzymes and formation of microfilaments that help sperm reach the egg.

acrosome A packet of enzymes found on the head of a sperm cell that help dissolve the zona pellucida or jelly layer.

actin A globular protein that can be polymerized to form actin filaments, which are involved in cell movement.

actin filament A long, thin fiber composed of two intertwined strands of polymerized actin. Involved in cell movement. Also called a microfilament.

action potential A rapid, temporary change in electrical potential across a membrane, from negative to positive and back to negative.

action spectrum The range of wavelengths of light that can drive photosynthesis.

activation energy The amount of energy required to cause a chemical reaction; specifically, the energy required to reach the transition state.

active site The location on an enzyme where substrates (reactant molecules) bind and react.

active transport The movement of ions or molecules across a plasma membrane against a concentration gradient or an electrochemical gradient. Requires energy, such as ATP.

adaptation A heritable trait that increases the fitness of an individual with that trait compared with individuals without it, in a certain environment.

adaptive management A conservation strategy in which information from environmental monitoring is continuously used to review and modify ongoing conservation programs.

adaptive radiation Rapid evolutionary diversification within one lineage, producing numerous descendant species with a wide range of adaptive forms.

addiction Compulsive use of a substance despite recognition of its undesirable consequences, coupled with increasing resistance to the substance (tolerance) and withdrawal symptoms upon cessation of use.

adenosine diphosphate (ADP) A molecule consisting of adenine, a sugar, and two phosphate groups. Addition of a third phosphate group produces adenosine triphosphate (ATP).

adenosine triphosphate (ATP) A molecule consisting of adenine, a sugar, and three phosphate groups that can be hydrolyzed to release energy. Universally used by cells to store and transfer energy.

adenylyl cyclase An enzyme that can catalyze the formation of cyclic AMP (cAMP) from ATP. Involved in control of the gene transcription of various operons in prokaryotes.

adhesion The tendency of certain dissimilar molecules to cling together due to attractive forces.

adipocyte A fat cell.

adipose tissue A connective tissue whose cells store fats.

ADP See adenosine diphosphate (ADP).

adrenal glands Small glands that sit above each kidney. The outer portion (cortex) secretes several steroid hormones; the inner portion (medulla) secretes epinephrine and norepinephrine.

adrenaline A catecholamine hormone from the adrenal medulla. Triggers rapid responses of the fight-or-flight response. Also called epinephrine.

adrenocorticotropic hormone (ACTH) A peptide hormone from the anterior pituitary gland that stimulates release of cortisol, corticosterone, and aldosterone from the adrenal cortex.

adventitious root A root that develops from a plant's shoot system instead of from the plant's root system.

aerobe An organism that uses aerobic respiration (i.e., uses oxygen as a final electron acceptor).

aerobic respiration Cellular respiration using oxygen as the electron acceptor, typically using the Krebs cycle and the electron transport chain.

afferent division The part of the nervous system that transmits sensory information to the central nervous system. Consists mainly of sensory neurons.

agar A gelatinous mix of polysaccharides that is commonly used in labs to grow cell cultures.

age class A group of individuals of a specific age.

agent A nonliving infectious entity, such as a virus. Also called a particle.

age pyramid A graph of a population's age structure, with horizontal bars representing the numbers of males and females, stacked against age on the vertical axis.

age-specific fecundity The average number of female offspring produced by a female in a certain age class.

age structure The proportion of individuals in a population that are of each possible age.

agglutination Clumping together of cells, typically caused by antibodies.

agonist A compound that can bind to and activate a receptor such as a hormone receptor or neurotransmitter receptor.

AIDS See acquired immune deficiency syndrome (AIDS).

air sacs Thin-walled sacs of air that extend throughout most of a bird's body and connect to the lungs. Function as bellows that continuously feed air to the lungs.

albumen A solution of water and protein (particularly albumins), found in amniotic eggs, that nourishes the growing embryo. Also called egg white.

albumin A class of large proteins found in plants and animals, particularly in the albumen of eggs and in blood plasma.

alcohol fermentation Fermentation (production of ATP without oxygen) in which the end product of glycolysis is converted to ethanol.

aldosterone A hormone from the adrenal cortex that stimulates the kidney to conserve salt and water and promotes retention of sodium.

aleurone layer In a seed, a layer that releases the starch-digesting enzyme α-amylase during germination.

alkaptonuria A medical condition characterized by an accumulation of homogentisic acid in the body, caused by a defect in the enzyme that metabolizes homogentisic acid.

allantois In an amniotic egg, the membrane-bound sac that holds waste materials.

allele A particular version of a gene.

allergen Any molecule that can trigger an allergic response.

allergy An abnormal immune response to a non-pathogenic substance, usually involving production of the IgE class of antibodies.

allometry A disproportionate relationship of body size to another feature, such as limb size or heart rate. Also, the general study of the effects of body size on biological processes.

allopatric speciation The divergence of populations into different species by physical isolation of populations in different geographic areas.

allopatry Living in an area different from the area in which some other population lives.

allopolyploid Polyploid (having multiple copies of chromosomes) due to hybridization between different species.

allosteric regulation Regulation of enzymatic catalysis via a change in the enzyme's shape, usually caused by a regulatory molecule binding at a specific site.

α-amylase A plant enzyme that digests the starch in a seed so that the plant embryo can use the resulting sugars for growth.

alpha chain (α chain) One of the two polypeptide chains that make up a T-cell receptor protein, enabling the T cells of the immune system to bind to antigens on other cells.

alternation of generations A life cycle involving alternation of a multicellular haploid stage (gametophyte) with a multicellular diploid stage (sporophyte). Occurs in most plants and some protists.

alternative splicing In eukaryotes, the splicing of the same primary RNA transcript in different ways to produce different mature mRNAs and thus different proteins.

altruism Any behavior that has a cost to the individual (such as lowered survival or reproduction) and a benefit to the recipient.

alveolus (plural: alveoli) One of the tiny air-filled sacs of a mammalian lung.

ambisense virus A virus whose genome contains both positive-sense and negative-sense sequences.

amino acid A small organic molecule with a central carbon atom bonded to an amino group (–NH₃), a carboxyl group (–COOH), a hydrogen atom, and a side group. Proteins are polymers of 20 common amino acids.

aminoacyl tRNA A transfer RNA molecule that is covalently bound to an amino acid.

aminoacyl tRNA synthetases Enzymes responsible for catalyzing the addition of amino acids to tRNA molecules.

ammonia The molecule NH_3, a toxin produced by the breakdown of proteins and nucleic acids.

ammonium ion The ion NH_4^+.

amnion The membrane in an amniotic egg that surrounds the embryo and encloses it in a protective pool of fluid (amniotic fluid).

Amniota A major lineage of vertebrates that reproduce with amniotic eggs. Includes all reptiles, birds, and mammals.

amniotic egg An egg containing a membrane-bound supply of water (the amnion), a membrane-bound supply of food (yolk sac), and a waste sac (allantois), all encased in a leathery or hard shell.

amniotic fluid The fluid inside the amnion of a developing mammalian embryo.

amoeba Any unicellular protist that lacks a cell wall and is extremely flexible in shape.

amphipathic Containing hydrophilic and hydrophobic elements.

amylase Any enzyme that can break down starch by catalyzing hydrolysis of the glycosidic linkages between the glucose monomers.

amyloplast Dense, starch-storing organelles that settle to the bottom of plant cells and that may be used as gravity detectors.

anabolic pathway A set of chemical reactions that synthesizes larger molecules from smaller ones.

anadromous Having a life cycle in which adults live in the ocean (or large lakes) but migrate up freshwater streams to breed and lay eggs.

anaerobic respiration Cellular respiration using an electron acceptor other than oxygen, such as nitrate or sulfate.

analogous trait Similarity between different species due to adaptation to a similar way of life rather than inheritance from a common ancestor. Often the result of convergent evolution.

anal pore An organelle in some protists used to expel wastes from the cell.

anaphase A stage in nuclear division (mitosis or meiosis) during which chromosomes are moved to opposite ends of the cell.

anaphylactic shock A life-threatening condition in which an allergic response constricts respiratory passages and dilates blood vessels, reducing blood pressure to dangerously low levels.

anatomy The study of the physical structure of organisms.

androgens A class of steroid hormones that generally promote male-like traits (although females have some androgens, too). Secreted primarily by the gonads and adrenal glands.

aneuploidy The state of having an abnormal number of copies of a certain chromosome.

angiosperm A flowering plant; a member of the lineage of plants that produces seeds within mature ovaries (fruits).

animal hemisphere The upper half of an amphibian egg cell, containing little of the yolk. Gives rise to most of the animal's body.

animal model Any disease that occurs in a non-human animal and has many parallels to a similar disease of humans. Studied by medical researchers in hope that findings may apply to human disease.

animals A major lineage of eukaryotes. Typically have a complex, large, multicellular body, eat other organisms, and are mobile.

anion A negatively charged ion.

annual plant A plant whose life cycle normally lasts only one growing season—less than one year.

anoxic For an environment, lacking oxygen.

anoxygenic photosynthesis Photosynthesis that does not include photosystem II and hence does not split water or produce oxygen.

antagonistic muscle groups Sets of muscles whose actions oppose each other and whose control must be coordinated. Often a pair of muscles that move a body part back and forth.

antenna (plural: antennae) A long appendage that is used to touch or smell.

antenna complex An array of chlorophyll molecules that receives energy from light and directs the energy to a reaction center. Part of a photosystem of the light-dependent reactions of photosynthesis.

anterior Toward an animal's head and away from its tail. The opposite of posterior.

anterior pituitary The anterior part of the pituitary gland, containing endocrine cells that release a variety of peptide hormones in response to other hormones from the hypothalamus.

anther The pollen-producing structure at the end of a flower stamen of an angiosperm plant.

antheridium (plural: antheridia) The sperm-producing structure in most land plants except angiosperms.

anthropoids One of the two major lineages of primates, including apes, humans, and all monkeys, but not prosimians.

antibiotic Any substance, such as penicillin, that can kill or inhibit the growth of bacteria.

antibiotic resistance The ability of a microorganism to grow and reproduce effectively in the presence of a certain antibiotic.

antibody A Y-shaped immunoglobulin protein, secreted from B cells, that can bind to a specific part of an antigen, tagging it for attack by the immune system.

anticodon The three bases of a transfer RNA molecule that can bind to a messenger RNA codon with a complementary sequence.

antidiuretic hormone (ADH) A peptide hormone from the posterior pituitary gland that stimulates water retention by the kidney.

antigen Any foreign molecule, often a protein, that can stimulate a specific response by the immune system.

antigen presentation Display of an antigen on the surface of an immune system cell to recruit other immune system cells.

antihistamines Drugs that oppose histamine release from mast cells, reducing inflammatory and allergic symptoms such as swelling and itching.

antiporter A membrane protein that allows an ion to diffuse down a concentration gradient, using the energy of that process to transport some other substance against its concentration gradient and in the opposite direction.

anus In a multicellular animal, the end of the digestive tract where wastes are expelled.

aorta In terrestrial vertebrates, the major artery carrying oxygenated blood away from the heart.

aphotic zone Deep water receiving no sunlight.

apical Toward the top. In plants, at the tip of a branch. In animals, on the side of an epithelial layer that faces the environment and not other body tissues.

apical bud A bud at the tip of a stem, where growth occurs to lengthen the stem.

apical dominance Inhibition of lateral bud growth by the apical meristem at the tip of the plant branch.

apical meristem A group of undifferentiated cells at the tip of a stem or root of a vascular plant. Responsible for primary growth.

apoplast A transport pathway in plant roots through the porous cell walls of adjacent cells.

apoptosis Programmed cell death. Occurs frequently during embryonic development; may occur later in response to infections or cell damage.

aquaporin A membrane protein through which water can move by osmosis.

arbuscular mycorrhizal fungi (AMF) Fungi whose hyphae enter the root cells of their host plants.

Archaea A domain consisting of unicellular prokaryotes distinguished by cell walls made of certain polysaccharides, unique plasma membranes but ribosomes and RNA polymerase similar to those of eukaryotes.

archegonium (plural: archegonia) The egg-producing structure in most land plants except angiosperms.

arteriole Tiny arteries that deliver blood to capillaries. Has muscular walls.

artery Any blood vessel that carries blood from the heart to capillaries. Has thick muscular walls that withstand and control blood pressure.

articulation A movable joint of a skeleton.

artificial selection Deliberate manipulation by humans, as in animal and plant breeding, of the hereditary traits of a population by allowing only certain individuals to reproduce.

asci (singular: ascus) Specialized spore-producing cells found at the ends of hyphae in sac fungi.

ascocarp A large, cup-shaped reproductive structure produced by some ascomycete fungi. Contains many microscopic asci, which produce spores.

asexual reproduction Any form of reproduction that results in offspring that are genetically identical to the parent.

A site The site in a ribosome where an aminoacyl transfer RNA pairs with a messenger RNA codon, in preparation for adding an amino acid to the growing peptide chain.

assimilation hypothesis The hypothesis that modern humans (*Homo sapiens*) evolved in Africa but acquired modern traits after interbreeding with other *Homo* species in Europe and Asia.

asthma A chronic lung condition characterized by periodic episodes of difficulty in breathing, due to smooth-muscle contractions in the respiratory passages.

astrobiology The study of biological molecules that occur in space (e.g., on meteors) and of possible extraterrestrial life.

asymmetric competition Ecological competition between two species in which one species suffers a much greater fitness decline than the other.

atherosclerosis A disease in which lipids accumulate on artery walls, eventually resulting in blockage of the artery.

atom The smallest unit of a chemical element that retains the characteristics of that element.

atomic mass unit (amu) A unit of mass equal to 1/12 the mass of one carbon-12 atom; about the mass of 1 proton or 1 neutron. Also called a dalton.

atomic number The number of protons in the nucleus of an atom, giving the atom its identity as a certain chemical element.

ATP See **adenosine triphosphate (ATP)**.

ATP synthase A membrane-bound enzyme in chloroplasts and mitochondria that uses the energy of protons flowing through it to synthesize ATP.

atrial natriuretic hormone A hormone from the heart that stimulates the kidney to excrete sodium, lowering blood pressure.

atrioventricular (AV) node A location in the heart between the atria and ventricles where electrical signals from the atria are slowed before spreading to the ventricles, allowing them to fill with blood before contracting.

atrium (plural: atria) A thin-walled chamber of the heart that receives blood from veins and pumps it to a neighboring chamber (the ventricle).

attenuated virus A virus that is functional but that has been rendered nonvirulent for a certain species, usually by culturing in cells of a different species. Used for vaccines.

attenuation Gradual reduction in strength. In bacterial genetics, fine-tuning of gene expression such that genes produce only as much of their proteins as required.

australopithecines Species of early hominins that appear in the fossil record shortly after the split from chimpanzees. Generally bipedal, but retained a chimpanzee-size brain.

autocrine signal A chemical signal that affects the same cell that released it.

autoimmunity A pathological condition in which the immune system attacks part of the body.

autonomic nervous system The part of the peripheral nervous system that controls internal organs and involuntary processes, such as stomach contraction, hormone release, and heart rate. Includes parasympathetic and sympathetic nerves.

autophagy The process by which damaged organelles are surrounded by a membrane and delivered to a lysosome to be destroyed.

autopolyploid Polyploid due to a mutation that doubled the chromosome number.

autosomal inheritance The inheritance patterns that occur when genes are located on autosomes rather than on sex chromosomes.

autosome One of any pair of chromosomes that do not carry the gene(s) that determine gender.

autotroph Any organism that can synthesize its own food—complex organic compounds—from simple inorganic sources such as CO_2 or CH_4, including most plants and some bacteria and archaea.

auxin Indoleacetic acid, a plant hormone that stimulates phototropism and some other responses.

avirulence (*avr*) loci Genes in pathogens that trigger a defense response in plants.

axon A long projection of a neuron that can propagate an action potential.

axoneme An arrangement of two central microtubules surrounded by nine doublet microtubules. Found in eukaryotic cilia and flagella and responsible for their motion.

axon hillock The location at which an axon joins the cell body of a neuron; the site at which action potentials are first triggered.

BAC (bacterial artificial chromosome) A lab-created loop of DNA that can be inserted into a living bacterial cell, which can then be grown in a cell culture to create many copies of the BAC.

background extinction The average rate of extinction that occurs normally, as opposed to the higher rate of extinction that occurs during mass extinction events.

BAC library A set of bacterial colonies containing BACs (bacterial artificial chromosomes) that each have a section of the DNA being studied. Commonly used in genome sequencing.

Bacteria A taxonomic domain that includes all bacteria—prokaryotes that have cell walls made of peptidoglycan, distinct forms of ribosomes and RNA polymerase, and certain other traits.

bacteriophage Any virus that infects bacteria.

baculum A bone inside the penis. Usually present in mammals that do not have erectile tissue.

bark The protective outer layer around woody plants, containing cork cells, cork cambium, and secondary phloem.

baroreceptors Specialized nerve cells in the walls of the heart and certain major arteries that detect changes in blood pressure and trigger appropriate responses by the brain.

basal body A structure of nine pairs of microtubules arranged in a circle at the base of eukaryotic cilia and flagella where they attach to the cell.

basal cell The lower cell produced by an angiosperm zygote. Gives rise to a structure that transfers nutrients from the parent plant to the developing embryo.

basal lamina A thick, collagen-rich extracellular matrix found in animal skin.

basal meristem A group of undifferentiated plant cells located below ground, from which the plant can regenerate the shoot system if necessary.

basal metabolic rate (BMR) The total energy consumption by an organism at rest in a comfortable environment. For aerobes, often measured as the amount of oxygen consumed per hour.

basal transcription complex A multi-protein structure that initiates transcription of eukaryotic genes.

basal transcription factors Proteins that bind to eukaryotic promoters and help initiate transcription.

base Any compound that acquires protons or gives up electrons during a chemical reaction or accepts hydrogen ions when dissolved in water.

baseline A measurement that represents normal values before a change or disturbance and that serves as a reference point for future comparisons.

basidia (singular: basidium) Specialized spore-producing cells at the ends of hyphae in club fungi.

basilar membrane The membrane in the vertebrate cochlea on which the hair cells sit.

basolateral Toward the bottom and sides. In animals, the side of an epithelial layer that faces other body tissues and not the environment.

basophil A circulating leukocyte involved in the inflammatory response and in allergies.

Batesian mimicry A type of mimicry in which a harmless species resembles a harmful species.

B cells A class of leukocytes that can produce antibodies. Produced by the bursa of birds and the bone marrow of mammals. Also called B lymphocytes.

B-cell receptor (BCR) A Y-shaped immunoglobulin protein found on the surfaces of B cells and to which antigens bind.

behavior Any action by an organism.

benign tumor A mass of abnormal tissue that grows slowly or not at all, does not disrupt surrounding tissues, and does not metastasize to other organs. Benign tumors are not cancers.

benthic Living at the bottom of an aquatic environment.

benthic zone The area along the bottom of an aquatic environment.

beta chain (β chain) One of the two polypeptide chains that make up a T-cell receptor protein, enabling the T cells of the immune system to bind to antigens on other cells.

bilateral symmetry An animal body pattern in which there is one plane of symmetry dividing the body into a left side and a right side. Typically, the body is long and narrow, with a distinct head end and tail end.

Bilateria A major lineage of animals that are bilaterally symmetrical at some point in their life cycle, have three embryonic germ layers, and have coeloms. Includes protostomes and deuterostomes.

bile A fluid produced by the liver, stored in the gall bladder, and secreted into the intestine, where it emulsifies fats during digestion.

bile salts Small lipids in bile that can emulsify fats.

binary fission Division of a bacterial cell to produce two daughter cells. Similar to mitosis of eukaryotic cells.

binomial nomenclature A system of naming species by using two-part Latinized names with a genus name and a species name. Always italicized, with genus name capitalized.

biodiversity The variety and relative abundance of species present in a certain area.

biofilm A tough, polysaccharide-rich substance secreted by certain prokaryotes. Encases and protects cells and attaches them to the organism's surface.

biogeochemical cycle The pattern of circulation of an element or molecule among living organisms and the environment.

biogeography The study of how species and populations are distributed geographically.

bioinformatics The field of study concerned with managing, analyzing, and interpreting biological information, particularly DNA sequences.

biological species concept The concept that species are best identified as groups that are reproductively isolated from each other; thus, different species cannot crossbreed in nature to produce viable and fertile hybrid offspring.

bioluminescence The emission of light by a living organism.

biomass The total mass of all organisms in a given population or geographical area; usually expressed as total dry weight.

biome A major category of ecosystem, characterized by a distinct type of vegetation and climate.

bioprospecting The search for naturally occurring compounds that can be useful to humans (e.g., as drugs, fragrances, or insecticides). Also called chemical prospecting.

bioremediation The use of living organisms, usually bacteria and archaea, to degrade environmental pollutants.

biotechnology The application of biological techniques and discoveries to medicine, industry, and agriculture.

biotic Living, or produced by a living organism.

bipedal Walking primarily on two legs.

bipolar cell Cells in the vertebrate retina that receive information from one or more photoreceptors and pass it to other bipolar cells and ganglion cells.

bivalves A lineage of mollusks that have two shells, such as clams and mussels.

bladder A mammalian organ that holds urine until it can be excreted.

blade The wide, flat part of a plant leaf.

blastomere The small cells created by cleavage divisions in early animal embryos.

blastopore A small pore in the surface of an early vertebrate embryo, through which cells move during gastrulation.

blastula An early stage of embryonic development in vertebrates, consisting of a ball of cells (a blastomere) enclosing a fluid-filled space. Immediately precedes gastrulation.

blood A type of liquid connective tissue consisting of red blood cells and leukocytes suspended in the fluid plasma.

blood doping The practice of giving an athlete a transfusion of his or her own red blood cells (removed and stored earlier) to increase aerobic capacity for a competition.

body mass index A mathematical relationship of weight and height used to assess obesity in humans. Calculated as weight (kg) divided by the square of height (m^2).

body plan The basic architecture of an animal's body, including the number and arrangement of limbs, body segments, and major tissue layers.

bog A wetland that has no or almost no water flow, resulting in very low oxygen levels and acidic conditions.

Bohr shift The rightward shift of the oxygen-hemoglobin dissociation curve that occurs with decreasing pH. Results in hemoglobin being more likely to release oxygen in the acidic environment of exercising muscle.

bone A vertebrate tissue consisting of living cells and blood vessels within a hard extracellular matrix of calcium phosphate $(CaPO_4)$ with small amounts of calcium carbonate $(CaCO_3)$ and protein fibers.

bone marrow The soft tissue filling the inside of long bones. Produces red blood cells and leukocytes and also contains fat.

Bowman's capsule The hollow, sphere-like end of a kidney nephron that surrounds a glomerulus.

braincase A bony, cartilaginous, or fibrous case that encloses and protects the brain of vertebrates. Forms part of the skull. Also called the cranium.

branch (1) A part of a phylogenetic tree that represents populations. (2) Any extension of a plant's shoot system.

bronchi (singular: bronchus) In mammals, the large tubes that lead from the trachea to each lung.

bronchioles The small tubes in mammalian lungs that carry air from the bronchi to the alveoli.

brown adipose tissue A specialized form of fat tissue whose cells have a high density of mitochondria as well as stored fats and that can produce extra body heat. Found in some mammals.

bryophytes Several phyla of green plants that lack vascular tissue. Includes liverworts, hornworts, and mosses. Also called nonvascular plants.

budding Asexual reproduction via growth of a small individual from part of the parent, eventually breaking free as an independent individual.

buffer A substance that, in solution, acts to minimize changes in the pH of that solution.

bulbourethral glands Small, paired glands at the base of the urethra in male mammals that secrete pre-ejaculatory fluid for lubrication during copulation. In humans, also called Cowper's glands.

bulk flow The directional movement of a substantial volume of fluid due to pressure differences, such as movement of water through plant phloem and movement of blood in animals.

bundle-sheath cell A type of cell that is located around the vascular tissue in the interior of plant leaves.

Burgess shale fauna A characteristic set of soft-bodied animal fossils found in several early Cambrian rock formations (525–515 million years old), particularly the Burgess Shale in Canada and the Chengjiang deposits in China.

bursa An immune system organ of birds that produces B cells.

C3 photosynthesis A common form of photosynthesis in which atmospheric carbon is used to form 3-phosphoglycerate, a three-carbon compound.

C4 photosynthesis A variant of photosynthesis in which carbon from CO_2 is first fixed into four-carbon compounds. Enhances photosynthetic efficiency in hot, dry environments by reducing loss of oxygen due to photorespiration.

cadherins Cell-surface proteins involved in cell adhesion and important for coordinating movements of cells during embryological development.

calcitonin A hormone from the thyroid gland that lowers blood calcium by preventing calcium and phosphorus withdrawal from bone.

callus A mass of undifferentiated plant cells that can generate roots and other tissues necessary to create a mature plant.

Calorie A unit of energy, often used to measure energy content of food. Also called a kilocalorie.

Calvin cycle In photosynthesis, the set of reactions that use the NADPH and ATP formed earlier (in the light-dependent reactions) to drive the reduction of atmospheric CO_2, ultimately producing sugars. Also called light-independent reactions.

CAM See **crassulacean acid metabolism (CAM)**.

cambium (plural: cambia) A layer of undifferentiated plant cells found in woody plants. Responsible for secondary growth. Also called secondary meristem or lateral meristem.

Cambrian explosion The rapid diversification of animal body types that began about 543 million years ago, during approximately 40 million years.

camera eye A type of eye in vertebrates and cephalopods, consisting of a hollow chamber with a hole at one end (through which light enters) and a sheet of light-sensitive cells against the opposite wall.

cAMP See **cyclic AMP (cAMP)**.

canopy The uppermost layers of branches in a forest (i.e., those fully exposed to the Sun).

5′ cap A structure added to the 5′ end of newly transcribed messenger RNA molecules. Consists of 7-methylguanylate and three phosphate groups.

CAP binding site A DNA sequence upstream of certain prokaryotic operons, to which catabolite activator protein can bind, increasing gene transcription.

capillarity The tendency of water to move up a narrow tube due to surface tension, adhesion, and cohesion.

capillary The smallest blood vessel, where gases and other molecules are exchanged between blood and tissues.

capillary bed A thick network of capillaries.

capsid A shell of protein enclosing the genome of a virus particle.

carapace In crustaceans, a large platelike section of the exoskeleton that covers and protects the cephalothorax (e.g., a crab's "shell").

carbohydrates A class of molecules, including sugars, starches, glycogen, and cellulose, that have a carbonyl group, several hydroxyl groups, and several to many carbon-hydrogen bonds.

carbonic anhydrase An enzyme that catalyzes the formation of carbonic acid (H_2CO_3) from carbon dioxide and water.

carbon sink A reservoir that holds carbon for very long time periods; e.g., fossil fuels, sedimentary rocks.

carcinogen Any cancer-causing agent.

cardiac cycle One complete heartbeat cycle, including systole and diastole.

cardiac muscle The muscle tissue of the vertebrate heart. Consists of long branched fibers that are electrically connected and that initiate their own contractions; not under voluntary control.

cardiac output The total volume of blood leaving the left ventricle per minute. Calculated as stroke volume times heart rate.

carnivore (1) Any animal whose diet consists predominantly of other animals. (2) Any member of the mammalian taxon Carnivora. (Most members of the Carnivora are meat eaters.)

carotenoids A class of plant pigments that absorb wavelengths not absorbed by chlorophyll. Typically appear yellow, orange, or red. Includes carotenes and xanthophylls.

carpel In a flower, the reproductive structure that produces female gametophytes; consists of the stigma, the style, and the ovary.

carrier (1) A heterozygous individual carrying a normal allele and a recessive allele for an inherited condition; does not display the phenotype of the condition but can pass the recessive gene to offspring. (2) A transmembrane protein that assists with facilitated diffusion by binding a specific ion or molecule and transporting it across the membrane.

carrying capacity The maximum population size of a certain species that a given habitat can support. Symbolized by K.

cartilage A type of connective tissue in the skeletons of vertebrates, consisting of relatively few cells scattered in a stiff matrix of polysaccharides and protein fibers.

Casparian strip A waxy layer that prevents movement of water through the walls in plant roots.

catabolic pathway Any set of chemical reactions that breaks down larger, complex molecules into smaller ones, releasing energy in the process.

catabolite The end product of a catabolic pathway; the final breakdown product of a substance.

catabolite activitor protein (CAP) A protein that can bind to the CAP binding site upstream of certain prokaryotic operons, facilitating binding of RNA polymerase and stimulating gene expression.

catabolite repression A type of inhibition of gene transcription in which a gene codes for an enzyme in a catabolic pathway and the end product of that pathway inhibits further transcription of that gene.

catalysis Acceleration of the rate of a chemical reaction, produced by lowering the potential energy of the transition state.

catalyst Any substance that increases the rate of a chemical reaction without itself undergoing any permanent chemical change. Functions by lowering the activation energy of the reaction.

catecholamines A class of small compounds, derived from the amino acid tyrosine, that are used as hormones or neurotransmitters. Includes epinephrine, norepinephrine, and dopamine.

cation A positively charged ion.

cation exchange The release of cations such as magnesium and calcium from soil particles, due to displacement by protons in acidic soil water.

CD4 A membrane protein on the surface of some T lymphocytes of the human immune system; these CD4$^+$ T cells can give rise to helper T cells.

CD8 A membrane protein on the surface of some T lymphocytes of the human immune system; these CD8$^+$ T cells can give rise to cytotoxic T cells.

Cdk See cyclin-dependent kinase (Cdk).

cDNA (complementary DNA) DNA created in a lab from an RNA transcript, using the enzyme reverse transcriptase. Corresponds to a certain gene but lacks introns. Is also created naturally by some viruses (retroviruses).

cDNA library A set of cDNAs from a particular organism or cell type. Usually exists as a set of bacteria colonies, each with a plasmid containing a particular cDNA sequence.

cecum A blind sac between the small intestine and the colon. Used in some species as a fermentation vat for digestion of cellulose.

cell A highly organized living entity that is bounded by a thin, flexible structure called a plasma membrane and that contains concentrated chemicals in an aqueous (watery) solution.

cell body The part of a neuron that contains the nucleus and where incoming signals are integrated. Also called the soma.

cell-cell signal A molecule that is secreted from one cell and affects the activity of a nearby cell.

cell crawling A form of cellular movement in which the cell produces bulges (pseudopodia) that stick to the substrate and pull the cell forward.

cell cycle The sequence of stages that a dividing eukaryotic cell goes through from the time it is created (by division of a parent cell) to the time it undergoes mitosis.

cell-cycle checkpoint A regulated point at which the cell cycle can be stopped.

cell division Creation of new cells by division of pre-existing cells (i.e., by mitosis or meiosis).

cell enlargement Growth in plants in which individual cells expand in size in a certain direction.

cell extract or cell homogenate A solution created by breaking cells apart. Cell extracts contain organelles, free macromolecules, and small membrane-bound vesicles.

cell-mediated response Defense mounted by cytotoxic T lymphocytes against infections.

cell plate A double layer of new plasma membrane that appears in the middle of a dividing plant cell; ultimately divides the cytoplasm into two separate cells.

cell sap An aqueous solution found in the vacuoles of plant cells.

cell theory The theory that all organisms are made of cells and that all cells come from pre-existing cells.

cellular respiration A common pathway for production of ATP, involving transfer of electrons from compounds with high potential energy (often NADH and FADH$_2$) to an electron transport chain and ultimately to an electron acceptor (often oxygen).

cellulases Enzymes that can digest cellulose.

cellulose A polysaccharide that is the major component of plant cell walls. Consists of β-glucose monomers joined with β-1,4-glycosidic linkages in long, straight chains.

cell wall A protective layer located outside the plasma membrane and usually composed of polysaccharides. Found in algae, plants, bacteria, fungi, and some other groups.

central dogma The long-accepted hypothesis that information in cells flows in one direction: DNA codes for RNA, which codes for proteins. Exceptions are now known (e.g., retroviruses).

central nervous system (CNS) The brain and spinal cord of vertebrate animals.

centrifugation A lab technique for separating substances by density and size by spinning them rapidly in a circle, so that larger or denser particles are flung to the outside of the sample container.

centriole One of two small cylindrical structures found together near the nucleus of a eukaryotic cell. Collectively called the centrosome, they serve as a microtubule hub for the cell's cytoskeleton.

centromere The structure that joins two sister chromatids during meiosis.

centrosome Structure in animal and fungal cells, consisting of two centrioles together near the nucleus. Serves as a microtubule organizing center for the cell's cytoskeleton and for the mitotic spindle during cell division.

cephalization The formation of a distinct head—an anterior region where sense organs and a mouth are clustered.

cephalochordates A lineage of small, mobile chordates also called lancelets or amphioxi.

cephalopods A lineage of mollusks including the squid, octopuses, and nautiluses. Distinguished by large brains, excellent vision, tentacles, and a reduced or absent shell.

cerebellum A posterior section of the vertebrate brain, involved in coordination of complex muscle movements, such as in locomotion and maintaining balance.

cerebrum The anteriormost section of the vertebrate brain. Divided into left and right hemispheres and involved in memory, interpretation of information, decision making, and (in humans) conscious thought.

cervix The narrow passageway between the vagina and the uterus of female mammals.

channel protein, or channel A membrane protein that forms a pore that selectively allows a specific ion or molecule to cross the membrane via diffusion. Channel proteins provide passive transport (facilitated diffusion) and do not require energy.

chelicerae Appendages found around the mouth of the chelicerates (spiders, mites, and allies).

chemical bond A strong attractive force binding two atoms together. Covalent bonds, ionic bonds, and hydrogen bonds are types of chemical bonds.

chemical carcinogen Any chemical that can cause cancer.

chemical energy The potential energy stored in covalent bonds between atoms.

chemical equilibrium A dynamic but stable state of a reversible chemical reaction in which the forward reaction and reverse reactions proceed at the same rate, so that the concentrations of reactants and products remain constant.

chemical evolution The hypothesis that simple chemical compounds in the ancient atmosphere and ocean combined by spontaneous chemical reactions to form larger, more complex substances, eventually leading to the origin of life and the start of biological evolution.

chemical reaction An event in which one compound or element is combined with others or is broken down.

chemiosmosis The production of ATP via proton movement, through ATP synthase, across a membrane, driven by a proton gradient.

chemiosmotic hypothesis The hypothesis that ATP synthesis in mitochondria and chloroplasts occurs indirectly via proton movement across a membrane.

chemokine A chemical signal that attracts leukocytes to a site of tissue injury or infection.

chemoreceptor A sensory cell or organ specialized for detection of specific molecules or classes of molecules.

chemotaxis Movement toward or away from a certain chemical.

chemotherapy Treatment with anticancer chemicals or drugs. Most chemotherapy drugs kill dividing cells or stop the cell cycle.

chiasma (plural: chiasmata) The X-shaped structure formed during meiosis by crossing over between adjacent chromatids of a pair of homologous chromosomes.

chitin A polysaccharide consisting of monomers of N-acetylglucosamine joined end to end in long, straight chains. Found in cell walls of fungi and many algae, and in external skeletons of insects and crustaceans.

chitons Marine mollusks that have a protective shell formed of eight calcium carbonate plates.

chloride cells In the gills of marine fish with bony skeletons, cells that excrete excess salt and maintain electrolyte balance.

chlorophyll A green pigment molecule that absorbs light energy to power photosynthesis. Found in plant cells and in photosynthetic protists.

chloroplast In plants, a chlorophyll-containing organelle in which photosynthesis occurs. Also, the location of amino-acid, fatty-acid, purine, and pyrimidine synthesis.

chloroplast DNA DNA found within a chloroplast of a eukaryotic plant cell. Consists of a single circular chromosome.

choanocytes The specialized feeding cells found in sponges.

choanoflagellates A phylum of protists thought to be the closest living relatives of animals.

cholecystokinin A peptide hormone from the small intestine that stimulates the secretion of digestive enzymes from the pancreas and bile from the liver and gallbladder.

cholera A human infectious disease characterized by watery diarrhea that can lead to dehydration and death. Caused by the bacterium *Vibrio cholerae*.

cholesterol A steroid that is a major component of plasma membranes. Required for synthesis of steroid hormones. In excess, can contribute to atherosclerosis.

chordate An animal of the phylum Chordata, distinguished by such traits as a dorsal hollow nerve cord, pharyngeal gill slits, and a post-anal tail. Includes vertebrates.

chorion In an amniotic egg, a highly vascularized membrane across which gas exchange occurs with the environment.

chromatid One of the daughter strands of a chromosome that has recently been copied (prior to mitosis or meiosis) and that is still connected to the other daughter strand.

chromatin The material that makes up eukaryotic chromosomes: a DNA molecule complexed with histone proteins. Can be highly compact (heterochromatin) or filamentous (euchromatin).

chromatin remodeling The process by which the DNA in chromatin is unwound from its associated proteins to allow transcription or replication.

chromatin-remodeling complex A protein involved in restructuring chromatin (DNA packed with proteins) in eukaryotic cells. The remodeling requires ATP.

chromosome A single long molecule of DNA and any associated proteins (e.g., histones).

chromosome inversion A mutation in which a segment of a chromosome breaks from the rest of the chromosome, flips, and rejoins with the opposite orientation as before.

chromosome painting A technique for producing high-resolution karyotypes by "painting" chromosomes with fluorescent tags that bind to particular regions of certain chromosomes. Also called spectral karyotyping (SKY).

chromosome theory of inheritance The theory that Mendel's rules of inheritance can be explained by the independent segregation of homologous chromosomes during meiosis.

chylomicron A ball of protein-coated lipids, used to transport the lipids through the bloodstream.

cilia (singular: cilium) Short, numerous, filamentous projections of some eukaryotic cells, used to move the cell and/or to move fluid or particles along a stationary cell.

circadian Lasting approximately one day.

circadian clock An internal mechanism found in most organisms that regulates many body processes (sleep-wake cycles, hormonal patterns, etc.) in a roughly 24-hour cycle.

circadian rhythm Any biological process that occurs on an approximately 24-hour cycle and that is controlled by an internal circadian clock.

circulation In physiology, mass movement of blood throughout the body.

cisterna (plural: cisternae) A flattened, membrane-bound compartments of the Golgi apparatus.

clade An evolutionary unit that includes an ancestral population, all of its descendants, and only its descendants. Also called a monophyletic group or a lineage.

cladistic approach A method for constructing a phylogenetic tree that is based on identifying the unique traits of each monophyletic group.

class A classic taxonomic rank above the order level and below the phylum level.

Class I MHC protein A type of MHC protein that is present on the plasma membrane of every body cell, marking the cells as "self" to the immune system.

Class II MHC protein A type of MHC protein, present only on the plasma membranes of macrophages, B cells, and T cells, that helps them join together during B-cell activation.

classical conditioning A type of learning in which an animal learns to associate two stimuli, so that a response originally given to just one stimulus can be evoked by the second stimulus as well.

cleavage Rapid cell division without production of new cytoplasm. Seen only in early embryonic development in animals. Cleavage transforms a zygote into a blastula.

cleavage furrow A pinching-in of the plasma membrane that occurs as the cytoplasm of an animal cell begins to divide; one of the final events of cell division in animals, fungi, and slime molds.

climate The prevailing long-term weather conditions in a particular region.

climax community The stable, final stage of an ecological community that develops from ecological succession.

clitoris A small rod of erectile tissue in the external genitalia of female mammals. Forms from the same embryonic tissue as the male penis and has a similar function in sexual arousal.

clonal expansion Rapid cell division by a particular T cell or B cell of the immune system in response to a particular antigen. Produces a large population of descendant cells that can attack the antigen.

clonal-selection theory The dominant theory of the development of acquired immunity in vertebrates, proposing that the immune system retains a vast pool of inactive lymphocytes, each with a unique receptor for a unique antigen. Lymphocytes that encounter their antigens are stimulated to divide (selected and cloned), producing daughter cells that combat infection and confer immunity.

clone An individual that is genetically identical to another individual.

closed circulatory system A circulatory system in which the circulating fluid (blood) is confined to blood vessels and flows in a continuous circuit.

clot A mass of red blood cells, platelets, and protein fibers that forms to plug a hole in the circulatory system and stop blood loss.

clutch size The number of eggs laid by a female in a single nest or a single breeding effort.

cnidocyte A specialized stinging cell found in cnidarians.

coactivators A class of regulatory proteins that help initiate transcription by bringing together the necessary transcription factor proteins.

cochlea A coiled, fluid-filled tube in the inner ear of mammals, birds, and crocodilians. Contains nerve cells that detect sounds of different pitches.

codominance An inheritance pattern in which heterozygotes exhibit both of the traits seen in either kind of homozygous individual.

codon A sequence of three nucleotides of DNA or RNA that codes for a certain amino acid or that initiates or terminates protein synthesis.

coefficient of relatedness A measurement of how closely two individuals are related. Calculated as the probability that an allele in two individuals is inherited from the same ancestor. Symbolized as r.

coelom An internal, usually fluid-filled, body cavity that forms within the mesoderm.

coelomate Possessing a coelom.

coenocytic Containing many nuclei and a continuous cytoplasm through a filamentous body, without the body being divided into distinct cells. Some fungi are coenocytic.

coenzyme Any non-protein molecule or ion that is a required cofactor for an enzyme-catalyzed reaction. Often transfers or receives electrons or functional groups.

coenzyme A (CoA) A coenzyme that is required for many cellular reactions involving transfer of acetyl groups ($-COCH_3$). In glycolysis, pyruvate reacts with CoA to produce acetyl CoA, which can then enter the Krebs cycle.

coenzyme Q A molecule that transfers electrons in the electron transport chain of cellular respiration. Also called ubiquinone or Q.

coevolutionary arms race A pattern of evolution in which one species evolves a defense against another species (e.g., a predator or parasite), which then evolves a counterdefense, and so on.

cofactor A non-protein molecule or ion that is required for an enzyme to function normally. May be a metal ion or a coenzyme.

cognition The mental processes involved in recognition and manipulation of facts about the world, particularly to form novel associations or insights.

cohesion A phenomenon seen in some liquids, such as water, in which attractive forces between molecules cause the liquid molecules to cling together and resist disruption.

cohesion-tension theory The theory that water movement upward through plant vascular tissues is due to transpiration (loss of water from leaves), which pulls a cohesive column of water upward.

cohort A group of individuals that are the same age and can be followed through time.

coleoptile A modified leaf that covers and protects the stems and leaves of young grasses.

collagen A fibrous, cable-like protein secreted by animal cells into the extracellular matrix. Forms a pliable, strong substance in which the cells sit.

collenchyma cell An elongated type of plant cell with cell walls thickened at the corners that provides support to growing plant parts; usually found in strands along leaf veins and stalks.

colonial growth Growth in which unicellular or multicellular individuals join to form a structure in which the multiple individuals are not specialized and each retains the ability to reproduce.

colony An assemblage of individuals. May refer to an assemblage of semi-independent cells or to a breeding population of multicellular organisms.

combination therapy Medical therapy that involves dosing an infected patient with several drugs simultaneously, to lessen the chances of the pathogen evolving resistance.

commensal A species that is dependent on another species but does not harm that species.

commensalism A species interaction in which one species benefits and the other is not harmed.

communication Any process in which a signal from one individual modifies the behavior of another individual.

community All of the species that interact with each other in a certain area.

companion cell Cell in phloem tissue that provides materials and nutrients to sieve-tube members.

compass orientation Motion in a specific compass direction.

competition The effect of two species or two individuals trying to use the same limited resource.

competitive exclusion principle The principle that two species cannot coexist in the same ecological niche in the same area because one species will out-compete the other.

competitive inhibition Inhibition of an enzyme's ability to catalyze a chemical reaction via a nonreactant molecule that competes with the substrate(s) for access to the active site.

complementary base pairing The specific pairing that occurs between nitrogenous bases of nucleic acids: Adenine pairs only with thymine (in DNA) or uracil (in RNA), and guanine pairs only with cytosine. Allows accurate replication of DNA and RNA sequences.

complementary DNA (cDNA) DNA created in the lab from an RNA transcript, using reverse transcriptase; corresponds to a particular gene but lacks introns. Created naturally by retroviruses.

complementary strand A new strand of RNA or DNA that has a base sequence complementary to that of the template strand.

complement system A class of proteins that circulate in the bloodstream and attack plasma membranes of bacteria.

compound eye An arthropod eye formed of many independent light-sensing columns (ommatidia).

concentration gradient Variation across space in the concentration of a dissolved substance, from a region of high concentration to a region of low concentration.

condensation reaction A type of chemical reaction involving the bonding together of two molecules by removal of an $-OH$ from one and an $-H$ from another to form water. Also called a dehydration reaction. The reverse of hydrolysis.

conditional strategy A behavioral response that varies with the current environmental and social conditions.

conduction Direct transfer of heat between two objects that are in physical contact.

cone (1) A photoreceptor cell with a cone-shaped outer portion that is particularly sensitive to bright light of a certain color. Found in eyes of vertebrates and some other animals. (2) The reproductive structure found in conifers.

confocal microscopy A technique for obtaining a focused image of a certain plane within a live cell.

conformational homeostasis Homeostasis (steady internal body conditions) that is achieved by the body's passively matching the conditions of a stable external environment.

conjugation The process by which DNA is exchanged between unicellular individuals. Occurs in bacteria, archaea, and some protists.

conjugation tube A connection between two prokaryotes that are in the process of transferring plasmids to each other.

connective tissue A class of animal tissue consisting of scattered cells in a liquid, jellylike, or solid extracellular matrix. Includes bone, cartilage, tendons, ligaments, and blood.

conservation biology The effort to study, preserve, and restore threated populations, communities, and ecosystems.

conservative replication A now-disproven hypothesis of DNA replication that proposed that one of the daughter molecules retains both original strands of DNA, while the other daughter molecule is built with two new strands.

constant (C) region A section of the light chains of antibodies that has the same amino acid sequence in every B cell of an individual.

constitutive Always occurring; always present (as in enzymes that are synthesized continuously).

constitutive defense A defensive trait that is always present, regardless of need. Also called standing defense.

constitutive mutant A mutant in which certain genetic loci are constantly transcribed due to flaws in gene regulation.

consumer An organism that consumes food created by other organisms; not a primary producer.

continental drift The motion of continents over large periods of time due to plate tectonics.

continental shelf The shallow, gently sloping portion of the ocean floor near continents.

continuous strand In DNA replication, the strand of new DNA synthesized in one continuous piece, with nucleotides added to the 3' end of the growing molecule. Also called leading strand.

contraception Any method to prevent pregnancy.

control In an experiment, a group of organisms or samples that do not receive the experimental treatment but are otherwise identical to the group that does.

convection Transfer of heat by movement of large volumes of a gas or liquid.

convergent evolution Evolution of similar traits in distantly related organisms due to adaptation to similar environments and a similar way of life. Often produces analogous traits.

convergent trait Similarity between different species that is due to adaptation to a similar way of life rather than inheritance from a common ancestor. Often the result of convergent evolution. Also called an analogous trait.

cooperative binding The tendency of the protein subunits of hemoglobin to affect each other's oxygen binding such that each bound oxygen molecule increases the likelihood of further oxygen binding.

copulation The act of transferring sperm from a male directly into a female's reproductive tract.

coral reef A large assemblage of colonial marine corals that usually serves as shallow-water, sunlit habitat for many other species as well.

co-receptor Any membrane protein that acts with some other membrane protein in a cell interaction or cell response.

core enzyme The part of a holoenzyme that contains the active site for catalysis.

cork cambium A ring of undifferentiated plant cells found just under the cork layer of woody plants. Produces new cork cells on its outer side.

cork cells The waxy cells in the protective outermost layer of a woody plant.

corm A rounded, thick underground stem that can produce new plants via asexual reproduction.

cornea The transparent sheet of connective tissue at the very front of the eye in vertebrates and some other animals. Protects the eye and helps focus light.

corolla All of the petals of a flower.

corona The cluster of cilia at the anterior end of a rotifer.

coronary heart disease Progressive weakening of the heart muscle due to chronic oxygen deprivation caused by blocked coronary arteries.

corpus callosum A thick band of neurons that connects the two cerebral hemispheres of the mammalian brain.

corpus luteum A yellowish structure in an ovary, formed from a follicle that has recently ovulated. Secretes progesterone.

cortex (1) The outermost region of an organ, such as the kidney or adrenal gland. (2) In plants, a layer of ground tissue found outside the vascular bundles and pith of a plant stem.

cortical granules Small enzyme-filled vesicles in the cortex of an egg cell. Involved in formation of the fertilization envelope after fertilization.

corticosterone The major glucocorticoid hormone released by the cortex of the adrenal gland in most reptiles, birds, and many mammals. Increases blood glucose and prepares the body for stress.

corticotropin-releasing hormone (CRH) A peptide hormone from the hypothalamus that stimulates the anterior pituitary gland to release ACTH.

cortisol The major glucocorticoid hormone released by the cortex of the adrenal gland in some mammals. Increases blood glucose and prepares the body for stress. Also called hydrocortisone.

cotransporter A membrane protein that allows an ion to diffuse down a previously established concentration gradient and uses the energy of that process to transport some other substance in the same direction. Also called a symporter.

cotyledon The first leaf, or seed leaf, of a plant embryo. Used for storing and digesting nutrients and/or for early photosynthesis.

countercurrent exchange A particularly efficient mechanism for the exchange of heat or a soluble substance, based on transfer between parallel tubes carrying fluids in opposite directions.

countercurrent heat exchanger A specialized network of blood vessels that recirculates body heat within a certain part of the body. A type of countercurrent exchanger.

covalent bond A type of molecular bond in which two atoms share one pair of electrons.

cranium A bony, cartilaginous, or fibrous case that encloses and protects the brain of vertebrates. Forms part of the skull. Also called braincase.

crassulacean acid metabolism (CAM) A variant of photosynthesis in which CO_2 is stored in organic compounds at night when stomata are open and released to enter the Calvin cycle during the day when stomata are closed. Helps reduce loss of water and oxygen by photorespiration in hot, dry environments.

crista (plural: cristae) Membranous sacs that articulate with the inner membrane of a mitochondrion. Location of the electron transport chain and ATP synthase.

critical period A short time span in a young animal's life during which the animal can learn certain things, such as song, language, or imprinting. Also called the sensitive period.

Cro-Magnon A prehistoric European population of modern humans (*Homo sapiens*) known from fossils, paintings, sculptures, and other artifacts.

crossing over The exchange of segments of non-sister chromatids between a pair of homologous chromosomes that occurs during meiosis.

cross-pollination Pollination of a flower by pollen from another individual, rather than by self-fertilization.

cryptochromes A class of plant photoreceptors that detect blue light and affect stem growth and flowering in shady conditions.

cud A partially digested package of food and symbiotic bacteria from the rumen that ruminants regurgitate for further chewing.

culture A collection of cells growing under controlled conditions in a lab, usually in suspension or on the surface of a dish on solid growth medium.

cup fungi A monophyletic lineage of fungi that produce large, often cup-shaped reproductive structures that contain spore-producing asci. Also called sac fungi.

current A flow of electrical charge past a point. Also called electric current.

Cushing's disease A human endocrine disorder caused by loss of feedback inhibition of cortisol on ACTH secretion. Characterized by high ACTH and cortisol levels and wasting of body protein reserves.

cuticle A protective coating secreted by the outermost layer of cells of an animal or a plant.

cyanobacteria A lineage of photosynthetic bacteria formerly known as blue-green algae. Likely the first life-forms to evolve by oxygenic photosynthesis.

cyclic AMP (cAMP) Cyclic adenosine monophosphate; a small molecule, derived from ATP, that is widely used in cells for signaling and regulation (e.g., in gene transcription, enzyme control, and hormone signal transduction).

cyclic photophosphorylation An alternative pathway in the light-dependent reactions of photosynthesis, in which excited electrons from photosystem I are transferred back to the electron transport chain of photosystem II to increase ATP generation.

cyclin-dependent kinase (Cdk) A protein kinase that is active only when bound to a cyclin. Involved in control of the cell cycle.

cyclins A class of proteins whose concentrations fluctuate cyclically, following the cell cycle.

cyst In some species, a protective structure containing a diploid cell in a resting state.

cystic fibrosis A human disease caused by a defective chloride channel. Causes thickened mucus in the respiratory tract and deterioration of the gastrointestinal and reproductive tracts.

cytochrome C A protein that helps transfer electrons between the parts of the electron transport chain in mitochondria.

cytokine Generally, any substance that stimulates cell division. Many cytokines are secreted by macrophages and helper T cells during an immune response, stimulating leukocyte production, tissue repair, and fever.

cytokinesis Division of the cytoplasm to form two daughter cells. Typically occurs immediately after division of the nucleus by mitosis or meiosis.

cytokinins A group of plant hormones that stimulate cell division.

cytoplasm All of the contents of a cell, excluding the nucleus.

cytoplasmic determinant A regulatory molecule that is distributed unevenly in the cytoplasm of the egg cells of many animals and that directs the differentiation of embryonic cells.

cytoplasmic streaming The directed flow of cytosol and organelles around the interior of a plant or fungal cell. Occurs along actin filaments and is powered by myosin.

cytoskeleton A network of protein fibers embedded in the cytoplasm of eukaryotic cells. Involved in cell shape, support, locomotion, and transport of materials within the cell. Includes microtubules, intermediate filaments, and actin filaments.

cytosol The fluid portion of the cytoplasm in a cell (i.e., not including the organelles).

cytotoxic T lymphocyte (CTL) A T cell that destroys infected cells and cancer cells. Cytotoxic T lymphocytes are descendants of an activated CD8$^+$ T cell. Also called killer T cell.

Darwinian fitness The ability of an organism to produce surviving fertile offspring. Also called fitness.

day-neutral plant A plant whose flowering time is not affected by photoperiod (relative length of day and night).

dead space Portions of the air passages that are not involved in gas exchange with the blood, such as the trachea and bronchi.

deciduous Shedding leaves annually.

decomposer A species that feeds on the dead bodies of other organisms. Decomposers include various bacteria, fungi, and protists.

decomposer food web An ecological network of detritus, decomposers that eat detritus, and predators and parasites of the decomposers.

definitive host The host species in which a parasite reproduces sexually.

dehydration reaction A type of chemical reaction involving the bonding of two molecules by removal of an –OH from one subunit and an –H from another to form water. Also called a condensation reaction. The reverse reaction of hydrolysis.

deleterious allele Any allele that reduces an individual's fitness.

deleterious mutation Any mutation that reduces an individual's Darwinian fitness.

demography The study of factors that determine the size and structure of populations through time.

denatured For a protein's three-dimensional structure, unfolded; usually due to breakage of hydrogen bonds and disulfide bonds.

dendrite A short extension from a neuron's cell body that receives neurotransmitters from other neurons.

dendritic cell A type of leukocyte that ingests foreign antigens, moves to a lymph node, and presents the antigens on its membrane to other immune system cells.

density-dependent factors Factors that affect birth rates or death rates differently, depending on population size.

density-dependent growth Population growth that is limited by increasing population size.

density-independent factors Factors that change birthrates and death rates irrespective of population size.

density-independent growth Population growth that is not affected by the population size.

dental plaque A biofilm of bacteria that can form on mammalian teeth.

deoxynucleoside triphosphate (dNTP) A monomer that can be polymerized to form DNA. Consists of deoxyribose, a base (A, T, G, or C), and three phosphate groups; similar to a nucleotide, but with two more phosphate groups.

deoxyribonucleic acid (DNA) A polymer consisting of two strings of deoxyribonucleotides, wound together in a double helix. Contains the genetic information of a cell.

deoxyribonucleotide A nucleotide consisting of the sugar deoxyribose, a phosphate group, and one of four nitrogen-containing bases (adenine, cytosine, guanine, or thymine). A subunit of deoxyribonucleic acid (DNA).

dephosphorylation Removal of a phosphate group from a molecule. A common mechanism of controlling protein shape or function.

depolarization A change in membrane potential from its resting negative state to a less negative or a positive state.

deposit feeder An organism that eats its way through a substrate.

dermal tissue Tissue forming the outer layer of an organism. In plants, also called epidermis; in animals, forms two distinct layers: dermis and epidermis.

descent with modification The phrase used by Darwin to describe his hypothesis of evolution by natural selection.

desmosome A complex physical connection between two animal cells, consisting of proteins that bind the cells' cytoskeletons together. Found where cells are strongly attached to each other.

detergent An amphipathic molecule that forms micelles in water and that can cleanse by suspending hydrophobic molecules (such as oily dirt) in water.

determination The irreversible commitment during development of a cell to becoming a particular cell type (e.g., liver cell, brain cell).

detritivore An organism whose diet consists mainly of detritus.

detritus A layer of dead organic matter that accumulates at ground level or on seafloors and lake bottoms.

deuterostomes A major lineage of animals that share a pattern of embryological development, including radial cleavage, formation of the anus earlier than the mouth, and formation of the coelom by pinching off of layers of mesoderm from the gut. Includes echinoderms and chordates.

developmental homology A similarity in embryonic form, or in the fate of embryonic tissues, that is due to inheritance from a common ancestor.

diabetes insipidus A human disease caused by defects in the kidney's system for conserving water. Characterized by production of large amounts of dilute urine.

diabetes mellitus A human disease caused by defects in insulin production or response. Characterized by abnormally high blood glucose levels and large amounts of glucose-containing urine.

diaphragm An elastic, sheetlike structure. In mammalian ventilation, the muscular sheet that separates the chest from the abdominal cavities and that can expand the chest cavity by moving downward.

diastole The portion of the heartbeat cycle during which the atria or ventricles of the heart are relaxed.

diastolic blood pressure Blood pressure in arteries during relaxation of the heart's left ventricle.

dicot A dicotyledonous plant (i.e., any plant that has two cotyledons upon germination).

dideoxy sequencing A lab technique for determining the exact nucleotide sequence of DNA. Relies on the use of dideoxynucleotide triphosphates (ddNTPs), which terminate DNA replication.

differential centrifugation Separation of cellular components by spinning a cell homogenate in a series of centrifuge runs at progressively higher velocities.

differential gene expression Expression of different genes in different cell types.

differentiation The process by which a cell becomes a particular cell type (e.g., liver cell, brain cell) by differential gene expression.

diffusion Spontaneous movement of molecules and ions along a concentration gradient, from an area of high concentration to one of low concentration.

digestion The breakdown of food into pieces that are small enough to be absorbed.

digestive tract The long tube that begins at the mouth and ends at the anus. Also called alimentary canal, gastrointestinal tract, or the gut.

dihybrid cross A mating between two parents that are heterozygous for both of the two genes being studied.

dimer An association of two identical molecules.

dioecious Having either male flowers or female flowers but not both.

diploblast An animal whose body develops from two basic embryonic cell layers—ectoderm and endoderm; includes cnidarians and ctenophores.

diploid With two sets of chromosomes. Most animals and many plants are diploid.

directional selection A pattern of natural selection in which individuals with a particular extreme phenotype have higher fitness than individuals with average phenotypes or with the other extreme of the phenotype.

direct sequencing A lab technique for discovery and study of unknown microscopic organisms that will not grow easily in the lab. Relies on detecting and amplifying copies of their DNA.

disaccharide A carbohydrate consisting of two monosaccharides (sugar subunits) linked together.

discontinuous replication In DNA replication, the process by which the lagging strand is synthesized in separate pieces (Okazaki fragments) that are joined together later.

discontinuous strand In DNA replication, the strand of new DNA that is synthesized discontinuously in a series of short pieces that are later joined together. Also called lagging strand.

discrete trait A phenotypic trait that exhibits distinct forms rather than the continuous variation seen in quantitative traits such as body height.

dispersal The movement of individuals from their place of origin (birth, hatching) to a new location.

dispersive replication A now-disproven hypothesis of DNA replication, in which each strand in the daughter molecules was proposed to consist of a mixture of old and new segments of DNA.

disruptive selection A pattern of natural selection in which individuals with extreme phenotypes (at either end of the range of phenotypic variation) have higher fitness than do individuals with an average phenotype.

dissociation curve The graphical relationship that depicts the percentage of hemoglobin in the blood that will bind to oxygen at various partial pressures of oxygen.

distal Away from the center of the body; toward the furthest tip of an appendage. The opposite of proximal.

disturbance Any event that removes some individuals or biomass from a community.

disturbance regime The characteristic disturbances that affect a given ecological community.

disulfide bond A covalent bond between two sulfur atoms, typically in the side groups of some amino acids (e.g., cysteine). Often contributes to tertiary structure of proteins.

DNA See deoxyribonucleic acid (DNA).

DNA fingerprinting Any of several methods for identifying individuals by unique features of their genomes. Commonly involves using PCR to create many copies of certain simple sequence repeats and then analyzing their lengths.

DNA footprinting A technique used to find and sequence stretches of DNA that are bound by particular regulatory proteins.

DNA ligase An enzyme that can connect pieces of DNA by catalyzing formation of a phosphodiester bond between the different pieces.

DNA microarray A lab tool involving the use of cDNAs to investigate whether any of several thousand genes are expressed in a certain cell or tissue.

DNA polymerase Any enzyme that catalyzes synthesis of DNA from deoxyribonucleotides.

domain (1) A section of a protein that has a distinctive tertiary structure and function. (2) A fundamental taxonomic group of organisms sharing similarities in basic cellular biochemistry, such as Bacteria, Archaea, and Eukarya.

dominant An allele that determines the phenotype of a heterozygous individual (i.e., one that can hide the presence of a recessive allele).

domino effect Progressive loss of species diversity in a fragmented habitat, in which the loss of one species causes the loss of further species.

dopamine A catecholamine neurotransmitter that functions mainly in a part of the mammal brain involved with muscle control.

dormancy A temporary state of greatly reduced, or no, metabolic activity.

dorsal Toward an animal's back and away from its belly. The opposite of ventral.

dorsal hollow nerve cord A bundle of nerves extending from the brain along the dorsal (back) side of a chordate animal, with cerebrospinal fluid inside a hollow central channel. One of the defining features of chordates.

double fertilization An unusual form of reproduction seen in flowering plants, in which one sperm nucleus fuses with an egg to form a zygote and the other sperm nucleus fuses with two polar nuclei to form the triploid endosperm.

double helix The three-dimensional shape of a molecule of DNA, consisting of two antiparallel DNA strands wound around each other.

Doushantuo fossils A characteristic assemblage of microscopic fossils found in pre-Cambrian rocks in China from about 580 million years ago, including evidence of sponges, different cell types, and multicellular embryos.

downstream In genetics, the direction in which RNA polymerase moves along a DNA strand.

Down syndrome A human developmental disorder caused by trisomy of chromosome 21.

duct A thin tube through which an exocrine gland secretes some substance.

dyad symmetry A type of symmetry in which an object can be superimposed on itself if rotated 180°. Occurs in some regulatory sequences of DNA. Also called two-fold rotational symmetry.

dynein A motor protein that produces movement of cilia and flagella. Dynein bridges use the chemical energy of ATP to "walk" along the adjacent microtubule doublets.

early endosome A membrane-bound vesicle, formed by endocytosis, that is an early stage in the process of becoming a lysosome.

ecdysone An insect hormone that triggers either molting (to a larger larval form) or metamorphosis (to the adult form), depending on the level of juvenile hormone.

Ecdysozoa A lineage of protostomes that grow by shedding their external skeletons and expanding their bodies, rather than by increasing the length of their skeletons. Includes arthropods and nematodes.

echolocation The use of echoes from vocalizations to obtain information about locations of objects in the environment.

ecology The study of how organisms interact with each other and with their surrounding environment.

ecosystem All organisms that live in a geographic area, together with abiotic components that affect or exchange materials with the organisms.

ecosystem diversity The variety of biotic components in a region along with abiotic components, such as soil, water, and nutrients.

ecosystem services Alteration of the physical components of an ecosystem by living organisms, especially beneficial changes in the quality of the atmosphere, soil, water, etc.

ectoderm One of the three basic embryonic cell layers of a triploblast animal. Forms the outer covering and nervous system.

ectomycorrhizal fungi (EMF) Fungi whose hyphae form a dense network that covers their host plant's roots but do not enter the root cells.

ectoparasite A parasite that lives on the outer surface of the host's body.

ectotherm An animal that does not use internally generated heat to regulate its body temperature.

Ediacaran fauna A characteristic set of animal fossils found in various pre-Cambrian rocks around 565–544 million years old, containing sponges, jellyfish, comb jellies, and other filter-feeding, shallow-water marine animals.

effector Any structure, cell, or organ with which an animal can respond to external or internal stimuli. Usually under control of the nervous system.

effector T cells T lymphocytes—descendants of activated T cells—that are actively involved in combating an infection. Includes helper T cells and cytotoxic T lymphocytes.

efferent division The part of the nervous system that carries commands from the central nervous system to the body. Consists primarily of motor neurons.

egg A mature female gamete and any associated external layers (such as a shell). Larger and less mobile than the male gamete. In animals, also called an ovum.

ejaculation The release of semen from the copulatory organ of a male animal.

ejaculatory duct A short duct connecting the vas deferens to the urethra, through which sperm move during ejaculation.

elasticity The ability to stretch and then spring back to the original shape.

electrical potential Potential energy created by a separation of electric charges between two points. Also called voltage.

electric current A flow of electrical charge past a point. Also called current.

electrocardiogram (EKG) A recording of the electrical activity of the heart, as measured through electrodes on the skin.

electrochemical gradient The combined effect of a concentration gradient and an electrical gradient. Affects the movement of ions across plasma membranes.

electrolyte Any compound that dissociates into ions when dissolved in water. In nutrition, refers to the major ions necessary for normal cell function.

electromagnetic spectrum The full range of wavelengths of electromagnetic radiation (light).

electron An extremely tiny, negatively charged particle that usually occupies an orbital around an atomic nucleus. Exhibits wave as well as particle characteristics.

electron acceptor A reactant that gains an electron and is reduced in a reduction-oxidation reaction.

electron carrier A molecule that readily donates electrons to other molecules.

electron donor A reactant that loses an electron and is oxidized in a reduction-oxidation reaction.

electronegativity The tendency of an atom to attract electrons toward itself.

electron microscope A microscope that uses beams of electrons instead of light to produce images of specimens. Can magnify hundreds of thousands of times.

electron shell Atomic orbitals of similar energies. Arranged in roughly concentric layers around the nucleus of an atom, with electrons in outer shells having more energy than those in inner shells.

electron transport chain (ETC) A set of molecules involved in a coordinated series of redox reactions in which the potential energy lost by electrons is used to pump protons from one side of a membrane to the other.

electroreceptor A sensory cell or organ specialized to detect electric fields.

element A fundamental chemical entity consisting of atoms with a specific number of protons. Elements preserve their identity in chemical changes.

elemental ion An ion that is a single element, such as K^+ or Cl^-, and not a molecule.

ELISA (enzyme-linked immunosorbent assay) A lab technique that can measure the concentration of a substances present in very small amounts.

elongation (1) The process by which messenger RNA lengthens during transcription. (2) The process by which polypeptide chains lengthen during translation.

elongation factors Proteins involved in the elongation phase of translation, assisting ribosomes in the synthesis of the growing peptide chain.

elongation phase The phase of DNA transcription in which RNA polymerase moves along the DNA molecule and synthesizes messenger RNA.

embryo A young developing organism; the stage after fertilization and zygote formation.

embryogenesis The process by which a single-celled zygote becomes a multicellular embryo.

embryophyte A plant, including a land plant, that nourishes its embryos inside its own body.

emergent vegetation Any plants in an aquatic habitat that extend above the surface of the water.

emerging disease Any infectious disease that suddenly afflicts significant numbers of individuals, often due to changes in host species or host population movements.

emerging virus Any of several pathogenic viruses that that suddenly afflict significant numbers of individuals, often due to changes in host species or host population movements.

emigration The movement of individuals from one population into another.

emphysema A lung disease caused by breakdown of alveoli and loss of elasticity of the lungs.

emulsified Broken up; the result of mixing of fat into an aqueous solution, usually with the aid of an emulsifying agent (an amphipathic substance such as a detergent) that can break large fat globules into microscopic fat droplets.

endemic species A species that occurs only in one limited area.

endergonic Said of a chemical reaction that will not occur spontaneously but requires an input of energy. For such a reaction, ΔG (Gibbs free-energy change) > 0.

endocrine Relating to hormones.

endocrine gland Any organ that secretes hormones into the blood.

endocrine system All of the glands that produce and secrete hormones into the bloodstream.

endocytosis Uptake of extracellular material by engulfing and pinching-off the plasma membrane to form a small membrane-bound vesicle in the cell.

endoderm One of the three basic embryonic cell layers of a triploblast animal. Forms the digestive tract and organs that connect to it (liver, lungs, etc.).

endodermis In plant roots, a cylindrical layer of cells that separates the cortex from the vascular tissue.

endomembrane system A system of organelles in eukaryotic cells that performs most protein and lipid synthesis. Includes the endoplasmic reticulum (ER), Golgi apparatus, and lysosomes.

endoparasite Any parasite that lives inside the host's body.

endophytic Living inside of a plant.

endoplasmic reticulum (ER) A network of interconnected membranous sacs and tubules found inside eukaryotic cells. Either rough ER (with ribosomes attached) or smooth ER.

endoskeleton An internal skeleton, such as that found in vertebrates.

endosome A membrane-bound vesicle created by endocytosis. Gradually transformed into a lysosome.

endosperm A triploid tissue in the seed of a flowering plant. Serves as food for the plant embryo.

endosymbiont An organism that lives in a symbiotic relationship inside the body of its host.

endosymbiosis A type of symbiotic relationship in which individuals of one species live inside the bodies of individuals of another species.

endosymbiosis theory The theory that mitochondria and chloroplasts evolved from prokaryotes that were engulfed by host cells and took up a symbiotic existence within those cells.

endotherm An animal whose primary source of body heat is internally generated heat.

endothermic (1) Said of a chemical reaction that absorbs heat. (2) Able to maintain a high body temperature using internally generated heat.

energy The capacity to do work or to supply heat. May be stored (potential energy) or available in the form of motion (kinetic energy).

enhancer In eukaryotes, a regulatory DNA sequence to which certain proteins can bind, enhancing the transcription of certain genes.

enrichment culture A method of growing cells in the lab that involves providing cells from the environment with a very specific set of conditions and isolating those that grow rapidly in response.

enterokinase An intestinal enzyme that converts trypsinogen (from the pancreas) to active trypsin, which then activates protein-digesting enzymes.

entomology The study of insects.

entropy The amount of disorder in any system, such as a group of molecules. Commonly symbolized by S.

envelope A membrane-like covering that encloses some viruses and their capsid coats, shielding them from attack by the host's immune system.

enveloped In a virus, having an envelope surrounding its capsid coat.

environmental variation The proportion of phenotypic variation in a trait that is due to environmental influences rather than genetic influences, in a certain population in a certain environment.

enzyme A protein catalyst used by living organisms to speed up and control biological reactions.

epicotyl In some embryonic plants, a portion of the embryonic stem that extends above the cotyledons.

epidemic The spread of an infectious disease throughout a large population in a short time.

epidermis The outermost layer of cells of any multicellular organism.

epididymis A coiled tube wrapped around the testis in reptiles, birds, and mammals. The site of the final stages of sperm maturation and storage.

epinephrine A catecholamine hormone from the adrenal medulla. Triggers rapid responses relating to the fight-or-flight response. Also called adrenaline.

epiphyte A plant that grows on trees or other solid objects and is not rooted in soil. Not parasitic—they do not harm the host.

epistasis An interaction of independently inherited genes, such that alleles at one locus alter the phenotypic effect of alleles at another locus.

epithelial tissue, or epithelium A class of animal tissues consisting of layers of tightly packed cells that line an organ, a duct, or a body surface. Also called epithelium (plural: epithelia).

epitope The unique region of a particular antigen to which antibodies or lymphocytes bind.

equilibrium A state of balance between forward and reverse processes, such as forward and reverse chemical reactions, or between diffusion rates from one side to the other of a selectively permeable membrane.

equilibrium potential The membrane potential at which there is no net movement of a particular ion into or out of a cell.

ER signal sequence A specific sequence of 20 amino acids, found at the beginning of certain newly synthesized proteins, that marks them for transport to the endoplasmic reticulum.

erythropoietin (EPO) A peptide hormone, released by the kidney in response to low blood oxygen levels, that stimulates the bone marrow to produce more red blood cells.

E site The site in a ribosome where an unattached transfer RNA is shunted after its amino acid is added to the growing polypeptide chain.

esophagus The muscular tube that connects the mouth to the stomach.

essential amino acid An amino acid that an animal cannot synthesize and must obtain from the diet. May refer specifically to one of the eight essential amino acids of adult humans: isoleucine, leucine, lysine, methionine, phenylalanine, threonine, tryptophan, and valine.

essential element A chemical element that an organism must obtain from its environment.

essential nutrient Any chemical element or compound required for normal growth, reproduction, and maintenance of a living organism.

ester linkage The covalent bond that joins a fatty acid to glycerol to form a fat or phospholipid.

estradiol The major estrogen produced by the ovaries of female mammals. Stimulates development of the female reproductive tract, growth of ovarian follicles, and growth of breast tissue.

estrogens A class of steroid hormones, including estradiol, estrone, estriol, and others, that generally promote female-like traits. Secreted by the gonads, fat tissue, and some other organs.

estuary An environment of brackish (partly salty) water where a river meets the ocean.

ethnobotanists Plant biologists who study how humans use plants, often focusing on indigenous cultures' knowledge of medical uses for plants.

ethylene A gaseous plant hormone that induces fruit to ripen, flowers to fade, and leaves to drop.

euchromatin Chromatin (a eukaryotic chromosome and its histone proteins) that is unwound in a long, filamentous structure.

eudicot A member of the lineage of angiosperms that includes complex flowering plants and trees. Eudicots include some of the plants classically called dicots.

Eukarya The taxonomic domain that includes all eukaryotes (protists, fungi, plants, animals, etc.), which share a cell nucleus, a cytoskeleton, and other features.

eukaryote A member of the domain Eukarya; an organism with complex cells with distinctive traits such as a nucleus, membrane-bound organelles, a cytoskeleton, and the presence of introns in genes. May be unicellular or multicellular.

eumetazoans Animals whose cells are organized into distinct tissues. All animals except sponges.

eutherians A lineage of mammals whose young develop in the uterus and are not housed in an abdominal pouch. Also called placental mammals.

eutrophication Deoxygenation of an aquatic ecosytem due to a bloom of photosynthetic algae that produces a bloom of decomposers, which use up all the oxygen.

evaporation The energy-absorbing phase change from a liquid state to a gaseous state. Many organisms evaporate water as a means of heat loss.

evo-devo The study of the developmental and molecular causes of major evolutionary changes such as novel body parts.

evolution (1) The theory that all organisms on Earth are related by common ancestry and that they have changed over time, predominantly via natural selection. (2) Any change in the genetic characteristics of a population over time; especially, a change in allele frequencies.

excision repair system A set of enzymes that identify and repair damaged sections of DNA.

excitable membrane A plasma membrane that is capable of generating an action potential.

excitatory postsynaptic potential (EPSP) A change in membrane potential at a neuron dendrite that makes an action potential more likely. Usually a depolarization.

exergonic Said of a chemical reaction that will occur spontaneously, releasing heat and/or increasing entropy. For such reactions, ΔG (Gibbs free-energy change) < 0.

exocrine gland A gland that secretes some substance through a duct into a space other than the circulatory system, such as the digestive tract or the skin surface.

exocytosis Secretion of cellular contents to the outside of the cell by fusion of vesicles to the plasma membrane.

exon A region of a eukaryotic gene that is translated into a peptide or protein.

exonuclease Any enzyme that can remove nucleotides from the end of a strand of DNA or RNA.

exoskeleton A hard covering secreted on the outside of the body, used for body support, protection, and muscle attachment.

exothermic Said of a chemical reaction that releases heat.

exotic From a different area. May refer specifically to exotic species introduced into a new area.

expansins A class of plant proteins that actively increase the length of the cell wall when the pH of the wall falls below 4.5.

exponential growth A constantly accelerating increase in population size that occurs when growth rate is constant (not affected by population size).

exportins A class of intracellular proteins whose function is to transport certain large molecules out of the nucleus.

extensor A muscle that pulls two bones further apart from each other, as in the extension of a limb or the spine.

extinct Said of a species that has died out.

extracellular digestion Digestion that takes place outside of an organism, as occurs in many fungi.

extracellular matrix (ECM) The substance that animal cells secrete and in which they are embedded; often has a fiber-composite structure with protein fibers (e.g., collagen, elastin).

extremophiles Any of several groups of bacteria and archaea that thrive in "extreme" (e.g., high-salt, high-temperature, low-temperature, or low-pressure) environments.

F_1 generation First filial generation. The first generation of offspring produced from a mating (i.e., the offspring of the parental generation).

facilitated diffusion Diffusion of a substance across a plasma membrane down a concentration gradient with the assistance of carrier proteins.

facilitation Ecological succession in which early-arriving species make conditions more favorable for later-arriving species.

facultative aerobe Any organism that can perform aerobic respiration when oxygen is available to serve as an electron acceptor but can switch to fermentation when it is not.

fallopian tube A narrow tube connecting the uterus to the ovary in humans, through which the egg travels after ovulation. Site of fertilization and cleavage. In nonhuman animals, called oviduct.

family A classic taxonomic rank above genus and below order. In animals, usually ends in the suffix -idae.

fast-twitch fibers A class of muscle fibers that contract rapidly and powerfully but fatigue quickly.

fate The likely developmental path that a certain embryonic cell will follow (i.e., which tissue types it will give rise to).

fate map A diagram of an embryo, showing the eventual fate of cells in that embryo—the ultimate location and the tissues each cell will give rise to.

fats A class of lipids consisting of three fatty acids joined to a glycerol molecule. Also called triacylglycerols or triglycerides.

fatty acid A type of lipid consisting of a hydrocarbon chain bonded to a carboxyl group (–COOH) at one end. Used by many organisms to store chemical energy; a major component of animal and plant fats.

fatty-acid binding protein A membrane protein of intestinal cells that binds to lipids from food and takes them into the cell.

feather A specialized skin outgrowth in all birds and only in birds. Composed of β-keratin. Used for flight, insulation, display, and other purposes.

feces The waste products of digestion.

fecundity The average number of female offspring produced by a single female per unit time.

feedback inhibition Inhibition of a process by high concentrations of the product of that process.

female gametophyte A multicellular haploid structure that can produce haploid eggs, in a species that exhibits alternation of generations. In flowering plants, consists of a sac (the embryo sac) containing haploid nuclei.

fermentation Any of several metabolic pathways that allow continued production of ATP via glycolysis by transferring electrons from a reduced compound such as glucose to an electron acceptor other than oxygen.

ferredoxin A molecule involved in the electron transport chain of photosystem I of photosynthesis. Passes electrons to the enzyme NADP$^+$ reductase, which catalyzes formation of NADPH.

fertilization Fusion of the nuclei of two haploid gametes to form a zygote with a diploid nucleus.

fertilization envelope A physical barrier that forms around a fertilized egg in amphibians and some other animals. Formed by an influx of water under the vitelline membrane.

fetal alcohol syndrome A condition thought to be caused by exposure to high blood alcohol concentrations during embryonic development.

fetus A later developmental stage of an embryo, usually developed sufficiently to be recognizable as a certain species.

fever A sustained, regulated elevation of body temperature in an endotherm, typically as part of a response to infection.

fiber In botany, a type of elongated sclerenchyma cell that provides support to vascular tissue.

fibronectins A class of proteins in the extracellular matrix to which cells bind to stay in place.

Fick's law of diffusion A law that gives the rate of diffusion of a gas into a liquid as a function of gas solubility, temperature, surface area, difference in partial pressures, and the thickness of the diffusion barrier.

fight-or-flight response Rapid physiological changes that prepare the body for emergencies. Includes increased heart rate, increased blood pressure, and decreased digestion. Triggered by catecholamines.

filament Any thin, threadlike structure, particularly (1) the threadlike extensions of a fish's gills or (2) the slender stalk that bears the anthers in a flower.

filter feeder Any organism that obtains food by filtering small particles or small organisms out of water or air. Also called suspension feeder.

filtrate Any fluid produced by filtration, such as the fluid in kidney nephrons.

filtration A process of removing large components from a fluid by forcing it through a filter, such as that in a renal corpuscle of the kidney.

finite rate of increase The rate of increase of a population over a given period of time. Calculated as the ending population size divided by the starting population size. Symbolized by lambda (λ).

first law of thermodynamics A fundamental principle of physics stating that energy is conserved; it cannot be created or destroyed.

fission The splitting of an organism into two daughter organisms.

fitness The ability of an organism to produce surviving fertile offspring. Also called Darwinian fitness.

fixed action patterns (FAPs) Highly stereotyped behavior patterns that occur in a certain invariant way in a certain species. A form of innate behavior.

flaccid Limp, as a result of low internal pressure; e.g., a wilted plant leaf.

flagellum (plural: flagella) A long, cellular projection that undulates (in eukaryotes) or rotates (in prokaryotes) to move the cell through an aqueous environment.

flavonoids A class of accessory pigments, found in many plants, that can absorb ultraviolet (UV) radiation and hence protect cells from damage by UV light.

flexor A muscle that pulls two bones closer together, as in the flexing of a limb or the spine.

floral meristem A group of undifferentiated plant cells that can produce flowers.

florigen A hypothesized hormone that may stimulate flowering but that has not yet been isolated.

flower The reproductive structure of angiosperm plants, containing microsporangia and/or megasporangia. Typically includes a calyx, a corolla, and one or more stamens or carpels.

flowering plant An angiosperm; a member of the lineage of plants that produces seeds within mature ovaries (fruits).

fluid-mosaic model The widely accepted hypothesis that plasma membranes consist of proteins embedded in a phospholipid bilayer and that these components move fluidly around the membrane.

fluorescence The spontaneous emission of light from an excited electron falling back to its normal (ground) state.

follicle An egg cell and its surrounding ring of supportive cells in a mammalian ovary.

follicle-stimulating hormone (FSH) A peptide hormone from the anterior pituitary that stimulates (in females) growth of eggs and follicles in the ovaries or (in males) sperm production in the testes.

follicular phase The first major phase of a menstrual cycle, when follicles are growing and estrogen levels are increasing. Ends with ovulation.

food A digestible material that contains nutrients.

food chain A simple pathway of energy through a few species in an ecosystem; e.g., a primary producer, a primary consumer, a secondary consumer, and a decomposer.

food vacuole A membrane-bound organelle containing food engulfed by the cell.

food web Any complex pathway along which energy moves among many species at different trophic levels of an ecosystem.

foot A muscular appendage of mollusks, used for movement and/or burrowing into sediment.

forb Herbaceous flowering plants such as daisies and coneflowers.

forensic science The use of scientific analyses to help solve crimes.

formed elements Cells and cell fragments found in blood, including red blood cells, leukocytes and platelets.

fossil Any trace of an organism that existed in the past. Includes tracks, burrows, fossilized bones, casts, etc.

fossil record All of the fossils that have been found anywhere on Earth and that have been formally described in the scientific literature.

founder effect A change in allele frequencies that often occurs when a new population is established from a small group of individuals, due to chance variations in gene frequency in the small group.

founder event The establishment of a new population from a small group of individuals.

fovea The small region of the vertebrate retina in which the photoreceptors are very tightly packed, producing the most acute vision.

F-plasmid A particular type of bacterial plasmid that gives the bacterium the ability to initiate conjugation with another bacterium.

fragile-X syndrome The most common form of inherited mental retardation in humans. Caused by an increase in the number of copies of a CGG codon at a certain location on the X chromosome.

free radical An atom with an unpaired electron. Unstable and highly reactive.

freeze-fracture electron microscopy A technique in which plasma membranes are frozen and split to obtain a highly magnified view of both the outer surface and the interior of the bilayer.

frequency The number of wave crests per second traveling past a stationary point. Determines the pitch of sound and the color of light.

fronds The large leaves of ferns.

frontal lobe The anteriormost region of each cerebral hemisphere of the mammal brain. In humans, involved in complex decision making.

fruit A mature, ripened plant ovary (or group of ovaries), along with the seeds it contains and any adjacent fused parts.

fruiting body A structure formed in some fungi and prokaryotes for spore dispersal, usually consisting of a base, a stalk, and a mass of spores at the top.

functional genomics The study of how a genome works (how the genes identified in genome sequencing interact to produce a functional organism).

functional group A small group of atoms bonded together in a precise configuration. Each group has particular chemical properties that it imparts to any organic molecule in which it occurs.

fundamental asymmetry of sex The fact that eggs are larger and more expensive to produce than sperm, and the resulting evolutionary consequences for males and females.

fundamental niche The ecological space that a species occupies in its habitat in the absence of competitors.

fungi A major lineage of eukaryotes that typically have a filamentous body (mycelium) and obtain nutrients by absorption.

fungicide Any substance that can kill fungi or slow their growth.

fur An insulative layer that covers most mammals' bodies, consisting of longer rain-shedding hairs covering a shorter layer of soft underfur.

G$_1$ phase The phase of a cell cycle that constitutes the first part of interphase before DNA synthesis (S phase).

G$_2$ phase The phase of a cell cycle between synthesis of DNA (S phase) and mitosis (M phase); the last part of interphase.

gall A tumorlike growth that forms on plants that are infected with certain bacteria or parasites.

gallbladder A small pouch that stores bile from the liver and releases it to the small intestine during digestion for emulsification of fats.

gametangium (plural: gametangia) (1) The gamete-forming structure found in all land plants except angiosperms. Contains a sperm-producing antheridium and an egg-producing archegonium. (2) The gamete-forming structure of chytrid fungi.

gamete A haploid reproductive cell that can fuse with another haploid cell to form a zygote. Most multicellular eukaryotes have two distinct forms of gametes: egg cells (ova) and sperm cells.

gametogenesis The production of gametes (eggs or sperm).

gametophyte The multicellular haploid phase in a species that exhibits alternation of generations. Arises from a single spore and produces gametes.

ganglion cells Neurons in the vertebrate retina that collect visual information from one or several bipolar cells and send it to the brain via the optic nerve.

Gap Analysis Program (GAP) An analysis aimed at identifying the gaps between geographic areas that are rich in biodiversity and areas that are being protected or managed for conservation.

gap genes A class of fruit-fly segmentation genes that organize embryonic cells into major groups of segments. Active in broad regions of the embryo.

gap junction A direct physical connection between two animal cells, consisting of a gap in the ECM and plasma membranes lined by specialized proteins that allow passage of water, ions and small molecules between the cells.

gastrin A hormone produced by the stomach in response to the arrival of food or to a signal via nerves from the brain. Stimulates other stomach cells to release hydrochloric acid.

gastropods The slugs and snails; mollusks distinguished by a large muscular foot and a unique feeding structure, the radula.

gastrula A vertebrate embryo just after gastrulation, containing the three embryonic germ layers (ectoderm, mesoderm, and endoderm) but with no nerve cord yet.

gastrulation The process by which some cells on the outside of a young embryo move to the interior of the embryo, resulting in the three distinct cell layers of endoderm, mesoderm, and ectoderm.

gated channel A channel protein that opens and closes in response to a certain stimulus, such as the binding of a particular molecule or a change in the electrical charge on the outside of the membrane.

gel electrophoresis A technique for separating molecules on the basis of size and electric charge, which affect their differing rates of movement through a gelatinous substance in an electric field.

gemma (plural: gemmae) A small reproductive structure produced in some liverworts by a gametophyte and can grow into another gametophyte.

gene A section of DNA (or RNA, for some viruses) that encodes information for building a polypeptide or a functional molecule of RNA.

gene duplication The creation of an additional copy of a gene, typically by misalignment of chromosomes during crossing over.

gene expression The transcription and translation of a gene, producing a protein.

gene family A set of genetic loci whose DNA sequences are extremely similar. Thought to have arisen by duplication of a single ancestral gene.

gene flow The movement of alleles between populations.

gene-for-gene hypothesis The hypothesis that there is a one-to-one correspondence between the resistance (*R*) loci of plants and the avirulence (*avr*) loci of pathogenic fungi; particularly that *R* genes produce receptors and *avr* genes produce molecules that bind to those receptors.

gene pool All of the alleles of all of the genes in a certain population.

generation The average time between a mother's first offspring and her daughter's first offspring.

generative cell A small haploid cell within the male gametophyte of a flowering plant. Gives rise to two haploid sperm cells.

gene therapy The treatment of an inherited disease by introducing normal alleles.

genetic bottleneck A reduction in allelic diversity via genetic drift due to a population bottleneck.

genetic cloning A technique for producing many identical copies of a certain gene, usually by inserting a cDNA version of the gene into a bacterial plasmid and growing the bacteria.

genetic code The set of all 64 codons of DNA and the particular amino acids that each specifies.

genetic correlation A type of evolutionary constraint in which selection on one trait causes a change in another trait as well, because the same gene(s) affect both traits.

genetic diversity The diversity of alleles in a population, species, or group of species.

genetic drift Any change in allele frequencies due to random events. Causes allele frequencies to drift up and down randomly over time, and eventually can lead to the fixation or loss of alleles.

genetic engineering The field of study of the manipulation of DNA sequences in living organisms.

genetic homology Similarities among certain organisms in DNA sequences or amino acid sequences of proteins that are due to inheritance from a common ancestor.

genetic map A map of the relative locations of specific genes on a certain chromosome. Also called a linkage map or meiotic map.

genetic marker A genetic locus that can be identified and traced in populations by lab techniques or by a distinctive phenotype. Includes reporter genes.

genetic model A set of hypotheses that explain how a certain trait is inherited.

genetic recombination A change in the combination of genes or alleles on a given chromosome or in a given individual. Also called recombination.

genetics The field of study concerned with the inheritance of traits.

genetic screen Any of several techniques for identifying individuals with a particular type of mutation. Also called a screen.

genetic variation (1) The number and relative frequency of alleles present in a particular population. (2) The proportion of phenotypic variation in a trait that is due to genetic rather than environmental influences in a certain population in a certain environment.

genitalia External copulatory organs.

genome All of the hereditary information in an organism, including not only genes but also other non-gene stretches of DNA.

genomic library A set of all the DNA sequences in a particular genome, split into small segments, each inserted into a vector for further study.

genomics The field of study concerned with sequencing, interpreting, and comparing whole genomes.

genotype All of the alleles of every gene present in a given individual. May refer specifically to the alleles of a particular set of genes under study.

genus (plural: genera) A taxonomic category of closely related species. Always italicized and capitalized to indicate that it is a recognized scientific genus.

geologic time scale The sequence of eons, epochs, and periods used to describe the geologic history of Earth.

geometric isomer A molecule that shares the same molecular formula as another molecule but differs in the arrangement of atoms or groups on either side of a double bond or ring structure.

germination The process by which a seed becomes a young plant.

germ layer In animals, one of the three basic types of tissues of early embryonic development that give rise to all other tissues: endoderm, mesoderm, or ectoderm.

germ theory of disease The theory that infectious diseases are caused by bacteria, viruses, and other microorganisms.

gestation The duration of embryonic development from fertilization to birth, in those species that have live birth.

gibberellic acid (GA) A plant hormone that stimulates elongation of shoots; a gibberellin.

gibberellins A class of plant hormones that stimulate growth.

Gibbs free-energy change A measure of the change in potential energy and entropy that occurs in a given chemical reaction. Calculated as $\Delta G = \Delta H - T \Delta S$, where ΔH is the change in potential energy, T is the temperature in kelvins, and ΔS is the change in entropy. Determines whether a reaction will be spontaneous.

gill Any organ in aquatic animals that exchanges gases and other dissolved substances between the blood and the surrounding water. Typically, a filamentous outgrowth of a body surface.

gill arch In aquatic vertebrates, curved regions of tissue between the gills. Gills are suspended from the gill arches.

gill filament The thin, pink strands of fish gills that extend from the gill arches into the water and across which gas exchange occurs.

gill lamellae (singular: lamella) Tiny, crescent-shaped flaps on the gill filaments of fish that serve to increase surface area for gas exchange.

gland An organ whose primary function is to secrete some substance, either into the blood (endocrine gland) or into some other space such as the gut or skin (exocrine gland).

glia Several types of cells in nervous tissue that are not neurons and do not conduct electrical signals but provide support, nourishment, and electrical insulation and perform other functions.

global carbon cycle The movement of carbon among terrestrial ecosystems, the oceans, and the atmosphere.

global nitrogen cycle The movement of nitrogen among terrestrial ecosystems, the oceans, and the atmosphere.

global warming A sustained increase in Earth's average surface temperature.

global water cycle The movement of water among terrestrial ecosystems, the oceans, and the atmosphere.

glomerulus (1) A ball-like cluster of capillaries at the beginning of a kidney nephron. Surrounded by Bowman's capsule. (2) The ball-shaped clusters of neurons in the olfactory bulb of the brain.

glucagon A peptide hormone produced by the pancreas in response to low blood glucose. Raises blood glucose by triggering breakdown of glycogen and stimulating gluconeogenesis.

glucocorticoids A class of steroid hormones released from the adrenal cortex that increase blood glucose and prepare the body for stress.

gluconeogenesis Synthesis of new glucose from non-carbohydrate sources, such as proteins and fatty acids. Occurs in the liver in response to low insulin levels and high glucagon levels.

glyceraldehyde-3-phosphate (G3P) A phosphorylated sugar produced during the Calvin cycle of photosynthesis.

glycerol A three-carbon molecule that forms the "backbone" of phospholipids and most fats.

glycogen A polysaccharide that is the major form of stored carbohydrate in animals. Consists of α-glucose monomers joined end to end in highly branched chains.

glycolipid Any lipid molecule that is covalently bonded to a carbohydrate group.

glycolysis A series of 10 chemical reactions that oxidize glucose to produce pyruvate and ATP. Used by all organisms as part of fermentation or cellular respiration.

glycoprotein Any protein with one or more covalently bonded carbohydrate groups.

glycosidic linkage The covalent bond between two sugar subunits of a polysaccharide.

glycosylation Addition of a carbohydrate group to a molecule.

glyoxisomes Specialized peroxisomes found in plant cells and that contain enzymes for processing the products of photosynthesis.

goblet cells The cells in the stomach lining that secrete mucus.

goiter A pronounced swelling of the thyroid gland in the neck, usually caused by a deficiency of iodine in the diet.

Golgi apparatus A stack of flattened membranous sacs (cisternae) in eukaryotic cells; processes proteins and lipids that will be secreted or directed to other organelles.

gonad An organ that produces reproductive cells; e.g., a testis or an ovary.

gonadotropin-releasing hormone (GnRH) A peptide hormone from the hypothalamus that stimulates release of FSH and LH from the anterior pituitary.

G proteins A class of peripheral membrane proteins that are important in signal transduction, typically by binding to GTP and activating a second messenger.

gracile australopithecines Several species of slender, lightly built hominins that appear in the fossil record shortly after the split from chimpanzees.

grade A group of species that share a position in an inferred evolutionary sequence of lineages but that are not a monophyletic group.

gram-negative bacteria Bacteria that look pink when treated with a Gram stain. Have a cell wall composed of a thin layer of peptidoglycan and an outer membrane.

gram-positive bacteria Bacteria that look purple when treated with a Gram stain. Have thick cell walls containing peptidoglycan.

Gram stain A dye that stains different types of bacteria different colors.

granum (plural: grana) A stack of the flattened, membrane-bound thylakoid disks inside plant chloroplasts.

gravitropism The growth or movement of a plant in a particular direction in response to gravity.

gray crescent A region of an amphibian zygote that becomes visible shortly after fertilization, opposite the point of sperm entry. Eventually gives rise to the blastopore and the organizer, which determines major body axes.

grazing food web The ecological network of herbivores and the predators and parasites that consume them.

great apes The hominids; members of the family Hominidae, including humans and extinct related forms, chimpanzees, gorillas, and orangutans. Distinguished by large body size, no tail, an exceptionally large brain, and a tendency toward bipedalism.

green fluorescent protein (GFP) A jellyfish protein that spontaneously emits green light after stimulation. Widely used in lab research to mark the location of certain molecules or cells.

greenhouse gas An atmospheric gas that absorbs and reflects infrared radiation, so that heat radiated from Earth is retained in the atmosphere instead of being lost to space.

green plants A lineage of eukaryotes that includes green algae and land plants.

gross photosynthetic efficiency The efficiency with which all the plants in a given area use the light energy available to them to produce sugars.

gross primary productivity The total amount of carbon fixed by photosynthesis, including that used for cellular respiration, in a given area over a given time period.

ground meristem The middle layer of a young plant embryo. Gives rise to the ground tissue.

ground tissue A plant tissue consisting of all cells beneath the outer protective layers of epidermis and cork, except for vascular tissue.

groundwater Any water below the land surface.

growth factor Any of several compounds that are secreted by certain cells and that stimulate other cells to divide or to differentiate.

growth hormone A peptide hormone produced by the mammalian pituitary gland. Involved in lengthening the long bones during childhood and in muscle growth, tissue repair, and lactation in adults.

growth-hormone-inhibiting-hormone (GHIH) A peptide hormone from the hypothalamus that inhibits release of growth hormone from the anterior pituitary.

guanosine triphosphate (GTP) A molecule consisting of guanine, a sugar, and three phosphate groups. Can be hydrolyzed to release free energy. Commonly used in many cellular reactions.

guard cell A specialized, crescent-shaped cell forming the border of a plant stoma. Changes shape to open or close the stoma.

gustation The perception of taste.

guttation Excretion of water droplets from plant leaves in the early morning, due to root pressure.

gymnosperms Four lineages of green plants that have vascular tissue and make seeds but that do not produce flowers. Includes cycads, ginkgoes, conifers, and gnetophytes.

H1 The histone protein associated with DNA in the "linker" stretches between the nucleosomes.

habitat destruction Human-caused destruction of a natural habitat with replacement by an urban, suburban, or agricultural landscape.

habitat fragmentation The breakup of a large region of a habitat into many smaller regions, separated from others by a different type of habitat.

Hadley cell An atmospheric cycle of large-scale air movement in which warm equatorial air rises, moves north or south, and then descends at approximately 30°N or 30°S latitude.

hair Mammalian fur that lacks an insulative layer of underfur; or a single strand of hair or fur consisting of keratin, dead cells, and pigments.

hair cell A pressure-detecting sensory cell that has tiny "hairs" (stereocilia) jutting from its surface.

hairpin A stable loop formed in an RNA molecule by hydrogen bonding between purine and pyrimidine bases on the same strand.

half-life The characteristic time taken for half of any amount of a particular radioactive isotope to decay.

halophile A bacterium or archaean that thrives in high-salt environments.

halophyte A plant that thrives in salty habitats.

Hamilton's rule The proposition that an allele for altruistic behavior will be favored by natural selection only if $Br > C$, where B = the fitness benefit to the recipient, C = the fitness cost to the actor, and r = the coefficient of relatedness between recipient and actor.

haploid Having one set of chromosomes. Bacteria, archaea, animal gametes, plant gametophytes, and many algae are haploid.

haploid number The number of different types of chromosomes in a cell. Symbolized as n.

Hardy-Weinberg principle A principle of population genetics stating that, if mutation, migration, genetic drift, random mating, and selection do not occur, then genotype frequencies will not occur in predictable ratios.

harmful algal bloom An extreme increase in the abundance of a toxin-producing protist in a particular aquatic environment.

HDL See high-density lipoprotein (HDL).

head A distinct anterior region of an organism's body, usually bearing a cluster of sensory organs and/or a mouth.

heart A muscular pump that circulates blood throughout the body.

heart attack An episode of cramping or death of heart muscle due to oxygen deprivation, usually caused by a blood clot in one or more partially blocked coronary arteries. Also called a myocardial infarction (MI).

heart murmur A distinctive sound caused by backflow of blood through a defective heart valve.

heartwood The older xylem in the center of an older stem or root, containing protective compounds and no longer functioning in water transport.

heat Thermal energy that is transferred between two objects.

heat of vaporization The energy required to vaporize 1 gram of a liquid into a gas.

heavy chain One of two identical polypeptides that together form the base of immunoglobulin proteins. Differences in heavy chains determine the different classes of immunoglobulins (IgA, IgE, etc.).

helicase An enzyme that catalyzes the breaking of hydrogen bonds between nucleotides of DNA, "unzipping" a double-stranded DNA molecule.

helix-turn-helix motif A motif seen in many repressor proteins in prokaryotes, consisting of two α-helices connected by a short stretch of amino acids that form a turn.

helper T cell A T cell that assists with the activation of other lymphocytes. Helper T cells are descendants of an activated CD4⁺ T cell.

hemagglutinin A protein that juts out from the surface of influenza viruses, enabling them to bind to host cells.

heme group A small molecule containing an iron atom that can bind to oxygen. Myoglobin and hemoglobin contain heme groups held within large specialized heme-carrying proteins (globins).

hemimetabolous metamorphosis A type of metamorphosis in which the animal increases in size from one stage to the next, but does not dramatically change its body form.

hemocoel A body cavity of arthropods and some mollusks, containing a pool of circulatory fluid (hemolymph) bathing the internal organs.

hemoglobin An oxygen-binding protein consisting of four polypeptide subunits, each containing a heme group. The major oxygen carrier of mammalian blood.

hemolymph The circulatory fluid of animals with open circulatory systems (e.g., insects) in which the fluid is not confined to blood vessels.

hemophilia A human disease characterized by defects in the blood-clotting system. Caused by an X-linked recessive allele.

herb A seed plant that lacks wood and has a relatively short-lived stem.

herbaceous Said of a plant that is not woody.

herbivore An animal that eats primarily plants and rarely or never eats meat.

herbivory The practice of eating plant tissues.

hereditary nonpolyposis colorectal cancer (HNPCC) A type of colon cancer that is associated with a mutated version of a mismatch repair gene on chromosome 2.

heredity The transmission of traits from parents to offspring via genetic information.

heritable Refers to traits that are influenced by hereditary genetic material (DNA, or RNA for some viruses).

hermaphroditic Producing both eggs and sperm.

heterochromatin Chromatin (a eukaryotic chromosome and its histone proteins) that is highly compact and supercoiled.

heterokaryotic Containing nuclei that are genetically distinct. Occurs naturally in many fungi.

heterospory The production of two distinct types of spore-producing structures and thus two distinct types of spores. Occurs in seed plants, which produce both microspores (which become the male gametophyte) and megaspores (which become the female gametophyte).

heterotherm An animal whose body temperature varies markedly with environmental conditions.

heterotroph Any organism that cannot synthesize reduced organic compounds from inorganic sources and that must obtain them by eating other organisms. Some bacteria, some archaea, and virtually all fungi and animals are heterotrophs.

heterozygote advantage A pattern of natural selection in which heterozygotes have greater fitness than either parental homozygote. Also called heterozygote superiority.

heterozygous Having two different alleles of a certain gene.

hexose A monosaccharide (simple sugar) with six carbons.

Hfr strain Any strain of bacteria in which an F-plasmid has been incorporated into the main chromosome. These bacteria have a high frequency of recombination.

hibernation A period of torpor (including a decreased body temperature) that continues for weeks or months.

high-density lipoprotein (HDL) Balls of protein and fat that transport cholesterol from body tissues to the liver for breakdown. Associated with decreased risk of heart disease.

hindgut The posterior portion of the digestive tract of an animal. Often functions to reabsorb water from wastes.

hinge helix A motif found in many repressor proteins in bacteria. Involved in locking the helix-turn-helix motif onto DNA.

histone A globular protein that is tightly associated with DNA in eukaryotic cells.

histone acetyl transferases In eukaryotes, a class of enzymes that loosens chromatin structure by adding acetyl groups to histone proteins.

histone deacetylases In eukaryotes, a class of enzymes that recondense chromatin by removing acetyl groups from histone proteins.

HIV See **human immunodeficiency virus (HIV)**.

holoenzyme A multipart enzyme consisting of a core enzyme (containing the active site for catalysis) along with other required proteins.

holometabolous metamorphosis A type of metamorphosis in which the animal completely changes its form.

homeobox A 180-base-pair sequence present in the *HOM/Hox* genes of animals. Codes for a helix-turn-helix motif in the proteins produced by these genes.

homeosis The occurrence of extra, fully formed segments or appendages normally found elsewhere in the body, replacing the structures usually formed at that location.

homeostasis The maintenance of a relatively constant physical and chemical environment within an organism.

homeotherm An animal that has a constant or relatively constant body temperature.

homeotic complex (*HOM-C*) A set of eight fruit-fly genes, closely linked on the same chromosome, that are active in different regions of the fly embryo, specifying the identity of body segments.

homeotic genes A class of genes that specify a location within an embryo, leading to the development of structures appropriate for that location.

homeotic mutant An individual with a mutation in a homeotic gene, causing the development of extra body parts or body parts in the wrong places.

hominids The great apes; members of the family Hominidae, which includes humans and extinct related forms; chimpanzees, gorillas, and orangutans. Distinguished by large body size, no tail, and an exceptionally large brain.

hominins Humans and extinct related forms; species in the lineage that branched off from chimpanzees and eventually led to humans.

homologous chromosomes In a diploid organism, chromosomes that are similar in size, shape, and gene content. Also called homologs.

homologous trait Any trait showing marked similarity between different species, due to inheritance from a common ancestor.

homology Similarity between organisms or DNA sequences that is due to inheritance from a common ancestor.

homospory The production of just one type of spore. Occurs in the seedless vascular plants, in contrast to the heterospory of seed plants.

homozygous Having two identical alleles of a certain gene.

hormone A signaling molecule that circulates throughout the body in blood or other body fluids; can trigger pronounced responses in distant target cells at very low concentrations.

hormone-response elements Sites on DNA to which a steroid hormone-receptor complex can bind and affect gene transcription.

host An individual or a species in or on which a parasite lives.

host cell A cell that has been invaded by an organism such as a parasite or a virus.

Hox **genes** A class of homeotic genes found in several animal phyla, including vertebrates. Controls pattern formation in early embryos.

human Any member of the genus *Homo*. Includes modern humans (*Homo sapiens*) and several extinct species.

human chorionic gonadotropin (hCG) A glycoprotein hormone produced by the human placenta from about week 3 to week 14 of pregnancy. Maintains the corpus luteum, which produces hormones that preserve the uterine lining.

Human Genome Project The multinational research project that sequenced the human genome.

human immunodeficiency virus (HIV) A retrovirus that causes AIDS (acquired immune deficiency syndrome) in humans.

humoral response Defense against infections and cancers via antibodies produced by B cells.

humus The completely decayed organic matter in soils.

Huntington's disease A degenerative brain disease of humans, caused by an autosomal dominant allele.

hybrid The offspring of parents from two different strains, populations, or species.

hybridoma A mass of cells produced in a lab from a myeloma cell that fused with an antibody-producing B cell. Used to produce large amounts of a monoclonal antibody.

hybrid zone A geographic area where interbreeding occurs between two species, sometimes producing fertile hybrid offspring.

hydrocarbon A molecule that contains only hydrogen and carbon, usually with the carbon atoms bonded covalently to form chains or rings.

hydrogen bond A weak interaction between two molecules, due to attraction between a hydrogen atom with a partial positive charge on one molecule and another atom (usually O or N) with a partial negative charge on the other molecule.

hydrogen ion A proton (H^+); typically, one that has dissolved in solution or that is being transferred from one atom to another in a chemical reaction.

hydrolysis A type of chemical reaction in which a compound reacts with water to break down into smaller molecules. In biology, most hydrolysis reactions involve polymers breaking down into monomers. The reverse of a condensation, or dehydration, reaction.

hydrophilic Mixing readily with water. Hydrophilic compounds are typically polar compounds with charged or electronegative atoms.

hydrophobic Not mixing readily with water. Hydrophobic compounds are typically nonpolar compounds without charged or electronegative atoms and often contain many C–C and C–H bonds.

hydroponic growth   Growth of plants in liquid cultures instead of soil.

hydrostatic skeleton   A system of body support involving fluid-filled compartments that can change in shape but cannot easily be compressed.

hydroxide ion   An oxygen atom bonded to a hydrogen atom and carrying a negative charge (OH⁻).

hyperpolarization   A change in membrane potential from its resting negative state to an even more negative state.

hypersensitive reaction   An allergic immune response in which previously sensitized mast cells and basophils produce large quantities of histamine, cytokines, etc., in response to a small amount of allergen. Triggers symptoms of allergies.

hypersensitive response   The rapid death of a plant cell that has been infected by a pathogen. Thought to be a strategic mechanism that protects the rest of the plant.

hypertension   Abnormally high blood pressure.

hypertonic   Having a greater solute concentration, and therefore a lower water concentration, relative to another solution.

hypha   (plural: hyphae) One of the strands of a fungal mycelium (the meshlike body of a fungus). Also found in some protists.

hypocotyl   The stem of a very young plant; the region between the cotyledon (embryonic leaf) and the radicle (embryonic root).

hypothalamic-pituitary axis   The combination of the hypothalamus and the pituitary gland, which together regulate most of the other endocrine glands in the body.

hypothalamus   A part of the brain that regulates the body's internal physiological state, such as the autonomic nervous system and endocrine system.

hypothesis   A proposed explanation for a phenomenon or for a set of observations.

hypotonic   Having a lower solute concentration, and therefore a higher water concentration, relative to another solution.

immigration   The movement of individuals into a certain population from some other population.

immune system   In vertebrates, the system of tissues and organs whose primary function is to defend the body against pathogens. Includes lymphocytes, lymph nodes, and other small organs.

immunity   Protection against infection by disease-causing pathogens.

immunization   The conferring of immunity to a particular disease.

immunoglobulins (Ig)   A class of Y-shaped proteins capable of binding to specific antigens. Responsible for acquired immunity.

immunological memory   The ability of the immune system to "remember" an antigen—i.e., to mount a rapid, effective response to a pathogen encountered years or decades earlier.

impact hypothesis   The hypothesis that the mass extinction at the end of the Cretaceous period, in which dinosaurs died out, was due to an asteroid impact.

imperfect   For a flower, containing male parts (stamens) or female parts (carpels) but not both.

implantation   The process by which an embryo buries itself in the uterine wall and forms a placenta. Occurs in mammals and a few other vertebrates.

importins   A class of intracellular proteins whose function is to transport certain large molecules into the nucleus.

imprinting   A type of rapid, irreversible learning in which young animals learn the distinctive appearance of the individual caring for them.

inactivated virus   A virus that has been deliberately damaged to render it incapable of causing infection, for use in vaccines.

inbreeding   Mating between closely related individuals or within a genetically homozygous strain.

inbreeding depression   A loss of fitness that occurs when homozygosity increases in a population.

incomplete dominance   An inheritance pattern in which the heterozygote phenotype is a blend or combination of both homozygote phenotypes.

independent assortment   The inheritance of the alleles of one gene independently of other genes. True only for genes on different chromosomes.

indeterminate growth   A pattern of growth in which an organism continues to increase its overall body size throughout its life.

indicator plate   A laboratory technique for detecting mutant cells by observing color changes due to cells cleaving (or not cleaving) the bonds in a pigmented substance.

induced defense   A defensive structure or compound produced by plants only in direct response to attack by pathogens or herbivores.

induced fit   The phenomenon whereby initial weak binding of a substrate to the active site of an enzyme causes the enzyme to change shape so as to bind the substrate more tightly.

inducer   A molecule that triggers transcription of a specific gene.

inducible defense   A defensive trait that is produced only when needed—i.e., in response to the presence of a predator or pathogen.

induction   The process by which one embryonic cell, or group of cells, alters the differentiation of neighboring cells.

infection thread   An invagination of a root hair membrane, through which beneficial nitrogen-fixing bacteria enter the roots of their host plants, legumes.

infectious disease   Any disease that can be transmitted from infected individuals to uninfected individuals.

infertile   Unable to produce offspring.

inflammatory response   An aspect of the innate immune response, seen in most cases of infection or tissue injury, in which the affected tissue becomes swollen, red, warm, and painful.

infrared light   Light with a wavelength longer than visible red light.

infrasound   Sound frequencies that are too low for humans to hear—lower than about 20 hertz.

ingestion   Taking nutrients or other substances into the body; feeding.

inheritance of acquired characters   The now-disproven theory that traits acquired during the lifetime of an organism due to noninherited causes will be passed genetically to its offspring.

inhibition   Ecological succession in which early-arriving species make conditions less favorable for the establishment of certain other species.

inhibitory hormone   A hormone that inhibits the release of some other hormone.

inhibitory postsynaptic potential (IPSP)   A change in membrane potential at a neuron dendrite that makes an action potential less likely. Usually a hyperpolarization.

initiation   The first stage of a molecular reaction, in which the necessary molecules are brought together. (1) In an enzyme-catalyzed reaction, the stage during which enzymes orient reactants precisely as they bind at specific locations within the enzyme's active site. (2) In DNA transcription, the binding of RNA polymerase to the promotor sequence. (3) In RNA translation, the binding of the ribosome to the mRNA molecule.

initiation factors   A class of proteins that assist ribosomes in binding to a messenger RNA molecule to initiate translation. Occur both in prokaryotes and in eukaryotes.

innate behavior   Behavior that is inherited genetically and does not have to be learned.

innate immune response   The body's nonspecific response to pathogens; the action of leukocytes.

innate immunity   A set of nonspecific defenses against pathogens that occurs even without previous exposure to the pathogen and that does not involve antibodies. Includes responses by mast cells, neutrophils, and macrophages; typically results in an inflammatory response.

inner ear   The innermost portion of the mammalian ear, consisting of a fluid-filled system of tubes that includes the cochlea (which receives sound vibrations from the middle ear) and the semicircular canals (which function in balance).

inoculation   Introduction of some substance into the body to increase immunity to disease. Usually involves vaccination.

in situ hybridization   A lab technique for revealing the locations of specific mRNAs in organisms (i.e., revealing which tissues transcribe certain genes).

insulin   A peptide hormone produced by the pancreas in response to high levels of glucose (or amino acids) in blood. Enables cells to absorb glucose and coordinates synthesis of fats, proteins, and glycogen.

integral membrane protein   Any membrane protein that spans the entire membrane. Also called transmembrane protein.

integration   Processing of information from many sources.

integrator   A component of an animal's nervous system that evaluates sensory information and triggers appropriate responses.

integrins   A class of transmembrane proteins that binds to fibronectins in the extracellular matrix, thus holding cells in place.

intercalated discs   The physical connections between adjacent heart muscle cells. Contain gap junctions to allow electrical signals to pass between the cells.

intermediate disturbance hypothesis   The hypothesis that moderate ecological disturbance causes high species diversity.

intermediate filament   Any of various long, thin cellular fibers composed of thin filaments of various protein polymers wound into thicker cables. Forms part of a cell's cytoskeleton.

intermediate host   The host species in which a parasite reproduces asexually.

interneuron   A neuron that passes information between two other neurons.

internode   The section of a plant stem between two nodes (sites where leaves attach).

interphase   The part of the cell cycle during which no cell division occurs. Includes the G_1 phase, the S phase, and the G_2 phase.

interspecific competition Competition between members of different species for the same limited resource.

interstitial fluid The plasma-like fluid found in the spaces between cells.

interstitial space Any area between two cells. Usually filled with fluid called interstitial fluid.

intertidal zone The region between the low-tide and high-tide marks on a seashore.

intracytoplasmic sperm injection (ICSI) A technique used to help infertile couples bear children, by injecting a single sperm directly into the cytoplasm of the egg.

intraspecific competition Competition between members of the same species for the same limited resource.

intrinsic rate of increase The rate at which a population will grow under optimal conditions (i.e., when birthrates are as high as possible and death rates are as low as possible). Symbolized by r_{max}.

intron A region of a eukaryotic gene that is transcribed into RNA but is later excised from the mRNA transcript before translation into protein.

invasive species An exotic species that, upon introduction to a new area, spreads rapidly and competes with native species.

invertebrates All multicellular animals that are not vertebrates. This is a paraphyletic group.

in vitro Outside the living body. Refers to a process or experiment done "in glass" (i.e., in a dish or test tube rather than in an intact living body).

in vitro fertilization (IVF) A technique for fertilizing mammalian eggs with sperm in a lab dish.

in vivo In a living body. Refers to a process or experiment done "in life."

ion An atom or a molecule that has gained or lost electrons and carries an electric charge.

ion channels A class of membrane proteins that allow certain ions to diffuse across a plasma membrane, via passive transport.

ionic bond A bond between atoms that is formed when an electron is completely transferred from one atom to another so that the atoms remain associated due to their opposite electric charges.

ionophore Any compound that increases membrane permeability for a certain ion. Many ionophores are membrane proteins that bind to the ion and carry it through the membrane.

iris A ring of pigmented muscle just inside the vertebrate eye that contracts or expands to control the amount of light entering the eye.

islets of Langerhans Clusters of cells in the pancreas that secrete insulin and glucagon directly into the blood.

isomer A molecule that has the same molecular formula as another molecule but differs from it in three-dimensional structure.

isometric (1) Characterized by scaling that is proportionate to some other feature, such as body size. (2) A type of muscle contraction in which the muscle exerts force without shortening.

isotonic Having the same solute concentration and water concentration relative to another solution.

isotope Any of several forms of an element that have the same number of protons but differ in the number of neutrons.

jelly layer A gelatinous layer that encloses the egg cells of some vertebrates.

jet propulsion Movement accomplished by forcibly ejecting water in the opposite direction. Used by various aquatic invertebrates.

joint A place where two pieces (bones, cartilages, etc.) of a skeleton meet. May be movable (an articulated joint) or immovable (e.g., skull sutures).

junctional diversity Genetic diversity created by variations in the joining of gene segments; occurs during gene recombination in lymphocytes of the immune system.

juvenile hormone An insect hormone that prevents larvae from metamorphosing into adults.

karyogamy Fusion of two haploid nuclei to form a diploid nucleus. Occurs in many fungi, and in animals and plants during fertilization of gametes.

karyotype The distinctive appearance of all of the chromosomes in an individual, including the number of chromosomes and their length and banding patterns (after staining with dyes).

keystone species A species that has an exceptionally great impact on the other species in its ecosystem relative to its abundance.

kidneys Paired organs situated at the back of the abdominal cavity that filter the blood, produce urine, and secrete several hormones.

kilocalorie (kcal) A unit of energy, often used to measure energy content of food. Also called a Calorie.

kinesin A motor protein that uses the chemical energy of ATP to transport vesicles, particles, or chromosomes along microtubules.

kinetic energy The energy of motion, as opposed to potential energy (stored in position or shape).

kinetochore The structure on chromatids where spindle fibers attach. Contains motor proteins that move the chromosome along the microtubule.

kingdom A classic taxonomic rank that is above the phylum level and below the domain level.

kinocilium A single cilium that juts from the surface of many hair cells and functions in detection of sound or pressure.

kin selection A form of natural selection that favors traits that increase survival or reproduction of an individual's kin at the expense of the individual.

Klinefelter syndrome A syndrome seen in humans who have an XXY karyotype. People with this syndrome have male sex organs, may have some female traits, and are sterile.

knock-out mutant A mutant allele that does not function at all, or an organism homozygous for such a mutation. Also called null mutant or loss-of-function mutant.

Koch's postulates Four criteria used to determine whether a putative infectious agent causes a particular disease.

Krebs cycle A series of chemical reactions, during which acetyl CoA (from glycolysis) is broken down to CO_2, producing ATP and reduced compounds for the electron transport chain. Occurs in the mitochondria.

labia majora (plural: labium majus) One of two outer folds of skin that protect the labia minora, clitoris, and vaginal opening of female mammals.

labia minora (plural: labium minus) One of two inner folds of skin that protect the opening of the urethra and vagina.

labor The strong muscular contractions of the uterus that expel the fetus during birth.

lac **operon** The operon in *E. coli* that includes genes responsible for metabolism of lactose.

lactation Production of milk from mammary glands of mammals.

lacteal Lymphatic vessels in the center of the villi of the small intestine. Receive chylomicrons containing fat absorbed from food and send them into the lymph system.

lactic acid fermentation Production of ATP without an electron transport chain, and the end product of glycolysis (pyruvate) is converted to lactic acid.

lagging strand In DNA replication, the strand of new DNA that is synthesized discontinuously in a series of short pieces that are later joined together. Also called discontinuous strand.

lamellae (singular: lamella) Any set of parallel platelike structures (e.g., the crescent-shaped flaps on the gill filaments of fish gills that serve to increase surface area for gas exchange).

large intestine The posterior part of the intestine, between the small intestine and the rectum.

larva (plural: larvae) An immature stage of a species in which the immature and adult stages have different body forms.

late endosome A membrane-bound vesicle, formed by endocytosis, that is a late stage in the process of becoming a lysosome.

lateral bud A bud on a plant stem that is capable of producing a new side branch.

lateral gene transfer Transfer of DNA between two different species, especially distantly related species. Commonly occurs among bacteria and archaea via plasmid exchange; also can occur in eukaryotes via viruses and some other mechanisms.

lateral meristem A layer of undifferentiated plant cells found in older stems and roots. Responsible for secondary growth. Also called cambium or secondary meristem.

lateral root A plant root extending from another, older root.

LDL See low-density lipoprotein (LDL).

leaching Loss of nutrients from soil via percolating water.

leading strand In DNA replication, the strand of new DNA that is synthesized in one continuous piece, with nucleotides added to the 3′ end of the growing molecule. Also called continuous strand.

leaf primordium (plural: leaf primordia) Small protuberances that form around an apical meristem of a plant shoot and that will develop into leaves.

leak channels The potassium channels in the membranes of nerve cells, which allow potassium ion to leak out of the cell in its resting state.

learning An enduring change in an individual's behavior that results from specific experience(s).

leghemoglobin An iron-containing protein similar to hemoglobin. Found in root nodules of legume plants where it binds oxygen, preventing it from poisoning a bacterial enzyme needed for nitrogen fixation.

legumes Members of the pea plant family. Form symbiotic associations with nitrogen-fixing bacteria in their roots.

lens A transparent, crystalline structure that focuses incoming light onto a retina or other light-sensing apparatus of an eye.

leptin A hormone produced by fat cells (adipocytes) that signals how much body fat is stored. Inhibits appetite.

leukocytes Immune system cells, including neutrophils, macrophages, B cells, and T cells, that circulate in blood or lymph and function in defense against disease. Also called white blood cells.

lichen A symbiotic association of a fungus and a photosynthetic alga.

life cycle The sequence of developmental events and phases that occurs during the life span of an organism, from fertilization to offspring production.

life history The sequence of events in an individual's life, from birth to reproduction to death. Also, the study of how the organism allocates resources and energy to these different activities.

life table A data set that summarizes the probability that an individual in a certain population will survive and reproduce in any given year over the course of its lifetime.

ligand Any molecule that binds to a specific site on a receptor molecule.

ligand-gated channel An ion channel that opens or closes in response to binding by a certain molecule.

light chain One of two identical short polypeptides forming part of an immunoglobulin protein (antibody, B-cell receptor, etc.). The light chains determine which antigen the molecule will bind to.

light-dependent reactions In photosynthesis, the set of reactions that use the energy of sunlight to split water, producing ATP, NADPH, and oxygen.

light-independent reactions In photosynthesis, the set of reactions that use the NADPH and ATP formed earlier (in the light-dependent reactions) to drive the reduction of atmospheric CO_2, ultimately producing sugars. Also called the Calvin cycle.

lignin An extremely strong polymer in the secondary cell walls of plant cells in woody plant parts. Composed of six-carbon rings joined.

lignin peroxidase A fungal enzyme that can digest lignin.

limiting nutrient An essential nutrient whose scarcity in the environment significantly affects growth and reproduction.

limnetic zone Open water (not near shore) that receives enough light to support photosynthesis.

lineage An evolutionary unit that includes an ancestral population, all of its descendants, and only its descendants. Also called a monophyletic group or a clade.

linkage A physical association between two genes because they are on the same chromosome; the inheritance patterns resulting from this association.

linkage map A map of the relative locations of specific genes on a certain chromosome. Also called a genetic map or meiotic map.

lipase Any enzyme that can digest fats.

lipid A carbon-containing subtance that is hydrophobic and thus does not dissolve in water, but dissolves well in nonpolar organic solvents. Lipids include fats, oils, phospholipids, and waxes.

lipid bilayer A double layer of phospholipid molecules, with hydrophobic tails oriented toward the inside and hydrophilic heads toward the outside. The fundamental component of membranes.

liposome An artificially formed tiny membrane-bound structure composed of a phospholipid bilayer membrane.

lithotroph An organism that produces ATP by oxidizing inorganic molecules with high potential energy, such as ammonia (NH_3) or methane (CH_4).

littoral zone Shallow waters near shore that receive enough sunlight to support photosynthesis. May be marine or freshwater.

liver An abdominal organ of vertebrates that performs many biochemical processes, including storage of glycogen, processing and conversion of food and wastes, and production of bile.

live virus vaccine A vaccine containing intact viruses that have been rendered nonvirulent for a certain species. Used for vaccines. Also called an attenuated virus vaccine.

lobe-finned fishes A lineage of fishes with fins supported by an arrangement of bones and muscles similar to that seen in tetrapod limbs. Includes two living groups: coelacanths and lungfishes.

lock-and-key model The hypothesis that enzymes have a precise three-dimensional structure into which substrates (reactant molecules) fit. Modified to reflect knowledge that enzymes are not rigid and may change shape.

locomotion Movement of an organism under its own power.

locus (plural: loci) A gene's physical location on a chromosome.

logistic growth equation The mathematical equation that defines how fast a population will grow over time, given a certain carrying capacity.

logistic population growth Changes in growth rate that occur as a function of population size.

long-day plant A plant that blooms in midsummer, in response to short nights.

long interspersed nuclear element (LINE) A type of parasitic DNA sequence commonly found in eukaryotic genomes. Contains genes for reverse transcriptase, integrase, and a promoter and can create copies of itself, inserted elsewhere in the genome.

loop of Henle A long loop of the nephrons in mammalian kidneys. Functions to set up a concentration gradient that allows reabsorption of water from a subsequent section of the nephron.

loose connective tissue A type of connective tissue consisting of fibrous proteins in a soft matrix. Often functions as padding for organs.

lophophore A specialized feeding structure found in three phyla of the Lophotrochozoa and used in filter feeding.

Lophotrochozoa The lineage of protostomes that includes mollusks and annelids. Many phyla have lophophore feeding structures, have trochophore larvae, and grow by extending the size of their skeletons rather than by molting.

loss-of-function mutant A mutant allele that does not function at all, or an organism homozygous for such a mutation. Also called knock-out mutant or null mutant.

low-density lipoprotein (LDL) Balls of protein and fat that transport cholesterol from the liver to the rest of the body for storage and use. Associated with increased risk of heart disease.

lumen The interior space of any hollow structure (e.g., the rough ER) or organ (e.g., the stomach).

lung Any respiratory organ used for gas exchange between blood and air.

luteal phase The second major phase of a menstrual cycle, after ovulation, when the progesterone levels are high and the body is preparing for a possible pregnancy.

luteinizing hormone (LH) A peptide hormone from the anterior pituitary that stimulates estrogen production, ovulation, and formation of the corpus luteum in females and testosterone production in males.

lymph The mixture of fluid and lymphocytes that circulates through the ducts and lymph nodes of the lymphatic system in vertebrates.

lymphatic duct One of the thin-walled tubes that collect excess fluid from body tissues and return it to the circulatory system, passing through lymph nodes along the way.

lymphatic system In vertebrates, a network of thin-walled tubes that collects excess fluid from body tissues and returns it to the veins of the circulatory system. Includes lymph nodes.

lymph nodes Small oval structures through which lymph ducts run. Filter the lymph and screen it for infection.

lymphocyte A type of leukocyte that circulates through the bloodstream and lymphatic system and that is responsible for the development of acquired immunity. Includes B cells and T cells.

lysogeny A type of viral replication in which the viral DNA is inserted into the host's chromosome, remaining there indefinitely and passively replicating whenever the host cell divides.

lysosome A small organelle in an animal cell containing acids and enzymes that catalyze hydrolysis reactions and can digest large molecules. In plants, fungi and some other groups, may be called vacuoles.

lysozyme An enzyme that acts as an antibiotic by digesting bacterial cell walls. Occurs in saliva, tears, mucus, and egg white.

lytic replication cycle A type of viral replication in which new virus particles are made inside a host cell and eventually burst out of the cell, killing it.

macromolecule Any very large molecule, usually made up of smaller molecules joined together. Include proteins, nucleic acids, and polysaccharides.

macronutrient An essential nutrient required in large quantities. Usually a major component of many organic molecules.

macrophage A type of leukocyte, capable of amoeboid movement through body tissues, that engulfs and digests pathogens and other foreign particles. Also plays roles in secreting cytokines and presenting foreign antigens to other immune system cells.

MADS box A DNA sequence found in some genes of fungi, animals, and plants (genes that control pattern formation in flowers). Codes for a stretch of 58 amino acids that can bind to DNA.

major histocompatibility proteins (MHC proteins) Mammalian cell-surface glycoproteins involved in immunity and in marking cells as "self" to the immune system.

maladaptive For a trait, reducing the fitness of individuals.

malaria A human disease caused by four species of the protist *Plasmodium*, which is passed to humans by mosquitoes.

male gametophyte A multicellular haploid structure that can produce haploid sperm cells in a species that exhibits alternation of generations. In flowering plants, consists of two sperm cells within a larger tube cell.

malignant tumor A tumor that is actively growing and disrupting local tissues and/or is spreading to other organs. A cancer consists of one or more malignant tumors.

Malpighian tubules A major excretory organ of insects, consisting of blind-ended tubes that extend from the gut into the hemocoel. Filter hemolymph to form pre-urine and then send it to the hindgut for further processing.

mammary glands Specialized skin glands that can produce milk for nursing offspring. A diagnostic feature of mammals.

mandibles Any chewing mouthparts. In vertebrates, the lower jaw. In insects, crustaceans, and myriapods, the first pair of mouthparts.

mantle In mollusks, the thick outer tissue that protects the visceral mass and that may secrete a calcium carbonate shell.

Marfan syndrome A human syndrome involving increased height, long limbs and fingers, an abnormally shaped chest, and heart disorders.

mark-recapture study A method of estimating population size involving release of marked individuals into a population and an assessment of how many individuals captured later are marked vs. unmarked.

marsh A wetland that lacks trees and usually has a slow but steady rate of water flow.

marsupial A member of the taxon Marsupiala, a lineage of mammals that nourish their young in an abdominal pouch after a very short period of development in the uterus.

mass extinction Rapid extinction of an unusually large number of diverse evolutionary groups across a wide geographic area. May occur due to sudden and extraordinary environmental changes.

mass number The total number of protons and neutrons in an atom.

mast cell A type of leukocyte that is stationary (embedded in tissue) and that helps trigger the inflammatory response to infection or injury, including secretion of histamine. Particularly important in allergic responses and defense against parasites.

master plate In replica plating, the original plate of bacteria that is used as the source for producing replica plates for further study.

maternal chromosome A chromosome inherited from the mother.

maternal effect inheritance A pattern of inheritance in which an individual's phenotype is determined by its mother's genotype. Common in embryological development, during which egg components made by the mother can influence development of the offspring.

mating type A form of fungal hyphae that carries certain alleles for the genes involved in mating and mates only with hyphae of a different mating type.

matrix A general term for any liquid or semisolid that fills or surrounds some structure and that has some role in maintaining the shape or function of organelles, cells, or tissues.

medulla The innermost part of an organ (e.g., kidney or adrenal gland). In the brain, the posteriormost portion, responsible for rhythmic body functions such as heart rate, respiration, and digestion.

medullary respiratory center An area at the base of the brain (in the region known as the medulla oblongata) that stimulates breathing.

medusa (plural: medusae) The free-floating stage of a cnidarian life cycle.

megapascal (MPa) A unit of pressure (force per unit area), equivalent to 1 million pascals (Pa).

megaphyll A leaf of a fern, horsetail, or vascular plant.

megasporangium (plural: megasporangia) In seed plants, a spore-producing structure that produces megaspores, which can grow to become the female gametophytes.

megaspore A type of small spore produced by the megasporangia of seed plants and that can grow to become a female gametophyte.

megasporocyte A diploid cell contained in the megasporangium of a flower ovule. Undergoes meioiss to produce four megaspores.

meiosis A type of cell division in which one diploid parent cell produces four haploid reproductive cells (gametes). In meiosis, chromosome pairs synapse and can exchange genes via crossing over.

meiosis I The first cell division of meiosis, in which synapsis and crossing over occur, and homologous chromosomes are separated from each other, producing daughter cells with half as many chromosomes as the parent cell.

meiosis II The second cell division of meiosis, in which sister chromatids are separated from each other. Similar to mitosis.

meiotic map A map of the relative locations of specific genes on a certain chromosome. Also called a genetic map or linkage map.

membrane potential A difference in electric charge across a cell membrane; a form of potential energy. Also called membrane voltage.

membrane protein Any protein found in a cell membrane. May span the entire membrane (transmembrane or integral membrane proteins) or may be found on only one side of the membrane (peripheral membrane proteins).

memory Retention of learned information.

memory cell A type of lymphocyte responsible for maintenance of immunity for years or decades after an infection. Descendants of B or T cells activated during a previous infection.

meniscus (plural: menisci) The concave boundary layer formed at most air-water interfaces, due to surface tension.

menstrual cycle A female reproductive cycle seen in Old World monkeys and apes (including humans), consisting of a follicular phase, ovulation, a luteal phase, and then (if no pregnancy occurs) menstruation.

menstrual synchrony The phenomenon in which human women living in close proximity experience synchronization of their menstrual cycles.

menstruation The periodic shedding of the uterine lining through the vagina that occurs in females of Old World monkeys and apes, including humans.

meristem A group of undifferentiated plant cells that can produce cells that differentiate into specific adult tissues.

mesoderm One of three basic embryonic cell layers of a triploblast animal. Forms the middle tissues between skin and gut: muscles, bones, blood, and some internal organs (kidney, spleen, etc.).

mesoglea A gelatinous material with scattered ectodermal cells, found in cnidarians between ectoderm and endoderm.

mesophyll cell A type of cell found near the surfaces of plant leaves and where most photosynthesis occurs.

messenger RNA (mRNA) An RNA molecule that carries encoded information, transcribed from DNA, and that can be used for the synthesis of one or more proteins.

meta-analysis An analysis that combines and compares the results of many smaller, previously published studies.

metabolic pathway A series of distinct chemical reactions that build up or break down a particular molecule. Often, each reaction is catalyzed by a different enzyme.

metabolic rate The total energy consumption of all the cells of an individual. For aerobic organisms, often measured as the amount of oxygen consumed per hour.

metabolic water The water that is produced as a by-product of cellular respiration.

metabolism All the chemical reactions occurring in a living cell or organism.

metallothioneins Small plant proteins that bind to and prevent excess metal ions from acting as toxins.

metamorphosis A dramatic change from the larval to the adult form of an animal.

metaphase A stage in cell division (mitosis or meiosis) during which chromosomes line up in the middle of the cell.

metaphase plate The plane along which chromosomes line up during metaphase of cell division (mitosis or meiosis).

metapopulation A population of a single species that is divided into many smaller populations.

metastasis The process by which cancerous cells leave the primary tumor and establish additional tumors elsewhere in the body.

methanogen A group of archaea that produce methane (CH_4; natural gas) as a by-product of cellular respiration.

methanotroph An organism that uses methane (CH_4; natural gas) as its primary electron donor and source of carbon.

methylation The addition of a methyl ($-CH_3$) group to a molecule.

MHC See **major histocompatibility complex proteins (MHC proteins).**

micelle A small droplet of similar molecules clumped together in a solution.

microbe Any microscopic organism. Includes bacteria, archaea, and various tiny eukaryotes.

microfilament A long, thin fiber composed of two intertwined strands of polymerized actin. Involved in cell movement. Also called actin filament.

micrograph A photograph of an image produced by a microscope.

micronutrient An essential nutrient required in very small quantities—usually an enzyme cofactor.

microphyll A type of small leaf on lycopods.

micropyle The tiny pore in a plant ovule through which the pollen tube reaches the embryo sac.

microsatellite A noncoding stretch of eukaryotic DNA that contains a repeating sequence one to five base pairs long. A type of simple sequence repeat.

microsporangium (plural: microsporangia) In seed plants, a spore-producing structure that produces microspores, which can grow to become the male gametophytes.

microspore A small haploid spore produced by the microsporangia of seed plants and that grow to become a male gametophyte.

microsporidians A lineage of single-celled, parasitic eukaryotes that are closely related to fungi.

microsporocytes Diploid cells contained within the microsporangium of a flower anther. Undergo meiosis to produce microspores.

microtubule A long, tubular polymer of protein subunits α-tubulin and β-tubulin. Involved in cell movement and transport of materials within the cell.

microtubule organizing center Any structure that organizes microtubules.

microvilli (singular: microvillus) Tiny protrusions from the surface of an epithelial cell that increase the surface area for absorption of substances.

middle ear The air-filled middle portion of the mammal ear, connecting to the throat via the Eustachian tube. Transmits and amplifies sound from the tympanic membrane to the inner ear.

middle lamella A central layer of gelatinous pectins between the primary cell walls of adjacent plant cells. Helps hold the cells together.

migration (1) A cyclical movement of large numbers of organisms from one geographic location or habitat to another. (2) In population genetics, movement of individuals from one population to another.

millivolt (mV) A unit of voltage equal to 1/1000 of a volt.

mimicry A phenomenon in which one species has evolved (or learns) to look or sound like another species.

minisatellite A noncoding stretch of eukaryotic DNA that contains a repeating sequence 6 to 500 base pairs long. A type of simple sequence repeat.

mismatch repair The process by which mismatched base pairs in DNA are fixed.

missense mutation A point mutation (change in a single base pair) that causes a change in the amino acid sequence of a protein. Also called replacement mutation.

mitochondrial DNA DNA found inside the mitochondria of eukaryotic cells.

mitochondrial matrix The solution inside the inner membrane of a mitochondrion. Contains the enzymes of the Krebs cycle.

mitochondrion (plural: mitochondria) A eukaryotic organelle that is the site of aerobic respiration.

mitosis Nuclear division of a eukaryotic cell producing two daughter nuclei that are genetically identical to the parent.

mitosis-promoting factor (MPF) A complex of two proteins (cyclin and cyclin-dependent kinase) that causes eukaryotic cells to initiate mitosis.

mitotic phase (M phase) The phase of the cell cycle during which cell division occurs. Includes mitosis and cytokinesis.

mitotic spindle An array of microtubules that moves chromosomes to opposite sides of the cell during cell division (mitosis or meiosis).

model organism An organism selected for intensive scientific study based on features that make it easy to work with (e.g., body size, life span), in hope that the findings will apply to other species.

molarity The number of moles of a dissolved solute in 1 liter of solution; a unit of concentration.

mole An amount of any substance containing 6.022×10^{23} atoms, ions, or molecules. This number of molecules of a compound will have a mass equal to the molecular weight of that compound expressed in grams.

molecular chaperones A class of proteins that facilitate the three-dimensional folding of newly synthesized proteins.

molecular clock The hypothesis that certain types of mutations tend to reach fixation in populations at a steady rate over large spans of time. As a result, comparisons of DNA sequences can be used to infer the timing of evolutionary divergences.

molecular formula A notation that indicates the numbers and types of atoms in a molecule, such as H_2O for the water molecule.

molecular ion An ion consisting of a group of several different atoms (rather than one element).

molecular weight The sum of the mass numbers of all of the atoms in a molecule; roughly, the total number of protons and neutrons in the molecule.

molecule Two or more atoms held together by covalent bonds.

molting A method of body growth, used by ecdysozoans, that involves the shedding of an external protective cuticle or skeleton, expansion of the soft body, and growth of a new external layer.

monoclonal antibody An antibody produced in the lab from a hybridoma derived from a single B cell. Such an antibody has a unique amino acid sequence and thus the ability to bind to a specific site of a particular antigen.

monocot A plant that has a single cotyledon (embryonic leaf) upon germination.

monoecious Having both male and female flowers.

monohybrid cross A mating between two parents that are both heterozygous for a given gene.

monomer A small molecular subunit that can bond to other subunits to form long macromolecules, or polymers.

monophyletic group An evolutionary unit that includes an ancestral population, all of its descendants, and only its descendants. Also called a clade or a lineage.

monosaccharide A single sugar monomer, such as glucose. Formally, a small carbohydrate of the chemical formula $(CH_2O)_n$ that cannot be hydrolyzed to form any smaller carbohydrates.

monosomy Having only one copy of a particular type of chromosome.

monotremes A member of the Monotremata, a lineage of mammals that lay eggs and then nourish the young with milk. Includes just three living species: the platypus and two species of echidna.

morphogenesis A process of embryologic development during which cells become organized into recognizable tissues, organs, and other structures.

morphology The shape and appearance of an organism's body and its component parts.

morphospecies concept The concept that species are best identified as groups that have measurably different anatomical features.

motif A repeating theme. In molecular biology, a domain (a section of a protein with a distinctive tertiary structure) in many different proteins.

motile Not sessile (not permanently attached to a substrate); capable of moving to another location.

motor neuron A nerve cell that carries signals from the central nervous system (brain and spinal cord) to an effector, such as a muscle or gland.

motor protein A protein whose major function is to convert the chemical energy of ATP into motion.

mRNA See **messenger RNA (RNA)**.

mucigel A slimy substance secreted by plant root caps to ease passage of the growing root through the soil.

mucins Glycoproteins, produced by salivary glands, that form mucus when mixed with water.

mucus A slimy mixture of glycoproteins and water, secreted by many organs for lubrication.

Müllerian inhibitory substance A peptide hormone secreted by the embryonic testis that causes regression (withering away) of the female reproductive ducts.

Müllerian mimicry A type of mimicry in which two (or more) harmful species resemble each other.

multicellular growth Growth in which individual cells join together to form a multicelled body and cells differentiate to perform specialized roles.

multicellularity The condition whereby an organism's body contains more than one cell and only certain cells pass genes to the next generation.

multiple allelism The occurrence of more than two alleles of a gene in a given population.

multiple sclerosis (MS) A human autoimmune disease caused by the immune system attacking the myelin sheaths that insulate nerve axons.

muscle fiber A single muscle cell.

muscle tissue A class of animal tissue consisting of bundles of long, thin contractile cells (muscle fibers).

mutagen Anything that can increase the rate of mutation.

mutant An individual that carries a mutation, particularly a new or rare mutation.

mutation Any change in the hereditary material of an organism (DNA in most organisms, RNA in some viruses).

mutualism A relationship between two species that benefits both species.

mutualist An organism that lives in a close relationship with a host and that benefits its host.

Myb proteins DNA-binding proteins that turn gene transcription on or off, acting as transcription activators or transcription repressors, respectively.

mycelium (plural: mycelia) A mass of underground filaments (hyphae) that form the body of a fungus. Also found in some protists and bacteria.

mycorrhiza (plural: mycorrhizae) A mutualistic association between certain fungi and most vascular plants, sometimes visible as nodules or nets in or around plant roots.

myelin sheath Multiple layers of myelin, a lipid, that are wrapped around the axons of neurons to provide electrical insulation.

myeloma A cancer or tumor of the cells of bone marrow. Myelomas of B cells are used in lab production of monoclonal antibodies.

MyoD A regulatory protein involved in differentiation of muscle cells during embryological development. Enhances transcription of muscle-specific genes.

myofibril A bundle of strands of contractile proteins organized into repeating units (sarcomeres) in vertebrate heart muscle and striated muscle.

myoglobin An oxygen-binding muscle protein consisting of a single globin and one heme group.

myosin A eukaryotic protein that can be polymerized to form thick filaments that are used in muscle contraction and intracellular movement.

natural history The branch of biology concerned with describing what exists in nature (i.e., primarily observational and not experimental).

natural selection The process by which individuals with certain heritable traits tend to produce more surviving offspring than do individuals without those traits, resulting in a change in the genetic makeup of the population. A major mechanism of evolution.

nauplius A distinct planktonic larval stage seen in many crustaceans.

Neanderthal A recently extinct European species of hominid, *Homo neanderthalensis*, closely related to but distinct from modern humans.

nectar The sugary fluid produced by flowers to attract and reward pollinating animals.

nectary The nectar-producing gland at the base of a flower.

negative control A type of gene regulation in which a transcription can occur only when a certain substance—typically a repressor protein that binds to a control sequence in the DNA and prevents transcription—is removed.

negative feedback A self-limiting, corrective response in which a deviation in some variable (e.g., body temperature, blood pH) triggers responses aimed at returning the variable to normal.

negative pressure ventilation Ventilation of the lungs that is accomplished by "pulling" air into the lungs by expansion of the rib cage.

negative result An experimental result that fails to show a predicted difference between two groups or conditions.

negative-sense virus A virus whose genome contains sequences complementary to those in the mRNA required to produce viral proteins.

nephron One of the tiny tubes within the vertebrate kidney that filter blood and concentrate salts to produce urine.

neritic zone Shallow marine waters beyond the intertidal zone, extending down to about 200 m, where the continental shelf ends.

Nernst equation A formula that converts the energy of a concentration gradient to the energy of an electrical potential, for a particular ion or ions.

nerve A long, tough strand of nervous tissue typically containing thousands of axons that carry information to or from the central nervous system.

nervous tissue A class of animal tissue consisting of nerve cells (neurons) and various supporting cells, and functioning in rapid transmission of complex information.

net primary productivity (NPP) In ecology, the amount of primary productivity that is stored in new biomass (not used for cellular respiration).

net reproductive rate The growth rate of a population per generation; equivalent to the average number of female offspring that each female produces over her lifetime. Symbolized by R_0.

neural Relating to nerve cells.

neural tube A folded tube of ectoderm that forms along the dorsal side of a young vertebrate embryo and that will give rise to the brain and spinal cord.

neuroendocrine Refers to nerve cells that release hormones into the blood.

neuron A nerve cell; a cell that is specialized for the transmission of nerve impulses. Typically has dendrites, a cell body, and a long axon that forms synapses with other neurons.

neurosecretory cell A neuron that secretes hormones into the blood; a neuroendocrine cell.

neurotoxin Any poison that specifically affects neuron function.

neurotransmitter A molecule that conveys information from one neuron to another or from a neuron to a muscle or gland. Released from the end of an axon and diffuse a very short distance to the next cell, where they can trigger an action potential.

neutral Any mutation that has no effect on an individual's fitness.

neutron An uncharged particle found in atomic nuclei. Variations in neutron number (with no change in proton number) produce different isotopes of the same element.

neutrophil A type of leukocyte, capable of amoeboid movement through body tissues, that engulfs and digests pathogens and other foreign particles and secretes various compounds that attack bacteria and fungi.

niche The particular set of habitat requirements of a certain species and the role that species plays in its ecosystem.

niche differentiation The tendency of competing species to use different ecological niches because of competition.

nitrogen fixation The incorporation of atmospheric nitrogen (N_2) into forms such as ammonia (NH_3) or nitrate (NO_3^-), which can be used to make many organic compounds. Occurs in only a few lineages of bacteria and archaea.

nociceptor A sensory cell or organ specialized to detect tissue damage, usually producing the sensation of pain.

node (1) Any small thickening (e.g., a lymph node); (2) The part of a stem where leaves or leaf buds are attached. (3) The point on a phylogenetic tree where two branches diverge, representing the point in time when an ancestral group split into two or more descendant groups.

node of Ranvier A point on a neuron's axon between sections of myelin sheath, where an action potential can be regenerated.

Nod factor Molecules produced by nitrogen-fixing bacteria that help them recognize and bind to legume roots.

nodules The lumplike structures in roots of the pea family, containing symbiotic nitrogen-fixing bacteria.

nondisjunction An error that can occur during meiosis or mitosis in which both homologous chromosomes of a pair move to the same side of the dividing cell. One daughter cell receives two copies of the chromosome, and the other daughter cell receives none.

nonenveloped For a virus, lacking an envelope surrounding its capsid coat.

nonpolar covalent bond A symmetrical covalent bond (i.e., one in which electrons are equally shared between the two atomic nuclei).

non-self Property of a molecule or cell whereby immune system cells will recognize it as "different" and attack it.

nonsense mutation A point mutation (change in a single base pair) that results in an early stop codon, resulting in a truncated polypeptide.

non-sister chromatids The chromosome copies of homologous chromosomes. Crossing over occurs between non-sister chromatids.

non-template strand The strand of DNA that is not transcribed to create RNA.

nonvascular plants Several phyla of green plants that lack vascular tissue. Includes liverworts, hornworts, and mosses. Also called bryophytes.

norepinephrine A catecholamine used as a neurotransmitter in the sympathetic nervous system and also released as a hormone from the adrenal medulla. Stimulates increased heart rate, increases blood pressure, decreases digestion, and produces other effects.

norm of reaction The range of phenotypes that are possible for an individual of a given genotype.

Northern blotting A lab technique for identifying the RNA produced by a particular gene. Involves separating RNAs by gel electrophoresis, transferring to a filter paper, and hybridizing to a labeled DNA probe.

notochord A long, gelatinous, supportive rod down the back of a chordate embryo, below the developing spinal cord. Replaced by vertebrae in adult vertebrates. A defining feature of chordates.

nuclear envelope The double-layered membrane enclosing the nucleus of a eukaryotic cell.

nuclear lamina A lattice-like sheet of fibrous nuclear lamin proteins that line the inner membrane of the nuclear envelope. Stiffens the envelope and helps organize the chromosomes.

nuclear lamins A class of fibers that form a dense mesh (the nuclear lamina) just inside the nuclear envelope. A form of intermediate filament.

nuclear localization signal (NLS) A certain sequence of amino acids that tags a protein for delivery to the nucleus by importins.

nuclear pore An opening in the nuclear envelope that connects the inside of the nucleus with the cytoplasm and through which molecules such as mRNA and some enzymes can pass.

nuclear pore complex A large complex of dozens of proteins lining a nuclear pore, defining its shape and transporting substances through the pore.

nuclease Any enzyme that can break down RNA and DNA molecules.

nucleic acids Polymers consisting of a chain of nucleotides. Generally used by cells to store or transmit hereditary information. Includes ribonucleic acid and deoxyribonucleic acid.

nucleoid The region of a prokaryotic cell that contains chromosomes.

nucleolus The structure in a eukaryotic nucleus where ribosomal RNA processing occurs and ribosomal subunits are assembled.

nucleoside A purine or pyrimidine base attached to a five-carbon sugar (ribose or deoxyribose).

nucleosome A repeating, bead-like structure of a eukaryotic chromosome, consisting of about 200 nucleotides of DNA wrapped twice around eight histone proteins.

nucleotide A monomer that can be polymerized to form the nucleic acid DNA or RNA. Consists of a five-carbon sugar (ribose or deoxyribose), a phosphate group, and one of several nitrogen-containing bases. Equivalent to a nucleoside plus one phosphate group.

nucleus (1) The center of an atom, containing protons and neutrons. (2) In eukaryotic cells, the membrane-bound organelle that contains DNA. (3) A discrete clump of neuron cell bodies in the brain, usually sharing a distinct function.

null hypothesis A hypothesis that specifies what the results of an experiment will be if the main hypothesis being tested is wrong. Often states that there will be no difference between experimental groups.

null mutant A mutant allele that does not function at all; or an organism homozygous for such a mutation. Also called knock-out mutant or loss-of-function mutant.

nutrient A substance that an organism requires for normal growth, maintenance, or reproduction.

nutritional balance A state in which an organism is taking in enough nutrients to maintain normal health and activity.

nymph An immature stage of an insect species in which the immature form looks like a miniature adult, such as that in dragonflies.

obesity The condition of having extremely and abnormally high reserves of body fat.

occipital lobe In the mammal brain, the posteriormost region of each cerebral hemisphere. Receives and interprets visual information.

oceanic zone The waters of the open ocean beyond the continental shelf.

oil A lipid that is liquid at room temperature.

Okazaki fragments Short fragments of DNA produced during DNA replication. Now known to be pieces of the lagging strand.

olfaction The perception of odors.

olfactory bulb A bulb-shaped projection of the brain just above the nose. Receives and interprets odor information from the nose.

oligopeptide A polypeptide with fewer than 50 amino acids. May also be referred to simply as "peptides."

ommatidium (plural: ommatidia) A light-sensing column of an arthropod's compound eye.

omnivore An animal whose diet regularly includes both meat and plants.

oncogene An allele that has mutated so as to stimulate cell growth at all times and thus promotes cancer development.

one-gene, one-enzyme hypothesis The hypothesis that each gene is responsible for making one (and only one) particular protein, in most cases an enzyme that catalyzes a specific reaction. Many exceptions to this hypothesis are now known.

oocyte A cell in the ovary that can undergo meiosis to produce an ovum.

oogenesis The production of egg cells (ova).

oogonia (singular: oogonium) The diploid cells in an ovary that can divide by mitosis to create more oogonia and primary oocytes, which can undergo meiosis.

open circulatory system A circulatory system in which the circulating fluid (hemolymph) is not confined to blood vessels.

open reading frame (ORF) Any DNA sequence that is suspected to be a functional gene because it has a start codon and a stop codon separated by a long stretch of DNA (and sometimes has other characteristic features, such as promoters).

operator The DNA binding site for a repressor protein in a prokaryotic operon.

operculum The stiff flap of tissue that covers the gills of teleost fishes.

operon A region of bacterial DNA that codes for a series of functionally related genes.

opsin One of several proteins involved in animal vision. An opsin joins with retinal to form the light-detecting pigment rhodopsin in rod cells.

optical isomer A molecule that shares the same molecular formula as another molecule but differs in the arrangement of atoms or groups around a carbon atom; left-handed or right-handed form of a molecule.

optic nerve A nerve that carries information from the eye to the brain. Vertebrates have two.

orbital The region around an atomic nucleus in which an electron orbits.

order A classic taxonomic rank above the family level and below the class level.

organ A group of tissues organized into a functional and structural unit.

organelle Any discrete, membrane-bound structure in the cytoplasm of a cell (e.g., mitochondrion).

organic For a compound, containing carbon and hydrogen and usually containing carbon-carbon bonds. Organic compounds are widely used by living organisms.

organism Any living entity that contains one or more cells.

organizer A region of an amphibian embryo (around the upper side of the blastopore) that can organize the development of the entire embryo.

organogenesis A stage of embryonic development just after gastrulation in vertebrate embryos, during which major organs develop from the three embryonic cell layers.

organotroph An organism that produces ATP, oxidizing organic molecules with high potential energy, such as sugars.

orientation A deliberate movement that results in a change in position relative to some external cue, such as toward the Sun or away from a sound.

origin of replication The place on a chromosome at which DNA replication begins.

osmoconformer An animal that does not actively regulate the osmolarity of its tissues but conforms to the osmolarity of the surrounding environment.

osmolarity The concentration of dissolved substances in a solution, measured in moles per liter.

osmoregulation The process by which a living organism controls the concentration of water and salts in its body.

osmoregulator An animal that actively regulates the osmolarity of its tissues.

osmosis Diffusion of water across a selectively permeable membrane from areas of high water concentration (low solute concentration) to areas of low water concentration (high solute concentration).

osmotic stress A condition in which there are abnormal concentrations of water and salts in an organism's cells or tissues.

ouabain A plant toxin that poisons the sodium-potassium pumps of animals.

outcrossing Reproduction by fusion of the gametes of different individuals, rather than self-fertilization. Typically refers to plants.

outer ear The outermost portion of the mammal ear, consisting of the pinna (ear flap) and the ear canal. Funnels sound to the tympanic membrane.

outgroup A taxon known to have diverged earliest from all the other taxa under study. Used to determine the root (most ancient node) of a phylogenetic tree.

out-of-Africa hypothesis The hypothesis that modern humans (*Homo sapiens*) evolved in Africa and spread to other continents, replacing other *Homo* species without interbreeding with them.

oval window A membrane separating the fluid-filled cochlea from the air-filled middle ear. The stapes, a middle ear bone, transmits sound vibrations to the cochlea by vibrating on the oval window.

ovary The egg-producing organ of a female animal, or the seed-producing structure of the female part of a flower.

oviduct A narrow tube that connects the uterus to the ovary, and through which the egg travels after ovulation. Fertilization and cleavage occur in the oviduct. In humans, called the fallopian tube.

oviparous Reproducing by laying eggs, rather than giving live birth.

ovoviviparous Reproducing by retaining eggs inside the body until they are ready to hatch. The embryos are nourished by egg yolk, not via a placenta.

ovulation The process by which an ovum is released from the ovary of a female vertebrate.

ovule A structure inside a flower ovary that produces the female gametophyte and eventually (if fertilized) becomes a seed.

ovum (plural: ova) An egg cell; a mature female gamete and any associated external layers. Larger and less mobile than the male gamete.

oxidation The loss of electrons from an atom during a redox reaction, either by donation of an electron to another atom or by the shared elec-

trons in covalent bonds moving farther from the atomic nucleus.

oxidative phosphorylation Production of ATP molecules from the redox reactions of an electron transport chain, starting with reduced compounds (such as NADH or FADH$_2$) and ending with oxygen as the final electron acceptor.

oxygenic photosynthesis Photosynthesis that involves photosystem II, which catalyzes the splitting of water and produces oxygen. Occurs in cyanobacteria, algae, and plants.

oxytocin A peptide hormone from the posterior pituitary that triggers labor and milk production in females and that stimulates pair bonding, parental care, and affiliative behavior in both sexes.

p53 A tumor-suppressor protein that responds to DNA damage by stopping the cell cycle and/or triggering apoptosis. Codes for a protein with a molecular weight of 53 kilodaltons.

pacemaker cells Cells with an inherent rhythm that can set a rhythm for other cells.

pair-rule genes A class of fruit-fly segmentation genes that organize embryonic cells into particular segments. Active in alternating segments.

paleontology The study of organisms that lived in the distant past.

palisade mesophyll Elongated parenchyma cells found in the ground tissue of leaves. Contain many chloroplasts and perform most photosynthesis.

pancreas A gland attached to the small intestine that secretes digestive enzymes into the intestine and several digestion-related hormones (notably, insulin and glucagon) into the bloodstream.

pancreatic amylase A carbohydrate-digesting enzyme secreted by the pancreas into the small intestine.

pancreatic lipase A fat-digesting enzyme secreted by the pancreas into the small intestine.

paper chromatography A technique for separating molecules on the basis of their size and solubility, by wicking them through a piece of filter paper using a certain solvent.

parabiosis An experimental technique for determining whether a certain physiological phenomenon is regulated by a hormone, by surgically uniting two individuals so that hormones can pass between them.

parabronchi (singular: parabronchus) The tiny parallel air tubes that run through a bird's lung.

paracrine signal A chemical signal that is released by one cell and affects neighboring cells.

parafollicular cells The cells of the thyroid gland that release calcitonin.

paraphyletic group A group of organisms that is not monophyletic; i.e, a group that includes some but not all descendants of the last common ancestor of the group. Paraphyletic groups are not meaningful evolutionary groups.

parasite An organism that lives on or in a host species and that damages its host.

parasitism A long-term relationship between two organisms that is beneficial to one organism (the parasite) but detrimental to the other (the host).

parasitoid An organism that has a parasitic larval stage and a free-living adult stage. Most parasitoids are insects that lay eggs in the bodies of other insects.

parasympathetic nervous system The part of the autonomic nervous system that stimulates activities of relaxation, repair, and rebuilding, such as reduced heart rate and increased digestion.

parathyroid glands Four small glands that are near or embedded in the thyroid gland of vertebrates. Secrete parathyroid hormone, which increases blood calcium.

parathyroid hormone (PTH) A peptide hormone from the parathyroid glands that increases blood calcium.

parazoans Animals whose cells are not organized into distinct tissues. Contains one living group: the sponges.

parenchyma cell A thin-walled type of plant cell found in leaves, the centers of stems and roots, and fruits. Involved in photosynthesis, starch storage, and new growth.

parental care Any action by which an animal expends energy or assumes risks to benefit its offspring (e.g., nest-building, feeding of young, defense).

parental generation The adult organisms used in the first experimental cross in a formal breeding experiment.

parietal cells The cells in the stomach lining that secrete hydrochloric acid.

parietal lobe In the mammal brain, the region of each cerebral hemisphere that is behind and above the frontal lobe. In humans, involved in integrating sensory and motor control.

Parkinson's disease A human neurological disorder that causes progressive deterioration of motor function. Due to the inactivation or destruction of dopamine-secreting neurons at the base of the brain.

parsimony The principle that the phylogenetic tree most likely to be correct is the one that requires the fewest evolutionary changes.

parthenogenesis Development of offspring from unfertilized eggs. A form of asexual reproduction.

partial pressure The pressure of one particular gas in a mixture; the contribution of that gas to the overall pressure.

particle A nonliving infectious entity, such as a virus. Also called an agent.

pascal (Pa) A unit of pressure (force per unit area).

passive transport Diffusion of a substance across a cell membrane down a concentration gradient. When this occurs with the assistance of carrier proteins, it is also called facilitated diffusion.

patch clamping A lab technique for studying the electrical currents that flow through individual ion channels, by sucking a tiny patch of membrane to the hollow tip of a microelectrode.

paternal chromosome A chromosome inherited from the father.

pathogen Any entity capable of causing disease, such as a microbe, virus, or prion.

pathogenic Capable of causing disease.

pattern formation The series of events that determines the spatial organization of an embryo, including alignment of the major body axes and orientation of the limbs.

pattern-recognition receptor Leukocyte membrane proteins that bind to molecules in many bacteria. Part of the innate immune response.

PCR See **polymerase chain reaction (PCR)**.

peat Semidecayed organic matter that accumulates in moist, low-oxygen environments such as bogs.

pectins A class of gelatinous polysaccharides found in the primary cell walls of plant cells. Attract and hold water, forming a gel that helps keep the cell wall moist.

pedigree A family tree of parents and offspring, showing inheritance of particular traits of interest.

pellet Any solid material that collects at the bottom of a test tube below a layer of liquid (the supernatant) during centrifugation.

penis The copulatory organ of male mammals, used to insert sperm into a female.

pentaradial symmetry A form of radial symmetry, found in adult echinoderms, in which the body has exactly five planes of symmetry. Typically, five (or multiples of five) body parts radiate from a central hub.

pentose A monosaccharide (simple sugar) with five carbons.

PEP carboxylase An enzyme that catalyze addition of CO_2 to three-carbon compounds, forming four-carbon compounds. Occurs in mesophyll cells of plants that perform C_4 photosynthesis.

pepsin A protein-digesting enzyme produced in the stomach.

pepsinogen The precursor of the digestive enzyme pepsin. Converted to pepsin by the acidic environment of the stomach.

peptide bond The C–N bond between two amino acid residues in a peptide or protein.

peptide hormone A hormone that is a chain of two or more amino acids.

peptidoglycan A polysaccharide found in bacterial cell walls.

per-capita rate of increase The growth rate of a population, expressed per individual. Calculated as the per-capita birthrate minus the per-capita death rate and symbolized r. Also called per-capita growth rate.

perennial plant A plant that normally lives for more than one year.

perfect For a flower, containing both male parts (stamens) and female parts (carpels).

perforations In plants, small holes in the primary and secondary cell walls of vessel elements that allow passage of water.

pericarp The part of a fruit that surrounds the seeds, formed from the ovary wall. The flesh of most edible fruits; the hard shells of most nuts.

pericycle In plant roots, a layer of cells that give rise to lateral roots.

peripheral membrane protein Any membrane protein that is found only on one side of the membrane, rather than spanning the entire membrane.

peripheral nervous system (PNS) All the components of the nervous system that are outside the central nervous system (the brain and spinal cord). Includes the somatic nervous system and the autonomic nervous system.

peristalsis Rhythmic waves of muscular contraction that push food along the digestive tract.

permafrost A permanently frozen layer of icy soil found in most tundra and some taiga.

permeability The tendency of a structure, such as a membrane, to allow a given substance to diffuse across it.

peroxisome A eukaryotic organelle that contains oxidative enzymes, usually for degrading fatty acids and amino acids and the resulting hydrogen peroxide.

petal One of the modified leaves arranged around the reproductive structures of a flower. Often colored to attract pollinators.

petiole The stalk of a leaf.

phagocytosis The engulfment and uptake of a small particle or cell by an extension of another cell's plasma membrane.

pharyngeal gill slits A set of parallel openings from the throat through the neck to the outside. One of the diagnostic traits of chordates.

pharyngeal jaw A secondary jaw in the back of the mouth, found in some fishes. Derived from modified gill arches.

phenetic approach A method for constructing a phylogenetic tree by computing a statistic that summarizes the overall similarity among populations, based on the available data.

phenotype Any of the observable traits of an individual. Commonly includes physical, physiological, and behavioral traits.

phenotypic variation The total observable variation in a particular trait in a certain population in a certain environment.

phenylketonuria (PKU) A human genetic disease caused by lack of the enzyme that converts the amino acid phenylalanine to tyrosine.

pheophytin A molecule that acts as an electron acceptor in the light-dependent reactions of photosynthesis, accepting excited electrons from chlorophyll and passing them to an electron transport chain.

pheromone A chemical signal, released by one individual into the external environment, that can trigger responses in a different individual.

phloem A plant vascular tissue that conducts sugars. Contains sieve-tube members and companion cells.

phloem loading The movement of sugars into plant phloem.

phloem sap The sugary fluid found in phloem tissue of plants.

phloem unloading The movement of sugars out of plant phloem.

phonotaxis Orientation toward or away from sound.

phosphate The functional group $-OPO_3^{2-}$. Breaking the O–P bonds between adjacent phosphate groups releases large amounts of energy.

phosphodiester bond The type of bond that links the nucleotides in DNA or RNA. Joins the phosphate group of one nucleotide to the hydroxyl group on the sugar of another nucleotide.

phosphofructokinase The enzyme that catalyzes the synthesis of fructose-1,6-bisphosphate from fructose-6-phosphate, a key reaction (step 3) in glycolysis.

phospholipid A type of lipid having a hydrophilic head (a phosphate group) and a hydrophobic tail (one or more fatty acids), often linked by a glycerol molecule. Major components of plasma membranes.

phosphorylase An enzyme that breaks down glycogen, by catalyzing hydrolysis of the α-glycosidic linkages between the glucose monomers.

phosphorylation The addition of a phosphate group to a molecule. Commonly, refers to phosphorylation of proteins to control protein shape or function.

phosphorylation cascade A sequence of events in which one enzyme phosphorylates other enzymes, which in turn phosphorylate many more, leading to phosphorylation of thousands of proteins. Commonly used in signal transduction of hormone messages; also called signal transduction cascade.

photic zone In an aquatic habitat, water that is shallow enough to receive some sunlight (whether or not it is enough to support photosynthesis).

photon   A discrete packet of light energy; a particle of light.

photoperiodism   Any response by an organism to photoperiod—the relative lengths of day and night.

photophosphorylation   Production of ATP molecules by using the energy of light to excite electrons, which are then passed down an electron transport chain. Occurs during photosynthesis.

photoreceptor   A molecule, a cell, or an organ that is specialized to detect light.

photorespiration   A series of light-driven chemical reactions that consumes oxygen and releases carbon dioxide, basically reversing photosynthesis. Usually occurs when there are high O_2 and low CO_2 concentrations inside plant cells, often in bright, hot, dry environments when stomata must be kept closed.

photoreversibility   A change in conformation that occurs in certain plant pigments when they are exposed to their preferred wavelengths of light and that triggers responses by the plant.

photosynthesis   A series of chemical reactions and electron transfer events that converts the energy of light into chemical energy stored in glucose.

photosynthesis-transpiration compromise   The balance that plants must strike between maximizing photosynthesis and conserving water.

photosystem   A system of 200–300 chlorophyll molecules, accessory pigments, and proteins, found in plant chloroplasts and involved in the light-dependent reactions of photosynthesis.

photosystem I   A system of molecules and enzymes in chloroplasts that absorbs light energy, using it to produce NADPH.

photosystem II   A system of molecules and enzymes in plant chloroplasts that absorbs light energy, using it to produce ATP and to split water into protons and oxygen.

phototaxis   Orientation toward or away from light.

phototroph   An organism that produces ATP through photosynthesis.

phototropins   A class of plant photoreceptors that detect blue light and initiate phototropic responses.

phototropism   Growth or movement in a particular direction in response to light.

pH scale   A measure of the concentration of protons in a solution and thus of acidity or alkalinity. Defined as the negative of the base-10 logarithm of the proton concentration: $pH = -\log[H^+]$.

phylogenetic species concept   The concept that species are best identified as the smallest monophyletic group in a phylogenetic tree.

phylogenetic tree   A diagram that depicts the evolutionary history of a group of organisms.

phylogeny   The evolutionary history of a group of organisms.

phylum   (plural: phyla) A classic taxonomic rank above the class level and below the kingdom level. (In plants, sometimes called a division.)

physical map   A map of a chromosome that shows the number of base pairs between various genetic markers.

physiology   The study of how an organism's body functions.

phytoalexin   Any small plant compound produced to combat an infection (usually a fungal infection).

phytochromes   A class of light-sensitive plant proteins involved in detecting light and timing certain physiological processes, such as flowering and germination.

phytoplankton   Plankton (small drifting aquatic organisms) that are photosynthetic.

phytoremediation   The use of plants to clean contaminated soils.

pigment   Any molecule that absorbs only certain wavelengths of visible light and reflects or transmits other wavelengths.

piloting   Finding one's way by using familiar landmarks (i.e., without a specific compass direction).

pinocytosis   Uptake of extracellular fluid by endocytosis (i.e., by pinching off the plasma membrane to form small membrane-bound vesicles).

pioneering species   Species that are often the first to appear in a recently disturbed area.

pitch   The sensation produced by a particular frequency of sound. Low frequencies are perceived as low pitches, high frequencies as high pitches.

pith   The center of a plant stem.

pits   In plants, small holes in the secondary cell walls of tracheids that allow passage of water.

pituitary dwarfism   Dwarfism (abnormally small body size) in mammals that is caused by defects in production of growth hormone by the pituitary gland.

pituitary gland   A small gland directly under the brain, close to the hypothalamus. Releases hormones that affect many other glands and organs.

placenta   An organ formed by a union of maternal and fetal tissues. Exchanges nutrients and wastes between mother and fetus, anchors the fetus to the uterine wall, and produces some hormones. Occurs in most mammals and in a few other vertebrates.

placental mammals   Members of the Eutheria, a major lineage of mammals whose young develop in the uterus and are not housed in an abdominal pouch. Also called eutherians.

planar bilayer   A lipid bilayer (double-layered membrane) constructed across a hole in a glass or plastic wall separating two aqueous solutions.

plankton   Any small organism that drifts near the surface of oceans or lakes and swims little if at all.

plant   A red alga, green alga, glaucophyte alga, or green plant.

plant-defense hypothesis   The hypothesis that rates of herbivory are limited by plant defenses such as toxins and spines.

plantlet   A small plant, particularly one that forms on a parent plant via asexual reproduction and drops, becoming an independent individual.

plasma   The fluid portion of blood that remains when red blood cells, leukocytes, and platelets are removed. (Equivalent to serum plus clotting factors.)

plasma cell   A type of leukocyte that produces large quantities of antibodies to combat an ongoing infection. A descendant of an activated B cell.

plasma membrane   A membrane that surrounds a cell, separating it from the external environment and selectively regulating passage of molecules and ions into and out of the cell.

plasmid   A small, usually circular, supercoiled DNA molecule independent of the cell's main chromosome(s) in prokaryotes and some eukaryotes.

plasmodesmata   (singular: plasmodesma) Physical connections between two plant cells, consisting of gaps in the cell walls through which the two cells' plasma membranes, cytoplasm, and smooth ER can connect directly.

plasmogamy   Fusion of the cytoplasm of two individuals. Occurs in many fungi.

plastocyanin   A small protein that shuttles electrons from photosystem II to photosystem I during photosynthesis.

plastoquinone   The molecule involved in the light-dependent reactions of photosynthesis that receives excited electrons from pheophytin and passes them to more electronegative molecules in the chain. Also carries protons to the lumen side of the thylakoid membrane.

platelet   A small membrane-bound cell fragment in vertebrate blood, important in blood clotting.

platelet-derived growth factor (PDGF)   A protein secreted by platelets and some other cells at the site of an injury. Promotes wound healing.

plate tectonics   The theory that Earth's crust is made up of separate plates that have moved throughout geologic history.

pleiotropy   A pattern of genetic expression in which one gene affects more than one phenotypic trait.

ploidy   The number of each type of chromosome present. Haploid cells have a ploidy of 1; diploid cells have a ploidy of 2.

pneumonia   Inflammation of the lungs.

podium   (plural: podia) The part of the tube foot of an echinoderm that extends outside of the body and makes contact with the substrate.

point mutation   A mutation that results in a change in a single nucleotide pair of DNA.

polar   (1) Asymmetrical or unidirectional. (2) Carrying a partial positive charge on one side of a molecule and a partial negative charge on the other. Polar molecules are generally hydrophilic.

polar bodies   The tiny, nonfunctional cells produced during meiosis of a primary oocyte, due to most of the cytoplasm going to the ovum.

polar covalent bond   A covalent bond that is asymmetrical, such that the electrons spend more time near one atomic nucleus than the other. Often results in the molecule being polar.

polar nuclei   (singular: polar nucleus) The nuclei in the female gametophyte of a flowering plant that will fuse with one sperm nucleus to produce the endosperm. Most species have two.

pollen grain   In flowering plants, a male gametophyte enclosed within a protective coat.

pollen tube   In flowering plants, a tube that grows out of a pollen grain and toward the ovule and through which the two sperm nuclei move.

pollination   The process by which pollen reaches the carpel of a flower (in flowering plants) or reaches the ovule directly (in conifers and their relatives).

poly (A) tail   In eukaryotes, a long sequence of 100–250 adenine nucleotides added to the 3' end of newly transcribed messenger RNA molecules.

polycistronic   For an mRNA molecule, containing more than one protein-coding segment, each with its own start and stop codons and each coding for a different protein. Common in prokaryotes.

polyclonal antibody   A mix of several different antibodies that all bind to the same antigen, typically produced by a living animal after injection with the antigen.

polygenic inheritance   The inheritance patterns that result when many genes influence one trait.

polymer   Any long molecule composed of small repeating subunits (monomers) bonded together.

polymerase chain reaction (PCR) A lab technique for rapidly generating millions of identical copies of a specific stretch of DNA. Involves incubating the original DNA sequence with primers, free nucleotides, and DNA polymerase.

polymerization The process by which one monomer (a small subunit molecule) is bound to others to form a polymer (a long chain molecule).

polymorphism (1) The occurrence of more than one allele at a certain genetic locus in a population. (2) The occurrence of more than two distinct phenotypes of a trait in a population.

polyp The sessile stage of a cnidarian life cycle.

polypeptide A peptide of three or more amino acids linked together in a chain. Very large polypeptides may be called proteins.

polyploid Having more than two chromosome sets.

polyribosome A structure consisting of one messenger RNA molecule along with many attached ribosomes and their growing peptide strands. Occurs in prokaryotes and eukaryotes.

polysaccharide A large carbohydrate polymer consisting of many monosaccharides linked together in a chain.

polyspermy Fertilization of an egg by multiple sperm. This is usually an abnormal situation.

polytomy A multibranched node on a phylogenetic tree. Represents a time when an ancestral population split into descendant populations.

pons A small region of the brain that relays information to the cerebellum and is also involved in control of breathing.

poor-nutrition hypothesis The hypothesis that herbivore populations are limited by the poor nutritional content of plants, especially low nitrogen.

population A group of individuals of the same species living in the same area at the same time.

population bottleneck An extreme reduction in population size, followed by re-expansion to a larger population size. May cause genetic bottleneck.

population cycles Regular fluctuations in size exhibited by certain populations.

population density The number of individuals of a population per unit area.

population dynamics Changes in population size through time.

population ecology The study of how and why the number of individuals in a population changes over time.

population viability analysis (PVA) A method of estimating the likelihood that a population will avoid extinction for a given time period.

pore Any small opening, such as the small opening in the stoma of a plant leaf or in the septa of fungal filaments.

positive control A type of gene regulation in which transcription can occur only when a certain substance—typically an activator protein that binds to a control sequence in the DNA and promotes transcription—is present.

positive feedback Stimulation of a reaction or process by the end result of that process. Tends to accelerate processes rapidly.

positive pressure ventilation Ventilation of the lungs that is accomplished by "pushing" air into the lungs by positive pressure in the mouth.

positive-sense virus A virus whose genome contains the same sequences as the mRNA required to produce viral proteins.

posterior Toward an animal's tail and away from its head. The opposite of anterior.

posterior pituitary The posterior part of the pituitary gland, consisting of the ends of neurosecretory cells from the hypothalamus, which secrete oxytocin and antidiuretic hormone.

postsynaptic neuron A neuron that receives neurotransmitters from another neuron at a particular synapse.

post-translational control Regulation of gene expression by modification of proteins after translation.

post-translational modification Any chemical alteration of a protein after it has been synthesized; includes cleavage of side chains, phosphorylation.

postzygotic isolation Result of mechanisms that prevent gene flow between different species even if mating occurs between them, typically due to death of hybrid embryos or reduced fitness of hybrids.

potential energy Energy stored in matter through its position or its shape, as opposed to kinetic energy (the energy of motion).

power stroke During muscle contraction, the motion of a myosin "head" that pulls myosin and actin filaments further along each other.

prairie An extensive grassland. Typically found in the dry interiors of continents in the temperate latitudes. Also called a steppe.

prebiotic soup A hypothetical solution of sugars, amino acids, nitrogenous bases, and other building blocks of larger molecules that may have formed in shallow waters or deep-ocean vents of ancient Earth and given rise to larger biological molecules.

predation The killing and eating of one organism by another.

predator Any organism that kills other organisms for food.

presentation Display of an antigen on the surface of an immune system cell to recruit other immune system cells.

pressure-flow hypothesis The hypothesis that sugar movement through phloem tissue is due to differences in the turgor pressure of phloem sap.

pressure potential Potential energy of water caused by pressure differences. Equals the sum of all the types of pressure that affect water, such as atmospheric pressure, wall pressure, and tension.

presynaptic neuron A neuron that releases neurotransmitters to another neuron at a synapse.

pre-urine The fluid in kidney nephrons that is formed by filtration of blood. Upon further processing, becomes urine. Also called filtrate.

prey A species that is commonly attacked and eaten by a predator species.

prezygotic isolation Anything that prevents individuals of two different species from mating.

primary cell wall The outermost layer of a plant cell wall, made of cellulose fibers with gelatinous polysaccharides, that defines the shape of the plant cell and withstands the turgor pressure of the plasma membrane.

primary consumer An herbivore; an organism that eats plants, algae, or other primary producers.

primary decomposer A decomposer that consumes detritus from plants.

primary growth Plant growth that results in an increase in length of stems and roots. Produced by apical meristems.

primary immune response An immune response to a pathogen that the immune system has not encountered before.

primary oocyte The large diploid cell in an ovarian follicle that can initiate meiosis to produce a haploid ovum.

primary producer An autotroph; a species that creates its own food through photosynthesis or from reduced inorganic compounds and that is a source of food for other species in its ecosystem.

primary productivity The total amount of carbon fixed by photosynthesis per unit area per year, including that used for cellular respiration.

primary RNA transcript In eukaryotes, a newly transcribed messenger RNA molecule that has not yet been processed—i.e., that still contains introns and has not received a 5′ cap or a poly (A) tail.

primary sex determination The process by which an embryonic gonad becomes either a testis or an ovary.

primary spermatocyte A diploid cell in the testis that can initiate meiosis I to produce two secondary spermatocytes.

primary structure The sequence of amino acids in a peptide or protein; also the sequence of nucleotides in a nucleic acid.

primary succession The gradual colonization of a habitat of bare rock or gravel, usually after an environmental disturbance that removes all soil and previous organisms.

primase An enzyme that synthesizes a short stretch of RNA to use as a primer during DNA replication.

primates The lineage of mammals that includes prosimians (lemurs, lorises, etc.), monkeys, and apes (including humans).

primer A short, single-stranded sequence of RNA that enables the start of replication of a DNA sequence.

principle of independent assortment The concept that each pair of hereditary elements (alleles of the same gene) behaves independently of other genes during meiosis. One of Mendel's two principles of genetics.

principle of segregation The concept that each pair of hereditary elements (alleles of the same gene) separate from each other during the formation of offspring (i.e., during meiosis). One of Mendel's two principles of genetics.

prion An infectious protein that is thought to cause disease by inducing other proteins to assume an abnormal three-dimentional structure. Likely cause of spongiform encephalopathies, such as mad cow disease.

probe A single-stranded fragment of a labeled, known DNA sequence that will bind to a complementary sequence in the sample being analyzed.

proboscis A long, narrow feeding appendage through which food can be obtained.

procambium A group of cells in the center of a young plant embryo that will give rise to the vascular tissue.

product The final atoms or molecules after a chemical reaction has occurred.

productivity The amount of energy used by a certain component of an ecosystem (e.g., a community or a trophic level) in a given area over a given time period.

progesterone A steroid hormone produced, along with estrogens, in the ovaries. Secreted by the corpus luteum after ovulation; causes the uterine lining to thicken.

prokaryote A member of the domain Bacteria or Archaea; a unicellular organism lacking certain complex cell features, such as a membrane-bound nucleus and gene introns.

prolactin A peptide hormone from the pituitary gland that promotes milk production in female mammals and that has a variety of effects on parental behavior and seasonal reproduction in other vertebrates.

prolactin-inhibiting hormone (PIH) A peptide hormone from the hypothalamus that inhibits release of prolactin from the anterior pituitary.

prometaphase A stage of cell division (mitosis or meiosis), during which the nuclear envelope breaks down and spindle fibers attach to chromatids.

promoter A short sequence of DNA that facilitates binding of RNA polymerase to enable transcription of downstream genes.

promoter-proximal elements In eukaryotes, regulatory sequences that are close to a promoter and that can bind regulatory transcription factors.

prophase The first stage of cell division (mitosis or meiosis) during which chromosomes become visible and the mitotic spindle forms. Synapsis and crossing over occur during prophase of meiosis I.

proplastid A colorless organelle found in undifferentiated plant cells that matures to become a plastid.

prosimians One of the two major lineages of primates, including lemurs, tarsiers, pottos, and lorises but not monkeys or apes.

protease An enzyme that can break apart proteins, by cleaving the peptide bonds between amino acids.

protein A long chain of 50 or more amino acids linked together; a large polypeptide.

proteinase inhibitors Defensive compounds produced by plants that block the enzymes in animal digestive tracts responsible for digesting proteins.

protein kinase An enzyme that catalyzes the addition of a phosphate group to another protein, typically activating or inactivating the other protein.

proteoglycans A type of glycoprotein commonly found in the extracellular matrix. Proteoglycans have a greater carbohydrate content and are larger than most other glycoproteins.

proteomics The study of the three-dimensional structure and function of proteins.

protist Any microscopic eukaryote that is not a green plant, animal, or fungus.

protoderm The exterior layer of a young plant embryo. Gives rise to the epidermis.

proton A small, positively charged particle in atomic nuclei. The number of protons in an atom gives that atom its characteristics as an element.

proton-motive force The combined effect of a proton gradient and an electric potential gradient across a membrane, which can drive protons across the membrane. Used by mitochondria and chloroplasts to power ATP synthesis.

proton pump A membrane protein that uses the energy of ATP to transport protons across the membrane against an electrochemical gradient.

proto-oncogene Any gene that encourages cell growth, typically by triggering specific phases in the cell cycle.

protostomes A major lineage of animals that share a pattern of embryological development, including spiral cleavage, formation of the mouth earlier than the anus, and formation of the coelom by splitting of a block of mesoderm. Includes arthropods, mollusks, and annelids.

proximal Toward or from the center of the body; away from the furthest tip of an appendage. The opposite of distal.

proximal tubule The convoluted section of a kidney nephron into which filtrate moves from Bowman's capsule. Involved in active reabsorption of certain solutes and water.

proximate causation In biology, the immediate, mechanistic cause of a phenomenon (how it happens), as opposed to the ultimate cause (why it evolved). Also called proximate explanation.

pseudocoelom A body cavity that forms between the endoderm and mesoderm layers. Occurs in roundworms (nematodes) and rotifers.

pseudogene A DNA sequence that closely resembles a working gene but is not transcribed. Thought to have arisen by duplication of the working gene followed by accidental inactivation due to a mutation.

pseudopodium (plural: pseudopodia) A mobile, outward bulging of a cell's plasma membrane, used in cell crawling or ingestion of food.

P site The site in a ribosome where peptide bonds are formed between amino acids.

puberty The process by which an immature animal attains reproductive maturity.

pulmonary artery A short, thick-walled artery that carries oxygen-poor blood from the heart to the lungs.

pulmonary circulation The part of the circulatory system that sends oxygen-poor blood to the lungs. In many vertebrates, the pulmonary circulation is separate from the rest of the circulatory system (the systemic circulation).

pulmonary vein A short, thin-walled vein that carries oxygen-rich blood from the lungs to the heart.

pulse-chase experiment A lab technique for marking a population of cells or molecules at a particular moment in time by means of a labeled molelcule and then following their fate over time.

Punnett square A diagram that depicts the genotypes and phenotypes that should appear in offspring of a certain cross.

pupa A metamorphosing insect that is enclosed in a protective case.

pupation A developmental stage of many insects, in which the body metamorphoses from the larval form to the adult form while enclosed in a protective case.

pupil The hole in the center of the iris through which light enters a vertebrate or cephalopod eye.

pure line In animal or plant breeding, a strain of individuals that produce offspring identical to themselves when self-pollinated or crossed to another member of the same population. Pure lines are homozygous for most, if not all, genetic loci.

purines A class of small, nitrogen-containing, double-ringed bases (guanine, adenine) found in nucleotides.

pyramid of productivity The characteristic pattern of productivity in ecosystems, in which productivity declines with each higher trophic level.

pyrimidines A class of small, nitrogen-containing, single-ringed bases (cytosine, uracil, thymine) found in nucleotides.

pyruvate dehydrogenase A large enzyme complex, located in the inner mitochondrial membrane, that is responsible for conversion of pyruvate to acetyl CoA during glycolysis.

Q A molecule in mitochondria that helps transfer electrons between the complexes of the electron transport chain during cellular respiration. Also called coenzyme Q or ubiquinone.

quadrat A small rectangular plot set up in a habitat to mark an area under intensive study.

quantitative trait A phenotypic trait that exhibits variation along a smooth, continuous scale of measurement (for example, human height), rather than the distinct forms seen in discrete traits.

quantitative variation Variation that exhibits differences in degree across a smooth, continuous scale of measurement.

quaternary structure The three-dimensional shape of several polypeptide chains and sometimes other small functional groups, arranged together to form a large, multiunit protein.

race A population that has different characteristics from another population of the same species, whether or not there are significant genetic differences between them.

radial cleavage The pattern of embryonic cleavage seen in protostomes, in which cells divide at right angles to each other to form tiers.

radial symmetry An animal body pattern in which there are least two planes of symmetry. Typically, the body is in the form of a cylinder or disk, with body parts radiating from a central hub.

radiation Production of electromagnetic energy (light). One of the mechanisms of heat transfer between organisms and the environment.

radicle The root of a plant embryo.

radioactive decay A spontaneous change in the mass number of an atomic nucleus via emission of radiation or a particle.

radioactive isotope An isotope that spontaneously emits radiation (gamma rays) and/or subatomic particles. In the latter case, the isotope decays to form a different isotope or element.

radiometric dating A technique for determining the age of a rock by measuring the amount of radioactive decay that has occurred since the rock solidified, usually by measuring the amount of new daughter element that has formed.

radula A rasping feeding appendage in gastropods.

rain shadow The dry region on the side of a mountain range away from the prevailing wind.

rays Lines of parenchyma cells that extend laterally through the xylem of plant wood.

reactant The atoms or molecules in a chemical reaction, in their starting states.

reaction center A central area in a photosystem of the light-dependent reactions of photosynthesis. Surrounded by chlorophyll molecules of the antenna complex and receives energy from them.

reactive oxygen intermediates (ROIs) Oxygen-containing compounds that are highly reactive and that are used in plant and animal cells to kill infected cells and for other purposes.

reading frame The division of a sequence of DNA or RNA into a particular series of three-nucleotide codons. There are three possible reading frames for any sequence.

realized niche The ecological niche that a species occupies in the presence of competitors.

receptor (1) A molecule, cell, or group of cells specialized for detecting environmental signals. (2) A molecule that binds to a particular chemical (e.g., hormone, sperm protein) and triggers a cellular response. (3) A cell-surface molecule necessary for a virus to gain entry to a cell.

receptor-mediated endocytosis Endocytosis triggered by the binding of certain macromolecules outside the cell to membrane proteins.

receptor tyrosine kinases Transmembrane proteins involved in signal transduction. Typically bind to a signalling molecule, triggering a phosphorylation cascade of other proteins inside the cell.

recessive Property of an allele whereby its influence on phenotype can be entirely hidden by the presence of another, dominant allele.

reciprocal altruism Altruistic behavior that is exchanged between a pair of individuals at different points in time (i.e., sometimes individual A helps individual B, and sometimes B helps A).

reciprocal cross A breeding experiment in which the mother's and father's phenotypes are the reverse of that examined in a previous breeding experiment.

recognition sequence A sequence of amino acids that binds to a specific sequence of DNA and that is found within the helix-turn-helix motif of many repressor proteins.

recognition site The specific sequence of DNA bases cut by a certain restriction endonuclease.

recombinant Possessing a new combination of alleles. May refer to a single chromosome or an entire organism.

recombinant DNA Any DNA altered by exchange with, or inclusion of, foreign DNA. May be produced via meiosis, viruses, or lab manipulation.

recombinant DNA technology A variety of lab techniques for isolating specific DNA fragments and introducing them into different regions of DNA and/or a different host organism. Also called biotechnology or genetic engineering.

recombination A change in the combination of genes or alleles on a given chromosome or in a given individual. Also called genetic recombination.

rectal gland A salt-excreting gland in the digestive system of sharks, skates, and rays.

rectum The last part of the digestive tract; a short tube that holds feces until they are expelled.

red blood cells Hemoglobin-containing cells that circulate in the blood and deliver oxygen from the lungs to the tissues.

redox reactions A class of chemical reactions that involve the loss and gain of electrons. Also called reduction-oxidation reactions.

reduction An atom's gain of electrons during a redox reaction, either by acceptance of an electron from another atom or by the electrons in covalent bonds moving closer to the atomic nucleus.

reduction-oxidation reactions A class of chemical reactions that involve the loss and gain of electrons. Also called redox reactions.

redundant code A code in which different sequences can represent the same information. The genetic code is redundant because some amino acids are coded for by two or three different codons.

reflex An involuntary response to environmental stimulation. May involve the brain (e.g., conditioned reflex) or not (e.g., spinal reflex).

refractory No longer responding to stimuli that previously elicited a response. For example, the tendency of voltage-gated sodium channels to remain closed immediately after an action potential.

regeneration Growth of a new body part to replace a lost body part.

regulatory cascade In embryonic development, a progressive series of interactions among genes and/or cytoplasmic determinants that organizes the body plan of the embryo.

regulatory evolution The evolution of new body parts and patterns via changes in the regulatory genes that affect pattern formation in embryos.

regulatory homeostasis Steady internal body conditions achieved by active physiological processes.

regulatory protein Any protein that affects gene transcription by binding to specific enhancers, silencers, or other sites in DNA.

regulatory sequence Any section of DNA that is involved in controlling the activity of other genes.

regulatory site A site on an enzyme to which a regulatory molecule can bind and affect the enzyme's activity, separate from the active site where catalysis occurs.

regulatory transcription factors Proteins that bind to eukaryotic enhancers, silencers, and promoter-proximal elements, but not to the promoter itself.

reinforcement Natural selection for traits that prevent interbreeding between recently diverged species.

release factors Proteins that can trigger termination of RNA translation when a ribosome reaches a stop codon.

releaser A simple stimulus that elicits an invariant, stereotyped behavioral response (fixed action pattern) from an animal. Also called a sign stimulus.

releasing hormone A hormone that stimulates release of some other hormone.

renal corpuscle The ball-like structure at the beginning of a kidney nephron, consisting of Bowman's capsule surrounding a glomerulus.

replacement mutation A point mutation (change in a single base pair) that causes a change in the amino acid sequence of a protein. Also called missense mutation.

replacement rate The reproductive rate at which each female produces two surviving offspring over her entire life—enough to exactly replace herself and her mate, resulting in zero population growth.

replica plate In replica plating, a copy of the master plate produced by transferring bacteria from it with a velvet-covered block.

replica plating A method of identifying bacterial colonies that have certain mutations by observing their growth on a plate that is exposed to different conditions.

replicated chromosome A chromosome that has been copied; consists of two chromatids joined at the centromere.

replication The exact copying of something—e.g., DNA replication.

replication fork The Y-shaped site at which a double-stranded molecule of DNA is separated into two single strands for replication.

repolarization A return to a normal membrane potential after a depolarization.

repressor Any regulatory protein that inhibits transcription of certain genes, typically by binding to a silencer upstream of the promoter.

reservoir In biogeochemical cycles, a location in the environment where elements are stored for a time.

resilience A measure of how quickly a community recovers following a disturbance.

resistance (1) The ability of an organism to defend itself against drugs, pathogens, or parasites. (2) A measure of how much a community is affected by a disturbance.

resistance (R) loci Genes associated with disease resistance in plants.

respiration The biochemical pathways that produce ATP from compounds with high potential energy, via an electron transport chain and using an inorganic final electron acceptor. Also called cellular respiration.

respiratory distress syndrome A syndrome in which premature infants can suffocate due to insufficient surfactant in their lungs.

resting potential The membrane potential of a cell in its resting, or normal, state.

restriction endonucleases Bacterial endonucleases that cut DNA at a specific base-pair sequence (recognition site). Also called restriction enzymes.

restriction fragment length polymorphisms (RFLPs) Variations in the size of DNA fragments that are produced by restriction endonucleases, due to differences in the DNA sequences at recognition sites.

retina A thin layer of light-sensitive cells (rods or cones) and neurons at the back of a camera-type eye, such as that of cephalopods or vertebrates.

retinal A carotenoid pigment derived from vitamin A that, with opsin, forms rhodopsin, the light-detecting pigment in rods and cones of animal eyes.

retrovirus A virus with an RNA genome that reproduces by transcribing its RNA into a DNA sequence and then inserting that DNA into the host's genome for replication.

reverse transcriptase A enzyme of retroviruses (RNA viruses) that can synthesize double-stranded DNA from a single-stranded RNA template.

rhizobia Members of the bacterial genus *Rhizobia*, nitrogen-fixing bacteria that live in root nodules of members of the pea family.

rhizoid The hairlike structure that anchors a bryophyte (nonvascular plant) to the substrate.

rhizome A plant stem extending horizontally underground.

rhodopsin A combination of two molecules (retinal and one of various opsins) instrumental in detection of light by rods and cones of vertebrate eyes.

ribonucleic acid (RNA) A polymer of ribonucleotides, usually single stranded. RNAs function as structural components of ribosomes (rRNA), transporters of amino acids (tRNA), and translaters of the message of the DNA code (mRNA).

ribonucleotide A nucleotide consisting of the five-carbon sugar ribose, a phosphate group, and one of several nitrogen-containing bases (adenine, guanine, thymine, or uracil). Can be polymerized to form ribonucleic acid (RNA).

ribosomal RNAs (rRNAs) A class of RNA molecules that form part of the structure of a ribosome.

ribosome A molecular machine that synthesizes proteins by using the genetic information encoded in messenger RNA strands. Consists of two subunits, each composed of ribosomal RNA and proteins.

ribosome binding site In bacteria, the sequence at the beginning of an mRNA molecule to which a ribosome binds to initiate translation. Also called the Shine-Dalgarno sequence.

ribozyme Any RNA molecule that can act as a catalyst for a chemical reaction.

ribulose bisphosphate (RuBP) A five-carbon compound that is the initial reactant in the Calvin cycle of photosynthesis.

rickets A human disorder characterized by malformed, soft bones. Usually caused by environmental factors (e.g., inadequate vitamin D intake), but also may be caused by an X-linked dominant allele.

RNA See **ribonucleic acid (RNA)**.

RNA polymerases A class of enzymes that catalyze synthesis of RNA from ribonucleotides, using a DNA template. Also called RNA pols.

RNA processing In eukaryotes, the changes that a primary RNA transcript undergoes to become a mature mRNA molecule. Includes splicing and the addition of 5′ caps and poly (A) tails.

RNA replicase A viral enzyme that can synthesize RNA from an RNA template.

RNA world hypothesis The hypothesis that life on Earth began as a polymer of ribonucleic acid.

robust australopithecines Several species of comparatively large, strong hominins that appear in the fossil record shortly after the split from chimpanzees.

rod A type of photoreceptor cell with a rod-shaped outer portion. Found in vertebrate retinas. Particularly sensitive to dim light, but not used to distinguish colors.

root (1) An underground appendage of a plant that anchors the plant and absorbs water and nutrients. (2) In a phylogenetic tree, the bottom, most ancient node.

root apical meristem A group of undifferentiated plant cells at the tip of a plant root.

root cap A small group of cells that covers and protects the tip of a plant root. Senses gravity and determines the direction of root growth.

rooted For a phylogenetic tree, oriented so that its bottom-most node is the most ancient node. Usually, the root is identified through the use of an outgroup.

root hair A long, thin outgrowth of the epidermal cells of plant roots, providing increased surface area for absorption of water and nutrients.

root pressure Positive pressure that is generated in plant roots during the night, due to accumulation of ions from the soil and subsequent movement of water into the root.

root system The belowground part of a plant.

rosette In plants, a compact growth form in which leaves pile up on each other in whorls.

rough ER (rough endoplasmic reticulum) A type of endoplasmic reticulum dotted with ribosomes. Involved in processing of membrane proteins and secretory proteins.

rRNAs See **ribosomal RNAs (rRNAs)**.

rubisco The enzyme that catalyzes the first step of the Calvin cycle of photosynthesis: the addition of a molecule of CO_2 to ribulose bisphosphate. Also called ribulose 1,5-bisphosphate carboxylase/oxygenase.

rumen The largest chamber of a ruminant's stomach, containing a large vat of symbiotic cellulose-digesting bacteria. Creates cud, which is sent back to the mouth for further chewing.

ruminants A group of hoofed mammals that have a four-chambered stomach specialized for digestion of plant cellulose, with one chamber containing symbiotic cellulose-digesting bacteria.

sac fungi A monophyletic lineage of fungi that produce large, often cup-shaped reproductive structures that contain asci. Also called cup fungi.

sage-steppe An arid, shrub-dominated habitat with some characteristics of deserts and grasslands.

salicylic acid A compound thought to play a role in the systemic acquired resistance (SAR) defensive mechanism of plants; a component of aspirin.

salivary glands Mammalian glands that secrete saliva (a mixture of water, mucins, and digestive enzymes) into the mouth.

sampling error The accidental selection of a nonrepresentative sample from some larger population, due to chance.

saprophyte An organism that feeds primarily on dead plant material. Usually, a fungus or plant.

sapwood The younger xylem in the outer layer of wood of a stem or root, functioning primarily in water transport.

sarcomere A single contractile unit of a skeletal muscle cell.

sarcoplasmic reticulum Sheets of smooth endoplasmic reticulum in a muscle cell. Contains high concentrations of calcium, which can be released into the cytoplasm to trigger contraction.

SARS See **severe acute respiratory syndrome (SARS)**.

saturated fat A fat that contains the maximum number of hydrogen atoms because all the carbon atoms in its fatty acid chains are joined by single bonds. Such fats have relatively high melting points.

scanning electron microscope (SEM) A microscope that produces images of the surfaces of objects by reflecting electrons from a specimen coated with a layer of metal atoms.

scarify Scraping, rasping, or other damage to the coat of a seed. Necessary in some species to trigger germination.

schizophrenia A human psychological disorder characterized by delusions, hallucinations, social withdrawal, and other symptoms.

Schwann cells Specialized cells that wrap around axons of neurons outside the brain and spinal cord, providing electrical insulation.

sclereid A type of plant sclerenchyma cell that usually functions in protection, such as in seed coats and nutshells.

sclerenchyma cells A thick-walled class of plant cells that provide support and typically contain the tough structural polymer lignin. Usually dead at maturity.

screen Any of several techniques for identifying individuals with a particular type of mutation. Also called genetic screen.

scrotum A sac of skin, containing the testes, suspended just outside the abdominal body cavity of many male mammals.

secondary antibody An antibody that will bind to another antibody. Used in ELISAs and in some other lab tests.

secondary cell wall An inner layer of the cell wall, formed by certain types of plant cells as they mature.

secondary consumer A carnivore; an organism that eats herbivores.

secondary endosymbiosis The presence of an organelle that originated with ingestion of a cell containing that organelle, which in turn was originally derived from ingestion and symbiosis.

secondary growth Plant growth that results in an increase in width of stems and roots.

secondary immune response The immune response, using memory cells, to an infection that the immune system has encountered before.

secondary meristem A layer of undifferentiated plant cells found in older stems and roots. Responsible for secondary growth (increase in width). Also called cambium or lateral meristem.

secondary metabolite Any poison produced by a plant that is synthesized by a variation of a biosynthetic pathway used for other purposes.

secondary phloem Phloem tissue produced by a lateral meristem rather than by an apical meristem.

secondary production The total amount of new body tissue produced by animals that eat plants. May involve growth and/or reproduction.

secondary spermatocyte A cell produced by meiosis I of a primary spermatocyte in the testis. Can undergo meiosis II to produce spermatids.

secondary structure A type of protein structure created by hydrogen bonding between C=O and N–H groups of the polypeptide backbone; most notably, the α-helix and β-pleated sheet structures.

secondary succession Gradual colonization of a habitat after an environmental disturbance (e.g., fire, windstorm, logging) that removes some or all previous organisms but leaves the soil intact.

secondary xylem Xylem tissue produced by a lateral meristem, rather than by an apical meristem.

second law of thermodynamics A fundamental principle of physics stating that, in an isolated system, entropy always increases during any chemical reaction.

second-male advantage The reproductive advantage of a male who mates with a female last, after other males have mated with her.

second messenger A nonprotein signaling molecule produced or activated inside a cell in response to stimulation at the cell surface. Commonly used to relay the message of a protein hormone.

secretin A peptide hormone produced by the small intestine in response to the arrival of food from the stomach. Stimulates secretion of bicarbonate (HCO_3^-) from the pancreas.

sedimentary rock A type of rock formed by gradual accumulation of sediment, as in riverbeds and on the ocean floor. Most fossils are found in sedimentary rocks.

seed A plant embryo with nutritive tissue (endosperm) to fuel its early growth, surrounded by an outer protective layer (seed coat). In angiosperms, forms from the fertilized ovule of a flower.

seed coat A protective layer around a seed that encases both the embryo and the endosperm.

seed dormancy A state of suspended development of a plant seed. Can be terminated by cues indicating that favorable environmental conditions have arrived or that unfavorable ones have passed.

seedless vascular plants Several phyla of green plants that have vascular tissue but do not make seeds. Include horsetails, ferns, lycophytes, and whisk ferns.

seedling A young plant that has emerged from a seed.

seed plants A group of several phyla of green plants that have vascular tissue and make seeds. Includes gymnosperms and angiosperms.

segment A well-defined region of the body along the anterior-posterior body axis, containing similar structures as other, nearby segments.

segmentation A body plan involving division of the body into many similar segments that bear similar or identical structures.

segmentation genes Genes that affect body segmentation in embryonic development. Includes gap genes, pair-rule genes, and segment polarity genes.

segment polarity genes A class of fruit-fly segmentation genes that establish the anterior-posterior orientation of each embryonic segment. Active in particular regions of each segment.

selective adhesion The tendency of cells of one tissue type to adhere to other cells of the same type.

selectively permeable membrane A membrane that some solutes can cross more readily than other solutes can.

selective permeability The property of a structure, such as a membrane, that allows some substances to diffuse across it much more readily than other substances.

self Property of a molecule or cell such that immune system cells do not attack it, due to certain molecular similarities to other body cells.

self-fertilization The fusion of two gametes from the same individual to form a diploid offspring. Also called selfing.

self-incompatible Incapable of self-fertilization.

selfing The fusion of two gametes from the same individual to form a diploid offspring. Also called self-fertilization.

selfish genes DNA sequences that survive and reproduce but reduce the fitness of the host genome.

self-pollination Pollination in which pollen from a certain individual lands on a flower stigma of that same individual.

semen The combination of sperm and accessory fluids that is released by male mammals and reptiles during ejaculation.

semiconservative replication The type of replication used by cells to copy DNA, in which each daughter DNA molecule is composed of one old strand and one new strand.

seminal vesicles Paired reproductive glands that, in mammals, secrete an alkaline fluid into semen to counteract the acidic environment of the vagina. In other vertebrates and invertebrates, often stores sperm.

senescence The process of aging.

sensitive period A short time span in a young animal's life during which the animal can learn certain things, such as song, language, or imprinting. Also called the critical period.

sensor A cell, organ, or structure that senses some aspect of the external or internal environment.

sensory neuron A nerve cell that carries sensory information to the central nervous system.

sepal One of the protective leaflike structures enclosing a flower bud and later supporting the blooming flower.

septa (singular: septum) Any wall-like structure. In fungi, cross-walls that divide fungul filaments into cell-like compartments.

serotonin A neurotransmitter involved in many brain functions, including sleep, pleasure, and mood.

serum The liquid that remains when clotted cells are removed from blood; plasma without clotting factors. Contains water, dissolved gases, hormones, food molecules, and other soluble substances.

sessile Permanently attached to a substrate; not capable of moving to another location.

set point A normal or target value for a regulated internal factor, such as body heat or blood pH.

severe acute respiratory syndrome (SARS) A human disease characterized by sudden and intense flu-like symptoms.

severe combined immune deficiency (SCID) A human disease characterized by an extremely high vulnerability to infectious disease, due to a genetic defect in the immune system.

sex chromosome One of the pair of chromosomes carrying the gene(s) that determine sex.

sex-linked inheritance Inheritance patterns observed in genes carried on sex chromosomes, so females and males have different numbers of alleles of a gene and may pass its trait only to one sex of offspring. Also called sex-linkage.

sexual dimorphism Any trait that differs between males and females.

sexual reproduction Reproduction in which genes from two parents are combined via fusion of gametes, producing offspring that are genetically distinct from both parents.

sexual selection The process by which individuals with certain heritable traits leave more offspring than other individuals specifically due to superiority in competing for mating opportunities.

Shannon-Weaver index A common measurement of species diversity, calculated as $H' = -\Sigma p_i \log(p_i)$, where p_i is the proportion of individuals in the community that belong to species i.

shell A hard protective outer structure. In protists, also called a test.

Shine-Dalgarno sequence In bacteria, the sequence at the beginning of an mRNA molecule to which a ribosome binds to initiate translation. Also called the ribosome binding sequence.

shoot The aboveground portion of a young plant, including stem and leaves.

shoot apical meristem A group of undifferentiated plant cells at the tip of a plant stem.

shoot system The aboveground part of a plant.

short-day plant A plant that blooms in response to long nights.

shotgun sequencing A method of sequencing genomes that is based on breaking the genome into small pieces, sequencing each piece separately, and then figuring out how the pieces are connected.

sieve plates The pore-containing structure at one end of a sieve-tube member in plant phloem tissue.

sieve-tube member An elongated sugar-conducting cell in phloem. Has sieve plates at both ends, allowing sap to flow to adjacent cells.

sigma A detachable protein subunit of RNA polymerase that binds to the −35 and −10 boxes to initiate transcription of DNA in prokaryotes.

signal Any information-containing behavior.

signal 1 The first step in activation of a T cell, in which the T cell engulfs an antigen presented by a dendritic cell.

signal 2 The second step in the activation of a T cell, in which additional receptors on the T cell bind to additional MHC-antigen complexes on a dendritic cell.

signal hypothesis The hypothesis that proteins destined for secretion are directed to the rough ER by a certain amino acid sequence at the beginning of the proteins.

signal receptor Any cellular protein that binds to a particular signaling molecule (such as a hormone or neurotransmitter) and triggers a response by the cell, usually by changing conformation or activity upon binding.

signal recognition particle (SRP) A complex of RNA and protein that transports certain newly synthesized proteins to the endoplasmic reticulum.

signal transducers and activators of transcription (STATs) In mammals, a group of regulatory transcription factors that, upon phosphorylation, can activate transcription of certain genes.

signal transduction The process by which a stimulus (e.g., a hormone, a neurotransmitter, or sensory information) outside a cell is translated into a response by the cell.

signal transduction cascade A sequence of events in which one enzyme phosphorylates other enzymes, which in turn phosphorylate many more, ultimately leading to phosphorylation of thousands of proteins. Commonly used in signal transduction of hormone messages. Also called phosphorylation cascade.

signal transduction pathway The exact sequence of molecular events by which signal transduction occurs.

sign stimulus A simple stimulus that elicits an invariant, stereotyped behavioral response (fixed action pattern) from an animal. Also called a releaser.

silencer A regulatory DNA sequence in eukaryotes to which repressor proteins can bind, inhibiting transcription of certain genes.

silent mutation A mutation that does not detectably affect the phenotype of the organism. Typically, a point mutation in the third position of certain codons that does not alter the amino acid coded for.

simian immunodeficiency viruses (SIVs) A family of lentiviruses that infect monkeys and apes and that are thought to be closely related to human immunodeficiency virus (HIV).

simple eye An eye with only one light-collecting apparatus (e.g., one lens), as in vertebrates.

simple sequence repeat A stretch of eukaryotic DNA consisting of repeats of a short, simple sequence that does not code for any protein or RNA.

single nucleotide polymorphism (SNP) A genetic locus where a single base pair varies between individuals of a certain species. Used as a genetic marker to help track the inheritance of nearby genes.

single-strand DNA-binding proteins A class of proteins that attach to separated strands of DNA during replication or transcription, preventing them from re-forming a double helix.

sink A location where an element or a molecule is consumed or taken out of circulation.

sinoatrial node (SA node) A cluster of heart muscle cells that initiates the heartbeat and determines the heart rate. In the wall of the right atrium.

siphon A tubelike appendage of many mollusks. Often used for feeding or propulsion.

sister chromatids The paired strands of a recently replicated chromosome that has not yet divided.

sister groups or sister taxa Closely related taxa.

skeletal muscle The muscle tissue attached to the bones of the vertebrate skeleton. Consists of long, unbranched muscle fibers with a characteristic striped (striated) appearance; controlled voluntarily. Also called striated muscle.

sliding clamp A doughnut-shaped structure that holds the enzyme DNA polymerase in place during DNA replication.

sliding-filament model The hypothesis that the contraction of muscle cells is caused by filaments of actin and myosin sliding past each other.

slow-twitch fibers A class of muscle fibers that contract relatively slowly, but do not fatigue easily.

slug (1) A member of a certain lineage of terrestrial gastropods, closely related to snails but lacking a shell. (2) A mobile aggregation of cells of a cellular slime mold.

small intestine The first section of the intestine, immediately after the stomach. The site of the final stages of digestion and of most nutrient absorption.

smooth ER (endoplasmic reticulum) Endoplasmic reticulum that does not have ribosomes attached to it. Involved in synthesis and secretion of lipids.

smooth muscle The unstriated muscle tissue that lines the intestine, blood vessels, and some other organs. Consists of tapered, unbranched cells that can sustain long contractions. Not voluntarily controlled.

snRNPs (small nuclear ribonucleoproteins) A complex of proteins and small RNA molecules that catalyze splicing (removal of introns from mRNA). Components of spliceosomes.

sodium-potassium pump A membrane protein that uses the energy of ATP to move sodium ions out of the cell and potassium ions in. Formally known as Na^+/K^+-ATPase.

soil erosion The removal of soil from an area by wind or water.

soil organic matter A mixture of partially and completely decomposed detritus.

solute Any substance that is dissolved in a liquid.

solute potential The potential energy of water caused by a difference in solute concentrations at two locations. Also called osmotic potential.

solution A liquid containing one or more dissolved solids or gases in a homogeneous mixture.

solvent Any liquid in which some substance will dissolve.

soma The part of a neuron that contains the nucleus and where incoming signals are integrated. Also called the cell body.

somatic cell Any type of cell that does not pass its genes on to the next generation. In a multicellular organism, all cells except eggs, sperm, and their parent cells (oogonia and spermatogonia) are somatic cells.

somatic hypermutation Rapid DNA mutation that occurs in a somatic cell. Occurs in certain cells of the immune system, such as memory cells that are fine-tuning antibody performance.

somatic nervous system The part of the peripheral nervous system (outside the brain and spinal cord) that controls skeletal muscles and is under voluntary control.

somatostatin A hormone secreted by many organs including the pancreas and hypothalamus, with a wide variety of effects.

somite A block of mesoderm on both sides of the developing spinal cord in a vertebrate embryo. Gives rise to muscle tissue, vertebrae, ribs, limbs, etc.

soredia (singular: soredium) Small reproductive structures produced by lichen. Contain both symbionts of the lichen (the fungus and green alga).

sorus (plural: sori) One of the small dots on the underside of fern fronds. Consists of many sporangia, each of which contains spores.

source A location where a substance is produced or enters circulation (e.g., in plants, the tissue where sugar enters the phloem).

Southern blotting A lab technique for the isolation and analysis of pieces of DNA. Involves cleavage with restriction enzymes, separation with gel electrophoresis, hybridization to a labeled probe, and visualization with autoradiography.

speciation The evolution of two or more distinct species from a single ancestral species.

species A distinct, identifiable group of populations that is thought to be evolutionarily independ-

ent of other populations. Generally distinct from other species in appearance, behavior, habitat, ecology, genetic characteristics, etc.

species-area relationship The mathematical relationship between the area of a certain habitat and the number of species that it can support.

species diversity The variety and relative abundance of the species present in a given ecological community.

species richness The number of species present in a given ecological community.

specific heat The amount of energy required to raise the temperature of 1 gram of a substance by 1°C; a measure of the capacity of a substance to absorb energy.

spectral karyotyping (SKY) A technique for producing high-resolution karyotypes by "painting" chromosomes with fluorescent tags that bind to particular regions of certain chromosomes. Also called chromosome painting.

spectrophotometer A lab instrument used to measure the wavelengths of light that are absorbed by a particular pigment.

sperm A mature male gamete, smaller and more mobile than the female gamete.

spermatid An immature sperm cell.

spermatogenesis The production of sperm.

spermatogonia (singular: spermatogonium) The diploid cells in a testis that can give rise to primary spermatocytes.

spermatophore A gelatinous package of sperm cells that is produced by males of species that have internal fertilization without copulation.

sperm competition Competition between sperm from different males to fertilize the eggs of the same female.

sperm nuclei The two haploid nuclei within the pollen grain of a flowering plant. One sperm nucleus fuses with the egg cell to form a zygote; the other fuses with two other nuclei to form triploid endosperm.

sphincter A muscular valve that can close off a tube, as in a blood vessel or a part of the digestive tract.

spicules Stiff spikes of silica or calcium carbonate found in the bodies of sponges.

spindle fibers Groups of microtubules that attach to chromosomes and pull them to opposite sides of the cell during cell division.

spine A modified plant leaf that functions as a sharp protective structure.

spiracle In insects, the small openings that connect air-filled tracheae to the external environment.

spiral cleavage The pattern of embryonic cleavage seen in deuterostomes, in which cells divide at oblique angles to form a spiral coil of cells.

spleen A dark red organ, found near the stomach of most vertebrates, that filters blood, stores extra red blood cells in case of emergency, and plays a role in immunity.

spliceosome In eukaryotes, an organized complex of snRNPs (small nuclear ribonucleoproteins) that catalyzes removal of introns from primary RNA transcripts.

splicing The process by which introns are removed from messenger RNA molecules and the remaining exons are connected together.

spongy mesophyll Rounded parenchyma cells found in the ground tissue of leaves near stomata. The site of most gas exchange.

spontaneous For a chemical reaction, occurring on its own, without any continuous external influences such as added energy.

spontaneous generation The disproven hypothesis that living organisms can develop spontaneously and rapidly from nonliving, noncellular materials under certain conditions.

sporangium (plural: sporangia) A spore-producing structure found in some plants, such as liverworts, and in some fungi, such as chytrids.

spore A single cell produced by mitosis or meiosis (not by cell fusion) that is capable of developing into an adult organism.

sporophyte The multicellular diploid phase of a species that exhibits alternation of generations; arises from two fused gametes; produces spores.

sporopollenin A watertight material that encases spores and pollen of modern land plants.

stabilizing selection A pattern of natural selection in which individuals with an average phenotype have higher fitness than those with extreme phenotypes.

stamen The male reproductive structure of a flower. Consists of an anther, which produces pollen grains, and a filament, which supports the anther.

standing defense Any defensive mechanism that is always present, regardless of need. Also called constitutive defense.

stapes A stirrup-shaped bone in the middle ear of vertebrates. Receives vibrations from the tympanic membrane and passes them to the cochlea.

starch A mixture of polysaccharides amylose and amylopectin; used primarily for food storage in plants. Both are helical polymers of α-glucose subunits, branched in amylopectin.

start codon The mRNA sequence AUG, which induces the beginning of protein synthesis and codes for the amino acid methionine.

statocyst A sensory organ of many arthropods that detects the animal's orientation in space (i.e., whether the animal is flipped upside down).

statolith A tiny stone or dense particle that can be used to sense gravity.

statolith hypothesis The hypothesis that amyloplasts (dense, starch-storing plant organelles) serve as statoliths (i.e., gravity detectors).

STATs See **signal transducers and activators of transcription (STATs)**.

stem A vertical aboveground part of a plant, usually bearing leaves, fruit, or flowers.

stem cell Undifferentiated cells that have the potential to give rise to any tissue type.

steppe An extensive grassland. Typically found in the dry interiors of continents in the temperate latitudes. Also called a prairie.

stereocilia (singular: stereocilium) Stiff outgrowths from the surface of a hair cell that is involved in detection of sound by terrestrial vertebrates or of waterborne vibrations by fishes.

steroids A class of lipids with a characteristic four-ring structure.

sticky ends The short, single-stranded ends of a DNA molecule cut by a restriction endonuclease. Tend to form hydrogen bonds with other sticky ends that have complementary sequences.

stigma The moist tip at the end of a flower carpel, to which pollen grains adhere.

stolon Modified stems that run horizontally over the soil surface.

stomata (singular: stoma) Generally, pores or openings. In plants, microscopic pores on the surface of a leaf or stem, through which gas exchange occurs for photosynthesis.

stomach A tough, muscular pouch in the digestive tract that breaks up food and delivers it to the intestine.

stop codon One of three messenger RNA triplets (UAG, UGA, or UAA) that cause termination of protein synthesis. Also called a termination codon.

strain A population of genetically similar or identical individuals.

striated muscle The muscle tissue attached to the bones of the vertebrate skeleton. Consists of long, unbranched muscle fibers with a characteristically striped (striated) appearance; controlled voluntarily. Also called skeletal muscle.

stroke An episode of oxygen deprivation to part of the brain that results in death of some neurons; usually due to a blood clot that blocks blood vessels of the brain.

stroke volume The volume of blood ejected from the left ventricle of the heart in one contraction.

stroma The fluid between a chloroplast's outer membrane and its thylakoid disks.

structural formula A graphical representation of the atoms and bonds in a molecule, often with covalent bonds represented by straight lines.

structural gene A stretch of DNA that codes for a functional protein or functional RNA molecule—i.e., not a promoter, enhancer, etc.

structural homology Similarities in organismal structures (e.g., limbs, shells, flowers) that are due to inheritance from a common ancestor.

structural isomer A molecule that shares the same molecular formula as another molecule but differs in the order in which covalently bonded atoms are attached.

style The slender stalk of a flower carpel, connecting the stigma and the ovary.

suberin A water-repellent compound that is a major component of the waxy Casparian strip in plant roots.

subspecies A population that has distinctive traits and some genetic differences relative to other populations of the same species but that is not distinct enough to be called a separate species.

substrate (1) A reactant that interacts with an enzyme in a chemical reaction. (2) A surface on which a cell or organism sits.

substrate-level phosphorylation Production of ATP molecules via transfer of a phosphate group from an intermediate substrate directly to ADP, unmediated by an electron transport chain. Occurs in glycolysis and in the Krebs cycle.

subzonal injection (SUZI) A technique used to help infertile couples bear children by selecting a few healthy sperm and injecting them directly into the zona pellucida of the woman's egg.

succession Gradual colonization of a habitat after an environmental disturbance (e.g., fire, flood), usually by a series of species assemblages.

successional pathway The specific sequence of species that appears over a period of time in an environment undergoing primary succession or secondary succession.

sugars A class of small, water-soluble organic compounds containing a carbonyl (–C=O) group and several hydroxyl (–OH) groups. Include monosaccharides and disaccharides.

sulfate-reducers A group of archaea that produce hydrogen sulfide (H_2S) as a by-product of cellular respiration.

summation The additive effect of different postsynaptic potentials at a nerve or muscle cell, such that several subthreshold stimulations can cause an action potential.

supernatant The liquid above a layer of solid particles (the pellet) in a tube after centrifugation.

surface tension The attractive force between liquid molecules that causes the liquid to form a rounded surface at an air-liquid interface.

surfactant A mixture of phospholipids and proteins produced by lung cells that reduces surface tension, allowing the lungs to expand more.

survivorship The average proportion of offspring produced that survive to a particular age in a certain population.

survivorship curve A graph depicting the percentage of a population that survives to different ages.

suspension culture A population of cells grown in the lab in a liquid flask of nutrients that is rotated continuously to keep the cells suspended and the nutrients mixed.

suspension feeder Any organism that obtains food by filtering small particles or small organisms out of water or air. Also called filter feeder.

sustainability The planned use of environmental resources at a rate no faster than the rate at which they are naturally replaced.

sustainable agriculture Agricultural techniques that are designed to maintain long-term soil quality and productivity.

sustainable development Economic development based on sustainability (i.e., that uses natural resources no faster than they are naturally replaced).

swamp A wetland that has a steady rate of water flow and is dominated by trees and shrubs.

swim bladder A gas-filled organ of many ray-finned fishes that regulates buoyancy.

symbiosis Any close and prolonged physical relationship between individuals of two different species. May be mutualistic, parasitic, or commensal.

symmetric competition Ecological competition between two species in which both suffer similar declines in fitness.

sympathetic nervous system The part of the autonomic nervous system that stimulates fight-or-flight responses, such as increased heart rate, increased blood pressure, and decreased digestion.

sympatric speciation Speciation that occurs while the two diverging species are living in the same area.

sympatry Living in the same geographic area as some other population.

symplast A pathway for water transport in plants roots through the cytoplasm of adjacent cells that are connected by plasmodesmata.

synapomorphy A shared, derived (altered from the ancestral state) character that can be used to infer evolutionary relationships.

synapse The connection between two neurons, or between a neuron and a muscle cell: a tiny space into which neurotransmitters are released.

synapsis The physical pairing of two homologous chromosomes during prophase I of meiosis. Crossing over occurs during synapsis.

synaptic cleft The gap of a synapse; the space between two communicating nerve cells, across which neurotransmitters diffuse.

synaptic plasticity Long-term changes in the responsiveness or physical structure of a synapse that can occur after particular stimulation patterns. Thought to be the basis of learning and memory.

synaptic vesicles Tiny neurotransmitter-containing vesicles at the end of an axon. Can fuse with the axon membrane to release a neurotransmitter into a synapse, stimulating the next neuron or effector cell.

synaptonemal complex A network of proteins that hold non-sister chromatids together during synapsis in meiosis.

syndrome A group of medical symptoms that often occur together and that are suspected to have the same underlying cause.

syngamy Fusion of two gametes to create a diploid zygote. In multicellular species, also called fertilization.

synthesis phase (S phase) The phase of the cell cycle during which DNA is synthesized and chromosomes are duplicated.

system A set of interacting elements, such as (1) a set of reactants and products at chemical equilibrium and (2) a group of organs that work together to perform a function.

systemic acquired resistance (SAR) A set of events through which plant tissues prepare to combat a possible infection.

systemic circulation The part of the circulatory system that sends oxygen-rich blood from the lungs out to the rest of the body. In mammals and birds, separate from the pulmonary circulation.

systemin A peptide hormone, produced by plant cells damaged by herbivores, that initiates a protective response in undamaged cells.

systole The portion of the heartbeat cycle during which the heart muscles are contracting.

systolic blood pressure Blood pressure in arteries during ventricular systole (heart contraction).

taiga A vast forest biome throughout subarctic regions, consisting primarily of short conifer trees. Characterized by intensely cold winters, short summers, and high annual variation in temperature.

taproot A large vertical main root of a plant.

***Taq* polymerase** A DNA polymerase commonly used in PCR due to its stability at high temperatures. Derived from *Thermus aquaticus*, a bacterium found in hot springs.

taste buds Sensory structures, found chiefly in the mammalian tongue, that are responsible for the sense of taste.

taste cells Spindle-shaped cells found in a taste bud of the mammalian tongue. Respond to certain chemical stimuli.

TATA box A DNA sequence in many eukaryotic promoters 30 base pairs upstream from the transcription start site. Recognized by RNA polymerase II.

taxis Movement toward or away from some external cue.

taxon (plural: taxa) Any named group of species at any level of a classification system.

taxonomy The branch of biology concerned with the classification and naming of organisms.

TBP (TATA-binding protein) A protein that binds to eukaryotic promoters and helps initiate transcription.

T-cell receptor (TCR) A T-cell membrane protein that can bind to antigens displayed on the surfaces of other cells.

T cells Lymphocytes that mature in the thymus and are involved in acquired immunity in vertebrates.

May be involved in activation of other lymphocytes (helper T cells) or destruction of infected cells (cytotoxic T cells). Also called T lymphocytes.

tectorial membrane A membrane in the vertebrate cochlea that takes part in the transduction of sound by bending the stereocilia of hair cells in response to sonic vibrations.

telomerase An enzyme that replicates telomeres by catalyzing DNA synthesis from an RNA template.

telomere The region at the end of a linear chromosome.

telophase The final stage in cell division (mitosis or meiosis), during which chromosomes finish moving and new nuclear envelopes begin to form around each set of daughter chromosomes.

temperate Having a climate with pronounced annual fluctuations in temperature (i.e., warm summers and cold winters) but typically neither as hot as the tropics nor as cold as the poles.

template strand (1) The strand of DNA that is transcribed by RNA polymerase to create RNA. (2) An original strand of RNA used to make a complementary strand of RNA.

temporal lobe In the mammalian brain, the region of each cerebral hemisphere near the ears. Functions in memory, interpretation of information from the ears, and, in humans, language.

tendon A band of tough, fibrous connective tissue that connects a muscle to a bone.

tension Any pulling force; the opposite of compression.

tentacle A long, thin, muscular appendage of gastropod mollusks.

terminal cell The upper cell produced by an angiosperm zygote. Gives rise to the plant embryo.

termination (1) In enzyme-catalyzed reactions, the final stage in which the enzyme returns to its original conformation and products are released. (2) In DNA transcription, the dissociation of RNA polymerase from DNA. (3) In RNA translation, the dissociation of ribosomes from mRNA.

territory An area that is actively defended by an animal from others of its species.

tertiary structure The overall three-dimensional shape of a single polypeptide chain, created by a variety of interactions among R-groups and the peptide backbone.

test A hard protective outer structure seen in some protists. Also called a shell.

testcross The breeding of an individual of unknown genotype with an individual having only recessive alleles for the traits of interest, in order to infer the unknown genotype from the phenotypic ratios seen in offspring.

testis (plural: testes) The sperm-producing organ of a male animal.

testosterone A steroid hormone from the testes that stimulates sperm production and various male traits and reproductive behaviors.

tetrad A pair of synapsed homologous chromosomes, each containing two chromatids.

tetraploid With four sets of chromosomes ($n = 4$).

tetrapod Any member of the taxon Tetrapoda, including all descendants of the first four-footed animals to move onto the land.

texture In soil biology, the proportions of different-sized particles present in soil.

theory A proposed explanation for a very general class of phenomena or observations.

thermal energy The kinetic energy of molecular motion.

thermocline A gradient (cline) in environmental temperature across a large geographic area.

thermophiles Bacteria or archaea that thrive in very hot environments.

thermoreceptor A sensory cell or an organ specialized for detection of changes in temperature.

thermoregulation Regulation of body temperature.

thick filaments Strands of myosin found in the middle of a muscle sarcomere. Bind to thin filaments (actin) and cause muscle contraction.

thin filaments Strands of actin found at the two ends of a muscle sarcomere. Bind to thick filaments (myosin) during muscle contraction.

thorax In mammals, the anterior region of the torso, containing the lungs. In insects, the middle of the three major body regions.

thorn A modified plant stem shaped as a sharp protective structure.

threshold The membrane potential at which a neuron's voltage-gated sodium channels will trigger an action potential.

thylakoid A flattened, membrane-bound disk inside a plant chloroplast. A stack of thylakoids is a granum.

thymus An organ, located in the anterior chest or neck of vertebrates, that, in young animals, processes T cells and sends them to lymph nodes.

thyroid gland A gland in the neck that releases thyroid hormone (which increases metabolic rate) and calcitonin (which lowers blood calcium).

thyroid-stimulating hormone (TSH) A peptide hormone from the pituitary gland that stimulates release of thyroid hormones from the thyroid gland.

thyroxine An iodine-containing peptide hormone from the thyroid gland that increases metabolic rate, both directly and via conversion to the more active hormone triiodothyronine.

tight junction A physical attachment between two adjacent animal cells, consisting of proteins that "stitch" the cells' plasma membranes together.

tip The end of a branch on a phylogenetic tree. Represents a specific population or species that has not (yet) produced descendants—either a group living today or a group that ended in extinction.

Ti plasmid A plasmid carried by *Agrobacterium* (a bacterium that infects plants) that can integrate into the plant cell's chromosomes and induce formation of a gall.

tissue A group of similar cells that function as a unit, such as muscle tissue or epithelial tissue.

tissue culture A collection of cells of a certain tissue type grown in a lab, typically in liquid suspension or in a petri dish on a solid food medium.

tolerance Ecological succession in which early-arriving species do not affect the probability that subsequent species will become established.

tonoplast The membrane surrounding a plant vacuole.

top-down control The limitation of herbivore population size by predation or disease rather than by limited or toxic nutritional resources.

topoisomerase An enzyme that cuts and rejoins DNA downstream of the replication fork, to ease the twisting that would otherwise occur as the DNA "unzips."

torpor A regulated, "deliberate" decrease in metabolic rate and body temperature in an endotherm.

totipotent Capable of dividing and developing to form a complete, mature organism.

tracheae (singular: trachea) A system of small air-filled tubes that extends throughout an insect's body and functions in gas exchange.

tracheid An elongated water-conducting plant cell that has gaps (pits) in its secondary cell wall to allow water movement from one cell to the next.

trade-off An inescapable compromise between two traits that cannot be optimized simultaneously.

trait Any characteristic of an individual.

transcription The process by which messenger RNA is made from a DNA template.

transcriptional control Regulation of gene expression via changes in the rate at which genes are transcribed to form messenger RNA.

transcription termination signal A DNA sequence that can terminate messenger RNA synthesis, usually because the RNA transcribed from this sequence forms a hairpin that prevents further transcription.

transduction Conversion of information from one mode to another. For example, the process by which a stimulus outside a cell is translated into a response by the cell.

transfer cells Cells that transfer nutrients from a parent plant to a developing plant seed. Occur in land plants.

transfer RNAs (tRNAs) A class of RNA molecules with an anticodon at one end and an amino acid binding site at the other. Match amino acids to messenger RNA codons during translation.

transformation Incorporation of DNA obtained directly from the environment into the genome. Occurs naturally in some bacteria; can be induced in the lab by certain processes.

transgenic Containing DNA that has been modified by genetic engineering.

transitional form A fossil species or population with traits that are intermediate between older and younger species.

transition state A high-energy intermediate state occurring during a chemical reaction that determines the activation energy necessary for the reaction to proceed.

translation The process by which proteins and peptides are synthesized from messenger RNA.

translational control The regulation of gene expression by altering the life span of messenger RNA or the efficiency of translation. A type of post-transcriptional regulation.

translocation (1) The movement of sugars through a plant by bulk flow. (2) A type of mutation in which a piece of a chromosome moves to a nonhomologous chromosome. (3) The process by which a ribosome moves down a messenger RNA molecule during translation.

transmembrane domain A structure that anchors a B-cell receptor molecule to the plasma membrane of the B cell.

transmembrane protein Any membrane protein that spans the entire membrane. Also called integral membrane protein.

transmission electron microscope (TEM) A microscope that forms an image from electrons that pass through a specimen.

transmission genetics The study of the genotypic and phenotypic patterns that occur as genes pass from one generation to the next.

transpiration Water loss from aboveground plant parts. Occurs primarily through the stomata.

transport protein Any membrane protein that enables specific molecules to cross plasma membranes, sometimes by causing a conformational change in the protein. Also called transporter.

transposable elements Any of several kinds of parasitic DNA sequences that are capable of moving themselves, or copies of themselves, to other locations in the genome.

tree of life A diagram depicting the genealogical relationships of all living organisms on Earth, with a single ancestral species at the base.

triacylglycerols Lipids consisting of three fatty acids joined to a glycerol molecule. Also called triglycerides or fats.

trichomes Protective hairlike appendages of the epidermal cells of some plants.

triglycerides Lipids consisting of three fatty acids joined to a glycerol molecule. Also called triacylglycerols or fats.

triiodothyronine An iodine-containing peptide hormone, secreted by the thyroid gland, that increases metabolic rate. Has a stronger effect than does the related hormone thyroxine.

trilobite A member of an extinct lineage of arthropods that were abundant from 550–440 million years ago.

trimester In humans, one of three major stages of gestation, each three months long.

triose A monosaccharide (simple sugar) that has three carbons.

triplet code A code in which a "word" of three letters encodes one piece of information. The genetic code is a triplet code because a codon is three nucleotides long and encodes one amino acid.

triploblast An animal whose body develops from three basic embryonic cell layers: ectoderm, mesoderm, and endoderm. All animals except sponges, cnidarians, and ctenophores are triploblasts.

triploid With three sets of chromosomes ($n = 3$).

trisomy The state of having three copies of one particular type of chromosome.

tRNAs See **transfer RNAs (tRNAs)**.

trochophore A type of larva that has a ring of cilia around its middle. Occurs in several phyla of the Lophotrochozoa lineage of protostomes.

trophic level A feeding level in an ecosystem.

tropomyosin A muscle protein that blocks the myosin-binding sites on actin filaments, preventing muscle contraction. Can be moved out of the way by troponin when intracellular calcium is high.

troponin A muscle protein that can trigger muscle contraction by moving tropomyosin off the myosin-binding sites on actin filaments. Activated by high intracellular calcium.

trp operon An *E. coli* operon that includes five cotranscribed genes involved in the synthesis of the amino acid tryptophan.

true navigation Navigation by which an animal can reach a specific point on Earth's surface.

trypsin A protein-digesting enzyme produced by the pancreas, secreted into the intestine, and activated by enterokinase. Trypsin in turn activates several other protein-digesting enzymes.

trypsinogen The precursor of protein-digesting enzyme trypsin. Secreted by the pancreas and activated by the intestinal enzyme enterokinase.

T tubules Membranous tubes that extend into the interior of muscle fibers. Propagate action potentials throughout the muscle cell and trigger release of calcium from the sarcoplasmic reticulum.

tube cell A large cell in a male gametophyte that will give rise to the pollen tube.

tube feet Small, mobile, fluid-filled extensions of the water vascular system of echinoderms. Used in locomotion.

tuber Plant rhizomes that are modified to function as carbohydrate-storage organs.

tuberculosis (TB) A highly contagious human disease caused by the bacterium *Mycobacterium tuberculosis*.

tube-within-a-tube Describing the basic body plan of all triploblast animals (i.e., an inner tube of endoderm within an outer tube of ectoderm, with mesoderm between the two tubes).

tumor A mass of cells formed by uncontrolled cell division. Can be benign or malignant.

tumor suppressor A gene that prevents cell division, particularly when the cell has DNA damage. Mutated forms are associated with cancer. Also, the protein produced by such a gene.

tundra The treeless biome in polar and alpine regions, characterized by short, slow-growing vegetation, permafrost, and a climate of long, intensely cold winters and very short summers.

turgid Refers to a plant cell that is firm (i.e., containing enough water for the cell cytoplasm to press against the cell wall).

turgor pressure The outward pressure exerted by the fluid contents of a plant cell against its cell wall.

Turner syndrome A human syndrome caused by the presence of only one X chromosome and no Y chromosome ("XO"). Individuals with this condition are female but sterile.

turnover In lake ecology, the complete mixing of upper and lower layers of water that occurs each spring and fall in temperate-zone lakes.

two-fold rotational symmetry A type of symmetry in which an object can be superimposed on itself if rotated 180°. Occurs in some regulatory sequences of DNA. Also called dyad symmetry.

tympanic membrane The eardrum; a membrane separating the middle ear from the outer ear in terrestrial vertebrates, or similar structures in insects.

type I diabetes mellitus Diabetes mellitus that is caused by insufficient secretion of insulin from the pancreas.

type II diabetes mellitus Diabetes mellitus that is caused primarily by insufficient responsiveness of tissues to insulin, despite normal secretion of insulin from the pancreas.

typhoid fever A human disease caused by the bacterium *Salmonella typhi*.

ubiquinone A molecule that transfers electrons in the electron transport chain of cellular respiration in mitochondria. Also called coenzyme Q or Q.

ulcer A hole or thin spot in the stomach or intestinal wall.

ultimate causation In biology, the reason that a trait or phenomenon is thought to have evolved; the adaptive advantage of that trait. Also called ultimate explanation.

ultrasound Sound frequencies that are too high for humans to hear—higher than about 20,000 Hz (20 kHz).

ultraviolet light Light with a wavelength shorter than visible blue light.

umami The taste of glutamate, responsible for the "meaty" taste of most proteins and of monosodium glutamate.

umbilical cord The cord that connects a developing mammal embryo or fetus to the placenta and through which the embryo or fetus receives oxygen and nutrients.

undershoot The brief phase after an action potential when a cell's membrane potential temporarily becomes more negative than the resting potential.

unequal crossover An error in crossing over during meiosis I in which the two chromatids match up at different sites. Results in gene duplication and gene loss in the two resulting chromatids.

uniformitarianism The concept that the laws of the physical universe are constant throughout time and space.

unsaturated fat A fat with one or more carbon-carbon double bonds in its fatty acid chains and thus fewer than the maximum number of hydrogen atoms. Double bonds produce kinks in the fatty acid chains and decrease the compound's melting point.

upstream Opposite to the direction in which RNA polymerase moves along a DNA strand.

urea A water-soluble excretory product of mammals and sharks, used to excrete excess nitrogen from amino acids.

ureter In mammals, a tube that transports urine from one kidney to the bladder.

urethra The tube that drains urine from the bladder to the outside environment. In male vertebrates, also used for passage of sperm during ejaculation.

uric acid A whitish excretory product of birds, reptiles, and terrestrial arthropods, used to get rid of excess nitrogen derived from amino acids.

urochordates The tunicates or sea squirts. A lineage of sessile, filter-feeding chordates.

uterus The organ in which developing embryos are housed in those vertebrates that give live birth. Common in most mammals and in some lizards, sharks, and other vertebrates.

vaccination Injection with weakened, killed, or altered pathogens to stimulate development of immunity against those pathogens.

vaccine Preparation containing pieces of a pathogen, or entire killed or weakened pathogens, administered to stimulate immunity without causing illness.

vacuole An organelle usually used for bulk storage of substances such as pigments, oils, carbohydrates, water, or toxins. In animal cells, these organelles are smaller and are called lysosomes.

vagina The birth canal of female mammals; a muscular tube that extends from the uterus through the pelvis to the exterior.

valence The number of electrons in the valence (outermost) electron shell of an atom; determines how many covalent bonds the atom can form.

valence electron An electron in the valence (outermost) electron shell of an atom. Valence electrons tend to be involved in chemical bonding.

valence shell The outermost electron shell of an atom.

valves In circulatory systems, flaps of tissue that prevent backward flow of blood, particularly in veins and between the chambers of the heart.

van der Waals interaction A weak electrical attraction between two hydrophobic side chains. Often contributes to tertiary structure in proteins.

variable (V) region A section of the light chains of antibodies that has a highly variable amino acid sequence, unique to each B cell.

vasa recta A network of blood vessels that runs alongside the loop of Henle of a kidney nephron, reabsorbing water and solutes.

vascular bundle A cluster of xylem and phloem strands in a plant stem.

vascular cambium A ring of undifferentiated plant cells inside the cork cambium of woody plants. Produces secondary xylem and secondary phloem.

vascular tissue In plants, a tissue that is involved in conducting water or solutes from one part of a plant to another, or that gives rise to such tissues.

vas deferens A pair of muscular tubes that store and transport semen from the epididymus to the ejaculatory duct. In nonhuman animals, called the ductus deferens.

vector (1) A plasmid or other vehicle used to transfer recombinant genes to a new host. (2) A biting insect or other organism that transfers pathogens between two other species.

vegetal hemisphere The lower half of an amphibian egg cell, containing most of the yolk. This portion eventually becomes part of the gut.

vein (1) Any blood vessel that carries blood from capillaries to the heart (oxygenated or not). Has thinner walls and lower blood pressure than an artery. (2) A strip of vascular tissue in a plant leaf. (3) A supporting filament of an insect wing.

veliger A unique form of larva in bivalves.

venae cavae (singular: vena cava) Large veins that return oxygen-poor blood to the heart.

ventilation Movement of air or water through the lungs or gills.

ventral Toward an animal's belly and away from its back. The opposite of dorsal.

ventricle (1) A thick-walled chamber of the heart that receives blood from an atrium and pumps it to the body or to the lungs. (2) One of several small fluid-filled chambers in the vertebrate brain.

vertebra (plural: vertebrae) One of the cartilaginous or bony elements that form the spine of vertebrate animals.

vertebrate An animal in the lineage Vertebrata, characterized by a skull, usually a spinal column, usually an endoskeleton of bone, and other traits.

vessel element An elongated water-conducting plant cell found in the xylem of certain advanced plants. Has gaps through both the primary and secondary cell walls, allowing unimpeded passage of water from one cell to the next.

vestigial trait Any rudimentary structure of unknown or minimal function that is homologous to functioning structures in other species. Vestigial traits are thought to reflect evolutionary history.

vicariance The physical splitting of a population into smaller, isolated populations by a geographic barrier.

villi (singular: villus) Small, fingerlike projections of the lining of the digestive tract. Function to increase surface area for absorption.

virion A single mature virus particle.

virulence The ability of a pathogen or parasite to cause disease and death.

virulent Tending to cause severe disease rather than mild disease.

virus A tiny infectious parasitic entity consisting of DNA or RNA enclosed in a protective covering (a capsid and sometimes an envelope). The DNA or RNA contain the necessary instructions to make more viruses but must use the machinery of a host cell to do so.

visceral mass The part of a mollusk containing most of the internal organs and external gill.

visible light The range of wavelengths of electromagnetic radiation that humans can see, from about 400 to 700 nanometers.

vitamin Any micronutrient that is a carbon-containing compound rather than a single chemical element. Usually functions as a coenzyme.

vitelline envelope A fibrous sheet of glycoproteins that surrounds a mature egg cell. Found in many vertebrates.

viviparous For animals, reproducing by live birth rather than by laying eggs.

volt A unit of voltage (electrical potential).

voltage Potential energy created by a separation of electric charges between two points. Also called electrical potential.

voltage clamping A lab technique for imposing a certain constant membrane potential on a cell. Widely used to investigate ion channels.

voltage-gated channel An ion channel that opens or closes in response to changes in membrane voltage.

wall pressure The inward pressure exerted by a cell wall against the fluid contents of a plant cell.

water potential The potential energy of water in a certain environment. In living organisms, the sum of solute potential and pressure potential. Determines movement of water into or out of cells.

water potential gradient A difference in water potential in one location compared with another. Determines the movement of water through plant tissues.

watershed The area drained by a single stream or river.

water table The upper limit of the underground layer of soil that is saturated with water.

water vascular system A system of fluid-filled tubes and chambers in echinoderms. Functions as a hydrostatic skeleton.

wavelength The distance between two successive wave crests in any regular wave, such as light waves, sound waves, or waves in water.

wax A type of lipid with long hydrocarbon tails, usually combinations of long-chain alcohols with fatty acids. Harder and less greasy than fats.

weather The specific short-term atmospheric conditions of temperature, moisture, sunlight, and wind in a certain area.

weathering The gradual wearing down of large rocks by rain, running water, and wind; one of the processes that transform rocks into soil.

weed A plant that is adapted for growth in disturbed soils.

Western blotting A lab technique for identifying proteins that will bind to a certain antibody. Involves separation of proteins with gel electrophoresis, transfer to a filter, and probing with a labeled antibody.

wetland A shallow-water habitat where the soil is saturated with water for at least part of the year.

white blood cells Immune system cells that circulate in the blood or lymph and function in the defense against disease. Also called leukocytes.

wild type The most common phenotype seen in a population; especially the most common phenotype in wild populations compared with inbred lab strains of the same species.

wilt To lose turgor, in a plant tissue.

wobble hypothesis The hypothesis that some tRNA molecules can pair with more than one codon of mRNA, tolerating some variation in the third base, as long as the first and second bases are correctly matched.

wood The secondary xylem of older woody stems and roots.

xanthophylls Carotenoid pigments found in many algae and some plants, typically appearing yellow.

xenotransplantation or xenografting The transplantation of tissues from one species to another.

xeroderma pigmentosum (XP) A human disease characterized by extreme sensitivity to ultraviolet light. Caused by an autosomal recessive allele that results in a defective DNA repair system.

X-linked inheritance Inheritance resulting from a gene being located on the mammalian X chromosome. Also called X-linkage.

X-ray crystallography A lab technique used to infer the three-dimensional structure of a molecule, on the basis of the diffraction patterns produced by X-rays beamed at the crystallized molecule.

xylem A plant vascular tissue that conducts water and ions; contains tracheids and/or vessel elements.

xylem sap The watery fluid found in xylem tissue of plants.

yeast Any fungus growing as a single-celled form. Also, a specific lineage of ascomycetes.

Y-linked inheritance Inheritance resulting from a gene being located on the mammalian Y chromosome. Also called Y-linkage.

yolk The nutrient-rich cytoplasm inside an egg cell, used as food for the growing embryo.

yolk sac In an amniotic egg, the membrane-bound sac that contains the yolk.

zeaxanthin A carotenoid pigment of plants that initiates the opening of stomata in response to blue light so that carbon dioxide can diffuse into photosynthesizing cells.

zero population growth (ZPG) A state of stable population size due to fertility staying at the replacement rate for at least a generation.

zona pellucida The gelatinous layer around the egg cell of a mammal.

zone of cellular division A group of plant cells just behind the root cap. Contains the apical meristem, where cells are actively dividing.

zone of cellular elongation A group of plant cells behind the apical meristem in plant roots. Consists of young cells that are increasing in length.

zone of cellular maturation A group of plant cells several centimeters behind the root cap that are differentiating into mature tissues.

Z scheme A widely accepted model for the passage of electrons from photosystem II to photosystem I during photosynthesis. Electrons on a graph of energy level versus location trace a Z shape.

zygosporangium (plural: zygosporangia) The spore-producing structure in fungi that are members of the Zygomycota.

zygote The diploid cell formed by the union of two haploid gametes. Capable of undergoing embryological development to form an adult.

Image Credits

Frontmatter
xv Jacob Halaska/Index Stock Imagery, Inc. **xvii** ©E.H. Newcomb & W.P. Wergin/Biological Photo Service **xviii** Dr. Gopal Murti/Science Photo Library/Photo Researchers, Inc. **xxi** Kent & Donna Dannen/Photo Researchers, Inc. **xxii** ©Beth Donidow/Visuals Unlimited **xxv** David Nunuk/Science Photo Library/Photo Researchers, Inc. **xxvi** ©Rob Nunnington/Foto Natura/Minden Pictures **xxix** ©Thomas Mangelsen/Minden Pictures **xxx** ©J.P. Ferrero/Peter Arnold **53.1** ©Christian Ziegler

Chapter 1
Opener ©Frans Lanting/Minden Pictures **1.1a** Burndy Library/Omikron/Photo Researchers, Inc. **1.1b** M.I. Walker/Photo Researchers, Inc. **1.3a** Photo by Kelly Buono/Courtesy of Dr. Richard Amasino, University of Wisconsin **1.3c** Bruce Forster/Getty Images Inc. - Stone Allstock **1.5a** Samuel F. Conti and Thomas D. Brock **1.5b** Kwangshin Kim/Photo Researchers, Inc. **1.6/1** ©Dr. David Phillips/Visuals Unlimited **1.6/2** Dennis Kunkel/Dennis Kunkel Microscopy, Inc. **1.6/3** Kolar, Richard/Animals Animals/Earth Scenes **1.6/4** Biophoto Associates/Photo Researchers, Inc. **1.6/5** Darwin Dale/Photo Researchers, Inc. **1.9b** Michael Hughes/Aurora & Quanta Productions Inc. **1.10a** Joshua J. Tewksbury **1.10b** William Weber/Visuals Unlimited **1.10c** Robert Dobbs **1.11** Data from Tewksbury and Nabhan. 2001. *Nature* 412: 403–404. Fig. 1a. **1.12** Data from Tewksbury and Nabhan. 2001. *Nature* 412: 403–404. Fig. 1b.

Unit 1
©Eric Meola/Image Bank/Getty Images

Chapter 2
Opener Science Photo Library/Photo Researchers, Inc. **2.1b** Mitchell Layton/Duomo Photography Incorporated **2.2** US Department of Energy/SPL/Photo Researchers, Inc. **2.4b** Minik Rosing, Geological Museum, Copenhagen **2.9c** Albert Copley/Visuals Unlimited **2.14a** P.W. Lipman/U.S. Geological Survey/U.S. Department of the Interior **2.14b** Richard Megna/Fundamental Photographs **2.17** Don Farrall/Getty Images Inc. - PhotoDisc **2.24** Robert and Beth Plowes Photography **2.27c** Geostock/Getty Images Inc. - PhotoDisc **2.29a** TSADO/NASA/Tom Stack & Associates, Inc. **2.29b** Jet Propulsion Laboratory/NASA Headquarters **T2.2** Source: Table 8.1, p. 312, in John McMurry and Robert C. Fay, *Chemistry*, 4th Edition, ©2004. Reprinted by permission of Pearson Education, Inc., Upper Saddle River, NJ

Chapter 3
Opener Jacob Halaska/Index Stock Imagery, Inc. **3.2** NOAA Vents Program **3.3** Photo courtesy R. Kempton, New England Meteoritical Services **3.6** Martin Bough/Fundamental Photographs **3.7b** Clare Sansom, Birkbeck College, University of London, London, England **3.12** Clare Sansom, Birkbeck College, University of London, London, England **3.13a** ©Microworks/Phototake **3.13b** ©Walter Reinhart/Phototake **3.15b** Clare Sansom, Birkbeck College, University of London, London, England **3.16** Clare Sansom, Birkbeck College, University of London, London, England **3.22** T.A. Steitz, Yale University **3.25** T.A. Steitz, Yale University **3.27a** Data from Nawani and Kapadnis. 2001. *Journal of Applied Microbiology* 90: 803–808. Fig. 3. Data also from Nawani et al. 2002. *Journal of Applied Microbiology* 93: 965–975. Fig. 7. **3.27b** Data from Hansen et al. 2002. *FEMS Microbiology Letters* 216: 249–253. Fig. 1. **3.29** ©Paul Fievez/Hulton Archive/Getty Images

Chapter 4
Opener Micheal Simpson/Getty Images, Inc. - Taxi **4.6** Reproduced by permission from J.P. Ferris et al., Synthesis of long prebiotic oligomers on mineral surfaces. Nature 381:59–61 (1996), Fig. 2. Copyright ©1996 Macmillan Magazines Limited. Image courtesy of James P. Ferris, Rensselaer Polytechnic Institute. **4.7** Omikron/Photo Researchers, Inc. **4.8** A. Barrington Brown/Science Source/Photo Researchers, Inc. **4.15** Reprinted with permission from *Science* 292: 1319–1325 Fig. 4B (2001) by Wendy K. Johnston, Peter J. Unrau, Michael S. Lawrence, Margaret E. Glasner, David P. Bartel "RNA-Catalyzed RNA Polymerization: Accurate and General RNA-Templated Primer Extension."

Chapter 5
Opener Dr. Jeremy Burgess/Photo Researchers, Inc. **5.8a** ©Biophoto Associates/ Photo Researchers, Inc. **5.8b** Dr. Jacob S. Ishay, and Dr. Eyal Rosenzweig. ©2000 E. Rosenzweig **5.8c** M. Jericho, Dept of Physics, Dalhousie University, Halifax, Nova Scotia and T.J. Beveridge, Dept of Microbiology, University of Guelph, Guelph, Ontario

Chapter 6
Opener Kit Pogliano and Marc Sharp, University of California at San Diego **6.1aL** JEOL USA Inc. **6.1aR** Photo courtesy of Peter M. O'Day, Juan Bacigalupo, Joan E. Haab, and Cecilia Vergara. The Journal of Neuroscience, October 1, 2000. 20(19): 7193–7198, Fig. 1C. ©2004 by the Society for Neuroscience. **6.1bL** SEM photo courtesy of JEOL-USA, Peabody, MA **6.1bR** ©Dennis Kunkel/Phototake **6.2a** Alec D. Bangham, M.D., F.R.S. **6.2b** Fred Hossler/Visuals Unlimited **6.8aL** James J. Cheetham, Carleton University **6.12a** ©Dorling Kindersley **6.12b** Clive Streeter ©Dorling Kindersley **6.12c** Phil Degginger/Color-Pic, Inc. **6.20** Don W. Fawcett/Photo Researchers, Inc. **6.25b** Kovacs, F., Quine, J. & Cross, T.A. (1999) "Validation of the Single-Stranded Channel Formation of Gramicidin A by Solid-State NMR" *Proc. Natl. Acad. Sci. U.S.A.* 96:7910–7915. **6.26L** Andrew Syred/Getty Images Inc. - Stone Allstock **6.26M** ©Dr. David Phillips/Visuals Unlimited **6.26R** Joseph F. Hoffman, Yale University School of Medicine **6.32** ©Science Photo Library/Photo Researchers, Inc.

Unit 2
©Yorgos Nikas/Stone/Getty Images

Chapter 7
Opener ©Albert Tousson/Phototake **7.1** ©Dr. T.J. Beveridge/Visuals Unlimited **7.2** ©Stanley C. Holt/Biological Photo Service **7.3** ©Dr. Gopal Murti/Visuals Unlimited **7.4** COVER (upper left image), from *Science* vol 285, no. 5436, September 1999. Image: J.H. Cate, M.M. Yusupov, G. Zh. Yusupova, T.N. Earnest, H.F. Noller. Copyright 1999 American Association for the Advancement of Science. **7.5** Wanner/Eye of Science/Photo Researchers, Inc. **7.7** ©Fawcett/Photo Researchers, Inc. **7.8** Photo Researchers, Inc. **7.9** ©Dr. Don Fawcett/Photo Researchers, Inc. **7.10** ©Biophoto Associates/Photo Researchers, Inc. **7.11** ©Don W. Fawcett/Photo Researchers, Inc. **7.12** ©Dr. Don Fawcett, Daniel Friend, and Richard Wood/Photo Researchers, Inc. **7.13** ©Dr. Gopal Murti/Visuals Unlimited **7.15** E.H. Newcomb & W.P. Wergin/Biological Photo Service/Getty Images Inc. - Stone Allstock **7.16** K.R. Porter/Photo Researchers, Inc. **7.17** W.P. Wergin/Biological Photo Service **7.18** ©E. H. Newcomb/Biological Photo Service **7.19a** ©Dr. Don Fawcett/S. Ito & A. Like/Photo Researchers, Inc. **7.19b** ©Dr. Don Fawcett/Photo Researchers, Inc. **7.19c** Courtesy of Dr. Julian R. Thorpe, Electron Microscope Division, The Sussex Centre for Advanced Microscopy, School of Life Sciences, University of Sussex. **7.19d** ©Jan Robert Factor/Photo Researchers, Inc. **7.21** ©Michael W. Davidson, Florida State University/Molecular Expressions (http://www.microscopy.fsu.edu) **7.22b** Don W. Fawcett/Photo Researchers, Inc. **7.27** Jamieson, J.D. and Palade, G.E. (1967). Intracellular transport of secretory proteins in the pancreatic exocrine cell. II. Transport to condensing vacuoles and zymogen granules. Reproduced from *The Journal of Cell Biology*, 1967, v. 34 , pp. 597–615. **7.33a** K.G. Murti/Visuals Unlimited **7.33b** Peter Dawson/Science Photo Library/Photo Researchers, Inc. **7.33c** ©Dr. Gopal Murti/Science Photo Library/Photo Researchers, Inc. **7.35** ©Dr. Conly L. Rieder/Biological Photo Service **7.36** Reproduced by permission of the American Society for Cell Biology from *Molecular Biology of the Cell* 9(12), December 1998, cover. Copyright ©1999 by the American Society for Cell Biology. Image courtesy of Bruce J. Schnapp, Oregon Health Sciences. **7.37a** John E. Heuser, M.D., Washington University School of Medicine, St. Louis, Missouri **7.38** Dennis Kunkel/Phototake NYC **7.39a** Dr. Gopal Murti/Science Photo Library/Photo Researchers, Inc. **7.40** Reproduced from C.J. Brokaw, Microtubule sliding in swimming sperm flagella: direct and indirect measurements on sea urchin and tunicate spermatozoa. *Journal of Cell Biology* 114:1201–1215 (1991), cover. Reproduced by copyright permission of The Rockefeller University Press.

Chapter 8
Opener ©E.H. Newcomb & W.P. Wergin/Biological Photo Service **8.2a** ©Spencer Grant, Photo Researcher, Inc. **8.2b** ©Biophoto Associates/Photo Researchers, Inc. **8.3a** Ken Eward/Science Source/Photo Researchers, Inc. **8.3b** Barry King, University of California, Davis/Biological Photo Service **8.5** C..T. Huang, Karen Xu, Gordon McFeters, and Philip S. Stewart **8.6** ©K.R. Porter/Photo Researchers, Inc. **8.8** Biophoto Associates/Science Source/Photo Researchers, Inc. **8.9a** ©Dr. Don Fawcett/Photo Researchers **8.10a** ©Dr. Don Fawcett/Gida Matoltsy/Photo Researchers, Inc. **8.13a** ©E. H. Newcomb & W.P. Wergin/Biological Photo Service **8.13b** ©Dr. Don Fawcett/Photo Researchers, Inc.

Chapter 9
Opener ©2004 Richard Megna/Fundamental Photographs **9.1/1** Peter Anderson ©Dorling Kindersley **9.1/2** ©Dorling Kindersley **9.1/3** ©Darren McCollester/CORBIS **9.2c** Clare Sansom, Birkbeck College, University of London, London, England **9.9** Clare Sansom, Birkbeck College, University of London, London, England. **9.19b** Modified from Wang and Oster. 1998. *Nature* 396: 279. Fig. 1.

Chapter 10
Opener J.P. Nacivet/Photo Researchers, Inc. **10.2a** John Durham/Science Photo Library/Photo Researchers, Inc. **10.2b** ©Dr. J. Burgess, Science Photo Library/Photo Researchers, Inc. **10.3L** Biophoto Associates/Photo Researchers, Inc. **10.3M** ©Dr. George Chapman/Visuals Unlimited **10.3R** Richard Green/Photo Researchers, Inc. **10.5b** Sinclair Stammers/Science Photo Library/Photo Researchers, Inc. **10.10** David Newman/Visuals Unlimited **10.19** James A. Bassham, Lawrence Berkeley Laboratory, UCB (retired) **10.21** Andersson I. Journal of Molecular Biology, May 1996, vol. 259, iss. 1, pp. 160–174(15). Courtesy of Rolf Bergmann, University of Hamburg **10.23a** Dr. Jeremy Burgess/Science Photo Library/Photo Researchers, Inc. **10.26a** Adam Hart-Davis, Science Photo Library/Photo Researchers, Inc. **10.26b** ©David Muench/CORBIS **10.28** ©David Woodfall/DRK Photo

Chapter 11
Opener Professor G. Schaten/Science Photo Library/Photo Researchers, Inc. **11.2a** Originally published in Walter Flemming, Zellsubstanz, Kern, und Zelltheilung. Leipzig: Verlag von F.C.W. Vogel, 1882. Image courtesy of Conly Rieder, Wadsworth Center, New York State Department of Health. **11.2b** ©Photo Researchers, Inc. **11.3** Mark E. Warchol, Central Institute for the Deaf, Washington University, and Jeffrey T. Corwin, University of Virginia School of Medicine **11.10a** ©Dr. Richard Kessel/Visuals Unlimited **11.10b** R. Calentine/Visuals Unlimited **11.11** Micrographs by Conly L. Rieder, Division of Molecular Medicine, Wadsworth Center, Albany, New York 12201–0509. **11.17** Source: American Cancer Society's *Cancer Facts and Figures – 2003*. Reprinted with permission. **11.18** Dr. E. Walker/Science Photo Library/Photo Researchers, Inc.

Unit 3
©Eye of Science/Photo Researchers, Inc.

Chapter 12
Opener David Phillips/The Population Council/Photo Researchers, Inc. **12.2a** Applied Image Corp., San Jose, CA **12.2b** ©Addenbrookes Hospital/Photo Researchers, Inc. **12.2c** ©Wessex Reg. Genetics Centre /Wellcome Photo Library **12.7** David A. Jones **12.10** Doug Sokell/Visuals Unlimited **12.16** ©Robert and Beth Plowes Photography

Chapter 13
Opener Margaret H. Peaslee, Vice President for Academic Affairs and Professor of Biology, University of Pittsburgh at Titusville. **13.7** Dr. Madan K. Bhattacharyya **13.10a** Robert Calentine/Visuals Unlimited **13.10b** Carolina Biological Supply Company/Phototake NYC **13.11a** ©Addenbrookes Hospital/Photo Researchers, Inc. **13.17a** Robert Calentine/Visuals Unlimited **13.18a** David Cavagnaro/Peter Arnold, Inc. **13.19a** Albert F. Blakeslee, *Journal of Heredity*, 1 Fig., 1914. Reproduced by permission of Oxford University Press. **13.23** ©Bettmann/CORBIS

Chapter 14

Opener Dr. Gopal Murti/Science Photo Library/Photo Researchers, Inc. **14.1** Reproduced from O.T. Avery, C.M. MacLeod, and M. McCarty, Studies on the chemical nature of the substance inducing transformation of pneumococcal types, *The Journal of Experimental Medicine*, 1944, 79:137–157, plate 1, by copyright permission of The Rockefeller University Press. **14.4b** Oliver Meckes/Max Planck Institut-Tubingen/Photo Researchers, Inc. **14.11a** Dr. Gopal Murti/Science Photo Library/Photo Researchers, Inc. **14.20** Vesalius Studios

Chapter 15

Opener Halaska, Jacob/Index Stock Imagery, Inc. **15.11** Omikron/Photo Researchers, Inc.

Chapter 16

Opener Oscar Miller/Science Photo Library/Photo Researchers, Inc. **16.2** Reproduced with permission from Structural basis of transcription initiation: an RNA polymerase holoenzyme-DNA complex. Murakami KS, Masuda S, Campbell EA, Muzzin O, Darst SA. *Science*. 2002 May 17;296(5571):1285–90. Fig 2a pg 1287. Copyright ©2004 **16.4** Hans Reinhard/Bruce Coleman Inc. **16.5** Bert W. O'Malley, M.D., Baylor College of Medicine **16.13** Reprinted from *Cell Press*, "Ribosome Structure and the Mechanism of Translation" by V. Ramakrishnan from Vol. 108, 557–572, Fig 1b, February 22, 2002, with permission from Elsevier. **16.16T** ©E.V. Kiseleva and Donald Fawcett/*Visuals Unlimited* **16.19b** Bill Longcore/Photo Researchers, Inc. **16.21a** "Structural basis for the interaction of antibiotics with peptidyl transferase centre in eubacteria: by Schluenzen et al." Figure 5 in *Nature* (2001) 413: 814–821 by Schluenzen et al.

Chapter 17

Opener EM Unit, VLA/Science Photo Library/Photo Researchers, Inc. **17.11** Reproduced by permission from A. Schmitz and D.J. Galas, The interaction of RNA polymerase and lac repressor with the lac control region. *Nucleic Acids Research* 6:111–137 (1979), fig. 2b. Copyright ©1979 by Oxford University Press. Image courtesy of David J. Galas, Keck Graduate Institute. **17.15** ©Dr. Dennis Kunkel/Visuals Unlimited **p. 381** Michael Gabridge/Visuals Unlimited

Chapter 18

Opener Prof. Oscar L. Miller/Science Photo Library/Photo Researchers, Inc. **18.2a** Ada Olins/Don Fawcett/Photo Researchers, Inc. **18.4b** Copyright ©2002 from *Molecular Biology of the Cell* 4/e fig. 4.23 by Bruce Alberts et al. Reproduced by Permission of Garland Science Taylor & Francis Books, Inc., and Dr. Barbara Hamkalo. **18.4c** Copyright ©2002 from *Molecular Biology of the Cell* 4/e fig. 4.23 by Bruce Alberts et al. Reproduced by Permission of Dr. Victoria E. Foe, and Garland Science/Taylor & Francis Books, Inc. **18.16** Walter J. Gehring

Chapter 19

Opener Work of Atsushi Miyawaki, Qing Xiong, Varda Lev-Ram, Paul Steinbach, and Roger Y. Tsien at the University of California, San Diego. **19.2** ©Dennis Kunkel/Phototake **19.12** Baylor College of Medicine/Peter Arnold, Inc. **19.14** ©Wayne P. Armstrong, Palomar College **19.16a** Brad Mogen/Visuals Unlimited **19.17b** Peter Beyer, University of Freiburg, Germany

Chapter 20

Opener ©Sanger Institute/Wellcome Photo Library **20.7** ©David Parker/Photo Researchers **20.8** AP Wide World Photos **20.10** Camilla M. Kao and Patrick O. Brown, Stanford University.

Unit 4

©Ted Levin/Animals Animals

Chapter 21

Opener Photo courtesy Dr. Andrew Ewald **21.2a** Microworks/Phototake NYC **21.2b** Cabisco/Visuals Unlimited **21.4a** Gregory Ochocki/Photo Researchers, Inc. **21.6a** Michael Whitaker/Science Photo Library/Photo Researchers, Inc. **21.6b** Victor D. Vacquier, Scripps Institution of Oceanography, University of California at San Diego **21.7b** Biodisc/Visuals Unlimited **21.8c** David M. Phillips/Visuals Unlimited **21.10a** David S. Addison/Visuals Unlimited **21.11a** Michael V. Danilchik, Oregon Health Sciences University **21.11b** Douglas A. Melton, Harvard University **21.12** Douglas A. Melton, Harvard University **21.15** Richard Hutchings/Photo Researchers, Inc. **21.18a** Holt Studios International/Photo Researchers, Inc. **21.24** ©Jochen Tack, Das Fotoarchiv/Peter Arnold

Chapter 22

Opener Gary C. Schoenwolf, University of Utah School of Medicine **22.1** F. Rudolf Turner, Indiana University **22.3a** Christiane Nusslein.Volhard, Max Planck Institute of Developmental Biology, Tubingen, Germany **22.3b** Wolfgang Driever, University of Freiburg, Germany **22.4a** Jim Langeland, Stephen Paddock, and Sean Carroll, University of Wisconsin at Madison **22.4b** Stephen J. Small, New York University **22.4c** Jim Langeland, Stephen Paddock, and Sean Carroll, University of Wisconsin at Madison **22.6** F. Rudolf Turner, Indiana University **22.7a/1** ©Oliver Meckes/Photo Researchers, Inc. **22.7a/2** Copyright ©Carolina Biological Supply Company/ Phototake **22.7b/1** ©Oliver Meckes/Photo Researchers, Inc. **22.7b/2** Edward B. Lewis, California Institute of Technology **22.11** John L. Bowman, University of California at Davis **22.12** Reprinted by permission from *Nature* (Vol 424, July 24, 2003, pg 439, by Kim et al) copyright (2004) Macmillan Publishers Ltd. Photograph supplied by Dr. Neelima Sinah. **22.13b** Roslin Institute/PA Photos Ltd **22.15** Reproduced by permission of Elsevier Science from K. Kuida et al., Reduced apoptosis and cytochrome c.mediated caspase activation in mice lacking caspase 9. *Cell* 94:325–337 (1998), figs. 2E and 2F. Copyright ©1998 by Elsevier Science Ltd. Image courtesy of Keisuke Kuida, Vertex Pharmaceuticals. **22.16a** Kathryn W. Tosney, University of Michigan

Unit 5

©Gary Meszaros/Photo Researchers

Chapter 23

Opener Kent & Donna Dannen/Photo Researchers, Inc. **23.1a** ©Bettmann/Corbis **23.1b** Alfred Russel Wallace by unknown artist, after a photograph by Thomas Sims, fl. 1860s. Reg. No.: 1765. National Portrait Gallery, London. **23.2a** ©Ken Lucas/Visuals Unlimited **23.2b** ©T. A. Wiewandt/DRK PHOTO **23.2c** Courtesy photographer, Knut Finstermeier, MPI.EVA Leipzig **23.3aL** Based on Gingerich et al. 2001. *Science* 293: 2239–2242 and Thewissen et al. 2001. *Nature* 413: 277–281. **23.3aR** Robert

Lubeck/Animals Animals/Earth Scenes **23.4aL** Vincent Zuber/Custom Medical Stock Photo, Inc. **23.4aR** CMCD/Getty Images Inc./PhotoDisc **23.4bL** Custom Medical Stock Photo, Inc. **23.4bR** Mary Beth Angelo/Photo Researchers, Inc. **23.5aTL** Marie Read/Animals Animals/Earth Scenes **23.5aTR** Tui De Roy/Bruce Coleman Inc. **23.5aBL** George D. Lepp/Photo Researchers, Inc. **23.5aBR** Mickey Gibson/Animals Animals/Earth Scenes **23.7L** Photo by Michael K. Richardson, reproduced by permission from *Anatomy and Embryology* 305, Fig. 7. Copyright ©Springer-Verlag GmbH & Co KG, Heidelberg, Germany. **23.7M** Photo k by Professor R. O'Rahilly. Photos a–j by Dr. Michael K. Richardson. Reproduced by permission *from Anatomy and Embryology* 305, Fig. 8. Copyright ©Springer-Verlag GmbH & Co KG, Heidelberg, Germany. **23.7R** From: Richardson, M. K., et al. *Science*. Vol 280: Pg 983c, Issue # 5366, May 15, 1998. Embryo from Professor R. O'Rahilly. National Museum of Health and Medicine/Armed Forces Institute of Pathology **23.10** Arthur C. Aufderheide, M.D., University of Minnesota School of Medicine, Duluth **23.12a** Frans Lanting/Minden Pictures **23.11** Modified from Grant and Grant. 2001. *Science* 296: 707–711. Fig. 1.

Chapter 24

Opener ©Wayne Lynch/DRK Photo **24.5** Otorohanga Kiwi House, New Zealand **24.6** Reprinted with permission from Fig. 1B, C from B Faivre et al., *Science* 300:103 (2003). Copyright 2004 American Association for the Advancement of Science. Photo courtesy of Bruno Faivre. **24.7a** Cyril Laubscher ©Dorling Kindersley **24.7b** Data from Blount et al. 2003. *Science* 300: 125–127. Fig. 1a. **24.8** David H. Funk **24.9a** Marc Moritsch/National Geographic Image Collection **24.10aT&B** Robert & Linda Mitchell/Robert & Linda Mitchell Photography **24.10bT** ©B. Schorre/VIREO **24.10bB** ©R. & A. Simpson/VIREO **24.10cT** Jeremy Woodhouse/Getty Images, Inc./Photodisc. **24.10cB** Van Os, Joseph/Getty Images Inc. - Image Bank **24.11a** ©James Hughes/Visuals Unlimited **24.14** Data from Johnston, M. 1992. *Evolution* 46: 688–702.

Chapter 25

Opener ©Wayne Lankinen/DRK PHOTO **25.2** Credit is to Don McGranaghan in Tattersall, I. 1995. *The Fossil Trail*. (Oxford: Oxford University Press) **25.4** Diane Pierce/NGS Image Collection **25.7a** Joseph T. Collins/Photo Researchers, Inc. **25.7b** Alvin E. Staffan/National Audubon Society/Photo Researchers, Inc. **25.12** H. Douglas Pratt/NGS Image Collection **25.13** Jason Rick/Loren H. Rieseberg **25.15a** ©Anthony Bannister; Gallo Images/CORBIS **25.15b** ©Barbara Cushman Rowell/DRK PHOTO **25.15c** ©Bob Krist/CORBIS **25.15d** ©Nik Wheeler/CORBIS **25.15e** ©Matthew McKee; Eye Ubiquitous/CORBIS **25.15f** ©Richard Cummins/CORBIS

Chapter 26

Opener ©1985 David L. Brill **26.7a** Reproduced by permission from P.S. Herendeen, W.L. Crepet, and K.C. Nixon, Chloranthus like stamens from the Upper Cretaceous of New Jersey, *American Journal of Botany* 80(8):865–871. ©Botanical Society of America. Micrograph courtesy of William L. Crepet, Cornell University. **26.7b** Martin Land/Science Photo Library/Photo Researchers, Inc. **26.7c** Monte Hieb & Harrison Hieb/Geocraft/www.geocraft.com **26.7d** John Gerlach/DRK Photo **26.8B** Neg. no. 325097, shot from Sanford Bird Hall. Courtesy Department of Library Services, American Museum of Natural History. **26.12b** Reproduced by permission from S. Xiao, Y. Zhang, and A.H. Knoll, Three-dimensional preservation of algae and animal embryos in a Neoproterozoic phosphorite. *Nature* 391:553–558 (February 5, 1998), fig. 5. Copyright ©1998 Macmillan Magazines Limited. **26.12cT** Simon Conway Morris, University of Cambridge, Cambridge, United Kingdom **26.12cB** Ken Lucas/Visuals Unlimited **26.12dT** B. Miller/Biological Photo Service **26.12dB** ©Ed Reschke/Peter Arnold, Inc. **26.14** Denis Duboule, University of Geneva, Geneva, Switzerland **26.16a** Jonathan B. Losos, Washington University in St. Louis **26.17a** Colin Keates ©Dorling Kindersley, Courtesy of the Natural History Museum, London **26.17b** ©Tim Fitzharris/Minden Pictures **26.17d** ©Joe Tucciarone/SPL/Photo Researchers, Inc. **26.19b** Glen A. Izett/U.S. Geological Survey, Denver **26.19c** Peter H. Schultz, Brown University; Steven D'Hondt, University of Rhode Island Graduate School of Oceanography

Unit 6

©John Gerlach /DRK Photo

Chapter 27

Opener ©Beth Donidow/Visuals Unlimited **27.2** ©Science VU/Visuals Unlimited **27.3** J. Robert Waaland/Biological Photo Service **27.4** ©Richard L. Carlton/Visuals Unlimited **27.6a** ©T. Bannor/Custom Medical Stock Photo **27.6b** ©Carolina Biological/Visuals Unlimited **27.7** Reprinted with permission of the American Association for the Advancement of Science from S.V. Liu et al., Thermophilic Fe(III)-reducing bacteria from the deep subsurface. *Science* 277:1106–1109 (1997), fig. 2. Copyright©1997 American Association for the Advancement of Science. Image courtesy of Yul Roh, Oak Ridge National Laboratory. **27.11aL** CNRI/Science Photo Library/Photo Researchers, Inc. **27.11aR** Reprinted with permission from H.N. Schulz, et al., Dense populations of a giant sulfur bacterium in Namibian shelf sediments. *Science* 284:493–495, Fig. 1b, August 4, 1999. Copyright 1999 American Association for the Advancement of Science. Image courtesy of Dr. Heide Schulz, Max-Planck-Institute for Marine Microbiology, Bremen, Germany. **27.11bL** ©Gary Gaugler/Visuals Unlimited **27.11bR** David M. Phillips/Visuals Unlimited **27.11cL** Linda Stannard, University of Cape Town/Science Photo Library/Photo Researchers, Inc. **27.11cR** Richard W. Castenholz, University of Oregon **27.12c** ©Jack Bostrack/Visuals Unlimited **27.16** ©Eric Grave/Phototake **27.17** ©National Cancer Institute/SPL/Photo Researchers, Inc. **27.18** ©Michael Abbey/Visuals Unlimited **27.19** ©R. Calentine/Visuals Unlimited **27.20** ©J. C. Revy/Phototake **27.21a** Photo by Yves B. Brun **27.21b** Hans Reichenbach, Gesellschaft fur Biotechnologische Forschung mbH, Braunschweig, Germany **27.23** Photograph courtesy Dr. Kenneth M. Stedman/Portland State University **27.23 (inset)** ©Corale Brierley/Visuals Unlimited **27.24** Marli Miller **27.24 (inset)** Photograph courtesy Dr. Bonnie K. Baxter/Westminster College

Chapter 28

Opener Dr. Ann Smith/Photo Researchers, Inc. **28.1a** Norman T. Nicoll **28.1b** Gregory Ochocki/Photo Researchers, Inc. **28.1c** Tom and Therisa Stack/Tom Stack & Associates, Inc. **28.1R** John Anderson/Animals Animals/Earth Scenes **28.2** Dr. Gopal Murti/Photo Researchers, Inc. **28.3** Sanford Berry/Visuals Unlimited **28.3 (inset)** J. Robert Waaland/Biological Photo Service **28.6** Biophoto Associates/Photo Researchers, Inc. **28.12L** M.I. Walker/Photo Researchers, Inc. **28.13** David L. Kirk, Washington University **28.14a–b** Andrew Syred/Science Photo Library/Photo Researchers, Inc. **28.14c** David M. Phillips/Visuals Unlimited **28.15a** Biophoto Associates/Photo Researchers, Inc. **28.15b** Bruce Coleman Collection **28.16** Eric Grave/Photo Researchers,

Index

Probe, 406
Proboscis, 658, 659, 732
Procambium, 923
Process, in scientific theory, 494–495
Prochlorococcus, 438, 583
Prochloron, 616
Producers, 750, 1248
 primary, 1244, 1250
Production, secondary, 1252
Productivity
 global patterns in, 1244–1246
 gross primary, 1251
 human effects on, 1261
 limiting in marine ecosystems, 1246–1247
 limiting in terrestrial ecosystems, 1246
 net primary. *See* Net primary productivity
 nitrogen and, 1258
 primary, 1238
 pyramid of, 1250
 species richness and, 1238
 tropical wet forests and, 1151
Product of chemical reaction, 30
Product rule, Punnett squares and, 275
Products, of chemical reaction, 32
Progesterone, menstrual cycle and, 1111–1113
Prokaryotes, 7
 electron transport chain and, 182
 gene number, 439
 See also Archaea; Bacteria
Prokaryotic cells, 129–131
 eukaryotic *vs.*, 131, 133
Prokaryotic genomes, history of, 431–432
Prolactin, 1081
Prolactin-inhibiting hormone (PIH), 1091
Prolactin (PRL), 1090, 1091
Proline, 50, 51, 58, 163
Prometaphase, 233, 234, 236
Promoter-proximal elements, 386
Promoters, 340, 345, 383, 386
Prophase, 233, 234, 236
 early, 254, 255
 late, 254, 255
 meiosis, 257–258
 meiosis I, 257
Prophase II, 255
Propionibacterium acnes, 584
Propionic acid, 597
Proplastids, 205, 812
Prop roots, 807
Propulsion, seed dispersal via, 925
Prosimians, 773
Prostaglandins, 171, 1079
Prostate gland, 1106, 1107
Protease, 450, 789
Protease inhibitors, 789–790, 792
Protected areas, 1279
Protein, 53
Proteinase inhibitors, 905
Protein kinase, 240
Protein kinase A, 1095
Proteins
 allosteric regulation and, 369
 amino acids and, 48–56
 catabolism of, 197–198
 in cell membranes, 116–123
 channel, 118–120
 complement, 63
 composition of, 47
 contractile, 63

 digestion of, 985–986, 987–988
 pancreatic enzymes and, 989–990
 DNA-binding, 374–378
 endomembrane system and, 145–149
 as first living entity, 70
 function of, 62–70
 catalysis, 63–64, 67–68
 in cells, 62–63
 enzymes, 64–70
 genes and, 305, 307
 heat-shock, 60
 integral membrane, 117–118
 involved in DNA synthesis, 315
 peptide bonds and, 54–55
 peripheral membrane, 117
 polymerization and, 53–56
 post-translational modifications of, 354–355
 prebiotic soup and, 46, 48
 production of viral, 788–790
 proteomics and, 440–441
 RDA, 978
 receptor, 63
 regulatory, 386–390
 ribosomes and synthesis of, 346–347
 RNA as intermediary with genes, 329–330
 structural, 63
 structure of, 56–62
 folding and function, 60–62
 primary, 57, 61
 quaternary, 59–60, 61
 secondary, 57–58, 61
 tertiary, 58–59, 61
 transmembrane, 117–118
 transport, 63, 120–122
Proteobacteria, 592, 601
Proteoglycans, 1121
Proteomics, 440–441
Proterozoic eon, 563
Protists (Protista), 8
 cell structure, 612
 direct sequencing, 617
 diversification of, 617–627
 ecological importance of, 610–612
 evolution of multicellularity in, 619–620
 feeding in, 621–623
 global warming and, 610–612
 green algae and, 661
 impacts on human health/welfare, 608–610
 key lineages, 627–633
 Alveolata, 629–630
 Ameobozoa, 633
 Cercozoa, 632
 Discricristata, 628
 Escavata, 627–728
 Plantae, 632–633
 Stramenopila, 630–631
 model organism, 618
 molecular phylogenies, 612–613
 morphological diversity of, 607, 619–621
 movement in, 623–624
 organelles, 619
 origin of chloroplasts, 615–617
 origin of mitochondria, 615–617
 overview, 607–608
 photosynthesis in, 622–623
 as primary producers, 610, 611
 reasons to study, 608–612
 reproduction in, 624–627

 research techniques, 612–617
 support structures, 620–621
Protoderm, 923
Proton gradient, establishing, 859–860
Proton-motive force, 192
Proton pumps, 846–848
 electrical signaling and, 880–882
 guard cell change and, 902
 location of, 846–848
 nutrient uptake and, 859–860
Protons, 21–22, 24
 in acid-base reactions, 41–42
 ATP production and, 193
Proto-oncogenes, 395
Proto-planets, 21
Protostomes, 701
 diversification of, 731–734
 ecdysozoans, 740–745
 feeding in, 732–733
 fossil record, 730
 life cycles, 734
 lophotrochozoans, 734–740
 molecular phylogenies, 708–709, 730–731
 morphological traits, 728–730
 movement in, 733–734
 overview, 724–725
 pattern of development, 704, 705
 reasons to study, 725–728
 reproduction, 734
 research techniques, 728–731
 water-to-land transition, 731
Proximal, 458
Proximal tubule, 967, 968–969
Proximate causation, 1167
Proximate explanation, 248
Prusiner, Stanley, 62
Pseudoceratina crassa, 719
Pseudoceros ferrugineus, 736
Pseudocoelom, 703, 728–729
Pseudogenes, 436, 498, 529
Pseudomonas aeruginosa, 164, 584
 genome, 431, 433
Pseudopodia, 150–151, 621, 623, 624, 632
Psilotophyta (whisk ferns), 644, 645, 647, 666
P site, 351, 354
PTC (phenylthiocarbamide), 1065
Pteridophyta, 667
Pterosaurs, 759, 770
Pterourus species, 744
PTH (parathyroid hormone), 1081, 1085
Puberty
 changes taking place during, 1109
 hormones and, 1084
 mammalian, 1109–1111
Puffballs, 690, 693
Puffer fish, 1035
Puffins, 1
Pulmonary artery, 1018
Pulmonary circulation, 1017
Pulmonary veins, 1018
Pulse-chase experiment, 145–146, 217–218, 221, 1008
Punnett square, 274, 275
Pupa, 464, 716
Pupation, 464, 716
Pupil, 1060
Pure lines, 271–272, 279
 cross between, 274–275
 for two traits, 276–277
Pure science, 924

Purines, 75, 76
 origin of, 76
 pyrimidine pairing, 81
Puromycin, 358
Purple loosestrife, 1161, 1164, 1274
Purple-spotted nudibranch, 739
Purple sulfur bacteria, 212
Putrescine, 597
PVA (population viability analysis), 1210
Pycnopodia helianthoides, 764
Pyramid of productivity, 1250
Pyrethroids, 813
Pyrimidines, 75, 76
 origin of, 76
 purine pairing, 81
Pyrophosphatase, 56
Pyruvate, 180, 184, 185
 and acetyl CoA, 187–188
 fermentation and, 196–197
Pyruvate dehydrogenase, 188–189

Q (coenzyme Q), 191, 192
Qori Kalis glacier, 1243
Quadrats, 1208
Quaking aspen trees, 261
Quantitative traits, 291–292
Quantitative variation, 287
Quaternary structure, 59–60, 61
Quiescent virus, 787
Quinidine, 642
Quinine, 642
Quinones, 213
Quinton, P. M., 124

Rabies virus, 781, 798
Races, human, 553
Racker, Efraim, 194
Radial cleavage, 453, 704
Radial symmetry, 702, 703, 751
Radiation, 950
Radicle, 923, 924, 928
Radioactive decay, 22
Radioactive isotopes, 22
Radiometric dating, 15, 21, 22–23
Radish, 871
Radula, 711, 739
Rafflesia arnoldii, 911
Ragged stunt virus, 798
Rain shadows, 1149
Ramón y Cajal, Santiago, 1028
Ram ventilation, 1004
Ran, 144
Rancher ants, 699, 1227
Random mating, Hardy-Weinberg principle and, 518, 520
Range, 1160
Raspberry, 926
Ratites, 544–545
Ray-finned fishes, 750
Rayment, Ivan, 1070
Rays, 766, 823
RDAs. *See* Recommended Dietary Allowances
Reabsorption, in kidney, 968–969
Reactants, 30, 32
Reaction center, 211
Reactive oxygen intermediates (ROIs), 883–884, 1122
Reading frames, 330–332
 open, 430
Realized niche, 1217
Receded traits, 272–273
Receptacle, 916
Receptor-mediated endocytosis, 136